712-2412

A First Course in Differential Equations
with Applications

A First Course in Differential Equations

with Applications
Third Edition

William R. Derrick

Stanley I. Grossman
University of Montana

West Publishing Company

St. Paul New York Los Angeles San Francisco

Production: Del Mar Associates
Copy Editor: John Thomas
Technical Art: Kristi Paulson, Pamela Becraft
Composition: Science Typographers

Copyright © 1976 by Addison-Wesley Publishing Company

Copyright © 1981 by Addison-Wesley Publishing Company

Copyright © 1987 by West Publishing Company
50 W. Kellogg Boulevard
P.O. Box 64526
St. Paul, MN 55164-1003

All rights reserved

Printed in the United States of America

Library of Congress Cataloging-in-Publication Data

Derrick, William R.
   Introduction to differential equations.

   Rev. ed. of: Elementary differential equations with applications. 2nd ed. c1981.
   Includes index.
   1. Differential equations.   I. Grossman, Stanley I.   II. Derrick, William R. Elementary differential equations with applications.   III. Title.
QA371.D435   1987   515.3'5   86-24755
ISBN 0-314-26895-2

Credits

Pages 21 and 50 (biographical sketches) 105–112, A1–A10, and A29–A38 from Grossman, S. I., *Calculus*, 3rd edition, © 1984 by Academic Press, Inc. Used by permission
Pages 37–39 (Perspective) from Scott, D., *The American Math Monthly*, Vol. 92, © 1985. Pages 422–423. Used by permission.
Page 39 (table) from Pearl, R., *Quarterly Review of Biology*, Vol. 2, © 1927. Figure 1. Used with permission of the publisher.
Page 53 (Problem 41) from *The American Math Monthly*, Vol. 61, © 1954. Page 545, problem 3. Used by permission.
Page 65 (figure and problem) from Shearer, Murphy, and Richardson, *Systems Dynamics*, © 1967 by Addison-Wesley Publishing Company, Inc., Reading, Massachusetts. Page 141 (figure). Reprinted with permission.
Page 150 (Figure 4.8), A/P Wide World Photos.
Pages 196 and 210 (biographical sketches), from Eves, *An Introduction to the History of Mathematics*, 3rd edition, © 1983 by Saunders College Publishing. Used by permission.
Pages 376–378 and A39–A46 from Grossman, S. I., *Elementary Linear Algebra*, 3rd edition, © 1987, 1984, 1980 by Wadsworth, Inc. Used by permission.

To the memory of
Benjamin Grossman
and
John K. Long

Contents

	Preface	xi
1	**Introduction to Differential Equations**	1
1.1	Introduction	1
1.2	Classification of Differential Equations and Direction Fields	11
1.3	An Excursion: Numerical Solutions of Differential Equations	20
	Review Exercises	26
2	**First-Order Equations**	28
2.1	Separation of Variables	28
2.2	Exact Equations and Integrating Factors	40
2.3	Linear First-Order Differential Equations	46
2.4	Homogeneous Equations and Substitution Techniques	54
2.5	Simple Electric Circuits	61
2.6	Further Applications: Compartmental Analysis and Curves of Pursuit	65
2.7	Successive Approximations: If Time Permits	74
2.8	An Excursion: First-Order Difference Equations	77
	Review Exercises	81
3	**Second- and Higher-Order Linear Differential Equations**	83
3.1	Theory of Linear Differential Equations	83
3.2	Using One Solution to Find Another: Reduction of Order	97
3.3	Homogeneous Equations with Constant Coefficients: Real Roots	100
3.4	Homogeneous Equations with Constant Coefficients: Complex Roots	103
3.5	Nonhomogeneous Equations: Undetermined Coefficients	105
3.6	Nonhomogeneous Equations: Variation of Parameters	112
3.7	Euler Equations	118
3.8	Higher-Order Linear Differential Equations	122
3.9	An Excursion: Second-Order Difference Equations	128
	Review Exercises	135
4	**Applications of Linear Differential Equations**	137
4.1	Vibrational Motion: Simple Harmonic Motion	137
4.2	Vibrational Motion: Damped Vibrations	143
4.3	Vibrational Motion: Forced Vibrations	146
4.4	Electric Circuits	152
	Review Exercises	155

vii

5	**Power Series Solutions of Differential Equations**	157
5.1	The Power Series Method	157
5.2	Ordinary and Singular Points	166
5.3	The Method of Frobenius: The Indicial Equation	173
5.4	Bessel Functions	186
5.5	Legendre Polynomials	195
5.6	An Excursion: Oscillatory Solutions of Linear Second-Order Differential Equations with Variable Coefficients	201
	Review Exercises	207
6	**Laplace Transforms**	209
6.1	Introduction: Definition and Basic Properties of the Laplace Transform	209
6.2	Solving Initial-Value Problems by Laplace Transform Methods	221
6.3	Step Functions, Impulse Functions, and Periodic Functions	234
6.4	Some Differential Equations with Discontinuous Forcing Functions: Applications to Electric Circuits (Optional)	247
6.5	The Transform of Convolution Integrals	252
	Review Exercises	258
7	**Systems of Linear Differential Equations**	259
7.1	Systems of First-Order Equations	260
7.2	Linear Systems: Theory	266
	Nonmatrix Methods	271
7.3	The Method of Elimination and the Method of Determinants for Linear Systems with Constant Coefficients	271
7.4	Laplace Transform Methods for Systems	280
	Matrix Methods	285
7.5	Review of Matrices	285
7.6	Matrices and Linear Systems of Differential Equations	299
7.7	Fundamental and Principal Matrix Solutions of a Homogeneous System of Differential Equations	304
7.8	Eigenvalues and Eigenvectors	315
7.9	Computation of the Principal Matrix Solution	327
7.10	Nonhomogeneous Systems	337
7.11	An Excursion: The Matrix Exponential	344
	Review Exercises	352
8	**Applications of Systems of Differential Equations**	354
8.1	Electric Circuits with Several Loops	354
8.2	Chemical Mixture and Population Biology Problems	360
8.3	Mechanical Systems	366
8.4	An Excursion: A Model for Epidemics	369
	Review Exercises	374
9	**Numerical Methods**	376
9.1	Error Analysis	376
9.2	Euler Methods for First-Order Differential Equations	379
9.3	The Runge-Kutta Method	386
9.4	Predictor-Corrector Formulas	393

9.5	Error Analysis for the Euler Method: If Time Permits	398
9.6	An Excursion: Numerical Instability Caused by the Propagation of Round-Off Error	402
9.7	Numerical Solution of Systems and Boundary Value Problems	404
	Review Exercises	410
10	**Fourier Series and Partial Differential Equations**	411
10.1	Introduction to Trigonometric Series	411
10.2	Orthogonal Sets of Functions	418
10.3	Fourier Series	423
10.4	Half-Range Expansions	432
10.5	Classification of Linear Second-Order Equations	438
10.6	The Vibrating String: d'Alembert's Method	444
10.7	Separation of Variables: The Wave Equation	450
10.8	Heat Flow and the Heat Equation	455
10.9	Two-Dimensional Heat Flow and Laplace's Equation	459
10.10	Laplace's Equation in Polar and Spherical Coordinates	463
	Review Exercises	472
	Appendixes	
1	Integral Tables	A1
2	Table of Laplace Transforms	A11
3	The Existence and Uniqueness of Solutions	A14
4	Determinants	A29
5	Complex Numbers	A39
	Answers to Odd-Numbered Problems	A47
	Index	I1

Preface

Since the time of Newton, differential equations have been of fundamental importance in the application of mathematics to the physical sciences. Lately, differential equations have gained increasing importance in the biological and social sciences. New information concerning the theory of differential equations and its application to all sciences appears at an increasing rate, leading to a continued interest in its study by students in all scientific disciplines.

Today, the subject of differential equations is well understood. There is a large and mathematically elegant body of theory that enables us to study differential equations in a very systematic way. This provides the writers of textbooks in the subject with a challenge: to provide the theoretical tools needed to understand fully how various topics fit together and, yet, not to lose sight of the important applications that led to the development of the theory.

In this book, the theory is not slighted. Every theorem that can be proved using the mathematics available to the student who has studied calculus is proved in the text. Occasionally, more difficult tools, the notion of uniform convergence, for example, are used to prove results that are central to the study of differential equations. However, the primary goal of this book is to teach students how to *use* differential equations in applied areas.

Here are some features that illustrate how our dual goals have been met:

Examples As students, we learned how to solve differential equations by reading examples and solving problems. *A First Course in Differential Equations with Applications, Third Edition*, contains 281 examples—many more than are commonly found in standard differential equations texts. Most examples contain all the algebraic and calculus steps needed to complete the solution. It is infuriating to be told that "it easily follows that . . ." when it is not at all obvious. Students have a right to see the "whole hand," so to speak, so that they always know how to get from A to B. In many instances, explanations are highlighted in color to make a step easier to follow.

Exercises The text contains over 1840 exercises—including both drill- and applied-type problems. More difficult problems are marked with an asterisk (*), and a few especially difficult ones are marked with a double asterisk (**).

The exercises provide the most important learning tool in any undergraduate mathematics textbook. We stress to our students that no matter how well they think they understand our lectures or the textbook, they do not really know the material until they have worked problems. A vast difference exists between understanding someone else's solution and solving a new problem by yourself. Learning mathematics without doing problems is about as easy as learning to ski without going to the slopes.

Chapter Review Exercises At the end of each chapter, we have provided a collection of review exercises. Any student who can do these exercises can feel confident that he or she understands the material in the chapter.

Applications Differential equations *is* applied mathematics. Consequently, this book includes many applied examples and exercises. The majority of the examples are drawn from the physical sciences. However, a wide range of applications in other areas is provided as well. Here is a partial list of applications:

- Computation of escape velocity and reentry into the atmosphere (Section 2.1)
- Electrical circuits (Sections 2.5, 4.4, 6.2, 6.4, 8.1)
- Harmonic and forced harmonic motion (Sections 4.1, 4.2, 4.3, 8.3)
- The collapse of the Tacoma Narrows Bridge (page 150)
- Chemical mixture problems (Sections 2.6, 7.3, 8.2)
- A model of epidemics (Section 8.4)
- Quality control in business (Section 3.9)

Perspectives In six sections of the book, we digress into short "perspectives" that explore an important topic from a different point of view. For example, every differential equations text discusses separable differential equations. We go further: After solving a number of differential equations in Section 2.1 by separating the variables, we ask (and answer) the question, "When is a differential equation separable?" (see the perspective on page 37).

Excursions The text contains seven section-long "excursions" into topics not ordinarily discussed in elementary differential equations texts. These are found at the end of chapters, so as not to interfere with standard topics. Two of these (Sections 2.8 and 3.9) discuss first- and second-order linear difference equations.

Numerical Methods Most differential equations cannot be solved in terms of elementary functions. For this reason, it is important that a differential equations textbook include a discussion of numerical methods for obtaining solutions. We provide this material in three ways:

- In Section 1.3, before a student has solved many differential equations, we discuss Euler's method. This provides the reader with a simple way to

approximate the solution to a first-order initial-value problem at a given point.
- Later, in Chapter 9, we provide a number of techniques for solving a variety of differential equations.
- Also in Chapter 9, we discuss the error in numerical approximation. Many books omit this important topic. To describe numerical techniques without discussing how errors can propagate is a bit like learning how to drive without learning the rules of the road.

Existence and Uniqueness Theory Proofs of the existence and uniqueness of solutions involve mathematical tools beyond those taught in a first calculus course. For that reason, we break our discussion of this topic into two parts. First, we discuss successive approximations in Section 2.7. Second, we prove both local and global existence-uniqueness theorems for scalar equations and systems of equations in Appendix 3. In that appendix, we discuss uniform convergence and related topics.

Systems of Differential Equations In order to study systems of differential equations in the most general setting, it is useful to know something about matrices and their eigenvalues and eigenvectors. Many students do not have this background. Therefore, in order to make this topic accessible to all readers, we use, in Chapter 7, three approaches.

- The **method of elimination** and its allied **method of determinants** (Section 7.3) provide efficient and elementary procedures for solving systems of two first-order linear differential equations in two unknown functions.
- The **method of Laplace transforms** (Section 7.4) converts a system of n linear first-order differential equations into a system of n linear equations, which may then be solved by the techniques of linear algebra.
- The rest of the chapter (Sections 7.5–7.11) provides a discussion of **matrix methods** for the solution of systems of linear differential equations. A brief review of matrices is given in Section 7.5 and a complete introduction to eigenvalues and eigenvectors is given in Section 7.8.

Determinants and Complex Numbers In certain sections of the book, students need to know something about determinants and complex numbers. Everything they need to know is contained in Appendices 4 and 5.

Integral Tables In the real world, most nontrivial integrals are computed with the use of an integral table. The table in Appendix 1 of this book contains 220 entries. Every integral that comes up in this text can be found there.

Biographical Sketches Mathematics becomes more interesting if one knows something about the historical development of the subject. We chose four mathematicians who contributed greatly to the understanding of differential equations. Full-page biographies of Euler, Bernoulli, Legendre, and Laplace appear on pages 21, 50, 196, and 210.

Numbering and Notation in the Text Numbering in the book is fairly standard. Within each section, examples, problems, theorems, and equations are numbered consecutively, starting with 1. Reference to an example, problem, theorem, or equation outside the section in which it appears is by chapter, section, and number. Thus, Example 4 in Section 2.3 is called, simply, Example 4 in that section, but outside the section it is referred to as Example 2.3.4. As already mentioned, the more difficult problems are marked (*) or occasionally (**). Finally, the ends of examples or groups of examples are marked with a line and the ends of proofs are marked with a ■.

Chapter Interdependence The following diagram indicates interdependence—that is, which sections and chapters depend on earlier material.

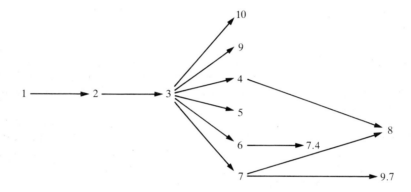

ACKNOWLEDGMENTS

We are grateful to the Wadsworth Publishing Company for permission to use, in Chapter 7, material from the book *Elementary Linear Algebra, Third Edition* (1987), by Stanley Grossman. Some of the material for the biographical sketches came from the excellent book *An Introduction to the History of Mathematics, Fifth Edition* (1983), by Howard Eves. We are grateful to W. B. Saunders Company for permission to use this information.

Some of the examples and exercises in this book first appeared in *Mathematics for the Biological Sciences* by Stanley Grossman and James E. Turner, originally published by Macmillan in 1974. We are grateful to Professor Turner for permission to use this material.

A number of reviewers provided useful suggestions for improving the quality of this text. In particular, we would like to thank the following:

Gerald Bradley
Claremont Mckenna College

Tom Cromer
NASA

J. M. Cushing
University of Arizona

Jack Eidswick
University of Nebraska

David Ellis
San Francisco State University

Garrett Etgen
University of Houston

G. Elton Graves
Rose-Hulman Institute

B. J. Harris
Northern Illinois University

Terry Herdman
Virginia Tech

Clark Kimberling
University of Evansville

M. K. Kwong
Louisiana State University

Deborah Lockhart
Michigan Technological University

Stanley Lukaweicki
Clemson University

Greg Passty
Southwest Texas State University

Nancy Poxon
California State University Sacramento

Robert Pruitt
San Jose State University

Kenneth Stoll
University of Louisville

Ted Suffridge
University of Kentucky

Wescott Vayo
University of Toledo

Professor Leon Gerber, at St. John's University in New York City, prepared the Student's and Instructor's manuals; he made many useful suggestions for the improvement of the problem sets. This book is a better teaching tool because of him.

Putting together a book like this is always a difficult process. Our task was made easier by the highly professional production work done by Nancy Sjoberg and her staff at Del Mar Associates in Del Mar, California. We are grateful for efforts that resulted not only in a nicely produced book, but also a nicely produced book that came out on schedule.

Finally, we owe a considerable debt to Rich Jones, mathematics editor at West Publishing Company. His insight and prodding (which was sometimes not too gentle) helped us immeasurably in the development of this third, greatly improved edition.

William R. Derrick
Stanley I. Grossman
January 1987

1

Introduction to Differential Equations

1.1 Introduction

Many of the basic laws of the physical sciences and, more recently, of the biological and social sciences are formulated in terms of mathematical equations involving certain known and unknown quantities and their derivatives. Such equations are called **differential equations**. In this section, we show how some differential equations arise and how their solutions are obtained.

Before citing any examples, however, we should emphasize that in the study of differential equations the most difficult problem often is to describe a real situation quantitatively. To do this, it is usually necessary to make simplifying assumptions that can be expressed in mathematical terms. Thus, for example, we initially describe the motion of a mass in space by assuming that (a) it is a point and (b) there is no friction or air resistance. These assumptions are not realistic, but the scientist can often glean valuable information from even highly idealized models that, once understood, can be modified to take other observable factors into account.

Example 1 **Free Fall** Newton's law of gravitation states that the magnitude of the gravitational force exerted by the earth on a body of mass m is proportional to its mass and inversely proportional to the square of its distance $r = r(t)$ from the center of the earth. Thus, if k denotes the constant of proportionality, then

$$F = \frac{km}{r^2}.$$

By Newton's second law of motion

$$F = ma = m\frac{d^2r}{dt^2},$$

where $a = a(r)$ is the acceleration of the body when the distance to the center of the earth is r. Hence, equating both forces and dividing by m, we get

$$\frac{d^2r}{dt^2} = \frac{k}{r^2}. \qquad (1)$$

Note[†] The velocity $v = dr/dt$ is negative because, as the object falls, its distance from the earth decreases. Moreover, the acceleration $a = dv/dt = d^2r/dt^2$ is also negative because, as the object falls, its velocity gets more and more negative; that is, v *decreases*. Thus the constant k is negative.

We denote the acceleration of gravity at the surface of the earth (when $r = R$: the mean radius of the earth) by $a(R) = -g$. Then equation (1) leads to

$$-g = a(R) = \frac{k}{R^2},$$

so that $k = -gR^2$. Hence we obtain the equation of motion

$$\frac{d^2r}{dt^2} = -g\frac{R^2}{r^2}, \qquad (2)$$

where g is approximately 9.81 m/sec² ($= 32.2$ ft/sec²). If we substitute $r = R + h$, where $h = h(t)$ is the height of the body from the surface of the earth, then $dr/dt = dh/dt$ and equation (2) becomes

$$\frac{d^2h}{dt^2} = -g\frac{R^2}{(R+h)^2}.$$

When the height h is very small in comparison to the radius of the earth R, the ratio $R/(R+h)$ is very close to 1, so the differential equation is well approximated by the usual equation found in calculus books:

$$\frac{d^2h}{dt^2} = -g. \qquad (3)$$

[†] In this problem the force is directed toward the center of the earth. In what follows we only consider vectors directed toward the earth's center or in the opposite direction. These can be distinguished by signs: minus for vectors toward the center of the earth, plus in the opposite direction.

Integrating† both sides of equation (3) with respect to t, we obtain

$$h'(t) = -gt + C_1.$$

The constant C_1 can be determined by setting $t = 0$. We obtain $C_1 = h'(0)$, the initial velocity. Thus the velocity of the body at any time t is given by the differential equation

$$h'(t) = -gt + h'(0). \tag{4}$$

Integrating once more with respect to t, we have

$$h(t) = -\frac{gt^2}{2} + h'(0)t + C_2.$$

The constant C_2 can also be found by setting $t = 0$. We obtain $C_2 = h(0)$, the initial height. Thus the height of the body at any time t is

$$h(t) = -\frac{gt^2}{2} + h'(0)t + h(0). \tag{5}$$

For example, if a ball is dropped from the top of a building 45 meters high, its initial velocity is $h'(0) = 0$ and its initial height is $h(0) = 4500$ centimeters. If we wish to find the time it takes for the ball to strike the ground, we substitute these values in equation (5) to obtain

$$0 = h(t) = -\frac{981t^2}{2} + 4500,$$

since the height at impact is zero. Solving for t, we have

$$490.5t^2 = 4500 \quad \text{or} \quad t^2 \approx 9.17.$$

Thus $t \approx \pm 3.03$. Since $t = -3.03$ has no physical significance, the answer is $t \approx 3.03$ seconds.

The differential equation in Example 1 was solved directly by integration. If it were always possible to do this, then differential equations would be a direct application of integral calculus and there would be no need for separate books on differential equations. However, most solvable differential equations can only be solved by other techniques. One type of differential equation that we can solve is often discussed in a beginning calculus class.

Example 2 Solve the differential equation

$$\frac{dy}{dx} = \alpha y, \quad y(0) = 10, \tag{6}$$

where α is a constant.

Solution We rewrite (6) as

$$\frac{dy}{y} = \alpha\, dx$$

---

†A table of integrals is given in Appendix 1.

where y is a function of x, and then integrate to obtain

$$\int \frac{dy}{y} = \int \alpha\, dx$$

or

$$\ln|y| = \alpha x + C.$$

Hence

(If $\ln a = b$, then $a = e^b$.)
↓

$$|y| = e^{\alpha x + C} = e^{\alpha x} e^C,$$

or

$$y = k e^{\alpha x}, \quad \text{where } k = \pm e^C. \tag{7}$$

We check our answer:

If $y(x) = ke^{\alpha x}$, then $dy/dx = k(\alpha e^{\alpha x}) = \alpha(ke^{\alpha x}) = \alpha y$, so $y = ke^{\alpha x}$ satisfies the differential equation in (6). Thus the differential equation has an infinite number of solutions, one for each real number k.

We obtain a unique solution if we use the **initial condition** $y(0) = 10$. Then

$$10 = ke^{\alpha \cdot 0} = ke^0 = k,$$

and $y = 10e^{\alpha x}$ is the unique solution to the **initial-value problem** (6).

The method used to solve (6) in Example 2 is called **separation of variables**, since it involves placing the dependent and independent variables on different sides of the equation. We will study this method in greater detail in Section 2.1.

If $\alpha > 0$, we say that $e^{\alpha x}$ is **growing exponentially**. If $\alpha < 0$, it is **decaying exponentially** (see Figure 1.1). Of course if $\alpha = 0$ there is no growth, and $y = e^0 = 1$ remains constant.

For a physical problem it would not make sense to have an infinite number of solutions. We can usually get around this difficulty, as in Example 2, by specifying the value of y for one particular value of x, say, $y(x_0) = y_0$. This value is called an **initial condition**, and it gives a unique solution to the problem. We see this illustrated in the examples that follow.

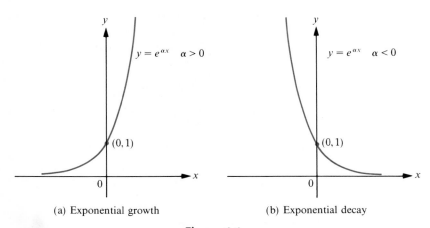

(a) Exponential growth   (b) Exponential decay

Figure 1.1

Example 3

a. Find all solutions to $dy/dx = 3y$.

b. Find the solution that satisfies the initial condition $y(0) = 2$.

Solution

a. Since $\alpha = 3$, by Example 2 all solutions are of the form $y = ce^{3x}$, where c is an arbitrary real constant.

b. $2 = y(0) = ce^{3 \cdot 0} = c \cdot 1 = c$, so $c = 2$ and the unique solution is $y = 2e^{3x}$.

This is illustrated in Figure 1.2. We can see that while there are indeed an infinite number of solutions, there is only one that passes through the point $(0, 2)$.

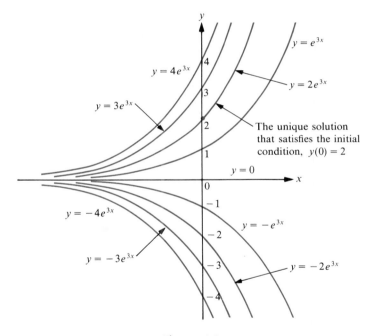

Figure 1.2

Example 4 In 1798, Thomas R. Malthus (1766–1834)[†] observed that the human population was growing at a rate proportional to the population size. If $N(t)$ is the number of individuals present at time t, then the growth of the population is given by

$$\frac{dN}{dt} = \alpha N,$$

where α is the constant of proportionality. This equation, called the **Malthu-**

[†] T. R. Malthus, "An Essay on the Principle of Population as it Affects the Future Improvement of Society," 1798.

sian law of population growth, has the solution (see Example 2)

$$N(t) = ke^{\alpha t}.$$

If $\alpha > 0$, the population grows exponentially. Malthus concluded that eventually the population would exceed the available resources, leading to famine and poverty.

Example 5 **Newton's law of cooling** states that the rate of change of the temperature of an object is proportional to the temperature difference between the body and its surrounding medium. Let $T(t)$ denote the temperature of the object at time t and let T_S denote the temperature of the surroundings (T_S is assumed to be constant throughout). Since a rate of change is expressed mathematically by a derivative, we may translate Newton's law of cooling into the equation

$$\frac{dT}{dt} = \alpha(T - T_S), \tag{8}$$

where α is the constant of proportionality.

Separating variables, we have

$$\int \frac{dT}{T - T_S} = \alpha \int dt + C$$

or

$$\ln|T - T_S| = \alpha t + C.$$

Then

$$|T - T_S| = e^{\ln|T-T_S|} = e^{\alpha t + C} = e^C e^{\alpha t},$$

so

$$T(t) - T_S = C_1 e^{\alpha t}, \tag{9}$$

where $C_1 = \pm e^C$. Equation (9) is the **general solution** to the differential equation (8), since every solution to (8) is of this form. The constant C_1 can be determined by setting $t = 0$:

$$C_1 = T(0) - T_S,$$

where $T(0) = T_0$ is the initial temperature of the object. Hence, from (9),

$$T(t) - T_S = (T_0 - T_S)e^{\alpha t},$$

or

$$T(t) = T_S + (T_0 - T_S)e^{\alpha t}. \tag{10}$$

As a practical example, suppose that a pot of boiling water (100°C) is removed from the fire and allowed to cool at 20°C room temperature. Two minutes later, the temperature of the water in the pot is 80°C. What will be the temperature of the water five minutes after it has been removed from the fire?

Here the initial temperature $T_0 = 100$ and $T_S = 20$, so $T_0 - T_S = 80$ and equation (10) becomes

$$T(t) = 20 + 80e^{\alpha t}. \tag{11}$$

When $t = 2$ minutes, the temperature is $T(2) = 80$, so if we substitute $t = 2$ into equation (11), we obtain
$$80 = 20 + 80e^{2\alpha},$$
$$80e^{2\alpha} = 60,$$
$$e^{2\alpha} = \tfrac{60}{80} = \tfrac{3}{4}.$$
Taking the natural logarithm of both sides of this equation, we get
$$2\alpha = \ln\left(\tfrac{3}{4}\right),$$
$$\alpha = \tfrac{1}{2}\ln\left(\tfrac{3}{4}\right).$$
Note that α is negative (≈ -0.1438). This makes sense because the temperature difference must decrease as time increases.

Substituting this value of α into equation (11) and recalling that $a \ln b = \ln(b^a)$, we obtain
$$T(t) = 20 + 80e^{\ln(3/4)^{t/2}}$$
or, since $e^{\ln x} = x$,
$$T(t) = 20 + 80\left(\tfrac{3}{4}\right)^{t/2}. \tag{12}$$

Equation (12) is a **particular solution** of (8) since it is uniquely determined by the conditions specified for this particular situation.

Finally, to find the temperature of the water in the pot at $t = 5$ minutes, we set $t = 5$ in equation (12):
$$T(5) = 20 + 80\left(\tfrac{3}{4}\right)^{5/2} \approx 20 + 38.97 = 58.97.$$

Thus the temperature of the water in the pot is approximately $58.97\,°C$ five minutes after it has been removed from the fire.

Remark The term **general solution** will be used for a family of solutions of the differential equation containing one or more arbitrary constants. A **particular solution** will be any solution that is free of arbitrary constants.

Example 6 **Carbon dating** is a technique used by archaeologists, geologists, and others who want to estimate the ages of certain artifacts and fossils. The technique is based on certain properties of the carbon atom. In its natural state the nucleus of the carbon atom ^{12}C has six protons and six neutrons. An **isotope** of carbon ^{12}C is ^{14}C, which has two additional neutrons in its nucleus. ^{14}C is **radioactive**; that is, it emits electrons until it reaches the stable state ^{14}N. We make the assumption that the ratio of ^{14}C to ^{12}C in the atmosphere is constant. This assumption has been shown experimentally to be approximately correct, for although ^{14}C is constantly being lost through **radioactive decay** (as this process is often termed), new ^{14}C is constantly being produced by the cosmic bombardment of nitrogen in the upper atmosphere. ^{12}C and ^{14}C are not distinguished in the biochemical processes of plants and animals, so at the time of death the ratio of ^{12}C to ^{14}C in an organism is the same as the ratio in the atmosphere. However, this ratio changes after death, because ^{14}C is converted to ^{14}N but no further ^{14}C is taken in.

It has been observed that ^{14}C decays at a rate proportional to its mass and that its **half-life** is approximately 5580 years; that is, if a substance starts with

1 gram of ^{14}C, then 5580 years later it will have 0.5 grams of ^{14}C, the other 0.5 grams having been converted to ^{14}N.[†]

We may now pose a question typically asked by an archaeologist. A fossil is unearthed and it is determined that the amount of ^{14}C present is 40 percent of what it would be for a living organism of similar size. What is the approximate age of the fossil?

Solution Let $M(t)$ denote the mass of ^{14}C present in the fossil. Since ^{14}C decays at a rate proportional to its mass, we have

$$\frac{dM}{dt} = -\alpha M,$$

where α is the constant of proportionality. Thus $M(t) = ce^{-\alpha t}$, where $c = M_0$, the initial amount of ^{14}C present. When $t = 0$, $M(0) = M_0$; when $t = 5580$ years, $M(5580) = \frac{1}{2}M_0$, since half the original amount of ^{14}C has been converted to ^{14}N. We can use this fact to eliminate α, since we have

$$\tfrac{1}{2}M_0 = M_0 e^{-\alpha \cdot 5580}, \quad \text{or} \quad e^{-5580\alpha} = \tfrac{1}{2}.$$

Thus

$$\left(e^{-\alpha}\right)^{5580} = \tfrac{1}{2}, \quad \text{or} \quad e^{-\alpha} = \left(\tfrac{1}{2}\right)^{1/5580}, \quad \text{and} \quad e^{-\alpha t} = \left(\tfrac{1}{2}\right)^{t/5580},$$

so

$$M(t) = M_0 \left(\tfrac{1}{2}\right)^{t/5580}.$$

Now we are told that after t years (from the death of the fossilized organism to the present) $M(t) = 0.4 M_0$, and we are asked to determine t. Then

$$0.4 M_0 = M_0 \left(\tfrac{1}{2}\right)^{t/5580},$$

and taking natural logarithms (after dividing by M_0), we obtain

$$\ln 0.4 = \frac{t}{5580} \ln\left(\frac{1}{2}\right), \quad \text{or} \quad t = \frac{5580 \ln(0.4)}{\ln\left(\frac{1}{2}\right)} \approx 7376 \text{ yr.}$$

Carbon dating has been used successfully on numerous occasions. It was this technique that established that the Dead Sea Scrolls were prepared and buried about two thousand years ago.

In Examples 2 through 6 we solved a very simple differential equation. Other, more elaborate differential equations can be solved by the method of separation of variables. We will do this in Section 2.1. However, it should be apparent from these examples and the problems that follow that a large variety of applied problems can be solved by considering a simple differential

---

[†] This number was first determined in 1941 by the American chemist W. S. Libby, who based his calculations on the wood from sequoia trees, whose ages were determined by rings marking years of growth. Libby's method has come to be regarded as the archaeologist's most important measuring scale. But in truth, this scale is flawed. Libby used the assumption that the atmosphere had at all times a constant amount of ^{14}C. However, the American chemist C. W. Ferguson of the University of Arizona deduced from his study of tree rings in 4000-year-old American giant trees that before 1500 B.C. the radiocarbon content of the atmosphere was considerably higher than it was later. This result implied that objects from the pre-1500 B.C. era were much older than previously believed; Libby's "clock" allowed for a smaller amount of ^{14}C than was actually present. For example, a find dated at 1800 B.C. was in fact from 2500 B.C. This fact has had a considerable impact on the study of prehistoric times. For a fascinating discussion of this subject, see Gerhard Herm, *The Celts* (New York: St. Martin's Press, 1975), pp. 90–92.

equation. It is the purpose of this book to show you how to deal with other, more complicated differential equations arising in physical applications.

The physical examples given in this section are such that in each case we know that a solution exists. However, there is an inherent danger of confusing physical reality with the mathematical model given by the differential equation we use to represent the real problem. It may well be that our modeling is faulty, in which case the equations obtained may bear no connection with reality. Then solutions to the equations need not exist. We should also note that not all differential equations have solutions. For example, the equation

$$\left(\frac{dy}{dx}\right)^2 + 3 = 0$$

has no real-valued solutions, since $(dy/dx)^2 + 3 \geq 3$. On the other hand, the equation

$$\left(\frac{dy}{dx}\right)^2 + y^2 = 0$$

has zero as its only solution, whereas the equation

$$\frac{dy}{dx} + y = 0$$

has an infinite set of solutions $y = ce^{-x}$ for any constant c.

PROBLEMS 1.1[†]

1. The growth rate of a bacteria population is proportional to its size. Initially the population is 10,000, and after 10 days it is 25,000. What is the population size after 20 days? After 30 days?

2. Suppose that in Problem 1 the population after 10 days is 6000. What is the population after 20 days? After 30 days?

3. The population of a certain city grows 6 percent a year. If the population in 1970 is 250,000, what is the population in 1980? In 2000?

4. When the air temperature is $70°F$, an object cools from $170°F$ to $140°F$ in 0.5 hour.

    a. What is the temperature after 1 hour?

    b. When does the temperature reach $90°F$? [*Hint:* Use Newton's law of cooling.]

5. A hot coal (temperature $150°C$) is immersed in ice water (temperature $0°C$). After 30 seconds the temperature of the coal is $60°C$. Assume that the ice water is kept at $0°C$.

    a. What is the temperature of the coal after 2 minutes?

    b. When does the temperature of the coal reach $10°C$?

**6. The president and vice-president sit down for coffee. They are each served a cup of hot black coffee (at the same temperature). The president immediately adds cream to his coffee, stirs it, and waits. The vice-president waits 10 minutes and then adds the same amount of cream (which has been kept cool) to her coffee and stirs it in. Then they both drink. Assuming that the temperature of the cream is lower than that of the air, who drinks the hotter coffee? [*Hint:* Use Newton's law of cooling. It is necessary to treat each case separately and to keep track of the volumes of coffee, cream, and the coffee-cream mixture].[‡]

7. A fossilized leaf contains 70 percent of a "normal" amount of ^{14}C. How old is the fossil?

8. Forty percent of a radioactive substance disappears in 100 years.

    a. What is its half-life?

[†] To complete most of these problems, you will need to use a hand calculator with ln and e^x function keys ([INV] [ln] gives e^x on some calculators).

[‡] This is a famous old problem that keeps popping up (with an ever-changing pair of characters) in books on games and puzzles in mathematics. The problem is hard and has stymied many a mathematician. Do not get frustrated if you cannot solve it. The trick is to write everything down and to keep track of all the variables. The fact that the air is warmer than the cream is critical. It should also be noted that guessing the correct answer is fairly easy; proving that your guess is correct is what makes the problem difficult.

b. After how many years will 90 percent be gone?

9. Salt decomposes in water into sodium [Na^+] and chloride [Cl^-] ions at a rate proportional to its mass. Suppose there are 25 kilograms of salt initially and 15 kilograms after 10 hours.

a. How much salt is left after one day?

b. After how many hours is there less than 0.5 kilograms of salt left?

10. X rays are absorbed into a uniform, partially opaque body as a function not of time but of penetration distance. The rate of change of the intensity $I(x)$ of the X ray is proportional to the intensity. Here x measures the distance of penetration. The more the X ray penetrates, the lower the intensity is. The constant of proportionality is the density D of the medium being penetrated.

a. Formulate a differential equation describing this phenomenon.

b. Solve for $I(x)$ in terms of x, D, and the initial (surface) intensity $I(0)$.

11. Radioactive beryllium is sometimes used to date fossils found in deep-sea sediment. The decay of beryllium satisfies the equation

$$\frac{dA}{dt} = -\alpha A, \quad \text{where } \alpha = 1.5 \times 10^{-7},$$

where t is measured in years. What is the half-life of beryllium?

†12. In a certain medical treatment a tracer dye is injected into the pancreas to measure its function rate. A normally active pancreas secretes 4 percent of the dye each minute. A physician injects 0.3 gram of the dye, and 30 minutes later 0.1 gram remains. How much dye would remain if the pancreas were functioning normally?

13. Atmospheric pressure is a function of altitude above sea level and is given by $dP/da = \beta P$, where β is a constant. The pressure is measured in millibars (mbar). At sea level ($a = 0$), $P(0)$ is 1013.25 mbar which means that the atmosphere at sea level will support a column of mercury 1013.25 millimeters high at a standard temperature of 15°C. At an altitude of $a = 1500$ meters, the pressure is 845.6 mbar.

a. What is the pressure at $a = 4000$ meters?

b. What is the pressure at 10 kilometers?

---

† This and similar mathematical models in medicine are discussed by J. S. Rustagi, "Mathematical Models in Medicine," *International Journal of Mathematical Education in Science and Technology* 2 (1971): 193–203.

c. In California the highest and lowest points are Mount Whitney (4418 meters) and Death Valley (86 meters below sea level). What is the difference in their atmospheric pressures?

d. What is the atmospheric pressure at the top of Mount Everest (elevation 8848 meters)?

e. At what elevation is the atmospheric pressure equal to 1 mbar?

14. A bacteria population is known to grow exponentially. The following data are collected:

Number of Days	Number of Bacteria
5	936
10	2,190
20	11,986

a. What is the initial population?

b. If the present growth rate continues, what is the population after 60 days?

15. A bacteria population is declining exponentially. The following data are collected:

Number of Hours	Number of Bacteria
12	5969
24	3563
48	1269

a. What is the initial population?

b. How many bacteria are left after one week?

c. When will there be no bacteria left (i.e., when is $P(t) < 1$)?

16. A ball is thrown upward with an initial velocity v_0 meters per second from the top of a building h_0 meters high. Find how high the ball travels and determine when it hits the ground for the following choices of v_0 and h_0 (we neglect air resistance and let $g = 9.8$ m/s^2).

a. $v_0 = 49$ m/s, $h_0 = 539$ m

b. $v_0 = 14$ m/s, $h_0 = 21$ m

c. $v_0 = 21$ m/s, $h_0 = 175$ m

d. $v_0 = 7$ m/s, $h_0 = 56$ m

e. $v_0 = 7.7$ m/s, $h_0 = 42$ m

In Problems 17–20, use Newton's law of cooling to determine how long to bake a cake at the given oven temperature, assuming that it takes exactly 30 minutes to change 70°F dough into a 170°F cake in a 350°F oven.

17. 250°F   **18.** 400°F

19. 300°F   **20.** 200°F

1.2 Classification of Differential Equations and Direction Fields

It should be apparent, if only from reading the examples in the previous section, that a great variety of types of differential equations can arise in the study of familiar phenomena. It is clearly necessary (and expedient) to study, independently, more restricted classes of these equations.

The most obvious classification is based on the nature of the derivative(s) in the equation. A differential equation involving only ordinary derivatives (derivatives of functions of one variable) is called an **ordinary differential equation**, whereas one containing partial derivatives is called a **partial differential equation**. We postpone the further classification of partial differential equations until the last chapter.

DEFINITION 1: ORDER

The **order** of a differential equation is defined as the order of the highest derivative appearing in the equation.

Example 1 The following are examples of differential equations with indicated orders.

a. $dy/dx = ay$ (first order)

b. $x''(t) - 3x'(t) + x(t) = \cos t$ (second order)

c. $(y^{(4)})^{3/5} - 2y'' = \cos x$ (fourth order)

Much of this book is concerned with the solutions of differential equations. Thus we need to explain what we mean by a *solution*. First we note that any nth-order differential equation can be written in the form

$$F(x, y, y', \ldots, y^{(n)}) = 0. \qquad (1)$$

For example, $y' = f(x, y)$ can be written $y' - f(x, y) = 0$. Here $F(x, y, y') = y' - f(x, y)$. The second-order equation $y'' + a(x)y' + b(x)y = f(x)$ can be written $F(x, y, y', y'') = y'' + a(x)y' + b(x)y - f(x) = 0$.

DEFINITION 2: SOLUTION TO AN nth-ORDER DIFFERENTIAL EQUATION

A **solution** to an nth-order differential equation is a function that is n times differentiable and that satisfies the differential equation.

Symbolically, this means that a solution of differential equation (1) is a function $y(x)$ whose derivatives $y'(x), y''(x), \ldots, y^{(n)}(x)$ exist and that satisfies the equation

$$F(x, y(x), y'(x), \ldots, y^{(n)}(x)) = 0$$

for all values of the independent variable x in some interval where $F(x, y(x), y'(x), \ldots, y^{(n)}(x))$ is defined.

Example 2 The differential equation in Example 1.1.2 can be rewritten in the form $y' - \alpha y = 0$. It is easy to check that $y(x) = ke^{\alpha x}$ is a solution for all x

and any real number k:

$$\frac{d}{dx}e^{\alpha x} = e^{\alpha x}\frac{d}{dx}\alpha x = \alpha e^{\alpha x}$$

$$y'(x) - \alpha y(x) = k(e^{\alpha x})' - \alpha(ke^{\alpha x}) = \alpha k e^{\alpha x} - \alpha k e^{\alpha x} = 0.$$

As we saw in Section 1.1, we are often interested in solving a first-order differential equation

$$\frac{dy}{dx} = f(x, y)$$

subject to the condition that $y = y_0$ when $x = x_0$, or

$$y(x_0) = y_0.$$

This is an example of an **initial-value problem**. The condition $y(x_0) = y_0$ is called an **initial condition** and x_0 is called the **initial point**. More generally, we have the following:

DEFINITION 3: INITIAL-VALUE PROBLEM

An **initial-value problem** consists of a differential equation (of any order) together with a collection of initial conditions that must be satisfied by the solution of the differential equation and its derivatives at the initial point.

Example 3 The following are examples of initial-value problems.

a. $dy/dx = 2y - 3x$, $y(0) = 2$. (Here $x = 0$ is the initial point.)

b. $x''(t) + 5x'(t) + \sin tx(t) = 0$, $x(1) = 0$, $x'(1) = 7$. (Here $t = 1$ is the initial point.)

DEFINITION 4: SOLUTION TO AN INITIAL-VALUE PROBLEM

We define a **solution** to an nth-order initial-value problem as a function that is n times differentiable, satisfies the given differential equation, and satisfies the given initial conditions.

Example 4 The function $y(x) = 2e^{3x}$ is a solution of the initial-value problem

$$\frac{dy}{dx} = 3y, \qquad y(0) = 2,$$

because $y(0) = 2e^{3 \cdot 0} = 2e^0 = 2$ and

$$\frac{dy}{dx} = 2\frac{d}{dx}(e^{3x}) = 6e^{3x} = 3y.$$

DEFINITION 5: BOUNDARY VALUE PROBLEM

A **boundary value problem** consists of a differential equation and a collection of values that must be satisfied by the solution of the differential equation or its derivatives at no fewer than two different points.

Example 5 The following are examples of boundary value problems:

a. $d^2y/dx^2 + 5xy = \cos x$, $y(0) = 0$, $y'(1) = 2$.

b. $dy/dx + 5xy = 0$, $y(0)y(1) = 2$.

We postpone further consideration and discussion of boundary value problems until a later chapter.

In each of Examples 1.1.1–1.1.6, we see that once initial conditions are satisfied, the resulting initial-value problem has a unique solution (note that two initial conditions were required in Example 1.1.1, p. 1: the initial height $h(0)$ and the initial velocity $h'(0)$). It is reasonable to ask whether every initial-value problem has a unique solution. Essentially we are asking two questions:

a. Is there a solution to the problem?

b. If there is a solution, is it the only one?

As we see in the next two examples, the answer may be *no* to each question.

Example 6 The initial-value problem

$$\left(\frac{dy}{dx}\right)^2 + y^2 + 1 = 0, \qquad y(0) = 1$$

has no real-valued solutions, since the left-hand side is always positive for real-valued functions.

Example 7 The initial-value problem

$$\frac{dy}{dx} = xy^{1/3}, \qquad y(0) = 0 \qquad (2)$$

has at least two solutions in the interval $-\infty < x < \infty$. Note that the functions

$$y = 0 \quad \text{and} \quad y = \frac{1}{3\sqrt{3}}x^3$$

both satisfy the initial condition and the differential equation in (2).

Because of examples such as these, it is important to obtain theorems that guarantee the existence of a unique solution. The following result, originally developed by Liouville and proved in its most general form by Picard, is very popular because of the ease with which its hypotheses are checked.

THEOREM 1: EXISTENCE-UNIQUENESS THEOREM

Let f and $\partial f/\partial y$ be continuous in a rectangle R given by $a < x < b$, $c < y < d$ that contains the point (x_0, y_0) (see Figure 1.3). Then, in an interval $x_0 - h < x < x_0 + h$ contained in $a < x < b$, there is a unique solution $y = y(x)$ of the initial-value problem

$$\frac{dy}{dx} = f(x, y), \qquad y(x_0) = y_0.$$

Theorem 1 is called an **existence-uniqueness theorem**, because it provides criteria guaranteeing the existence of a unique solution. It requires that we

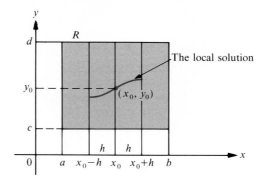

Figure 1.3

check that f and $\partial f/\partial y$ are continuous functions in a rectangle R containing the initial point (in many of our examples, f and $\partial f/\partial y$ are continuous everywhere). When these criteria hold, the theorem guarantees that a unique solution exists in some region containing the initial point (see Figure 1.3). Since the guarantee is only for a (possibly) small region around the initial point, we call such a theorem a **local** existence-uniqueness theorem. We now illustrate how to use Theorem 1, but we defer its proof to Section 2.7 and Appendix 3, where we also prove a **global** existence-uniqueness theorem for linear initial-value problems.

Example 8   Does the initial-value problem

$$\frac{dy}{dx} = x^2 + y^3, \qquad y(0) = 1$$

have a unique solution in a region around its initial point $(0,1)$?

Solution   Since $f(x, y) = x^2 + y^3$ and $\partial f/\partial y = 3y^2$ are continuous everywhere, they are certainly continuous in any rectangle R containing the initial point $(0,1)$. Hence a unique local solution to the initial-value problem exists.

Example 9   If we look again at equation (2) in Example 7 $[dy/dx = xy^{1/3}, y(0) = 0]$, we see that $\partial f/\partial y = x/3y^{2/3}$, which is not defined (and therefore not continuous) at the initial point $(0,0)$. Thus Theorem 1 does not apply to (2).

Theorem 1 is only one of many known existence-uniqueness theorems. One can relax its conditions while still retaining its conclusions. We will refer to a variety of existence-uniqueness theorems in subsequent sections. We will show in Appendix 3 that, under certain easily stated conditions that apply to a wide variety of problems, the following principle holds:

> There is a unique solution to an nth-order differential equation if the value of the unknown function and all its derivatives up to the $(n-1)$st are specified at a given point.

Although the major portion of this and the next five chapters concerns methods for solving differential equations, a basic fact remains: many differential equations cannot be solved; that is, for many differential equations it is impossible to express a solution in terms of elementary functions.[†]

There are many ways to deal with this vexing problem. One is to look for numerical solutions; in other words, rather than looking for a function $y(x)$ that solves the problem for every value of the independent variable x, we try to find an approximation to the function at one or more values. This approach will be discussed further in Section 1.3 and in Chapter 9.

Another approach is to try to describe, without solving the equation, how solutions "behave." Typical questions that can be asked in such a *qualitative* approach are

a. Do solutions grow without bound as x increases?

b. Do solutions tend to zero or some other value?

c. Do solutions oscillate between certain values?

Much of the modern research on differential equations centers on these questions. In the rest of this section, we describe one relatively simple way to obtain information about the solution to a differential equation. We will say more about these questions in Chapter 3.

Direction Fields

Consider the first-order differential equation

$$y' = f(x, y). \tag{3}$$

Equation (3) contains a great deal of information. Under certain conditions, the differential equation has a unique solution if we specify an initial condition; that is, there is a unique function $y(x)$ satisfying $y'(x) = f(x, y)$ and $y(x_0) = y_0$ for arbitrarily chosen numbers x_0 and y_0. The function $y(x)$ is a curve in the xy-plane. Even though we may not be able to find $y(x)$, *we do know the slope of y at every point on the curve*. If the solution $y(x)$ passes through the point (x, y), since $y' = f(x, y)$, we conclude the following:

> The slope of the tangent line to the curve $y(x)$ at the point (x,y) is given by $f(x, y)$.

Thus we know the direction of the solution curve $y(x)$ through any point in the xy-plane. The set of all these directions in the plane is called the **direction field** of the differential equation $y' = f(x, y)$. In many cases, we can use the direction field to sketch the solution to a differential equation without actually solving the differential equation.

Example 10 Consider the initial-value problem

$$y' = 2xy, \qquad y(0) = 1. \tag{4}$$

[†] The elementary functions are algebraic functions, exponential functions, logarithmic functions and trigonometric functions.

16 CHAPTER 1/INTRODUCTION TO DIFFERENTIAL EQUATIONS

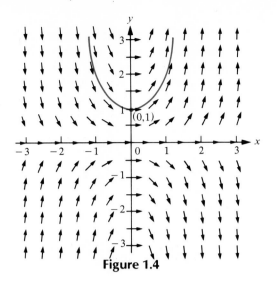

Figure 1.4

We have

$$y'(x) > 0, \quad \text{if } xy > 0$$
$$y'(x) < 0, \quad \text{if } xy < 0.$$

Thus $y'(x) > 0$ in the first and third quadrants and $y'(x) < 0$ in the second and fourth quadrants. The direction field is sketched in Figure 1.4. This figure consists of arrows with ends at a grid of points (x, y) and slopes $f(x, y) = 2xy$, where x and y are the coordinates of each grid point. Note that since x and y are positive in the first quadrant, the slopes of the tangent lines to any solution curve are positive, so the solution curves increase and become steeper as x and y become larger. Along the axes, the solution is flat (has a horizontal tangent) because the derivative y' is zero. In the second quadrant, the slopes of the tangent lines are negative, since $y' < 0$. Similar conditions apply in the third and fourth quadrants.

For the particular initial-value problem we are considering in equation (4), we know that the solution curve must satisfy the initial condition $y(0) = 1$. Thus the solution curve must pass through the point $(0, 1)$. Since that point is on the y-axis, the curve is initially flat and moves into the first quadrant with increasing values of x. As x increases, the solution curve begins to rise as $y'(x) > 0$ in the first quadrant. Hence $y > 1$ for all values of $x > 0$. On the other hand, if we allow x to be negative, the solution curve extends into the second quadrant. Since the slope is negative in this quadrant, the solution curve decreases as the curve moves to the right. Because xy becomes larger in absolute value as we move away from the y-axis, the curve becomes steeper. Plotting this information, we obtain the curve in Figure 1.5, which is also superimposed on Figure 1.4.

In this case the differential equation can be solved as follows:

$$\int \frac{dy}{y} = 2 \int x \, dx + C,$$

so

$$\ln|y| = x^2 + C,$$

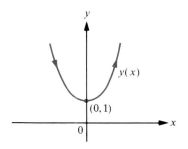

Figure 1.5

or
$$y(x) = C_1 e^{x^2}.$$

Finally, since $y(0) = 1$, it follows that $C_1 = 1$; the solution we find by using the direction field is thus

$$y = e^{x^2}.$$

Example 11 Consider the differential equation
$$y'(x) = y(y+1). \qquad (5)$$

The following facts are evident:

$$y' > 0, \quad \text{if} \quad y < -1,$$
$$y' = 0, \quad \text{if} \quad y = -1,$$
$$y' < 0, \quad \text{if} \quad -1 < y < 0,$$
$$y' = 0, \quad \text{if} \quad y = 0,$$
$$y' > 0, \quad \text{if} \quad y > 0.$$

Since there is no x-term in the right-hand side of equation (5), the direction field depends only on the values of y. The direction field, together with eight possible solutions for eight different initial values $(0, y_0)$, is given in Figure 1.6.

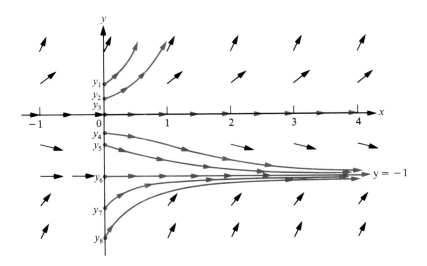

Figure 1.6

Note that the curve $y(x) = 0$ is the unique solution for the initial value $(0, 0)$, and the curve $y(x) = -1$ is the unique solution passing through the initial value $(0, -1)$. Note, too, that if $y(0) > 0$, then $y(x)$ increases without bound as x increases. On the other hand, if $y(0) < 0$, then all solutions approach the line $y(x) = -1$. This is a substantial amount of information obtained with very little work.

In Section 2.1 we will discuss a method for solving this differential equation. You should verify that

$$y = \frac{1}{Ce^{-x} - 1}. \tag{6}$$

is a solution for every constant C. Another solution is $y \equiv 0$ [this satisfies $y(0) = 0$].

If $C < 1$, then the denominator in (6) is always negative and tends to -1 as $x \to \infty$. This shows that if $y(0) < 0$, the solution curve approaches -1 as x becomes arbitrarily large. If $C > 1$, so that $y(0) > 0$, there is a positive value for x such that $Ce^{-x} = 1$ and the denominator of (6) is zero. This occurs when $x = \ln C$, and since $y(0) = 1/(C - 1)$ by (6), $C = 1/y(0) + 1 = (y(0) + 1)/y(0)$. Thus the solution curve tends to infinity as x approaches $\ln[(y(0) + 1)/y(0)]$. This confirms the shape of our rough sketches in Figure 1.6.

Of course, if we can solve the differential equation directly, we need not plot the direction field. Nevertheless, direction fields provide a quick and useful, if crude, tool for getting an idea of the shape of the solution. And if a solution is not readily obtainable, direction fields provide an important first step in analyzing the behavior of solutions.

PROBLEMS 1.2

In Problems 1–7 state the order of each of the following differential equations.

1. $y' + ay = \sin^2 x$

2. $\left(\dfrac{d^2 x}{dt^2}\right)^3 - 3x \dfrac{dx}{dt} = 4 \cos t$

3. $s'''(t) - s''(t) = 0$

4. $\dfrac{d^5 y}{dx^5} = 0$

5. $y'' + y = 0$

6. $\left(\dfrac{dx}{dt}\right)^3 = x^5$

7. $x' - x^2 = 3x'''$

In Problems 8–13 state whether each of the following differential equations is an initial-value problem or a boundary value problem.

8. $y'' + \omega^2 y = 0$, $y(0) = 0$, $y(1) = 1$ (here ω^2 is a constant)

9. $y'' + \omega^2 y = 0$, $y(0) = 0$, $y'(0) = 1$ (ω^2 constant)

10. $y'' + \omega^2 y = 0$, $y(0) = 0$, $y'(1) = 1$ (ω^2 constant)

11. $\left(\dfrac{dx}{dt}\right)^3 - 4x^2 = \sin t$, $x(0) = 3$

12. $y''' + 3y'' - (y')^2 + e^y = \sin x$,
    $y(0) = 0$, $y'(0) = 3$, $y''(0) = 5$

13. $y''' + 3y'' - (y')^2 + e^y = \sin x$,
    $y(0) = y(1) = 0$, $y'(0) = 2$

In Problems 14–18 verify that the given function or functions are solutions to the given differential equation.

14. $y'' + y = 0$; $y_1 = 2 \sin x$, $y_2 = -5 \cos x$

15. $y''' - y'' + y' - y = x$; $y_1 = e^x - x - 1$,
    $y_2 = 3 \cos x - x - 1$,
    $y_3 = \cos x + \sin x + e^x - x - 1$

16. $x^2 y'' - 2xy' + 2y = 0$; $y_1 = x$, $y_2 = x^2$,
    $y_3 = 2x - 3x^2$

17. $y'' - y = e^x$; $y_1 = \dfrac{x}{2} e^x$,
    $y_2 = \left(4 + \dfrac{x}{2}\right) e^x + 3e^{-x}$

18. $x^2 y'' + 5xy' + 4y = 0;$

$$y_1 = \frac{4 \ln x}{x^2}, \quad y_2 = \frac{-6}{x^2} \quad (x > 0)$$

19. By "guessing" that there is a solution to the equation
$$y'' - 4y' + 5y = 0$$
of the form $y = e^{ax} \cos bx$, find this solution. Try to "guess" a second solution.

20. By "guessing" that there is a solution to
$$y'' - 3y' - 4y = 0$$
of the form $y = e^{ax}$ for some constant a, find two solutions of the equation.

21. Given that $y_1(x)$ and $y_2(x)$ are two solutions of the equation in Problem 20, check to see that $y_3(x) = c_1 y_1(x) + c_2 y_2(x)$ is also a solution, where c_1 and c_2 are arbitrary constants, by substituting y_3 into the differential equation.

22. Determine $\phi(x)$ so that the functions $\sin \ln x$ and $\cos \ln x$ $(x > 0)$ are solutions of the differential equation
$$[\phi(x) y']' + \frac{y}{x} = 0.$$

23. Show that $\sin(1/x)$ and $\cos(1/x)$ are solutions of the differential equation
$$\frac{d}{dx}\left(x^2 \frac{dy}{dx} \right) + \frac{y}{x^2} = 0.$$

24. Verify that $y_1 = \sinh x$ and $y_2 = \cosh x$ are solutions of the differential equation $y'' - y = 0$. [*Hint:* $\cosh x = \frac{1}{2}(e^x + e^{-x})$; $\sinh x = \frac{1}{2}(e^x - e^{-x})$.]

25. Suppose that $\phi(x)$ is a solution of the initial-value problem $y'' + yy' = x^3$, with $y(-1) = 1$, $y'(-1) = 2$. Find $\phi''(-1)$ and $\phi'''(-1)$.

26. Let $\phi(x)$ be a solution to $y' = x^2 + y^2$ with $y(1) = 2$. Find $\phi'(1)$, $\phi''(1)$, and $\phi'''(1)$.

DIRECTION FIELDS

27. a. Plot the direction field for the differential equation
$$y' = y^{4/5}.$$
b. Plot the solution that satisfies $y(0) = 2$.
c. Plot the solution that satisfies $y(0) = -1$.

28. a. Plot the direction field for the equation
$$y' = y^{3/5}.$$
b. Plot the solution that satisfies $y(0) = 3$.
c. Plot the solution that satisfies $y(0) = -2$.

29. a. Plot the direction field for the equation
$$y' = 2y(3 - y).$$
b. Plot the solution that satisfies $y(0) = -1$.

c. Plot the solution that satisfies $y(0) = 2$.
d. Plot the solution that satisfies $y(0) = 4$.

30. a. Plot the direction field for the equation
$$y' = (y^2 - 4)(y - 4).$$
b. Plot the solution that satisfies $y(0) = -3$.
c. Plot the solution that satisfies $y(0) = -1$.
d. Plot the solution that satisfies $y(0) = 1$.
e. Plot the solution that satisfies $y(0) = 3$.
f. Plot the solution that satisfies $y(0) = 5$.

31. a. Plot the direction field for the equation
$$y' = x^2 + y^2.$$
b. Plot the solution that satisfies $y(1) = 2$.

32. a. Plot the direction field of the equation
$$y' = \frac{2xy}{1 + y^2}.$$
b. Plot the solution that satisfies $y(1) = 1$.
c. Plot the solution that satisfies $y(1) = -1$.

33. Plot the direction field of the equation
$$y' = \frac{4y - 5x}{y + x}.$$

34. a. Plot the direction field of the equation
$$y' = 1 + x + y.$$
b. Plot the solution that satisfies $y(0) = 1$.

35. a. Plot the direction field of the equation
$$y' = e^{xy} - 1.$$
b. Discuss the solution that passes through the origin.
c. Compare the solution through $(0, 1)$ to that through $(0, -1)$.

In Problems 36–39 determine the region in the xy-plane for which the existence of a unique solution through one of its points is guaranteed by the existence-uniqueness theorem (Theorem 1).

36. $y' = x - y^2$ **37.** $y' = \dfrac{x - y}{x + y}$

38. $y' = \dfrac{1}{(x - y)^2}$ **39.** $y' = \sqrt{1 - y^2}$

40. Show that the initial-value problem $y' = y^{1/5}$, $y(0) = 0$ has three solutions: $y \equiv 0$, $y = (\frac{4}{5}x)^{5/4}$, and $y = -(\frac{4}{5}x)^{5/4}$. Does this contradict the existence-uniqueness theorem? Explain.

41. Show that $y = (1 - x)^{-1}$ is a solution of the initial-value problem
$$y' = y^2, \quad y(0) = 1.$$
Where is this solution valid? [This illustrates why Theorem 1 is a local result.]

1.3 An Excursion: Numerical Solutions of Differential Equations

In Section 1.2 we discussed an elementary graphic technique for determining the solution of a differential equation. In this section we present an elementary numerical method for "solving" differential equations.

Before presenting this numerical technique, it is useful to discuss the situations in which numerical methods could or should be employed. Such methods are used frequently when other methods are not applicable. Even when other methods do apply, there may be an advantage in having a numerical solution; solutions in terms of more exotic special functions are sometimes difficult to interpret. There may also be computational advantages: the exact solution may be extremely tedious to obtain. Finally, there are situations (see, for example, Section 2.2) for which we can solve a differential equation but for which the solution is given implicitly; in these cases a numerical method can tell us more about the solution.

On the other hand, care must always be exercised in using any numerical scheme, as the accuracy of the solution depends, not only on the "correctness" of the numerical method being used, but also on the precision of the device (hand calculator or computer) used for the computations.

We assume that the initial-value problem

$$\frac{dy}{dx} = f(x, y), \qquad y(x_0) = y_0 \tag{1}$$

has a unique solution $y(x)$ (see Theorem 1.2.1, p. 13). The technique we describe below approximates this solution $y(x)$ only at a finite number of points

$$x_0, \quad x_1 = x_0 + h, \quad x_2 = x_0 + 2h, \ldots, \quad x_n = x_0 + nh,$$

where h is some (nonzero) real number. The method provides a value y_k that is an approximation of the exact value $y(x_k)$ for $k = 0, 1, \ldots, n$.

THE EULER METHOD[†]

This procedure is crude but very simple. The idea is to approximate $y(x_1) = y_1$ by assuming that $f(x, y)$ varies so little on the interval $x_0 \leq x \leq x_1$ that only a very small error is made by replacing it by the constant value $f(x_0, y_0)$. Integrating

$$\frac{dy}{dx} = f(x, y)$$

from x_0 to x_1, we obtain

$$y_1 - y_0 = y(x_1) - y(x_0)$$

$$= \int_{x_0}^{x_1} \frac{dy}{dx} dx$$

$$= \int_{x_0}^{x_1} f(x, y)\, dx$$

$$\approx f(x_0, y_0)(x_1 - x_0). \tag{2}$$

[†] See the accompanying biographical sketch.

LEONHARD EULER (1707–1783)

Leonhard Euler (pronounced "Oiler") was born in Basel, Switzerland. His father, a clergyman, planned that his son follow him into the ministry. The father was gifted at mathematics and, together with Johann Bernoulli, instructed young Leonhard in that subject as well as in theology, astronomy, physics, medicine, and several Eastern languages.

In the early eighteenth century, Catherine I of Russia, wife of Peter the Great, founded the St. Petersburg Academy. In 1727 Euler applied and was accepted for a chair in the faculty of medicine and physiology at the academy. However, Catherine died the day Euler arrived in Russia and the academy was plunged into turmoil. By 1730 Euler found himself in the chair of natural philosophy, from which he pursued his mathematical career. In 1741 he went to Berlin to head the Prussian Academy, accepting an invitation from Frederick the Great. Twenty-five years later he returned to St. Petersburg, where he died in 1783 at the age of 76.

*Leonhard Euler
(Bettmann Archive)*

Euler was the most prolific writer in the history of mathematics. He found results in virtually every branch of pure and applied mathematics. Although German was his native language, he wrote mostly in Latin, and occasionally in French. His amazing productivity did not decline even when, in 1766, he became totally blind. During his lifetime Euler published 530 books and papers. When he died, he left so many unpublished manuscripts that the St. Petersburg Academy was still publishing his work in its *Proceedings* almost half a century later. Euler enriched such diverse areas as hydraulics, celestial mechanics, lunar theory, and the theory of music, as well as mathematics.

Euler had a phenomenal memory. As a young man he memorized the entire Aeneid by Virgil (in Latin), and many years later he could still recite the entire work. He was able to solve astonishingly complex mathematical problems in his head, including, it is said, problems in astronomy that stymied Newton. The French academician François Arago once commented that Euler could calculate without effort "just as men breathe, as eagles sustain themselves in the air."

Euler wrote in a mathematical language that is largely in use today. The following symbols, among many others, were first used by him:

$f(x)$ for functional notation
e for the base of the natural logarithm
Σ for the summation sign
i to denote the imaginary unit

Euler's textbooks are models of clarity. His texts include *Introductio in analysis infinitorum* (1748), *Institutiones calculi differentialis* (1755), and the three-volume *Institutiones calculi integralis* (1768–1774). These and others of his works served as models for many of today's mathematics textbooks.

It is said that Euler did for mathematical analysis what Euclid did for geometry. It is no wonder that so many mathematicians who followed express the debt they owe him.

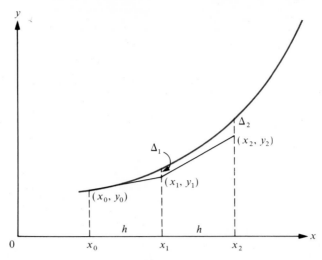

Figure 1.7

Since $h = x_1 - x_0$,

$$y_1 = y_0 + hf(x_0, y_0).$$

Repeating the process with (x_1, y_1) to obtain y_2, and so on, we obtain the **difference equation**

$$\boxed{y_{n+1} = y_n + hf(x_n, y_n),} \tag{3}$$

where $y_n = y(x_0 + nh)$, $n = 0, 1, 2, \ldots$. We solve equation (3) iteratively, that is, by first finding y_1, then using it to find y_2, and so on.

The geometric meaning of equation (3) is easily seen by considering the direction field of differential equation (1): we are simply following the tangent to the solution curve passing through (x_n, y_n) for a small horizontal distance. Looking at Figure 1.7, where the smooth curve is the unknown exact solution to the initial-value problem (1), we see how equation (3) approximates the exact solution. Since $f(x_0, y_0)$ is the slope of the exact solution at (x_0, y_0), we follow this line to the point (x_1, y_1). Some solution to the differential equation passes through this point. We follow its tangent line at this point to reach (x_2, y_2), and so on. The differences Δ_k are errors at the kth stage in the process.

Example 1 Use the Euler method to estimate $y(1)$ where $y(x)$ satisfies the initial-value problem

$$\frac{dy}{dx} = y, \quad y(0) = 1.$$

Solution From Section 1.1 we know that the solution to this problem is $y(x) = e^x$. Thus $y(1) = e^1 = e \approx 2.71828$. Let us apply the Euler method with $h = 1/5$ to this familiar problem to see if this answer is obtained.

We begin by dividing the interval $[0,1]$ into five subintervals. Then $h = 1/5 = 0.2$ and $f(x_n, y_n) = y_n$. Thus we obtain, successively,

$$y_0 = y(0) = 1$$
$$y_1 = y_0 + 0.2\, y_0 = (1 + 0.2)\, y_0 = (1.2)\, y_0 = 1.2$$
$$y_2 = y_1 + 0.2\, y_1 = (1.2)\, y_1 = (1.2)^2$$
$$y_3 = y_2 + 0.2\, y_2 = (1.2)\, y_2 = (1.2)^3$$
$$y_4 = y_3 + 0.2\, y_3 = (1.2)\, y_3 = (1.2)^4$$
$$y_5 = y_4 + 0.2\, y_4 = (1.2)\, y_4 = (1.2)^5 \approx 2.48832.$$

Thus, with $h = 0.2$, the Euler method yields $y(1) \approx 2.48832$. Since we know that $y(1) = e \approx 2.71828$, our error at $x = 1$ is given by

$$\text{error} = 2.71828 - 2.48832 = 0.22996.$$

What happens if we double the number of subintervals? If $n = 10$, then $h = 0.1$, and, computing as before, we obtain $y(1) \approx y_{10} = (1.1)^{10} \approx 2.59374$. The error is given by

$$\text{error} = 2.71828 - 2.59374 = 0.12454.$$

In general, if $h = 1/n$, then

$$y_1 = y_0 + hy_0 = (1+h)\,y_0 = 1 + h = 1 + \frac{1}{n}$$

$$y_2 = y_1\left(1 + \frac{1}{n}\right) = \left(1 + \frac{1}{n}\right)^2$$

$$y_3 = y_2\left(1 + \frac{1}{n}\right) = \left(1 + \frac{1}{n}\right)^3$$

$$\vdots$$

$$y_n = \left(1 + \frac{1}{n}\right)^n.$$

Thus

$$y(1) \approx \left(1 + \frac{1}{n}\right)^n.$$

Different values of $(1 + 1/n)^n$ are given in Table 1.1 (to five decimal places of accuracy).

The numbers in Table 1.1 should not be surprising. In fact, in elementary calculus the number e is *defined* by

$$e = \lim_{n \to \infty} \left(1 + \frac{1}{n}\right)^n.$$

Thus we have shown that $e = \lim_{n \to \infty} y_n$, where y_n is the nth iterate in Euler's method with $h = 1/n$; that is, y_n approximates $y(1)$ with better and better accuracy as n increases.

TABLE 1.1

n	$\left(1 + \dfrac{1}{n}\right)^n$
1	2
2	2.25
5	2.48832
10	2.59374
100	2.70481
1,000	2.71692
10,000	2.71815
100,000	2.71827
1,000,000	2.71828

Example 2 Find an approximate value for $y(1)$ if $y(x)$ satisfies the initial-value problem

$$\frac{dy}{dx} = y + x^2, \qquad y(0) = 1. \tag{4}$$

Use five subintervals in your approximation.

Solution Here $h = 1/n = 1/5 = 0.2$ and we wish to find $y(1)$ by approximating the solution at $x = 0.0, 0.2, 0.4, 0.6, 0.8,$ and 1.0. We see that $f(x_n, y_n) = y_n + x_n^2$, and the Euler method [equation (3)] yields

$$y_{n+1} = y_n + h \cdot f(x_n, y_n) = y_n + h(y_n + x_n^2).$$

Since $y_0 = y(0) = 1$, we obtain

$$y_1 = y_0 + h \cdot (y_0 + x_0^2) = 1 + 0.2(1 + 0^2) = 1.2$$

$$y_2 = y_1 + h \cdot (y_1 + x_1^2) = 1.2 + 0.2[1.2 + (0.2)^2] = 1.448 \approx 1.45$$

$$y_3 = y_2 + h(y_2 + x_2^2) = 1.45 + 0.2[1.45 + (0.4)^2] \approx 1.77$$

$$y_4 = y_3 + h(y_3 + x_3^2) = 1.77 + 0.2[1.77 + (0.6)^2] \approx 2.20$$

$$y_5 = y_4 + h(y_4 + x_4^2) = 2.20 + 0.2[2.20 + (0.8)^2] \approx 2.77.$$

We arrange our work as shown in Table 1.2. The value $y_5 = 2.77$, corresponding to $x_5 = 1.0$, is our approximate value for $y(1)$. Equation (4) has the exact solution $y = 3e^x - x^2 - 2x - 2$ (check this), so that $y(1) = 3e - 5 \approx 3.154$. Thus the Euler method estimate was off by about 12 percent.[†] This is not surprising, because we treated the derivative as a constant over intervals of length 0.2 units. The error that arises in this way is called **discretization error**, because the "discrete" function $f(x_n, y_n)$ is substituted for the "continuously valued" function $f(x, y)$. It is usually true that if we reduce the step size h, we can improve the accuracy of our answer, since then the "discretized" function $f(x_n, y_n)$ is closer to the true value of $f(x, y)$ over the interval $[0,1]$. This is illustrated in Figure 1.8 with $h = 0.2$ and $h = 0.1$. Indeed, carrying out similar calculations with $h = 0.1$ yields an approximation of $y(1)$ of 3.07, which is a good deal more accurate (an error of less than 3 percent).[‡]

TABLE 1.2

x_n	y_n	$f(x_n, y_n) = y_n + x_n^2$	$y_{n+1} = y_n + hf(x_n, y_n)$
0.0	1.00	1.00	1.20
0.2	1.20	1.24	1.45
0.4	1.45	1.61	1.77
0.6	1.77	2.13	2.20
0.8	2.20	2.84	2.77
1.0	2.77		

[†] $y(1) - y_5 \approx 3.15 - 2.77 = 0.38$ and $0.38/3.15 \approx 0.12 = 12\%$.

[‡] $y(1) - y_{10} \approx 3.15 - 3.07 = 0.08$ and $0.08/3.15 \approx 0.0254 \approx 2\frac{1}{2}\%$.

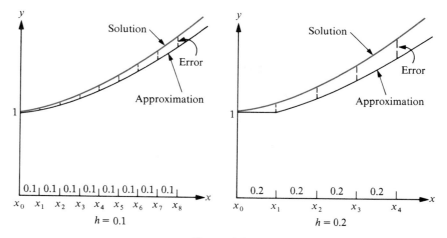

Figure 1.8

In general, reducing step size improves accuracy. However, a warning must be attached to this. Reducing step size increases the amount of work that must be done. Moreover, at every stage of the computation **round-off errors** are introduced. For example, in our calculations with $h = 0.2$, we rounded off the exact value 1.448 to the value 1.45 (correct to two decimal places). The rounded-off value was then used to calculate further values of y_n. It is not unusual for a computer solution of a more complicated differential equation to take several thousand individual computations, thus having several thousand round-off errors. In some problems the accumulated round-off error can be so large that the resulting computed solution is sufficiently inaccurate to invalidate the result. Fortunately, this usually does not occur, since round-off errors can be positive or negative and tend to cancel one another out. The statement that reducing step size improves accuracy is made under the assumption (usually true) that the average of the round-off errors is zero. In any event, it should be clear that reducing the step size, thereby increasing the number of computations, is a procedure that should be carried out carefully. In general, each problem has an optimal step size, and a smaller than optimal step size yields a greater error due to accumulated round-off errors.

The material in this section is intended as a brief introduction to the numerical solution of differential equations. A far more extensive discussion of this topic will be given in Chapter 9.

PROBLEMS 1.3

Solve Problems 1–10 by using the Euler method and the indicated value of h.

1. $\dfrac{dy}{dx} = x + y$, $y(0) = 1$. Find $y(1)$ with $h = 0.2$.

2. $\dfrac{dy}{dx} = x - y$, $y(1) = 2$. Find $y(3)$ with $h = 0.4$.

3. $\dfrac{dy}{dx} = \dfrac{x - y}{x + y}$, $y(2) = 1$. Find $y(1)$ with $h = -0.2$.

4. $\dfrac{dy}{dx} = \dfrac{y}{x} + \left(\dfrac{y}{x}\right)^2$, $y(1) = 1$. Find $y(2)$ with $h = 0.2$.

5. $\dfrac{dy}{dx} = x\sqrt{1+y^2}$, $y(1) = 0$. Find $y(3)$ with $h = 0.4$.

6. $\dfrac{dy}{dx} = x\sqrt{1-y^2}$, $y(1) = 0$. Find $y(2)$ with $h = 0.125$.

7. $\dfrac{dy}{dx} = \dfrac{y}{x} - \dfrac{5}{2}x^2 y^3$, $y(1) = \dfrac{1}{\sqrt{2}}$. Find $y(2)$ with $h = 0.125$.

8. $\dfrac{dy}{dx} = \dfrac{-y}{x} + x^2 y^2$, $y(1) = \dfrac{2}{9}$. Find $y(3)$ with $h = \tfrac{1}{3}$.

9. $\dfrac{dy}{dx} = ye^x$, $y(0) = 2$. Find $y(2)$ with $h = 0.2$.

10. $\dfrac{dy}{dx} = xe^y$, $y(0) = 0$. Find $y(1)$ with $h = 0.1$.

In Problems 11–20, use the Euler method to graph approximately the solution of the given initial-value problem by plotting the points (x_k, y_k) over the indicated range, where $x_k = x_0 + kh$.

11. $y' = xy^2 + y^3$, $y(0) = 1$, $h = 0.02$, $0 \le x \le 0.1$

12. $y' = x + \sin(\pi y)$, $y(1) = 0$, $h = 0.2$, $1 \le x \le 2$

13. $y' = x + \cos(\pi y)$, $y(0) = 0$, $h = 0.4$, $0 \le x \le 2$

14. $y' = \cos(xy)$, $y(0) = 0$, $h = \pi/4$, $0 \le x \le \pi$

15. $y' = \sin(xy)$, $y(0) = 1$, $h = \pi/4$, $0 \le x \le 2\pi$

16. $y' = \sqrt{x^2 + y^2}$, $y(0) = 1$, $h = 0.5$, $0 \le x \le 5$

17. $y' = \sqrt{y^2 - x^2}$, $y(0) = 1$, $h = 0.1$, $0 \le x \le 1$

18. $y' = \sqrt{x + y^2}$, $y(0) = 1$, $h = 0.2$, $0 \le x \le 1$

19. $y' = \sqrt{x + y^2}$, $y(1) = 2$, $h = -0.2$, $0 \le x \le 1$

20. $y' = \sqrt{x^2 + y^2}$, $y(1) = 5$, $h = -0.2$, $0 \le x \le 1$

Review Exercises for Chapter 1

In Exercises 1–4 find all solutions to the given differential equation. When an initial condition is given, find the particular solution that satisfies that condition.

1. $\dfrac{dy}{dx} = 3x$

2. $\dfrac{dx}{dt} = -2t$, $x(0) = 4$

3. $\dfrac{dx}{dt} = \dfrac{x}{2}$, $x(0) = -3$

4. $\dfrac{dy}{dx} = 100y$

5. The relative annual rate of growth of a population is 15 percent. If the initial population is 10,000, what is the population after 5 years? After 10 years?

6. In Exercise 5, how long does it take for the population to double?

7. When a cake is taken out of the oven, its temperature is 125° C. Room temperature is 23° C. The temperature of the cake is 80° C after 10 minutes.
   a. What is its temperature after 20 minutes?
   b. How long does the cake take to cool to 25° C?

8. A fossil contains 35 percent of the normal amount of ^{14}C. What is its approximate age?

9. What is the half-life of an exponentially decaying substance that loses 20 percent of its mass in one week?

10. How long does it take the substance in Exercise 9 to lose 75 percent of its mass? 95 percent of its mass?

11. a. Plot the direction field for the differential equation $y' = y^{2/5}$.
    b. Plot the solution that satisfies $y(0) = 5$.
    c. Plot the solution that satisfies $y(0) = -2$.

12. a. Plot the direction field for the differential equation $y' = -4xy/(1 + y^2)$.
    b. Plot the solution that satisfies $y(2) = -1$.
    c. Plot the solution that satisfies $y(2) = 3$.

In Exercises 13–18 use the Euler method and the given value of h to obtain an approximate solution at the indicated value of x.

13. $\dfrac{dy}{dx} = \dfrac{e^x}{y}$, $y(0) = 2$. Find $y(3)$ with $h = \tfrac{1}{2}$.

14. $\dfrac{dy}{dx} = \dfrac{e^y}{x}$, $y(1) = 0$. Find $y(\tfrac{1}{2})$ with $h = -0.1$.

15. $\dfrac{dy}{dx} = \dfrac{y}{\sqrt{1 + x^2}}$, $y(0) = 1$. Find $y(3)$ with $h = \tfrac{1}{2}$.

16. $xy\dfrac{dy}{dx} = y^2 - x^2$, $y(1) = 2$. Find $y(3)$ with $h = \tfrac{1}{2}$.

17. $\dfrac{dy}{dx} = y - xy^3$, $y(0) = 1$. Find $y(3)$ with $h = \tfrac{1}{2}$.

18. $\dfrac{dy}{dx} = \dfrac{2xy}{3x^2 - y^2}$, $y(-\tfrac{3}{8}) = -\tfrac{3}{4}$. Find $y(6)$ with $h = \tfrac{3}{8}$.

19. **a.** Apply the Euler method with $h = 0.2$ and estimate $y(1)$ for the initial-value problem

$$y' = y^3,\ y(0) = 1.$$

b. Show that $y = (1 - x)^{-1/2}$ is a local solution to the initial-value problem in part (a).

c. Explain the apparent contradiction.

2

First-Order Equations

In this chapter we provide a variety of techniques for solving first-order ordinary differential equations. In Sections 2.5 and 2.6 we give some interesting applications using these methods. For each method we assume that f and $\partial f/\partial y$ are continuous in some rectangle containing the initial point (x_0, y_0). According to Theorem 1.2.1 (p. 13) the differential equation has a unique solution in the interval $(x_0 - h, x_0 + h)$ for some number $h > 0$.

2.1 Separation of Variables

In Section 1.1 we solved the differential equation $dy/dx = \alpha y$ by separating the variables; that is, we put all the x's on one side, all the y's on the other side, and then integrated. This technique can be used to solve a variety of first-order differential equations.

Consider the differential equation

$$\frac{dy}{dx} = f(x, y) \tag{1}$$

and suppose that the function $f(x, y)$ can be factored into a product,

$$f(x, y) = g(x)h(y), \tag{2}$$

where $g(x)$ and $h(y)$ are each functions of only one variable. When this occurs, equation (1) can be solved by the method of **separation of variables**. To solve the equation, we substitute the product (2) into (1) to obtain

$$\frac{dy}{dx} = g(x)h(y),$$

or

$$\frac{1}{h(y)}\frac{dy}{dx} = g(x). \qquad (3)$$

Integrating both sides of equation (3) with respect to x, we have

$$\int \frac{1}{h(y)}\frac{dy}{dx}\,dx = \int g(x)\,dx + C,$$

and by the change of variables procedure for integrals we derive

$$\int \frac{dy}{h(y)} = \int g(x)\,dx + C.^\dagger \qquad (4)$$

If both integrals in equation (4) can be evaluated, a solution to the differential equation (1) is obtained.[‡] We illustrate this method with several examples.

Example 1 Solve the differential equation $dx/dt = t\sqrt{1-x^2}$.

Solution We have

$$\frac{dx}{\sqrt{1-x^2}} = t\,dt \quad \text{or} \quad \int \frac{dx}{\sqrt{1-x^2}} = \int t\,dt + C.$$

Integration yields

$$\sin^{-1}x = \frac{t^2}{2} + C \quad \text{or} \quad x = \sin\left(\frac{t^2}{2} + C\right).$$

There are an infinite number of solutions, one for each value of C when $|C| \leq \pi/2$. (What happens when $|C| > \pi/2$?) For certain initial conditions, such as $x(0) = 1/2$, there is a unique solution:

$$C = \sin^{-1}\frac{1}{2} = \frac{\pi}{6} \quad \text{and} \quad x(t) = \sin\left(\frac{t^2}{2} + \frac{\pi}{6}\right).$$

However, there is no solution for the initial condition $x(0) = 2$, since the sine function takes values in the interval $[-1, 1]$.

Example 2 Solve the initial-value problem

$$\frac{dy}{dx} = y(y-2), \qquad y(0) = 3.$$

[†] There is no need to use two constants of integration,

$$\int \frac{dy}{h(y)} + C_1 = \int g(x)\,dx + C_2,$$

since a single arbitrary constant $C = C_2 - C_1$ serves the same purpose.

[‡] Equation (4) may give the solution implicitly. Sometimes it is very difficult or impossible to write y explicitly as a function of x. In these cases it is necessary to use an approximation technique, like the Euler method (discussed in Section 1.3), to obtain solutions at certain values of x.

Solution Separating the variables, we have

$$\frac{dy}{y(y-2)} = dx,$$

or

$$\int \frac{dy}{y(y-2)} = x + C.$$

To integrate, we use partial fractions:

$$\frac{1}{y(y-2)} = \frac{A}{y} + \frac{B}{y-2},$$

for some numbers A and B. Cross-multiplying, we obtain

$$\frac{1}{y(y-2)} = \frac{A(y-2) + By}{y(y-2)} = \frac{(A+B)y - 2A}{y(y-2)}.$$

Equating the numerators on the left and right, we find that

$$A + B = 0,$$
$$-2A = 1,$$

so that $A = -1/2$, $B = 1/2$, and

$$\frac{1}{y(y-2)} = \frac{1}{2}\left(-\frac{1}{y} + \frac{1}{y-2}\right).$$

Thus

$$x + C = \int \frac{dy}{y(y-2)} = \frac{1}{2}\int\left(-\frac{1}{y} + \frac{1}{y-2}\right)dy = \frac{1}{2}(-\ln|y| + \ln|y-2|)$$

$$\ln\frac{b}{a} = \ln b - \ln a$$

$$= \frac{1}{2}\ln\left|\frac{y-2}{y}\right|.$$

Then

$$\ln\left|\frac{y-2}{y}\right| = 2x + C_1, \qquad C_1 = 2C$$

$$\left|\frac{y-2}{y}\right| = e^{2x + C_1} = e^{C_1}e^{2x},$$

and

$$\frac{y-2}{y} = ke^{2x}. \qquad k = \pm e^{C_1}$$

Since $y(0) = 3$, we have

$$\frac{3-2}{3} = ke^0 = k = \frac{1}{3}.$$

Finally,
$$\frac{y-2}{y} = \frac{1}{3}e^{2x},$$
$$y - 2 = \frac{1}{3}ye^{2x},$$
$$y\left(1 - \frac{1}{3}e^{2x}\right) = 2,$$
$$y = \frac{2}{1 - \frac{1}{3}e^{2x}} = \frac{6}{3 - e^{2x}}.$$

We have avoided a technical detail in this and Example 1 in order to emphasize the solution procedure used in solving a problem by separation of variables: *care should be taken to make certain that divisors are not zero*. In this case, when we divide both sides of the differential equation
$$\frac{dy}{dx} = y(y-2)$$
by $y(y-2)$, we disallow the solution y from being equal to 0 or 2. Observe that
$$y \equiv 0 \quad \text{and} \quad y \equiv 2$$
are both solutions of the differential equation (check!). Had the initial condition been $y(0) = 0$, we would have been unable to proceed beyond the equation
$$\frac{y-2}{y} = ke^{2x},$$
because division by zero is not allowed. Indeed, the constant solution $y \equiv 0$ cannot be obtained from this equation.† However it is interesting to note that the constant solution $y \equiv 2$ can be recovered by setting $k = 0$.

Example 3 **Escape Velocity** In Example 1.1.1 we studied the motion of a body falling freely subject to the gravitational force of the earth. In that example we assumed that the distance the body fell was small in comparison to the radius R of the earth.‡ However, if we wish to study the equation of motion for a communications satellite or an interplanetary vehicle, the distance r of the object from the center of the earth may be considerably larger than R. In such a case the approximation we made in obtaining equation (1.1.3) (p. 2) is no longer valid. Returning to equation (1.1.2),
$$\frac{d^2r}{dt^2} = -g\frac{R^2}{r^2}, \tag{5}$$
and setting $v = dr/dt$, we see by the chain rule that
$$\frac{d^2r}{dt^2} = \frac{dv}{dt} = \frac{dv}{dr}\frac{dr}{dt} = \frac{v\,dv}{dr}. \tag{6}$$

---

† The $y \equiv 0$ solution can be obtained by replacing the arbitrary constant k by $1/C$, so that $y/(y-2) = Ce^{-2x}$, and setting $C = 0$ yields $y \equiv 0$.
‡ The radius of the earth is approximately 6378 kilometers (3963 miles) at the equator and 6357 kilometers (3950 miles) at the poles.

Hence equation (5) can be rewritten as

$$\frac{v\,dv}{dr} = -g\frac{R^2}{r^2},$$

where g (≈ 9.81 m/s^2) and R are constant. Separating variables and integrating, we have

$$\int v\,dv = -gR^2 \int \frac{dr}{r^2} + C,$$

or

$$\tfrac{1}{2}v^2 = \frac{gR^2}{r} + C.$$

Assuming that the object is at the surface of the earth when $t = 0$, we get

$$\tfrac{1}{2}v(0)^2 = g\frac{R^2}{R} + C,$$

or

$$C = \tfrac{1}{2}v(0)^2 - gR.$$

Thus

$$v^2 = 2g\frac{R^2}{r} + v(0)^2 - 2gR. \tag{7}$$

For the object to escape the gravitational force of the earth, it is necessary that $v > 0$ for all time t. If we select $v(0) = \sqrt{2gR}$, the last two terms in equation (7) cancel, so that $v^2 > 0$ for all r. Observe that any smaller choice for $v(0)$ allows the right side of equation (7) to be zero for some sufficiently large value of r. Thus $v(0) = \sqrt{2gR} \approx 11.2$ kilometers per second is the initial velocity an object needs to escape the gravitational attraction of the earth. This is called the **escape velocity**.

The substitution we used in equation (6) can always be used to reduce an equation involving a second derivative to an equation involving only first derivatives, provided that the independent variable does not appear explicitly in the equation.

Example 4 **Reentry into Atmosphere** One of the problems facing space vehicles is their reentry into the earth's atmosphere. For simplicity, assume that gravity has a negligible effect in determining the maximum deceleration during reentry of a vehicle heading toward (a flat) earth at a constant angle α and speed V. Let $s(t)$ denote the distance to impact of the reentry vehicle (see Figure 2.1). Then ds/dt is its velocity and d^2s/dt^2 its deceleration. Assume that the density of the earth's atmosphere at height z is

$$\rho(z) = \rho_0 e^{-z/k_0},$$

where ρ_0 and k_0 are constants, and that the **drag force** of the atmosphere is proportional to the product of the air density and the square of the velocity of the vehicle. By Newton's second law of motion,

$$m\frac{d^2s}{dt^2} = \beta\rho_0 e^{-z/k_0}\left(\frac{ds}{dt}\right)^2,$$

where m is the mass of the vehicle and β is the constant of proportionality.

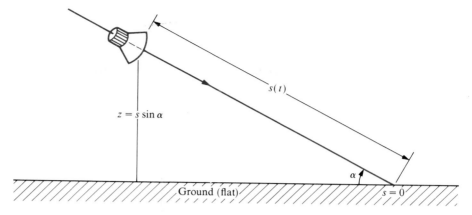

Figure 2.1
Reentry into atmosphere.

Dividing both sides by m and substituting $N = \beta \rho_0/m$ and $k = k_0/\sin \alpha$, we get the equation of motion $[-z/k_0 = -s \sin \alpha/k_0 = -s/(k_0/\sin \alpha) = -s/k]$

$$\frac{d^2s}{dt^2} = Ne^{-s/k}\left(\frac{ds}{dt}\right)^2. \tag{8}$$

Letting $v = ds/dt$ and using the same technique as in equation (6), we have

$$\frac{d^2s}{dt^2} = \frac{dv}{dt} = \frac{dv}{ds}\frac{ds}{dt} = v\frac{dv}{ds},$$

or

$$v\frac{dv}{ds} = Ne^{-s/k}v^2. \tag{9}$$

Separating variables, we get

$$\int \frac{dv}{v} = \int Ne^{-s/k}\,ds + C_1,$$

or

$$\ln|v| = -kNe^{-s/k} + C_1,$$

so

$$v = Ce^{-kNe^{-s/k}}.$$

The arbitrary constant C can now be determined by noting that the vehicle is traveling at velocity V when it is very far from earth. Thus, letting s tend to ∞, we have

$$V = Ce^0 = C,$$

or

$$\frac{ds}{dt} = v(s) = Ve^{-kNe^{-s/k}}. \tag{10}$$

Combining equations (8) and (10), we get

$$\frac{d^2s}{dt^2} = Ne^{-s/k}(Ve^{-kNe^{-s/k}})^2. \tag{11}$$

To find the *maximum* deceleration, we must determine the value of s that maximizes the right side $f(s)$ of equation (11). This can be found by differentiating the right side of equation (11) with respect to s by logarithmic differentiation and using the first derivative test of calculus:

$$f'(s) = f(s)\left(-\frac{1}{k} + 2Ne^{-s/k}\right) = 0. \tag{12}$$

Solving for s, we get $e^{-s/k} = 1/2kN$, or $s = k\ln(2kN)$. This value is easily seen to be a maximum, since $f > 0$ and the term in parentheses in (12) is decreasing. Substituting $s_{\max} = k\ln(2kN)$ into equation (11), we have

$$\left(\frac{d^2s}{dt^2}\right)\bigg|_{s_{\max}} = Ne^{-\ln 2kN}\left(Ve^{-kNe^{-\ln 2kN}}\right)^2$$

$$= \frac{N}{2kN}\left(Ve^{-(kN/2kN)}\right)^2$$

$$\overset{k = k_0/\sin\alpha}{= \frac{V^2}{2ek} \;\downarrow\; = \frac{V^2\sin\alpha}{2ek_0},}$$

which is independent of the drag coefficient $N = \beta\rho_0/m$.

Example 5 Let $P(t)$ denote the population of a species at time t. The **growth rate** per individual of the population is defined as the growth of the population divided by the size of the population; for example, if the birth rate is 3.2 per hundred and the death rate is 1.8 per hundred, then the growth rate is $3.2 - 1.8 = 1.4$ per hundred $= 1.4/100 = 0.014$. We then write $dP/dt = 0.014P$.

Suppose that in a given population the average birth rate is a positive constant β. It is reasonable to assume that the average death rate is proportional to the number of individuals in the population. Greater populations mean greater crowding and more competition for food and territory. We call this constant of proportionality δ (which is greater than zero). Since dP/dt is the growth rate of the population, the growth rate per individual of the population is

$$\frac{1}{P}\frac{dP}{dt}.$$

The differential equation that governs the growth of this population is

$$\frac{1}{P}\frac{dP}{dt} = \beta - \delta P.$$

Multiplying both sides of this equation by P, we have

$$\frac{dP}{dt} = P(\beta - \delta P), \tag{13}$$

which is called the **logistic equation**.[†] The growth shown by this equation is called **logistic growth**.

While equation (13) can be solved by separation of variables, it is useful to consider its direction field. Assuming that the constants β and δ are positive,

---

[†] The logistic equation is sometimes referred to as **Verhulst's equation** in honor of P. F. Verhulst (1804–1849), a Belgian mathematician who proposed this model for human population growth in 1838.

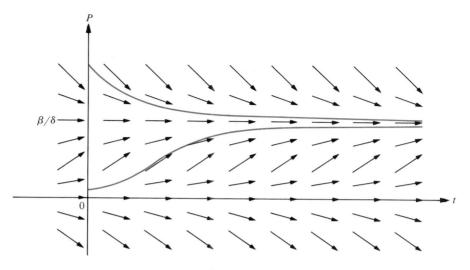

Figure 2.2
Direction field for $P' = P(\beta - \delta P)$.

the following facts are evident:

$$\begin{aligned}
P' &< 0, & &\text{if } P < 0, \\
P' &= 0, & &\text{if } P = 0, \\
P' &> 0, & &\text{if } 0 < P < \beta/\delta, \\
P' &= 0, & &\text{if } P = \beta/\delta, \\
P' &< 0, & &\text{if } P > \beta/\delta.
\end{aligned}$$

Drawing arrows at the points (t, P) to indicate the slope of the tangent line of the solution through that point, we obtain the direction field shown in Figure 2.2. In particular, any solution for which $P(0) > 0$ must approach the value of β/δ, called the **carrying capacity** of the environment. We now obtain a formula for these solutions and show that this qualitative behavior occurs.

Separating the variables, we have

$$\int \frac{dP}{P(\beta - \delta P)} = \int dt + C. \tag{14}$$

Using partial fractions, it is easy to verify that

$$\frac{1}{P(\beta - \delta P)} = \frac{1}{\beta P} + \frac{\delta}{\beta(\beta - \delta P)}.$$

Substituting the right-hand side of this equation into equation (14) and integrating, we obtain

$$\frac{1}{\beta} \ln|P| - \frac{1}{\beta} \ln|\beta - \delta P| = t + C,$$

or

$$\ln \left| \frac{P}{\beta - \delta P} \right|^{1/\beta} = t + C. \tag{15}$$

Exponentiating both sides of this equation and denoting the arbitrary constant

$\pm e^{\beta C}$ by C_1, we have

$$\frac{P}{\beta - \delta P} = C_1 e^{\beta t}. \tag{16}$$

[**Warning** We will frequently make such changes of constant without further notice.] Setting $t = 0$, we find that

$$\frac{P(0)}{\beta - \delta P(0)} = C_1,$$

and substituting this value of C_1 into equation (16) yields

$$\frac{P(t)}{\beta - \delta P(t)} = \frac{P(0)}{\beta - \delta P(0)} e^{\beta t}.$$

Cross-multiplying and solving for $P(t)$, we obtain

$$P(t)[\beta - \delta P(0)] = P(0)[\beta - \delta P(t)] e^{\beta t},$$

or

$$\beta P(t) - \delta P(t)P(0) = \beta P(0) e^{\beta t} - \delta P(0) P(t) e^{\beta t}.$$

Thus

$$P(t)[\beta - \delta P(0) + \delta P(0) e^{\beta t}] = \beta P(0) e^{\beta t},$$

and dividing top and bottom by $P(0)e^{\beta t}$, we have

$$P(t) = \frac{\beta P(0) e^{\beta t}}{\beta - \delta P(0) + \delta P(0) e^{\beta t}}$$

$$= \frac{\beta}{\delta + \left(\dfrac{\beta}{P(0)} - \delta\right) e^{-\beta t}}. \tag{17}$$

Observe that since $\beta > 0$, the term $e^{-\beta t}$ approaches zero as t increases. Thus the population approaches a limiting value of β/δ beyond which it cannot increase, since setting $P = \beta/\delta$ in equation (13) yields $dP/dt = 0$.

Warning Do not do the following:

There is a common error that students make when they try to separate variables. Consider the following problem, *done incorrectly*:

$$\frac{dy}{dx} = y + x$$

$$dy = y \, dx + x \, dx$$

$$\int dy = \int y \, dx + \int x \, dx$$

$$y = \frac{y^2}{2} + \frac{x^2}{2} + C$$

What is wrong? *It is not correct* that

$$\int y \, dx = \frac{y^2}{2},$$

because y is a function of x. For example, if $y = \cos x$, then
$$\int y\,dx = \int \cos x\,dx = \sin x + C \neq \frac{y^2}{2} + C = \frac{\cos^2 x}{2} + C.$$
Of course
$$\int y\,dy = \frac{y^2}{2} + C.$$
So, be careful. (Using the techniques of Section 2.3, we can show that the solutions to $dy/dx = y + x$ are $y = -x - 1 + Ce^x$ for every real number C.)

A Perspective

WHEN IS A DIFFERENTIAL EQUATION SEPARABLE?

In this section we saw how to solve a first-order differential equation when the equation was separable. However, it is not always clear that an equation is separable. For example, it is obvious that $f(x, y) = e^x \cos y$ is separable, but it is not so obvious that $f(x, y) = 2x^2 + y - x^2 y + xy - 2x - 2$ is separable.[†] In this perspective, we give conditions that ensure that an equation is separable.[‡]

THEOREM 1

Suppose that $f(x, y) = g(x)h(y)$, where both g and h are differentiable. Then

$$\boxed{f(x,y)f_{xy}(x,y) = f_x(x,y)f_y(x,y).} \tag{18}$$

Here the subscripts denote partial derivatives.

Proof Observe that
$$f_x(x, y) = g'(x)h(y)$$
$$f_y(x, y) = g(x)h'(y)$$
$$f_{xy}(x, y) = g'(x)h'(y).$$

Hence
$$f(x,y)f_{xy}(x,y) = g(x)h(y)g'(x)h'(y) = [g'(x)h(y)][g(x)h'(y)]$$
$$= f_x(x,y)f_y(x,y). \quad \blacksquare$$

It turns out that, under further conditions, if equation (18) holds, then $f(x, y)$ is separable. For what follows, D denotes an open disk in the xy-plane; that is, $D = \{(x, y): (x - a)^2 + (y - b)^2 < r^2\}$, where a, b, and r are real numbers and $r > 0$.

THEOREM 2

Suppose that in D, f, f_x, f_y, and f_{xy} exist and are continuous, $f(x, y) \neq 0$, and equation (18) holds. Then there are continuously differentiable functions $g(x)$ and $h(y)$ such that, for every $(x, y) \in D$,

$$f(x, y) = g(x)h(y). \tag{19}$$

[†] $2x^2 + y - x^2 y + xy - 2x - 2 = (1 + x - x^2)(y - 2)$.

[‡] The results here are based on the paper "When is an Ordinary Differential Equation Separable?" by David Scott in *American Mathematical Monthly* 92 (1985): 422–423.

Proof Since $f(x, y) \neq 0$ and f is continuous on D, f has the same sign on D. Assume $f(x, y) > 0$ for $(x, y) \in D$. A similar proof works if $f(x, y) < 0$ if f is replaced by $-f$. Now, from the quotient rule of differentiation,

equation (18)
$$\frac{\partial}{\partial y} \frac{f_x(x, y)}{f(x, y)} = \frac{f(x, y) f_{xy}(x, y) - f_x(x, y) f_y(x, y)}{f^2(x, y)} = 0.$$

When the partial derivative with respect to y of a function of x and y is zero, then that function must be a function of x only. Thus there is a function $\alpha(x)$ such that

$$\frac{f_x(x, y)}{f(x, y)} = \alpha(x).$$

Also, since $f(x, y) > 0$, $\ln f(x, y)$ is defined and

$$\frac{\partial}{\partial x} \ln f(x, y) = \frac{f_x(x, y)}{f(x, y)} = \alpha(x).$$

The function $\alpha(x)$ is continuous in D because it is the quotient of continuous functions and the function in the denominator is nonzero. Let $\beta(x) = \int \alpha(x)\, dx$. Then

$$\ln f(x, y) = \int \left[\frac{\partial}{\partial x} \ln f(x, y) \right] dx = \int \alpha(x)\, dx = \beta(x) + \gamma(y),$$

where γ is a function of y only. (The partial derivative with respect to x of a function of y only is zero. Thus $\gamma(y)$ represents the most general constant of integration.) Finally, let $g(x) = e^{\beta(x)}$ and $h(y) = e^{\gamma(y)}$. Then

$$f(x, y) = e^{\ln f(x, y)} = e^{\beta(x) + \gamma(y)} = e^{\beta(x)} e^{\gamma(y)} = g(x) h(y). \quad \blacksquare$$

Example 1 Let $f(x, y) = 2x^2 + y - x^2 y + xy - 2x - 2$. Then

$$f_x(x, y) = 4x - 2xy + y - 2$$
$$f_y(x, y) = 1 - x^2 + x$$
$$f_{xy}(x, y) = -2x + 1$$
$$f(x, y) f_{xy}(x, y) = (2x^2 + y - x^2 y + xy - 2x - 2)(-2x + 1)$$
$$= -4x^3 - xy + 2x^3 y - 3x^2 y + 6x^2 + 2x + y - 2$$

and
$$f_x(x, y) f_y(x, y) = (4x - 2xy + y - 2)(1 - x^2 + x)$$
$$= 2x - xy + y - 2 - 4x^3 + 2x^3 y - 3x^2 y + 6x^2.$$

Since the last two expressions are equal, we conclude by Theorem 2 that $f(x, y)$ is separable.

Example 2 Let $f(x, y) = 1 + xy$. Then

$$f_x(x, y) = y$$
$$f_y(x, y) = x$$
$$f_{xy}(x, y) = 1$$
$$f(x, y) f_{xy}(x, y) = 1 + xy$$

and
$$f_x(x, y) f_y(x, y) = xy.$$

PROBLEMS 2.1

In Problems 1–24 find the general solution by separating variables. If a specific condition is given, find the particular solution that satisfies that condition.

1. $\dfrac{dy}{dx} = \dfrac{e^x}{2y}$

2. $xy' = 3y,\ y(2) = 5$

3. $\dfrac{dy}{dx} = \dfrac{e^y x}{e^y + x^2 e^y}$

4. $\dfrac{dx}{dy} = x \cos y,\ x\left(\dfrac{\pi}{2}\right) = 1$

5. $\dfrac{dz}{dr} = r^2(1 + z^2)$

6. $\dfrac{dy}{dx} + y = y(xe^{x^2} + 1),\ y(0) = 1$

7. $\dfrac{dP}{dQ} = P(\cos Q + \sin Q)$

8. $\dfrac{dy}{dx} = y^2(1 + x^2),\ y(0) = 1$

9. $\dfrac{ds}{dt} + 2s = st^2,\ s(0) = 1$

10. $\dfrac{dy}{dx} = \sqrt{1 - y^2}$

11. $(1 + x)\dfrac{dy}{dx} = -3y,\ y(6) = 7$

12. $\dfrac{dx}{dt} + (\cos t)e^x = 0$

13. $\cot x \dfrac{dy}{dx} + y + 3 = 0$

14. $\dfrac{dx}{dt} = x(1 - \sin t),\ x(0) = 1$

15. $\dfrac{dy}{dx} + \sqrt{\dfrac{1 - y^2}{1 - x^2}} = 0$

16. $(\tan y)\dfrac{dy}{dx} - \tan x = 0,\ y(0) = 0$

17. $x^2 \dfrac{dy}{dx} + y^2 = 0,\ y(1) = 3$

18. $\dfrac{dy}{dx} = \dfrac{y^3 + 2y}{x^2 + 3x},\ y(1) = 1$

19. $e^x\left(\dfrac{dx}{dt} + 1\right) = 1,\ x(0) = 1$

20. $\dfrac{ds}{dr} = \dfrac{s^2 + s - 2}{r^2 - 2r - 8},\ s(0) = 0$

21. $yy' = e^x$

22. $y' + y = y(xe^x + 1)$

23. $\dfrac{dy}{dx} = \dfrac{x^2 - xy - x + y}{xy - y^2}$

24. $\dfrac{dy}{dx} = \dfrac{x}{y} - \dfrac{x}{1 + y},\ y(0) = 1$

25. The table below shows data for the growth of yeast in a culture. Use equation (17), with $\beta = 0.55$ and $\delta = 8.3 \times 10^{-4}$, to calculate the predicted growth, and find the percentage error between observed and predicted values at $t = 0$, 9, and 18 hours.

Time in hours	Yeast biomass	Time in hours	Yeast biomass
0	9.6	10	513.3
1	18.3	11	559.7
2	29.0	12	594.8
3	47.2	13	629.4
4	71.1	14	640.8
5	119.1	15	651.1
6	174.6	16	655.9
7	257.3	17	659.6
8	350.7	18	661.8
9	441.0		

Data from R. Pearl, "The Growth of Population," *Quarterly Review of Biology* 2 (1927): 532–548.

26. **Obsidian dating**[†] is a technique used by archaeologists which allows the dating of certain artifacts well beyond the reliable 40,000-year range of carbon dating. Obsidian, a glassy volcanic rock, absorbs water from the atmosphere. The water forms a hydration layer—a compound of water molecules and obsidian molecules. The depth of the hydration layer $x(t)$ beneath the surface of an obsidian artifact is a function of the time t that has elapsed since the manufacture of that artifact. The velocity at

---

[†] We wish to thank H. W. Vayo for bringing to our attention the article by I. Friedman and R. L. Smith, "A New Dating Method Using Obsidian," *American Antiquity* 25 (1960): 476–522.

which the hydration layer grows is inversely proportional to its depth. Find the depth of the layer for all time t.

27. A rocket is launched from an initial position (x_0, y_0) with an initial speed v_0 and with an angle θ $(0 \leq \theta \leq \pi/2)$. Find its horizontal and vertical coordinates $x(t)$ and $y(t)$ as functions of time. Assume that there is no air resistance, and that the force of gravity g is constant.

28. The half-life of a radioactive substance is defined as the time required to decompose 50 percent of the substance. If $r(t)$ denotes the amount of the radioactive substance present after t years, $r(0) = r_0$, and the half-life is H years, what is a differential equation for $r(t)$ taking all side conditions into account?

29. A bacteria population is known to double every 3 hours. If the population initially consists of 1000 bacteria, how long does it take for the population to reach 10,000?

In Problems 30–32 assume that the population of the given country is growing at a rate proportional to the size of the population. Calculate each country's population in the year 2000 A.D. given the two census figures shown.

30. Australia: 1968 census, 12,100,000; 1973 census, 13,268,000

31. Colombia: 1968 census, 19,825,000; 1973 census, 23,210,000

32. India: 1968 census, 523,893,000; 1973 census, 574,220,000

33. Assume India has resources that will provide only enough food for 750,000,000 humans. Using the information given in Problem 32, find India's population in the year 2000 A.D. [*Hint:* Use the logistic equation.]

34. The economist Vilfredo Pareto (1848–1923) discovered that the rate of decrease of the number of people y in a stable economy having an income of at least x dollars is directly proportional to the number of such people and inversely proportional to their income. Obtain an expression (**Pareto's law**) for y in terms of x.

*35. A pond is shaped like a cone, with radius r and depth d. Water flows into the pond at a constant rate i and is lost through evaporation at a rate proportional to the surface area.

  a. Show that the volume $V(t)$ of water at time t satisfies the differential equation
  $$V' = i - k\pi \left(\frac{3rV}{\pi d}\right)^{2/3}.$$
  b. Solve this equation.
  c. What condition must be satisfied so that the pond will not overflow?

*36. A large open cistern filled with water has the shape of a hemisphere with radius 25 feet. The bowl has a circular hole of radius 1 foot in the bottom. By **Torricelli's law**,[†] water flows out of the hole with the same speed it would attain in falling freely from the level of the water to the hole. How long does it take for all the water to flow from the cistern?

*37. In Problem 37, find the shape of the cistern that ensures that the water level drops at a constant rate.[‡]

**38. On a certain day it begins to snow early in the morning, and the snow continues to fall at a constant rate. The velocity at which a snowplow is able to clear a road is inversely proportional to the height of the accumulated snow. The snowplow starts at 11 A.M. and clears four miles by 2 P.M. By 5 P.M. it clears another two miles. When did it start snowing?[§]

**39. (Snowplow chase).[‖] If a second snowplow starts at noon along the same path as the snowplow in Problem 38, when does it catch up to the first snowplow?

**40. (Snowplow collision).[#] Suppose three identical snowplows start clearing the same road at 10 A.M., 11 A.M., and noon. If all three collide sometime after noon, when did it start snowing?

[†] Evangelista Torricelli (1608–1647) was an Italian physicist.

[‡] The ancient Egyptians (ca. 1380 B.C.) used water clocks based on this principle to tell time.

[§] Based on problem E275 of the Otto Dunkel Memorial Problem Book, *American Mathematical Monthly* 64 (1957): 54.

[‖] This problem was first proposed by Fred Wan, *Applied Mathematics Notes*, (January 1975): 6–11.

[#] This problem was first proposed by M. S. Klamkin, *American Mathematical Monthly* 59 (1952): 42 (problem E963).

2.2 Exact Equations and Integrating Factors

The **total differential** dg of a function of two variables $g(x, y)$ is defined by

$$dg = \frac{\partial g}{\partial x} dx + \frac{\partial g}{\partial y} dy.$$

Example 1 Let $g(x, y) = x^2 y^3 + e^{4x} \sin y$. Compute the total differential dg.

Solution

$$\frac{\partial g}{\partial x} = 2xy^3 + 4e^{4x} \sin y,$$

and

$$\frac{\partial g}{\partial y} = 3x^2 y^2 + e^{4x} \cos y.$$

Hence

$$dg = (2xy^3 + 4e^{4x} \sin y) \, dx + (3x^2 y^2 + e^{4x} \cos y) \, dy.$$

We now use partial derivatives to solve ordinary differential equations. Suppose that we take the total differential of the equation $g(x, y) = c$:

$$dg = \frac{\partial g}{\partial x} dx + \frac{\partial g}{\partial y} dy = 0. \qquad (1)$$

For example, the equation $xy = c$ has the total differential $y \, dx + x \, dy = 0$, which may be rewritten as the differential equation $y' = -y/x$. Reversing the situation, suppose that we start with the differential equation

$$M(x, y) \, dx + N(x, y) \, dy = 0. \qquad (2)$$

If we can find a function $g(x, y)$ such that

$$\frac{\partial g}{\partial x} = M \quad \text{and} \quad \frac{\partial g}{\partial y} = N,$$

then equation (2) becomes $dg = 0$, so that $g(x, y) = c$ is the general solution of equation (2). In this case $M \, dx + N \, dy$ is said to be an **exact differential**, and equation (2) is called an **exact differential equation**.

It is very easy to determine whether a differential equation is exact by using the **cross-derivative test**: The equation $M(x, y) \, dx + N(x, y) \, dy = 0$ is exact if and only if

$$\boxed{\frac{\partial M}{\partial y} = \frac{\partial N}{\partial x}.} \qquad (3)$$

If $M \, dx + N \, dy$ is exact, then we can solve the differential equation (2) by finding the function g given above. The procedure for doing this is illustrated in Example 2.

Example 2 Solve the equation

$$(1 - \sin x \tan y) \, dx + (\cos x \sec^2 y) \, dy = 0. \qquad (4)$$

Solution Letting $M(x, y) = 1 - \sin x \tan y$ and $N(x, y) = \cos x \sec^2 y$, we have

$$\frac{\partial M}{\partial y} = -\sin x \sec^2 y = \frac{\partial N}{\partial x},$$

so the equation is exact. We now seek a function g of two variables such that $\partial g/\partial x = M$ and $\partial g/\partial y = N$. But if $\partial g/\partial x = M$, then

$$g(x, y) = \int M\, dx = \int (1 - \sin x \tan y)\, dx$$
$$= x + \cos x \tan y + h(y). \tag{5}$$

The *constant of integration* $h(y)$ occurring in (5) is an arbitrary function of y since we must introduce the most general term that vanishes under partial differentiation with respect to x. But

$$\cos x \sec^2 y = N(x, y) = \frac{\partial g}{\partial y} = \underbrace{\cos x \sec^2 y + h'(y)}_{\text{differentiating (5) with respect to } y}.$$

This means that $h'(y) = 0$, so $h(y) = k$, a constant. Thus the general solution to (5) is

$$g(x, y) = x + \cos x \tan y + k = C, \qquad \text{(another constant)}$$

or

$$x + \cos x \tan y = C_1.$$

MULTIPLYING BY AN INTEGRATING FACTOR TO MAKE AN EQUATION EXACT

It should be apparent that exact equations are comparatively rare, since the condition in equation (3) requires a precise balance of the functions M and N. For example,

$$(3x + 2y)\, dx + x\, dy = 0$$

is not exact. However, if we multiply the equation by x, the new equation

$$(3x^2 + 2xy)\, dx + x^2\, dy = 0$$

is exact. The question we now must ask is, If

$$M(x, y)\, dx + N(x, y)\, dy = 0 \tag{6}$$

Integrating Factor is not exact, under what conditions can we multiply both sides of (6) by an **integrating factor** $\mu(x, y)$ so that

$$\mu M\, dx + \mu N\, dy = 0$$

is exact? The surprising answer is: Whenever equation (6) has a general solution $g(x, y) = c$. To see this, we solve (6) for dy/dx: By the chain rule,

$$dg = \frac{\partial g}{\partial x}\, dx + \frac{\partial g}{\partial y}\, dy = \mu M\, dx + \mu N\, dy = 0.$$

Thus

$$\frac{dy}{dx} = -\frac{M}{N} = -\frac{\partial g/\partial x}{\partial g/\partial y},$$

from which it follows that

$$\frac{\partial g/\partial x}{M} = \frac{\partial g/\partial y}{N}.$$

Denote either side of the equation above by $\mu(x, y)$. Then

$$\frac{\partial g}{\partial x} = \mu M, \qquad \frac{\partial g}{\partial y} = \mu N, \tag{7}$$

and (6) has at least one integrating factor μ. However, finding integrating factors is in general very difficult. Here is one procedure that is sometimes successful. Since equation (7) indicates that $\mu M\, dx + \mu N\, dy = 0$ is exact, by equation (3) we have

$$\mu \frac{\partial M}{\partial y} + M \frac{\partial \mu}{\partial y} = \frac{\partial}{\partial y}(\mu M) \stackrel{(3)}{=} \frac{\partial}{\partial x}(\mu N) = \mu \frac{\partial N}{\partial x} + N \frac{\partial \mu}{dx};$$

therefore

$$\frac{1}{\mu}\left(N\frac{\partial \mu}{\partial x} - M\frac{\partial \mu}{\partial y}\right) = \frac{\partial M}{\partial y} - \frac{\partial N}{\partial x}. \tag{8}$$

In case the integrating factor μ depends only on x, equation (8) becomes

$$\frac{1}{\mu}\frac{d\mu}{dx} = \frac{\partial M/\partial y - \partial N/\partial x}{N} = k(x, y). \tag{9}$$

Since the left-hand side of equation (9) consists only of functions of x, k *must* also be a function of x. If this is indeed true, then μ can be found by separating the variables: $\mu(x) = e^{\int k(x)\,dx}$. A similar result holds if μ is a function of y alone, in which case

$$K = \frac{\partial M/\partial y - \partial N/\partial x}{-M}$$

is also a function of y. In this case, $\mu(y) = e^{\int K(y)\,dy}$ is the integrating factor.

Example 3 Solve the equation

$$(3x^2 - y^2)\, dy - 2xy\, dx = 0.$$

Solution In this problem, $M = -2xy$ and $N = 3x^2 - y^2$, so

$$\frac{\partial M}{\partial y} = -2x \quad \text{and} \quad \frac{\partial N}{\partial x} = 6x.$$

Then

$$K = \frac{\partial M/\partial y - \partial N/\partial x}{-M} = \frac{-4}{y}$$

and

$$\mu = e^{-4\int y^{-1}\,dy} = e^{-4\ln y} = y^{-4}.$$

Multiplying the differential equations by y^{-4}, we obtain the exact equation

$$-\frac{2x}{y^3}\, dx + \left(\frac{3x^2 - y^2}{y^4}\right) dy = 0.$$

Integrating $M = -2x/y^3$ with respect to x and $N = (3x^2 - y^2)/y^4$ with respect to y, we get

$$-\int \frac{2x}{y^3} dx + h(y) = \int \frac{3x^2 - y^2}{y^4} dy + k(x),$$

or

$$-\frac{x^2}{y^3} + h(y) = -\frac{x^2}{y^3} + \frac{1}{y} + k(x).$$

Then $k(x) = 0$, $h(y) = 1/y$, and we obtain the general solution

$$g(x, y) = \frac{1}{y} - \frac{x^2}{y^3} = c,$$

or

$$cy^3 - y^2 + x^2 = 0.$$

A Perspective

SINGULAR SOLUTIONS

In describing the solution of a first-order differential equation, we have used the term **general solution** for a *family* of solutions of the differential equation containing one arbitrary constant. The term **particular solution** has been used for a solution that is free of arbitrary constants, usually as the result of requiring that the solution satisfy some initial condition. In most of the examples that we have considered, the particular solution was obtained by choosing a specific value for the arbitrary constant in the general solution.

Sometimes a differential equation has a particular solution that cannot be obtained by selecting a specific value for the arbitrary constant in the general solution. Consider the initial-value problem

$$\frac{dy}{dx} = xy^{1/3}, \qquad y(0) = 0. \tag{10}$$

Separating variables and integrating, we have

$$\int y^{-1/3} dy = \int x \, dx + C$$

and

$$\frac{3}{2} y^{2/3} = \frac{x^2}{2} + C,$$

with the general solution

$$y = \pm \left(\frac{x^2}{3} + C \right)^{3/2}.$$

Thus *two* particular solutions of the initial-value problem (10) (obtained by setting $C = 0$) are $y = \pm x^3/3\sqrt{3}$. However, $y \equiv 0$ is *also* a solution of the initial-value problem, and this solution cannot be obtained from the general solution, no matter what value we assign to C. Solutions that cannot be obtained from the general solution are called **singular solutions**.

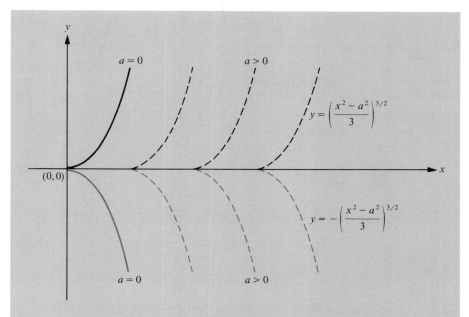

Figure 2.3

The reason the solution $y \equiv 0$ is lost is that the initial step of separating variables involves dividing by $y^{1/3}$, and division by zero is not allowed.

One final point of interest: These are not all of the solutions of the initial-value problem (10). It possesses infinitely many more solutions, defined by the piecewise functions (graphed in Figure 2.3)

$$y = \begin{cases} 0, & x < a, \\ \pm \left(\dfrac{x^2 - a^2}{3} \right)^{3/2}, & x \geq a, \end{cases}$$

for any choice of the parameter $a \geq 0$. (Check.)

PROBLEMS 2.2

In Problems 1–11 verify that each given differential equation is exact and find the general solution. Find a particular solution when an initial condition is given.

1. $2xy \, dx + (x^2 + 1) \, dy = 0$
2. $[x \cos(x + y) + \sin(x + y)] \, dx + x \cos(x + y) \, dy = 0$, $y(1) = \pi/2 - 1$
3. $\left(4x^3 y^3 + \dfrac{1}{x} \right) dx + \left(3x^4 y^2 - \dfrac{1}{y} \right) dy = 0$, $x(e) = 1$
4. $\left(\dfrac{\ln(\ln y)}{x} + \dfrac{2}{3} xy^3 \right) dx + \left(\dfrac{\ln x}{y \ln y} + x^2 y^2 \right) dy = 0$
5. $(x - y \cos x) \, dx - \sin x \, dy = 0$, $y(\pi/2) = 1$
6. $\cosh 2x \cosh 2y \, dx + \sinh 2x \sinh 2y \, dy = 0$
7. $(ye^{xy} + 4y^3) \, dx + (xe^{xy} + 12xy^2 - 2y) \, dy = 0$, $y(0) = 2$
8. $(3x^2 \ln x + x^2 - y) \, dx - x \, dy = 0$, $y(1) = 5$
9. $(2xy + e^y) \, dx + (x^2 + xe^y) \, dy = 0$

10. $(x^2 + y^2)\,dx + 2xy\,dy = 0$, $y(1) = 1$

11. $\left(\dfrac{1}{x} - \dfrac{y}{x^2 + y^2}\right) dx + \left(\dfrac{x}{x^2 + y^2} - \dfrac{1}{y}\right) dy = 0$

In Problems 12–16 find an integrating factor for each differential equation and obtain the general solution.

12. $y\,dx + (y - x)\,dy = 0$
13. $(x^2 + y^2 + x)\,dx + y\,dy = 0$
14. $2y^2\,dx + (2x + 3xy)\,dy = 0$
15. $(x^2 + 2y)\,dx - x\,dy = 0$
16. $(x^2 + y^2)\,dx + (3xy)\,dy = 0$

17. Solve $xy\,dx + (x^2 + 2y^2 + 2)\,dy = 0$.

18. Let $M = yF(xy)$ and $N = xG(xy)$. Show that $1/(xM - yN)$ is an integrating factor for
$$M\,dx + N\,dy = 0.$$

19. Use the result of Problem 18 to solve the equation
$$2x^2 y^3\,dx + x^3 y^2\,dy = 0.$$

20. Solve $(x^2 + y^2 + 1)\,dx - (xy + y)\,dy = 0$. [*Hint:* Try an integrating factor of the form $\mu(x, y) = (x + 1)^n$.]

2.3 Linear First-Order Differential Equations

An nth-order differential equation is **linear** if it can be written in the form

$$\frac{d^n y}{dx^n} + a_{n-1}(x)\frac{d^{n-1} y}{dx^{n-1}} + \cdots + a_1(x)\frac{dy}{dx} + a_0(x) y = f(x).$$

Hence a first-order linear equation has the form

$$\frac{dy}{dx} + a(x) y = f(x), \tag{1}$$

and a second-order linear equation can be written as

$$\frac{d^2 y}{dx^2} + a(x)\frac{dy}{dx} + b(x) y = f(x).$$

The notation indicates that $a(x)$, $b(x)$, $f(x)$, and so on are functions of x alone. If the function $f(x)$ is the zero function, the linear differential equation is said to be **homogeneous**. Otherwise, we say that the linear differential equation is **nonhomogeneous**. Any differential equation that cannot be written in the form above is said to be **nonlinear**. For example,

$$\frac{dy}{dx} = y^2$$

is nonlinear.

Before dealing with the nonhomogeneous first-order linear equation (1), it is important to discuss the solution of the homogeneous equation

$$\frac{dy}{dx} + a(x) y = 0, \tag{2}$$

or

$$\frac{dy}{dx} = -a(x) y.$$

Separating variables, we have

$$\int \frac{dy}{y} = -\int a(x)\,dx + C.$$

Integrating, we have

$$\ln|y| = -\int a(x)\,dx + C$$

and
$$y = C_1 e^{-\int a(x)\,dx}. \tag{3}$$

This is the general solution to equation (2); it indicates that a solution is obtainable whenever the antiderivative can be found. We illustrate this situation with two examples.

Example 1   Solve the homogeneous differential equation
$$y' + 3y = 0.$$

Solution   Rewriting the equation as
$$y' = -3y$$
and separating variables, we have
$$\int \frac{dy}{y} = -3 \int dx + C$$
$$\ln|y| = -3x + C$$
$$y = C_1 e^{-3x}.$$

Example 2   Solve the equation
$$\frac{dy}{dx} = -2xy, \qquad y(1) = 1 \tag{4}$$

Solution   Separating variables, we have
$$\int \frac{dy}{y} = -2 \int x\,dx + C$$
$$\ln|y| = -x^2 + C$$
$$y = C_1 e^{-x^2}.$$

Since $y(1) = 1 = C_1 e^{-1}$, it follows that $C_1 = e$; the particular solution to the initial-value problem (4) is therefore
$$y = e^{1-x^2}.$$

We see that equation (3) can be written in the form
$$y e^{\int a(x)\,dx} = C \tag{5}$$

by multiplying both sides by $e^{\int a(x)\,dx}$. If we differentiate both sides of equation (5), we obtain
$$y' e^{\int a(x)\,dx} + a(x) y e^{\int a(x)\,dx} = [y' + a(x)y] e^{\int a(x)\,dx} = 0, \tag{6}$$
$$\left(\frac{d}{dx} e^{\int a(x)\,dx} = e^{\int a(x)\,dx} \frac{d}{dx} \int a(x)\,dx = a(x) e^{\int a(x)\,dx} \right)$$

since the derivative of an indefinite integral is the integrand. Notice that the expression in brackets is the left side of the original differential equation (2). We call the exponential

Integrating Factor
$$e^{\int a(x)\,dx} \tag{7}$$

an **integrating factor** for the linear equation (2).

We can use the integrating factor to obtain the solution of the nonhomogeneous linear equation

$$\frac{dy}{dx} + a(x)y = f(x). \tag{8}$$

We multiply both sides of equation (8) by the integrating factor $e^{\int a(x)\,dx}$ to obtain

$$[y' + a(x)y]e^{\int a(x)\,dx} = f(x)e^{\int a(x)\,dx}. \tag{9}$$

But as we have seen in going from equation (5) to equation (6), the left side of equation (9) is the derivative of $ye^{\int a(x)\,dx}$. Thus

$$\frac{d}{dx}\left[ye^{\int a(x)\,dx}\right] = f(x)e^{\int a(x)\,dx}. \tag{10}$$

Integrating both sides of equation (10) with respect to x, we have

$$\int \frac{d}{dx}\left[ye^{\int a(x)\,dx}\right]dx = \int f(x)e^{\int a(x)\,dx}\,dx + C,$$

or

$$ye^{\int a(x)\,dx} = \int f(x)e^{\int a(x)\,dx}\,dx + C.$$

This can be written as

$$\boxed{y = \left[\int f(x)e^{\int a(x)\,dx}\,dx + C\right]e^{-\int a(x)\,dx}.} \tag{11}$$

Equation (11) provides an expression for the general solution of the first-order nonhomogeneous linear differential equation (8). It is usually better to go through the process of multiplying both sides of (8) by the integrating factor to obtain the solution than to try to memorize equation (11). We illustrate this process with several examples.

Example 3 Solve the nonhomogeneous linear equation

$$y' = y + x^2, \qquad y(0) = 1. \tag{12}$$

Solution Rewriting equation (12) in the form

$$y' - y = x^2, \tag{13}$$

we see that $a(x) = -1$; thus the integrating factor is

$$e^{-\int dx} = e^{-x}.$$

Multiplying both sides of equation (13) by e^{-x}, we get

$$e^{-x}(y' - y) = x^2 e^{-x},$$

or

$$(ye^{-x})' = x^2 e^{-x}.$$

Integrating both sides, we have

$$ye^{-x} = \int x^2 e^{-x}\,dx + C$$
$$= C - (x^2 + 2x + 2)e^{-x},$$
↑
integrate by parts twice

so
$$y = Ce^x - (x^2 + 2x + 2).$$

Finally, setting $x = 0$, we get
$$1 = y(0) = C - 2,$$
so $C = 3$ and the solution of the initial-value problem is
$$y = 3e^x - (x^2 + 2x + 2).$$

Example 4 Consider the equation $dy/dx = x^3 - 2xy$, where $y = 1$ when $x = 1$. Rewriting the equation as $dy/dx + 2xy = x^3$, we see that $a(x) = 2x$ and the integrating factor is $e^{\int a(x)\,dx} = e^{x^2}$. Multiplying both sides by e^{x^2} and integrating, we have

$$e^{x^2} y = \int x^3 e^{x^2}\,dx + C,$$

so
$$y = e^{-x^2}\left(\int x^3 e^{x^2}\,dx + C\right).$$

We can integrate the integral by parts as follows:
$$\int x^3 e^{x^2}\,dx = \int x^2 (xe^{x^2})\,dx = \frac{x^2 e^{x^2}}{2} - \int xe^{x^2}\,dx = e^{x^2}\left(\frac{x^2 - 1}{2}\right).$$

Replacing this term for the integral above, we have
$$y = e^{-x^2}\left[e^{x^2}\left(\frac{x^2 - 1}{2}\right) + C\right] = \frac{x^2 - 1}{2} + Ce^{-x^2}.$$

Setting $x = 1$ yields
$$1 = y(1) = Ce^{-1}.$$

Thus $C = e$ and the solution to the problem is
$$y = \tfrac{1}{2}(x^2 - 1) + e^{1-x^2}.$$

BERNOULLI'S EQUATION[†]

Certain nonlinear first-order equations can be reduced to linear equations by a suitable change of variables. The equation

$$\frac{dy}{dx} + a(x)y = f(x)y^n, \qquad (14)$$

[†] See the accompanying biographical sketch.

JAKOB BERNOULLI (1654–1705)

*Jakob Bernoulli
(David Smith Collection)*

One of the most distinguished families in the history of mathematics and science are the Bernoullis of Switzerland. The family record starts in the seventeenth century with two brothers, Jakob and Johann. These two men gave up earlier vocational interests and became mathematicians when Leibniz's papers began to appear in the *Acta eruditorum*. Among the first mathematicians to realize the surprising power of the calculus, they applied the tool to a great diversity of problems. From 1687 until his death, Jakob occupied the mathematics chair at the University of Basel. The two brothers, often bitter rivals, maintained an almost constant exchange of ideas with Leibniz and with each other.

Among Jakob Bernoulli's contributions to mathematics are the early use of polar coordinates, the derivation in both rectangular and polar coordinates of the formula for the radius of curvature of a plane curve, the study of the catenary curve with extensions to strings of variable density and strings under the action of a central force, the study of a number of other higher-plane curves, the discovery of the so-called **isochrone**—or curve along which a body will fall with uniform vertical velocity (it turned out to be a semicubical parabola with a vertical cusptangent), the determination of the form taken by an elastic rod fixed at one end and carrying a weight at the other, the form assumed by a flexible rectangular sheet having two opposite edges held horizontally fixed at the same height and loaded with a heavy liquid, and the shape of a rectangular sail filled with wind. He also proposed and discussed the problem of isoperimetric figures (planar closed paths of given species and fixed perimeter which include a maximum area) and was one of the first mathematicians to work in the calculus of variations. He was also one of the early students of mathematical probability; his book in this field, the *Ars conjectandi*, was published posthumously in 1713. There are several mathematical entities that now bear Jakob Bernoulli's name; among them are the *Bernoulli distribution* and *Bernoulli theorem* of statistics and probability theory, the *Bernoulli equation* met by every student of a first course in differential equations, the *Bernoulli numbers* and *Bernoulli polynomials* of number-theory interest, and the *lemniscate of Bernoulli* encountered in a first course in the calculus. In Jakob Bernoulli's solution to the problem of the isochrone curve, which was published in the *Acta eruditorum* in 1690, we meet for the first time the word *integral* in a calculus sense.

Jakob Bernoulli was struck by the way the equiangular spiral reproduces itself under a variety of transformations and asked, in imitation of Archimedes, that such a spiral be engraved on his tombstone along with the inscription "Eadem mutata resurgo" (I shall arise the same, though changed).

which is known as **Bernoulli's equation**, is of this type. Set $z = y^{1-n}$. Then $z' = (1-n)y^{-n}y'$, so if we multiply both sides of equation (14) by $(1-n)y^{-n}$, we obtain

$$(1-n)y^{-n}y' + (1-n)a(x)y^{1-n} = (1-n)f(x),$$

or

$$\frac{dz}{dx} + (1-n)a(x)z = (1-n)f(x).$$

The equation is now linear and may be solved as before.

Example 5 Solve

$$\frac{dy}{dx} - \frac{y}{x} = -\frac{5}{2}x^2 y^3. \tag{15}$$

Solution Here $n = 3$, so we let $z = y^{-2}$, let $z' = -2y^{-3}y'$, and multiply both sides of equation (15) by $-2y^{-3}$ to obtain

$$-2y^{-3}y' + \frac{2}{x}y^{-2} = 5x^2,$$

or

$$z' + \frac{2z}{x} = 5x^2. \tag{16}$$

The integrating factor for this linear equation is

$$e^{2\int \frac{dx}{x}} = e^{2\ln x} = e^{\ln x^2} = x^2.$$

Multiplying both sides of equation (16) by x^2, we have

$$x^2 z' + 2xz = 5x^4$$
$$(x^2 z)' = 5x^4.$$

Thus

$$x^2 z = 5\int x^4 \, dx + C = x^5 + C.$$

Hence

$$y^{-2} = z = x^3 + Cx^{-2},$$

or

$$y = (x^3 + Cx^{-2})^{-1/2}.$$

A similar procedure can be used to solve

$$\frac{dy}{dx} + a(x)y = f(x) y \ln y. \tag{17}$$

Let $z = \ln y$. Then $z' = y'/y$, so dividing equation (17) by y, we obtain the linear equation

$$\frac{dz}{dx} + a(x) = f(x)z.$$

PROBLEMS 2.3

In Problems 1–11 find the general solution for each equation. When an initial condition is given, find the particular solution that satisfies the condition.

1. $\dfrac{dx}{dt} = 3x$

2. $\dfrac{dy}{dx} + 22y = 0$, $y(1) = 2$

3. $\dfrac{dx}{dt} = x + 1$, $x(0) = 1$

4. $\dfrac{dy}{dx} + y = \sin x$, $y(0) = 0$

★ 5. $\dfrac{dx}{dy} - x \ln y = y^y$

6. $\dfrac{dy}{dx} + y = \dfrac{1}{1 + e^{2x}}$

7. $\dfrac{dy}{dx} - \dfrac{3}{x}y = x^3, \quad y(1) = 4$

8. $\dfrac{dx}{dt} + x \cot t = 2t \csc t$

9. $x' - 2x = t^2 e^{2t}$

10. $y' + \dfrac{2}{x}y = \dfrac{\cos x}{x^2}, \quad y(\pi) = 0$

11. $\dfrac{ds}{du} + s = ue^{-u} + 1$

12. Solve the equation
$$y - x\dfrac{dy}{dx} = \dfrac{dy}{dx} y^2 e^y$$
by reversing the roles of x and y (that is, treat x as the dependent variable).

13. Use the method shown in Problem 12 to solve
$$\dfrac{dy}{dx} = \dfrac{1}{e^{-y} - x}.$$

14. Find the solution of $dy/dx = 2(2x - y)$ that passes through the point $(0, -1)$.

15. In a study[†] on the rate at which education is being forgotten or made obsolete, the following linear first-order differential equation was used:
$$x' = 1 - kx,$$
where $x(t)$ denotes the education of an individual at time t and k is a constant given by the rate at which that education is being lost. Obtain an equation for x at time t.

16. Data collected in a botanical experiment[‡] led to the differential equation
$$\dfrac{dI}{dw} = 0.088(2.4 - I).$$
Find the value of I as $w \to \infty$.

17. Assume that there exists an upper bound B for the size y of a crop in a given field. E. A. Mitscherlich proposed in 1939 the use of the linear differential equation
$$\dfrac{dy}{dt} = k(B - y)$$
as a model for agricultural growth. Find the general solution of this equation.

18. Suppose a population is growing at a rate proportional to its size and, in addition, individuals are immigrating into the population at a constant rate. (a) Find the linear differential equation governing this situation. (b) Find its general solution.

19. Use the method we have developed in this section for the solution of Bernoulli's equation (14) to solve the logistic equation (see Example 2.1.5)
$$\dfrac{dP}{dt} = P(\beta - \delta P).$$

*§20. Let $N(t)$ be the biomass of a fish species in a given area of the ocean and suppose the rate of change of the biomass is governed by the logistic equation
$$\dfrac{dN}{dt} = rN\left(1 - \dfrac{N}{K}\right),$$
where the net proportional growth rate r is a constant and K is the carrying capacity (see p. 35) for that species in that area. Assume that the rate at which fish are caught depends on the biomass. If E is the constant effort expended to harvest that species, then

a. find the resulting growth rate of the biomass;[‖]

b. solve the differential equation in part (a) using the Bernoulli method.

[Note: Effort can be measured in man-hours per year, tonnage of the fishing fleet, or volume seined by the fleet's nets.]

*21. Suppose fish are harvested at a constant rate h independent of their biomass. Answer parts (a) and (b) of Problem 20 for this situation.

22. Find the effort E that maximizes the sustainable yield in Problem 20. [This is the limit of the yield as $t \to \infty$.]

23. Show that if $h > rK/4$, in Problem 21, the species will become extinct regardless of the initial size of the biomass.

24. The differential equation governing the velocity v of an object of mass m subject to air resistance proportional to the instantaneous velocity is
$$m\dfrac{dv}{dt} = -mg - kv.$$
Solve the equation and determine the limiting velocity of the object as $t \to \infty$.

25. Repeat Problem 24 if air resistance is proportional to the square of the instantaneous velocity.

26. An infectious disease is introduced into a large population. The proportion of people exposed to

---

[†] L. Southwick and S. Zionts, "An Optimal-Control-Theory Approach to the Education Investment Decision." *Operations Research* 22 (1974): 1156–1174.

[‡] R. L. Specht, "Dark Island Heath," *Australian Journal of Botany* 5 (1957): 137–172.

[§] A number of other examples of models of renewable resources are given in C. W. Clark, *Mathematical Bioeconomics* (New York: Wiley-Interscience, 1976).

[‖] This is called the **Schaefer model**, after the biologist M. B. Schaefer.

the disease increases with time. Suppose that $P(t)$ is the proportion of people exposed to the disease within t years of its introduction. If $P'(t) = [1 - P(t)]/3$ and $P(0) = 0$, after how many years does the proportion increase to 90 percent?

In Problems 27–32 find the general solution for each equation and a particular solution when an initial condition is given.

27. $\dfrac{dy}{dx} = -\dfrac{(6y^2 - x - 1)y}{2x}$

28. $y' = -y^3 x e^{-2x} + y$

29. $x\dfrac{dy}{dx} + y = x^4 y^3$, $y(1) = 1$

30. $tx^2 \dfrac{dx}{dt} + x^3 = t \cos t$

31. $\dfrac{dy}{dx} + \dfrac{3}{x} y = x^2 y^2$, $y(1) = 2$

32. $xyy' - y^2 + x^2 = 0$

33. We have seen that the equation
$$y' + f(x) y = 0$$
has the general solution
$$y = c e^{-\int f(x)\,dx}.$$
This fact prompted J. L. Lagrange (1736–1813) to seek a solution of the equation
$$y' + f(x) y = g(x) \qquad (i)$$
of the form
$$y = c(x) e^{-\int f(x)\,dx},$$
where $c = c(x)$ is a function of x.[†]
a. Show that $c'(x) = g(x) e^{\int f(x)\,dx}$;
b. Integrate part (a) to obtain the general solution to equation (i).

34. Use the method in Problem 33 to solve the equation
$$y' + \dfrac{1}{x} y = e^x.$$

35. Consider the second-order linear equation
$$y'' + 5y' + 6y = 0. \qquad (ii)$$
a. Let $z = y' + 2y$. Show that equation (ii) reduces to
$$z' + 3z = 0. \qquad (iii)$$
b. Solve equation (iii), substitute z in the equation $y' + 2y = z$, and use the methods given in this section to obtain the solution to equation (ii).

36. Use the procedure outlined in Problem 35 to find the general solution to
$$y'' + (a + b) y' + aby = 0,$$
where a and b are constants and $a \neq b$. [*Hint:* Let $z = y' + ay$.]

37. Use the method given in Problem 35 to find the general solution to
$$y'' + 2ay' + a^2 y = 0,$$
where a is a constant.

In Problems 38 and 39 the function f is given by
$$f(x) = \begin{cases} 1, & \text{if } 0 \leq x \leq 1, \\ 0, & \text{if } x > 1. \end{cases}$$

38. Solve the initial-value problem
$$y' + y = f(x), \qquad y(0) = 0.$$

39. Solve the initial-value problem
$$y' + f(x) y = 0, \qquad y(0) = 1.$$
[*Hint:* Solve separately on $0 \leq x \leq 1$ and $x > 1$ and match at $x = 1$.]

**‡ 40. Show that the general solution of the differential equation
$$y' - 2xy = x^2$$
is of the form $y = f(x) + ce^{x^2}$, where
$$\left| f(x) + \dfrac{x}{2} \right| \leq \dfrac{1}{4x}, \quad \text{for } x \geq 2.$$

**§ 41. Show that if the family of solutions of
$$y' + a(x) y = f(x), \qquad a(x) f(x) \neq 0,$$
all cross the line $x = c$, the tangent lines to the solution curves at the points of intersection are **concurrent** (have a point in common).

**42. Daniel Bernoulli obtained the following differential equation (in 1760),
$$S' = -pS + (S/N) N' + pS^2/mN,$$
in studying the effects of smallpox. Here $N(t)$ is the number of individuals that survive at age t, $S(t)$ is the number that are susceptible to smallpox at age t (the disease imparts lifetime immunity if survived), p is the probability of a susceptible individual getting the disease, and $1/m$ is the proportion of those who die from the disease. Let $y = N/S$ and find a solution of the resulting equation.

---

‡ R. C. Buck, "On 'Solving' Differential Equations," *American Mathematical Monthly* 63 (1956): 414.

§ Problem 3 of the William Lowell Putnam Mathematical Competition, *American Mathematical Monthly* 61 (1954): 545.

† This technique, called the method of **variation of constants**, or **variation of parameters**, can be extended to equations of higher order (see Section 3.6).

2.4 Homogeneous Equations and Substitution Techniques

Consider the differential equation

$$\frac{dy}{dx} = f(x, y). \qquad (1)$$

Homogeneous Equation

If $f(tx, ty) = t^n f(x, y)$, for some real number n, then the function f is said to be **homogeneous of degree n**. In this case we call the differential equation (1) a **homogeneous differential equation**.

Example 1

a. $f(x, y) = x + \sqrt[3]{x^3 - y^3}$ is homogeneous of degree 1 because

$$f(tx, ty) = tx + \sqrt[3]{(tx)^3 - (ty)^3}$$

$$= tx + \sqrt[3]{t^3(x^3 - y^3)}$$

$$= t\left[x + \sqrt[3]{x^3 - y^3}\right] = tf(x, y).$$

b. $f(x, y) = \dfrac{x^{12} + y^{12}}{x^3 - y^3}$ is homogeneous of degree 9 because

$$f(tx, ty) = \frac{t^{12}x^{12} + t^{12}y^{12}}{t^3 x^3 - t^3 y^3} = \frac{t^{12}}{t^3}\left(\frac{x^{12} + y^{12}}{x^3 - y^3}\right)$$

$$= t^9 f(x, y).$$

Note that if $f(x, y)$ is a homogeneous function of degree n, then we can set $t = 1/x$ and obtain

$$f\left(1, \frac{y}{x}\right) = f(tx, ty) = t^n f(x, y) = \frac{f(x, y)}{x^n},$$

or

$$f(x, y) = x^n f\left(1, \frac{y}{x}\right).$$

Homogeneous of Degree Zero

In particular, if $f(x, y)$ is **homogeneous of degree zero** (that is, if $f(tx, ty) = f(x, y)$), then

$$f(x, y) = f\left(1, \frac{y}{x}\right) = F\left(\frac{y}{x}\right).$$

Equation (1) now becomes

$$\frac{dy}{dx} = F\left(\frac{y}{x}\right) \qquad (2)$$

and the right-hand side of the equation can be written as a function of the variable y/x. It is then natural to try the substitution $z = y/x$. Since the function y depends on x, so does the function z. Differentiating $y = xz$ with respect to x, we have

$$\frac{dy}{dx} = z + x\frac{dz}{dx}. \tag{3}$$

Substituting $z = y/x$ and replacing the left-hand side of equation (2) by the right-hand side of equation (3), we obtain

$$z + x\frac{dz}{dx} = F(z).$$

The variables in this equation can be separated, since

$$x\frac{dz}{dx} = F(z) - z,$$

and so

$$\frac{dz}{F(z) - z} = \frac{dx}{x}.$$

A complete solution can now be obtained by integrating both sides of this equation and replacing each z by y/x. We illustrate this procedure in the next two examples.

Example 2 Solve the equation

$$\frac{dy}{dx} = \frac{x-y}{x+y}.$$

Solution If $f(x, y) = (x - y)/(x + y)$, then
$f(tx, ty) = (tx - ty)/(tx + ty) = t(x - y)/t(x + y) = (x - y)/(x + y) = f(x, y);$
so f is homogeneous of degree zero.

Dividing the numerator and denominator of the right-hand side of this equation by x yields

$$\frac{dy}{dx} = \frac{1 - (y/x)}{1 + (y/x)} = \frac{1 - z}{1 + z} = F(z).$$

Replacing the left-hand side by $z + x(dz/dx)$ and separating variables, we have (after some algebra)

$$\left(\frac{1+z}{1 - 2z - z^2}\right) dz = \frac{dx}{x}.$$

After integrating, we obtain

$$\ln(1 - 2z - z^2) = -2\ln x + c = \ln(cx^{-2});$$

exponentiating and replacing z by y/x leads to the implicit solution

$$1 - \frac{2y}{x} - \frac{y^2}{x^2} = \frac{c}{x^2}.$$

Finally, multiplying both sides by x^2 yields $x^2 - 2xy - y^2 = c$.

Example 3 Find the shape of a curved mirror such that light from a source at the origin is reflected in a beam parallel to the x-axis.

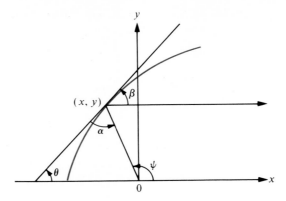

Figure 2.4

Solution By symmetry, the mirror has the shape of a surface of revolution obtained by revolving a curve about the x-axis. Let (x, y) be any point on the cross section in the xy-plane (see Figure 2.4). The law of reflection states that the angle of incidence α must equal the angle of reflection β; thus $\alpha = \beta = \theta$. Since the interior angles of a triangle add up to $180°$, we have $\psi = \alpha + \theta = 2\theta$. What is important about the angles θ and ψ is that

$$y' = \tan\theta \quad \text{and} \quad \frac{y}{x} = \tan\psi.$$

Using the trigonometric formula for the tangent of a double angle, we have

$$\frac{y}{x} = \tan\psi = \tan 2\theta = \frac{2\tan\theta}{1 - \tan^2\theta} = \frac{2y'}{1 - (y')^2}.$$

Solving for y', we obtain the quadratic equation

$$y(y')^2 + 2xy' - y = 0,$$

and by the quadratic formula we find that

$$y' = \frac{-x \pm \sqrt{x^2 + y^2}}{y}. \tag{4}$$

Since $y' > 0$, we need only consider the $+$ sign of the $\pm$ in equation (4), because $\sqrt{x^2 + y^2} > |x|$, yielding the nonlinear differential equation

$$y' = \frac{-x + \sqrt{x^2 + y^2}}{y}. \tag{5}$$

Dividing the numerator and denominator of equation (5) by x, and remembering that $\sqrt{x^2 + y^2}/x = \sqrt{(x^2 + y^2)/x^2}$, we obtain

$$\frac{dy}{dx} = \frac{-1 + \sqrt{1 + (y/x)^2}}{y/x}. \tag{6}$$

Substituting $z = y/x$ into equation (6) and using equation (3), we have

$$z + x\frac{dz}{dx} = \frac{-1 + \sqrt{1 + z^2}}{z},$$

or, subtracting z from both sides and simplifying the right-hand side, we have
$$x\frac{dz}{dx} = \frac{\sqrt{1+z^2} - (1+z^2)}{z}.$$
Separating variables, we obtain
$$\frac{z\,dz}{\sqrt{1+z^2}\,(1-\sqrt{1+z^2})} = \frac{dx}{x}. \tag{7}$$
To integrate the left-hand side of equation (7) we substitute
$$u = 1 - \sqrt{1+z^2} \quad \text{and} \quad du = -\frac{z\,dz}{\sqrt{1+z^2}},$$
obtaining
$$-\int \frac{du}{u} = \int \frac{dx}{x},$$
or
$$-\ln|u| = \ln|x| + c.$$
Exponentiating both sides, we have
$$\frac{1}{u} = cx;$$
inverting both sides leads to
$$1 - \sqrt{1+z^2} = -\frac{c}{x}.$$
Replacing z by y/x and simplifying, we have
$$1 + \frac{c}{x} = \sqrt{1 + \left(\frac{y}{x}\right)^2},$$
which may be squared to obtain
$$1 + \frac{2c}{x} + \frac{c^2}{x^2} = 1 + \frac{y^2}{x^2}. \tag{8}$$
Canceling the ones and multiplying both sides of equation (8) by x^2, we get
$$y^2 = 2cx + c^2,$$
which is the equation of the family of all parabolas with foci at the origin that are symmetric with respect to the x-axis.

Another useful technique applies to equations of the form
$$\frac{dy}{dx} = F(ax + by + c), \tag{9}$$
where a, b, and c are real constants. If the substitution $z = ax + by + c$ is made in equation (9), we obtain
$$\frac{z' - a}{b} = F(z),$$
since $z' = a + by'$. This equation has separable variables.

Example 4 Solve
$$\frac{dy}{dx} = (x+y+1)^2 - 2.$$

Solution Letting $z = x + y + 1$, we have $z' = 1 + y'$. Thus
$$z' - 1 = y' = (x+y+1)^2 - 2$$
$$= z^2 - 2,$$
or
$$z' = z^2 - 1.$$
We now separate variables to obtain
$$\int \frac{dz}{z^2 - 1} = \int dx. \tag{10}$$
We can integrate the left-hand side of equation (10) by partial fractions, using
$$\frac{1}{z^2 - 1} = \frac{1}{2}\left(\frac{1}{z-1} - \frac{1}{z+1}\right),$$
so that equation (10) yields
$$\tfrac{1}{2}\left[\ln|z-1| - \ln|z+1|\right] = x + c,$$
or
$$\ln\left|\frac{z-1}{z+1}\right| = 2x + c.$$
Exponentiating both sides, we obtain
$$\frac{z-1}{z+1} = ce^{2x},$$
or
$$(z-1) = ce^{2x}(z+1).$$
Gathering all terms involving z on one side and the remaining terms on the other, we get
$$z = \frac{1 + ce^{2x}}{1 - ce^{2x}}.$$
Finally, we replace z by $x + y + 1$, which yields
$$y = \frac{1 + ce^{2x}}{1 - ce^{2x}} - x - 1.$$

As these examples indicate, we may make *any* substitution we want in trying to solve a differential equation. Of course, *there is no guarantee of success*. Moreover, there may be more than one substitution that yields the answer, as illustrated in the following example.

Example 5 Here we reconsider Example 3, in which we were trying to find the shape of a curved mirror that reflects light from a source at the origin in a beam parallel to the *x*-axis. Recall that the geometric considerations in

Example 3 led to the differential equation [see equation (5)]

$$y' = \frac{-x + \sqrt{x^2 + y^2}}{y}. \tag{11}$$

Now, instead of proceeding as we did in Example 3, suppose we substitute $z = x^2 + y^2$. Then

$$\frac{dz}{dx} = 2x + 2y\frac{dy}{dx},$$

and replacing dy/dx by the right-hand side of equation (11), we have

$$\frac{dz}{dx} = 2x + 2y\left(\frac{-x + \sqrt{x^2 + y^2}}{y}\right)$$

$$= 2x - 2x + 2\sqrt{x^2 + y^2},$$

or

$$\frac{dz}{dx} = 2\sqrt{z}.$$

Separating variables gives us

$$\frac{dz}{2\sqrt{z}} = dx,$$

and integrating, we get

$$\sqrt{z} = x + c.$$

Replacing z by $x^2 + y^2$ and squaring both sides yields

$$x^2 + y^2 = x^2 + 2cx + c^2,$$

or

$$y^2 = 2cx + c^2.$$

Since c is arbitrary, we have the same family of parabolas as before.

PROBLEMS 2.4

Find the general solution in Problems 1–10 by using the method given in Example 2. When an initial condition is given, find the particular solution that satisfies that condition.

1. $xy' - y = \sqrt{xy}$
2. $\dfrac{dx}{dt} = \dfrac{x}{t} + \cosh\dfrac{x}{t}$
3. $x\dfrac{dy}{dx} = xe^{y/x} + y$, $y(1) = 0$
4. $2xyy' = x^2 + y^2$, $y(-1) = 1$
5. $y' = \dfrac{y^2 + xy}{x^2}$, $y(1) = 1$
6. $(x + v)\dfrac{dx}{dv} = x$

7. $\dfrac{dy}{dx} = \dfrac{2xy - y^2}{2xy - x^2}$, $y(1) = 2$
8. $3xyy' + x^2 + y^2 = 0$
9. $xy' - y = \sqrt{x^2 + y^2}$
10. $\dfrac{dy}{dx} = \dfrac{x^2 y + 2xy^2 - y^3}{2y^3 - xy^2 + x^3}$

Use the method given in Example 4 to solve Problems 11–16.

11. $2y' = x^2 + 4xy + 4y^2 + 3$
12. $9y' + (x + y - 1)^2 = 0$
13. $(x + y)y' = 2x + 2y - 3$
14. $y' + 1 = \sqrt{x + y + 2}$, $y(0) = 2$
15. $y' + \sin^2(x + y) = 0$

16. $y' = \dfrac{e^y}{e^x} - 1$

Find a substitution that provides a solution to each of Problems 17–22.

17. $xy' + y = (xy)^3$
18. $xy' = \sqrt{1 - x^2 y^2} - y$
19. $xy' = e^{xy} - y$, $y(1) = 1$
20. $y' = y^2 e^x - y$
21. $y' = (x + y)\ln(x + y) - 1$
22. $y' = y(\ln y + x^2 - 2x)$
23. Show, if $ad - bc \neq 0$, that there are constants h and k such that the substitutions $x = u + h$ and $y = v + k$ convert the quotient
$$\dfrac{ax + by + m}{cx + dy + n}$$
into the quotient
$$\dfrac{au + bv}{cu + dv}.$$

Use Problem 23 and the method given in Example 2 to find a solution for each of Problems 24–29.

24. $\dfrac{dy}{dx} = \dfrac{x - y - 5}{x + y - 1}$
25. $y' = \dfrac{x + 2y + 2}{y - 2x}$
26. $y' = \dfrac{y - x - 1}{y + x}$
27. $\dfrac{dx}{dt} = \dfrac{x + 1}{x + t}$
28. $\dfrac{dy}{dx} = \dfrac{2x - 3y + 4}{4x + y - 6}$
29. $\dfrac{dy}{dx} = \dfrac{3x - y - 3}{x + y - 5}$

30. Solve the equation
$$\dfrac{dy}{dx} = \dfrac{1 - xy^2}{2x^2 y}$$
by making the substitution $v = y/x^n$ for an appropriate value of n.

31. Use the method of Problem 30 to solve
$$\dfrac{dx}{dt} = \dfrac{x - tx^2}{t + xt^2}.$$

Another equation that can be solved directly is **Clairaut's**[†] **equation**:
$$y = xy' + f(y'). \tag{i}$$

We can differentiate both sides with respect to x to obtain
$$y' = y' + xy'' + f'(y')y'',$$
where the last term is a result of the chain rule. Canceling like terms, we get
$$[x + f'(y')]y'' = 0.$$
Since one of the factors must vanish, two different solutions arise:

a. If $y'' = 0$, then $y' = c$, and substituting this value into (i) produces the *general solution*
$$y = cx + f(c),$$
which is a collection of straight lines.

b. If $x + f'(y') = 0$, then $x = -f'(y')$, and (i) may be rewritten as
$$y = f(y') - y'f'(y').$$
Here x and y are both expressed in terms of functions of y', so if we substitute $y' = t$, we obtain the parametrized curve
$$x = -f'(t), \qquad y = f(t) - tf'(t).$$
This curve is also a solution, called the **singular solution**, to Clairaut's equation. For example, the equation
$$y = xy' - e^{y'}$$
has the general solution $y = cx - e^c$ and the parametrized curve
$$x = e^t, \qquad y = e^t(1 - t),$$
as a singular solution. Generally we can proceed no further, but in this case we can eliminate the parameter t by letting $t = \ln x$, so that the singular solution is given by
$$y = x(1 - \ln x), \qquad x > 0.$$

In Problems 32–36 find the general and singular solutions for each Clairaut equation.

32. $y = xy' + \ln y'$
33. $y = xy' + (y')^3$
34. $y = xy' - \sqrt{y'}$
35. $y = x\dfrac{dy}{dx} + \dfrac{1}{4}\left(\dfrac{dy}{dx}\right)^4$
36. $(y - xy')^2 - (y')^2 = 1$

[†] Named for the French mathematician Alexis Claude Clairaut (1713–1765). Clairaut, a mathematical prodigy, wrote a well-received paper on third-order curves at the age of eleven. This paper helped him win a seat in the French Academy of Sciences at the ineligible age of eighteen. In 1736 he traveled to Lapland with Pierre de Maupertius to measure the length of a degree of one of the earth's meridians. This measurement helped to confirm a belief of Newton and Huygens that the earth was flattened at the poles. Clairaut made many other contributions to mathematics and physics. Among them was his computation of the 1759 return of Halley's comet with an error of about a month.

2.5 Simple Electric Circuits

In this section we consider simple electric circuits containing a resistor and an inductor or capacitor in series with a source of electromotive force (emf). Such circuits are shown in Figure 2.5. Their action can be understood very easily without any special knowledge of electricity.

a. An electromotive force E (volts), usually a battery or generator, drives an electric charge Q (coulombs) and produces a current I (amperes). The current is defined as the rate of flow of the charge, and we can write

$$I = \frac{dQ}{dt}. \tag{1}$$

b. A resistor of resistance R (ohms) is a component of the circuit that opposes the current, dissipating the energy in the form of heat. It produces a drop in voltage given by **Ohm's law**:

$$E_R = RI. \tag{2}$$

c. An inductor of inductance L (henrys) opposes any change in current by producing a voltage drop of

$$E_L = L\frac{dI}{dt}. \tag{3}$$

d. A capacitor of capacitance C (farads) stores charge. In so doing, it resists the flow of further charge, causing a drop in the voltage of

$$E_C = \frac{Q}{C}. \tag{4}$$

The quantities R, L, and C are usually constants associated with the particular component in the circuit; E may be a constant or a function of time. The fundamental principle guiding such circuits is given by **Kirchhoff's voltage law**:

> The algebraic sum of all voltage drops around a closed circuit is zero.

In the circuit of Figure 2.5(a), the resistor and the inductor cause voltage drops of E_R and E_L, respectively. The emf, however, *provides* a voltage of E

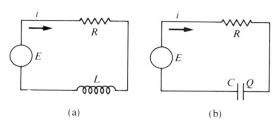

Figure 2.5

(that is, a voltage drop of $-E$). Thus Kirchhoff's voltage law yields
$$E_R + E_L - E = 0.$$

Transposing E to the other side of the equation and using equations (2) and (3) to replace E_R and E_L, we have

$$L\frac{dI}{dt} + RI = E. \tag{5}$$

The following two examples illustrate the use of equation (5) in analyzing the circuit shown in Figure 2.5(a).

Example 1[†] An inductance of 2 henrys and a resistance of 10 ohms are connected in series with an emf of 100 volts. If the current is zero when $t = 0$, what is the current at the end of 0.1 second?

Solution Since $L = 2$, $R = 10$, and $E = 100$, equation (5) and the initial current yield the initial-value problem:

$$2\frac{dI}{dt} + 10I = 100, \qquad I(0) = 0. \tag{6}$$

Dividing both sides of equation (6) by 2, we note that the resulting linear first-order equation has e^{5t} as an integrating factor; that is,

$$\frac{d}{dt}(e^{5t}I) = e^{5t}\left(\frac{dI}{dt} + 5I\right) = 50e^{5t}. \tag{7}$$

Integrating both ends of equation (7), we get
$$e^{5t}I(t) = 10e^{5t} + c,$$

or
$$I(t) = 10 + ce^{-5t}. \tag{8}$$

Setting $t = 0$ in equation (8) and using the initial condition $I(0) = 0$, we have
$$0 = I(0) = 10 + c,$$

which implies that $c = -10$. Substituting this value into equation (8), we obtain an equation for the current at all times t:
$$I(t) = 10(1 - e^{-5t}).$$

Thus, when $t = 0.1$, we have
$$I(0.1) = 10(1 - e^{-0.5}) \approx 3.93 \text{ amperes}.$$

Example 2 Suppose that the emf $E = 100 \sin 60t$ volts but all other values remain the same as those given in Example 1. In this case equation (5) yields

$$2\frac{dI}{dt} + 10I = 100 \sin 60t, \qquad I(0) = 0. \tag{9}$$

Again dividing by 2 and multiplying both sides by the integrating factor e^{5t},

---

[†] This example is typical in electrical engineering. It is, for example, very similar to Exercise 12a in C. A. Desoer and E. S. Kuh, *Basic Circuit Theory* (New York: McGraw Hill, 1969), p. 169.

we have

$$\frac{d}{dt}(e^{5t}I) = e^{5t}\left(\frac{dI}{dt} + 5I\right) = 50e^{5t}\sin 60t. \tag{10}$$

Integrating both ends of equation (10) and using Formula 168 of the integral table in Appendix 1, we obtain

$$I(t) = e^{-5t}\left[50\int(\sin 60t)e^{5t}\,dt + c\right]$$

$$= e^{-5t}\left[50e^{5t}\left(\frac{5\sin 60t - 60\cos 60t}{3625}\right) + c\right]$$

$$= \frac{2\sin 60t - 24\cos 60t}{29} + ce^{-5t}.$$

Setting $t = 0$, we find that $c = 24/29$ and

$$I(t) = \frac{2\sin 60t - 24\cos 60t}{29} + \frac{24}{29}e^{-5t}.$$

Thus

$$I(0.1) = \frac{2\sin 6 - 24\cos 6}{29} + \frac{24}{29}e^{-0.5} \approx -0.31 \text{ amperes.}$$

Transient and Steady-State Current

In the previous example the term $24e^{-5t}/29$ is called the **transient current** because it approaches zero as t increases without bound. The other part of the current, $(2\sin 60t - 24\cos 60t)/29$, is called the **steady-state current**.

For the circuit in Figure 2.5(b) we have $E_R + E_C - E = 0$, or

$$RI + \frac{Q}{C} = E.$$

Using the fact that $I = dQ/dt$, we obtain the linear first-order equation

$$R\frac{dQ}{dt} + \frac{Q}{C} = E. \tag{11}$$

The next example illustrates how to use equation (11).

Example 3   If a resistance of 2000 ohms and a capacitance of 5×10^{-6} farad are connected in series with an emf of 100 volts, what is the current at $t = 0.1$ second if $I(0) = 0.01$ ampere?

Solution   Setting $R = 2000$, $C = 5 \times 10^{-6}$, and $E = 100$ in equation (11), we have

$$2000\left(\frac{dQ}{dt} + 100Q\right) = 100,$$

or

$$\frac{dQ}{dt} + 100Q = \frac{1}{20}, \tag{12}$$

from which we can determine $Q(0)$ since

$$\frac{1}{20} = Q'(0) + 100Q(0) = I(0) + 100Q(0).$$

Thus

$$Q(0) = \frac{1}{100}\left[\frac{1}{20} - I(0)\right] = \frac{1}{100}\left(\frac{1}{20} - \frac{1}{100}\right)$$

$$= \frac{1}{100}\left(\frac{4}{100}\right) = 4 \times 10^{-4} \text{ coulombs.} \tag{13}$$

Multiplying both sides of equation (12) by the integrating factor e^{100t}, we get

$$\frac{d}{dt}(e^{100t}Q) = \frac{e^{100t}}{20},$$

and integrating this equation yields

$$e^{100t}Q = \frac{e^{100t}}{2000} + c.$$

Dividing both sides by e^{100t} gives us

$$Q(t) = \frac{1}{2000} + ce^{-100t},$$

and setting $t = 0$, we find that

$$c = Q(0) - \frac{1}{2000} = (4 \times 10^{-4}) - (5 \times 10^{-4}) = -10^{-4}.$$

Thus the charge at all times t is

$$Q(t) = (5 - e^{-100t})/10^4,$$

and the current is

$$I(t) = Q'(t) = \frac{1}{100}e^{-100t}.$$

Thus $I(0.1) = 10^{-2}e^{-10} \approx 4.54 \times 10^{-7}$ amperes.

PROBLEMS 2.5

In Problems 1–5 assume that the *RL* circuit shown in Figure 2.5(a) has the given resistance (ohms, Ω), inductance (henrys, H), emf (volts, V), and initial current (amperes, amp). Find an expression for the current at all times t and calculate the current after 1 second.

1. $R = 10\ \Omega$, $L = 1$ H, $E = 12$ V, $I(0) = 0$ amp
2. $R = 8\ \Omega$, $L = 1$ H, $E = 6$ V, $I(0) = 1$ amp
3. $R = 50\ \Omega$, $L = 2$ H, $E = 100$ V, $I(0) = 0$ amp
4. $R = 10\ \Omega$, $L = 5$ H, $E = 10\sin t$ V, $I(0) = 1$ amp
5. $R = 10\ \Omega$, $L = 10$ H, $E = e^t$ V, $I(0) = 0$ amp

In Problems 6–10 use the given resistance, capacitance (farads, f), emf, and initial charge (coulombs) in the *RC* circuit shown in Figure 2.5(b). Find an expression for the charge at all time t.

6. $R = 1\ \Omega$, $C = 1$ f, $E = 12$ V, $Q(0) = 0$ coulomb
7. $R = 10\ \Omega$, $C = 0.001$ f, $E = 10\cos 60t$ V, $Q(0) = 0$ coulomb
8. $R = 1\ \Omega$, $C = 0.01$ f, $E = \sin 60t$ V, $Q(0) = 0$ coulomb
9. $R = 100\ \Omega$, $C = 10^{-4}$ f, $E = 100$ V, $Q(0) = 1$ coulomb

10. $R = 200\ \Omega$, $C = 5 \times 10^{-5}$ f, $E = 1000$ V, $Q(0) = 1$ coulomb

†11. The capacitor C in the circuit illustrated below is charged to 10 volts when the switch is closed. Obtain a differential equation for the capacitor voltage and find the voltage for all times t given that $R = 1000$ ohms and $C = 10^{-6}$ farad.

12. An inductance of 1 henry and a resistance of 2 ohms are connected in series with a battery of $6e^{-0.0001t}$ volt. No current is flowing initially. When does the current measure 0.5 ampere?

13. A variable resistance $R = 1/(5 + t)$ ohms and a capacitance of 5×10^{-6} farad are connected in series with an emf of 100 volts. If $Q(0) = 0$, what is the charge on the capacitor after 1 minute?

---

† This example is Exercise 5.28 in Shearer et al., *System Dynamics* (Reading, Mass.: Addison-Wesley, 1971), p. 141. Reprinted with permission of Addison-Wesley Publishing Co.

14. In the RC circuit [Figure 2.5(b)] with constant voltage E, how long will it take the current to decrease to one-half its original value?

15. Suppose that the voltage in an RC circuit is $E(t) = E_0 \cos \omega t$, where $2\pi/\omega$ is the period of the cycle. Assuming that the initial charge is zero, what are the charge and current as functions of R, C, ω, and t?

16. Show that the current in Problem 15 consists of two parts: a steady-state term that has a period of $2\pi/\omega$ and a transient term that tends to zero as t increases.

17. Show that if R in Problem 16 is small, then the transient term can be quite large for small values of t. [This is why fuses can blow when a switch is flipped.]

18. Find the steady-state current, given that a resistance of 2000 ohms and a capacitance of 3×10^{-6} farad are connected in series with an alternating emf of $120 \cos 2t$ volts.

19. Find an expression for the current of a series RL circuit, where $R = 100$ ohms, $L = 2$ henrys, $I(0) = 0$ amp, and the emf voltage satisfies

$$E = \begin{cases} 6, & \text{for } 0 \leq t \leq 10, \\ 7 - e^{10-t}, & \text{for } t \geq 10. \end{cases}$$

20. Repeat Problem 19 with $R = 100/(1 + t)$, all other values remaining the same.

2.6 Further Applications: Compartmental Analysis and Curves of Pursuit

In this section we use the techniques developed earlier in the chapter to solve two interesting types of problems.

COMPARTMENTAL ANALYSIS

A complicated physical or biological process can often be divided into several distinct stages. The entire process can then be described by the interactions between the individual stages. Each such stage is called a **compartment** or pool, and the contents of each compartment are assumed to be well mixed. Material from one compartment is transferred to another and is immediately incorporated into the latter. Because of the name we have given to the stages, the entire process is called a **compartmental system**.‡ An **open** system is one in which there are inputs to or outputs from the system through one or more compartments. A system that is not open is said to be **closed**.

In this section we investigate only the simplest such system: the one-compartment system. Additional work on more complicated systems will be found in later chapters.

---

‡ This name is frequently used in mathematical biology. Engineers refer to such systems as **block diagrams**.

Figure 2.6

Figure 2.6 illustrates a one-compartment system consisting of a quantity $x(t)$ of material in the compartment, an input rate $i(t)$ at which material is being introduced to the system, and a **fractional transfer coefficient** k indicating the fraction of the material in the compartment that is being removed from the system per unit time. It is clear that the rate at which the quantity x is changing depends on the difference between the input and output at any time t, leading to the differential equation

$$\frac{dx}{dt} = i(t) - kx(t). \tag{1}$$

As we saw in Section 2.3 [equation (11), p. 48], this linear equation has the solution

$$x(t) = e^{-kt}\left(\int i(t)e^{kt}\,dt + c\right). \tag{2}$$

This simple model applies to many different problems, as we illustrate below.

Example 1   Strontium 90 (^{90}Sr) has a half-life of 25 years. If 10 grams of ^{90}Sr are initially placed in a sealed container, how many grams remain after 10 years?

Solution   Let $x(t)$ be the number of grams of ^{90}Sr at time t (years). Since the number of atoms present is very large, the number decaying per unit time is directly proportional to the number present at that time. The constant of proportionality k is the fractional transfer coefficient. Since there is no input, the equation involved is

$$\frac{dx}{dt} = -kx(t). \tag{3}$$

Equation (3) has the solution $x(t) = x_0 e^{-kt}$, where $x_0 = 10$ grams. To find k, we set $t = 25$ to obtain

$$5 = 10e^{-25k},$$

from which we find, after taking logarithms, that $k = (\ln 2)/25$. Thus

$$x(10) = 10e^{-(10\ln 2)/25}$$
$$= 10(2)^{-2/5} \approx 7.579 \text{ grams}.$$

Example 2   Consider a tank holding 100 gallons of water in which are dissolved 50 pounds of salt. Suppose that 2 gallons of brine, each containing 3 pounds of dissolved salt, run into the tank per minute, and the mixture, kept uniform by high-speed stirring, runs out of the tank at the rate of 2 gallons per minute. Find the amount of salt in the tank at any time t.

Solution Let $x(t)$ be the number of pounds of salt at the end of t minutes. Since each gallon of brine that enters the compartment (tank) contains 3 pounds of salt, we know that $i(t) = 6$. On the other hand, $k = 2/100$, since 2 of the 100 gallons in the tank are being removed each minute. Thus equation (1) becomes

$$\frac{dx}{dt} = 6 - \frac{2}{100}x,$$

or

$$\frac{dx}{dt} + \frac{1}{50}x = 6.$$

Multiplying both sides by the integrating factor $e^{t/50}$, we get

$$(e^{t/50}x)' = 6e^{t/50},$$

which has the solution

$$x(t) = e^{-t/50}\left(6\int e^{t/50}\,dt + c\right),$$
$$= 300 + ce^{-t/50}.$$

At $t = 0$ we have

$$50 = x(0) = 300 + c,$$

so that

$$x(t) = 300 - 250e^{-t/50}.$$

Observe that x increases and the ratio of salt to water in the tank approaches the ratio of salt to water in the input stream (3 lb/gal) as time increases.

The fractional transfer coefficient k may be a function of time, as we see in the following example.

Example 3 Suppose that, in Example 2, 3 gallons of brine, each containing 1 pound of salt, run into the tank each minute, and that all other facts are the same. Now $i(t) = 3$, but since the quantity of brine in the tank increases with time, the fraction that is being transferred is $k = 2/(100 + t)$. The numerator of k is the number of gallons being removed, and $100 + t$ is the number of gallons in the tank at time t. The equation describing the system is

$$\frac{dx}{dt} = 3 - \frac{2x}{100 + t}, \qquad (4)$$

or

$$\frac{dx}{dt} + \frac{2x}{100 + t} = 3.$$

The integrating factor in this case is

$$e^{\int \frac{2\,dt}{100+t}} = e^{\ln(100+t)^2} = (100 + t)^2,$$

so we have, successively,
$$[(100+t)^2 x]' = 3(100+t)^2,$$
$$(100+t)^2 x = (100+t)^3 + c,$$
$$x(t) = (100+t) + c(100+t)^{-2}.$$

Setting $t = 0$, we find that $c = -50(100)^2$, so
$$x(t) = 100 + t - 50(1 + t/100)^{-2}.$$

After 100 minutes, we have
$$x(100) = 200 - 50/4 = 187.5 \text{ pounds}$$
of salt in the tank.

The input function $i(t)$ may depend not only on time but also on the quantity present.

Example 4 Systems with periodic inputs and fractional transfer coefficients often occur in biological processes because of diurnal periods of activity. For example, ACTH (adrenocorticotropic hormone) secretion by the anterior pituitary follows a 24-hour cycle that drives the secretion of adrenal steroids in such a way that the levels of these steroids in the blood plasma peak near 8:00 A.M. and are at a minimum near 8:00 P.M. Let $k(t) = A + B \sin \omega t$, with $A > B$, in equation (1), which leads to

$$\frac{dx}{dt} = i(t) - (A + B \sin \omega t)x. \tag{5}$$

Since
$$\int (A + B \sin \omega t) \, dt = At - \frac{B}{\omega} \cos \omega t + c,$$

we may use the integrating factor $e^{At + (B/\omega)(1 - \cos \omega t)}$ on both sides of equation (5):

$$\frac{d}{dt}\left[e^{At+(B/\omega)(1-\cos \omega t)}x(t)\right] = e^{At+(B/\omega)(1-\cos \omega t)}[x' + (A + B \sin \omega t)x]$$
$$= e^{At+(B/\omega)(1-\cos \omega t)} i(t). \tag{6}$$

Integrating both sides of equation (6) from 0 to t, we have

$$e^{At+(B/\omega)(1-\cos \omega t)}x(t)\Big|_0^t = \int_0^t i(t) e^{At+(B/\omega)(1-\cos \omega t)} \, dt,$$

or

$$x(t) = e^{-At-(B/\omega)(1-\cos \omega t)}\left(x(0) + \int_0^t i(t) e^{At+(B/\omega)(1-\cos \omega t)} \, dt\right). \tag{7}$$

Since $1 - \cos \omega t = 2 \sin^2(\omega t/2)$, we can write equation (7) as

$$x(t) = e^{-At - 2B \sin^2(\omega t/2)/\omega}\left(x(0) + \int_0^t i(t) e^{At + 2B \sin^2(\omega t/2)/\omega} \, dt\right). \tag{8}$$

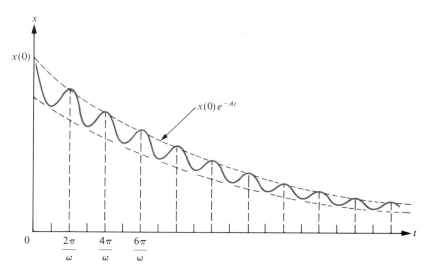

Figure 2.7

If $i(t) = 0$, then $x(t)$ behaves as shown in Figure 2.7, where $x(0)e^{-At}$ is an upper bound, and the factor $e^{-2B\sin^2(\omega t/2)/\omega}$ oscillates between $e^{-2B/\omega}$ and 1.

CURVES OF PURSUIT[†]

Many interesting differential equations arise in studying the path of a pursuer in tracking prey.

Example 5 Suppose that a hawk P at the point $(a, 0)$ spots a pigeon Q at the origin flying along the y-axis at a speed v. The hawk immediately flies toward the pigeon at a speed w. What is the flight path of the hawk?

Solution Let time $t = 0$ at the instant the hawk starts flying toward the pigeon. After t seconds the pigeon is at the point $Q = (0, vt)$ and the hawk at $P = (x, y)$. Since the hawk flies toward the pigeon, the line PQ is tangent to the path (see Figure 2.8). We find that its slope is given by $y' = (y - vt)/x$, so that

$$xy' - y = -vt. \qquad (9)$$

On the other hand, the length of the path traveled by the hawk can be computed by the formula for arc length of basic calculus,

$$wt = \int ds = \int_x^a \sqrt{1 + (y')^2}\, dx. \qquad (10)$$

Solving equations (9) and (10) for t and equating them, we have

$$\frac{y - xy'}{v} = \frac{1}{w} \int_x^a \sqrt{1 + (y')^2}\, dx. \qquad (11)$$

[†] An interesting discussion of this topic is contained in A. Bernhart, "Curves of Pursuit II," *Scripta Mathematica* 23 (1957): 49–66.

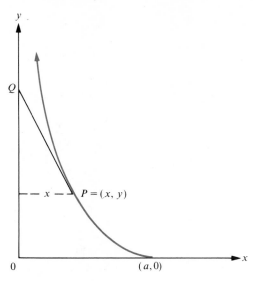

Figure 2.8

Differentiating both sides of equation (11) with respect to x yields

$$xy'' = \frac{v}{w}\sqrt{1 + (y')^2}. \tag{12}$$

Setting $p = y'$, we find that equation (12) becomes

$$xp' = \frac{v}{w}\sqrt{1 + p^2},$$

and we can separate the variables to obtain

$$\frac{dp}{\sqrt{1 + p^2}} = \frac{v}{w}\frac{dx}{x}.$$

Integrating both sides of this equation (see Formula 67 of the integral table in Appendix 1), we have

$$\ln\left(p + \sqrt{1 + p^2}\right) = \frac{v}{w}\ln x - c.$$

Since $p = y' = 0$ when $x = a$ (the slope of the line PQ at $t = 0$ is zero), it follows that $c = (v/w)\ln a$. Exponentiating both sides of this equation yields

$$p + \sqrt{1 + p^2} = \left(\frac{x}{a}\right)^{v/w},$$

which, after some algebra, yields

$$\frac{dy}{dx} = p = \frac{1}{2}\left[\left(\frac{x}{a}\right)^{v/w} - \left(\frac{x}{a}\right)^{-v/w}\right]. \tag{13}$$

If we assume that the hawk flies faster than the pigeon ($w > v$), we may integrate equation (13) to obtain

$$y = \frac{a}{2}\left[\frac{(x/a)^{1+v/w}}{1 + v/w} - \frac{(x/a)^{1-v/w}}{1 - v/w}\right] + c.$$

Since $y = 0$ when $x = a$, we have

$$c = -\frac{a}{2}\left[\frac{1}{1+v/w} - \frac{1}{1-v/w}\right] = \frac{avw}{w^2 - v^2}.$$

The hawk catches the pigeon at $x = 0$ and $y = c = avw/(w^2 - v^2)$. The situation in which the hawk flies no faster than the pigeon ($w \leq v$) is discussed in Problems 12 and 13.

Example 6 A destroyer is in a dense fog, which lifts for an instant, disclosing an enemy submarine on the surface 4 miles away. Suppose that the submarine dives immediately and proceeds at full speed in an unknown direction. What path should the destroyer select to be certain of passing directly over the submarine if its velocity v is three times that of the submarine?

Solution Suppose that the destroyer has traveled 3 miles toward the place where the submarine was originally spotted. Then the submarine lies on the circle of radius 1 mile centered at where it was when spotted (see Figure 2.9), since its velocity is one-third that of the destroyer. Since the location of the submarine can be described easily in polar coordinates, we assume that $r = f(\theta)$ is the path the destroyer must follow to be certain of passing over the submarine, regardless of the direction the latter chooses. The distance traveled by the submarine to the point where the paths cross is then $r - 1$, whereas that of the destroyer (which is three times longer) is given by the arc length formula in polar coordinates:

$$3(r-1) = \int_0^\theta ds = \int_0^\theta \sqrt{(dr)^2 + (r\,d\theta)^2}$$
$$= \int_0^\theta \sqrt{(dr/d\theta)^2 + r^2}\,d\theta. \qquad (14)$$

Differentiating both sides of equation (14) with respect to θ yields the differential equation

$$3r' = \sqrt{(r')^2 + r^2},$$

Figure 2.9

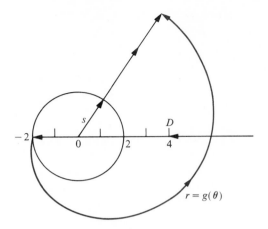

Figure 2.10

which simplifies to $8(r')^2 = r^2$. Taking the square roots of both sides and separating the variables, we have

$$\frac{dr}{r} = \frac{d\theta}{\sqrt{8}},$$

from which it follows that $\ln r = \theta/\sqrt{8} + c$, or

$$r = ce^{\theta/\sqrt{8}}. \quad (15)$$

Since $r = 1$ when $\theta = 0$, it follows that $c = 1$ and the path that the destroyer should follow is the spiral $r = e^{\theta/\sqrt{8}}$ after proceeding 3 miles toward where the submarine was spotted.

It should be noted that this path is not the only curve that the destroyer could follow. For example, suppose that the destroyer has gone 6 miles toward where the submarine was spotted (see Figure 2.10). At this point, we can again follow a path $r = g(\theta)$. Since by now the submarine is 2 miles from the origin, the distance traveled by the submarine to where the paths cross is $r - 2$, whereas the destroyer must go a distance

$$3(r - 2) = \int_{-\pi}^{\theta} \sqrt{(dr/d\theta)^2 + r^2} \, d\theta. \quad (16)$$

Equation (16) again leads to the general solution (15), but in this case $r = 2$ when $\theta = -\pi$, so that $c = 2e^{\pi/\sqrt{8}}$. Thus the spiral that the destroyer must follow is

$$r = 2e^{(\theta - \pi)/\sqrt{8}}.$$

Of course, the submarine captain can evade detection by not going at full speed or by following a curved trajectory.

PROBLEMS 2.6

1. Carbon 14 (^{14}C) has a half-life of 5700 years and is uniformly distributed in the atmosphere in the form of carbon dioxide. Living plants absorb carbon dioxide and maintain a fixed ratio of ^{14}C to the stable element ^{12}C. At death, the disintegration of ^{14}C changes this ratio. Compare the concentrations of ^{14}C in two identical pieces of wood, one of them freshly cut, the other 2000 years old.

2. Radioactive iodine ^{131}I is often used as a tracer in medicine. Suppose that a given dose Q_0 is

injected into the bloodstream at time $t = 0$ and is evenly distributed in the entire bloodstream before any loss occurs. If the daily removal rate of the iodine is k_1 percent by the kidney and k_2 percent by the thyroid gland, what percentage of the initial amount remains in the blood after one day?

3. Suppose that an infected individual is introduced into a population of size N, all of whom are susceptible to the disease. If we assume that the rate of infection is proportional to the product of the numbers of infectives and susceptibles present, what is the number of infections at any time t? Let k be the **specific infection rate**.

4. A tank initially contains 100 liters of fresh water. Brine containing 20 grams per liter of salt flows into the tank at the rate of 4 liters per minute, and the mixture, kept uniform by stirring, runs out at the same rate. How long does it take for the quantity of salt in the tank to become 1 kilogram?

5. Given the same data as in Problem 4, determine how long it takes for the quantity of salt in the tank to increase from 1 to 1.5 kilograms.

6. A tank contains 100 gallons of fresh water. Brine containing 2 pounds per gallon of salt runs into the tank at the rate of 4 gallons per minute, and the mixture, kept uniform by stirring, runs out at the rate of 2 gallons per minute. Find (a) the amount of salt present when the tank has 120 gallons of brine; (b) the concentration of salt in the tank at the end of 20 minutes.

7. A tank contains 50 liters of water. Brine containing x grams per liter of salt enters the tank at the rate of 1.5 liters per minute. The mixture, thoroughly stirred, leaves the tank at the rate of 1 liter per minute. If the concentration is to be 20 grams per liter at the end of 20 minutes, what is the value of x?

8. A tank holds 500 gallons of brine. Brine containing 2 pounds per gallon of salt flows into the tank at the rate of 5 gallons per minute, and the mixture, kept uniform, flows out at the rate of 10 gallons per minute. If the maximum amount of salt is found in the tank at the end of 20 minutes, what was the initial salt content of the tank?

9. Phosphate excretion in human metabolism is at a minimum at 3:00 A.M. and rises to a peak at 6:00 P.M. If the rate of excretion is

$$\frac{1}{3} - \frac{1}{6}\cos\frac{\pi}{12}(t-6)$$

grams per hour at time t hours ($0 \le t \le 24$), the body contains 400 grams of phosphate, and the patient is only allowed to drink water, what is the amount of phosphate in the patient's body at all times t for $0 \le t \le 24$?

10. Suppose that in Problem 9 the patient is allowed three meals during the day in such a way that the body takes in phosphate at a rate given by

$$i(t) = \begin{cases} 1/3 \text{ g/hr}, & 8 \le t \le 16, \\ 0 \text{ g/hr}, & \text{otherwise.} \end{cases}$$

Obtain a formula for the amount of phosphate in the patient's body at all times t. When is it at a maximum?

11. Given a one-compartment system with k constant and $i(t) = A + B \sin \omega t$, $A > B$, find a solution of the system. How does it differ from that of the system in which the input is constant and the fractional transfer coefficient is periodic [see equation (8)]?

*12. Suppose that $v = w$ in Example 5. Prove that

$$y = \frac{a}{2}\left\{\frac{1}{2}\left[\left(\frac{x}{a}\right)^2 - 1\right] - \ln\frac{x}{a}\right\},$$

so that the hawk never catches the pigeon. Using equations (10) and (13), show that the distance between the hawk and the pigeon is $(x^2 + a^2)/2a$ whenever the hawk is at the point (x, y) on the path—and thus that the hawk does not come as close as $a/2$ to the pigeon.

*13. Suppose that $v > w$ in Example 5. Show that

$$y = \frac{a}{2}\left[\frac{(x/a)^{1+(v/w)} - 1}{1 + (v/w)} + \frac{(a/x)^{(v/w)-1} - 1}{(v/w) - 1}\right],$$

so that the hawk never catches the pigeon. Find the distance between the hawk and the pigeon in terms of the variable x.

14. Let the y-axis and the line $x = b$ be the banks of a river with a current of velocity v (in the negative y-direction). A man is at the origin, and his dog is at the point $(b, 0)$. When the man calls, the dog enters the river, swimming toward the man at a constant velocity w (greater than v). Describe the path of the dog.

15. Where does the dog of Problem 14 reach the bank if $w = v$?

16. Show that the dog of Problem 14 never reaches the bank if $w < v$. Suppose that the man walks down river at velocity v while calling his dog. Is the dog now able to land?

17. Suppose the destroyer in Example 6 proceeds to where the submarine was sighted, then turns 90° left and proceeds 2 miles before beginning the spiral search pattern. What is the equation of the path the destroyer should now follow?

18. Suppose the destroyer in Example 6 is only twice as fast as the submarine and the submarine is spotted when it is 3 miles away. Find a path that guarantees that the destroyer passes over the submarine, assuming that both ships execute the same maneuvers as those given in the example.

19. Three snails at the corners of an equilateral triangle of side a begin to move with the same velocity, each toward the snail to its right. Centering the triangle at the origin, with one vertex along the polar axis, find an equation for the slime path left by the snail that started on the polar axis.

20. Consider the same problem with four snails at the corners of the square $[0, a] \times [0, a]$. How far do the snails travel before they meet?

*†21. A hawk is flying 100 feet below a sparrow that is 50 feet below an eagle. The sparrow flies straight forward horizontally, while both the eagle and hawk fly directly toward the sparrow. The hawk flies twice as fast as the sparrow and reaches it at the same instant the eagle does. How far does each bird fly and how fast does the eagle fly?

_____

† Based on *American Mathematical Monthly* 40 (1933): 436–437.

2.7 Successive Approximations: If Time Permits

In Sections 2.1–2.6 we discussed a number of techniques that can be used to solve first-order differential equations or initial-value problems. In this section we describe an iterative technique, due to Picard, that provides an alternate approach to solving initial-value problems. The main use of this method, however, is theoretical: it forms the basis for the proof of Picard's theorem (Theorem 1.2.1). The technique that we develop here will be used in Appendix 3 to prove Theorem 1.2.1.

Consider the initial-value problem

$$y' = f(x, y), \qquad y(x_0) = y_0. \tag{1}$$

If we integrate both sides of the differential equation in (1) from x_0 to x with respect to x, we have

$$\int_{x_0}^{x} y'(x) \, dx = \int_{x_0}^{x} f(x, y(x)) \, dx,$$

or

$$y(x) - y(x_0) = \int_{x_0}^{x} f(t, y(t)) \, dt. \tag{2}$$

(We have changed the variable of integration on the right-hand side to t to avoid confusion.)

Rewriting (2), we get (since $y(x_0) = y_0$)

$$y(x) = y_0 + \int_{x_0}^{x} f(t, y(t)) \, dt. \tag{3}$$

Equation (3) is an alternative way of writing initial-value problem (1): note that if we set $x = x_0$ in (3), then $y(x_0) = y_0$, since the integral is zero. Furthermore, if we differentiate both sides of (3), we obtain the differential equation

$$y'(x) = f(x, y(x)). \tag{4}$$

Thus (1) and (3) are equivalent ways of writing the same initial-value problem.

We now define a sequence of functions $\{y_n(x)\}$, called **Picard**[†] **iterations**, by successive formulas:

$$\begin{aligned} y_0(x) &= y_0 \\ y_1(x) &= y_0 + \int_{x_0}^{x} f[t, y_0(t)] \, dt \\ y_2(x) &= y_0 + \int_{x_0}^{x} f[t, y_1(t)] \, dt \\ &\vdots \\ y_n(x) &= y_0 + \int_{x_0}^{x} f[t, y_{n-1}(t)] \, dt \end{aligned} \tag{5}$$

We will show in Appendix 3 that under the conditions of Theorem 1.2.1 (p. 13) the Picard iterations defined by (5) converge uniformly to a solution of equation (2). We illustrate the process of this iteration by a simple example.

Example 1   Consider the initial-value problem

$$y'(x) = y(x), \qquad y(0) = 1. \tag{6}$$

As we know, equation (6) has the unique solution $y(x) = e^x$. In this case, the function $f(x, y)$ in equation (1) is given by $f(x, y(x)) = y(x)$, so the Picard iterations defined by (5) yield, successively,

$$y_0(x) = y_0 = 1,$$
$$y_1(x) = 1 + \int_0^x (1) \, dt = 1 + x,$$
$$y_2(x) = 1 + \int_0^x (1 + t) \, dt = 1 + x + \frac{x^2}{2},$$
$$y_3(x) = 1 + \int_0^x \left(1 + t + \frac{t^2}{2}\right) dt = 1 + x + \frac{x^2}{2!} + \frac{x^3}{3!},$$

and clearly,

$$y_n(x) = 1 + x + \frac{x^2}{2!} + \cdots + \frac{x^n}{n!} = \sum_{k=0}^{n} \frac{x^k}{k!}.$$

Hence

$$\lim_{n \to \infty} y_n(x) = \sum_{k=0}^{\infty} \frac{x^k}{k!} = e^x,$$

since the series is the Maclaurin series of e^x.

[†] Charles Emile Picard (1856–1941), one of the most eminent French mathematicians of the past century, made several outstanding contributions to mathematical analysis. Picard published the results we discuss in this section in 1890 and 1893: *Journal de Mathematique* (4) 6 (1890): 145–210; 9 (1893): 217–271.

You should not be fooled by the relative ease with which we obtained the solution in Example 1. In general, the Picard iterations quickly give rise to formidable integrals that are often difficult or impossible to integrate.

Example 2 If we apply Picard iterations to the initial-value problem

$$y' = e^{x^2}y, \qquad y(0) = 1,$$

we get

$$y_0(x) = y_0 = 1,$$
$$y_1(x) = 1 + \int_0^x e^{t^2}\, dt,$$

and this integral cannot be written in terms of elementary functions.

We will use Picard iterations in Appendix 3 to prove Theorem 1.2.1. However, we can give a simple existence-uniqueness proof for linear first-order equations without using this powerful tool.

THEOREM 1: UNIQUENESS THEOREM

Suppose that $a(x)$ and $f(x)$ are continuous functions. Then the linear initial-value problem

$$\frac{dy}{dx} + a(x)y = f(x), \qquad y(x_0) = y_0 \tag{7}$$

has at most one solution.

Remark This theorem does not claim that a solution exists. However, a global existence result for this equation has already been proved. One solution to equation (7) is given by equation (2.3.11) (p. 48) for an appropriate constant C. All integrals in equation (2.3.11) exist because $a(x)$ and $f(x)$ are assumed continuous.

Proof Suppose that both $y_1(x)$ and $y_2(x)$ satisfy (7). Let $y_3(x) = y_1(x) - y_2(x)$. We must show that $y_3(x) \equiv 0$. But, since y_1 and y_2 satisfy (7),

$$y_3' + a(x)y_3 = (y_1 - y_2)' + a(x)(y_1 - y_2)$$
$$= [y_1' + a(x)y_1] - [y_2' + a(x)y_2] = f(x) - f(x) = 0.$$

Also, $y_3(x_0) = y_1(x_0) - y_2(x_0) = y_0 - y_0 = 0$. Thus y_3 satisfies the initial-value problem

$$\frac{dy}{dx} + a(x)y = 0, \qquad y(x_0) = 0. \tag{8}$$

Multiplying both sides of the differential equation in (8) by the integrating factor

$$e^{\int_{x_0}^x a(t)\, dt}$$

yields

$$\left[e^{\int_{x_0}^x a(t)\, dt} y_3(x) \right]' = e^{\int_{x_0}^x a(t)\, dt} [y_3' + a(x)y_3] = 0,$$

so

$$e^{\int_{x_0}^{x} a(t)\, dt} y_3(x) = C \text{ (a constant)}.$$

Thus

$$y_3(x) = Ce^{-\int_{x_0}^{x} a(t)\, dt}, \qquad (9)$$

but by the initial condition in (8),

$$0 = y_3(x_0) = Ce^{-\int_{x_0}^{x_0} a(t)\, dt} = C, \qquad (10)$$

so that $C = 0$. Substituting this value of C in (9) proves that $y_3(x) \equiv 0$ is the only solution to (8). This completes the proof. ∎

PROBLEMS 2.7

Use Picard iterations to solve the following initial-value problems (in closed form).

1. $y' = -y$, $\quad y(0) = 1$
2. $y' = 2y$, $\quad y(0) = 5$
3. $y' = x + y$, $\quad y(0) = 0$
4. $y' = 2xy$, $\quad y(0) = 1$

2.8 An Excursion: First-Order Difference Equations

Differential equations arise in physics and biology when we consider the instantaneous rate of change of one variable with respect to another. However, in many situations it may be more meaningful to study a process in a sequence of well-defined stages or steps. For example, the available data in some experiment may be organized in fixed increments, such as the length to the nearest inch, age in weeks, or populations measured every year. Problems of this nature are often posed as **difference equations**, that is, as equations involving the differences between values of the variable at different stages.

Example 1 A patient in a hospital is suddenly administered oxygen. Let V be the volume of gas contained in the lungs after inspiration and V_D the amount present at the end of expiration (commonly called the *dead space*). Assuming uniform and complete mixing of the gases in the lungs, what is the concentration of nitrogen in the lungs at the end of the nth inspiration?

Solution The amount of nitrogen in the lungs at the end of the nth inspiration must equal the amount in the dead space at the end of the $(n-1)$st expiration. If x_n is the concentration of nitrogen in the lungs at the end of the nth inspiration, then

$$Vx_n = V_D x_{n-1}. \qquad (1)$$

Subtracting Vx_{n-1} from both sides of (1), we have

$$V(x_n - x_{n-1}) = (V_D - V)x_{n-1}. \qquad (2)$$

The difference $x_n - x_{n-1}$ is the discrete analogue of a derivative, since it measures the change in concentration of nitrogen in the lungs. Thus (2) is a discrete version of a first-order differential equation.

If we can find an expression (in n) for the concentrations x_n that satisfies (1) for all values of $n = 1, 2, 3, \ldots$, then we say we have a **solution** for difference equation (1). It is easy to solve (1), since

$$x_n = \frac{V_D}{V} x_{n-1} = \left(\frac{V_D}{V}\right)^2 x_{n-2} = \cdots = \left(\frac{V_D}{V}\right)^n x_0,$$

where x_0 is the concentration of nitrogen in the air.

Example 2 In modeling some fisheries, it is often assumed that each fish in the parent stock of the $(n-1)$st generation gives birth, on an average, before it dies to k fish that reach adulthood. If in each generation we harvest h fish from the adult stock, then the remainder (called the *escapement*) forms the parent stock of the nth generation. If we use P_n to denote the population in the nth generation, then

$$P_n = kP_{n-1} - h. \tag{3}$$

Since equation (3) holds for each generation,

$$P_{n-1} = kP_{n-2} - h.$$

Subtracting the two equations, we have the difference equation

$$P_n - P_{n-1} = (kP_{n-1} - h) - (kP_{n-2} - h) = k(P_{n-1} - P_{n-2}),$$

or

$$P_n = (k+1)P_{n-1} - kP_{n-2}, \tag{4}$$

independent of the harvest h.

Examples 1 and 2 provide instances of the use of difference equations. Formally, any equation that relates the values of a variable at different stages is called a difference equation.

If N_1 and N_2 are, respectively, the largest and smallest values of n that occur in the equation, then the **order** of the difference equation is $N_1 - N_2$.

Equation (4) is a difference equation of order $n - (n-2) = 2$, whereas equations (1), (2), and (3) are of the first order.

A difference equation is said to be **linear** if it can be written in the form

$$y_{n+k} + a_n y_{n+k-1} + \cdots + b_n y_{n+1} + c_n y_n = f_n, \tag{5}$$

consisting only of sums of the unknown terms $y_{n+k}, y_{n+k-1}, \ldots, y_n$. Here we consider only the first-order linear difference equation

$$y_{n+1} + a_n y_n = f_n. \tag{6}$$

In Section 3.9 we will consider linear difference equations of higher orders.

We say that Equation (6) is **homogeneous** if $f_n = 0$ for all n; otherwise (6) is said to be **nonhomogeneous**. In Examples 1 and 2, (1) is homogeneous, while (3) is nonhomogeneous.

In Example 1, we solved the given homogeneous linear first-order equation by successive iteration. In general, if $y_n - ay_{n-1} = 0$, or

$$y_n = ay_{n-1}, \tag{7}$$

then successive iteration gives

$$y_n = ay_{n-1} = a(ay_{n-2}) = \cdots = a^n y_0, \tag{8}$$

which is the general solution of equation (7). If we compare equation (8) with the general solution $y = y(0)e^{ax}$ of the first-order linear differential equation

$$y' - ay = 0,$$

it is easy to see that y_0 corresponds to $y(0)$ and the product a^{n+1} corresponds to e^{ax}. Since exponential functions were of fundamental importance in solving linear differential equations, it is not unreasonable to hope that products may be useful in solving linear difference equations. In particular, if

$$y_n = a_{n-1} y_{n-1}, \tag{9}$$

then

$$y_n = a_{n-1} y_{n-1} = a_{n-1}(a_{n-2} y_{n-2}) = \cdots = (a_{n-1} \cdot a_{n-2} \cdots a_0) y_0,$$

or, using a capital pi to denote a product,

$$y_n = y_0 \prod_{j=0}^{n-1} a_j. \tag{10}$$

Next we examine what happens if the equation is nonhomogeneous, by considering

$$y_n = a y_{n-1} + f_{n-1}, \tag{11}$$

with a constant, for known values of the sequence f_n. Using successive iterations, we have

$$y_1 = a y_0 + f_0$$
$$y_2 = a y_1 + f_1 = a^2 y_0 + a f_0 + f_1$$
$$y_3 = a y_2 + f_2 = a^3 y_0 + a^2 f_0 + a f_1 + f_2$$
$$\vdots$$
$$y_n = a^n y_0 + \sum_{k=0}^{n-1} a^{n-k} f_k. \tag{12}$$

Comparing equation (12) to the solution $y = e^{ax}[c + \int f(x) e^{-ax} dx]$ of

$$y' = ay + f(x),$$

we again see that a^n corresponds to e^{ax}, and

$$\sum_{k=0}^{n-1} a^{n-k} f_k = a^n \sum_{k=0}^{n-1} a^{-k} f_k$$

corresponds to $e^{ax} \int f(x) e^{-ax} dx$.

If we apply the method above to equation (3) in Example 2, we have

$$P_1 = k P_0 - h$$
$$P_2 = k P_1 - h = k^2 P_0 - kh - h$$
$$\vdots$$
$$P_n = k^n P_0 - h \sum_{j=0}^{n-1} k^j. \tag{13}$$

But the first $n-1$ terms of a geometric progression satisfy the identity

$$1 + k + k^2 + \cdots + k^{n-1} = \frac{k^n - 1}{k - 1}, \tag{14}$$

so

$$P_n = k^n P_0 - h\left(\frac{k^n - 1}{k - 1}\right). \tag{15}$$

Finally, we look at the general first-order linear equation

$$y_n = a_{n-1} y_{n-1} + f_{n-1}. \tag{16}$$

Proceeding inductively, we obtain

$$y_1 = a_0 y_0 + f_0, \qquad y_2 = a_1 y_1 + f_1 = a_1 a_0 y_0 + a_1 f_0 + f_1, \ldots,$$

and, in general

$$y_n = (a_{n-1} a_{n-2} \cdots a_1 a_0) y_0 + (a_{n-1} \cdots a_1) f_0 + (a_{n-1} \cdots a_2) f_1 + \cdots$$
$$+ a_{n-1} f_{n-2} + f_{n-1}$$
$$= y_0 \prod_{k=0}^{n-1} a_k + \sum_{k=0}^{n-1} \left(\prod_{j=k+1}^{n-1} a_j\right) f_k, \tag{17}$$

where we define the "empty" product by

$$\prod_{j=n}^{n-1} a_j = 1.$$

If we compare equation (17) to the solution

$$y = e^{\int a(x)\,dx}\left[c + \int f(x) e^{-\int a(x)\,dx}\,dx\right]$$

of the linear differential equation $y' = a(x)y + f(x)$, it is apparent that the product $\prod_{k=0}^{n-1} a_k$ corresponds to the exponential $e^{\int a(x)\,dx}$.

PROBLEMS 2.8

In Problems 1–9 find the general solution of each difference equation and a particular solution when an initial condition is specified.

1. $y_{n+1} - y_n = 2^{-n}$
2. $y_{n+1} = \dfrac{n+5}{n+3} y_n$
3. $y_{n+1} - 3y_n = 3y_{n+1} - y_n$
4. $2y_{n+1} = y_n$, $y_0 = 1$
5. $(n+1) y_{n+1} = (n+2) y_n$, $y_0 = 1$
6. $y_{n+1} = n y_n$
7. $y_{n+1} - 5 y_n = 2$, $y_0 = 2$
8. $y_{n+1} - n y_n = n!$, $y_0 = 5$
9. $y_{n+1} - e^{-2n} y_n = e^{-n^2}$
10. Radium transmutes at a rate of 1 percent every 25 years. Consider a sample of r_0 grams of radium. If r_n is the amount of radium remaining in the sample after $25n$ years, obtain a difference equation for r_n and find its solution. How much radium is left after 100 years?

11. A fair coin is marked "1" on one side and "2" on the other side. The coin is tossed repeatedly and a cumulative score of the outcomes is recorded. Define P_n to be the probability that at some time the cumulative score takes on the value n. Prove that $P_n = 1 - (1/2) P_{n-1}$. Assuming that $P_0 = 1$, derive the formula for P_n.

12. In constructing a mathematical model of a population, it is assumed that the probability P_n that a couple produces exactly n offspring satisfies the equation $P_n = 0.7 P_{n-1}$. Find P_n in terms of P_0 and determine P_0 from the fact that

$$P_0 + P_1 + P_2 + \cdots = 1.$$

13. An alternative model to Problem 12 is given by $P_n = (1/n)P_{n-1}$. For this model, find P_n in terms of P_0 and prove that $P_0 = 1/e$.

14. Let x_k denote the number of permutations of n objects taken k at a time. For every permutation of k objects, we can get a total of $n - k$ permutations of $k + 1$ objects by taking one of the remaining $n - k$ objects and placing it at the end. Thus $x_{k+1} = (n-k)x_k$. Prove that the number of permutations of n objects taken k at a time is $n!/(n-k)!$

15. If we let x_n in Problem 14 denote the number of combinations of the n objects (order does not count), then every permutation of $k + 1$ objects occurs (in different orders) $k + 1$ times. Thus

$$x_{k+1} = \frac{n-k}{k+1}x_k.$$

Prove that the number of combinations of n objects taken k at a time is

$$\frac{n!}{(n-k)!k!} \equiv \binom{n}{k}.$$

The expression on the right-hand side is called the **binomial coefficient**.

Review Exercises for Chapter 2

Find the general solution to each of Exercises 1–30. When an initial condition is given, find the particular solution that satisfies the condition.

1. $x\dfrac{dy}{dx} = y^2$, $y(1) = 1$

2. $\dfrac{dy}{dx} = y\sqrt{1-x}$

3. $\dfrac{dy}{dx} = \dfrac{\sqrt{1-y^2}}{x}$

4. $x\dfrac{dy}{dx} = \tan y$, $y(1) = \dfrac{\pi}{2}$

5. $\dfrac{dy}{dx} + x = x(y^2 + 1)$

6. $xy' = y(1 - 2y)$, $y(1) = 2$

7. $yy' = \cos x$, $y(\pi) = 0$

8. $y' = xy(2 - 3y)$

9. $y = xy' + \dfrac{1}{y'}$

10. $y - xy' = \sqrt{1 - (y')^2}$

11. $y' - xy = 0$, $y(1) = 1$

12. $xy' - y = x$, $y(1) = 1$

13. $y' - (\sin x)y = \sin x$

14. $y' - \dfrac{1}{x}y = e^x$

15. $(1 + x^2)y' + xy = \sqrt{1 + x^2}$

16. $y' - (\cos x)y = x^2$

17. $xy' - 2y = x^2$, $y(1) = 1$

18. $xy' + (1-x)y = xe^x$, $y(1) = e$

19. $y' - xy = \begin{cases} 1, & x \le 0 \\ 0, & x > 0 \end{cases}$

20. $y' + xy = \begin{cases} x, & x \le 1 \\ 1, & x > 1 \end{cases}$

21. $y' = \dfrac{x-y}{x+2y}$, $y(0) = 1$

22. $xy' = 2y + \sqrt{y^2 + x^2}$

23. $xy' = -y + \sqrt{xy+1}$

24. $xy' = \sqrt{x^2y^2 - 1} - y$, $y(1) = 2$

25. $y' = y + xy^2$

26. $y' - xy = e^x y^3$

27. $(y - e^y \sec^2 x)\,dx + (x - e^y \tan x)\,dy = 0$

28. $(2x^2y^3 - y^2)\,dx + (x^3y^2 - x)\,dy = 0$

29. $\dfrac{dy}{dx} = \dfrac{3y^4 + 3x^2y^3 - x^4y}{xy^3 - 2x^5}$

30. $\dfrac{dy}{dx} + \dfrac{x + y\sqrt{x^2 + y^2}}{y + x\sqrt{x^2 + y^2}} = 0$

31. A paper mill is located where a river enters a lake of volume 10^9 cubic meters. The river, which has a constant flow of 1000 cubic meters per second, is the lake's only inlet. Assume that at time $t = 0$, the paper mill begins pumping pollutants into the river at the rate of 1 m³/sec, and that the inflow and outflow of the lake are constant. How high a concentration of pollutant is there in the lake after 10 hours? after 100 hours? after one year?

32. Assume that the paper mill in Exercise 31 stops polluting the river at the end of 1 hour. Find an expression for the concentration of pollutant in the lake at all time t.

33. Assume the paper mill in Exercise 31 pollutes the river for 1 hour each day. Find an expression for the concentration of pollutant in the lake at all time t. What is the maximum concentration that the pollution reaches in the lake?

34. A $20' \times 12' \times 8'$ room contains five chain smokers who are playing poker. An exhaust fan is removing 10 ft^3/min of smoky air, which is replaced by pure air seeping in under the door. Each chain smoker is contributing 0.1 ft^3/min of smoke to the room. Find an expression for the concentration of smoke in the room at time t, assuming the room contains no smoke at time $t = 0$.

35. **Newton's law of cooling** can be used to estimate the time of death in a homicide.[†] Assume that at the time of death the body temperature is 98.6°F, when the corpse is discovered its temperature is 75°F, and 1 hour after discovery it is 70°F. If the constant ambient temperature is 68°F, how long before the corpse was discovered did the homicide occur?

36. **Stefan's law** of radiation states that the rate of change of temperature from a body at absolute temperature T is $T' = k(T^4 - T_0^4)$, where T_0 is the absolute temperature of the surrounding medium.

    a. Solve this differential equation.

    *b. Show that if $T - T_0$ is small compared to T_0, then Newton's law of cooling is a close approximation to Stefan's law.

37. Another equation that has been proposed to model population growth is the **Gompertz**[‡] equation,
$$\frac{dP}{dt} = P(a - b \ln P).$$
Find a solution to this equation and determine its behavior as $t \to \infty$.

38. A. G. W. Cameron,[§] in an article concerning the processes in the primitive solar nebula, obtained the first-order differential equation
$$\frac{dx}{dt} = \frac{ax^{5/6}}{(b - Bt)^{3/2}},$$
where a, b, and B are constants. Solve this equation.

39. A boat weighing 4000 pounds drifts away from a dock at 3 feet per second. What is the minimum distance it drifts if a 180-pound crewman exerts a force equal to his weight on a rope tied to the bow of the boat?

[†] D. A. Smith, "The Homicide Problem Revisited," *The Two Year College Mathematics Journal* 9 (1978): 141–145.

[‡] Named after Benjamin Gompertz (1779–1865), an English mathematician.

[§] "Accumulation Processes in the Primitive Solar Nebula," *Icarus* 18 (1973): 407–450.

40. L. L. Thurstone[||] used the separable differential equation
$$y' = \frac{2k}{\sqrt{m}} [y(1-y)]^{3/2}$$
to describe the state of a learner $y(t)$ at time t, where k and m are positive constants depending on the learner and the complexity of the task, respectively. Find a solution to this equation.

41. A 6-foot chain weighing 10 pounds per foot is placed on a frictionless table so that 1 foot of chain hangs over the edge of the table. Find an equation describing the amount of chain still on the table for all time $t \geq 0$. [*Hint:* The weight of the chain that is falling changes with time.]

*42. A power cable, hanging from fixed towers, has a weight of w pounds per foot.

   a. Let $y(x)$ be the position of the cable x feet horizontally away from a tower and T_H be the horizontal component of tension in the cable. Show that
$$y'(x + \Delta x) - y'(x) = \frac{w}{T_H} \Delta s,$$
where Δs is the length of the cable over the horizontal interval of length Δx.

   b. Use part (a), the Pythagorean theorem, and limits to deduce the differential equation
$$y'' = \frac{w}{T_H} \sqrt{1 + (y')^2}.$$

   c. Solve the differential equation in part (b) to obtain an expression for $y'(x)$, assuming $y'(0) = 0$.

   d. If $y(0) = 0$, find $y(x)$.

43. Suppose a constant capacitor is connected in series to an emf whose voltage is a sine wave. Show that the current is 90° out of phase with the voltage.

44. Repeat Exercise 43 with the capacitor replaced by an inductor. What can you say in this case?

In Exercises 45–47 use Picard iterations to solve the given initial-value problem.

45. $y' = y + 1$, $\quad y(0) = 0$
46. $y' = 3x^2 y$, $\quad y(0) = 1$
47. $y' = y/x$, $\quad y(1) = 1$
    [*Hint:* $x_0 = 1$ in equation (2.7.5).]

[||] "The Learning Function," *Journal of General Psychology* 3 (1930): 469–493.

3

Second- and Higher-Order Linear Differential Equations

3.1 Theory of Linear Differential Equations

Although there is no procedure for explicitly solving arbitrary differential equations, systematic methods do exist for certain classes of differential equations. In this chapter we study a class of problems for which there are always unique solutions and present some methods for calculating them.

Linear Equations

Recall that a differential equation is **linear** if it does not involve nonlinear functions (squares, exponentials, etc.) or products of the dependent variable and its derivatives. Thus $y'' + (x^3 \sin x)^5 y' + y = \cos x^3$ is linear, whereas $y'' + (y')^2 + y = 0$ is nonlinear. The most general second-order linear equation can be written

$$y''(x) + a(x)y'(x) + b(x)y(x) = f(x), \qquad (1)$$

whereas the most general third-order linear equation can be written

$$y'''(x) + a(x)y''(x) + b(x)y'(x) + c(x)y(x) = f(x), \qquad (2)$$

where $a(x)$, $b(x)$, $c(x)$, and $f(x)$ are functions of the independent variable x only. Equations (1) and (2) are special cases of the **general linear nth-order equation**:

$$y^{(n)}(x) + a_{n-1}(x)y^{(n-1)}(x) + \cdots + a_1(x)y'(x) + a_0(x)y(x) = f(x). \qquad (3)$$

Homogeneous and Nonhomogeneous Equations

If the function $f(x)$ is identically zero, we say that equations (1), (2), and (3) are **homogeneous**. Otherwise, they are **nonhomogeneous**.

Example 1

a. The equation $y'' + 2xy' + 3y = 0$ is homogeneous.

b. The equation $y'' + 2xy' + 3y = e^x$ is nonhomogeneous.

Constant and Variable Coefficients

If the coefficient functions $a(x)$ and $b(x)$ are constants, $a(x) = a$ and $b(x) = b$, the equation is said to have **constant coefficients**. (As we see below, linear differential equations with constant coefficients are the easiest to solve.) If either $a(x)$ or $b(x)$ is not constant, the equation is said to have **variable coefficients**.

Example 2

a. The equation $y'' + 3y' - 10y = 0$ has constant coefficients.

b. The equation $y'' + 3xy' - 10x^2 y = 0$ has variable coefficients.

In Section 2.3 we discussed the general first-order linear equation
$$y'(x) + a(x)y(x) = f(x).$$
If $a(x)$ and $f(x)$ are continuous, this equation has infinitely many solutions, since it involves an arbitrary constant [see equation (2.3.11), p. 48]. As we have seen, this arbitrary constant can be determined if one condition $y(x_0) = y_0$ is given. In this case, the equation has a unique solution. We can restate this basic fact as follows:

If $a(x)$ and $f(x)$ are continuous, then the equation
$$y'(x) + a(x)y(x) = f(x)$$
has one and only one solution that satisfies the initial condition $y(x_0) = y_0$.

This is a very nice result, for it tells us that every linear first-order equation with a given initial condition has a unique solution. We need only set about finding it. It turns out that this special property holds for linear initial-value problems of any order. The only difference is that in order to have a unique solution to a second-order equation, we must specify two initial conditions, for a third-order equation three conditions, and so on. One case of the following central theorem is proved in Appendix 3 (Theorem 8).

THEOREM 1   EXISTENCE-UNIQUENESS THEOREM FOR LINEAR INITIAL-VALUE PROBLEMS

Let $a_1(x), a_2(x), \ldots, a_n(x)$, and $f(x)$ be continuous functions on the interval $[x_1, x_2]$, and let $c_0, c_1, c_2, \ldots, c_{n-1}$ be n given constants. Then there exists a unique function $y(x)$ that satisfies the linear differential equation

$$y^{(n)}(x) + a_{n-1}(x)y^{(n-1)}(x) + a_{n-2}(x)y^{(n-2)}(x) + \cdots + a_0(x)y(x) = f(x)$$

(3)

on $[x_1, x_2]$ *and* the n initial conditions

$$y(x_0) = c_0, \quad y'(x_0) = c_1, \quad y''(x_0) = c_2, \ldots, y^{(n-1)}(x_0) = c_{n-1} \quad (4)$$

for some value x_0 in $[x_1, x_2]$.

Note The conditions given in (4) all involve evaluations of the unknown function y and its derivatives at the *same* point x_0. This is a *crucial* requirement for the existence and uniqueness of a solution.

There is another type of problem involving the differential equation (3) in which conditions at more than one point are given. For example, we might specify $y(x_1) = k_1$, and $y'(x_2) = k_2, \ldots, y^{(n-1)}(x_{n-1}) = k_{n-1}$. Conditions of this sort are called **boundary conditions**, and a differential equation together with a set of boundary conditions is called a **boundary value problem**. Boundary value problems are much more difficult to handle; they will be discussed in a later chapter.

It is important to note that the existence and uniqueness of a solution, guaranteed by Theorem 1 for initial-value problems, does not hold for boundary value problems.

Example 3   Observe that the boundary value problem

$$y'' + y = 0, \quad y(0) = y(\pi) = 0,$$

has infinitely many solutions:

$$y = c \sin x, \quad \text{for any constant } c.$$

To check, note that $\sin 0 = \sin \pi = 0$, so the boundary conditions are satisfied, and

$$(c \sin x)'' + c \sin x = -c \sin x + c \sin x = 0.$$

We will be able to show in Section 3.4 (see Problem 3.4.13) that the boundary value problem

$$y'' + y = 0, \quad y(0) = 0, \quad y(\pi) = 1$$

has *no* solution. On the other hand, the initial-value problem

$$y'' + y = 0, \quad y(0) = 0, \quad y'(0) = 1$$

has the unique solution $y = \sin x$; the initial conditions hold since $\sin 0 = 0$, $(\sin x)' = \cos x$, and $\cos 0 = 1$, and

$$(\sin x)'' + \sin x = -\sin x + \sin x = 0.$$

If we apply Theorem 1 to the second-order equation (1), we have the following result.

> If $a(x)$, $b(x)$, and $f(x)$ are continuous functions, then the equation
>
> $$y''(x) + a(x)y'(x) + b(x)y(x) = f(x) \quad (5)$$
>
> has a unique solution that satisfies the conditions
>
> $$y(x_0) = y_0, \quad y'(x_0) = y_1$$
>
> for any real numbers x_0, y_0, and y_1.

For simplicity we limit most of our discussion in this chapter to second-order linear equations (and associated systems). We emphasize, however, that *every* result we prove can be extended to higher-order linear equations (see Section 3.8).

The requirement that the functions $a(x)$, $b(x)$, and $f(x)$ in (5) be continuous is also an essential requirement for the existence of a unique solution, as the following example demonstrates.

Example 4   Verify that the function
$$y = cx^3 + x$$
is a solution (for any constant c) of the initial-value problem
$$x^2 y'' - 3xy' + 3y = 0, \quad y(0) = 0, \quad y'(0) = 1;$$
that is, the problem does not have a unique solution—it has infinitely many solutions.

Solution   Since $y' = 3cx^2 + 1$, the initial conditions are satisfied. Then $y'' = 6cx$, and
$$x^2(6cx) - 3x(3cx^2 + 1) + 3(cx^3 + x) = 6cx^3 - 9cx^3 - 3x + 3cx^3 + 3x = 0.$$

Theorem 1 does not apply here, because in order to obtain a differential equation of the form (5), we must divide $x^2 y'' - 3xy' + 3y = 0$ by x^2:
$$y'' - \frac{3}{x} y' + \frac{3}{x^2} y = 0.$$

Then $a(x) = -3/x$ and $b(x) = 3/x^2$. Theorem 1 requires that $a(x)$ and $b(x)$ be continuous in an interval containing x_0 (here $x_0 = 0$). However, there is *no* interval containing 0 such that $-3/x$ and $3/x^2$ are continuous (neither function is even defined at zero).

There are special techniques for handling some problems of this sort, which we will discuss in Chapter 5. In the remainder of this chapter we assume, unless otherwise stated, that all functions in each equation are continuous for all real values of x.

Before solving a second-order differential equation, it helps to know what we are seeking. A clue is provided by examining a first-order equation. Consider the equation
$$y' + 2y = 0. \tag{6}$$
In Section 1.1 we saw that one solution to this equation is $y = e^{-2x}$. In fact, $y = ce^{-2x}$ is a solution for any constant c, and every solution to (6) has the form ce^{-2x}. We can summarize this result by noting that once we have found one nonzero solution to (6), we have found all of the solutions, since every other solution is a constant multiple of this one solution.

It turns out that similar results hold for all homogeneous second-order equations:
$$y'' + a(x) y' + b(x) y = 0. \tag{7}$$
The major difference is that now we have to find *two* solutions to (7) where

Linear Combination and Linear Independence

neither solution is a multiple of the other. We now make these ideas more precise.

Let y_1 and y_2 be any two functions. By a **linear combination** of y_1 and y_2 we mean a function $y(x)$ that can be written in the form

$$y(x) = c_1 y_1(x) + c_2 y_2(x),$$

for some constants c_1 and c_2. Two functions are **linearly independent** on an interval $[x_1, x_2]$ whenever the relation

$$c_1 y_1(x) + c_2 y_2(x) = 0, \tag{8}$$

for all x in $[x_1, x_2]$, implies that $c_1 = c_2 = 0$. Otherwise they are **linearly dependent**.

Example 5 Verify that the functions $y_1 = 1$ and $y_2 = x$ are linearly independent on the interval $[0, 1]$.

Solution To determine linear independence or dependence we must consider equation (8):

$$c_1 y_1 + c_2 y_2 = c_1 \cdot 1 + c_2 \cdot x = 0. \tag{9}$$

This equation must hold for all x in $[0, 1]$. If $x = 0$, we have

$$c_1 \cdot 1 + c_2 \cdot 0 = c_1 = 0.$$

But then at $x = 1$ we get $c_2 \cdot 1 = 0$. Hence (9) holds for all x in $[0, 1]$ if and only if $c_1 = c_2 = 0$. This proves that $y_1 = 1$ and $y_2 = x$ are linearly independent.

The notions of linear combination, linear independence, and linear dependence extend easily to a collection of n functions $y_1(x), y_2(x), \ldots, y_n(x)$, with $n > 2$. A **linear combination** of these functions is any function of the form

$$y(x) = c_1 y_1(x) + c_2 y_2(x) + \cdots + c_n y_n(x),$$

where $c_1, c_2, \ldots, c_n$ are constants. We will say that the collection of functions $y_1(x), y_2(x), \ldots, y_n(x)$ are **linearly independent** on an interval $[x_1, x_2]$ if the equation

$$c_1 y_1(x) + c_2 y_2(x) + \cdots + c_n y_n(x) = 0$$

is only true for all x in $[x_1, x_2]$ when $c_1 = c_2 = \cdots = c_n = 0$. Otherwise, we say that the collection of functions is **linearly dependent** on $[x_1, x_2]$. We will say more about this in Section 3.8.

There is an easy way to see that two functions y_1 and y_2 are linearly dependent. If $c_1 y_1(x) + c_2 y_2(x) = 0$ (where not both c_1 and c_2 are zero), we may suppose that $c_1 \neq 0$. Dividing the above expression by c_1, we obtain

$$y_1(x) + \frac{c_2}{c_1} y_2(x) = 0,$$

or

$$y_1(x) = -\frac{c_2}{c_1} y_2(x) = c y_2(x).$$

This leads to a useful fact:

> Two functions are linearly dependent on the interval $[x_0, x_1]$ if and only if one of the functions is a constant multiple of the other. (10)

Example 5 (revisited) It is easy to see that the functions $y_1 = 1$ and $y_2 = x$ are linearly independent on $[0, 1]$, since x is not a *constant* multiple of 1.

The notions of linear combination and linear independence are central to the theory of linear homogeneous equations, as is illustrated by the results that follow.

THEOREM 2

Every homogeneous linear second-order differential equation
$$y'' + a(x)y' + b(x)y = 0 \qquad (11)$$
has two linearly independent solutions.

Proof The existence part of Theorem 1 guarantees that we can find a solution $y_1(x)$ to equation (11) satisfying
$$y_1(x_0) = 1 \quad \text{and} \quad y_1'(x_0) = 0.$$
Similarly, we can also find a solution $y_2(x)$ to (11) satisfying
$$y_2(x_0) = 0 \quad \text{and} \quad y_2'(x_0) = 1.$$
Now consider the equation
$$c_1 y_1(x) + c_2 y_2(x) = 0 \qquad (12)$$
and its derivative
$$c_1 y_1'(x) + c_2 y_2'(x) = 0. \qquad (13)$$
Setting $x = x_0$ in (12) yields $c_1 \cdot 1 + c_2 \cdot 0 = 0$, or $c_1 = 0$, whereas substituting $x = x_0$ in (13) gives $c_1 \cdot 0 + c_2 \cdot 1 = c_2 = 0$. Thus (12) holds only when $c_1 = c_2 = 0$, implying that solutions $y_1(x)$ and $y_2(x)$ are linearly independent. ■

Example 6 Verify that $\sin x$ and $\cos x$ are linearly independent solutions to $y'' + y = 0$.

Solution To check that $\sin x$ and $\cos x$ are solutions, we write
$$(\sin x)'' + \sin x = -\sin x + \sin x = 0,$$
$$(\cos x)'' + \cos x = -\cos x + \cos x = 0.$$
Now consider the equation
$$c_1 \sin x + c_2 \cos x = 0,$$
for all values x, and its derivative
$$c_1 \cos x - c_2 \sin x = 0.$$
If $x = 0$, the first equation yields $c_2 = 0$ whereas the second gives $c_1 = 0$. Thus

sin x and cos x are linearly independent. Alternatively, they are independent according to (10), because
$$\frac{\sin x}{\cos x} = \tan x \neq \text{constant}.$$

Every linear, homogeneous, second-order differential equation has two linearly independent solutions. The next two theorems show us that this is all the information we need; that is, once we have two linearly independent solutions, we can construct them all.

THEOREM 3

Let $y_1(x)$ and $y_2(x)$ be any two solutions of the homogeneous equation
$$y'' + a(x)y' + b(x)y = 0. \tag{14}$$
Then any linear combination of them is also a solution of (14).

Proof Let $y(x) = c_1 y_1(x) + c_2 y_2(x)$. Then
$$\begin{aligned} y'' + ay' + by &= c_1 y_1'' + c_2 y_2'' + c_1 a y_1' + c_2 a y_2' + c_1 b y_1 + c_2 b y_2 \\ &= c_1(y_1'' + ay_1' + by_1) + c_2(y_2'' + ay_2' + by_2) \\ &= c_1 \cdot 0 + c_2 \cdot 0 = 0, \end{aligned}$$
since y_1 and y_2 are solutions of the homogeneous equation (14). ∎

THEOREM 4

Let $y_1(x)$ and $y_2(x)$ be linearly independent solutions to (14) and let $y_3(x)$ be another solution. Then there exist unique constants c_1 and c_2 such that
$$y_3(x) = c_1 y_1(x) + c_2 y_2(x).$$

In other words, any solutions of (14) can be written as a linear combination of two given linearly independent solutions of (14).

We stress the importance of this theorem. It indicates that once we have found two linearly independent solutions y_1 and y_2 of equation (14), we have essentially found *all* the solutions of (14). (We delay the proof of this theorem until the end of the section.)

GENERAL SOLUTION TO A LINEAR, SECOND-ORDER EQUATION

The general solution of (14) is given by the linear combination
$$\boxed{y(x) = c_1 y_1(x) + c_2 y_2(x),}$$
where c_1 and c_2 are arbitrary constants and $y_1(x)$ and $y_2(x)$ are linearly independent solutions of (14).

Example 6 (revisited) The general solution to $y'' + y = 0$ is
$$y = c_1 \cos x + c_2 \sin x.$$

If we look at the proof of Theorem 2 and at Example 6, we see that equations (12) and (13) yield a system of equations

$$c_1 y_1(x) + c_2 y_2(x) = 0,$$
$$c_1 y_1'(x) + c_2 y_2'(x) = 0, \qquad (15)$$

that can be used to determine whether the solutions $y_1(x)$ and $y_2(x)$ are linearly independent. The procedure involves substituting a value $x = x_0$ in (15), which determines $y_1(x_0)$, $y_2(x_0)$, $y_1'(x_0)$, and $y_2'(x_0)$. We are then left with a system of two homogeneous† equations in the two unknowns c_1 and c_2, and this system‡ has the trivial solution $c_1 = c_2 = 0$ if and only if the determinant

$$\begin{vmatrix} y_1(x_0) & y_2(x_0) \\ y_1'(x_0) & y_2'(x_0) \end{vmatrix} = y_1(x_0) y_2'(x_0) - y_1'(x_0) y_2(x_0)$$

does not equal zero. This fact leads to the following very useful definition.

WRONSKIAN

Let $y_1(x)$ and $y_2(x)$ be any two solutions to the differential equation

$$y'' + a(x) y' + b(x) y = 0. \qquad (16)$$

The **Wronskian**§ of y_1 and y_2 is defined as

$$W(y_1, y_2)(x) = \begin{vmatrix} y_1(x) & y_2(x) \\ y_1'(x) & y_2'(x) \end{vmatrix} = y_1(x) y_2'(x) - y_1'(x) y_2(x).$$

---

† Don't be confused by the many uses of the word "homogeneous" in mathematics. This is our third encounter with this word in this text:

a. The first-order differential equation $y' = f(x, y)$ is *homogeneous* (of order zero) if $f(tx, ty) = f(x, y)$ (see p. 54). Example: $dy/dx = (x + y)/(x - y)$

b. The linear second-order differential equation $y'' + a(x) y' + b(x) y = f(x)$ is *homogeneous* if $f(x) = 0$. Example: $y'' + y' + y = 0$.

c. The system of two linear equations in two unknowns

$$a_{11} x + a_{12} y = b_1$$
$$a_{21} x + a_{22} y = b_2$$

is *homogeneous* if $b_1 = b_2 = 0$.

Each of these definitions is different. However, the meaning will always be clear from the context.

‡ A homogeneous system

$$ax + by = 0$$
$$cx + dy = 0$$

has nontrivial solutions for the unknowns x and y if and only if the determinant

$$\begin{vmatrix} a & b \\ c & d \end{vmatrix} = ad - bc$$

equals zero (see Theorem 3 in Appendix 4).

§ Named after the Polish philosopher Józef Maria Hoene-Wroński (1778–1853).

The Wronskian is defined at all points x at which $y_1(x)$ and $y_2(x)$ are differentiable.

Since $y_1(x)$ and $y_2(x)$ are solutions to (16), their second derivatives exist. Hence we can differentiate $W(y_1, y_2)(x)$ with respect to x:

$$W'(y_1, y_2)(x) = [y_1(x)y_2'(x) - y_1'(x)y_2(x)]'$$
$$= y_1'(x)y_2'(x) + y_1(x)y_2''(x) - y_1''(x)y_2(x)$$
$$\quad - y_1'(x)y_2'(x)$$
$$= y_1(x)y_2''(x) - y_1''(x)y_2(x).$$

Since y_1 and y_2 are solutions to (16),

$$y_1'' + ay_1' + by_1 = 0 \quad \text{and} \quad y_2'' + ay_2' + by_2 = 0.$$

Multiplying the first of these equations by y_2 and the second by y_1 and subtracting, we obtain

$$y_1 y_2'' - y_2 y_1'' + a(y_1 y_2' - y_2 y_1') = 0,$$

which is just

$$W' + aW = 0. \tag{17}$$

But (17) is a linear first-order equation similar to equation (2.3.2) (p. 46) with solution [see equation (2.3.3)]

$$\boxed{W(y_1, y_2)(x) = ce^{-\int a(x)\,dx}} \tag{18}$$

for some arbitrary constant c. Equation (18) is known as **Abel's formula**.

Since an exponential is never zero, we see that $W(y_1, y_2)(x)$ is either always zero (when $c = 0$) or never zero (when $c \neq 0$). The importance of this fact is given by the following theorem.

THEOREM 5

Two solutions $y_1(x)$ and $y_2(x)$ of equation (16) are linearly independent on $[x_0, x_1]$ if and only if $W(y_1, y_2)(x) \neq 0$.

Theorem 5 is useful in at least three ways. First, it provides us with an easy way to determine whether or not two solutions are linearly independent. Second, it greatly simplifies the proof of Theorem 4—as we see later in this section. Third, the Wronskian can be easily extended to third- and higher-order equations with similar results. It is not easy to verify directly that three solutions are linearly independent, but the task is made easy by use of the Wronskian.

Example 6 (revisited) We have seen that $y_1(x) = \sin x$ and $y_2(x) = \cos x$ are linearly independent solutions of $y'' + y = 0$. The linear independence is easily verified using the Wronskian:

$$W(y_1, y_2)(x) = \begin{vmatrix} \sin x & \cos x \\ \cos x & -\sin x \end{vmatrix} = -\sin^2 x - \cos^2 x = -1 \neq 0.$$

Example 7

a. Verify that $y_1(x) = e^{-5x}$ and $y_2(x) = e^{2x}$ are linearly independent solutions of the differential equation
$$y'' + 3y' - 10y = 0. \tag{19}$$

b. Use the general solution obtained from y_1 and y_2 to solve the initial-value problem
$$y'' + 3y' - 10y = 0, \quad y(0) = 3, \quad y'(0) = -1. \tag{20}$$

Solution

a. That $y_1(x)$ and $y_2(x)$ are solutions to the differential equation follows from
$$(e^{-5x})'' + 3(e^{-5x})' - 10(e^{-5x}) = 25e^{-5x} - 15e^{-5x} - 10e^{-5x} = 0,$$
$$(e^{2x})'' + 3(e^{2x})' - 10(e^{2x}) = 4e^{2x} + 6e^{2x} - 10e^{2x} = 0.$$

Also,
$$W(y_1, y_2)(x) = \begin{vmatrix} e^{-5x} & e^{2x} \\ -5e^{-5x} & 2e^{2x} \end{vmatrix} = 2e^{-3x} + 5e^{-3x} = 7e^{-3x} \neq 0,$$

so the solutions are linearly independent. (This is also true since $e^{-5x}/e^{2x} = e^{-7x} \neq$ constant.) Thus the general solution of (19) is
$$y(x) = c_1 y_1(x) + c_2 y_2(x) = c_1 e^{-5x} + c_2 e^{2x}, \tag{21}$$

where c_1 and c_2 are arbitrary constants.

b. To find the particular solution of initial-value problem (20), we must determine the constants c_1 and c_2 so that (21) satisfies the initial conditions in (20). Setting $x = 0$ in (21), we have
$$c_1 + c_2 = y(0) = 3.$$

Differentiating (21), we get
$$y'(x) = -5c_1 e^{-5x} + 2c_2 e^{2x},$$

and setting $x = 0$ gives
$$-5c_1 + 2c_2 = y'(0) = -1.$$

Hence
$$c_1 + c_2 = 3$$
$$-5c_1 + 2c_2 = -1.$$

The first equation implies that $c_1 = 3 - c_2$, so
$$-5c_1 + 2c_2 = -5(3 - c_2) + 2c_2 = -1,$$

or
$$-15 + 5c_2 + 2c_2 = -1,$$
$$7c_2 = 14,$$
$$c_2 = 2.$$

Hence $c_1 = 1$, and the particular solution of initial-value problem (20) is
$$y = e^{-5x} + 2e^{2x}.$$

Solution of Nonhomogeneous Equations

We now turn briefly to the nonhomogeneous equation
$$y'' + a(x)y' + b(x)y = f(x). \tag{22}$$
Let y_p be any particular solution to equation (22). If we know the general solution to the homogeneous equation
$$y'' + a(x)y' + b(x)y = 0, \tag{23}$$
we can find all solutions to (22).

THEOREM 6

Let $y_p(x)$ be a particular solution of (22) and let $y^*(x)$ be any other solution. Then $y^*(x) - y_p(x)$ is a solution of (23); that is,
$$y^*(x) = c_1 y_1(x) + c_2 y_2(x) + y_p(x),$$
for some constants c_1 and c_2, where y_1, y_2 are two linearly independent solutions of (22).

Thus in order to find all solutions to the nonhomogeneous equation, we need find only one solution to the nonhomogeneous equation and the general solution of the homogeneous equation.

Proof We have
$$(y^* - y_p)'' + a(y^* - y_p)' + b(y^* - y_p)$$
$$= (y^{*\prime\prime} + ay^{*\prime} + by^*) - (y_p'' + ay_p' + by_p)$$
$$= f - f = 0. \quad \blacksquare$$

General Solution to a Linear, Nonhomogeneous Second-Order Equation

Let $y_p(x)$ be one solution to the nonhomogeneous equation (22) and let $y_1(x)$ and $y_2(x)$ be two linearly independent solutions to the homogeneous equation (23). Then the general solution to (22) is given by

$$\boxed{y(x) = c_1 y_1(x) + c_2 y_2(x) + y_p(x),}$$

where c_1 and c_2 are arbitrary constants.

Example 8 It is not difficult to verify that $\frac{1}{2}te^t$ is a particular solution of $x'' - x = e^t$. Two linearly independent solutions of $x'' - x = 0$ are given by $x_1 = e^t$ and $x_2 = e^{-t}$. The general solution is therefore $x(t) = \frac{1}{2}te^t + c_1 e^t + c_2 e^{-t}$. Note that x_1 and x_2 are independent since $W(x_1, x_2)(t) = e^t(-e^{-t}) - e^{-t}(e^t) = -2 \neq 0$.

In the next three sections we present methods for finding the general solution of homogeneous equations. In Sections 3.5 and 3.6 techniques for obtaining a solution y_p of a nonhomogeneous equation are developed.

A Perspective

GENERAL SOLUTIONS

We have used the term **general solution** to describe a two-parameter (c_1 and c_2) family of solutions of either the homogeneous equation (23) or the nonhomogeneous equation (22). The general solution of an nth-order linear differential equation is a family of solutions involving n parameters (see Section 3.8). Any solution that is free of arbitrary parameters is called a **particular solution**.

There is another school of thought about the concept of a general solution. This viewpoint holds that a general solution must contain *all* solutions to the differential equation in some interval. No difficulties arise from this requirement when we are considering linear differential equations, since Theorems 4 and 6 guarantee that every solution is obtained by the linear combinations

$$c_1 y_1(x) + c_2 y_2(x) \quad \text{or} \quad c_1 y_1(x) + c_2 y_2(x) + y_p(x),$$

respectively. However, similar results are *not* known for nonlinear second- and higher-order equations. Even if we have a two-parameter family of solutions to a nonlinear second-order equation (that is, a solution that involves two arbitrary constants), it is certainly not easy to determine if this family includes all solutions to the equation. The perspective on page 44 provides an example of infinitely many singular solutions that are not included in a one-parameter family of solutions. Higher-order nonlinear equations often possess singular solutions.

Thus, on a practical level, the requirement that a "general solution" include all solutions presents an extra complication. Observe, however, that both definitions yield the same outcome for linear equations.

PROOFS OF THEOREMS 4 AND 5

Theorem 5 Two solutions $y_1(x)$ and $y_2(x)$ of

$$y'' + a(x)y' + b(x)y = 0$$

are linearly independent on $[x_0, x_1]$ if and only if $W(y_1, y_2)(x) \neq 0$.

Proof We first show that if $W(y_1, y_2)(x) = 0$, then y_1 and y_2 are linearly dependent. Let x_2 be a point in the interval $x_0 \leq x \leq x_1$. Consider the system of equations

$$\begin{aligned} c_1 y_1(x_2) + c_2 y_2(x_2) &= 0, \\ c_1 y_1'(x_2) + c_2 y_2'(x_2) &= 0. \end{aligned} \quad (24)$$

The determinant of this system is

$$y_1(x_2) y_2'(x_2) - y_2(x_2) y_1'(x_2) = W(y_1, y_2)(x_2) = 0.$$

Thus, according to the theory of determinants (see Theorem 3 in Appendix 4), there exists a solution (c_1, c_2) for (24) where c_1 and c_2 are not both equal to zero. Define $y(x) = c_1 y_1(x) + c_2 y_2(x)$. By Theorem 3, $y(x)$ is a solution of

$$y'' + a(x)y' + b(x)y = 0.$$

But since c_1 and c_2 satisfy (24),

$$y(x_2) = c_1 y_1(x_2) + c_2 y_2(x_2) = 0$$

and
$$y'(x_2) = c_1 y_1'(x_2) + c_2 y_2'(x_2) = 0.$$
Thus $y(x)$ solves the initial-value problem
$$y'' + a(x)y' + b(x)y = 0, \qquad y(x_2) = y'(x_2) = 0.$$
But this initial-value problem also has the solution $y_3(x) \equiv 0$ for all values of x in $x_0 \le x \le x_1$. By Theorem 1, the solution of this initial-value problem is unique, so necessarily $y(x) = y_3(x) \equiv 0$. Thus
$$y(x) = c_1 y_1(x) + c_2 y_2(x) = 0,$$
for all values of x in $x_0 \le x \le x_1$, which proves that y_1 and y_2 are linearly dependent.

We now assume that $W(y_1, y_2)(x) \ne 0$ in $[x_0, x_1]$ and prove that y_1 and y_2 are linearly independent. Consider the equation
$$c_1 y_1(x) + c_2 y_2(x) = 0,$$
for all x in $[x_0, x_1]$. Differentiating with respect to x, we have
$$c_1 y_1'(x) + c_2 y_2'(x) = 0,$$
for all x in $[x_0, x_1]$. Select any x_2 in $[x_0, x_1]$ and consider the homogeneous system
$$c_1 y_1(x_2) + c_2 y_2(x_2) = 0,$$
$$c_1 y_1'(x_2) + c_2 y_2'(x_2) = 0.$$
Its determinant,
$$\begin{vmatrix} y_1(x_2) & y_2(x_2) \\ y_1'(x_2) & y_2'(x_2) \end{vmatrix} = W(y_1, y_2)(x_2),$$
does not equal zero; thus $c_1 = c_2 = 0$. Hence $y_1(x)$ and $y_2(x)$ are linearly independent on $[x_0, x_1]$. ∎

Theorem 4 Let $y_1(x)$ and $y_2(x)$ be linearly independent solutions to
$$y'' + a(x)y' + b(x)y = 0$$
and let $y_3(x)$ be another solution. Then there exist unique constants c_1 and c_2 such that
$$y_3(x) = c_1 y_1(x) + c_2 y_2(x).$$

Proof Let $y_3(x_0) = a$ and $y_3'(x_0) = b$. Consider the linear system of equations in two unknowns c_1 and c_2:
$$y_1(x_0) c_1 + y_2(x_0) c_2 = a,$$
$$y_1'(x_0) c_1 + y_2'(x_0) c_2 = b. \tag{25}$$
As we saw earlier, the determinant of this system is $W(y_1, y_2)(x_0)$, which is nonzero since the solutions are linearly independent. Thus (by Theorem 5) there is a unique solution (c_1, c_2) to (25) and a solution $y^*(x) = c_1 y_1(x) + c_2 y_2(x)$ that satisfies the conditions $y^*(x_0) = a$ and $y^{*\prime}(x_0) = b$. Since every initial-value problem has a unique solution (by Theorem 1), it must follow that $y_3(x) = y^*(x)$ on the interval $x_0 \le x \le x_1$, and so the proof is complete. ∎

PROBLEMS 3.1

In Problems 1–10 determine whether the given equation is linear or nonlinear. If it is linear, state whether it is homogeneous or nonhomogeneous with constant or variable coefficients.

1. $y'' + 2x^3 y' + y = 0$
2. $y'' + 2y' + y^2 = x$
3. $y'' + 3y' + yy' = 0$
4. $y'' + 3y' + 4y = 0$
5. $y'' + 3y' + 4y = \sin x$
6. $y'' + y(2 + 3y) = e^x$
7. $y'' + 4xy' + 2x^3 y = e^{2x}$
8. $y'' + \sin(xe^x) y' + 4xy = 0$
9. $3y'' + 16y' + 2y = 0$
10. $yy'y'' = 1$

In Problems 11–14 use the Wronskian to determine how many of the given functions are linearly independent on $[0,1]$.

11. $y_0 = 1$, $y_1 = 1 + x$, $y_2 = x^2$,
    $y_3 = x(1 - x)$, $y_4 = x$
12. $y_0 = \sin^2 x$, $y_1 = 1$, $y_2 = \sin x \cos x$,
    $y_3 = \cos^2 x$, $y_4 = \sin 2x$
13. $y_0 = 1 + x$, $y_1 = 1 - x$, $y_2 = 1$,
    $y_3 = x^2$, $y_4 = 1 + x^2$
14. $y_0 = 2\cos^2 x - 1$, $y_1 = \cos^2 x - \sin^2 x$,
    $y_2 = 1 - 2\sin^2 x$, $y_3 = \cos 2x$

In Problems 15–18 test each of the functions 1, x, x^2, and x^3 to see which functions satisfy the given differential equation. Then construct the *general solution* to the equation by writing a linear combination of the linearly independent solutions you have found.

15. $y'' = 0$
16. $y''' = 0$
17. $xy'' - y' = 0$
18. $x^2 y'' - 2xy' + 2y = 0$

19. Let $y_1(x)$ be a solution of the homogeneous equation
$$y'' + a(x)y' + b(x)y = 0$$
on the interval $\alpha \leq x \leq \beta$. Suppose that the curve y_1 is tangent to the x-axis at some point of this interval. Prove that y_1 must be identically zero.

20. Let $y_1(x)$ and $y_2(x)$ be two nontrivial solutions of the homogeneous equation
$$y'' + a(x)y' + b(x)y = 0$$
on the interval $\alpha \leq x \leq \beta$. Suppose $y_1(x_0) = y_2(x_0) = 0$ for some point $\alpha \leq x_0 \leq \beta$. Show that y_2 is a constant multiple of y_1.

21. **a.** Show that x and x^3 are linearly independent on $|x| < 1$ even though the Wronskian $W(x, x^3) = 0$ at $x = 0$.
    b. Show that x and x^3 are solutions to
$$x^2 y'' - 3xy' + 3y = 0.$$
Does this contradict Theorem 5?

22. **a.** Show that $y_1(x) = \sin x^2$ and $y_2(x) = \cos x^2$ are linearly independent solutions of
$$xy'' - y' + 4x^3 y = 0.$$
    b. Calculate $W(y_1, y_2)(x)$ and show that it is zero when $x = 0$. Does this result contradict Theorem 5? [*Hint*: In Theorem 5, as elsewhere in this section, it is assumed that $a(x)$ and $b(x)$ are continuous.]

23. Show that
$$y_1(x) = \sin x$$
and
$$y_2(x) = 4\sin x - 2\cos x$$
are linearly independent solutions of $y'' + y = 0$. Write the solution $y_3(x) = \cos x$ as a linear combination of y_1 and y_2.

24. Prove that $e^x \sin x$ and $e^x \cos x$ are linearly independent solutions of the equation
$$y'' - 2y' + 2y = 0.$$
    a. Find a solution that satisfies the conditions $y(0) = 1$, $y'(0) = 4$.
    b. Find another pair of linearly independent solutions.

25. Assume that some nonzero solution of
$$y'' + a(x)y' + b(x)y = 0, \qquad y(0) = 0$$
vanishes at some point x_1, where $x_1 > 0$. Prove that any other solution vanishes at $x = x_1$.

26. Define the function $s(x)$ to be the unique solution of the initial-value problem
$$y'' + y = 0; \qquad y(0) = 0, \qquad y'(0) = 1,$$
and the function $c(x)$ as the solution of
$$y'' + y = 0; \qquad y(0) = 1, \qquad y'(0) = 0.$$
Without using trigonometry, prove that
    a. $\dfrac{ds}{dx} = c(x)$; **b.** $\dfrac{dc}{dx} = -s(x)$;
    c. $s^2 + c^2 = 1$.

27. **a.** Show that $y_1 = \sin \ln x^2$ and $y_2 = \cos \ln x^2$ are linearly independent solutions of
$$y'' + \frac{1}{x} y' + \frac{4}{x^2} y = 0 \quad (x > 0).$$
    b. Calculate $W(y_1, y_2)(x)$.

3.2 Using One Solution to Find Another: Reduction of Order

As we saw in Theorem 3.1.4, it is easy to write down the general solution of the homogeneous equation

$$y'' + a(x)y' + b(x)y = 0, \qquad (1)$$

provided we know two linearly independent solutions y_1 and y_2 of equation (1). The general solution is then given by

$$y = c_1 y_1 + c_2 y_2,$$

where c_1 and c_2 are arbitrary constants. Unfortunately, there is no general procedure for determining y_1 and y_2. However, a standard procedure does exist for finding y_2 when y_1 is known. This method is of considerable importance, since it is often possible to find one solution by inspecting the equation or by trial and error.

We assume that y_1 is a nonzero solution of (1) and seek another solution y_2 such that y_1 and y_2 are linearly independent. Suppose that y_2 can be found. Then, since y_1 and y_2 are linearly independent, $y_2 \neq k y_1$ for any constant k. So

$$\frac{y_2}{y_1} = v(x)$$

must be a nonconstant function of x, and $y_2 = v y_1$ must satisfy (1). Thus

$$(v y_1)'' + a(v y_1)' + b(v y_1) = 0. \qquad (2)$$

But

$$(v y_1)' = v y_1' + v' y_1 \qquad (3)$$

and

$$(v y_1)'' = (v y_1' + v' y_1)' = v y_1'' + v' y_1' + v' y_1' + v'' y_1$$
$$= v y_1'' + 2 v' y_1' + v'' y_1. \qquad (4)$$

Using (3) and (4) in (2), we have

$$(v y_1'' + 2 v' y_1' + v'' y_1) + a(v y_1' + v' y_1) + b v y_1 = 0,$$

or

$$v(y_1'' + a y_1' + b y_1) + v'(2 y_1' + a y_1) + v'' y_1 = 0. \qquad (5)$$

The first term in parentheses in (5) vanishes since y_1 is a solution of (1), so we obtain

$$v'' y_1 + v'(2 y_1' + a y_1) = 0.$$

Dividing by $v' y_1$, we can rewrite this equation in the form

$$\frac{v''}{v'} = -2 \frac{y_1'}{y_1} - a. \qquad (6)$$

We set $z = v'$ in (6). Then $z' = v''$ and (6) becomes

$$\frac{z'}{z} = -2 \frac{y_1'}{y_1} - a, \qquad (7)$$

a separable first-order equation. Thus we have *reduced the order* of our equation. The functions in both sides of (7) are functions of x, so we integrate with respect to x to obtain

$$\ln z = -2\ln y_1 - \int a(x)\,dx.$$

Exponentiating, we have

$$z = e^{\ln z} = e^{-2\ln y_1 - \int a(x)\,dx} = \frac{1}{y_1^2}e^{-\int a(x)\,dx}.$$

But, since $z = v'$, we have

$$v' = \frac{1}{y_1^2}e^{-\int a(x)\,dx}.$$

Since the exponential is never zero, v is nonconstant. To find v, we perform another integration and obtain

$$\boxed{y_2 = y_1 v = y_1(x)\int \frac{e^{-\int a(x)\,dx}}{y_1^2(x)}\,dx.} \qquad (8)$$

Remark It is not advisable to memorize formula (8). It is only necessary to remember the substitution $y_2 = vy_1$ and substitute this into the original differential equation. The following examples illustrate this procedure.

Example 1 Note that $y_1 = x$ is a solution of

$$x^2 y'' - xy' + y = 0, \quad x > 0. \qquad (9)$$

Setting $y_2 = y_1 v = xv(x)$, it follows that $y_2' = xv' + v$, $y_2'' = xv'' + 2v'$, and (9) becomes

$$x^2(xv'' + 2v') - x(xv' + v) + (xv) = 0,$$

or

$$x^3 v'' + x^2 v' = 0.$$

Setting $z = v'$ and separating variables, we have

$$\frac{dz}{z} = \frac{-dx}{x} \quad \text{or} \quad \ln|z| = -\ln x, \qquad \text{(since } x > 0\text{)}$$

from which we obtain by exponentiation

$$v' = z = \frac{1}{x}.$$

Thus $v(x) = \ln x$, so $y_2 = y_1 v = x\ln x$ and the general solution of (9) is

$$y = c_1 x + c_2 x \ln x, \quad x > 0.$$

Example 2 Consider the **Legendre equation of order one**:

$$(1 - x^2)y'' - 2xy' + 2y = 0, \quad -1 < x < 1. \qquad (10)$$

Again, it is not hard to verify that $y_1(x) = x$ is a solution. Setting $y_2 = xv$, we have

$$(1-x^2)(xv''+2v') - 2x(xv'+v) + 2xv = x(1-x^2)v'' + 2(1-2x^2)v' = 0;$$

when we substitute $z = v'$ and divide by $x(1-x)^2$, we can use partial fractions to obtain

$$\frac{dz}{dx} + \frac{2(1-2x^2)z}{x(1-x^2)} = \frac{dz}{dx} + \left(\frac{2}{x} - \frac{2x}{1-x^2}\right)z = 0.$$

Separating variables and integrating, we obtain

$$\ln|z| = \int \frac{dz}{z} = \int \left(\frac{2x}{1-x^2} - \frac{2}{x}\right) dx = -\ln(1-x^2) - \ln(x^2).$$

Thus, exponentiating and using partial fractions, we get

$$v' = \frac{1}{x^2(1-x^2)} = \frac{1}{x^2} + \frac{1}{2}\left(\frac{1}{1+x} + \frac{1}{1-x}\right).$$

Hence

$$y_2 = y_1 v = x \int \left[\frac{1}{x^2} + \frac{1}{2}\left(\frac{1}{1+x} + \frac{1}{1-x}\right)\right] dx = x\left[\frac{-1}{x} + \frac{1}{2}\ln\left(\frac{1+x}{1-x}\right)\right],$$

or

$$y_2 = \frac{x}{2}\ln\left(\frac{1+x}{1-x}\right) - 1, \quad |x| < 1.$$

PROBLEMS 3.2

In each of Problems 1–18 a second-order differential equation and one solution $y_1(x)$ are given. Verify that $y_1(x)$ is indeed a solution and find a second linearly independent solution.

1. $y'' - 2y' + y = 0$, $y_1(x) = e^x$
2. $y'' + 4y = 0$, $y_1(x) = \sin 2x$
3. $y'' - 4y = 0$, $y_1(x) = e^{2x}$
4. $y'' - 4y' + 4y = 0$, $y_1(x) = xe^{2x}$
5. $y'' + 5y' + 6y = 0$, $y_1(x) = e^{-2x}$
6. $y'' + 2y' + 2y = 0$, $y_1(x) = e^{-x}\cos x$
7. $y'' + y' + 7y = 0$, $y_1(x) = e^{-x/2}\cos\frac{3\sqrt{3}}{2}x$
8. $y'' + 10y' + 25y = 0$, $y_1(x) = xe^{-5x}$
9. $y'' - 30y' + 200y = 0$, $y_1(x) = e^{20x}$
10. $y'' + \left(\frac{3}{x}\right)y' = 0$, $y_1(x) = 1$
11. $x^2 y'' + xy' - 4y = 0$, $y_1(x) = x^2$
12. $y'' - 2xy' + 2y = 0$, $y_1(x) = x$
13. $x^2 y'' - 2xy' + (x^2 + 2)y = 0$, $(x > 0)$, $y_1 = x\sin x$
14. $xy'' + (2x-1)y' - 2y = 0$, $(x > 0)$, $y_1 = e^{-2x}$
15. $xy'' + (x-1)y' + (3-12x)y = 0$, $(x > 0)$, $y_1 = e^{3x}$
16. $xy'' - y' + 4x^3 y = 0$, $(x > 0)$, $y_1 = \sin(x^2)$
17. $xy'' - (x+n)y' + ny = 0$, (integer $n > 0$), $y_1(x) = e^x$
18. $x^{1/3} y'' + y' + \left(\frac{1}{4}x^{-1/3} - \frac{1}{6x} - 6x^{-5/3}\right)y = 0$, $y_1 = x^3 e^{-3x^{2/3}/4}$, $(x > 0)$

19. The **Bessel differential equation** is given by
$$x^2 y'' + xy' + (x^2 - p^2)y = 0.$$
For $p = \frac{1}{2}$, verify that $y_1(x) = (\sin x)/\sqrt{x}$ is a solution for $x > 0$. Find a second linearly independent solution.

20. Letting $p = 0$ in the equation of Problem 19, we obtain the **Bessel differential equation of index zero**, which we will study in Chapter 5. One solution is the **Bessel function of order zero** denoted by $J_0(x)$. In terms of $J_0(x)$, find a second linearly independent solution.

3.3 Homogeneous Equations with Constant Coefficients: Real Roots

In this section we present a simple procedure for finding the general solution to the linear homogeneous equation with constant coefficients

$$y'' + ay' + by = 0. \tag{1}$$

Recall that for the comparable first-order equation $y' + ay = 0$ the general solution is $y(x) = ce^{-ax}$. It is then not implausible to "guess" that there may be a solution to (1) of the form $y(x) = e^{\lambda x}$ for some number λ (real or complex). Setting $y(x) = e^{\lambda x}$, we obtain $y' = \lambda e^{\lambda x}$ and $y'' = \lambda^2 e^{\lambda x}$, so that (1) yields

$$\lambda^2 e^{\lambda x} + a\lambda e^{\lambda x} + be^{\lambda x} = 0.$$

Since $e^{\lambda x} \neq 0$, we can divide this equation by $e^{\lambda x}$ to obtain

Characteristic Equation

$$\boxed{\lambda^2 + a\lambda + b = 0,} \tag{2}$$

where a and b are real numbers. Equation (2) is called the **characteristic equation** of the differential equation (1). It is clear that if λ satisfies (2), then $y(x) = e^{\lambda x}$ is a solution to (1). As we saw in Section 3.1, we need only obtain two linearly independent solutions. Equation (2) has the roots

$$\lambda_1 = \frac{-a + \sqrt{a^2 - 4b}}{2} \quad \text{and} \quad \lambda_2 = \frac{-a - \sqrt{a^2 - 4b}}{2}. \tag{3}$$

There are three possibilities: $a^2 - 4b > 0$, $a^2 - 4b = 0$, $a^2 - 4b < 0$.

Case 1: Roots Real and Unequal If $a^2 - 4b > 0$, then λ_1 and λ_2 are distinct real numbers, given by (3), and $y_1(x) = e^{\lambda_1 x}$ and $y_2 = e^{\lambda_2 x}$ are distinct solutions.

These two solutions are linearly independent because

$$\frac{y_1}{y_2} = e^{(\lambda_1 - \lambda_2)x},$$

which is clearly not a constant when $\lambda_1 \neq \lambda_2$. Thus we have proved the following theorem:

THEOREM 1

If $a^2 - 4b > 0$, then the roots of the characteristic equation are real and unequal and the general solution to equation (1) is given by

$$\boxed{y(x) = c_1 e^{\lambda_1 x} + c_2 e^{\lambda_2 x},} \tag{4}$$

where c_1 and c_2 are arbitrary constants and λ_1 and λ_2 are the real roots of (2).

Example 1 Consider the equation
$$y'' + 3y' - 10y = 0.$$
The characteristic equation is $\lambda^2 + 3\lambda - 10 = 0$, $a^2 - 4b = 49$, and the roots are $\lambda_1 = 2$ and $\lambda_2 = -5$ (the order in which the roots are taken is irrelevant). The general solution is
$$y(x) = c_1 e^{2x} + c_2 e^{-5x}.$$
If we specify the initial conditions $y(0) = 1$ and $y'(0) = 3$, for example, then differentiating and substituting $x = 0$, we obtain the simultaneous equations
$$c_1 + c_2 = 1,$$
$$2c_1 - 5c_2 = 3,$$
which have the unique solution $c_1 = \frac{8}{7}$ and $c_2 = -\frac{1}{7}$. The unique solution to the initial-value problem is therefore
$$y(x) = \tfrac{1}{7}(8e^{2x} - e^{-5x}).$$

Case 2: Roots Real and Equal Suppose $a^2 - 4b = 0$. In this case (2) has the double root $\lambda_1 = \lambda_2 = -a/2$. Thus $y_1(x) = e^{-ax/2}$ is a solution of (1). To find the second solution y_2, we make use of equation (3.2.8), p. 98, since one solution is known:

$$y_2(x) = y_1(x) \int \frac{e^{-ax}}{y_1^2(x)} dx = e^{-ax/2} \int \frac{e^{-ax}}{(e^{-ax/2})^2} dx$$

$$= e^{-ax/2} \int \frac{e^{-ax}}{e^{-ax}} dx = e^{-ax/2} \int dx = xe^{-ax/2}.$$

Since $y_2/y_1 = x$, it follows that y_1 and y_2 are linearly independent. Hence we have the following result:

THEOREM 2

If $a^2 - 4b = 0$, then the roots of the characteristic equation are equal and the general solution to equation (1) is given by

$$\boxed{y(x) = c_1 e^{-ax/2} + c_2 x e^{-ax/2},} \qquad (5)$$

where c_1 and c_2 are arbitrary constants.

Example 2 Consider the equation
$$y'' - 6y' + 9 = 0.$$
The characteristic equation is $\lambda^2 - 6\lambda + 9 = 0$, and $a^2 - 4b = 0$, yielding the unique double root $\lambda_1 = -a/2 = 3$. The general solution is
$$y(x) = c_1 e^{3x} + c_2 x e^{3x}.$$

If we use the initial conditions $y(0) = 1$, $y'(0) = 7$, we obtain the simultaneous equations

$$c_1 = 1,$$
$$3c_1 + c_2 = 7,$$

which yield the unique solution (since $c_2 = 7 - 3c_1 = 4$)

$$y(x) = e^{3x} + 4xe^{3x} = e^{3x}(1 + 4x).$$

We will deal with the more complicated situation of complex roots ($a^2 - 4b < 0$) in Section 3.4.

PROBLEMS 3.3

In Problems 1–20 find the general solution of each equation. When initial conditions are specified, give the particular solution that satisfies them.

1. $y'' - 4y = 0$
2. $x'' + x' - 6x = 0$, $x(0) = 0$, $x'(0) = 5$
3. $y'' - 3y' + 2y = 0$
4. $y'' + 5y' + 6y = 0$, $y(0) = 1$, $y'(0) = 2$
5. $4x'' + 20x' + 25x = 0$, $x(0) = 1$, $x'(0) = 2$
6. $y'' + 6y' + 9y = 0$
7. $x'' - x' - 6x = 0$, $x(0) = -1$, $x'(0) = 1$
8. $y'' - 8y' + 16y = 0$, $y(0) = 2$, $y'(0) = -1$
9. $y'' - 5y' = 0$
10. $y'' + 17y' = 0$, $y(0) = 1$, $y'(0) = 0$
11. $y'' + 2\pi y' + \pi^2 y = 0$
12. $y'' - 13y' + 42y = 0$
13. $z'' + 2z' - 15z = 0$
14. $w'' + 8w' + 12w = 0$
15. $y'' - 8y' + 16y = 0$, $y(0) = 1$, $y'(0) = 6$
16. $y'' + 2y' + y = 0$, $y(1) = 2/e$, $y'(1) = -3/e$
17. $y'' - 2y = 0$
18. $y'' + 6y' + 5y = 0$
19. $y'' - 5y = 0$, $y(0) = 3$, $y'(0) = -\sqrt{5}$
20. $y'' - 2y' - 2y = 0$, $y(0) = 1$, $y'(0) = 1 + 3\sqrt{3}$

THE RICCATI EQUATION†

Linear second-order differential equations may also be used in finding the solution to the **Riccati** equation:

$$y' + y^2 + a(x)y + b(x) = 0. \quad \text{(i)}$$

This nonlinear first-order equation frequently occurs in physical applications. To change it into a linear second-order equation, let $y = z'/z$. Then $y' = (z''/z) - (z'/z)^2$, so (i) becomes

$$\frac{z''}{z} - \left(\frac{z'}{z}\right)^2 + \left(\frac{z'}{z}\right)^2 + a(x)\left(\frac{z'}{z}\right) + b(x) = 0.$$

Multiplying by z, we obtain the linear second-order equation

$$z'' + a(x)z' + b(x)z = 0. \quad \text{(ii)}$$

If the general solution to (ii) can be found, the quotient $y = z'/z$ is the general solution to (i).

21. Suppose $z = c_1 z_1 + c_2 z_2$ is the general solution to equation (ii). Explain why the quotient z'/z involves only one arbitrary constant. [*Hint*: Divide numerator and denominator by c_1.]

22. For arbitrary constants a, b, and c, find the substitution that changes the nonlinear equation

$$y' + ay^2 + by + c = 0$$

into a linear second-order equation with constant coefficients. What second-order equation is obtained?

Use the method above to find the general solution to the Riccati equations in Problems 23–28. If an initial condition is specified, give the particular solution that satisfies that condition.

23. $y' + y^2 - 1 = 0$, $y(0) = -\frac{1}{3}$
24. $\dfrac{dx}{dt} + x^2 + 1 = 0$

---

† Jacopo Francesco Riccati (1676–1754), an Italian mathematician, physicist, and philosopher, was responsible for bringing much of Newton's work on calculus to the attention of Italian mathematicians.

25. $y' + y^2 - 2y + 1 = 0$
26. $y' + y^2 + 3y + 2 = 0$, $y(0) = 1$
27. $y' + y^2 - y - 2 = 0$
28. $y' + y^2 + 2y + 1 = 0$, $y(1) = 0$
29. Suppose that the two roots λ_1 and λ_2 of the characteristic equation (2) are real and satisfy $\lambda_2 = \lambda_1 + h$.

   a. Verify that
   $$\phi_h(x) = \frac{e^{\lambda_2 x} - e^{\lambda_1 x}}{\lambda_2 - \lambda_1}$$
   is a solution of equation (1).
   b. Hold λ_1 fixed and evaluate $\phi_h(x)$, as $h \to 0$, with L'Hôpital's rule.
   c. Verify that parts (a) and (b) yield Theorem 2's conclusion.

3.4 Homogeneous Equations with Constant Coefficients: Complex Roots

The material in this section requires some familiarity with the basic properties of complex numbers. A review of these properties is given in Appendix 5.

One of the facts we use repeatedly in this section is the **Euler formula** (see Appendix 5):

$$e^{i\beta x} = \cos(\beta x) + i\sin(\beta x).$$

We return to our examination of the homogeneous equation with constant coefficients

$$y'' + ay' + by = 0, \qquad (1)$$

where $a^2 - 4b < 0$.

Case 3: Complex Conjugate Roots Suppose $a^2 - 4b < 0$. The roots of the characteristic equation to equation (1) are

$$\lambda_1 = \alpha + i\beta, \qquad \lambda_2 = \alpha - i\beta, \qquad (2)$$

where $\alpha = -a/2$ and $\beta = \sqrt{4b - a^2}/2$. Thus $y_1 = e^{\lambda_1 x}$ and $y_2 = e^{\lambda_2 x}$ are solutions to (1). However, in this case it is useful to recall that any linear combination of solutions is also a solution (see Theorem 3.1.3, p. 89) and to consider instead the solutions

$$y_1^* = \frac{e^{\lambda_1 x} + e^{\lambda_2 x}}{2} \quad \text{and} \quad y_2^* = \frac{e^{\lambda_1 x} - e^{\lambda_2 x}}{2i}.$$

Since $\cos(-\theta) = \cos\theta$ and $\sin(-\theta) = -\sin\theta$, we can rewrite y_1^* as

$$y_1^* = \frac{e^{(\alpha + i\beta)x} + e^{(\alpha - i\beta)x}}{2} = \frac{e^{\alpha x}}{2}(e^{i\beta x} + e^{-i\beta x})$$

$$\underset{\text{Euler formula}}{\downarrow}$$

$$= \frac{e^{\alpha x}}{2}[\cos\beta x + i\sin\beta x + \cos(-\beta x) + i\sin(-\beta x)]$$

$$= e^{\alpha x}\cos\beta x.$$

Similarly, $y_2^* = e^{\alpha x}\sin\beta x$, and the linear independence of y_1^* and y_2^* follows easily since

$$\frac{y_1^*}{y_2^*} = \cot \beta x, \quad \beta \neq 0,$$

which is not a constant. Alternatively, we can compute $W(y_1^*, y_2^*)(x)$:

$$W(y_1^*, y_2^*)(x) = \begin{vmatrix} e^{\alpha x}\cos \beta x & e^{\alpha x}\sin \beta x \\ e^{\alpha x}(\alpha \cos \beta x - \beta \sin \beta x) & e^{\alpha x}(\alpha \sin \beta x + \beta \cos \beta x) \end{vmatrix}$$

$$= e^{2\alpha x}(\alpha \cos \beta x \sin \beta x + \beta \cos^2 \beta x - \alpha \cos \beta x \sin \beta x + \beta \sin^2 \beta x)$$

$\beta(\cos^2 \beta x + \sin^2 \beta x) = \beta$

$$= \beta e^{2\alpha x} \neq 0.$$

Thus we have proved the following theorem:

THEOREM 1

If $a^2 - 4b < 0$, then the characteristic equation has complex conjugate roots, and the general solution to

$$y'' + ay' + by = 0$$

is given by

$$\boxed{y(x) = e^{\alpha x}(c_1 \cos \beta x + c_2 \sin \beta x),} \qquad (3)$$

where c_1 and c_2 are arbitrary constants and

$$\alpha = -\frac{a}{2}, \quad \beta = \frac{\sqrt{4b - a^2}}{2}.$$

Example 1 Let $y'' + y = 0$. Then the characteristic equation is $\lambda^2 + 1 = 0$ with roots $\pm i$. We have $\alpha = 0$ and $\beta = 1$, so the general solution is

$$y(x) = c_1 \cos x + c_2 \sin x. \qquad (4)$$

This is the **equation of harmonic motion** [see equation (4.1.2), p. 139].

Example 2 Consider the equation $y'' + y' + y = 0$, $y(0) = 1$, $y'(0) = 3$. We have $\lambda^2 + \lambda + 1 = 0$ with roots $\lambda_1 = (-1 + i\sqrt{3})/2$ and $\lambda_2 = (-1 - i\sqrt{3})/2$. Then $\alpha = -\frac{1}{2}$ and $\beta = \sqrt{3}/2$, so the general solution is

$$y(x) = e^{-x/2}\left(c_1 \cos \frac{\sqrt{3}}{2}x + c_2 \sin \frac{\sqrt{3}}{2}x\right).$$

To solve the initial-value problem, we differentiate, set $x = 0$, and solve the simultaneous equations

$$c_1 = 1,$$

$$\frac{\sqrt{3}}{2}c_2 - \frac{1}{2}c_1 = 3.$$

Thus $c_1 = 1$, $c_2 = 7/\sqrt{3}$, and
$$y(x) = e^{-x/2}\left(\cos\frac{\sqrt{3}}{2}x + \frac{7}{\sqrt{3}}\sin\frac{\sqrt{3}}{2}x\right).$$

PROBLEMS 3.4

In Problems 1–12 find the general solution of each equation. When initial conditions are specified, give the particular solution that satisfies them.

1. $y'' + 2y' + 2y = 0$
2. $8y'' + 4y' + y = 0$, $y(0) = 0$, $y'(0) = 1$
3. $x'' + x' + 7x = 0$
4. $y'' + y' + 2y = 0$
5. $\dfrac{d^2x}{d\theta^2} + 4x = 0$, $x\left(\dfrac{\pi}{4}\right) = 1$, $x'\left(\dfrac{\pi}{4}\right) = 3$
6. $y'' + y = 0$, $y(\pi) = 2$, $y'(\pi) = -1$
7. $y'' + \frac{1}{4}y = 0$, $y(\pi) = 1$, $y'(\pi) = -1$
8. $y'' + 6y' + 12y = 0$
9. $y'' + 2y' + 5y = 0$
10. $y'' + 2y' + 5y = 0$, $y(0) = 1$, $y'(0) = -3$
11. $y'' + 2y' + 2y = 0$, $y(\pi) = e^{-\pi}$, $y'(\pi) = -2e^{-\pi}$
12. $y'' + 2y' + 5y = 0$, $y(\pi) = e^{-\pi}$, $y'(\pi) = 3e^{-\pi}$
13. Show that the boundary value problem
$$y'' + y = 0, \quad y(0) = 0, \quad y(\pi) = 1$$
has no solutions.

3.5 Nonhomogeneous Equations: Undetermined Coefficients

In this and the following section we present methods for finding a particular solution to the nonhomogeneous linear equation

$$y'' + ay' + by = f(x). \tag{1}$$

First, however, we prove a very useful result concerning nonhomogeneous equations called the principle of superposition:

THEOREM 1: PRINCIPLE OF SUPERPOSITION

Suppose the function $f(x)$ in (1) is a sum of two functions $f_1(x)$ and $f_2(x)$:

$$f(x) = f_1(x) + f_2(x).$$

If $y_1(x)$ is a solution of the equation

$$y'' + ay' + by = f_1(x), \tag{2}$$

and $y_2(x)$ is a solution of the equation

$$y'' + ay' + by = f_2(x), \tag{3}$$

then $y = y_1 + y_2$ is a solution of equation (1); that is, the solution of (1) is obtained by superimposing the solution of equation (3) on that of equation (2).

Proof Substituting $y = y_1 + y_2$ in the left-hand side of equation (1), we have

$$y'' + ay' + by = (y_1'' + y_2'') + a(y_1' + y_2') + b(y_1 + y_2)$$
$$= (y_1'' + ay_1' + by_1) + (y_2'' + ay_2' + by_2)$$
$$= f_1 + f_2 = f,$$

since y_1 and y_2 are solutions of equations (2) and (3), respectively. ∎

Briefly, the principle of superposition tells us that if we can split the function $f(x)$ into a sum of two (or more) simpler expressions $f_k(x)$, then we can restrict our attention to solving the nonhomogeneous equations

$$y'' + ay' + by = f_k(x), \qquad k = 1, 2, \ldots, m, \tag{4}$$

because the solution to equation (1) is simply the sum of the solutions of these equations.

The method we present in this section *requires* that the function $f(x)$ in equation (1) be of *one* of the following three forms:

> i. $P_n(x)$,
> ii. $P_n(x)e^{ax}$,
> iii. $e^{ax}[P_n(x)\cos bx + Q_n(x)\sin bx]$,

where $P_n(x)$ and $Q_n(x)$ are polynomials in x of degree n $(n \geq 0)$. The method we present below can also be used if $f(x)$ is a sum of functions $f_k(x)$ of these three forms, since by the principle of superposition we can solve each of the equations in (4) and add the solutions together. However, if any term of $f(x)$ is not of one of these three forms, we cannot use the method of this section.[†]

Note that these three forms involve a multitude of situations. The following three functions are all of one of these forms:

a. $2e^{3x}$ (the polynomial is the constant 2)

b. $e^{4x}\cos x$ (here $P_n(x) = 1$ and $Q_n(x) = 0$ are both polynomials of degree zero)

c. $x \cos x + \sin x$ (here $a = 0$, $b = 1$, $P_n(x) = x + 0$, and $Q_n(x) = 0 \cdot x + 1$)

Indeed, forms (i) and (ii) are special cases of form (iii): set $b = 0$ in form (iii) to obtain form (ii), and then set $a = 0$ to get form (i).

The **method of undetermined coefficients** assumes that the solution to equation (1) is of exactly the same form as $f(x)$. The technique requires that we replace each dependent variable y in (1) with an expression of the same form as $f(x)$ having polynomial terms with **undetermined coefficients**. If we compare both sides of the resulting equation, it is then possible to "determine" the unknown coefficients. The method is illustrated by a number of examples.

Example 1   Solve the following equation:

$$y'' - y = x^2 \tag{5}$$

Solution   Since $f(x) = x^2$ is a polynomial of degree two, we "guess" that (5) has a solution $y_p(x)$ that is a polynomial of degree two. We try the most general polynomial of degree two:

$$y_p(x) = a + bx + cx^2.$$

[†] Instead we must use the method of variation of parameters, which will be described in Section 3.6.

3.5 / NONHOMOGENEOUS EQUATIONS: UNDETERMINED COEFFICIENTS

Then $y_p' = b + 2cx$ and $y_p'' = 2c$, so if we substitute y_p'' and y_p into (5) we obtain

$$2c - (a + bx + cx^2) = x^2.$$

Equating coefficients, we have

$$\underset{\substack{\uparrow \\ \text{Coefficient of constant term}}}{2c - a = 0,} \quad \underset{\substack{\uparrow \\ \text{Coefficient of } x}}{-b = 0,} \quad \underset{\substack{\uparrow \\ \text{Coefficient of } x^2}}{-c = 1,}$$

which immediately yields $a = -2$, $b = 0$, $c = -1$, and the particular solution

$$y_p(x) = -2 - x^2.$$

This particular solution is easily verified by substitution into (5). Finally, since the general solution of the homogeneous equation $y'' - y = 0$ is given by

$$y = c_1 e^x + c_2 e^{-x},$$

the general solution of (5) is

$$y = c_1 e^x + c_2 e^{-x} - 2 - x^2.$$

Example 2 Solve

$$y'' - 3y' + 2y = e^x \sin x. \qquad (6)$$

Solution Since $f(x)$ is of form (iii) with $P_n(x) = 0$ and $Q_n(x) = 1$, we "guess" that there is a solution to (6) of the form

$$y_p(x) = ae^x \sin x + be^x \cos x.$$

Then

$$y_p'(x) = (a - b)e^x \sin x + (a + b)e^x \cos x$$

and

$$y_p''(x) = 2ae^x \cos x - 2be^x \sin x.$$

Substituting these expressions into (6) we have

$$e^x(2a \cos x - 2b \sin x) - 3e^x[(a - b)\sin x + (a + b)\cos x]$$
$$+ 2e^x(a \sin x + b \cos x) = e^x \sin x.$$

Dividing both sides by e^x and equating the coefficients of $\sin x$ and $\cos x$, we have

$$2a - 3(a + b) + 2b = 0,$$
$$-2b - 3(a - b) + 2a = 1,$$

which yield $a = -\frac{1}{2}$ and $b = \frac{1}{2}$ so that

$$y_p = \frac{e^x}{2}(\cos x - \sin x).$$

Again this result is easily verified by substitution. Finally, the general solution of (6) is

$$y = c_1 e^{2x} + c_2 e^x + \frac{e^x}{2}(\cos x - \sin x).$$

Example 3 Solve $y'' + y = xe^{2x}$.

Solution Here $f(x)$ is of form (ii), where $P_n(x)$ is a polynomial of degree one, so we try a solution of the form
$$y_p(x) = e^{2x}(a + bx).$$
Then
$$y_p'(x) = e^{2x}(2a + b + 2bx), \quad y_p''(x) = e^{2x}(4a + 4b + 4bx),$$
and substitution yields
$$e^{2x}(4a + 4b + 4bx) + e^{2x}(a + bx) = xe^{2x}.$$
Dividing both sides by e^{2x} and equating like powers of x, we obtain the equations
$$5a + 4b = 0, \quad 5b = 1.$$
Thus $a = -\frac{4}{25}$, $b = \frac{1}{5}$, and a particular solution is
$$y_p(x) = \frac{e^{2x}}{25}(5x - 4).$$
Therefore the general solution of this example is (since $y'' + y = 0$ is the equation of the harmonic oscillator)
$$y(x) = c_1 \sin x + c_2 \cos x + \frac{e^{2x}}{25}(5x - 4).$$

Difficulties arise in connection with problems of this type whenever any term of the guessed solution is a solution of the homogeneous equation
$$y'' + ay' + by = 0. \tag{7}$$
For example, in the equation
$$y'' + y = (1 + x + x^2)\sin x, \tag{8}$$
the function $f(x)$ is the sum of three functions, one of which ($\sin x$) is a solution to the homogeneous equation $y'' + y = 0$. As another example, in
$$y'' + y = (x + x^2)\sin x \tag{9}$$
the guessed solution is $y_p = (a_0 + a_1 x + a_2 x^2)\sin x + (b_0 + b_1 x + b_2 x^2)\cos x$ and, $a_0 \sin x + b_0 \cos x$ is a solution to the homogeneous equation $y'' + y = 0$. When this situation occurs, the method of undetermined coefficients must be modified. To see why, consider the following example.

Example 4 Find the solution to the equation
$$y'' - y = 2e^x. \tag{10}$$

Solution The general solution of $y'' - y = 0$ is
$$y(x) = c_1 e^x + c_2 e^{-x}.$$
Here $f(x) = 2e^x$ is a solution to the homogeneous equation. If we try to find a solution of the form Ae^x we get nowhere, since Ae^x is a solution to the homogeneous equation for every constant A and, therefore, it cannot possibly be a solution to the nonhomogeneous equation.

What do we do? Recall that if λ is a double root of the characteristic equation for a homogeneous differential equation, then two solutions are $e^{\lambda x}$ and $xe^{\lambda x}$. This suggests that we try Axe^x instead of Ae^x as a possible solution to (10). Thus we consider a particular solution of the form

$$y_p = Axe^x.$$

Then $y_p' = Ae^x(x+1)$, $y_p''(x) = Ae^x(x+2)$, and

$$y_p'' - y_p = Ae^x(x+2) - Axe^x = 2Ae^x = 2e^x.$$

Hence $A = 1$ and $y_p = xe^x$. Thus the general solution is

$$y(x) = c_1 e^x + c_2 e^{-x} + xe^x.$$

The preceding example suggests the following rule:

MODIFICATIONS OF THE METHOD

If any term of the guessed solution $y_p(x)$ is a solution of the homogeneous equation (7), multiply $y_p(x)$ by x repeatedly until no term of the product $x^k y_p(x)$ is a solution of (7). Then use the product $x^k y_p(x)$ to solve equation (1).

Example 5 Find the solution to

$$y'' + y = \cos x \tag{11}$$

that satisfies $y(0) = 2$ and $y'(0) = -3$.

Solution The general solution to $y'' + y = 0$ is $y = c_1 \cos x + c_2 \sin x$. Since $y = \cos x$ is a solution, we must use the modification of the method to find a particular solution to (11). Ordinarily we would guess a solution of the form $y_p = A\cos x + B\sin x$. Instead we multiply by x and try a solution of the form

$$y_p = Ax\cos x + Bx\sin x.$$

Note that no term of y_p is a solution to $y'' + y = 0$. Then

$$y_p' = A\cos x - Ax\sin x + B\sin x + Bx\cos x$$

and

From (11)
↓

$$\cos x = y_p'' + y_p = (-2A\sin x - Ax\cos x + 2B\cos x - Bx\sin x)$$
$$+ (Ax\cos x + Bx\sin x)$$
$$= -2A\sin x + 2B\cos x.$$

Therefore

$$-2A = 0, \quad 2B = 1, \quad B = \tfrac{1}{2},$$

and

$$y_p = \tfrac{1}{2}x\sin x.$$

Thus the general solution to (11) is
$$y = c_1 \cos x + c_2 \sin x + \tfrac{1}{2}x \sin x.$$
We are not finished yet, as initial conditions were given. We have
$$y' = -c_1 \sin x + c_2 \cos x + \tfrac{1}{2}x \cos x + \tfrac{1}{2}\sin x.$$
Then
$$y(0) = c_1 = 2 \quad \text{and} \quad y'(0) = c_2 = -3,$$
which yields the unique solution
$$y(x) = 2 \cos x - 3 \sin x + \tfrac{1}{2}x \sin x.$$

Example 6 Find the general solution of
$$y'' - 4y' + 4y = e^{2x}.$$

Solution The homogeneous equation $y'' - 4y' + 4y = 0$ has the independent solutions e^{2x} and xe^{2x}. Thus, multiplying $f(x) = e^{2x}$ by x twice, we look for a particular solution of the form $y_p = ax^2 e^{2x}$. Then
$$y_p' = ae^{2x}(2x^2 + 2x)$$
and
$$y_p'' = ae^{2x}(4x^2 + 8x + 2),$$
so
$$y_p'' - 4y_p' + 4y_p = ae^{2x}(4x^2 + 8x + 2 - 8x^2 - 8x + 4x^2)$$
$$= 2ae^{2x} = e^{2x},$$
or $2a = 1$ and $a = \tfrac{1}{2}$. Thus $y_p = \tfrac{1}{2}x^2 e^{2x}$, and the general solution is
$$y(x) = c_1 e^{2x} + c_2 x e^{2x} + \tfrac{1}{2}x^2 e^{2x} = e^{2x}\left(c_1 + c_2 x + \tfrac{1}{2}x^2\right).$$

Example 7 Consider the equation
$$y'' - y = x^2 + 2e^x.$$
Using the results of Examples 1 and 4 and the principle of superposition, we find immediately that a particular solution is given by
$$y_p(x) = -2 - x^2 + xe^x.$$

Example 8 Find the general solution to
$$y'' + y = x \sin x.$$

Solution The guessed solution is $y_p = (Ax + B)\cos x + (Cx + D)\sin x$. Since $B \cos x + D \sin x$ solves $y'' + y = 0$, the modification is required. We therefore multiply by x and try a solution of the form
$$y_p = (Ax^2 + Bx)\cos x + (Cx^2 + Dx)\sin x.$$

Then
$$y_p' = [Cx^2 + (2A + D)x + B]\cos x + [-Ax^2 + (2C - B)x + D]\sin x,$$
$$y_p'' = [-Ax^2 + (4C - B)x + 2A + 2D]\cos x$$
$$+ [-Cx^2 - (4A + D)x + 2C - 2B]\sin x,$$
and

$$y_p'' + y_p = [4Cx + 2A + 2D]\cos x + [-4Ax + 2C - 2B]\sin x \overset{\text{given}}{=} x \sin x.$$

This yields $A = -\frac{1}{4}$, $B = 0$, $C = 0$, $D = \frac{1}{4}$, and the particular solution
$$y_p = -\frac{1}{4}x^2 \cos x + \frac{1}{4}x \sin x.$$

Thus the general solution is
$$y = \left(c_1 - \tfrac{1}{4}x^2\right)\cos x + \left(c_2 + \tfrac{1}{4}x\right)\sin x.$$

Let us now summarize the results of this section as follows: Consider the nonhomogeneous equation
$$y'' + ay' + by = f(x) \tag{12}$$
and the homogeneous equation
$$y'' + ay' + by = 0. \tag{13}$$

Case 1 *No term in the guessed solution $y_p(x)$ is a solution of equation (13)*. A particular solution of equation (12) has the form $y_p(x)$ given by the table below:

$f(x)$	$y_p(x)$
$P_n(x)$	$a_0 + a_1 x + a_2 x^2 + \cdots + a_n x^n$
$P_n(x)e^{ax}$	$(a_0 + a_1 x + a_2 x^2 + \cdots + a_n x^n)e^{ax}$
$P_n(x)e^{ax}\sin bx$ $+$ $Q_n(x)e^{ax}\cos bx$	$(a_0 + a_1 x + a_2 x^2 + \cdots + a_n x^n)e^{ax}\sin bx +$ $(c_0 + c_1 x + c_2 x^2 + \cdots + c_n x^n)e^{ax}\cos bx$

Case 2 *If any term of $y_p(x)$ is a solution of equation (13), then multiply the appropriate function $y_p(x)$ of Case 1 by x^k, where k is the smallest integer such that no term in $x^k y_p(x)$ is a solution of equation (13).*

PROBLEMS 3.5

In Problems 1–13 find the general solution of each differential equation. If initial conditions are given, then find the particular solution that satisfies them.

1. $y'' + 4y = 3 \sin x$
2. $y'' - y' - 6y = 20e^{-2x}$, $y(0) = 0$, $y'(0) = 6$
3. $y'' - 3y' + 2y = 6e^{3x}$
4. $y'' + y' = 3x^2$, $y(0) = 4$, $y'(0) = 0$
5. $y'' - 2y' + y = -4e^x$
6. $y'' - 4y' + 4y = 6xe^{2x}$, $y(0) = 0$, $y'(0) = 3$
7. $y'' - 7y' + 10y = 100x$, $y(0) = 0$, $y'(0) = 5$
8. $y'' + y = 1 + x + x^2$
9. $y'' + y' = x^3 - x^2$

10. $y'' + 4y = 16x \sin 2x$
11. $y'' - 4y' + 5y = 2e^{2x}\cos x$
12. $y'' - y' - 2y = x^2 + \cos x$
13. $y'' + 6y' + 9y = 10e^{-3x}$

Use the principle of superposition to find the general solution of each of the equations in Problems 14–17.

14. $y'' + y = 1 + 2\sin x$
15. $y'' - 2y' - 3y = x - x^2 + e^x$
16. $y'' + 4y = 3\cos 2x - 7x^2$
17. $y'' + 4y' + 4y = xe^x + \sin x$
18. Show by the methods of this section that a particular solution of

$$y'' + 2ay' + b^2 y = A \sin \omega x \quad (a, \omega > 0)$$

is given by

$$y = \frac{A \sin(\omega x - \alpha)}{\sqrt{(b^2 - \omega^2)^2 + 4\omega^2 a^2}},$$

where

$$\alpha = \tan^{-1} \frac{2a\omega}{(b^2 - \omega^2)}, \quad (0 < \alpha < \pi).$$

19. Let $f(x)$ be a polynomial of degree n. Show that, if $b \neq 0$, there is always a solution that is a polynomial of degree n for the equation
$y'' + ay' + by = f(x)$.
20. Use the method indicated in Problem 19 to find a particular solution of

$$y'' + 3y' + 2y = 9 + 2x - 2x^2.$$

In Problems 21–24 find particular solutions to the given differential equation.

21. $y'' + y = (x + x^2)\sin x$
22. $y'' - y' = x^2$
23. $y'' - 2y' + y = x^2 e^x$
24. $y'' - 4y' + 3y = x^3 e^{3x}$
25. In many physical problems (see, for example, Section 6.3), the nonhomogeneous term $f(x)$ is specified by different formulas in different intervals of x. Find
   a. a general solution of the equation

$$y'' + y = \begin{cases} x, & 0 \le x \le 1, \\ 1, & x \ge 1; \end{cases}$$

   [*Note:* This "solution" is not differentiable at $x = 1$.]
   b. a particular solution of part (a) that satisfies the initial conditions

$$y(0) = 0, \quad y'(0) = 1.$$

3.6 Nonhomogeneous Equations: Variation of Parameters[†]

In this section we consider a procedure, due to J. L. Lagrange (1736–1813), for finding a particular solution of any nonhomogeneous linear equation

$$y'' + a(x)y' + b(x)y = f(x), \tag{1}$$

where the functions $a(x)$, $b(x)$, and $f(x)$ are continuous. To use this method it is necessary to know the general solution $c_1 y_1(x) + c_2 y_2(x)$ of the homogeneous equation

$$y'' + a(x)y' + b(x)y = 0. \tag{2}$$

If $a(x)$ and $b(x)$ are constants, the general solution to (2) can always be obtained by the methods of Sections 3.3 and 3.4. If $a(x)$ and $b(x)$ are not both constants, it may be difficult to find this general solution; however, if one solution y_1 of (2) can be found, then the method of reduction of order (see Section 3.2) yields the general solution to (2).

Lagrange[‡] noticed that any particular solution y_p of (1) must have the property that y_p/y_1 and y_p/y_2 are not constants, suggesting that we look for a

---

[†] This procedure is also called the method of **variation of constants** or **Lagrange's method**.
[‡] The reason he made this observation is evident from Problem 2.3.33 (p. 53). It also follows from the method of reduction of order used in Section 3.2.

3.6 / NONHOMOGENEOUS EQUATIONS: VARIATION OF PARAMETERS

particular solution of (1) of the form

$$y(x) = c_1(x)y_1(x) + c_2(x)y_2(x). \tag{3}$$

This replacement of constants or parameters by variables gives the method its name. Differentiating (3), we obtain

$$y'(x) = c_1(x)y_1'(x) + c_2(x)y_2'(x) + c_1'(x)y_1(x) + c_2'(x)y_2(x).$$

To simplify this expression, it is convenient (but not necessary—see Problem 25) to set

$$c_1'(x)y_1(x) + c_2'(x)y_2(x) = 0. \tag{4}$$

Then

$$y'(x) = c_1(x)y_1'(x) + c_2(x)y_2'(x).$$

Differentiating once again, we obtain

$$y''(x) = c_1(x)y_1''(x) + c_2(x)y_2''(x) + c_1'(x)y_1'(x) + c_2'(x)y_2'(x).$$

Substitution of the expressions for $y(x)$, $y'(x)$, and $y''(x)$ into (1) yields

$$y'' + a(x)y' + b(x)y = c_1(x)(y_1'' + ay_1' + by_1) + c_2(x)(y_2'' + ay_2' + by_2)$$
$$+ c_1'y_1' + c_2'y_2'$$
$$= f(x).$$

But y_1 and y_2 are solutions to the homogeneous equation, so the equation above reduces to

$$c_1'y_1' + c_2'y_2' = f(x). \tag{5}$$

This gives a second equation relating $c_1'(x)$ and $c_2'(x)$, and we have the simultaneous equations

$$\boxed{\begin{aligned} y_1 c_1' + y_2 c_2' &= 0, \\ y_1' c_1' + y_2' c_2' &= f(x). \end{aligned}} \tag{6}$$

The determinant of system (6) is the Wronskian

$$\begin{vmatrix} y_1 & y_2 \\ y_1' & y_2' \end{vmatrix} = W(y_1, y_2)(x) \neq 0. \quad \text{since } y_1 \text{ and } y_2 \text{ are linearly independent}$$

Thus, for each value of x, $c_1'(x)$ and $c_2'(x)$ are uniquely determined and the problem has essentially been solved. We obtain, from (6),

$$y_1 y_2' c_1' + y_2 y_2' c_2' = 0, \quad \text{first equation multiplied by } y_2'$$
$$y_1' y_2 c_1' + y_2' y_2 c_2' = y_2 f(x). \quad \text{second equation multiplied by } y_2$$

So

$$(y_1 y_2' - y_1' y_2) c_1' = -y_2 f(x),$$

or

$$c_1' = \frac{-y_2 f(x)}{W(y_1, y_2)(x)}.$$

A similar calculation yields an expression for c_2'. Thus we obtain

$$c_1'(x) = \frac{-f(x)y_2(x)}{y_1(x)y_2'(x) - y_1'(x)y_2(x)} = \frac{-f(x)y_2(x)}{W(y_1, y_2)(x)}, \qquad (7)$$

$$c_2'(x) = \frac{f(x)y_1(x)}{y_1(x)y_2'(x) - y_1'(x)y_2(x)} = \frac{f(x)y_1(x)}{W(y_1, y_2)(x)}. \qquad (8)$$

Finally, if we can integrate c_1' and c_2', we can substitute c_1 and c_2 into (3) to obtain a particular solution to the nonhomogeneous equation; that is,

$$y_p(x) = c_1(x)y_1(x) + c_2(x)y_2(x), \qquad (9)$$

where

$$c_1(x) = \int c_1'(x)\,dx, \qquad c_2(x) = \int c_2'(x)\,dx,$$

with $c_1'(x)$ and $c_2'(x)$ given by (7) and (8).

Remark It is not advisable to try to memorize equations (7) and (8), particularly as it is almost always easier to solve system (6) directly. The key concept to this method is system (6). We illustrate with three examples.

Example 1 Solve $y'' - y = e^{2x}$ by the method of variation of parameters.

Solution The solutions to the homogeneous equation are $y_1 = e^{-x}$ and $y_2 = e^x$. Using these solutions, system (6) becomes

$$e^{-x}c_1' + e^x c_2' = 0,$$
$$-e^{-x}c_1' + e^x c_2' = e^{2x}.$$

Adding these two equations and then subtracting them, we get

$$2e^x c_2' = e^{2x} \qquad \text{and} \qquad 2e^{-x}c_1' = -e^{2x},$$

or

$$c_1' = \frac{-e^{3x}}{2} \qquad \text{and} \qquad c_2' = \frac{e^x}{2}.$$

Integrating these functions, we obtain $c_1(x) = -e^{3x}/6$ and $c_2(x) = e^x/2$.[†] A particular solution is therefore

$$c_1(x)y_1(x) + c_2(x)y_2(x) = \frac{-e^{2x}}{6} + \frac{e^{2x}}{2} = \frac{e^{2x}}{3},$$

and the general solution is

$$y(x) = c_1 e^x + c_2 e^{-x} + \frac{e^{2x}}{3}.$$

[†] It is not necessary or even desirable to include arbitrary constants of integration in the computation of $c_1(x)$ and $c_2(x)$. Recall that the task at hand is simply to find one particular solution.

Example 2 Determine the solution of
$$y'' + y = 4\sin x$$
that satisfies $y(0) = 3$ and $y'(0) = -1$.

Solution Here $y_1(x) = \cos x$, $y_2(x) = \sin x$, and (6) becomes
$$\cos x \cdot c_1' + \sin x \cdot c_2' = 0,$$
$$-\sin x \cdot c_1' + \cos x \cdot c_2' = 4\sin x.$$
Solving these equations simultaneously for c_1' and c_2', we have
$$c_1' = -4\sin^2 x \quad \text{and} \quad c_2' = 4\sin x \cos x.$$
Since
$$\int \sin^2 x \, dx = \tfrac{1}{2}(x - \sin x \cos x),$$
we see that
$$c_1 = 2(\sin x \cos x - x), \quad c_2 = 2\sin^2 x,$$
and a particular solution is
$$y_p(x) = c_1(x) y_1(x) + c_2(x) y_2(x)$$
$$= 2(\sin x \cos x - x)\cos x + 2\sin^2 x \sin x$$
$$= 2\sin x(\cos^2 x + \sin^2 x) - 2x\cos x = 2\sin x - 2x\cos x.$$
Thus the general solution is
$$y = c_1 \cos x + c_2 \sin x + 2\sin x - 2x \cos x$$
$$= c_1 \cos x + c_2^* \sin x - 2x \cos x,$$
where $c_2^* = c_2 + 2$. To solve the initial-value problem, we differentiate:
$$y' = -c_1 \sin x + c_2^* \cos x + 2x \sin x - 2\cos x.$$
But
$$3 = y(0) = c_1 \quad \text{and} \quad -1 = y'(0) = c_2^* - 2,$$
so $c_2^* = 1$ and the unique solution is
$$y = 3\cos x + \sin x - 2x\cos x = (3 - 2x)\cos x + \sin x.$$

Example 3 Solve $y'' + y = \tan x$.

Solution The solutions to the homogeneous equation are $y_1 = \cos x$ and $y_2 = \sin x$. System (6) becomes
$$\cos x \cdot c_1' + \sin x \cdot c_2' = 0,$$
$$-\sin x \cdot c_1' + \cos x \cdot c_2' = \tan x,$$
for which we obtain by elimination
$$c_1'(x) = -\tan x \sin x = -\frac{\sin^2 x}{\cos x} = \frac{\cos^2 x - 1}{\cos x} = \cos x - \sec x,$$
$$c_2'(x) = \tan x \cos x = \sin x.$$

Hence
$$c_1(x) = \sin x - \ln|\sec x + \tan x|$$
and
$$c_2(x) = -\cos x.$$
Thus the particular solution is
$$\begin{aligned}y_p(x) &= c_1(x)y_1(x) + c_2(x)y_2(x)\\ &= \cos x \sin x - \cos x \ln|\sec x + \tan x| - \sin x \cos x\\ &= -\cos x \ln|\sec x + \tan x|,\end{aligned}$$
and the general solution is
$$y(x) = c_1\cos x + c_2\sin x - \cos x \ln|\sec x + \tan x|.$$

Example 3 illustrates that there are instances in which we cannot apply the method of undetermined coefficients. (Try to "guess" a solution in this case.) As a rule, the method of undetermined coefficients is easier to use if the function $f(x)$ is in the right form. However, the method of variation of parameters is far more general, since it yields a solution whenever the functions c_1' and c_2' have known antiderivatives.

A Perspective

GREEN'S FUNCTIONS

We may rewrite equation (9) in integral form as
$$y_p(x) = c_1(x)y_1(x) + c_2(x)y_2(x)$$
$$= -y_1(x)\int^x \frac{y_2(t)f(t)}{W(y_1,y_2)(t)}\,dt + y_2(x)\int^x \frac{y_1(t)f(t)}{W(y_1,y_2)(t)}\,dt,$$
or
$$y_p(x) = \int^x \left[\frac{y_1(t)y_2(x) - y_1(x)y_2(t)}{W(y_1,y_2)(t)}\right] f(t)\,dt. \tag{10}$$

If we set the lower limit of integration of (10) to, for example, x_0, then
$$y_p(x) = \int_{x_0}^x \left[\frac{y_1(t)y_2(x) - y_1(x)y_2(t)}{W(y_1,y_2)(t)}\right] f(t)\,dt \tag{11}$$
is a particular solution to the initial-value problem
$$y'' + a(x)y' + b(x)y = f(x), \quad y(x_0) = 0, \quad y'(x_0) = 0.$$

This is so because it is a solution to the nonhomogeneous linear equation by the development earlier in this section, $y_p(x_0) = 0$ trivially from (11), and
$$y_p'(x) = \overbrace{\left[\frac{y_1(x)y_2(x) - y_1(x)y_2(x)}{W(y_1,y_2)(x)}\right]f(x)}^{=0}$$
$$+ \int_{x_0}^x \left[\frac{y_1(t)y_2'(x) - y_1'(x)y_2(t)}{W(y_1,y_2)(t)}\right] f(t)\,dt,$$
so that $y_p'(x_0) = 0$.

If we wish to solve the initial-value problem

$$y'' + a(x)y' + b(x)y = f(x), \quad y(x_0) = y_0, \quad y'(x_0) = y'_0, \quad (12)$$

we need only select linearly independent solutions y_1 and y_2 of the homogeneous equation that satisfy the conditions

$$y_1(x_0) = 1, \quad y'_1(x_0) = 0,$$

and

$$y_2(x_0) = 0, \quad y'_2(x_0) = 1.$$

Then

$$y(x) = y_0 y_1(x) + y'_0 y_2(x)$$

$$+ \int_{x_0}^{x} \left[\frac{y_1(t) y_2(x) - y_1(x) y_2(t)}{W(y_1, y_2)(t)} \right] f(t) \, dt \quad (13)$$

is the unique solution to initial-value problem (12). The function in brackets in (13) is called a **Green's function** for the differential equation in (12).

PROBLEMS 3.6

In Problems 1–20 find the general solution to each equation by the method of variation of parameters.

1. $y'' - 3y' + 2y = 10$
2. $y'' - y' - 6y = 20e^{-2x}$
3. $y'' + 4y = 3 \sin x$
4. $y'' + y' = 3x^2$
5. $y'' - 3y' + 2y = 6e^{3x}$
6. $y'' - 4y' + 4y = 6xe^{2x}$
7. $y'' - 2y' + y = -4e^x$
8. $y'' + y = 1 + x + x^2$
9. $y'' - 7y' + 10y = 100x$
10. $y'' + 4y = 16x \sin 2x$
11. $y'' - y' = \sec^2 x - \tan x$
12. $y'' + y = \cot x$
13. $y'' + 4y = \sec 2x$
14. $y'' + 4y = \sec x \tan x$
15. $y'' - 2y' + y = \dfrac{e^x}{(1-x)^2}$
16. $y'' - y = \sin^2 x$
17. $y'' - y = \dfrac{(2x-1)e^x}{x^2}$
18. $y'' - 3y' - 4y = \dfrac{e^{4x}(5x-2)}{x^3}$
19. $y'' - 4y' + 4y = \dfrac{e^{2x}}{(1+x)}$
20. $y'' + 2y' + y = e^{-x} \ln|x|$

21. Find a particular solution of

$$y'' + \frac{1}{x}y' - \frac{y}{x^2} = \frac{1}{x^2 + x^3}, \quad (x > 0),$$

given that two solutions of the associated homogeneous equations are $y_1 = x$ and $y_2 = 1/x$.

22. Find a particular solution of

$$y'' - \frac{2}{x}y' + \frac{2}{x^2}y = \frac{\ln|x|}{x}, \quad (x > 0),$$

given the two homogeneous solutions $y_1 = x$ and $y_2 = x^2$.

23. Verify that

$$y = \frac{1}{\omega} \int_0^x f(t) \sin \omega(x - t) \, dt$$

is a particular solution of $y'' + \omega^2 y = f(x)$.
[*Hint:* Use equation (10).]

24. Find a particular solution to the initial-value problem

$$y'' - \omega^2 y = f(x), \quad y(0) = y'(0) = 0.$$

*25. This problem shows why there is no loss in generality in equation (4) by setting

$$c'_1 y_1 + c'_2 y_2 = 0.$$

Suppose that we instead let $c'_1 y_1 + c'_2 y_2 = z(x)$, with $z(x)$ an undetermined function of x.

a. Show that we then obtain the system

$$c'_1 y_1 + c'_2 y_2 = z,$$
$$c'_1 y'_1 + c'_2 y'_2 = f - z' - az.$$

b. Show that the system in part (a) has the solution

$$c_1' = \frac{-y_2 f}{W(y_1, y_2)} + \frac{\left(e^{\int a(x)\,dx} z y_2\right)'}{e^{\int a(x)\,dx} W(y_1, y_2)},$$

$$c_2' = \frac{y_1 f}{W(y_1, y_2)} - \frac{\left(e^{\int a(x)\,dx} z y_1\right)'}{e^{\int a(x)\,dx} W(y_1, y_2)}.$$

c. Integrate by parts to show that

$$\int \frac{\left(e^{\int a(x)\,dx} z y_i\right)'}{e^{\int a(x)\,dx} W(y_1, y_2)}\,dx = \frac{z y_i}{W(y_1, y_2)}, \quad i = 1, 2.$$

d. Conclude that the particular solution obtained by letting $c_1' y_1 + c_2' y_2 = z$ is identical to that obtained by assuming equation (4).

*26. Suppose one solution $y_1(x)$ of the homogeneous counterpart of the linear differential equation

$$y'' + a(x) y' + b(x) y = f(x) \quad (i)$$

is known. Use the substitution $y_2 = v y_1$ of Section 3.2 to find the general solution of (i):

$$y = c_1 y_1(x) + c_2 y_1(x) \int^x \frac{h(t)}{y_1^2(t)}\,dt$$

$$+ y_1(x) \int^x \frac{h(t)}{y_1^2(t)} \left[\int^t \frac{y_1(s) f(s)}{h(s)}\,ds \right] dt,$$

where $h(t) = \exp(-\int^t a(u)\,du)$.

3.7 Euler Equations

For most linear second-order equations with variable coefficients it is impossible to write solutions in terms of elementary functions. In most cases it is necessary to use techniques such as the power series method (Chapter 5) to obtain information about solutions. However, there is one class of such equations that do arise in applications for which solutions in terms of elementary functions can be obtained:

Euler Equation An equation of the form

$$x^2 y'' + a x y' + b y = f(x), \quad x \neq 0 \quad (1)$$

is called an **Euler equation**.

Note Equation (1) can be written

$$y'' + \frac{a}{x} y' + \frac{b}{x^2} y = \frac{f(x)}{x^2},$$

which is not defined for $x = 0$. This is why we make the restriction that $x \neq 0$.

We begin by solving the homogeneous Euler equation

$$x^2 y'' + a x y' + b y = 0, \quad x \neq 0. \quad (2)$$

If we can find two linearly independent solutions to (2), we can solve (1) by the method of variation of parameters. There are two ways to solve equation (2); each one involves a certain trick. We give one method here and leave the other for the problem set (see Problem 18).

The first method involves guessing an appropriate solution to (2). We note that if $y = x^\lambda$ for some number λ, then $y' = \lambda x^{\lambda - 1}$ and $y'' = \lambda(\lambda - 1) x^{\lambda - 2}$. This is interesting because $x^2 y''$, $x y'$, and y all can be written as constant multiples of x^λ. Therefore we guess that there is a solution having the form $y = x^\lambda$. Substituting this into equation (2), we obtain

$$\lambda(\lambda - 1) x^\lambda + a \lambda x^\lambda + b x^\lambda = x^\lambda [\lambda(\lambda - 1) + a \lambda + b] = 0.$$

If $x \neq 0$, we can divide by x^λ to obtain the **characteristic equation**† for the **Euler equation**:

$$\lambda(\lambda - 1) + a\lambda + b = 0, \qquad (3)$$

or

$$\lambda^2 + (a - 1)\lambda + b = 0. \qquad (4)$$

As with constant-coefficient equations, there are three cases to consider.

Case 1 Characteristic equation (4) has two real, distinct roots.

Example 1 Find the general solution to
$$x^2 y'' + 2xy' - 12 y = 0, \qquad x \neq 0.$$

Solution The characteristic equation is
$$\lambda(\lambda - 1) + 2\lambda - 12 = \lambda^2 + \lambda - 12 = 0 = (\lambda + 4)(\lambda - 3),$$
with roots $\lambda_1 = -4$ and $\lambda_2 = 3$. Thus two solutions (that are linearly independent) are
$$y_1 = x^{-4} = \frac{1}{x^4} \qquad \text{and} \qquad y_2 = x^3,$$
and the general solution is
$$y(x) = \frac{c_1}{x^4} + c_2 x^3.$$

In general, we have the following result:

THEOREM 1

If λ_1 and λ_2 are real and distinct, then the general solution to equation (2) is

$$y(x) = c_1 x^{\lambda_1} + c_2 x^{\lambda_2}, \quad x \neq 0. \qquad (5)$$

Case 2 The roots are real and equal ($\lambda_1 = \lambda_2$).

Example 2 Find the general solution to
$$x^2 y'' - 3xy' + 4y = 0, \quad x > 0. \qquad (6)$$

---

† The term "characteristic equation" is generally reserved for linear equations with *constant* coefficients. The only reason we can use this term in this case is that the substitution in Problem 18 converts equation (1) into a second-order constant-coefficient equation with characteristic equation (4). *The method of characteristic equations is generally not applicable to equations with variable coefficients.*

Solution The characteristic equation is
$$\lambda^2 - 4\lambda + 4 = (\lambda - 2)^2 = 0,$$
with the single root $\lambda = 2$. Thus one solution is $y_1(x) = x^2$. To find a second solution we use the method of reduction of order (see Section 3.2). Let $y_2 = vy_1 = x^2v$. Then $y_2' = x^2v' + 2xv$ and $y_2'' = x^2v'' + 4xv' + 2v$, so (6) becomes
$$x^2(x^2v'' + 4xv' + 2v) - 3x(x^2v' + 2xv) + 4(x^2v) = 0,$$
or, with $z = v'$,
$$x^4v'' + x^3v' = x^4z' + x^3z = 0.$$
Separating variables, we have
$$\ln|z| = \int \frac{dz}{z} = -\int \frac{dx}{x} = -\ln x,$$
so that
$$v' = z = x^{-1}.$$
Then
$$y_2(x) = y_1v = x^2 \int \frac{dx}{x} = x^2 \ln x.$$
Thus the general solution to equation (6) is
$$y(x) = c_1 x^2 + c_2 x^2 \ln x = x^2(c_1 + c_2 \ln x).$$

THEOREM 2

If λ is the only root of characteristic equation (4), then the general solution to (2) is
$$y(x) = x^\lambda(c_1 + c_2 \ln|x|).$$

Case 3 The roots are complex conjugates ($\lambda_1 = \alpha + i\beta$, $\lambda_2 = \alpha - i\beta$).

Example 3 Find the general solution of
$$x^2 y'' + 5xy' + 13y = 0, \quad x > 0. \tag{7}$$

Solution The characteristic equation is
$$\lambda^2 + 4\lambda + 13 = 0,$$
and
$$\lambda = \frac{-4 \pm \sqrt{16 - 4(13)}}{2} = \frac{-4 \pm \sqrt{-36}}{2} = -2 \pm 3i.$$
Thus two linearly independent solutions are
$$y_1(x) = x^{-2+3i} \quad \text{and} \quad y_2(x) = x^{-2-3i}.$$
Using the material in Appendix 5, we can eliminate the imaginary exponents. First we note that
$$x^a = e^{\ln x^a} = e^{a \ln x}.$$

By the Euler formula ($e^{i\beta x} = \cos \beta x + i \sin \beta x$),
$$y_1(x) = (x^{-2})(x^{3i}) = x^{-2}e^{3i \ln x}$$
$$= x^{-2}[\cos(3 \ln x) + i \sin(3 \ln x)]$$
and
$$y_2(x) = (x^{-2})(x^{-3i}) = x^{-2}e^{-3i \ln x}$$
$$= x^{-2}[\cos(3 \ln x) - i \sin(3 \ln x)].$$

We now form two new solutions:
$$y_3(x) = \tfrac{1}{2}[y_1(x) + y_2(x)] = x^{-2}\cos(3 \ln x)$$
and
$$y_4(x) = \frac{1}{2i}[y_1(x) - y_2(x)] = x^{-2}\sin(3 \ln x).$$

These new solutions contain no complex numbers and are easier to work with. The general solution to (7) is
$$y(x) = x^{-2}[c_1 \cos(3 \ln x) + c_2 \sin(3 \ln x)].$$

THEOREM 3

If $\lambda_1 = \alpha + i\beta$ and $\lambda_2 = \alpha - i\beta$ are complex conjugate roots of characteristic equation (4), then the general solution to (2) is

$$\boxed{y(x) = x^\alpha [c_1 \cos(\beta \ln|x|) + c_2 \sin(\beta \ln|x|)].} \tag{8}$$

Example 4 Find the general solution to
$$x^2 y'' + 2xy' - 12y = \sqrt{x}, \quad x > 0. \tag{9}$$

Solution In Example 1 we found the homogeneous solutions
$$y_1 = x^{-4} \quad \text{and} \quad y_2 = x^3.$$

Dividing both sides of (9) by x^2, we obtain the standard form
$$y'' + \frac{2}{x}y' - \frac{12}{x^2}y = x^{-3/2}.$$

We now apply the method of variation of parameters, obtaining the system
$$x^{-4}c_1' + x^3 c_2' = 0,$$
$$-4x^{-5}c_1' + 3x^2 c_2' = x^{-3/2}.$$

Solving these equations simultaneously, we get
$$c_1' = \frac{-x^{7/2}}{7} \quad \text{and} \quad c_2' = \frac{x^{-7/2}}{7}.$$

Hence
$$c_1(x) = -\frac{1}{7} \cdot \frac{2}{9} x^{9/2}, \quad c_2(x) = -\frac{1}{7} \cdot \frac{2}{5} x^{-5/2},$$

so
$$y_p(x) = c_1(x)y_1(x) + c_2(x)y_2(x) = -\frac{1}{7}\left[\frac{2}{9}x^{9/2} \cdot x^{-4} + \frac{2}{5}x^{-5/2} \cdot x^3\right]$$
$$= \left(\frac{2}{9} + \frac{2}{5}\right)\frac{-x^{1/2}}{7} = -\frac{4}{45}x^{1/2}.$$

Thus the general solution is given by
$$y(x) = c_1 x^{-4} + c_2 x^3 - \frac{4}{45}x^{1/2}.$$

PROBLEMS 3.7

In Problems 1–17 find the general solution to the given Euler equation for $x > 0$. Find the unique solution when initial conditions are given.

1. $x^2 y'' + xy' - y = 0$
2. $x^2 y'' - 5xy' + 9y = 0$
3. $x^2 y'' - xy' + 2y = 0$
4. $x^2 y'' - 2y = 0$, $y(1) = 3$, $y'(1) = 1$
5. $4x^2 y'' - 4xy' + 3y = 0$, $y(1) = 0$, $y'(1) = 1$
6. $x^2 y'' + 3xy' + 2y = 0$
7. $x^2 y'' - 3xy' + 3y = 0$
8. $x^2 y'' + 5xy' + 4y = 0$, $y(1) = 1$, $y'(1) = 3$
9. $x^2 y'' + 5xy' + 5y = 0$
10. $4x^2 y'' - 8xy' + 8y = 0$
11. $x^2 y'' + 2xy' - 12y = 0$
12. $x^2 y'' + xy' + y = 0$
13. $x^2 y'' + 3xy' - 15y = \dfrac{1}{x}$
14. $x^2 y'' + 3xy' + y = 3x^6$
15. $x^2 y'' - 5xy' + 9y = x^3$
16. $x^2 y'' + 3xy' - 15y = x^2 e^x$
17. $x^2 y'' + xy' + y = 10$

*18. Show that the homogeneous Euler equation (2) can be transformed into the constant-coefficient equation $y'' + (a-1)y' + by = 0$ by making the substitution $x = e^t$ ($t = \ln x$). [Hint: By the chain rule,
$$\frac{dy}{dt} = \frac{dy}{dx}\frac{dx}{dt} = x\frac{dy}{dx}$$
and
$$\frac{d^2 y}{dt^2} = \frac{d}{dx}\left(x\frac{dy}{dx}\right)\frac{dx}{dt} = x^2\frac{d^2 y}{dx^2} + x\frac{dy}{dx}.]$$

Use the method of Problem 18 to solve Problems 19–22.

19. $x^2 y'' + 7xy' + 5y = x$
20. $x^2 y'' + 3xy' - 3y = 5x^2$
21. $x^2 y'' - 2y = \ln x$, $(x > 0)$
22. $4x^2 y'' - 4xy' + 3y = \sin \ln(-x)$ $(x < 0)$
23. The equation $xy'' + 4y' = 0$, $x > 0$, arises in astronomy.[†] Obtain its general solution.

[†] Z. Kopal, "Stress History of the Moon and of Terrestrial Planets," *Icarus* 2 (1963): 381.

3.8 Higher-Order Linear Differential Equations

In this section we extend the results of the chapter to linear differential equations of order higher than two. There is little theoretical difference between second- and higher-order systems, so we can be relatively brief. We state all theorems without proof.

The *general nonhomogeneous linear nth-order equation* is
$$y^{(n)}(x) + a_{n-1}(x)y^{(n-1)}(x) + \cdots + a_1(x)y'(x) + a_0(x)y(x) = f(x). \quad (1)$$

The associated homogeneous equation is
$$y^{(n)}(x) + a_{n-1}(x)y^{(n-1)}(x) + \cdots + a_1(x)y'(x) + a_0(x)y(x) = 0. \quad (2)$$

In Theorem 3.1.1 (p. 84) we stated that equation (1) has a unique solution provided that all the functions in the equation are continuous and n initial conditions are specified. Now we concern ourselves with finding the general solutions to equations (1) and (2). To do so we follow the procedures we have developed for solving second-order equations.

Linear Independence

We say that the functions $y_1, y_2, \ldots, y_n$ are **linearly independent** in $[x_0, x_1]$ if the following condition holds:

$$c_1 y_1(x) + c_2 y_2(x) + \cdots + c_n y_n(x) = 0 \quad \text{for all } x \in [x_0, x_1]$$

implies that $c_1 = c_2 = \cdots = c_n = 0$.

Linear Combination

Otherwise the functions are **linearly dependent**. The expression $c_1 y_1 + c_2 y_2 + \cdots + c_n y_n$ is called a **linear combination** of the functions $y_1, y_2, \ldots, y_n$.

Wronskian

The Wronskian of $y_1, y_2, \ldots, y_n$ is defined by

$$W(y_1, y_2, \ldots, y_n)(x) = \begin{vmatrix} y_1 & y_2 & \cdots & y_n \\ y_1' & y_2' & \cdots & y_n' \\ y_1'' & y_2'' & \cdots & y_n'' \\ \vdots & \vdots & & \vdots \\ y_1^{(n-1)} & y_2^{(n-1)} & \cdots & y_n^{(n-1)} \end{vmatrix}. \quad (3)$$

THEOREM 1

Let $a_0, a_1, \ldots, a_{n-1}$ be continuous in $[x_0, x_1]$ and let $y_1, y_2, \ldots, y_n$ be n solutions of equation (2). Then

a. $W(y_1, y_2, \ldots, y_n)(x)$ is zero either for all $x \in [x_0, x_1]$ or for no $x \in [x_0, x_1]$.

b. $y_1, y_2, \ldots, y_n$ are linearly independent if and only if $W(y_1, y_2, \ldots, y_n)(x) \neq 0$.

Example 1 The functions 1, x, and x^2 are solutions to the equation $y'''(x) = 0$. Determine whether they are linearly independent or dependent for all x.

Solution The easiest way to test for linear independence is to use the Wronskian:

$$W(y_1, y_2, y_3)(x) = \begin{vmatrix} 1 & x & x^2 \\ 0 & 1 & 2x \\ 0 & 0 & 2 \end{vmatrix} = 2 \neq 0,$$

so the functions are linearly independent. Alternatively, consider

$$c_1 \cdot 1 + c_2 x + c_3 x^2 = 0.$$

Setting $x = 0$, it follows that $c_1 = 0$. Setting $x = \pm 1$ yields the system

$$c_2 + c_3 = 0,$$
$$-c_2 + c_3 = 0,$$

so $c_2 = c_3 = 0$, implying that the functions are linearly independent.

CHAPTER 3/SECOND- AND HIGHER-ORDER LINEAR DIFFERENTIAL EQUATIONS

General Solution

The procedure for solving a linear nth-order equation is as follows:

Step 1 Find n linearly independent solutions, $y_1, y_2, \ldots, y_n$, to the homogeneous equation

$$y^{(n)} + a_{n-1}(x)y^{(n-1)} + \cdots + a_1(x)y' + a_0(x)y = 0.$$

The **general solution** to this equation is then

$$\boxed{y(x) = c_1 y_1(x) + c_2 y_2(x) + \cdots + c_n y_n(x).} \qquad (4)$$

Step 2 Find one solution, $y_p(x)$, to the nonhomogeneous equation

$$y^{(n)} + a_{n-1}(x)y^{(n-1)} + \cdots + a_1(x)y' + a_0(x)y = f(x).$$

The **general solution** to this equation is then given by

$$\boxed{y(x) = c_1 y_1(x) + c_2 y_2(x) + \cdots + c_n y_n(x) + y_p,} \qquad (5)$$

where $y_1, y_2, \ldots, y_n$ are the n linearly independent solutions of Step 1.

As in the case of second-order equations, we can generally obtain these solutions only when the coefficients $a_k(x)$ are all constants. In this case, equations (1) and (2) are said to have **constant coefficients**. We only deal with the case where these constants are real.

The general nth-order linear, homogeneous constant-coefficient equation is

$$y^{(n)}(x) + a_{n-1}y^{(n-1)}(x) + \cdots + a_1 y'(x) + a_0 y(x) = 0. \qquad (6)$$

Note that

$$\frac{d^n}{dx^n} e^{\lambda x} = \lambda^n e^{\lambda x}.$$

If we substitute $y = e^{\lambda x}$ into (6) and then divide by $e^{\lambda x}$, we obtain the **characteristic equation**

$$\boxed{\lambda^n + a_{n-1}\lambda^{n-1} + \cdots + a_1 \lambda + a_0 = 0.} \qquad (7)$$

Equation (7) has n roots $\lambda_1, \lambda_2, \ldots, \lambda_n$. Some of these roots may be real and distinct, real and equal, distinct complex conjugate pairs, or equal complex conjugate pairs. If a root λ_k (real or complex) occurs m times, we say that it has **multiplicity** m. The following rules tell us how to find the general solution to equation (6).

PROCEDURE FOR SOLVING LINEAR HOMOGENEOUS EQUATIONS WITH CONSTANT COEFFICIENTS

a. Obtain characteristic equation (7).

b. Find the roots $\lambda_1, \lambda_2, \ldots, \lambda_n$ of (7). (This is usually the most difficult step.)

c. For each real root λ_k of multiplicity 1 (*single root*), one solution to (6) is $y_k = e^{\lambda_k x}$.

d. For each real root λ_k of multiplicity $m > 1$, m solutions to (6) are
$$y_1 = e^{\lambda_k x},\ y_2 = xe^{\lambda_k x},\ldots,\ y_m = x^{m-1}e^{\lambda_k x}.$$

e. If $\alpha + i\beta$ and $\alpha - i\beta$ are simple roots, then two solutions to (6) are
$$y_1 = e^{\alpha x}\cos \beta x \quad \text{and} \quad y_2 = e^{\alpha x}\sin \beta x.$$

f. If $\alpha + i\beta$ and $\alpha - i\beta$ are roots of multiplicity $m > 1$, then $2m$ solutions to (6) are
$$y_1 = e^{\alpha x}\cos \beta x,\ y_2 = xe^{\alpha x}\cos \beta x,\ldots,\ y_m = x^{m-1}e^{\alpha x}\cos \beta x,$$
$$y_{m+1} = e^{\alpha x}\sin \beta x,\ y_{m+2} = xe^{\alpha x}\sin \beta x,\ldots,\ y_{2m} = x^{m-1}e^{\alpha x}\sin \beta x.$$

g. If $y_1, y_2, \ldots, y_n$ are the n solutions obtained in Steps c–f, then $y_1, y_2, \ldots, y_n$ are linearly independent and the general solution to (6) is given by
$$y(x) = c_1 y_1(x) + c_2 y_2(x) + \cdots + c_n y_n(x).$$

Example 2 Find the general solution of
$$y''' - 3y'' - 10y' + 24y = 0.$$

Solution The characteristic equation is
$$\lambda^3 - 3\lambda^2 - 10\lambda + 24 = (\lambda - 2)(\lambda + 3)(\lambda - 4) = 0,$$
with roots $\lambda_1 = 2$, $\lambda_2 = -3$, and $\lambda_3 = 4$. Since these roots are real and distinct, three linearly independent solutions are
$$y_1 = e^{2x}, \quad y_2 = e^{-3x}, \quad y_3 = e^{4x},$$
and the general solution is
$$y(x) = c_1 e^{2x} + c_2 e^{-3x} + c_3 e^{4x}.$$

Example 3 Find the general solution of
$$y^{(4)} - 4y''' + 6y'' - 4y' + y = 0.$$

Solution The characteristic equation is
$$\lambda^4 - 4\lambda^3 + 6\lambda^2 - 4\lambda + 1 = (\lambda - 1)^4 = 0,$$
with the single root $\lambda = 1$ of multiplicity 4. Thus four linearly independent solutions are
$$y_1 = e^x, \quad y_2 = xe^x, \quad y_3 = x^2 e^x, \quad y_4 = x^3 e^x,$$
and the general solution is
$$y(x) = e^x (c_1 + c_2 x + c_3 x^2 + c_4 x^3).$$

Example 4 Find the general solution of
$$y^{(5)} - 2y^{(4)} + 8y'' - 12y' + 8y = 0.$$

Solution The characteristic equation is
$$\lambda^5 - 2\lambda^4 + 8\lambda^2 - 12\lambda + 8 = 0,$$
which can be factored into
$$(\lambda + 2)(\lambda^2 - 2\lambda + 2)^2 = 0.$$
The solutions to $\lambda^2 - 2\lambda + 2 = 0$ are $\lambda = 1 \pm i$. Thus the roots are
$$\lambda_1 = -2 \text{ (simple)}, \quad \lambda_2 = 1 + i, \quad \lambda_3 = 1 - i,$$
with the complex roots λ_2 and λ_3 having multiplicity 2. Thus five linearly independent solutions are
$$y_1 = e^{-2x}, \quad y_2 = e^x \cos x, \quad y_3 = xe^x \cos x,$$
$$y_4 = e^x \sin x, \quad y_5 = xe^x \sin x,$$
and the general solution is
$$y(x) = c_1 e^{-2x} + (c_2 + c_3 x) e^x \cos x + (c_4 + c_5 x) e^x \sin x.$$

Remark In solving the last three characteristic equations we made the factoring look easy. Finding roots of a polynomial of degree greater than two is, in general, very difficult.

How do we find a particular solution to the nonhomogeneous equation (1)? As with second-order equations, there are two methods: undetermined coefficients and variation of parameters. The method of undetermined coefficients is identical to the technique we used for second-order equations. The method of variation of parameters is discussed in Problems 29 and 30.

Finally, certain equations with variable coefficients can be solved. The higher-order Euler equation is discussed in Problems 31–34.

PROBLEMS 3.8

In Problems 1–16 find the general solution to the given equation. If initial conditions are given, find the particular solutions that satisfy them.

1. $y^{(4)} + 2y'' + y = 0.$
2. $y''' - y'' - y' + y = 0$
3. $y''' - 3y'' + 3y' - y = 0, \ y(0) = 1,$
   $y'(0) = 2, \ y''(0) = 3$
4. $x''' + 5x'' - x' - 5x = 0$
5. $y''' - 9y' = 0, \ y(0) = 3, \ y'(0) = 0,$
   $y''(0) = 18$
6. $y''' - 6y'' + 3y' + 10y = 0$
7. $y^{(4)} = 0$
8. $y^{(4)} - 9y'' = 0$
9. $y^{(4)} - 5y'' + 4y = 0$
10. $y^{(5)} - 2y''' + y' = 0$
11. $y^{(4)} - 4y'' = 0, \ y(0) = 1, \ y'(0) = 3,$
    $y''(0) = 0, \ y'''(0) = 16$
12. $y^{(4)} - 4y''' - 7y'' + 22y' + 24y = 0$
13. $y''' - y'' + y' - y = 0$
14. $y''' - 3y'' + 4y' - 2y = 0, \ y(0) = 1,$
    $y'(0) = 2, \ y''(0) = 3$
15. $y''' - 27y = 0$
16. $y^{(5)} + 2y''' + y' = 0, \ y(\pi/2) = 0, \ y'(\pi/2) = 1,$
    $y''(\pi/2) = 0, \ y'''(\pi/2) = -3, \ y^{(4)}(\pi/2) = 0$
17. Show that the solutions y_1, y_2, and y_3 of the linear third-order differential equation
$$y''' + a_1(x) y'' + a_2(x) y' + a_3(x) y = 0$$

that satisfy the conditions

$$y_1(x_0) = 1, \quad y_1'(x_0) = 0, \quad y_1''(x_0) = 0,$$
$$y_2(x_0) = 0, \quad y_2'(x_0) = 1, \quad y_2''(x_0) = 0,$$
$$y_3(x_0) = 0, \quad y_3'(x_0) = 0, \quad y_3''(x_0) = 1,$$

respectively, are linearly independent.

18. Show that *any* solution of

$$y''' + a_1(x)y'' + a_2(x)y' + a_3(x)y = 0$$

can be expressed as a linear combination of the solutions y_1, y_2, y_3 given in Problem 17. [*Hint*: If $y(x_0) = c_1$, $y'(x_0) = c_2$, and $y''(x_0) = c_3$, consider the linear combination $c_1 y_1 + c_2 y_2 + c_3 y_3$.]

*19. Consider the third-order equation

$$y''' + a(x)y'' + b(x)y' + c(x)y = 0$$

and let $y_1(x)$ and $y_2(x)$ be two linearly independent solutions. Define $y_3(x) = v(x) y_1(x)$ and assume that $y_3(x)$ is a solution to the equation.

a. Find a second-order differential equation that is satisfied by v'.

b. Show that $(y_2/y_1)'$ is a solution of this equation.

c. Use the result of part (b) to find a second linearly independent solution of the equation derived in part (a).

*20. Consider the equation

$$y''' - \left(\frac{3}{x^2}\right)y' + \left(\frac{3}{x^3}\right)y = 0, \quad (x > 0).$$

a. Show that $y_1(x) = x$ and $y_2(x) = x^3$ are two linearly independent solutions.

b. Use the results of Problem 19 to get a third linearly independent solution.

21. Consider the third-order equation

$$y''' + a(x)y'' + b(x)y' + c(x)y = 0,$$

where a, b, and c are continuous functions of x in some interval I. Prove that if $y_1(x)$, $y_2(x)$, and $y_3(x)$ are solutions to the equation, then so is any linear combination of them.

*22. In Problem 21, let

$$W(y_1, y_2, y_3)(x) = \begin{vmatrix} y_1 & y_2 & y_3 \\ y_1' & y_2' & y_3' \\ y_1'' & y_2'' & y_3'' \end{vmatrix}.$$

a. Show that W satisfies the differential equation

$$W'(x) = -a(x)W.$$

b. Prove that $W(y_1, y_2, y_3)(x)$ is either always zero or never zero.

*23. a. Prove that the solutions $y_1(x), y_2(x), y_3(x)$ of the equation in Problem 21 are linearly independent on $[x_0, x_1]$ if and only if $W(y_1, y_2, y_3) \neq 0$.

b. Show that $\sin t$, $\cos t$, and e^t are linearly independent solutions of

$$y''' - y'' + y' - y = 0$$

on any interval (a, b) where $-\infty < a < b < \infty$.

24. Assume that $y_1(x)$ and $y_2(x)$ are two solutions to

$$y''' + a(x)y'' + b(x)y' + c(x)y = f(x).$$

Prove that $y_3(x) = y_1(x) - y_2(x)$ is a solution of the associated homogeneous equation.

In Problems 25–28 use the method of undetermined coefficients to find the general solution of the given equation.

25. $y''' - y'' - y' + y = e^x$
26. $y''' - y'' - y' + y = e^{-x}$
27. $y''' - 3y'' - 10y' + 24y = x + 3$
28. $y^{(4)} + 2y'' + y = 3\cos x$

*29. Consider the third-order equation

$$y''' + ay'' + by' + cy = f(x).$$

Let $y_1(x)$, $y_2(x)$, and $y_3(x)$ be three linearly independent solutions to the associated homogeneous equation. Assume that there is a solution of this equation of the form $y(x) = c_1(x) y_1(x) + c_2(x) y_2(x) + c_3(x) y_3(x)$.

a. Following the steps used in deriving the method of variation of parameters for second-order equations, derive a method for solving third-order equations.

b. Find a particular solution of the equation

$$y''' - 2y' - 4y = e^{-x} \tan x.$$

30. Use the method derived in Problem 29 to find a particular solution of

$$y''' + 5y'' + 9y' + 5y = 2e^{-2x} \sec x.$$

In Problems 31–33 guess that there is a solution of the form $y = x^\lambda$ to solve the given Euler equation.

31. $x^3 y''' + 2x^2 y'' - xy' + y = 0$
32. $x^3 y''' - 12xy' + 24y = 0$
33. $x^3 y''' + 4x^2 y'' + 3xy' + y = 0$

34. Show that the substitution $x = e^t$ can be used to solve the third-order Euler equation

$$x^3 y''' + x^2 y'' - 2xy' + 2y = 0.$$

3.9 An Excursion: Second-Order Difference Equations

In Section 2.8 we presented an excursion on linear first-order difference equations. Here we derive a method for the solution of linear second-order difference equations. The technique we give is easily extended to higher-order linear difference equations.

First consider the linear homogeneous second-order difference equation with constant coefficients

$$y_{n+2} + ay_{n+1} + by_n = 0, \quad b \neq 0. \tag{1}$$

We saw in Section 2.8 that the first-order equation $y_n = ay_{n-1}$ has the general solution $y_n = a^n y_0$. It is, then, not implausible to "guess" that there are solutions to equation (1) of the form $y_n = \lambda^n$ for some λ (real or complex). Substituting $y_n = \lambda^n$ into equation (1), we obtain

$$\lambda^{n+2} + a\lambda^{n+1} + b\lambda^n = 0.$$

Since this equation is true for all $n \geq 0$, it holds for $n = 0$, so

$$\boxed{\lambda^2 + a\lambda + b = 0.} \tag{2}$$

This is the **characteristic equation** for the difference equation (1) and is identical to the characteristic equation for the second-order differential equation derived in Section 3.3. As before, the roots are

$$\lambda_1 = \frac{-a + \sqrt{a^2 - 4b}}{2} \quad \text{and} \quad \lambda_2 = \frac{-a - \sqrt{a^2 - 4b}}{2}. \tag{3}$$

Again there are three cases.

Case 1 λ_1 and λ_2 are distinct real roots of (2). Then the general solution of equation (1) is given by

$$\boxed{y_n = c_1 \lambda_1^n + c_2 \lambda_2^n.} \tag{4}$$

Example 1 Consider the equation

$$x_{n+2} - 5x_{n+1} - 6x_n = 0.$$

The characteristic equation is $\lambda^2 - 5\lambda - 6 = 0$, with the roots $\lambda_1 = 6$ and $\lambda_2 = -1$. The general solution is given by $y_n = c_1(6)^n + c_2(-1)^n$. If we specify the initial conditions $y_0 = 3$, $y_1 = 11$, for example, we obtain the system

$$c_1 + c_2 = 3,$$
$$6c_1 - c_2 = 11,$$

which has the unique solution $c_1 = 2$, $c_2 = 1$, and the specific solution

$$y_n = 2 \cdot 6^n + (-1)^n.$$

Case 2 If $a^2 - 4b = 0$, the two roots of equation (2) are equal, and we have the solution $x_n = \lambda^n$ where $\lambda = -a/2$. A second linearly independent solution is $n\lambda^n$.

Check If $y_n = n\lambda^n$, then

$$y_{n+2} + ay_{n+1} + by_n = (n+2)\lambda^{n+2} + a(n+1)\lambda^{n+1} + bn\lambda^n$$
$$= \lambda^n\left[n(\lambda^2 + a\lambda + b) + 2\lambda(\lambda + a/2)\right] = 0,$$

because $\lambda^2 + a\lambda + b = 0$ and $\lambda = -a/2$. Thus the general solution to (1) is given by

$$\boxed{y_n = (c_1 + c_2 n)\lambda^n.} \tag{5}$$

Example 2 Consider the equation

$$y_{n+2} - 6y_{n+1} + 9y_n = 0, \tag{6}$$

with the initial conditions $y_0 = 5$, $y_1 = 12$. The characteristic equation is $\lambda^2 - 6\lambda + 9 = 0$ with the double root $\lambda = 3$. The general solution is therefore

$$y_n = c_1 3^n + c_2 n 3^n = 3^n(c_1 + nc_2).$$

Using the initial conditions, we obtain

$$c_1 = 5,$$
$$3c_1 + 3c_2 = 12.$$

The unique solution is $c_1 = 5$, $c_2 = -1$, and we have the specific solution to equation (6): $y_n = 5 \cdot 3^n - n \cdot 3^n = 3^n(5 - n)$.

Case 3 If $a^2 - 4b < 0$, the roots of $\lambda^2 + a\lambda + b = 0$ are the complex conjugates

$$\lambda_1 = \alpha + i\beta \quad \text{and} \quad \lambda_2 = \alpha - i\beta.$$

We write these numbers in polar form (see Appendix 5):

$$\lambda_1 = re^{i\theta} \quad \text{and} \quad \lambda_2 = re^{-i\theta},$$

where

$$r = \sqrt{\alpha^2 + \beta^2} \quad \text{and} \quad \theta = \tan^{-1}\frac{\beta}{\alpha}.$$

Equation (1) has two solutions,

$$x_n = r^n e^{in\theta} = r^n(\cos n\theta + i\sin n\theta)$$

and

$$y_n = r^n e^{-in\theta} = r^n(\cos n\theta - i\sin n\theta).$$

Two other solutions are

$$x_n^* = \tfrac{1}{2}(x_n + y_n) = r^n \cos n\theta$$

and
$$y_n^* = \frac{1}{2i}(x_n - y_n) = r^n \sin n\theta.$$

The general solution to (1) is then

$$y_n = r^n(c_1 \cos n\theta + c_2 \sin n\theta). \qquad (7)$$

Example 3 Let $y_{n+2} + y_n = 0$. The characteristic equation is $\lambda^2 + 1 = 0$, with the roots $\pm i$. Here $\alpha = 0$ and $\beta = 1$, so $r = 1$ and $\theta = \pi/2$. The general solution is given by

$$y_n = c_1 \cos \frac{n\pi}{2} + c_2 \sin \frac{n\pi}{2}.$$

This is the equation for **discrete harmonic motion**, which corresponds to the equation of ordinary harmonic motion as discussed in Example 3.4.1. If we specify $y_0 = 0$ and $y_1 = 1000$, we obtain

$$c_1 = 0, \qquad c_1 \cos \frac{\pi}{2} + c_2 \sin \frac{\pi}{2} = 1000,$$

with the solution $c_1 = 0$, $c_2 = 1000$, and the specific solution

$$y_n = 1000 \sin \frac{n\pi}{2}.$$

Example 4 Let P_n be the population in the nth generation of a species of bacteria. Assume that the population in the $(n+2)$nd generation depends on the two preceding generations:

$$P_{n+2} - rP_{n+1} - sP_n = 0, \qquad (8)$$

where r and s are measures of the relative importance of the preceding two generations.

The characteristic equation is $\lambda^2 - r\lambda - s = 0$, with the roots

$$\lambda_1 = \frac{r + \sqrt{r^2 + 4s}}{2}, \qquad \lambda_2 = \frac{r - \sqrt{r^2 + 4s}}{2}.$$

If $r^2 > -4s$, the roots are real and distinct, and the general solution is $P_n = c_1 \lambda_1^n + c_2 \lambda_2^n$. If $|\lambda_1| < 1$ and $|\lambda_2| < 1$, then $P_n \to 0$ as $n \to \infty$; but if either $|\lambda_1|$ or $|\lambda_2|$ is greater than 1, then $|P_n| \to \infty$. If $r^2 = -4s$ ($s < 0$), then the general solution is $P_n = c_1(r/2)^n + c_2 n(r/2)^{n-1}$, in which case $P_n \to 0$ if $|r| < 2$, but if $|r| \geq 2$, the solution tends to ∞. In the third case, $r^2 < -4s$, the general solution is

$$P_n = (-s)^{n/2}(c_1 \cos n\theta + c_2 \sin n\theta),$$

where $\theta = \tan^{-1}(\sqrt{-r^2 - 4s}/r)$. When $-s < 1$, the solution P_n tends to zero in an oscillatory manner. This is called **damped discrete harmonic motion** [see Figure 3.1(a)]. If $-s = 1$, we have the harmonic motion of the previous example. Finally, if $-s > 1$, the solution grows in an oscillatory motion, called **forced discrete harmonic motion** [see Figure 3.1(b)].

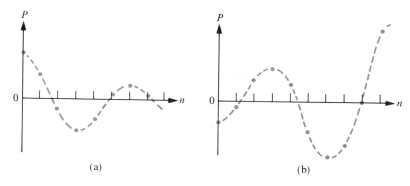

Figure 3.1

It is interesting to see that even in this very simple model the behavior of solutions depends critically on the relative importance of the populations in the preceding two generations.

The theory of linear difference equations is very similar to the theory of linear differential equations. The definitions of linear combination and of linear independence of solutions are the same as the definitions in Section 3.1. The general solution of a homogeneous second-order difference equation is given by $c_1 x_n + c_2 y_n$, where x_n and y_n are linearly independent solutions to (1); that is, if we know two solutions, we know them all. The general solution to (1) therefore always takes one of the forms (4), (5), or (7) depending on the nature of the roots of the characteristic equation.

We can also study the nonhomogeneous equation

$$y_{n+2} + a y_{n+1} + b y_n = f_n. \qquad (9)$$

It is not difficult to show that if y_n^* and y_n^{**} are two solutions to (9), then their difference $y_n^* - y_n^{**}$ is a solution to (1). Thus the general solution to the nonhomogeneous equation (9) is given by

$$y_n = c_1 x_n + c_2 x_n^* + y_n^*,$$

where x_n and x_n^* are linearly independent solutions to (1).

Finally, there is a method of variation of parameters that can be used to find one solution to (9). Rather than say more about nonhomogeneous equations in this brief excursion, we provide an interesting application of homogeneous difference equations.

AN APPLICATION TO GAMES AND QUALITY CONTROL

Suppose that two people, A and B, play a certain unspecified game several times in succession. One of the players must win on each play of the game (ties are ruled out). The probability that A wins is p ($\neq 0$), and the corresponding probability for B is q. Of course, $p + q = 1$. One dollar is bet by each competitor on each play of the game. Suppose that player A starts with k dollars and player B with j dollars. We let P_n denote the probability that A will bankrupt B (that is, win all B's money) when A has n dollars. Clearly $P_0 = 0$ (this is the probability that player A will wipe out player B when A has no money) and $P_{k+j} = 1$ (A has already eliminated B from further competition). To calculate P_n for other values of n between 0 and $k+j$, we have the following difference equation:

$$P_n = qP_{n-1} + pP_{n+1}. \tag{10}$$

To understand this equation, we note that if A has n dollars at one turn, his probability of having $n-1$ dollars at the next turn is q, and similarly for the other term in equation (10).

Equation (10) can be written in our usual format as

$$pP_{n+1} - P_n + qP_{n-1} = 0,$$

or

$$P_{n+2} - \frac{1}{p}P_{n+1} + \frac{q}{p}P_n = 0.$$

The characteristic equation is

$$\lambda^2 - \frac{1}{p}\lambda + \frac{q}{p} = 0,$$

which has the roots

$$\frac{1}{2p}\left(1 \pm \sqrt{1-4pq}\right).$$

Since

$$\sqrt{1-4pq} = \sqrt{1-4p(1-p)} = \sqrt{4p^2 - 4p + 1} = |1-2p|,$$

we discover that the roots of the characteristic equation are

$$\lambda_1 = \frac{1-p}{p} = \frac{q}{p}, \qquad \lambda_2 = 1.$$

The general solution to (10) is therefore

$$P_n = c_1\left(\frac{q}{p}\right)^n + c_2$$

when $p \neq q$, and

$$P_n = c_1 + c_2 n$$

when $p = q = \frac{1}{2}$ (by the rules for repeated roots). Keeping in mind the fact that $0 \leq P_n \leq 1$ for all n and using the boundary conditions $P_0 = 0$ and $P_{k+j} = 1$, we can finally obtain

$$P_n = \frac{1-(q/p)^n}{1-(q/p)^{k+j}}, \qquad \text{if } p \neq q, \tag{11}$$

and

$$P_n = \frac{n}{k+j}, \qquad \text{if } p = q = \frac{1}{2}, \tag{12}$$

for all n such that $0 \leq n \leq k+j$.

Example 5 In a game of roulette a gambler bets against a casino. Suppose that the gambler bets one dollar on red each time. On a typical roulette wheel there are eighteen reds, eighteen blacks, a zero, and a double zero. If the wheel is honest, the probability that the casino will win is $p = 20/38 \approx 0.5263$. Then $q \approx 0.4737$ and $q/p = 0.9$. Suppose that both the gambler and the casino start with k dollars. Then the probability that the gambler will be cleaned out is

$$P_k = \frac{1 - (q/p)^k}{1 - (q/p)^{2k}} = \frac{1}{1 + (q/p)^k} = \frac{1}{1 + (0.9)^k}.$$

The second column in Table 3.1 shows the probability that the gambler will lose all his money for different initial amounts k. Note that the more money both start with, the more likely it is that the gambler will lose all his money. It is evident that as the gambler continues to play, the probability that he will be wiped out approaches certainty. This is why even a very small advantage enables the casino, in almost every case, to break the gambler if he continues to bet.

TABLE 3.1

k	P_k	P_k^*
1	0.5263	0.5263
5	0.6287	0.8740
10	0.7415	0.9492
25	0.9330	0.9923
50	0.9949	0.9995
100	0.9999	0.9999

The third column in Table 3.1 gives the probability that the gambler will lose all his money if he has one dollar and the casino has k dollars:

$$P_k^* = \frac{1 - (q/p)^k}{1 - (q/p)^{k+1}}.$$

Thus, if the gambler has much less money than the casino, he is even more likely to lose it all. The combination of a small advantage, table limits, and huge financial resources makes running a casino a very profitable business indeed.

Example 6   The previous example can be used as a simple model of competition for resources between two species in a single ecological niche. Suppose that two species, A and B, control a combined territory of N resource units. The resource units may be such things as acres of grassland, number of trees, and so on. If species A controls k units, then species B controls the remaining $N - k$ units. Suppose that during each unit of time there is competition for one unit of resource between the two species. Let species A have the probability p of being successful. If we let P_n denote the probability that species B will lose all its resources when species A has n units of resource, then P_n is determined by equation (10), with initial conditions $P_0 = 0$ and $P_N = 1$. By equations (2) and (3), the solution to this equation is

$$P_n = \begin{cases} \dfrac{1-[(1-p)/p]^n}{1-[(1-p)/p]^N}, & \text{if } p \neq \dfrac{1}{2}, \\ \dfrac{n}{N}, & \text{if } p = \dfrac{1}{2}. \end{cases}$$

As in the previous example, even a small competitive advantage virtually ensures the extinction of the weaker species. For example, if $p = 0.55$ and species A has n of a total of 100 units, then we have the probabilities in Table 3.2.

Note that even if the species with the competitive advantage starts with as few as 4 of the 100 units of resource, it has a better than even chance to supplant the weaker species. With one quarter of the resources, it is virtually certain to do so.

TABLE 3.2

n	P_n
1	0.1818
2	0.3306
3	0.4523
4	0.5519
5	0.6334
10	0.8656
25	0.9934

A process similar to the one discussed here has been used with success to determine whether, in a manufacturing process, a batch of articles is satisfactory. Let us briefly describe this process here. More extensive details can be found in the paper by G. A. Barnard.[†]

To test whether a batch of articles is satisfactory, we introduce a scoring system. The score is initially set at N. If a randomly sampled item is found to be defective, we subtract k. If it is acceptable, we add 1. The procedure stops when the score reaches either $2N$ or becomes nonpositive. If $2N$, the batch is accepted; if nonpositive, it is rejected. Suppose that the probability of selecting an acceptable item is p, and $q = 1 - p$. Let P_n denote the probability that the batch will be rejected when the score is at n. Then after the next choice, the score will be either increased by 1 with probability p or decreased by k with probability q. Thus

$$P_n = pP_{n+1} + qP_{n-k},$$

which can be written as the $(k+1)$st-order difference equation

$$P_{n+k+1} - \frac{1}{p}P_{n+k} + \frac{q}{p}P_n = 0,$$

with boundary conditions

$$P_{1-k} = \cdots = P_0 = 1, \quad P_{2N} = 0.$$

The case of $k = 1$ reduces to Example 5, with the result that a batch is almost certain to be accepted or rejected depending on whether p is greater or less than $\frac{1}{2}$.

[†] G. A. Barnard, "Sequential Tests in Industrial Statistics," Supplement to the *Journal of the Royal Statistical Society* 8 (1) (1946).

PROBLEMS 3.9

In Problems 1–9 find the general solution to each given difference equation. When initial conditions are specified, find the unique solution that satisfies them.

1. $y_{n+2} - 3y_{n+1} - 4y_n = 0$
2. $y_{n+2} + 7y_{n+1} + 6y_n = 0$, $y_0 = 0$, $y_1 = 1$
3. $6y_{n+2} + 5y_{n+1} + y_n = 0$, $y_0 = 1$, $y_1 = 0$
4. $y_{n+2} + y_{n+1} - 6y_n = 0$, $y_0 = 1$, $y_1 = 2$
5. $y_{n+2} + 2y_{n+1} + y_n = 0$
6. $y_{n+2} - \sqrt{2}\,y_{n+1} + y_n = 0$
7. $y_{n+2} - 2y_{n+1} + 2y_n = 0$, $y_0 = 1$, $y_1 = 2$
8. $y_{n+2} + 8y_n = 0$
9. $y_{n+2} - 2y_{n+1} + 4y_n = 0$, $y_0 = 0$, $y_1 = 1$
10. The **Fibonacci numbers** are a sequence of numbers such that each one is the sum of its two predecessors. The first few Fibonacci numbers are $1, 1, 2, 3, 5, 8, 13, \ldots$.

    a. Formulate an initial-value difference equation that generates the Fibonacci numbers.

    b. Find the solution to this equation.

    c. Show that the ratio of successive Fibonacci numbers tends to $(1 + \sqrt{5})/2$ as $n \to \infty$. This ratio, known as the **golden ratio**, was often used in ancient Greek architecture whenever rectangular structures were constructed. It was believed that when the ratio of the sides of a rectangle was this number, the resulting structure was most pleasing to the eye.

11. In a study of infectious diseases, a record is kept of outbreaks of measles in a particular school. It is estimated that the probability of at least one new infection occurring in the nth week after an outbreak is $P_n = P_{n-1} - \frac{1}{5}P_{n-2}$. If $P_0 = 0$ and $P_1 = 1$, what is P_n? After how many weeks is the probability of the occurrence of a new case of measles less than 10 percent?

12. Two competing species of drosophila (fruit flies) are growing under favorable conditions. In each generation, species A increases its population by 60 percent and species B increases by 40 percent. If initially there are 1000 flies of each species, what is the total population after n generations?

*13. Snow geese mate in pairs in late spring of each year. Each female lays an average of five eggs, approximately 20 percent of which are claimed by predators and foul weather. The goslings, 60 percent of which are female, mature rapidly and are fully developed by the time the annual migration begins. Hunters and disease claim about 300,000 geese and 200,000 ganders annually. Are snow geese in danger of extinction?

14. In the discussion that introduces equation (10), assume that player A has a 10 percent competitive advantage. If player A starts with three dollars, how much money must player B start with to have a better than even chance to win all of A's money? To have an 80 percent chance?

15. Answer the questions in Problem 14 given that player A has

    a. a 2 percent advantage;

    b. a 20 percent advantage.

16. In the discussion that introduces equation (10), suppose that on each play of the game player A has the probability p to win one dollar, q to win two dollars, and r to lose one dollar, where $p + q + r = 1$. Let P_n be as before. Write a difference equation that determines P_n, assuming that each player starts with N dollars.

Review Exercises for Chapter 3

In Exercises 1–5 a second-order differential equation and one solution $y_1(x)$ are given. Verify that $y_1(x)$ is indeed a solution and find a second linearly independent solution.

1. $y'' + 4y = 0$; $y_1(x) = \sin 2x$
2. $y'' - 6y' + 9y = 0$; $y_1(x) = e^{3x}$
3. $x^2 y'' + xy' - 4y = 0$; $y_1(x) = x^2$
4. $y'' + \dfrac{1}{x} y' + \left(1 - \dfrac{1}{4x^2}\right) y = 0$;

   $y_1(x) = x^{-1/2} \sin x$.

5. $(1 - x^2) y'' - 2xy' + 2y = 0$; $y_1(x) = x$

In Exercises 6–24 find the general solution to the given equation. If initial conditions are given, find the particular solution that satisfies them.

6. $y'' - 9y' + 20y = 0$
7. $y'' - 9y' + 20y = 0$; $y(0) = 3$, $y'(0) = 2$
8. $y'' - 3y' + 4y = 0$
9. $y'' - 3y' + 4y = 0$; $y(0) = 0$, $y'(0) = 1$
10. $y'' = 0$
11. $4y'' + 4y' + y = 0$

12. $y'' - 11y = 0$
13. $y'' - 2y' + 7y = 0$
14. $y'' - y' - 2y = \sin 2x$
15. $y''' - 6y'' + 11y' - 6y = 0$
16. $y'' - 2y' + y = xe^x$
17. $y'' - 2y' + y = x^2 - 1$; $y(0) = 2$, $y'(0) = 1$
18. $y'' + y = \sec x$, $0 < x < \dfrac{\pi}{2}$
19. $y'' - 2y' + y = \dfrac{2e^x}{x^3}$
20. $y'' + 4y' + 4y = e^{-2x}/x^2$; $x > 0$
21. $x^2 y'' + 5xy' + 4y = 0$; $x > 0$
22. $x^2 y'' - 2xy' + 3y = 0$; $x > 0$
23. $y''' + y'' - 8y' - 12y = 0$
24. $y^{(4)} + 8y'' + 4y = 0$

*25. Prove that if y_1 and y_2 are linearly independent solutions of
$$y'' + a(x)y' + b(x)y = 0,$$
then they cannot have a common point of inflection unless $a(x)$ and $b(x)$ vanish simultaneously at that point.

*26. By setting $y = u(x)v(x)$, show that it is possible to select $v(x)$ so that
$$y'' + a(x)y' + b(x)y = 0$$
takes the form
$$u'' + c(x)u = 0.$$
[This is known as the **normal form** of a second-order linear differential equation.]

**27. If y_1 and y_2 are linearly independent solutions of
$$y'' + a(x)y' + b(x)y = 0,$$
show that between consecutive zeros of y_1 there is exactly one zero of y_2.

4

Applications of Linear Differential Equations

In Chapter 3 we developed several techniques for solving initial-value problems of the form

$$y'' + ay' + by = f(t), \qquad y(t_0) = y_0, \qquad y'(t_0) = y_0', \tag{1}$$

where a and b are constants and t is time. In this chapter we see that many physical problems, such as the motion of an object in a mechanical system or the flow of electric current in a simple series circuit, lead to a problem of form (1) when expressed mathematically. Thus it is only necessary to make proper interpretations of the terms in (1) to obtain solutions to physical problems in different disciplines.

4.1 Vibrational Motion: Simple Harmonic Motion

Differential equations were first studied in attempts to describe the motion of particles with mass subject to various forces. As a simple example, consider a spring, whose upper end is securely fastened, of natural (unstretched) length l_0 [see Figure 4.1(a)], to which we attach an object of mass m [see Figure 4.1(b)].[†]

[†] The most common systems of units are given in the table below.

Systems of units	Force	Length	Mass	Time
International (SI)	newton (N)	meter (m)	kilogram (kg)	second (s)
English	pound (lb)	foot (ft)	slug	second (sec)

$1 \text{ N} = 1 \text{ kg-m}/\text{s}^2 = 0.22481 \text{ lb}$ $\quad$ $1 \text{ kg} = 0.06852 \text{ slug}$
$1 \text{ m} = 3.28084 \text{ ft}$ $\quad$ $1 \text{ lb} = 1 \text{ slug-ft}/\text{s}^2 = 4.4482 \text{ N}$

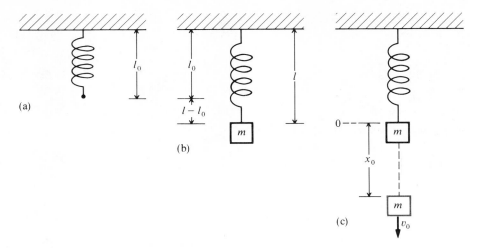

Figure 4.1

The addition of the mass m stretches the spring to length l. This static elongation of the spring is the result of two forces:

a. The force of gravity $F_g = mg$, acting downward.

b. The spring force F_s, acting upward.

In Figure 4.1(b), the object is in its equilibrium position, that is, at the point where the object remains at rest. For this to happen the two forces must be numerically equal. According to **Hooke's law**[†] the spring force exerted on the object is proportional to the difference between the length l of the spring and its natural or equilibrium length l_0. The positive constant of proportionality k is called the **spring constant**. Thus

$$F_s = k(l - l_0).$$

If the mass is at rest, then it is the force of gravity that pulls the spring beyond its natural length; this means that

$$mg = k(l - l_0)$$

as well. It is the case that $F_g = F_s$ only as long as the mass is left at rest.

In Figure 4.1(c) we denote the equilibrium position of the object on the spring, that is, the position where the spring has length l, by $x = 0$. Suppose that the object is given an initial displacement x_0 and an initial velocity v_0. Can we describe the movement of the object?

In answering the question we have just posed, we make three simplifying assumptions, the last two of which we remove in the next two sections.

[†] Robert Hooke (1638–1703), a British mathematician and physicist, was one of the first scientists to state the **inverse square law**: the force of gravitational attraction between two bodies is inversely proportional to the square of the distance between them.

> **a.** All motion is along a vertical line through the center of gravity of the object, which is treated as a point mass.
>
> **b.** There is no damping force F_d due to the medium in which the mass is moving (such as air resistance).
>
> **c.** No other forces (beyond the ones already mentioned) are applied to the mass.

Now suppose we have put the mass in Figure 4.1(c) in motion. What forces act on the mass? Clearly the gravitational force is constant, $F_g = mg$, but the spring force depends on the displacement x of the mass:

$$F_s = k(x + l - l_0).$$

As we increase x (moving downward in the same direction as the gravitational force), the spring force acting upward increases, so the total force on the mass is

$$F = F_g - F_s = mg - k(x + l - l_0)$$
$$= -kx,$$

since $mg = k(l - l_0)$.

Newton's second law of motion states that the force F acting on this mass moving with varying velocity v is equal to the time rate of change of the momentum mv; since the mass is constant,

$$F = \frac{d(mv)}{dt} = m\frac{d^2x}{dt^2}. \qquad (v = \frac{dx}{dt})$$

Equating the two forces, we have

$$m\frac{d^2x}{dt^2} = -kx. \qquad (1)$$

Two conditions were initially imposed on the mechanical system: at time $t = 0$ the mass has an initial displacement x_0 and an initial velocity v_0. Hence we have the initial-value problem

$$\frac{d^2x}{dt^2} + \frac{k}{m}x = 0, \qquad x(0) = x_0, \qquad x'(0) = v_0. \qquad (2)$$

To find the solution of (2), we note that the characteristic equation has the complex root $\pm i\omega_0$, where $\omega_0 = \sqrt{k/m}$, leading to the general solution

$$x(t) = c_1 \cos \omega_0 t + c_2 \sin \omega_0 t.$$

Using the initial conditions, we find that $c_1 = x_0$ and $c_2 = v_0/\omega_0$, so the solution of (2) is given by

$$x(t) = x_0 \cos \omega_0 t + (v_0/\omega_0)\sin \omega_0 t. \qquad (3)$$

Equation (3) describes the **free vibrations** or **free motion** of the mechanical system, since it is free of external influencing forces other than those imposed by gravity and the spring itself.

We would like to write $x(t)$ in the form
$$x(t) = A\sin(\omega_0 t + \phi),$$
so that we can graph (and understand) the superposition of the sine and cosine functions in (3). To do so we use the trigonometric formula for $\sin(x+y)$:
$$x(t) = A\sin(\omega_0 t + \phi) = A\sin\omega_0 t\cos\phi + A\cos\omega_0 t\sin\phi$$
$$\overset{\text{from (3)}}{=} x_0\cos\omega_0 t + \frac{v_0}{\omega_0}\sin\omega_0 t.$$

Equating coefficients of $\sin\omega_0 t$ and $\cos\omega_0 t$, we have
$$A\sin\phi = x_0, \qquad A\cos\phi = \frac{v_0}{\omega_0}$$
and
$$x_0^2 + \left(\frac{v_0}{\omega_0}\right)^2 = A^2\sin^2\phi + A^2\cos^2\phi = A^2(\sin^2\phi + \cos^2\phi) = A^2,$$
so
$$A = \sqrt{x_0^2 + (v_0/\omega_0)^2}.$$
Also,
$$\cos\phi = \frac{1}{A}\frac{v_0}{\omega_0} \qquad \text{and} \qquad \sin\phi = \frac{x_0}{A},$$
so
$$\tan\phi = \frac{\sin\phi}{\cos\phi} = \frac{x_0/A}{v_0/\omega_0 A} = \frac{x_0\omega_0}{v_0}.$$

Thus we may write equation (3) as
$$x(t) = A\sin(\omega_0 t + \phi), \tag{4}$$
with $A = \sqrt{x_0^2 + (v_0/\omega_0)^2}$ and
$$\phi = \tan^{-1}\frac{x_0\omega_0}{v_0} \qquad \text{or} \qquad \phi = \tan^{-1}\frac{x_0\omega_0}{v_0} + \pi.$$

(We have to be careful to determine which quadrant ϕ is in.)

The motion of the mass in (4) is called **simple harmonic motion**, since it is sinusoidal.

From this equation it is clear that the mass oscillates between the extreme positions $\pm A$; A is called the **amplitude** of the motion. Since the sine term has period $2\pi/\omega_0$, this is the time required for each complete oscillation. The **natural frequency** f of the motion is the number of complete oscillations per unit time:[†]
$$f = \frac{\omega_0}{2\pi}. \tag{5}$$

[†] Cycles/sec = hertz (Hz).

Note that although the amplitude depends on the initial conditions, the frequency does not.

Example 1 Suppose that $x_0 = 0.5$ meters, $k = 0.4$, $m = 10$ kilograms, and $v_0 = 0.25$ meters per second. Then $\omega_0 = \sqrt{k/m} = \sqrt{0.04} = 0.2$ and equation (3) becomes

$$x(t) = 0.5\cos 0.2t + \frac{0.25}{0.2}\sin 0.2t = 0.5\cos 0.2t + 1.25\sin 0.2t.$$

Now

$$A = \sqrt{0.5^2 + 1.25^2} = \sqrt{1.8125} \approx 1.3463$$

and

$$\phi = \tan^{-1}\frac{(0.5)(0.2)}{0.25} = \tan^{-1} 0.4 \approx 0.3805 \text{ radians } (\approx 21.8°),$$

so we may write

$$x(t) \approx 1.3463 \sin(0.2t + 0.3805) \text{ meters}.$$

Example 2 Consider a spring fixed at its upper end and supporting a weight of 10 pounds at its lower end. Suppose the 10-pound weight stretches the spring by 6 inches. Find the equation of motion of the weight if it is drawn to a position 4 inches below its equilibrium position and released.

Solution By Hooke's law, since a force of 10 pounds stretches the spring by $\frac{1}{2}$ foot, $10 = k(\frac{1}{2})$ or $k = 20$ (lb/ft). We are given the initial values $x_0 = \frac{1}{3}$(ft) and $v_0 = 0$, so by (3) and the identity† $k/m = gk/w \approx 64/\sec^2$, we obtain

$$x(t) = \tfrac{1}{3}\cos 8t \text{ feet}.$$

Thus the amplitude is $\frac{1}{3}$ foot ($= 4$ in.), and the frequency is $f \approx 4/\pi$ hertz.

---

† The identity $w = mg$ may be used to convert weight to mass. Keep in mind that pounds and newtons are units of weight (force) whereas slugs and kilograms are units of mass. The gravitational constant $g = 9.81$ m/s² $= 32.2$ ft/sec² (approximately).

PROBLEMS 4.1

In Problems 1–6 determine the equation of motion of a point mass m attached to a coiled spring with spring constant k initially displaced a distance x_0 from equilibrium and released with velocity v_0 subject to no damping or other external forces.

1. $m = 10$ kg, $k = 1000$ N/m, $x_0 = 1$ m, $v_0 = 0$
2. $m = 10$ kg, $k = 10$ N/m, $x_0 = 0$, $v_0 = 1$ m/s
3. $m = 10$ kg, $k = 10$ N/m, $x_0 = 3$ m, $v_0 = 4$ m/s
4. $m = 1$ kg, $k = 16$ N/m, $x_0 = 4$ m, $v_0 = 0$
5. $m = 1$ kg, $k = 25$ N/m, $x_0 = 0$, $v_0 = 3$ m/s
6. $m = 9$ kg, $k = 1$ N/m, $x_0 = 4$ m, $v_0 = 1$ m/s

7. One end of a rubber band is fixed at a point A. An object of 1-kilogram mass, attached to the other end, stretches the rubber band vertically to point B in such a way that the length AB is 16 centimeters greater than the natural length of the band. If the mass is further drawn to a position 8 centimeters below B and released, what is its velocity (if we neglect resistance) as it passes point B?

8. If the mass in Problem 7 is released at a position 8 centimeters above B, what is its velocity as it passes 1 centimeter above B?

*9. A cylindrical block of wood of radius and height 1 foot and weighing 124.8 pounds floats with its axis vertical in water (62.4 lb/ft^3). If it is depressed so that the surface of the water is tangent to the block and is then released, what is its period of vibration and equation of motion? Neglect resistance. [*Hint*: The upward force on the block is equal to the weight of the water displaced by the block.]

*10. A cubical block of wood, 1 foot on a side, is depressed so that its upper face lies along the surface of the water and is then released. The period of vibration is found to be 1 second. Neglecting resistance, what is the weight of the block of wood?

11. An ideal pendulum consists of a weightless rod of length l attached at one end to a frictionless hinge and supporting a body of mass m at the other end. Suppose the pendulum is displaced an angle θ_0 and released, as illustrated below. The tangential acceleration of the ideal pendulum is $l\theta''$ and must be proportional, by Newton's second law of motion, to the tangential component of gravitational force.

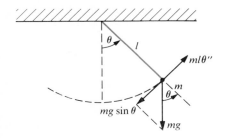

a. Neglecting air resistance, show that the ideal pendulum satisfies the nonlinear initial-value problem

$$l\frac{d^2\theta}{dt^2} = -g\sin\theta, \quad \theta(0) = \theta_0, \quad \theta'(0) = 0. \quad \text{(i)}$$

b. Assuming θ_0 is small, explain why equation (i) may be approximated by the linear initial-value problem

$$\frac{d^2\theta}{dt^2} + \frac{g}{l}\theta = 0, \quad \theta(0) = \theta_0, \quad \theta'(0) = 0. \quad \text{(ii)}$$

c. Solve equation (ii) assuming that the rod is 6 inches long and that the initial displacement θ_0 is 0.5 radian. What is the frequency of the pendulum?

12. A grandfather clock has a pendulum that is 1 meter long. The clock ticks each time the pendulum reaches the rightmost extent of its swing. Neglecting friction and air resistance, and assuming that the motion is small, determine how many times the clock ticks in 1 minute.

13. A turbine of unknown weight is placed on a spring-supported mounting platform of unknown spring constant. What is the natural frequency of the system if the turbine lowers the platform by 1/8 inch?

*14. A cubical block of wood 1 foot on a side and weighing 41.6 pounds floats in water (62.4 lb/ft^3). Find its equation of motion, neglecting resistance, if it is depressed so that its top just touches the surface of the water and is then released.

15. Assume that a straight tube has been bored through the center of the earth and that a particle of weight w pounds is dropped into the tube. If the radius of the earth is 3960 miles and the gravitational attraction is proportional to the distance from its center,† how long does it take (neglecting resistance) for the particle to drop halfway to the center? Does it pass through the tube to the other side?

16. A weight w is suspended by two springs, having spring constants k_1 and k_2, connected in series (see illustration). What is the **effective** spring constant k for such a system? [*Hint*: No derivatives are required.]

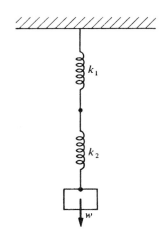

*17. A small ball of weight w is placed in the middle of a tightly stretched, vertical string of length $2L$ and tension T_0. Show that for small horizontal displacements the ball undergoes simple harmonic motion. What is its period? [*Note*: Neglect gravity.]

---

† This is not really contrary to Newton's law, since inside the earth the only gravitational effect is from that portion of the earth closer to the center than the particle.

4.2 Vibrational Motion: Damped Vibrations

Throughout the discussion in Section 4.1 we made the assumption that the only forces involved were gravity and the spring force. This assumption, however, is not very realistic. To take care of such things as friction in the spring and air resistance, we now assume that there is a damping force (that tends to slow things down), which can be thought of as the resultant of all other external forces acting on the object, which we still treat as a point mass. It is reasonable to assume that the magnitude of the damping force is proportional to the velocity of the particle (for example, the slower the movement, the smaller the air resistance).† Therefore we add the term $c(dx/dt)$, where c is the damping constant that depends on all external factors, to equation (4.1.1). The equation of motion then becomes

$$\frac{d^2x}{dt^2} = -\frac{k}{m}x - \frac{c}{m}\frac{dx}{dt}, \qquad x(0) = x_0, \qquad x'(0) = v_0, \qquad (1)$$

or

$$\frac{d^2x}{dt^2} + \frac{c}{m}\frac{dx}{dt} + \frac{k}{m}x = 0, \qquad x(0) = x_0, \qquad x'(0) = v_0. \qquad (2)$$

(Of course, since c depends on external factors, it may very well not be a constant at all but may vary with time and position. In that case c is really $c(t,x)$, and the equation becomes much harder to analyze than the constant-coefficient case.)

To study equation (2), we first find the roots of the characteristic equation:

$$\frac{-c \pm \sqrt{c^2 - 4mk}}{2m}. \qquad (3)$$

The nature of the general solution depends on the discriminant $\sqrt{c^2 - 4mk}$. If $c^2 > 4mk$, then both roots are negative, since $\sqrt{c^2 - 4mk} < c$. So in this case

$$x(t) = c_1 \exp\left(\frac{-c + \sqrt{c^2 - 4mk}}{2m}\right)t + c_2 \exp\left(\frac{-c - \sqrt{c^2 - 4mk}}{2m}\right)t \qquad (4)$$

becomes small as t becomes large whatever the initial conditions may be. Similarly, in the event that the discriminant vanishes,

$$x(t) = (c_1 + c_2 t)e^{(-c/2m)t}, \qquad (5)$$

and the solution has a similar behavior. Equations (4) or (5) give the **(free) damped vibrations** of the mechanical system.

Example 1 A spring fixed at its upper end is stretched 6 inches by a 10-pound weight attached at its lower end. The spring-mass system is suspended in a viscous medium (such as oil or water) so that the system is subjected to a damping force (pounds) of

$$5\frac{dx}{dt}.$$

Describe the motion of the system if the weight is drawn down an additional 4 inches and released.

---

† In hydrodynamics the damping force is proportional to velocity squared.

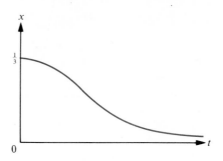

Figure 4.2

Solution The differential equation is given by

$$\frac{d^2x}{dt^2} + 16\frac{dx}{dt} + 64x = 0, \qquad (6)$$

since $k/m = gk/w \approx 32[10/(\frac{1}{2})]/10 = 64$ per second per second and $c/m = gc/w \approx 32(5)/10 = 16$ per second. The initial conditions are $x(0) = -1/3$ and $x'(0) = 0$. Hence the roots of the characteristic equation are

$$\frac{-16 \pm \sqrt{(16)^2 - 4(64)}}{2} = -8,$$

and (6) has the general solution

$$x(t) = e^{-8t}(c_1 + c_2 t).$$

Applying the initial conditions yields

$$x(t) = \tfrac{1}{3} e^{-8t}(1 + 8t) \text{ feet,}$$

which has the graph shown in Figure 4.2. We observe that the solution does not oscillate. This system is **overdamped, critically damped,** or **underdamped** accordingly as c^2 is greater than, equal to, or less than $4mk$.

If $c^2 < 4mk$, the general solution of equation (2) is

$$x(t) = e^{(-c/2m)t}\left(c_1 \cos\frac{\sqrt{4mk - c^2}}{2m}t + c_2 \sin\frac{\sqrt{4mk - c^2}}{2m}t\right), \qquad (7)$$

which shows an oscillation with frequency

$$f = \frac{\sqrt{4mk - c^2}}{4\pi m}.$$

The factor $e^{(-c/2m)t}$ is called the **damping factor.** Letting $c = 4$ pounds per foot per second in Example 1 leads to the general solution

$$x(t) = e^{-32t/5}\left(c_1 \cos\frac{24}{5}t + c_2 \sin\frac{24}{5}t\right) \text{ feet.}$$

Note that $e^{-32t/5} \to 0$ as $t \to \infty$, so the damped motion decays to zero as time increases.

Using the initial values, we find that $c_1 = 1/3$ and $c_2 = 4/9$. The motion is illustrated in Figure 4.3.

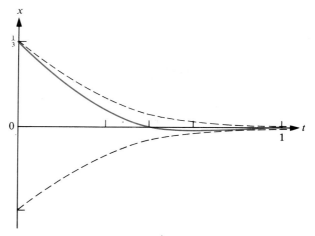

Figure 4.3

Damping forces also occur as the result of mechanical devices. A typical example is provided by automobile shock absorbers in which a piston is driven into a dashpot or damper (see Figure 4.4). Mechanical systems involving torsional motion, such as brakes on automobiles, also experience damped vibrations (see Figure 4.5). In this case the equation of motion is given by

$$I\frac{d^2\theta}{dt^2} + c\frac{d\theta}{dt} + k\theta = 0, \quad \theta(0) = \theta_0, \quad \theta'(0) = \theta_0', \tag{8}$$

Figure 4.4

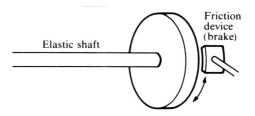

Figure 4.5

where I is the moment of inertia, θ is the angular displacement, $c\,d\theta/dt$ is the damping torque, and $k\theta$ is the elastic torque due to twisting of the shaft.

PROBLEMS 4.2

In Problems 1–6 determine the equation of motion of a point mass m attached to a coiled spring with spring constant k initially displaced a distance x_0 from equilibrium and released with initial velocity v_0, subject to a damping constant c.

1. $m = 10$ kg, $k = 1000$ N/m, $x_0 = 1$ m, $v_0 = 0$, $c = 200$ kg/s $= 200$ N/(m/s)
2. $m = 10$ kg, $k = 10$ N/m, $x_0 = 0$, $v_0 = 1$ m/s, $c = 20$ kg/s
3. $m = 10$ kg, $k = 10$ N/m, $x_0 = 3$ m, $v_0 = 4$ m/s, $c = 10\sqrt{5}$ kg/s
4. $m = 1$ kg, $k = 16$ N/m, $x_0 = 4$ m, $v_0 = 0$, $c = 10$ kg/s
5. $m = 1$ kg, $k = 25$ N/m, $x_0 = 0$, $v_0 = 3$ m/s, $c = 8$ kg/s
6. $m = 9$ kg, $k = 1$ N/m, $x_0 = 4$ m, $v_0 = 1$ m/s, $c = 10$ kg/s
7. An object of 10-gram mass is suspended from a vibrating spring, the resistance in kg being numerically equal to half the velocity (in m/s) at any instant. If the period of the motion is 8 seconds, what is the spring constant (in kg/s^2)?
8. A weight w (lb) is suspended from a spring whose constant is 10 pounds per foot. The motion of the weight is subject to a resistance (lb) numerically equal to half the velocity (ft/sec). If the motion is to have a 1-second period, what are the possible values of w?
9. A weight of 48 pounds hangs from a spring with spring constant 50 pounds per foot. The damping in the system is 30 percent of critical. Determine the motion of the weight if it is pulled down 2 inches from its equilibrium position and released with an upward velocity of 1 foot per second.
10. Answer Problem 9 for damping that is 90 percent of critical.
11. Answer Problem 9 for critical damping.
12. Answer Problem 9 for damping that is 150 percent of critical.
*13. Answer Problem 9 if 48 pounds of additional weight is suddenly applied to the system when it is at rest at equilibrium.
*14. A weight, hung on an ideal spring in a vacuum (assume no air resistance) and set vibrating, has a period of 1 second. When the spring-weight system is placed in a viscous medium such that the resistance is proportional to the velocity, its (damped) period is found to be 2 seconds. Determine the differential equation corresponding to the damped vibrations.
*15. A container weighing 2 pounds is half filled with 4 pounds of mercury. When hung on the end of a spring, it stretches it by 2 inches. The period of oscillation is found to be 0.3 seconds. If 4 pounds more mercury is added, the period becomes 0.4 seconds. Can the resistance be proportional to the velocity in this situation?
**16. The disk in Figure 4.5 has a radius of 12 inches and weighs 100 pounds. The observed frequency of torsional vibration is 2π radians per second. When a different body is attached to the same shaft, the observed frequency of torsional vibration is 2.4π radians per second. Find the moment of inertia of the second body with respect to the axis of the shaft. [See (8).]

4.3 Vibrational Motion: Forced Vibrations

The motion of the point mass considered in Sections 4.1 and 4.2 is determined by the inherent forces of the spring-weight system and the natural forces acting on the system. Accordingly, the vibrations are called **free** or **natural vibrations**. We now assume that the point mass is also subject to an external periodic force $F_0 \sin \omega t$, which is due to the motion of the object to which the upper end of the spring is attached (see Figure 4.6). In this case the mass undergoes **forced vibrations**.

We have seen in the previous sections that a spring-weight system, having a damping force proportional to the velocity, satisfies the initial-value problem

$$m\frac{d^2x}{dt^2} = -kx - c\frac{dx}{dt}, \qquad x(0) = x_0, \qquad x'(0) = v_0.$$

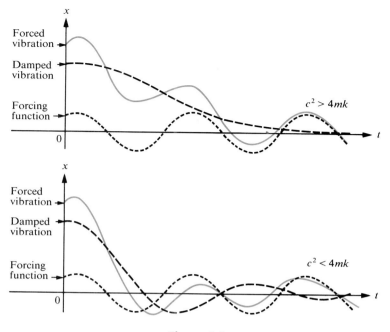

Figure 4.6

If we subject such a system to an additional periodic external force $F_0 \sin \omega t$, we obtain the nonhomogeneous second-order differential equation

$$m\frac{d^2x}{dt^2} = -kx - c\frac{dx}{dt} + F_0 \sin \omega t,$$

which we write in the form

$$\frac{d^2x}{dt^2} + \frac{c}{m}\frac{dx}{dt} + \frac{k}{m}x = \frac{F_0}{m}\sin \omega t. \tag{1}$$

By the method of undetermined coefficients, we know that $x(t)$ has a particular solution of the form

$$x_p(t) = b_1 \cos \omega t + b_2 \sin \omega t. \tag{2}$$

Substituting this function into equation (1) yields the simultaneous equations

$$\left(\omega_0^2 - \omega^2\right)b_1 + \frac{c\omega}{m}b_2 = 0,$$

$$-\frac{c\omega}{m}b_1 + \left(\omega_0^2 - \omega^2\right)b_2 = \frac{F_0}{m}, \tag{3}$$

where $\omega_0 = \sqrt{k/m}$, from which we obtain

$$b_1 = \frac{-F_0 c\omega}{m^2\left(\omega_0^2 - \omega^2\right)^2 + (c\omega)^2},$$

$$b_2 = \frac{F_0 m\left(\omega_0^2 - \omega^2\right)}{m^2\left(\omega_0^2 - \omega^2\right)^2 + (c\omega)^2}.$$

Using the same method we used to obtain equation (4.1.4), we have
$$x_p(t) = A\sin(\omega t + \phi), \tag{4}$$
where
$$A = \frac{F_0/k}{\sqrt{\left[1 - \left(\dfrac{\omega}{\omega_0}\right)^2\right]^2 + \left(2\dfrac{c}{c_0}\dfrac{\omega}{\omega_0}\right)^2}},$$
and
$$\tan\phi = \frac{2\dfrac{c}{c_0}\dfrac{\omega}{\omega_0}}{\left(\dfrac{\omega}{\omega_0}\right)^2 - 1},$$

with $c_0 = 2m\omega_0$. Here A is the amplitude of the motion, ϕ is the **phase angle**, c/c_0 is the **damping ratio**, and ω/ω_0 is the **frequency ratio** of the motion.

The general solution is found by superimposing the periodic function (4) on the general solution [equations (4.2.4), (4.2.5), or (4.2.7)] of the homogeneous equation (4.2.2). Since the solution of the homogeneous equation damps out as t increases, the general solution is very close to (4) for large values of t. Figure 4.6 illustrates two typical situations.

It is interesting to see what occurs if the damping constant c vanishes. There are two cases.

Case 1 If $\omega^2 \ne \omega_0^2$, we superimpose the periodic function (4) on the general solution of the homogeneous equation $x'' + \omega_0^2 x = 0$, obtaining
$$x(t) = c_1\cos\omega_0 t + c_2\sin\omega_0 t + \frac{F_0/k}{1 - (\omega/\omega_0)^2}\sin\omega t. \tag{5}$$

Using the initial conditions, we find that
$$c_1 = x_0 \quad \text{and} \quad c_2 = \frac{v_0}{\omega_0} - \frac{(F_0/k)(\omega/\omega_0)}{1 - (\omega/\omega_0)^2},$$
so
$$x(t) = A\sin(\omega_0 t + \phi) + \frac{F_0/k}{1 - (\omega/\omega_0)^2}\sin\omega t,$$
where
$$A = \sqrt{c_1^2 + c_2^2} \quad \text{and} \quad \tan\phi = c_1/c_2.$$

Hence the motion in this case is simply the sum of two sinusoidal curves, as illustrated in Figure 4.7.

Case 2 If $\omega^2 = \omega_0^2$, we must seek a particular solution of the form
$$x_p(t) = b_1 t\cos\omega t + b_2 t\sin\omega t, \tag{6}$$
since equation (2) is a solution of the homogeneous equation (see Form (iii) in

Figure 4.7

our discussion of undetermined coefficients, p. 106). Substituting equation (6) into

$$\frac{d^2x}{dt^2} + \frac{k}{m}x = \frac{F_0}{m}\sin \omega t$$

we get

$$b_1 = \frac{-F_0}{2m\omega} \quad \text{and} \quad b_2 = 0,$$

so the general solution has the form

$$x(t) = c_1\cos \omega t + c_2\sin \omega t - \frac{F_0}{2m\omega}t\cos \omega t. \tag{7}$$

Note that as t increases, the vibrations caused by the last term in equation (7) increase without bound. The external force is said to be in **resonance** with the vibrating mass. It is evident that the displacement here becomes so large that the elastic limit of the spring is exceeded, leading to fracture or to a permanent distortion in the spring.

Suppose that c is positive but very close to zero, while $\omega^2 = \omega_0^2$. Note that (3) yields $b_1 = -F_0/c\omega$ and $b_2 = 0$ when (2) is substituted into (1). Superimposing

$$x_p(t) = \frac{-F_0}{c\omega}\cos \omega t$$

on the solution of the homogeneous equation [see equation (4.2.7), since c^2 is very small] and letting $c_0 = 2m\omega_0$, we obtain

$$x(t) = e^{(-c/c_0)\omega_0 t}\left[c_1\cos \omega_0\sqrt{1 - \left(\frac{c}{c_0}\right)^2}\,t + c_2\sin \omega_0\sqrt{1 - \left(\frac{c}{c_0}\right)^2}\,t\right]$$

$$-\frac{F_0}{c\omega}\cos \omega t. \tag{8}$$

Since c/c_0 is very small, for small values of t we see that equation (8) can be approximated by

$$x(t) \approx c_1\cos \omega t + c_2\sin \omega t - \frac{F_0}{2m\omega}\left(\frac{2m}{c}\right)\cos \omega t,$$

which bears a marked resemblance to (7) *when (7) is evaluated at large values of t* (since $2m/c$ is large). Thus the *damped* spring problem approaches resonance. This phenomenon is extremely important in engineering, since resonance may produce undesirable effects such as metal fatigue and structural fracture, as well as desirable objectives such as sound and light amplification.

An example of resonance occurred when a column of soldiers marched across Broughton bridge, near Manchester, England, in 1831. Their rhythmic marching produced a periodic force of large amplitude closely approximating the natural frequency of the bridge, causing it to collapse. For that reason soldiers today are required to march out of step when crossing a bridge.

A more recent disaster was the collapse of the Tacoma (Washington) Narrows Bridge at Puget Sound on November 7, 1940 (see Figure 4.8). When it was completed, the bridge was the third largest suspension bridge in the world. From its opening on July 1, 1940, the bridge enjoyed enormous popularity because of its strange weaving motions and soon earned the nickname "Galloping Gertie." Many automobile passengers complained of motion sickness when crossing it.

On November 7, winds reached 42 miles per hour in the Tacoma area. At around 7:00 A.M. the bridge began to resonate violently. By 10 A.M. the oscillations were so great that one edge of the roadway alternated between being 28 feet higher and 28 feet lower than the other edge. Shortly after 11 A.M. the entire bridge crashed 190 feet into Puget Sound. The only vehicles on the bridge at the time of its collapse were a logging truck and the automobile of a reporter Mr. Leonard Coatsworth. Both drivers abandoned their vehicles

Figure 4.8

and crawled to safety. Mr. Coatsworth struggled over sections of the bridge, pitching at 45° angles, and made his way 500 yards to the support towers, 425 feet high. The only loss of life was Mr. Coatsworth's pet cocker spaniel, which went down with the car.

The bridge was built at a cost of $6.4 million and, unsurprisingly, many explanations and excuses were given following its collapse. Some construction engineers opined that the introduction of stiffening plate or web girders in place of the older lattice or open girders was responsible. Clark W. Eldridge, chief engineer of the bridge, said that Washington State highway engineers had protested against the design, but that it was built anyway in the interest of economy. He blamed the ill-fated design on eastern engineers. "The employment of eastern engineers was because of a requirement by the money-lending agencies that engineers of a national reputation be employed." Perhaps the clearest explanation of what happened was provided in a *New York Times* editorial on November 9, 1940.

> Like all suspension bridges, that at Tacoma both heaved and swayed with a high wind. It takes only a tap to start a pendulum swinging. Time successive taps correctly and soon the pendulum swings with its maximum amplitude. So with this bridge. What physicists call resonance was established, with the result that the swaying and heaving exceeded the limits of safety. Steel buckled like cardboard. The concrete roadway cracked like ice. The whole structure broke, with the exception of the towers.

None of this collapse was anticipated, probably because few of the individuals who helped design the bridge understood resonance. Even afterward, there was eagerness to repeat the same mistake. Shortly after the collapse of the bridge, Governor Clarence D. Martin of the state of Washington declared, "We are going to build the same exact [sic] bridge, exactly as before." A well-known engineer who had had nothing to do with the original construction and who did understand what had happened responded, "If you build the exact same bridge exactly as before, it will fall into the exact same river exactly as before."

PROBLEMS 4.3

In Problems 1–6 determine the equation of motion of a point mass m attached to a coiled spring with spring constant k initially displaced a distance x_0 from equilibrium and released with velocity v_0 subject to a damping constant c and an external force $F_0 \sin \omega t$ newtons.

1. $m = 10$ kg, $k = 1000$ N/m, $x_0 = 1$ m, $v_0 = 0$, $c = 200$ kg/s, $F_0 = 1$ N, $\omega = 10$ rad/s.
2. $m = 10$ kg, $k = 10$ N/m, $x_0 = 0$, $v_0 = 1$ m/s, $c = 20$ kg/s, $F_0 = 1$ N, $\omega = 1$ rad/s.
3. $m = 10$ kg, $k = 10$ N/m, $x_0 = 3$ m, $v_0 = 4$ m/s, $c = 10\sqrt{5}$ kg/s, $F_0 = 1$ N, $\omega = 1$ rad/s.
4. $m = 1$ kg, $k = 16$ N/m, $x_0 = 4$ m, $v_0 = 0$, $c = 10$ kg/s, $F_0 = 4$ N, $\omega = 4$ rad/s.
5. $m = 1$ kg, $k = 25$ N/m, $x_0 = 0$, $v_0 = 3$ m/s, $c = 8$ kg/s, $F_0 = 1$ N, $\omega = 3$ rad/s.
6. $m = 9$ kg, $k = 1$ N/m, $x_0 = 4$ m, $v_0 = 1$ m/s, $c = 10$ kg/s, $F_0 = 2$ N, $\omega = \frac{1}{3}$ rad/s.
7. An object of 1-gram mass is hanging at rest on a spring that is stretched 25 centimeters by the weight. The upper end of the spring is given the periodic force $0.01 \sin 2t$ newtons and air resistance has a magnitude (kg/s) 0.02162 times the velocity in meters per second. Find the equation of motion of the object.
8. A 100-pound weight is suspended from a spring with spring constant 20 pounds per inch. When the system is vibrating freely, we observe that in consecutive cycles the amplitude decreases by 40 percent. If a force of $20 \cos \omega t$ newtons acts on the system, find the amplitude and phase shift of the resulting steady-state motion if $\omega = 9$ radians per second.

9. Repeat Problem 8 for $\omega = 12$ radians per second.

10. A particle of weight w moves along the x-axis under the influence of a force $F = -kx$. Friction on the particle is proportional to the force between the particle and the surface on which it moves. Find the differential equation governing the motion of this particle.

*11. Consider the forced vibrations of an undamped mechanical spring-weight system, where the external force is $F_0 \sin \omega t$ newtons.

  a. Show that if $\omega \neq \omega_0$ $(= \sqrt{k/m})$, then the solution is given by
  $$x(t) = c_1 \cos \omega_0 t + c_2 \sin \omega_0 t + \frac{F_0}{k - m\omega^2} \sin \omega t.$$

  b. Discuss what happens if ω is close, but not equal, to ω_0. [*Hint*: Use the procedure for finding equation (4.1.4). The phenomenon that occurs is called **beats** (see illustration), which occurs whenever an impressed frequency is close to a natural frequency of a mechanical system.]

*12. A 48-pound weight is suspended from a spring with spring constant 50 pounds per inch. In 10 cycles we see that the maximum displacement decreases by 50 percent. Suppose an external force of $F_0(\sin 15t + \sin 16t)$ newtons is applied to the spring-weight system. Will beats occur? Why?

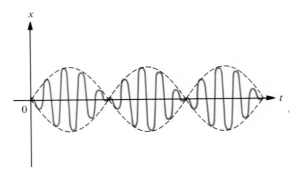

Figure for Problem 11
Beats. The dotted curve is called the *envelope* of the graph of $x(t)$

4.4 Electric Circuits

We make use of the concepts of electric circuitry developed in Section 2.5 and the methods of Chapter 3 to study a simple electric circuit containing a resistor, an inductor, and a capacitor in series with an electromotive force (Figure 4.9). Suppose that R, L, C, and E are constants. Applying Kirchhoff's law, we obtain

$$L\frac{dI}{dt} + RI + \frac{Q}{C} = E. \tag{1}$$

Since $dQ/dt = I$, we may differentiate (1) to get the second-order homogeneous differential equation

$$L\frac{d^2I}{dt^2} + R\frac{dI}{dt} + \frac{I}{C} = 0. \tag{2}$$

Figure 4.9

To solve this equation, we note that the characteristic equation

$$\lambda^2 + \frac{R}{L}\lambda + \frac{1}{CL} = 0$$

has the following roots:

$$\lambda_1 = \frac{-R + \sqrt{R^2 - 4L/C}}{2L}, \quad \lambda_2 = \frac{-R - \sqrt{R^2 - 4L/C}}{2L},$$

or, rewriting the radical in dimensionless units,[†] we have

$$\lambda_1 = \frac{R}{2L}\left(-1 + \sqrt{1 - \frac{4L}{CR^2}}\right), \quad \lambda_2 = \frac{R}{2L}\left(-1 - \sqrt{1 - \frac{4L}{CR^2}}\right). \quad (3)$$

Equation (2) may now be solved using the methods of Sections 3.3 and 3.4.

Remark Equation (1) can be turned into a second-order differential equation even if E is a nondifferentiable function (like a square wave). Setting $I = dQ/dt$ and $dI/dt = d^2Q/dt^2$ in (1), we obtain

$$L\frac{d^2Q}{dt^2} + R\frac{dQ}{dt} + \frac{Q}{c} = E. \quad (1')$$

Example 1 Let $L = 1$ henry (H), $R = 100$ ohms (Ω), $C = 10^{-4}$ farad (f), and $E = 1000$ volts (V) in the circuit shown in Figure 4.9. Suppose that no charge is present and no current is flowing at time $t = 0$ when E is applied. By equation (3) we see that the characteristic equation has the roots $\lambda_1 = -50 + 50\sqrt{3}\,i$ and $\lambda_2 = -50 - 50\sqrt{3}\,i$, since

$$R^2 - 4L/C = 10,000 - 4 \times 10^4 = -30,000$$

and $R/2L = 50$. Thus

$$I(t) = e^{-50t}\left(c_1 \cos 50\sqrt{3}\,t + c_2 \sin 50\sqrt{3}\,t\right).$$

Applying the initial condition $I(0) = 0$, we have $c_1 = 0$. Hence

$$I(t) = c_2 e^{-50t} \sin 50\sqrt{3}\,t,$$

and

$$I'(t) = 50 c_2 e^{-50t}\left(\sqrt{3}\cos 50\sqrt{3}\,t - \sin 50\sqrt{3}\,t\right).$$

To establish the value of c_2, we must make use of equation (1) and the initial condition $Q(0) = 0$. Since

$$Q(t) = C\left(E - L\frac{dI}{dt} - RI\right)$$

$$= 10^{-4}\left[1000 - 50 c_2 e^{-50t}\left(\sqrt{3}\cos 50\sqrt{3}\,t - \sin 50\sqrt{3}\,t + 2\sin 50\sqrt{3}\,t\right)\right]$$

$$= \frac{1}{10} - \frac{c_2}{200} e^{-50t}\left(\sin 50\sqrt{3}\,t + \sqrt{3}\cos 50\sqrt{3}\,t\right),$$

[†] 1 henry = 1 volt-sec/amp; 1 farad = 1 coulomb/volt; 1 ohm = 1 volt/amp; 1 coulomb = 1 amp-sec.

it follows that
$$Q(0) = \frac{1}{10} - \frac{c_2\sqrt{3}}{200} = 0 \quad \text{or} \quad c_2 = \frac{20}{\sqrt{3}}.$$

Hence
$$Q(t) = \frac{1}{10} - \frac{1}{10\sqrt{3}} e^{-50t}(\sin 50\sqrt{3}\, t + \sqrt{3}\cos 50\sqrt{3}\, t)$$

and
$$I(t) = \frac{20}{\sqrt{3}} e^{-50t} \sin 50\sqrt{3}\, t.$$

From these equations we observe that the current rapidly damps out and that the charge rapidly approaches its **steady-state value** of 0.1 coulomb. Here $I(t)$ is called the **transient current** (because its significance is brief).

Example 2   Let the inductance, resistance, and capacitance in Example 1 remain the same, but suppose that $E = 962 \sin 60t$ volts. By equation (1) we have

$$\frac{dI}{dt} + 100I + 10^4 Q = 962 \sin 60t, \tag{4}$$

and converting equation (4) so that all expressions are in terms of $Q(t)$, we obtain

$$\frac{d^2Q}{dt^2} + 100\frac{dQ}{dt} + 10^4 Q = 962 \sin 60t. \tag{5}$$

It is evident that (5) has a particular solution of the form
$$Q_p(t) = A_1 \sin 60t + A_2 \cos 60t. \tag{6}$$

To determine the values A_1 and A_2, we substitute (6) into (4), obtaining the simultaneous equations
$$6400 A_1 - 6000 A_2 = 962,$$
$$6000 A_1 + 6400 A_2 = 0.$$

Thus $A_1 = 2/25$, $A_2 = -3/40$, and since the general solution of the homogeneous equation is the same as that of (2), the general solution of (5) is

$$Q(t) = e^{-50t}(c_1 \cos 50\sqrt{3}\, t + c_2 \sin 50\sqrt{3}\, t) + \frac{2}{25}\sin 60t - \frac{3}{40}\cos 60t. \tag{7}$$

Differentiating (7), we obtain
$$I(t) = 50 e^{-50t}\left[(\sqrt{3}\, c_2 - c_1)\cos 50\sqrt{3}\, t - (c_2 + \sqrt{3}\, c_1)\sin 50\sqrt{3}\, t\right]$$
$$+ \frac{24}{5}\cos 60t + \frac{9}{2}\sin 60t.$$

Setting $t = 0$ and using the initial conditions, we obtain
$$Q(0) = c_1 - \frac{3}{40} = 0,$$
$$I(0) = 50(\sqrt{3}\, c_2 - c_1) + \frac{24}{5} = 0,$$

so $c_1 = 3/40$ and $c_2 = -21/1000\sqrt{3}$. Therefore

$$Q(t) = \frac{e^{-50t}}{1000}(75\cos 50\sqrt{3}\,t - 7\sqrt{3}\sin 50\sqrt{3}\,t) + \frac{80\sin 60t - 75\cos 60t}{1000},$$

$$I(t) = -\frac{e^{-50t}}{5}(24\cos 50\sqrt{3}\,t + 17\sqrt{3}\sin 50\sqrt{3}\,t) + \frac{48\cos 60t + 45\sin 60t}{10}.$$

PROBLEMS 4.4

1. Let $L = 10$ H, $R = 250$ Ω, $C = 10^{-3}$ f, and $E = 900$ V in Example 1. With the same assumptions, calculate the current and charge for all values of $t \geq 0$.

2. Suppose instead that $E = 50\cos 30t$ in Problem 1. Find $Q(t)$ for $t \geq 0$.

In Problems 3–6 find the steady-state current in the RLC circuit of Figure 4.9 for the given values.

3. $L = 5$ H, $R = 10$ Ω, $C = 0.1$ f, $E = 25\sin t$ V.
4. $L = 10$ H, $R = 40$ Ω, $C = 0.025$ f, $E = 100\cos 5t$ V.
5. $L = 1$ H, $R = 7$ Ω, $C = 0.1$ f, $E = 100\sin 10t$ V.
6. $L = 2.5$ H, $R = 10$ Ω, $C = 0.08$ f, $E = 100\cos 5t$ V.

Find the transient current in the RLC circuit of Figure 4.9 for Problems 7–12.

7. Values from Problem 3.
8. Values from Problem 4.
9. Values from Problem 5.
10. Values from Problem 6.
11. $L = 20$ H, $R = 40$ Ω, $C = 10^{-3}$ f, $E = 500\sin t$ V.
12. $L = 24$ H, $R = 48$ Ω, $C = 0.375$ f, $E = 900\cos 2t$ V.
13. Given that $L = 1$ H, $R = 1200$ Ω, $C = 10^{-6}$ f, $I(0) = Q(0) = 0$, and $E = 100\sin 600t$ V, determine the transient current and the steady-state current.
14. Find the ratio of the current flowing in the circuit of Problem 13 to that which would be flowing if there were no resistance, at $t = 0.001$ sec.

15. Consider the system governed by equation (1) for the case where the resistance is zero and $E = E_0\sin\omega t$. Show that the solution consists of two parts, a general solution with frequency $1/\sqrt{LC}$ and a particular solution with frequency ω. The frequency $1/\sqrt{LC}$ is called the **natural frequency** of the circuit. Note that if $\omega = 1/\sqrt{LC}$, the particular solution disappears.

16. To allow for different variations of the voltage, let us assume in equation (1) that $E = E_0 e^{it}$ ($= E_0\cos t + iE_0\sin t$). Assume also, as in Problem 15, that $R = 0$. Finally, for simplicity assume that $E_0 = L = C = 1$. Then $1 = \omega = 1/\sqrt{LC}$.

   a. Show that equation (2) becomes
   $$\frac{d^2 I}{dt^2} + I = e^{it}.$$
   b. Determine λ such that $I(t) = \lambda t e^{it}$ is a solution.
   c. Calculate the general solution and show that the magnitude of the current increases without bound as t increases. This phenomenon produces resonance.

17. Let an inductance of L henrys, a resistance of R ohms, and a capacitance of C farads be connected in series with an emf of $E_0\sin\omega t$ volts. Suppose $Q(0) = I(0) = 0$, and $4L > R^2 C$.

   a. Find the expressions for $Q(t)$ and $I(t)$.
   b. What value of ω produces resonance?

18. Solve Problem 17 for $4L = R^2 C$.
19. Solve Problem 17 for $4L < R^2 C$.

Review Exercises for Chapter 4

In Exercises 1–6 determine the equation of motion of a mass m attached to a coiled spring with spring constant k initially displaced a distance x_0 from equilibrium and released with velocity v_0 subject to

a. no damping or external forces,

b. a damping constant c but no external force,

c. an external force $F_0\sin\omega t$ but no damping,

d. both a damping constant c and external force $F_0\sin\omega t$.

1. $m = 20$ kg, $k = 1000$ N/m, $x_0 = 1$ m, $v_0 = 0$, $c = 200$ kg/s, $F_0 = 1$ N, $\omega = 10$ rad/s

2. $m = 25$ kg, $k = 40$ N/m, $x_0 = 0$, $v_0 = 1$ m/s, $c = 20$ kg/s, $F_0 = 1$ N, $\omega = 1$ rad/s

3. $m = 25$ kg, $k = 40$ N/m, $x_0 = 3$ m, $v_0 = 4$ m/s, $c = 10\sqrt{5}$ kg/s, $F_0 = 1$ N, $\omega = 1$ rad/s

4. $m = 1$ kg, $k = 36$ N/m, $x_0 = 4$ m, $v_0 = 0$, $c = 10$ kg/s, $F_0 = 4$ N, $\omega = 4$ rad/s

5. $m = 4$ kg, $k = 25$ N/m, $x_0 = 0$, $v_0 = 3$ m/s, $c = 8$ kg/s, $F_0 = 1$ N, $\omega = 3$ rad/s

6. $m = 9$ kg, $k = 81$ N/m, $x_0 = 4$ m, $v_0 = 1$ m/s, $c = 10$ kg/s, $F_0 = 2$ N, $\omega = \frac{1}{3}$ rad/s

7. Let an inductance of $L = 2$ henrys, a resistance of $R = 50$ ohms and a capacitance $C = 10^{-4}$ farad be connected in series with an emf of $E = 1000$ volts (see Figure 4.9). Suppose no charge is present and no current is flowing at time $t = 0$ when E is applied. Find the transient and steady-state solutions for the charge Q at all times t.

In Exercises 8–11 find the steady-state current in the RLC circuit of Figure 4.9 for the given values.

8. $L = 5$ H, $R = 20$ Ω, $C = 0.1$ f, $E = 25 \sin t$ V.

9. $L = 10$ H, $R = 240$ Ω, $C = 0.025$ f, $E = 100 \cos 5t$ V.

10. $L = 1$ H, $R = 9$ Ω, $C = 0.1$ f, $E = 100 \sin 10t$ V.

11. $L = 2.5$ H, $R = 20$ Ω, $C = 0.08$ f, $E = 100 \cos 5t$ V.

12. A 27-pound weight hangs from a spring of spring constant 18 pounds per inch. We find that during free motion the amplitude decreases to one tenth in six complete cycles of the motion. Find the equation describing the motion of the spring-weight system.

13. A spring fixed at its upper end supports a 10-pound weight that stretches the spring by 1/2 foot. An external periodic force of $2 \cos 8t$ pounds is applied to the spring-weight system. Describe the motion, neglecting resistance.

14. Repeat Exercise 13 with a damping constant of $c = 0.01$.

15. Repeat Exercise 14 with an external periodic force of $2 \sin \frac{65}{8} t$ pounds.

5

Power Series Solutions of Differential Equations

In Chapter 3 we studied several methods for solving second- and higher-order differential equations. With the exception of the Euler equation and a few equations in which one solution was easily guessed, the techniques applied only to linear differential equations with *constant coefficients*. The case of linear differential equations with *variable coefficients* is much more complicated. Unfortunately, many of the most important differential equations in applied mathematics—for example, the Bessel equation and the Legendre equation—are of this type. In this chapter we consider a method that is often successful for obtaining solutions to such equations. Since the solutions so obtained are in the form of power series, the procedure used is known as the **power series method**. For the method to apply, the variable coefficients must be expressed as power series. The last two sections of the chapter are an introduction to Bessel functions and Legendre polynomials. In those sections we consider some of the properties of these special functions and a few of the standard procedures used in working with them.

5.1 The Power Series Method

In this section we review some of the basic properties of power series before discussing the power series method. We take it for granted that most readers have received some background in power series in an earlier course in calculus. A **power series** in $(x - a)$ is an infinite series of the form

$$\sum_{n=0}^{\infty} c_n(x-a)^n = c_0 + c_1(x-a) + c_2(x-a)^2 + \cdots, \tag{1}$$

where $c_0, c_1, \ldots$ are constants, called the **coefficients** of the series, a is a constant called the **center** of the series, and x is an independent variable. In particular, a power series centered at zero ($a = 0$) has the form

$$\sum_{n=0}^{\infty} c_n x^n = c_0 + c_1 x + c_2 x^2 + c_3 x^3 + \cdots. \tag{2}$$

Note that polynomials are also power series, since they have this form.

A series of form (1) can always be reduced to form (2) by the substitution $X = x - a$. This substitution is merely a translation of the coordinate system. It is easy to see that the behavior of equation (2) near zero is exactly the same as the behavior of equation (1) near a. For this reason we need only study the properties of series of form (2).

Every power series (2) has an **interval of convergence** consisting of all values x for which the series converges. The interval of convergence includes the interval $|x| < R$, where R is the **radius of convergence** of the power series (2), and *may* include one or both of the endpoints $x = \pm R$. The radius of convergence is often obtained[†] from the **ratio test**:

$$R = \lim_{n \to \infty} \left| \frac{c_n}{c_{n+1}} \right|,$$

when this limit exists. The power series (2) **converges absolutely**, that is, $\sum_{n=0}^{\infty} |c_n| |x|^n$ converges, for all x in $|x| < R$, and (2) **diverges** for all x in $|x| > R$. When $R = 0$, the interval of convergence consists only of $x = 0$; when $R = \infty$, the power series (2) converges (absolutely) for all x.

Every power series (2) represents a differentiable function

$$f(x) = \sum_{n=0}^{\infty} c_n x^n \tag{3}$$

in the interval $|x| < R$. Moreover, *the derivative $f'(x)$ is obtained by term-by-term differentiation*:

$$f'(x) = \left(\sum_{n=0}^{\infty} c_n x^n \right)' = \sum_{n=1}^{\infty} n c_n x^{n-1}, \tag{4}$$

and the termwise derivative $f'(x)$ has the same radius of convergence as the original series (3). The series in (4) may again be differentiated term by term to obtain the second derivative $f''(x)$, and this process may be repeated infinitely many times. Observe that $f(0) = c_0$, $f'(0) = c_1$, $f''(0) = 2!c_2$, and, in general, $f^{(n)}(0) = n!c_n$. Thus $c_n = f^{(n)}(0)/n!$, so (3) becomes the **Maclaurin series**[‡] of $f(x)$:

$$f(x) = \sum_{n=0}^{\infty} \frac{f^{(n)}(0)}{n!} x^n. \tag{5}$$

Any function that has a Maclaurin series (5), or a **Taylor series**,[§]

$$f(x) = \sum_{n=0}^{\infty} \frac{f^{(n)}(a)}{n!} (x-a)^n, \tag{6}$$

[†] When the ratio test fails, the radius of convergence can be obtained using the **root test**; see R. C. Buck, *Advanced Calculus* (New York: McGraw-Hill, 1965), p. 197.

[‡] Named after the Scottish mathematician Colin Maclaurin (1698–1746).

[§] Named after the British mathematician Brook Taylor (1685–1731).

is said to be **analytic** in the disk $|x| < R$, or $|x - a| < R$, respectively. The following Maclaurin series, analytic for all x, should be familiar:

$$\left.\begin{aligned} e^x &= \sum_{n=0}^{\infty} \frac{x^n}{n!} = 1 + x + \frac{x^2}{2!} + \frac{x^3}{3!} + \cdots, \\ \sin x &= \sum_{n=0}^{\infty} \frac{(-1)^n x^{2n+1}}{(2n+1)!} = x - \frac{x^3}{3!} + \frac{x^5}{5!} - \frac{x^7}{7!} + \cdots, \\ \cos x &= \sum_{n=0}^{\infty} \frac{(-1)^n x^{2n}}{(2n)!} = 1 - \frac{x^2}{2!} + \frac{x^4}{4!} - \frac{x^6}{6!} + \cdots, \\ \sinh x &= \sum_{n=0}^{\infty} \frac{x^{2n+1}}{(2n+1)!} = x + \frac{x^3}{3!} + \frac{x^5}{5!} + \frac{x^7}{7!} + \cdots, \\ \cosh x &= \sum_{n=0}^{\infty} \frac{x^{2n}}{(2n)!} = 1 + \frac{x^2}{2!} + \frac{x^4}{4!} + \frac{x^6}{6!} + \cdots. \end{aligned}\right.$$

There are functions $f(x)$ that have derivatives of all orders at a given point a and yet are not analytic. In these cases, the remainder term $R_N(x - a)$ of **Taylor's formula**

$$f(x) = \left[\sum_{n=0}^{N} \frac{f^{(n)}(a)}{n!} (x - a)^n \right] + R_N(x - a)$$

does not tend to zero, for all x in $|x - a| < R$, as N tends to infinity. An example of such a function is given in Problem 17.

The fundamental assumption made in solving a differential equation $f(x, y, y', y'', \ldots) = 0$ by the power series method is that *the solution of the differential equation can be expressed in the form of a power series*, say,

$$y = \sum_{n=0}^{\infty} c_n x^n = c_0 + c_1 x + c_2 x^2 + \cdots. \tag{7}$$

Power series expansions for $y', y'', \ldots$ are then obtained by differentiating equation (7) term by term:

$$y' = \sum_{n=1}^{\infty} n c_n x^{n-1} = c_1 + 2c_2 x + 3c_3 x^2 + \cdots \tag{8}$$

$$y'' = \sum_{n=2}^{\infty} n(n-1) c_n x^{n-2} = 2c_2 + 3 \cdot 2 c_3 x + 4 \cdot 3 c_4 x^2 + \cdots \tag{9}$$

$$\vdots$$

and these series are substituted into the given differential equation. Collecting the terms involving like powers of x, we then obtain an expression of the form

$$k_0 + k_1 x + k_2 x^2 + \cdots = \sum_{n=0}^{\infty} k_n x^n = 0, \tag{10}$$

where the coefficients $k_0, k_1, k_2, \ldots$ are expressions involving the unknown coefficients $c_0, c_1, c_2, \ldots$. Since equation (10) must hold for all values of x in the interval of convergence, all the coefficients $k_0, k_1, k_2, \ldots$ must be zero. From the equations

$$k_0 = 0, \quad k_1 = 0, \quad k_2 = 0, \ldots$$

it is then possible to determine successively the coefficients $c_0, c_1, c_2, \ldots$. To illustrate that the power series method does provide the required solution, we solve three problems, two of which can be solved more easily by other methods.

Example 1 Solve the initial-value problem
$$y' = y + x^2, \qquad y(0) = 1. \tag{11}$$

Solution Inserting equations (7) and (8) into the differential equation, we have
$$c_1 + 2c_2 x + 3c_3 x^2 + 4c_4 x^3 + \cdots = (c_0 + c_1 x + c_2 x^2 + c_3 x^3 + \cdots) + x^2.$$
Collecting like powers of x yields
$$(c_1 - c_0) + (2c_2 - c_1)x + (3c_3 - c_2 - 1)x^2 + (4c_4 - c_3)x^3 + \cdots = 0.$$
Equating each of the coefficients to zero, we obtain the identities
$$c_1 - c_0 = 0, \quad 2c_2 - c_1 = 0, \quad 3c_3 - c_2 - 1 = 0, \quad 4c_4 - c_3 = 0, \ldots,$$
from which we find that
$$c_1 = c_0, \quad c_2 = \frac{c_1}{2} = \frac{c_0}{2!}, \quad c_3 = \frac{c_2 + 1}{3} = \frac{c_0 + 2}{3!}, \quad c_4 = \frac{c_3}{4} = \frac{c_0 + 2}{4!}, \ldots.$$
Substituting these values into equation (7), we obtain *the power series solution to the differential equation in* (11):
$$y = c_0 + c_0 x + \frac{c_0}{2!} x^2 + \frac{c_0 + 2}{3!} x^3 + \frac{c_0 + 2}{4!} x^4 + \frac{c_0 + 2}{5!} x^5 + \cdots.$$
Adding and subtracting $2[1 + x + x^2/2!]$, we can rewrite y as
$$y = (c_0 + 2)\left[1 + x + \frac{x^2}{2!} + \frac{x^3}{3!} + \frac{x^4}{4!} + \cdots\right] - 2\left[1 + x + \frac{x^2}{2!}\right].$$

To solve the initial-value problem, we set $x = 0$, obtaining
$$1 = y(0) = c_0 + 2 - 2 = c_0.$$
Thus the solution of the initial-value problem (11) is
$$y = 3\left[1 + x + \frac{x^2}{2!} + \frac{x^3}{3!} + \frac{x^4}{4!} + \cdots\right] - 2\left[1 + x + \frac{x^2}{2!}\right].$$

This example exhibits an unusual occurrence: looking carefully at the series in square brackets, we recognize the expansion for e^x, so we have the solution
$$y = 3e^x - x^2 - 2x - 2.$$
It is often difficult or impossible to recognize the power series solution of a differential equation.

Example 2 Solve the differential equation
$$y'' + y = 0. \tag{12}$$

Solution Using equations (7) and (9), we have
$$(2c_2 + 3\cdot 2c_3 x + 4\cdot 3c_4 x^2 + \cdots) + (c_0 + c_1 x + c_2 x^2 + \cdots) = 0.$$
Gathering like powers of x yields
$$(2c_2 + c_0) + (3\cdot 2c_3 + c_1)x + (4\cdot 3c_4 + c_2)x^2 + \cdots = 0.$$
Setting each of the coefficients to zero, we obtain
$$2c_2 + c_0 = 0, \quad 3\cdot 2c_3 + c_1 = 0, \quad 4\cdot 3c_4 + c_2 = 0, \quad 5\cdot 4c_5 + c_3 = 0, \ldots,$$
and
$$c_2 = -\frac{c_0}{2!}, \quad c_3 = -\frac{c_1}{3!}, \quad c_4 = -\frac{c_2}{4\cdot 3} = \frac{c_0}{4!}, \quad c_5 = -\frac{c_3}{5\cdot 4} = \frac{c_1}{5!}, \ldots.$$
Substituting these values into the power series (7) for y yields
$$y = c_0 + c_1 x - \frac{c_0}{2!}x^2 - \frac{c_1}{3!}x^3 + \frac{c_0}{4!}x^4 + \frac{c_1}{5!}x^5 + \cdots.$$
Splitting this series into two parts, we have the general power series solution to equation (12),
$$y = c_0\left(1 - \frac{x^2}{2!} + \frac{x^4}{4!} - \cdots\right) + c_1\left(x - \frac{x^3}{3!} + \frac{x^5}{5!} - \cdots\right).$$

Here, again, we can recognize the power series: using the Maclaurin series for $\sin x$ and $\cos x$, we obtain the familiar general solution
$$y = c_0 \cos x + c_1 \sin x$$
to equation (12). We observe that in this case the power series method produces two arbitrary constants c_0 and c_1.

So far we have considered only linear equations with constant coefficients. We turn now to linear equations with variable coefficients.

Example 3 Solve the differential equation
$$y'' + xy' + y = 0. \tag{13}$$

Solution Using the power series method, we obtain the equation
$$\sum_{n=2}^{\infty} n(n-1)c_n x^{n-2} + x\sum_{n=1}^{\infty} nc_n x^{n-1} + \sum_{n=0}^{\infty} c_n x^n = 0. \tag{14}$$
We use the summation notation in this example in order to develop the skill in manipulating power series that will be required later on. *In order to gather all three power series into a single one, we need to rewrite each of the sums in equation (14) so that the general term contains the same power of x.* Consider the first series:
$$T_1 = \sum_{n=2}^{\infty} n(n-1)c_n x^{n-2}.$$

To obtain the exponent k in place of $n - 2$, we make the substitution $k = n - 2$. Then $n = k + 2$, so every place we see an n, we replace it with $k + 2$. Since n ranges from 2 to ∞ and $k = n - 2$, k ranges from $2 - 2 = 0$ to

∞ ($\infty - 2 = \infty$; explain why). Thus we have

$$T_1 = \sum_{k=0}^{\infty} (k+2)(k+1)c_{k+2}x^k.$$

with $n = 2 \to k=0$, $n \to k$, $n-1 = (k+2)-1$, $n-2 \to$ exponent.

Next, we need to rewrite the second series so that the general term is x^k. Here

$$T_2 = x \sum_{n=1}^{\infty} nc_n x^{n-1} = \sum_{n=1}^{\infty} nc_n x \cdot x^{n-1} = \sum_{n=1}^{\infty} nc_n x^n = \sum_{k=1}^{\infty} kc_k x^k$$

if we set $k = n$. Similarly, setting $k = n$ in the third series, we have

$$T_3 = \sum_{n=0}^{\infty} c_n x^n = \sum_{k=0}^{\infty} c_k x^k.$$

Thus (14) becomes

$$\sum_{k=0}^{\infty} (k+2)(k+1)c_{k+2}x^k + \sum_{k=1}^{\infty} kc_k x^k + \sum_{k=0}^{\infty} c_k x^k = 0.$$

Note that the second sum can also be allowed to range from zero to infinity, since $kc_k x^k = 0$ when $k = 0$. (Had this not happened, we would have to treat the terms corresponding to the index $k = 0$ separately.)

Gathering like terms in x produces the equation

$$\sum_{k=0}^{\infty} [(k+2)(k+1)c_{k+2} + (k+1)c_k] x^k = 0.$$

Setting the coefficients equal to zero, we obtain the general recursion formula

$$(k+2)(k+1)c_{k+2} + (k+1)c_k = 0.$$

Therefore

$$(k+2)c_{k+2} = -c_k \quad \text{or} \quad c_{k+2} = -\frac{c_k}{k+2}$$

and

$$c_2 = -\frac{c_0}{2}, \quad c_3 = -\frac{c_1}{3}, \quad c_4 = -\frac{c_2}{4} = \frac{c_0}{2 \cdot 4},$$

$$c_5 = -\frac{c_3}{5} = \frac{c_1}{3 \cdot 5}, \quad c_6 = -\frac{c_4}{6} = -\frac{c_0}{2 \cdot 4 \cdot 6}, \ldots.$$

Hence the power series for y can be written in the form

$$y = c_0 + c_1 x - \frac{c_0}{2}x^2 - \frac{c_1}{3}x^3 + \frac{c_0}{2 \cdot 4}x^4 + \frac{c_1}{3 \cdot 5}x^5 - \cdots$$

$$= c_0\left(1 - \frac{x^2}{2} + \frac{x^4}{2 \cdot 4} - \frac{x^6}{2 \cdot 4 \cdot 6} + \cdots\right)$$

$$+ c_1\left(x - \frac{x^3}{3} + \frac{x^5}{3 \cdot 5} - \frac{x^7}{3 \cdot 5 \cdot 7} + \cdots\right)$$

by separating the terms that involve c_0 and c_1.

At this point we can try to see whether we recognize the two series that have been obtained by the power series method. Very frequently this is an unproductive task, and no further simplification of the solution is possible, but in this instance we are fortunate:

$$1 - \frac{x^2}{2} + \frac{x^4}{2 \cdot 4} - \frac{x^6}{2 \cdot 4 \cdot 6} + \cdots$$

$$= 1 + \left(-\frac{x^2}{2}\right) + \frac{1}{2!}\left(-\frac{x^2}{2}\right)^2 + \frac{1}{3!}\left(-\frac{x^2}{2}\right)^3 + \cdots = e^{-x^2/2},$$

so the first series is

$$y_1 = e^{-x^2/2}.$$

Although we cannot recognize the second series, knowing one solution allows us to determine the other. Substitute $y_2 = v(x)y_1$ into equation (13). This yields

$$v''y_1 + v'(2y_1' + xy_1) + v(y_1'' + xy_1' + y_1) = 0.$$

The term in the last parentheses is zero since y_1 is a solution to equation (13). Setting $z = v'$, we obtain the *first*-order differential equation

$$y_1 z' + (2y_1' + xy_1)z = e^{-x^2/2}(z' - xz) = 0.$$

Separating variables we get $z = e^{x^2/2}$, so the second solution to equation (13) is

$$y_2 = y_1 v = e^{-x^2/2} \int e^{x^2/2}\, dx.$$

Hence the general solution to equation (13) is given by

$$y = c_0 e^{-x^2/2} + c_1 e^{-x^2/2} \int e^{x^2/2}\, dx.$$

Note that this is as far as we can go, since $\int e^{x^2/2}\, dx$ cannot be expressed in terms of elementary functions (i.e., the rational, trigonometric, exponential, or logarithmic functions).

A Perspective

THE TAYLOR SERIES METHOD

The Taylor series and the derivatives of a differential equation can also be used to construct its power series solution.

Example 1 (revisited) Consider the initial-value problem
$$y' = y + x^2, \qquad y(0) = 1.$$
Differentiating both sides of the differential equation repeatedly and successively evaluating each derivative at the initial value of $x = 0$, we have
$$y'(0) = y + x^2\big|_{x=0} = y(0) + (0)^2 = 1,$$
$$y''(0) = y' + 2x\big|_{x=0} = y'(0) + 2(0) = 1,$$
$$y'''(0) = y'' + 2\big|_{x=0} = y''(0) + 2 = 3,$$
$$y^{(4)}(0) = y'''\big|_{x=0} = y'''(0) = 3, \ldots.$$
Substituting these derivatives in the Taylor series
$$y(x) = \sum_{n=0}^{\infty} \frac{y^{(n)}(a)}{n!}(x-a)^n$$
with $a = 0$, we have
$$y(x) = 1 + x + \frac{x^2}{2!} + \frac{3x^3}{3!} + \frac{3x^4}{4!} + \cdots = 3e^x - 2 - 2x - x^2,$$
which is the result that we obtained before.

Example 3 (revisited) From Theorem 3.1.2, we know that the homogeneous differential equation
$$y'' + xy' + y = 0 \tag{15}$$
has two linearly independent solutions y_1 and y_2 satisfying
$$y_1(0) = 1, \quad y_1'(0) = 0, \quad \text{and} \quad y_2(0) = 0, \quad y_2'(0) = 1.$$
Thus, differentiating repeatedly, we have
$$y'' = -xy' - y,$$
$$y''' = -xy'' - 2y',$$
$$y^{(4)} = -xy''' - 3y'',$$
$$\vdots$$
$$y^{(n+1)} = -xy^{(n)} - ny^{(n-1)}.$$
Using the initial conditions for the solution y_1, we can successively evaluate $y_1''(0), y_1'''(0), \ldots$:
$$y_1''(0) = -xy_1' - y_1\big|_{x=0} = -1,$$
$$y_1'''(0) = -xy_1'' - 2y_1'\big|_{x=0} = 0,$$
$$y_1^{(4)}(0) = -xy_1''' - 3y_1''\big|_{x=0} = 3,$$
$$y_1^{(5)}(0) = -xy_1^{(4)} - 4y_1'''\big|_{x=0} = 0,$$
$$y_1^{(6)}(0) = -xy_1^{(5)} - 5y_1^{(4)}\big|_{x=0} = -15, \ldots,$$

so the Taylor series (6) (with $a = 0$) becomes

$$y_1(x) = 1 - \frac{x^2}{2!} + \frac{3x^4}{4!} - \frac{15x^5}{6!} + \cdots = 1 - \frac{x^2}{2} + \frac{x^4}{2 \cdot 4} - \frac{x^6}{2 \cdot 4 \cdot 6} + \cdots.$$

Similarly, using the initial conditions for y_2, we have successively

$$y_2''(0) = -xy_2' - y_2|_{x=0} = 0,$$
$$y_2^{(3)}(0) = -xy_2'' - 2y_2'|_{x=0} = -2,$$
$$y_2^{(4)}(0) = -xy_2^{(3)} - 3y_2''|_{x=0} = 0,$$
$$y_2^{(5)}(0) = -xy_2^{(4)} - 4y_2^{(3)}|_{x=0} = 8, \ldots$$

yielding

$$y_2(x) = x - \frac{2x^3}{3!} + \frac{8x^5}{5!} - \cdots = x - \frac{x^3}{3} + \frac{x^5}{3 \cdot 5} - \cdots.$$

These are the same two linearly independent power series solutions that were obtained in Example 3 using the power series method.

PROBLEMS 5.1

In Problems 1–10 find the Taylor series centered at a and its corresponding radius of convergence for the given function.

1. $f(x) = e^x$, $a = 1$
2. $f(x) = e^{-x}$, $a = 0$
3. $f(x) = \cos x$, $a = \pi/4$
4. $f(x) = \sinh x$, $a = \ln 2$
5. $f(x) = e^{bx}$, $a = -1$
6. $f(x) = xe^x$, $a = 1$
7. $f(x) = x^2 e^{-x^2}$, $a = 0$
8. $f(x) = \begin{cases} \dfrac{\sin x}{x}, & x \neq 0, \\ 1, & x = 0 \end{cases}$ $a = 0$
9. $f(x) = (x-1) \ln x$, $a = 1$
10. $f(x) = \sin^2 x$, $a = 0$

In Problems 11–14 derive the given Taylor series.

11. $\ln(1 + x) = x - \dfrac{x^2}{2} + \dfrac{x^3}{3} - \dfrac{x^4}{4} + \cdots$, $|x| < 1$

12. $\sin^{-1} x = x + \dfrac{1}{2} \cdot \dfrac{x^3}{3} + \dfrac{1}{2} \cdot \dfrac{3}{4} \cdot \dfrac{x^5}{5}$
$+ \dfrac{1}{2} \cdot \dfrac{3}{4} \cdot \dfrac{5}{6} \cdot \dfrac{x^7}{7} + \cdots$, $|x| < 1$

13. $\ln x = (x-1) - \dfrac{(x-1)^2}{2} + \dfrac{(x-1)^3}{3}$
$- \dfrac{(x-1)^4}{4} + \cdots$, $0 < x < 2$

14. $\dfrac{1}{2-x} = 1 + (x-1) + (x-1)^2 + (x-1)^3 + \cdots$, $0 < x < 2$

15. Show that

$$\frac{1}{1+x} = 1 - x + x^2 - x^3 + \cdots, \quad |x| < 1.$$

Then prove that

a. $\ln(1+x) = x - \dfrac{x^2}{2} + \dfrac{x^3}{3} - \dfrac{x^4}{4} + \cdots,$
$|x| < 1.$

b. $\tan^{-1} x = x - \dfrac{x^3}{3} + \dfrac{x^5}{5} - \dfrac{x^7}{7} + \cdots,$
$|x| < 1.$

c. $\dfrac{1}{(1+x)^2} = 1 - 2x + 3x^2 - 4x^3 + \cdots,$
$|x| < 1.$

16. Show that the series

$$\sum_{n=1}^{\infty} \frac{x^n}{n} = x + \frac{x^2}{2} + \frac{x^3}{3} + \frac{x^4}{4} + \cdots$$

diverges at $x = 1$ by proving that the partial sums satisfy the inequality

$$S_{2^k}(1) = \sum_{n=1}^{2^k} \frac{1}{n} \geq 1 + k/2.$$

[This exercise shows that, even though the terms

in a series may tend to zero, the series itself may diverge, in this case at $x = 1$.]

17. Consider the function
$$f(x) = \begin{cases} e^{-1/x^2}, & x \neq 0, \\ 0, & x = 0. \end{cases}$$

   a. Show that f has derivatives of all orders at $x = 0$ and that
   $$f'(0) = f''(0) = \cdots = 0.$$
   [*Hint*: Use the limit $f'(0) = \lim_{x \to 0} \dfrac{f(x) - f(0)}{x}$.]

   b. Conclude that $f(x)$ does not have a Taylor series expansion at $x = 0$, even though it is infinitely differentiable there. Thus f is not analytic at $x = 0$ (why?).

18. Using Taylor's formula, prove the **binomial formula**
$$(1 + x)^p = 1 + px + \frac{p(p-1)}{1 \cdot 2} x^2 + \frac{p(p-1)(p-2)}{1 \cdot 2 \cdot 3} x^3 + \cdots.$$
   Where does it converge?

In Problems 19–34 find the general power series solution of each equation by the power series method. When initial conditions are specified, give the solution that satisfies them. Then try to recognize a solution in terms of elementary functions (this will *not* be possible in some problems).

19. $y' = y - x$, $y(0) = 2$
20. $y' = x^3 - 2xy$, $y(0) = 1$
21. $y'' + y = x$
22. $y'' + 4y = 0$, $y(0) = 1$, $y'(0) = 0$
23. $(1 + x^2) y'' + 2xy' - 2y = 0$
24. $xy'' - xy' + y = e^x$, $y(0) = 1$, $y'(0) = 2$
25. $xy'' - x^2 y' + (x^2 - 2) y = 0$, $y(0) = 0$, $y'(0) = 1$
26. $(1 - x) y'' - y' + xy = 0$, $y(0) = y'(0) = 1$
27. $y'' - 2xy' + 4y = 0$, $y(0) = 1$, $y'(0) = 0$
28. $(1 - x^2) y'' - xy' + y = 0$, $y(0) = 0$, $y'(0) = 1$
29. $y'' - xy' + y = -x \cos x$, $y(0) = 0$, $y'(0) = 2$
30. $y'' - xy' + xy = 0$, $y(0) = 2$, $y'(0) = 1$
31. $(1 - x)^2 y'' - (1 - x) y' - y = 0$, $y(0) = y'(0) = 1$
32. $y'' - 2xy' + 2y = 0$
33. $y'' - 2xy' - 2y = x$, $y(0) = 1$, $y'(0) = -\frac{1}{4}$
34. $y'' - x^2 y = 0$
35. The **Airy equation**,[†]
$$y'' - xy = 0,$$
has applications in the theory of diffraction. Find the general solution of this equation.
36. The **Hermite equation**,[‡]
$$y'' - 2xy' + 2py = 0,$$
where p is constant, arises in quantum mechanics in connection with the Schrödinger[§] equation for a harmonic oscillator. Show that if p is a positive integer, one of the two linearly independent solutions of the Hermite equation is a polynomial, called the **Hermite polynomial** $H_p(x)$.

Use the Taylor series method to find the general solution of the differential equation in Problems 37–42.

37. $y'' - 2xy' + 2y = 0$
38. $x^2 y'' - xy' + y = 0$, at $a = 1$
39. $x^2 y'' + xy' - 4y = 0$, at $a = 1$
40. $x^2 y'' - 2xy' + (x^2 + 2) y = 0$, at $a = 1$
41. $(1 - x^2) y'' - 2xy' + 6y = 0$
42. $(1 - x^2) y'' - 2xy' + 2y = 0$

[†] Sir George Biddell Airy (1801–1892) was Lucasian Professor of Mathematics, director of the observatory, and Plumian Professor of Astronomy at Cambridge University in England until 1835. Then he was appointed director of the Greenwich Observatory (Astronomer royal). He remained there until his retirement in 1881. He did much work in lunar and solar photography, planetary motion, optics, and other areas.

[‡] Charles Hermite (1822–1901) was a French mathematician known for his contributions in algebra and number theory.

[§] Erwin Schrödinger (1887–1961) was an Austrian physicist. He was awarded the Nobel prize in 1933 (jointly with P. A. M. Dirac) for his work in quantum mechanics.

5.2 Ordinary and Singular Points

The power series method sometimes fails to yield a solution for one equation while working very well for an apparently similar equation.

Example 1 Solve the equations
$$x^2 y'' + axy' + by = 0, \qquad (1)$$

where

i. $a = -2$, $b = 2$;

ii. $a = -1$, $b = 1$;

iii. $a = 1$, $b = 1$.

Solution Set

$$y = \sum_{n=0}^{\infty} c_n x^n, \qquad y' = \sum_{n=1}^{\infty} nc_n x^{n-1}, \qquad y'' = \sum_{n=2}^{\infty} n(n-1)c_n x^{n-2}.$$

Then, since $n(n-1) = 0$ at $n = 0$ and $n = 1$,

$$x^2 y'' + axy' + by = x^2 \sum_{n=0}^{\infty} n(n-1)c_n x^{n-2} + ax \sum_{n=0}^{\infty} nc_n x^{n-1} + b \sum_{n=0}^{\infty} c_n x^n$$

$$= \sum_{n=0}^{\infty} n(n-1)c_n x^n + \sum_{n=0}^{\infty} anc_n x^n + \sum_{n=0}^{\infty} bc_n x^n$$

$$= \sum_{n=0}^{\infty} [n(n-1) + an + b]c_n x^n = 0. \tag{2}$$

i. Substitute $a = -2$ and $b = 2$ into equation (2) to obtain

$$\sum_{n=0}^{\infty} (n^2 - 3n + 2)c_n x^n = 0.$$

Equating each of the coefficients to zero, we have

$$(n-2)(n-1)c_n = 0,$$

implying that $c_n = 0$ for all $n \neq 1$ or 2. Hence

$$y = c_1 x + c_2 x^2$$

is the general solution to equation (1) with $a = -2$ and $b = 2$.

ii. Equation (2) yields, with $a = -1$ and $b = 1$,

$$\sum_{n=0}^{\infty} (n^2 - 2n + 1)c_n x^n = 0,$$

so $(n-1)^2 c_n = 0$. Thus $c_n = 0$ for all $n \neq 1$, yielding the solution

$$y = c_1 x.$$

Since the general solution of a second-order linear differential equation involves *two* linearly independent solutions, the power series method has given us only half of the general solution. We can use the method in Example 5.1.3 (see Section 3.2) to find the other solution. Substitute $y = xv$ into equation (1) (with $a = -1$ and $b = 1$) to obtain

$$x^2(xv)'' - x(xv)' + (xv) = 0,$$

or

$$x^3 v'' + x^2 v' = 0.$$

Setting $z = v'$, we obtain the first-order separable differential equation

$$x^3 \frac{dz}{dx} + x^2 z = 0.$$

Thus
$$\frac{dz}{z} = -\frac{dx}{x},$$
so $z = c/x$ and $v = c \ln|x|$. Hence the general solution to (ii) is
$$y = Ax + Bx \ln|x|.$$

iii. Equation (2) gives us the series
$$\sum_{n=0}^{\infty} (n^2 + 1) c_n = 0.$$
Equating the coefficients to zero, we have $(n^2 + 1)c_n = 0$, so $c_n = 0$ for every n. Hence the power series method fails completely in helping us find the general solution
$$y = A \cos(\ln|x|) + B \sin(\ln|x|)$$
of the Euler equation (1) (check!).

It would be very useful to find the reason for this anomaly; once we know the reason for the failure of the power series method, a method for avoiding this difficulty may become apparent. The main clue to the puzzle can be obtained by making the coefficient of the highest-order derivative equal to 1 as required by the existence-uniqueness theorem for initial-value problems (Theorem 3.1.1). Thus, if we write each of the second-order homogeneous equations in Example 1 in the form
$$y'' + a(x)y' + b(x)y = 0, \tag{3}$$
the equations become
$$y'' + \frac{a}{x}y' + \frac{b}{x^2}y = 0.$$

Neither of the terms $a(x) = a/x$ or $b(x) = b/x^2$ is defined at $x = 0$, so we certainly cannot expect to find a power series representation for $a(x)$ or $b(x)$ that converges in an open interval containing $x = 0$.

When $a(x)$ and $b(x)$ can both be represented by Maclaurin series that converge in an open interval containing $x = 0$, we have the following result.

THEOREM 1

There is a unique Maclaurin series $y(x)$ satisfying the initial-value problem
$$y'' + a(x)y' + b(x)y = 0, \qquad y(0) = \alpha, \qquad y'(0) = \beta,$$
provided $a(x)$ and $b(x)$ can both be represented by Maclaurin series converging in an interval $|x| < R$. The power series $y(x)$ also converges in $|x| < R$.

The proof of this theorem is complicated and is not given in this book.[†]

[†] A proof may be found in E. A. Coddington, *An Introduction to Ordinary Differential Equations* (Englewood Cliffs, N.J.: Prentice-Hall, 1961), § 3.9.

5.2/ORDINARY AND SINGULAR POINTS

Ordinary Point

Note, however, that it guarantees the success of the power series method whenever $a(x)$ and $b(x)$ are analytic at $x = 0$. We call $x = 0$ an **ordinary point** of the differential equation

$$y'' + a(x)y' + b(x)y = 0$$

Singular Point

when both $a(x)$ and $b(x)$ are analytic at $x = 0$. If $x = 0$ is not an ordinary point, it is called a **singular point** of the differential equation. Hence $x = 0$ is a singular point of equation (1) in Example 1, and it is an ordinary point of the differential equations in Examples 5.1.1–5.1.3.

Example 2 Determine whether the power series method yields a solution to each of the following equations:

$$(1 - x^2)y'' - 2xy' + 2y = 0. \quad (4)$$

$$xy'' + 2y' + xy = 0. \quad (5)$$

Is $x = 0$ an ordinary or a singular point of each equation?

Solution

i. Rewriting equation (4) in the form

$$y'' - \frac{2x}{1-x^2}y' + \frac{2}{1-x^2}y = 0,$$

we observe that

$$a(x) = \frac{-2x}{1-x^2} = -2x(1 + x^2 + x^4 + \cdots)$$

and

$$b(x) = \frac{2}{1-x^2} = 2(1 + x^2 + x^4 + \cdots)$$

are Maclaurin series that converge in the interval $|x| < 1$. Hence $x = 0$ is an ordinary point of equation (4) and the power series method will yield the general solution. Setting

$$y = \sum_{n=0}^{\infty} c_n x^n$$

and multiplying the power series representations of y, y', and y'' by the appropriate coefficients of equation (4), we get

$$(1 - x^2) \sum_{n=2}^{\infty} n(n-1)c_n x^{n-2} - 2x \sum_{n=1}^{\infty} nc_n x^{n-1} + 2 \sum_{n=0}^{\infty} c_n x^n = 0.$$

But

$$(1-x^2)\sum_{n=2}^{\infty} n(n-1)c_n x^{n-2}$$

$$= \sum_{n=2}^{\infty} n(n-1)c_n x^{n-2} - \sum_{n=2}^{\infty} n(n-1)c_n x^n$$

$k = n-2$
$\downarrow$

$$= \sum_{k=0}^{\infty} (k+2)(k+1)c_{k+2} x^k - \sum_{n=2}^{\infty} n(n-1)c_n x^n$$

$n(n-1) = 0$
at $n=0$ and $n=1$
$\downarrow$

$$= \sum_{k=0}^{\infty} (k+2)(k+1)c_{k+2} x^k - \sum_{n=0}^{\infty} n(n-1)c_n x^n$$

$k = n$
$\downarrow$

$$= \sum_{n=0}^{\infty} (n+2)(n+1)c_{n+2} x^n - \sum_{n=0}^{\infty} n(n-1)c_n x^n.$$

Thus we obtain

$$\sum_{n=0}^{\infty} [(n+2)(n+1)c_{n+2} - n(n-1)c_n] x^n - \sum_{n=0}^{\infty} 2nc_n x^n + \sum_{n=0}^{\infty} 2c_n x^n = 0,$$

or

$$\sum_{n=0}^{\infty} [(n+2)(n+1)c_{n+2} + (-n(n-1) - 2n + 2)c_n] x^n$$

$$= \sum_{n=0}^{\infty} [(n+2)(n+1)c_{n+2} + (-n^2 - n + 2)c_n] x^n$$

$$= \sum_{n=0}^{\infty} [(n+2)(n+1)c_{n+2} - (n+2)(n-1)c_n] x^n$$

$$= \sum_{n=0}^{\infty} (n+2)[(n+1)c_{n+2} - (n-1)c_n] x^n = 0. \tag{6}$$

Every example involving power series involves algebraic manipulations like the ones above. Here we have provided every detail. In subsequent examples we leave some of the details to you.

We note that equation (6) holds only if, for each n,

$$(n+1)c_{n+2} = (n-1)c_n,$$

or

$$c_{n+2} = \frac{n-1}{n+1} c_n.$$

Setting $n = 1$, we immediately see that $c_3 = 0$. Hence $c_3 = c_5 = c_7 = \cdots = 0$. If even values of n are chosen, we have

$$c_2 = \frac{0-1}{0+1}c_0 = -c_0, \qquad c_4 = \frac{2-1}{2+1}c_2 = \frac{1}{3}c_2 = -\frac{1}{3}c_0,$$

$$c_6 = \frac{3}{5}c_4 = -\frac{3}{5}\cdot\frac{1}{3}c_0 = -\frac{1}{5}c_0, \qquad c_8 = \frac{5}{7}c_6 = -\frac{1}{7}c_0,\ldots,$$

and, choosing c_1 arbitrarily, we find the general solution

$$y = c_1 x + c_0\left(1 - x^2 - \frac{x^4}{3} - \frac{x^6}{5} - \cdots\right).$$

ii. Here we have

$$y'' + \frac{2}{x}y' + y = 0,$$

so $a(x)$ is not defined at $x=0$. Thus $x=0$ is a singular point of equation (5), and we should anticipate the possibility of trouble in using the power series method. We have $xy'' + 2y' + xy = 0$ and

$$x\sum_{n=2}^{\infty} n(n-1)c_n x^{n-2} + 2\sum_{n=1}^{\infty} nc_n x^{n-1} + x\sum_{n=0}^{\infty} c_n x^n = 0,$$

or

$$\sum_{n=2}^{\infty} n(n-1)c_n x^{n-1} + \sum_{n=1}^{\infty} 2nc_n x^{n-1} + \sum_{n=0}^{\infty} c_n x^{n+1} = 0. \qquad (7)$$

Setting $k = n - 1$, the first series is

$$\sum_{k=1}^{\infty} (k+1)kc_{k+1} x^k,$$

while the second is

taking out the first term
$$\sum_{k=0}^{\infty} 2(k+1)c_{k+1} x^k \overset{\downarrow}{=} 2c_1 + \sum_{k=1}^{\infty} 2(k+1)c_{k+1} x^k.$$

Finally, setting $k = n + 1$, the third series is

$$\sum_{k=1}^{\infty} c_{k-1} x^k$$

and (7) becomes

$$2c_1 + \sum_{k=1}^{\infty} \{[(k+1)k + 2(k+1)]c_{k+1} + c_{k-1}\} x^k$$

$$= 2c_1 + \sum_{k=1}^{\infty} [(k+1)(k+2)c_{k+1} + c_{k-1}] x^k = 0.$$

Therefore $c_1 = 0$ and

$$c_{k+1} = -\frac{c_{k-1}}{(k+1)(k+2)}.$$

Thus $c_3 = c_5 = c_7 = \cdots = 0$. If odd values of k are chosen, we have

$$c_2 = -\frac{c_0}{3 \cdot 2} = -\frac{c_0}{3!}, \quad c_4 = -\frac{c_2}{5 \cdot 4} = \frac{c_0}{5 \cdot 4 \cdot 3!} = \frac{c_0}{5!}, \ldots,$$

and we obtain

$$y = c_0\left(1 - \frac{x^2}{3!} + \frac{x^4}{5!} - \frac{x^6}{7!} + \cdots\right) = \frac{c_0}{x}\left(x - \frac{x^3}{3!} + \frac{x^5}{5!} - \cdots\right)$$

$$= c_0 \frac{\sin x}{x}.$$

Hence one solution to equation (5) is

$$y = c_0 \frac{\sin x}{x}.$$

However, the power series method does not yield the general solution. We can find that solution by the method in Example 5.1.3. Setting $y = (v \sin x)/x$, we have, after some algebra,

$$(\sin x)v'' + 2(\cos x)v' = 0,$$

or, letting $z = v'$,

$$\sin x \frac{dz}{dx} = -2z \cos x.$$

Hence

$$\frac{dz}{z} = -\frac{2 \cos x \, dx}{\sin x},$$

or

$$\ln|z| = -2\ln|\sin x| + c,$$

or

$$v' = \frac{k}{\sin^2 x} = k \csc^2 x.$$

Thus $v = -k \cot x$, so equation (5) becomes

$$y = \left(\frac{\sin x}{x}\right)\cot x = \left(\frac{\sin x}{x}\right)\frac{\cos x}{\sin x} = \frac{\cos x}{x}$$

and the general solution is

$$y = A\frac{\sin x}{x} + B\frac{\cos x}{x}.$$

PROBLEMS 5.2

In Problems 1–16 find two linearly independent power series about the ordinary point $x = 0$ that are solutions to the given differential equation.

1. $y'' - xy' + y = 0$
2. $y'' + xy' + y = 0$
3. $y'' - 3xy = 0$
4. $y'' - 2xy' + y = 0$
5. $y'' - xy' + xy = 0$
6. $y'' - x^2 y = 0$
7. $y'' + x^2 y' + 2xy = 0$

8. $y'' + x^2 y' + xy = 0$
9. $(1 + x^2) y'' + 2xy' - 2y = 0$
10. $(1 - x) y'' - y' + xy = 0$
11. $(1 - x)^2 y'' - (1 - x) y' - y = 0$
12. $(x^2 + 1) y'' - 6y = 0$
13. $(2x^2 + 1) y'' + 2xy' - 18y = 0$
14. $(x^2 + 2) y'' + 3xy' + y = 0$
15. $y'' - xy' + y = -x \cos x$
16. $y'' - 2xy' - 2y = x$
17. Find a solution to the initial-value problem
$$y'''' - xy = 0, \quad y(0) = 1, \quad y'(0) = 0,$$
$$y''(0) = 0.$$
18. Solve the **Airy equation**
$$y'' - xy = 0, \quad y(1) = 1, \quad y'(1) = 0.$$
19. Solve the initial-value problem
$$y'' - xy' - y = 0, \quad y(0) = 1, \quad y'(0) = 0.$$

In Problems 20–23 find the first four terms in the power series solution to the given initial-value problem.

20. $y'' + (\sin x) y = 0, \quad y(0) = 1, \quad y'(0) = 0$
21. $y'' - e^x y = 0, \quad y(0) = y'(0) = 1$
22. $y'' + (\cos x) y = 0, \quad y(0) = 1, \quad y'(0) = 0$
23. $y'' + (\cos x) y = 0, \quad y(0) = 0, \quad y'(0) = 1$

In Problems 24–32 use the power series method to obtain at least one solution about the singular point $x = 0$. Then use the method of reduction of order (Section 3.2) to find the general solution.

24. $x^2 y'' + 2xy' - 2y = 0$
25. $xy'' + (1 - 2x) y' - (1 - x) y = 0$
26. $xy'' + 2y' - xy = 0$
27. $x^2 y'' + x(x - 1) y' - (x - 1) y = 0$
28. $xy'' + (1 - x) y' - y = 0$
29. $xy'' - (1 - x) y' - 2y = 0$
30. $x(x - 1) y'' - (1 - 3x) y' + y = 0$
31. $x(x - 1) y'' + 3y' - 2y = 0$
32. $x^2 y'' - x(1 - x) y' + y = 0$
33. Does the power series method yield a solution to the equation
   a. $x^2 y' = y$?
   b. $x^3 y' = y$?
34. Show that the power series method fails for
$$x^2 y'' + x^2 y' + y = 0.$$
35. Show that the power series method fails for
$$x^3 y'' + xy' + y = 0.$$
36. Show that the power series method fails for
$$x^4 y'' + 2x^3 y' - y = 0.$$

5.3 The Method of Frobenius: The Indicial Equation

If we look at the solution of Example 5.2.2(ii) (p. 171), we see that the solution $(\cos x)/x$ was not obtained by the power series method. In fact, $(\cos x)/x$ cannot be written as a power series in x. However, it *can* be written as a power series in x times a power of x:

$$\frac{\cos x}{x} = x^{-1} \left(1 - \frac{x^2}{2!} + \frac{x^4}{4!} - \cdots \right).$$

This suggests that we should try to find solutions of the form

$$y = x^r \left(c_0 + c_1 x + c_2 x^2 + c_3 x^3 + \cdots \right), \tag{1}$$

where r is some real or complex number, whenever $x = 0$ is a singular point of the differential equation. For one class of singular points, this modification of the power series method does yield solutions.

Regular Singular Point

We call $x = 0$ a **regular singular point** of the differential equation

$$y'' + a(x) y' + b(x) y = 0 \tag{2}$$

if both the functions $xa(x)$ and $x^2b(x)$ have convergent Maclaurin series in an open interval containing $x = 0$. Observe that $x = 0$ is a regular singular point for the differential equation in Example 5.2.2(ii) because $xa(x) = 2$ and $x^2b(x) = x^2$, both of which are convergent Maclaurin series (for all x) with only one nonzero coefficient.

A singular point that is not regular is called **irregular**. For example, the point $x = 0$ is an irregular singular point of the two equations

$$y'' + \frac{1}{x^2}y' + y = 0$$

and

$$y'' + \frac{1}{x}y' + \frac{1}{x^3}y = 0$$

because in the first case $xa(x) = x^{-1}$, while in the second case $x^2b(x) = x^{-1}$. The function of x^{-1} is not defined at $x = 0$, so it cannot have a convergent Maclaurin series.

To simplify the explanation of the modified power series method, called the **method of Frobenius**,[†] we assume that $x = 0$ is a regular singular point of the equation

$$y'' + a(x)y' + b(x)y = 0 \tag{3}$$

and that equation (3) has a solution of the form

$$y = x^r(c_0 + c_1 x + c_2 x^2 + \cdots) = \sum_{n=0}^{\infty} c_n x^{r+n}, \quad x > 0, \tag{4}$$

where r is some real or complex number. We can assume that $c_0 = 1$, since any constant multiple of a solution is again a solution of the differential equation. In addition, the choice $c_0 = 1$ simplifies much of the following discussion. The restriction $x > 0$ is necessary to prevent difficulties for certain values of r, such as $r = \frac{1}{2}$ and $-\frac{1}{4}$, since we are not interested in imaginary solutions. [If we need to find a solution valid for $x < 0$, we can change variables by substituting $X = -x$ into equation (3) and solve the resulting equation for $X > 0$.]

Since

$$y' = \sum_{n=0}^{\infty} c_n(r+n)x^{r+n-1}$$

and

$$y'' = \sum_{n=0}^{\infty} c_n(r+n)(r+n-1)x^{r+n-2},$$

equation (3) can be rewritten as

$$\sum_{n=0}^{\infty} c_n(r+n)(r+n-1)x^{r+n-2} + a(x)\sum_{n=0}^{\infty} c_n(r+n)x^{r+n-1}$$

$$+ b(x)\sum_{n=0}^{\infty} c_n x^{r+n} = 0,$$

[†] Georg Ferdinand Frobenius (1848–1917) was a German mathematician.

or, factoring an x and an x^2 in the second and third series, respectively, we obtain

$$\sum_{n=0}^{\infty} c_n \left[(r+n)(r+n-1) + (r+n)xa(x) + x^2 b(x) \right] x^{r+n-2} = 0. \quad (5)$$

As $x = 0$ is a regular singular point, both $xa(x)$ and $x^2 b(x)$ can be expressed as convergent power series in x:

$$xa(x) = a_0 + a_1 x + a_2 x^2 + \cdots,$$
$$x^2 b(x) = b_0 + b_1 x + b_2 x^2 + \cdots.$$

But $n \geq 0$, so x^{r-2} is the smallest power of x in equation (5). Since the coefficients of a power series whose sum is zero must vanish, we have, for $n = 0$,

$$c_0 \left[r(r-1) + a_0 r + b_0 \right] = 0.$$

By hypothesis $c_0 = 1$, so we obtain the **indicial equation**

Indicial Equation

$$\boxed{r(r-1) + a_0 r + b_0 = 0,} \quad (6)$$

whose roots, r_1 and r_2, are called the **exponents** of the differential equation (3). In what follows we see that one of the solutions of equation (3) is always of form (4) and that there are three possible forms for the second linearly independent solution corresponding to the following cases:

Case 1: r_1 and r_2 differ but not by an integer.

Case 2: $r_1 = r_2$.

Case 3: r_1 and r_2 differ by a nonzero integer.

We consider the three cases separately.

Case 1: r_1 **and** r_2 **differ but not by an integer** This is the easiest case, since equation (3) has two solutions, for $x > 0$, of the forms

$$y_1(x) = x^{r_1}(c_0 + c_1 x + c_2 x^2 + \cdots), \qquad c_0 = 1,$$
$$y_2(x) = x^{r_2}(c_0^* + c_1^* x + c_2^* x^2 + \cdots), \qquad c_0^* = 1.$$

That y_1 and y_2 are linearly independent follows easily from the fact that y_1/y_2 cannot be constant, since if it were, the roots r_1 and r_2 would coincide. The coefficients $c_1, c_2, \ldots$ are obtained by setting the coefficients of each power of x equal to zero in equation (5). We find $c_1^*, c_2^*, \ldots$ in a similar manner. The procedure is demonstrated in the following two examples.

Example 1 Solve the Euler equation

$$y'' + \frac{1}{4x} y' + \frac{1}{8x^2} y = 0, \qquad x > 0. \quad (7)$$

Solution We substitute

$$y = x^r(c_0 + c_1 x + c_2 x^2 + \cdots) = x^r \sum_{n=0}^{\infty} c_n x^n = \sum_{n=0}^{\infty} c_n x^{r+n}$$

into equation (7) to obtain

$$\sum_{n=0}^{\infty} (r+n)(r+n-1) c_n x^{r+n-2} + \frac{1}{4x} \sum_{n=0}^{\infty} (r+n) c_n x^{r+n-1} + \frac{1}{8x^2} \sum_{n=0}^{\infty} c_n x^{r+n}$$

$$= \sum_{n=0}^{\infty} (r+n)(r+n-1) c_n x^{r+n-2} + \sum_{n=0}^{\infty} \tfrac{1}{4}(r+n) c_n x^{r+n-2}$$

$$+ \sum_{n=0}^{\infty} \tfrac{1}{8} c_n x^{r+n-2}$$

$$= \sum_{n=0}^{\infty} c_n \left[(r+n)(r+n-1) + \tfrac{1}{4}(r+n) + \tfrac{1}{8} \right] x^{r+n-2} = 0. \tag{8}$$

The indicial equation is obtained by setting the expression in brackets, for $n = 0$, equal to zero.

$$r(r-1) + \tfrac{1}{4}r + \tfrac{1}{8} = r^2 - \tfrac{3}{4}r + \tfrac{1}{8} = (r - \tfrac{1}{2})(r - \tfrac{1}{4}) = 0,$$

with roots $r = \tfrac{1}{4}$ and $r = \tfrac{1}{2}$ that do not differ by an integer. Assume a solution of the form

$$y = x^{1/4}(c_0 + c_1 x + c_2 x^2 + \cdots);$$

then equation (8) becomes (since $r = \tfrac{1}{4}$)

$$\sum_{n=0}^{\infty} c_n \left[(\tfrac{1}{4} + n)(\tfrac{1}{4} + n - 1) + (\tfrac{1}{4} + n)\tfrac{1}{4} + \tfrac{1}{8} \right] x^{(1/4)+n-2} = 0,$$

or, after a bit of algebra,

$$\sum_{n=0}^{\infty} c_n \left[n^2 - \frac{n}{4} \right] x^{n-7/4} = 0.$$

Equating *all* terms of this series to zero, we get $n(n - \tfrac{1}{4}) c_n = 0$, which holds only if $c_n = 0$, for $n > 0$. Thus $y_1(x) = c_0 x^{1/4} = x^{1/4}$.

To find the second solution, we set $r = \tfrac{1}{2}$ in equation (8), obtaining

$$\sum_{n=0}^{\infty} c_n^* \left[(\tfrac{1}{2} + n)(\tfrac{1}{2} + n - 1) + (\tfrac{1}{2} + n)\tfrac{1}{4} + \tfrac{1}{8} \right] x^{(1/2)+n-2} = 0,$$

from which we get $n(n + \tfrac{1}{4}) c_n^* = 0$. Thus $c_n^* = 0$ for $n > 0$, so $y_2(x) = c_0^* \sqrt{x}$. Hence the general solution to equation (7) is

$$y = Ax^{1/4} + Bx^{1/2}.$$

The roots of the indicial equation may also be complex, as the following example illustrates.

Example 2 Find the general solution of the equation

$$y'' + \frac{1}{x} y' + \frac{1}{x^2} y = 0, \qquad x > 0. \tag{9}$$

Solution Substitute $y = \sum_{n=0}^{\infty} c_n x^{r+n}$ into (9) to obtain

$$\sum_{n=0}^{\infty} c_n [(r+n)(r+n-1) + (r+n) + 1] x^{r+n-2} = 0. \qquad (10)$$

The indicial equation (obtained by setting $n = 0$ in the expression in brackets) is

$$r(r-1) + r + 1 = r^2 + 1 = 0,$$

with the roots $r_1 = i$, $r_2 = -i$, which do not differ by an integer. Setting $r = i$ in equation (10), we have

$$\sum_{n=0}^{\infty} c_n [(i+n)(i+n-1) + (i+n) + 1] x^{i+n-2} = 0.$$

Equating all coefficients of this series to zero, we have, after combining terms, and using the fact that $(i+n)^2 = i^2 + 2in + n^2 = n^2 + 2in - 1$,

$$0 = c_n \left[(i+n)^2 + 1 \right] = c_n (n^2 + 2in) = c_n n(n + 2i),$$

which holds only if $c_n = 0$ for $n > 0$. Note that

$$e^{\ln u} = u$$

and that

$$e^{iu} = \cos u + i \sin u,$$

for any $u > 0$. This suggests that

$$x^i = e^{\ln x^i} = e^{i \ln x} = \cos \ln x + i \sin \ln x.$$

Thus, setting $c_0 = 1$ (any constant gives us a solution),

$$y_1(x) = x^i = e^{i(\ln x)} = [\cos(\ln x) + i \sin(\ln x)].$$

Similarly, substituting $r = -i$ into equation (10) yields the series

$$\sum_{n=0}^{\infty} c_n^* [(-i+n)(-i+n-1) + (-i+n) + 1] x^{-i+n-2} = 0,$$

whose coefficients satisfy the condition

$$c_n^* [n^2 - 2in] = 0.$$

Thus $c_n^* = 0$ for $n > 0$. Hence, setting $c_0^* = 1$,

$$y_2(x) = x^{-i} = [\cos(\ln x) - i \sin(\ln x)].$$

Finally, since linear combinations of solutions are solutions, the real and imaginary parts of y_1 and y_2,

$$y_1^*(x) = \frac{1}{2}(y_1 + y_2) = \cos(\ln x),$$

$$y_2^*(x) = \frac{1}{2i}(y_1 - y_2) = \sin(\ln x)$$

are solutions of equation (9). That y_1^* and y_2^* are linearly independent follows since $y_2^*/y_1^* = \tan(\ln x)$, which is nonconstant. Hence equation (9) has the general solution (see Example 5.2.1(iii), p. 168)

$$y = A \cos(\ln x) + B \sin(\ln x), \qquad x > 0.$$

Case 2: $r_1 = r_2$ Here we set $r = r_1$ and determine the coefficients $c_1, c_2, \ldots$ as in case 1. We can then use the method of reduction of order to find the second linearly independent solution, since one solution is known. Consider the following example.

Example 3 Use the method of Frobenius to find the general solution of
$$x^2 y'' - xy' + y = 0, \qquad x > 0.$$

Solution Rewriting this equation in the form
$$y'' - \frac{1}{x} y' + \frac{1}{x^2} y = 0$$
and substituting $y = x^r(c_0 + c_1 x + c_2 x^2 + \cdots)$, we have
$$\sum_{n=0}^{\infty} c_n [(r+n)(r+n-1) - (r+n) + 1] x^{r+n-2} = 0. \tag{11}$$

Setting $n = 0$, we obtain the indicial equation
$$r(r-1) - r + 1 = r^2 - 2r + 1 = (r-1)^2 = 0,$$
with the double root $r = 1$. Then equation (11) yields (with $r = 1$)
$$\sum_{n=0}^{\infty} c_n [(n+1)n - (n+1) + 1] x^{n-1} = \sum_{n=0}^{\infty} n^2 c_n x^{n-1} = 0.$$

Hence $c_0 = 1$, $c_n = 0$ for $n > 0$, and one solution is $y_1 = x$. The second linearly independent solution, $y_2 = x \ln |x|$, is found as in Example 5.2.1(ii).

Example 4 Solve the equation
$$y'' + y' + \frac{1}{4x^2} y = 0, \qquad x > 0. \tag{12}$$

Solution If we substitute $y = \sum_{n=0}^{\infty} c_n x^{r+n}$ into (12) we obtain
$$\sum_{n=0}^{\infty} c_n (r+n)(r+n-1) x^{r+n-2}$$
$$+ \sum_{n=0}^{\infty} c_n (r+n) x^{r+n-1} + \sum_{n=0}^{\infty} \tfrac{1}{4} c_n x^{r+n-2} = 0.$$

We need to express all sums in terms of the same power of x. If $k = n + 1$, then $n = k - 1$ and $x^{r+n-1} = x^{r+k-2}$. Thus, with $k = n + 1$, the second sum above can be written
$$\sum_{k=1}^{\infty} c_{k-1}(r+k-1) x^{r+k-2} \overset{\text{setting } n=k}{=} \sum_{n=1}^{\infty} c_{n-1}(r+n-1) x^{r+n-2},$$
and we have (taking out the $n = 0$ terms from the first and third sums)
$$c_0 [r(r-1) + \tfrac{1}{4}] x^{r-2}$$
$$+ \sum_{n=1}^{\infty} \{ c_n [(r+n)(r+n-1) + \tfrac{1}{4}] + c_{n-1}(r+n-1) \} x^{r+n-2} = 0. \tag{13}$$

The indicial equation is
$$r(r-1) + \tfrac{1}{4} = r^2 - r + \tfrac{1}{4} = \left(r - \tfrac{1}{2}\right)^2 = 0,$$
which has the double root $r = \tfrac{1}{2}$. Substitute $r = \tfrac{1}{2}$ into (13) to obtain
$$\sum_{n=1}^{\infty} \left\{ c_n\left[\left(n + \tfrac{1}{2}\right)\left(n - \tfrac{1}{2}\right) + \tfrac{1}{4}\right] + c_{n-1}\left(n - \tfrac{1}{2}\right)\right\} x^{n-3/2}$$
$$= \sum_{n=1}^{\infty} \left[n^2 c_n + \left(n - \tfrac{1}{2}\right) c_{n-1}\right] x^{n-3/2} = 0.$$

This leads to the recurrence equation
$$c_n = -\frac{\left(n - \tfrac{1}{2}\right) c_{n-1}}{n^2}, \quad \text{for } n \geq 1.$$

Hence
$$c_1 = -\frac{c_0}{2}, \quad c_2 = -\frac{c_1\left(\tfrac{3}{2}\right)}{2^2} = \frac{3 c_0}{2^2 \cdot 2^2}, \quad c_3 = -\frac{5 c_2}{2 \cdot 3^2} = -\frac{3 \cdot 5 c_0}{2^3 \cdot 2^2 \cdot 3^2},$$
$$c_4 = -\frac{7 c_3}{2 \cdot 4^2} = \frac{3 \cdot 5 \cdot 7 c_0}{2^4 \cdot 2^2 \cdot 3^2 \cdot 4^2}, \ldots,$$

so that
$$y_1(x) = x^{1/2}\left(c_0 - \frac{c_0}{2} x + \frac{3 c_0}{2^2 \cdot 2^2} x^2 - \frac{3 \cdot 5 c_0}{2^3 \cdot 2^2 \cdot 3^2} x^3 + \frac{3 \cdot 5 \cdot 7 c_0}{2^4 \cdot 2^2 \cdot 3^2 \cdot 4^2} x^4 - \cdots\right)$$
$$= c_0 x^{1/2}\left[1 - \left(\frac{x}{2}\right) + \frac{3}{2^2}\left(\frac{x}{2}\right)^2 - \frac{3 \cdot 5}{2^2 \cdot 3^2}\left(\frac{x}{2}\right)^3 + \frac{3 \cdot 5 \cdot 7}{2^2 \cdot 3^2 \cdot 4^2}\left(\frac{x}{2}\right)^4 - \cdots\right]$$
$$\overset{c_0 = 1}{=} x^{1/2} \sum_{n=0}^{\infty} \frac{(2n)!}{(n!)^3} \left(\frac{-x}{4}\right)^n, \quad x > 0.$$

Without worrying whether this series converges, we can use the method of reduction of order to produce the second linearly independent solution y_2. Recall that to find y_2 we set $y_2 = v y_1$, where y_1 is the solution above. Hence
$$y_2'' + y_2' + \frac{1}{4x^2} y_2 = v'' y_1 + v'(2 y_1' + y_1) + v\left(y_1'' + y_1' + \frac{1}{4x^2} y_1\right)$$
$$= v'' y_1 + v'(2 y_1' + y_1) = 0$$
because y_1 satisfies the differential equation. Thus
$$\frac{v''}{v'} = -2 \frac{y_1'}{y_1} - 1 = \frac{-2\left(\frac{1}{2\sqrt{x}}\right)\left[1 - 3\left(\frac{x}{2}\right) + \frac{3 \cdot 5}{2^2}\left(\frac{x}{2}\right)^2 - \cdots\right]}{\sqrt{x}\left[1 - \left(\frac{x}{2}\right) + \frac{3}{2^2}\left(\frac{x}{2}\right)^2 + \cdots\right]} - 1.$$

After finding the first few terms by long division (carried out exactly as in the division of one polynomial by another), we have
$$\frac{v''}{v'} = \frac{-1}{x}\left(1 - x + \frac{x^2}{4} - \cdots\right) - 1 = \frac{-1}{x} - \frac{x}{4} + \cdots. \tag{14}$$

Integrating both sides of equation (14), we obtain

$$\ln v' = -\ln x - \frac{x^2}{8} + \cdots,$$

or

$$v' = \frac{1}{x} \exp\left(-\frac{x^2}{8} + \cdots\right)$$

$$= \frac{1}{x}\left[1 + \left(\frac{-x^2}{8} + \cdots\right) + \frac{1}{2!}\left(\frac{-x^2}{8} + \cdots\right)^2 + \cdots\right].$$

After expanding the exponential as a power series in x, we integrate once more and find that v has the form

$$v = \ln x - \frac{x^2}{16} + \cdots.$$

Then

$$y_2 = vy_1 = \left(\ln x - \frac{x^2}{16} + \cdots\right) \cdot \sqrt{x}\left[1 - \left(\frac{x}{2}\right) + \frac{3}{2^2}\left(\frac{x}{2}\right)^2 - \cdots\right]$$

$$= (\ln x) y_1 + \sqrt{x}\left(-\frac{x^2}{16} + \cdots\right),$$

and the general solution of equation (12) has the form

$$y(x) = \sqrt{x}\left\{(A + B \ln x)\left[1 - \left(\frac{x}{2}\right) + \cdots\right] + B\left(-\frac{x^2}{16} + \cdots\right)\right\}, \qquad x > 0.$$

Indeed, it is always true, when $r_1 = r_2$, that the general solution has the form

$$y(x) = x^r\left(\sum_{n=0}^{\infty} c_n x^n + \ln x \sum_{n=0}^{\infty} c_n^* x^n\right), \qquad x > 0.$$

Case 3: r_1 and r_2 differ by a nonzero integer Suppose that $r_1 > r_2$. Then one solution of equation (3) has the form

$$y_1 = x^{r_1}(c_0 + c_1 x + c_2 x^2 + \cdots), \qquad c_0 = 1, \quad x > 0,$$

as in case 1. In some instances it is not possible to determine y_2 as was done in case 1, because the procedure regenerates the same series expansion we obtained for y_1 (in this case, the first $r_1 - r_2$ coefficients c_n^* vanish). When this occurs, we proceed as in case 2. These two possibilities are illustrated in the following two examples.

Example 5 Solve the **Bessel equation of order one-half** (see Section 5.4):

$$x^2 y'' + xy' + \left[x^2 - \left(\tfrac{1}{2}\right)^2\right] y = 0. \qquad (15)$$

Solution We divide the Bessel equation by x^2:

$$y'' + \frac{1}{x} y' + \left(1 - \frac{1}{4x^2}\right) y = 0.$$

Then, if $y = \sum_{n=0}^{\infty} c_n x^{r+n}$, we obtain

$$\sum_{n=0}^{\infty} c_n(r+n)(r+n-1)x^{r+n-2} + \sum_{n=0}^{\infty} c_n(r+n)x^{r+n-2}$$

$$+ \sum_{n=0}^{\infty} \left(-\tfrac{1}{4}\right) c_n x^{r+n-2} + \sum_{n=0}^{\infty} c_n x^{r+n} = 0.$$

The last sum can be written as $\sum_{n=2}^{\infty} c_{n-2} x^{r+n-2}$ (let $k = n + 2$) and we have, after taking out the $n = 0$ and $n = 1$ terms from the first three sums,

$$c_0 \left[r(r-1) + r - \tfrac{1}{4}\right] x^{r-2} + c_1 \left[r(r+1) + (r+1) - \tfrac{1}{4}\right] x^{r-1}$$

$$+ \sum_{n=2}^{\infty} \left\{c_n\left[(r+n)(r+n-1) + (r+n) - \tfrac{1}{4}\right] + c_{n-2}\right\} x^{r+n-2} = 0. \quad (16)$$

The indicial equation is

$$r(r-1) + r - \tfrac{1}{4} = r^2 - \tfrac{1}{4} = \left(r - \tfrac{1}{2}\right)\left(r + \tfrac{1}{2}\right) = 0.$$

Here the roots differ by the integer 1. Substituting $r = \tfrac{1}{2}$ in (16), we get

$$2c_1 x^{-1/2} + \sum_{n=2}^{\infty} \left[c_n(n^2 + n) + c_{n-2}\right] x^{n-3/2} = 0.$$

Thus $c_1 = 0$ and

$$c_n = -\frac{c_{n-2}}{n(n+1)}, \quad \text{for } n \geq 2.$$

Clearly all the coefficients with odd-numbered subscripts are zero, and

$$c_2 = -\frac{c_0}{3!}, \quad c_4 = -\frac{c_2}{4 \cdot 5} = \frac{c_0}{5!}, \quad c_6 = -\frac{c_4}{6 \cdot 7} = -\frac{c_0}{7!}, \ldots.$$

We thus have the solution

$$y_1(x) = \sqrt{x}\left(c_0 - \frac{c_0}{3!}x^2 + \frac{c_0}{5!}x^4 - \frac{c_0}{7!}x^6 + \cdots\right)$$

$$= c_0\sqrt{x}\left(1 - \frac{x^2}{3!} + \frac{x^4}{5!} - \frac{x^6}{7!} + \cdots\right)$$

$$= \frac{1}{\sqrt{x}}\left(x - \frac{x^3}{3!} + \frac{x^5}{5!} - \frac{x^7}{7!} + \cdots\right) = \frac{\sin x}{\sqrt{x}},$$

since $c_0 = 1$. Now setting $r = -\tfrac{1}{2}$ in equation (16), we obtain (omitting some details)

$$\sum_{n=2}^{\infty} \left[c_n(n^2 - n) + c_{n-2}\right] x^{n-5/2} = 0,$$

from which we get the recurrence relation

$$c_n = -\frac{c_{n-2}}{n(n-1)}.$$

Hence

$$c_2 = -\frac{c_0}{2!}, \quad c_4 = -\frac{c_2}{3 \cdot 4} = \frac{c_0}{4!}, \quad c_6 = -\frac{c_4}{5 \cdot 6} = -\frac{c_0}{6!}, \ldots,$$

$$c_3 = -\frac{c_1}{3!}, \quad c_5 = -\frac{c_3}{4 \cdot 5} = \frac{c_1}{5!}, \quad c_7 = -\frac{c_5}{6 \cdot 7} = -\frac{c_1}{7!}, \ldots,$$

and
$$y_2(x) = x^{-1/2}\left(c_0 + c_1 x - c_0 \frac{x^2}{2!} - c_1 \frac{x^3}{3!} + c_0 \frac{x^4}{4!} + c_1 \frac{x^5}{5!} - \cdots\right)$$
$$= \frac{1}{\sqrt{x}}(c_0 \cos x + c_1 \sin x), \quad c_0 = 1.$$

Since the last term of y_2 is a multiple of y_1 and we are looking for linearly independent solutions, we may set $c_1 = 0$ to obtain
$$y_2 = \frac{\cos x}{\sqrt{x}}, \quad x > 0.$$

That y_1 and y_2 are linearly independent follows from the fact that $y_1/y_2 = \tan x$, which is nonconstant. Thus the general solution of equation (15) is
$$y = \frac{A}{\sqrt{x}} \cos x + \frac{B}{\sqrt{x}} \sin x, \quad x > 0.$$

Example 6 Solve the **Bessel equation of order one**:
$$x^2 y'' + xy' + (x^2 - 1)y = 0.$$

Solution Again, we first divide by x^2:
$$y'' + \frac{1}{x} y' + \left(1 - \frac{1}{x^2}\right) y = 0. \tag{17}$$

Substituting $y = \sum_{n=0}^{\infty} c_n x^{r+n}$ into (17) and changing the index of summation as in Example 5, we obtain
$$c_0[r(r-1) + r - 1]x^{r-2} + c_1[r(r+1) + (r+1) - 1]x^{r-1}$$
$$+ \sum_{n=2}^{\infty} \{c_n[(r+n)(r+n-1) + (r+n) - 1] + c_{n-2}\} x^{r+n-2} = 0. \tag{18}$$

The indicial equation is
$$r(r-1) + r - 1 = r^2 - 1 = 0,$$
with roots $r_1 = 1$ and $r_2 = -1$, which differ by the integer 2. Setting $r = 1$ in (18) leads to
$$3c_1 + \sum_{n=2}^{\infty} \left[c_n(n^2 + 2n) + c_{n-2}\right] x^{n-1} = 0,$$
so $c_1 = 0$ and
$$c_n = -\frac{c_{n-2}}{n(n+2)}, \quad \text{for } n \geq 2.$$

Thus all the coefficients with odd-numbered subscripts are zero, and
$$c_2 = -\frac{c_0}{2 \cdot 4}, \quad c_4 = -\frac{c_2}{4 \cdot 6} = \frac{c_0}{2 \cdot 4^2 \cdot 6}, \quad c_6 = -\frac{c_4}{6 \cdot 8} = -\frac{c_0}{2 \cdot 4^2 \cdot 6^2 \cdot 8}, \ldots.$$

Hence, since $c_0 = 1$,

$$y_1(x) = x\left(c_0 - \frac{c_0}{2\cdot 4}x^2 + \frac{c_0}{2\cdot 4^2\cdot 6}x^4 - \frac{c_0}{2\cdot 4^2\cdot 6^2\cdot 8}x^6 + \cdots\right)$$

$$= x\left(1 - \frac{1}{1!2!}\left(\frac{x}{2}\right)^2 + \frac{1}{2!3!}\left(\frac{x}{2}\right)^4 - \frac{1}{3!4!}\left(\frac{x}{2}\right)^6 + \cdots\right).$$

Now we can see that the method of Frobenius does not work for the second root $r_2 = -1$. Setting $r = -1$ in equation (18), we obtain

$$-c_1 x^{-2} + \sum_{n=2}^{\infty}\left[c_n(n^2 - 2n) + c_{n-2}\right]x^{n-3} = 0,$$

and so $c_1 = 0$ and $n(n-2)c_n = c_{n-2}$. Setting $n = 2$, we see that $c_0 = 0$, contradicting the assumption that $c_0 = 1$ (see p. 175). Note that all the coefficients are zero, so the method fails for the second root. However, we do have *one* solution, indicating that the other solution can be obtained by the method of Example 4:

$$\frac{v''}{v'} = -\frac{2y_1'}{y_1} - \frac{1}{x} = -\frac{3}{x} + \frac{x}{2} + \cdots.$$

Integrating both ends of this equation, we have

$$\ln v' = -3\ln x + \frac{x^2}{4} + \cdots,$$

or

$$v' = x^{-3}\exp\left(\frac{x^2}{4} + \cdots\right) = x^{-3} + \frac{1}{4}x^{-1} + \cdots.$$

Integrating once more, we get

$$v = -\tfrac{1}{2}x^{-2} + \tfrac{1}{4}\ln x + \cdots,$$

so

$$y_2 = vy_1 = \frac{1}{4}y_1 \ln x - \frac{1}{2}x^{-1} + \frac{x}{16} + \cdots, \qquad x > 0.$$

In general, if the method of case 1 fails for r_2, the procedure above yields the solution

$$y_2 = k_{-1}(\ln x)y_1 + x^{r_2}(k_0 + k_1 x + \cdots), \qquad x > 0,$$

where k_{-1} may equal zero.

We gather all the facts we have proved in this section in one theorem:

THEOREM 2

Let $x = 0$ be a regular singular point of the differential equation

$$y'' + a(x)y' + b(x)y = 0, \qquad x \neq 0, \qquad (19)$$

and let r_1 and r_2 be the roots of the indicial equation

$$r(r-1) + a_0 r + b_0 = 0,$$

where a_0 and b_0 are given by the power series expansions
$$xa(x) = a_0 + a_1 x + a_2 x^2 + \cdots,$$
$$x^2 b(x) = b_0 + b_1 x + b_2 x^2 + \cdots.$$

Then equation (19) has two linearly independent solutions y_1 and y_2 whose form depends on r_1 and r_2 as follows:

Case 1 If r_1 and r_2 differ but not by an integer, then
$$y_1(x) = |x|^{r_1}\left(\sum_{n=0}^{\infty} c_n x^n\right), \qquad c_0 = 1,$$
$$y_2(x) = |x|^{r_2}\left(\sum_{n=0}^{\infty} c_n^* x^n\right), \qquad c_0^* = 1.$$

(The absolute-value signs are needed to avoid the assumption that $x > 0$.)

Case 2 If $r_1 = r_2 = r$, then
$$y_1(x) = |x|^{r}\left(\sum_{n=0}^{\infty} c_n x^n\right), \qquad c_0 = 1,$$
$$y_2(x) = |x|^{r}\left(\sum_{n=1}^{\infty} c_n^* x^n\right) + y_1(x) \ln|x|.$$

Case 3 If $r_1 - r_2$ is a positive integer, then
$$y_1(x) = |x|^{r_1}\left(\sum_{n=0}^{\infty} c_n x^n\right), \qquad c_0 = 1,$$
$$y_2(x) = |x|^{r_2}\left(\sum_{n=0}^{\infty} c_n^* x^n\right) + c_{-1}^* y_1(x)\ln|x|, \qquad c_0^* = 1,$$

and c_{-1}^* may equal zero.

Furthermore, if the power series expansions for $xa(x)$ and $x^2 b(x)$ are valid for $|x| < R$, then the solutions y_1 and y_2 are valid for $0 < |x| < R$. The proof of this fact is left as an exercise for the reader (see Problems 29 and 30).

PROBLEMS 5.3

In Problems 1–23 find the general solution to the given differential equation by the method of Frobenius at $x = 0$.

1. $y'' + \dfrac{1}{2x} y' + \dfrac{1}{4x} y = 0$

2. $y'' + \dfrac{2(1 - 2x)}{x(1-x)} y' - \dfrac{2}{x(1-x)} y = 0$

3. $y'' + \dfrac{6}{x} y' + \left(\dfrac{6}{x^2} - 1\right) y = 0$

4. $y'' + \dfrac{4}{x} y' + \left(1 + \dfrac{2}{x^2}\right) y = 0$

5. $y'' + \dfrac{3}{x} y' + 4x^2 y = 0$

6. $(x - 1)y'' - \left(\dfrac{4x^2 - 3x + 1}{2x}\right) y' + \left(\dfrac{2x^2 - x + 2}{2x}\right) y = 0$

7. $y'' + \dfrac{2}{x} y' - \dfrac{2}{x^2} y = 0$

8. $4xy'' + 2y' + y = 0$

9. $xy'' + 2y' + xy = 0$

10. $xy'' - y' + 4x^3 y = 0$
11. $xy'' + (1 - 2x) y' - (1 - x) y = 0$
12. $x(x + 1)^2 y'' + (1 - x^2) y' - (1 - x) y = 0$
13. $y'' - 2y' + \left(1 + \dfrac{1}{4x^2}\right) y = 0$
14. $x^2 y'' + Axy' + By = 0$, A and B constants
15. $x(x - 1) y'' - (1 - 3x) y' + y = 0$
16. $x^2(x^2 - 1) y'' - x(x^2 + 1) y' + (x^2 + 1) y = 0$
17. $y'' + \dfrac{y'}{x} - y = 0$
18. $2xy'' - (x - 3) y' - y = 0$
19. $y'' + \dfrac{x + 1}{2x} y' + \dfrac{3}{2x} y = 0$
20. $y'' + \dfrac{1}{2x} y' - \dfrac{x + 1}{2x^2} y = 0$
21. $x^2 y'' + x(x - 1) y' - (x - 1) y = 0$
22. $xy'' - (3 + x) y' + 2y = 0$
23. $y'' + \dfrac{1}{4x^2} y = 0$

24. Prove that the second linearly independent solution of the equation
$$y'' + \dfrac{1}{x} y' + y = 0$$
has the form
$$y_2 = y_1 \ln x$$
$$+ \sum_{n=1}^{\infty} \dfrac{(-1)^{n+1}}{2^{2n}(n!)^2} \left(1 + \dfrac{1}{2} + \cdots + \dfrac{1}{n}\right) x^{2n}, \quad x > 0.$$

25. Consider the differential equation
$$y'' - \dfrac{1}{x^2} y' + \dfrac{1}{x^3} y = 0.$$

   a. Show that $x = 0$ is an irregular singular point of this equation.

   b. Use the fact that $y_1 = x$ is a solution to find a second independent solution.

   c. Show that the solution y_2 cannot be expressed as a series of form (4). Thus this solution cannot be found by the method of Frobenius.

26. The differential equation
$$x^2 y'' + (4x - 1) y' + 2y = 0$$
has $x = 0$ as an irregular singular point.

   a. Suppose that equation (4) is inserted into this equation. Show that $r = 0$ and that the corresponding "solution" by the method of Frobenius is
$$y = \sum_{n=0}^{\infty} (n + 1)! x^n.$$

   *b. Prove that the series in (a) has radius of convergence $R = 0$. Hence, even though a Frobenius series may formally satisfy a differential equation, it may not be a valid solution at an irregular singular point.

*27. If $r_1 = r_2$, verify that the technique in Example 4 always yields
$$v(x) = \ln x - (a_1 + 2c_1) x + \cdots,$$
where
$$xa(x) = a_0 + a_1 x + a_2 x^2 + a_3 x^3 + \cdots$$
and $y_2 = v y_1$ with
$$y_1 = x^r (c_0 + c_1 x + c_2 x^2 + \cdots), \quad c_0 = 1.$$

28. Show that the method of Frobenius fails for the equation
$$x^4 y'' + 2x^3 y' - y = 0, \quad x > 0,$$
which has $x = 0$ as an irregular singular point. Find a solution by assuming
$$y = \sum_{n=0}^{\infty} c_n x^{-n}.$$

*29. Let $x_0 = 0$ and suppose that the Maclaurin series
$$a(x) = \sum_{n=0}^{\infty} a_n x^n, \quad b(x) = \sum_{n=0}^{\infty} b_n x^n$$
are valid for $|x| < R$.

   a. Show that if equation (3) has a power series solution
$$y(x) = \sum_{n=0}^{\infty} c_n x^n, \quad c_0 = 1,$$
then the coefficients c_n satisfy the recursion formula
$$(n + 1)(n + 2) c_{n+2}$$
$$= - \sum_{k=0}^{n} [(k + 1) a_{n-k} c_{k+1} + b_{n-k} c_k].$$

   b. Use the root test formula for finding the radius of convergence to show that for any r such that $0 < r < R$ there is a constant $M > 0$ such that
$$(n + 1)(n + 2) |c_{n+2}|$$
$$\leq \dfrac{M}{r^n} \sum_{k=0}^{\infty} [(k + 1) |c_{k+1}| + |c_k|] r^k$$
$$+ M |c_{n+1}| r.$$

c. Use the ratio test to prove that the series

$$\sum_{n=0}^{\infty} |c_n| x^n$$

converges for $|x| < r$. Then by the comparison test for series, it follows that the series $y(x)$ converges for $|x| < r$. Since r is arbitrary, it follows that the series representation of the solution is valid for $|x| < R$.

*30. Modify the argument in Problem 29 to justify the last statement of Theorem 2.

5.4 Bessel Functions

The differential equation

$$x^2 y'' + x y' + (x^2 - p^2) y = 0, \qquad (1)$$

which is known as the **Bessel equation of order** p (≥ 0), is one of the most important differential equations in applied mathematics. The equation was first investigated in 1703 by Jakob Bernoulli (see p. 50) in connection with the oscillatory behavior of a hanging chain, and later by the German mathematician Friedrich Wilhelm Bessel (1784–1846) in his studies of planetary motion. Since then, the Bessel functions have been used in the studies of elasticity, fluid motion, potential theory, diffusion, and the propagation of waves. We will present a few applications of the Bessel functions in the last chapter when we study the solutions of partial differential equations.

Recall that in Section 5.3 we found the solution of the Bessel equation for $p = \frac{1}{2}$ and 1 (see Examples 5.3.5 and 5.3.6). In each of these examples, the method of Frobenius was an important tool, so we again anticipate the successful application of this procedure. By Theorem 5.3.2, a solution of the form

$$y(x) = \sum_{n=0}^{\infty} c_n x^{r+n}, \qquad x \neq 0, \qquad c_0 = 1, \qquad (2)$$

exists for the Bessel equation of order p. We divide equation (1) by x^2:

$$y'' + \frac{1}{x} y' + \left(1 - \frac{p^2}{x^2}\right) y = 0$$

and then substitute (2) into this equation to obtain (after setting $k = n + 2$ in $\sum_{n=0}^{\infty} c_n x^{r+n}$)

$$\sum_{n=0}^{\infty} c_n (r+n)(r+n-1) x^{r+n-2} + \sum_{n=0}^{\infty} c_n (r+n) x^{r+n-2}$$

$$+ \sum_{n=0}^{\infty} -p^2 c_n x^{r+n-2} + \sum_{n=2}^{\infty} c_{n-2} x^{r+n-2} = 0,$$

or

$$c_0 (r^2 - p^2) x^{r-2} + c_1 \left[(r+1)^2 - p^2 \right] x^{r-1}$$

$$+ \sum_{n=2}^{\infty} \left\{ c_n \left[(n+r)^2 - p^2 \right] + c_{n-2} \right\} x^{r+n-2} = 0. \qquad (3)$$

The indicial equation is $r^2 - p^2 = 0$, with roots $r_1 = p$ (≥ 0) and $r_2 = -p$. Setting $r = p$ in (3) yields

$$(1 + 2p) c_1 x^{p-1} + \sum_{n=2}^{\infty} \left[n(n+2p) c_n + c_{n-2} \right] x^{n+p-2} = 0,$$

indicating that $c_1 = 0$ and that

$$c_n = -\frac{c_{n-2}}{n(n+2p)}, \quad \text{for } n \geq 2. \tag{4}$$

Hence all the coefficients with odd-numbered subscripts c_{2j+1} are zero, since by equation (4) they can all be expressed as a multiple of c_1. Letting $n = 2j + 2$, we see that the coefficients with even-numbered subscripts satisfy the equation

$$c_{2(j+1)} = -\frac{c_{2j}}{2^2(j+1)(p+j+1)}, \quad \text{for } j \geq 0,$$

which yields

$$c_2 = -\frac{c_0}{2^2(p+1)},$$

$$c_4 = -\frac{c_2}{2^2 \cdot 2(p+2)} = \frac{c_0}{2^4 2!(p+1)(p+2)},$$

$$c_6 = -\frac{c_4}{2^2 \cdot 3(p+3)} = -\frac{c_0}{2^6 3!(p+1)(p+2)(p+3)}, \ldots$$

Hence series (2) becomes

$$y_1(x) = |x|^p \left[c_0 - \frac{c_0}{2^2(p+1)} x^2 + \frac{c_0}{2^4 2!(p+1)(p+2)} x^4 - \cdots \right]$$

$$= c_0 |x|^p \sum_{n=0}^{\infty} (-1)^n \frac{x^{2n}}{2^{2n} n!(p+1)(p+2)\cdots(p+n)}. \tag{5}$$

The Gamma Function

To write equation (5) in a more compact form, we define the **gamma function** for all values $p > -1$:

$$\Gamma(p+1) = \int_0^\infty e^{-t} t^p \, dt.$$

Integrating $\Gamma(p+1)$ by parts, we have

$$\Gamma(p+1) = \int_0^\infty e^{-t} t^p \, dt = -e^{-t} t^p \Big|_0^\infty + p \int_0^\infty e^{-t} t^{p-1} \, dt.$$

The first expression on the right is zero, and the integral on the right-hand side is $\Gamma(p)$. We thus have the basic property of gamma functions:

$$\Gamma(p+1) = p\Gamma(p).$$

Since

$$\Gamma(1) = \int_0^\infty e^{-t} \, dt = -e^{-t} \Big|_0^\infty = 1,$$

it follows that $\Gamma(2) = \Gamma(1) = 1!$, $\Gamma(3) = 2\Gamma(2) = 2!, \ldots,$ and in general, $\Gamma(n+1) = n!$. Thus the gamma function is the extension to real numbers $p > -1$ of the factorial function.

It is customary in equation (5) to let $c_0 = [2^p \Gamma(p+1)]^{-1}$. Then equation (5) becomes

$$J_p(x) = \left|\frac{x}{2}\right|^p \sum_{n=0}^{\infty} (-1)^n \frac{(x/2)^{2n}}{n! \Gamma(p+n+1)}, \quad x \neq 0, \tag{6}$$

which is known as the **Bessel function of the first kind of order** p. Thus $J_p(x)$ is the first solution of equation (1). It can be shown that the series $J_p(x)$ converges for all real x.

To find the second solution, we must consider the difference $r_1 - r_2 = 2p$. By case 1 of Section 5.3, if p is not a multiple of $\frac{1}{2}$, we can again apply the method of Frobenius with $r = -p$ to find the second solution. The results of Examples 5.3.5 and 5.3.6 make it appear likely that we obtain $\ln|x|$ terms only when p is an integer. Therefore we set $r = -p$ in equation (3) and assume that p is not an integer. Then we obtain

$$(1-2p)c_1 x^{-p-1} + \sum_{n=2}^{\infty} \left[n(n-2p)c_n + c_{n-2} \right] x^{n-p-2} = 0, \qquad (7)$$

indicating that $c_1 = 0$ if $p \neq \frac{1}{2}$ and that

$$c_n = -\frac{c_{n-2}}{n(n-2p)}. \qquad (8)$$

Note that when p is not a multiple of $\frac{1}{2}$, all the coefficients c_n with odd-numbered subscripts are zero. If $p = (2m+1)/2$, then c_{2m+1} is arbitrary and the recurrence relation (8) yields the coefficients

$$c_{2m+3} = -\frac{c_{2m+1}}{2^2(p+1)}, \qquad c_{2m+5} = \frac{c_{2m+1}}{2^4 2!(p+1)(p+2)}, \ldots$$

Hence the odd-numbered coefficients generate the series

$$|x|^{-p} \left[c_{2m+1} x^{2m+1} - \frac{c_{2m+1}}{2^2(p+1)} x^{2m+3} + \frac{c_{2m+1} x^{2m+5}}{2^4 2!(p+1)(p+2)} - \cdots \right]$$

$$= c_{2m+1} |x|^p \left[1 - \frac{x^2}{2^2(p+1)} + \frac{x^4}{2^4 2!(p+1)(p+2)} - \cdots \right],$$

which is a multiple of $J_p(x)$. Thus we can ignore the odd-numbered coefficients and concentrate on using equation (7) to calculate the coefficients with even-numbered subscripts. Then

$$c_2 = -\frac{c_0}{2^2(1-p)}, \qquad c_4 = -\frac{c_2}{2^2 \cdot 2(2-p)} = \frac{c_0}{2^4 2!(1-p)(2-p)}, \ldots,$$

and the second solution is the convergent series

$$J_{-p}(x) = \left|\frac{x}{2}\right|^{-p} \sum_{n=0}^{\infty} (-1)^n \frac{(x/2)^{2n}}{n!\Gamma(n-p+1)}. \qquad (9)$$

To see that equations (6) and (9) are linearly independent, we obtain by long division

$$\frac{J_p(x)}{J_{-p}(x)} = \frac{|x/2|^p/\Gamma(p+1) - |x/2|^{p+2}/2!\Gamma(p+3) + \cdots}{|x/2|^{-p}/\Gamma(1-p) - |x/2|^{2-p}/2!\Gamma(3-p) + \cdots}$$

$$= \frac{|x/2|^{2p}}{\Gamma(1+p)/\Gamma(1-p)} + \frac{3p|x/2|^{2p+2}}{\Gamma(3+p)/\Gamma(1-p)} + \cdots,$$

which clearly is not a constant function. We have shown the following:

THEOREM 1

> If p is not an integer, then
> $$y(x) = AJ_p(x) + BJ_{-p}(x)$$
> is the general solution of the Bessel equation for all values $x \neq 0$.

If p is an integer, then the term $(n - 2p)$ in the recurrence relation (8) is zero for the integer $n = 2p$. Hence c_{2p-2} is zero, and iterating equation (8) repeatedly, we see that $c_{2p-2} = c_{2p-4} = \cdots = c_2 = c_0 = 0$. But this contradicts the assumed form (2) of the solution. Thus the method of Frobenius cannot be used when p is a positive integer, and the second linearly independent solution of equation (1) must be calculated by the method in Example 5.3.6 (p. 182). After an extremely long (but straightforward) calculation, we obtain

$$y_2(x) = J_p(x) \ln|x| - \frac{1}{2}\left[\sum_{k=0}^{p-1} \frac{(p-k-1)!}{k!}\left(\frac{x}{2}\right)^{2k-p} + \frac{h_p}{p!}\left(\frac{x}{2}\right)^p \right.$$
$$\left. + \sum_{k=1}^{\infty} \frac{(-1)^k[h_k + h_{p+k}]}{k!(p+k)!}\left(\frac{x}{2}\right)^{2k+p}\right], \quad (10)$$

where

$$h_p = 1 + \frac{1}{2} + \frac{1}{3} + \cdots + \frac{1}{p} \quad (11)$$

and p is a positive integer.

It is customary to replace equation (10) by the linear combination of solutions

$$Y_p(x) = \frac{2}{\pi}\left[y_2(x) + (\gamma - \ln 2)J_p(x)\right], \quad p = 0, 1, 2, \ldots,$$

where

$$\gamma = \lim_{p \to \infty} (h_p - \ln p) = 0.5772156649\ldots$$

is the **Euler constant**. This particular solution is obviously independent of $J_p(x)$ and is called the **Bessel function of the second kind of order p**, or **Neumann's function of order p**.[†] It is defined by the formula

$$Y_p(x) = \frac{2}{\pi}J_p(x)\left(\ln\frac{x}{2} + \gamma\right)$$
$$- \frac{1}{\pi}\left[\sum_{k=0}^{p-1} \frac{(p-k-1)!}{k!}\left(\frac{x}{2}\right)^{2k-p} + \frac{h_p}{p!}\left(\frac{x}{2}\right)^p\right.$$
$$\left. + \sum_{k=1}^{\infty} (-1)^k \frac{[h_k + h_{p+k}]}{k!(p+k)!}\left(\frac{x}{2}\right)^{2k+p}\right],$$

for all integers $p = 0, 1, 2, \ldots$.

[†]Carl Gottfried Neumann (1832–1925) was a German mathematician.

The function Y_p may be extended to all real numbers $p \geq 0$ (see Problem 24) by letting

$$Y_p(x) = \frac{1}{\sin p\pi}\left[J_p(x)\cos p\pi - J_{-p}(x)\right], \qquad p \neq 0,1,2,\ldots.$$

Using this definition of Y_p, we have the following result:

THEOREM 2

> The general solution of the Bessel equation of order p is
> $$y(x) = AJ_p(x) + BY_p(x), \qquad x \neq 0.$$

Graphs of the functions J_0, J_1, J_2, Y_0, Y_1, and Y_2 are given in Figure 5.1.

PROPERTIES OF BESSEL FUNCTIONS

Now that we have the expansions for $J_p(x)$ and $Y_p(x)$, we can derive a number of important expressions involving Bessel functions and their derivatives. For simplicity, we assume $x > 0$. The first two identities are immediate consequences of equation (6):

$$\frac{d}{dx}\left[x^p J_p(x)\right] = x^p J_{p-1}(x), \tag{12}$$

$$\frac{d}{dx}\left[x^{-p} J_p(x)\right] = -x^{-p} J_{p+1}(x). \tag{13}$$

To prove (12), we differentiate the product $x^p J_p$ term by term:

$$\frac{d}{dx}\sum_{n=0}^{\infty}(-1)^n \frac{2^p(x/2)^{2n+2p}}{n!\Gamma(p+n+1)} = \sum_{n=0}^{\infty}(-1)^n \frac{2^{p-1}(x/2)^{2n+2p-1}2(n+p)}{n!\Gamma(p+n+1)}$$

$$= x^p \sum_{n=0}^{\infty}(-1)^n \frac{(x/2)^{2n+p-1}}{n!\Gamma(p+n)} = x^p J_{p-1}(x),$$

since $\Gamma(p+n+1) = (p+n)\Gamma(p+n)$. The proof of equation (13) is similar (see Problem 4). Expanding the left-hand sides of equations (12) and (13), we have

$$x^p J_p' + px^{p-1}J_p = x^p J_{p-1}$$

and

$$x^{-p}J_p' - px^{-p-1}J_p = -x^{-p}J_{p+1},$$

which may be simplified to yield the identities

$$xJ_p' = xJ_{p-1} - pJ_p, \tag{14}$$

$$xJ_p' = pJ_p - xJ_{p+1}. \tag{15}$$

Subtracting equation (15) from equation (14), we obtain the recursion relation

$$xJ_{p+1} - 2pJ_p + xJ_{p-1} = 0. \tag{16}$$

Adding the two together yields

$$2J_p' = J_{p-1} - J_{p+1}. \tag{17}$$

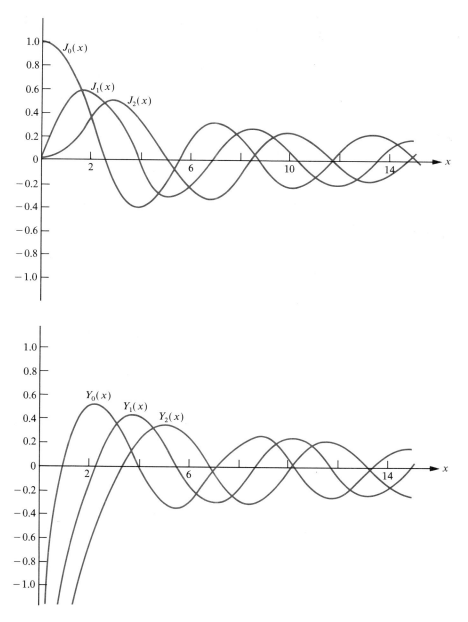

Figure 5.1

Equations (12) through (17) are extremely important in solving problems involving the Bessel functions, since they allow us to express Bessel functions of higher order in terms of lower-order functions.

Example 1 Express $J_3(x)$ in terms of $J_0(x)$ and $J_1(x)$.

Solution We let $p = 2$ in equation (16):
$$xJ_3 = 4J_2 - xJ_1.$$

If we apply (16) with $p = 1$ to J_2, we obtain
$$xJ_2 = 2J_1 - xJ_0.$$

Thus
$$J_3(x) = \frac{4}{x}J_2 - J_1 = \frac{4}{x^2}(2J_1 - xJ_0) - J_1 = \left(\frac{8}{x^2} - 1\right)J_1(x) - \frac{4}{x}J_0(x).$$

Example 2 Evaluate the integral
$$\int x^4 J_1(x)\, dx. \tag{18}$$

Solution Integrating (18) by parts, we have, by equation (12),

$$\int x^2 [x^2 J_1(x)]\, dx \stackrel{(x^2 J_2)' = x^2 J_1}{=} x^2(x^2 J_2) - \int x^2 J_2 \cdot 2x\, dx = x^4 J_2(x) - 2\int x^3 J_2\, dx.$$

Applying equation (12) to the last integral, we obtain
$$\int x^4 J_1(x)\, dx = x^4 J_2(x) - 2x^3 J_3(x) + C.$$

In general, an integral of the form
$$\int x^m J_n(x)\, dx,$$

where m and n are integers such that $m + n \geq 0$, can be completely integrated if $m + n$ is odd. But if $m + n$ is even, then the result depends on the residual integral $\int J_0(x)\, dx$. It is not possible to reduce $\int J_0(x)\, dx$ [or $\int Y_0(x)\, dx$] any further and for this reason the functions

$$\int_0^x J_0(x)\, dx$$

and
$$\int_0^x Y_0(x)\, dx$$

have been tabulated.[†]

Example 3 Express $J_{3/2}(x)$ in terms of $\sin x$ and $\cos x$.

Solution Recall from Example 5.3.5 that the general solution of the Bessel equation of order one-half can be written in terms of sines and cosines. Multiplying $y_1(x)$ and $y_2(x)$ by $c_0 = [\sqrt{2}\,\Gamma(\tfrac{3}{2})]^{-1}$, we have

$$J_{1/2}(x) = \frac{1}{\sqrt{2}\,\Gamma(\tfrac{3}{2})} \frac{\sin x}{\sqrt{x}} = \sqrt{\frac{2}{\pi x}} \sin x$$

[†]Milton Abramowitz and Irene A. Stegun, *Handbook of Mathematical Functions* (New York: Dover, 1965), Table 11.1, pp. 492–493.

and
$$J_{-1/2}(x) = \frac{1}{\sqrt{2}\,\Gamma(\frac{3}{2})} \frac{\cos x}{\sqrt{x}} = \sqrt{\frac{2}{\pi x}} \cos x,$$

since $\Gamma(\frac{3}{2}) = (\frac{1}{2})\Gamma(\frac{1}{2}) = \sqrt{\pi}/2$ (see Appendix 1, Formula 219). By equation (16),

$$J_{3/2}(x) = \frac{1}{x} J_{1/2} - J_{-1/2} = \sqrt{\frac{2}{\pi x}} \left(\frac{\sin x}{x} - \cos x \right).$$

Similar results hold for $Y_p(x)$ (see Problems 5, 6, and 7).

Many differential equations with variable coefficients can be reduced to Bessel equations.

Example 4 Solve the equation
$$y'' + k^2 xy = 0. \tag{19}$$

Solution The following substitution reduces equation (19) to a Bessel equation: Let $u = y/\sqrt{x}$ and $z = 2kx^{3/2}/3$. Then

$$\frac{du}{dz} = \frac{du/dx}{dz/dx} = \frac{y'}{kx} - \frac{y}{2kx^2}$$

and

$$\frac{d^2 u}{dz^2} = \frac{\frac{d}{dx}\left(\frac{du}{dz}\right)}{dz/dx} = \frac{y''}{k^2 x^{3/2}} - \frac{3 y'}{2 k^2 x^{5/2}} + \frac{y}{k^2 x^{7/2}}.$$

Hence
$$z^2 \frac{d^2 u}{dz^2} + z \frac{du}{dz} = \frac{4}{9} x^{3/2} y'' + \frac{1}{9} \frac{y}{\sqrt{x}},$$

and using equation (19) for y'', we obtain

$$z^2 \frac{d^2 u}{dz^2} + z \frac{du}{dz} = -\frac{4}{9} k^2 x^3 \left(\frac{y}{\sqrt{x}} \right) + \frac{1}{9} \frac{y}{\sqrt{x}} = -\left(z^2 - \frac{1}{9} \right) u,$$

or
$$z^2 \frac{d^2 u}{dz^2} + z \frac{du}{dz} + \left(z^2 - \frac{1}{9} \right) u = 0, \tag{20}$$

the Bessel equation of order one-third. Since equation (20) has the general solution
$$u(z) = A J_{1/3}(z) + B J_{-1/3}(z),$$

equation (19) has the solution
$$y(x) = \sqrt{x} \left[A J_{1/3}\left(\frac{2}{3} k x^{3/2} \right) + B J_{-1/3}\left(\frac{2}{3} k x^{3/2} \right) \right].$$

Additional facts concerning the zeros of Bessel functions will be developed in Section 5.6.

PROBLEMS 5.4

1. Express $J_5(x)$ in terms of $J_0(x)$ and $J_1(x)$.
2. Express $J_{5/2}(x)$ in terms of $\sin x$ and $\cos x$.
3. Show that
   a. $4J_p''(x) = J_{p+2}(x) - 2J_p(x) + J_{p-2}(x)$;
   b. $-8J_p'''(x) = J_{p+3}(x) - 3J_{p+1}(x) + 3J_{p-1}(x) - J_{p-3}(x)$.
4. Prove that $[x^{-p}J_p(x)]' = -x^{-p}J_{p+1}(x)$.
5. Show that $[x^p Y_p(x)]' = x^p Y_{p-1}(x)$.
6. Prove that $[x^{-p}Y_p(x)]' = -x^{-p}Y_{p+1}(x)$.
7. Using the equations of Problems 5 and 6, show that
   a. $x(Y_{p+1} + Y_{p-1}) = 2pY_p$;
   b. $2Y_p' = Y_{p-1} - Y_{p+1}$.
8. Prove the following identities:
   a. $\int J_1(x)\,dx = -J_0(x) + C$
   b. $\int x^2 J_1(x)\,dx = 2xJ_1(x) - x^2 J_0(x) + C$
   c. $\int x J_0(x)\,dx = xJ_1(x) + C$
   d. $\int x^3 J_0(x)\,dx = (x^3 - 4x)J_1(x) + 2x^2 J_0(x) + C$
   e. $\int J_0(x)\cos x\,dx = xJ_0(x)\cos x + xJ_1(x)\sin x + C$
   f. $\int J_0(x)\sin x\,dx = xJ_0(x)\sin x - xJ_1(x)\cos x + C$
   g. $\int J_1(x)\cos x\,dx = xJ_1(x)\cos x - (x\sin x + \cos x)J_0(x) + C$
   h. $\int J_1(x)\sin x\,dx = xJ_1(x)\sin x + (x\cos x - \sin x)J_0(x) + C$
9. Verify the following identities:
   a. $\int x^2 J_0(x)\,dx = x^2 J_1(x) + xJ_0(x) - \int J_0(x)\,dx + C$
   b. $\int x^{-1} J_1(x)\,dx = -J_1(x) + \int J_0(x)\,dx + C$
   c. $\int xJ_1(x)\,dx = -xJ_0(x) + \int J_0(x)\,dx + C$
   d. $\int x^3 J_1(x)\,dx = 3x^2 J_1(x) - (x^3 - 3x)J_0(x) - 3\int J_0(x)\,dx + C$
10. Show that $\int J_0(\sqrt{x})\,dx = 2\sqrt{x}\, J_1(\sqrt{x}) + C$.
11. Show that $\int xJ_0(x)\sin x\,dx = \frac{1}{3}\{x^2[J_0(x)\sin x - J_1(x)\cos x] + xJ_1(x)\sin x\} + C$.
12. Show that $\int xJ_1(x)\cos x\,dx = \frac{1}{3}\{x^2[J_1(x)\cos x - J_0(x)\sin x] + 2xJ_1(x)\sin x\} + C$.
13. Show that $\int J_0(x)\,dx = 2[J_1(x) + J_3(x) + J_5(x) + \cdots] + C$.

In Problems 14–21 reduce each given equation to a Bessel equation and solve it.

14. $x^2 y'' + xy' + (a^2 x^2 - p^2)y = 0$
15. $4x^2 y'' + 4xy' + (x^2 - p^2)y = 0$
16. $x^2 y'' + xy' + 4(x^4 - p^2)y = 0$
17. $xy'' - y' + xy = 0$
18. $x^2 y'' + (x^2 + \tfrac{1}{4})y = 0$
19. $xy'' + (1 + 2k)y' + xy = 0$
20. $y'' + k^2 x^2 y = 0$
21. $y'' + k^2 x^4 y = 0$
22. Prove that the second linearly independent solution of the equation
$$y'' + \frac{1}{x}y' + y = 0$$
has the form
$$y_2(x) = J_0(x)\ln x + \sum_{n=1}^{\infty} \frac{(-1)^{n+1}}{2^{2n}(n!)^2}\left(1 + \frac{1}{2} + \cdots + \frac{1}{n}\right)x^{2n},$$
$x > 0$.

23. Obtain equation (10) for $p = 1$ by the methods of this section.

*24. Prove that if $p \geq 0$ is an integer,
$$\lim_{q \to p} \frac{J_q(x)\cos q\pi - J_{-q}(x)}{\sin q\pi} = Y_p(x).$$
This is the extension of Y_p to all real values $p \geq 0$.

25. a. Expand $e^{(x/2)[t-(1/t)]}$ as a power series in t by multiplying the series for $e^{xt/2}$ and $e^{-x/2t}$.
    b. Show that the coefficient of t^n in the expansion obtained in part (a) is $J_n(x)$.
    c. Conclude that
$$e^{(x/2)[t-(1/t)]} = J_0(x) + \sum_{n=1}^{\infty} J_n(x)[t^n + (-t)^{-n}].$$
This function is called the **generating function** of the Bessel functions.
    d. Set $t = e^{i\theta}$ in the expression in part (c) and obtain the identities
$$\cos(x\sin\theta) = J_0(x) + 2\sum_{n=1}^{\infty} J_{2n}(x)\cos 2n\theta$$
and
$$\sin(x\sin\theta) = 2\sum_{n=1}^{\infty} J_{2n-1}(x)\sin(2n-1)\theta.$$

26. Show that $J_0(x) = (1/\pi)\int_0^\pi \cos(x\cos t)\,dt$.
27. Show that $J_n(x) = (1/\pi)\int_0^\pi \cos(nt - x\sin t)\,dt$.
28. **Modified Bessel functions**: The function $I_p(x) = i^{-p}J_p(ix)$, $i^2 = -1$ is called the **modified Bessel function of the first kind of order p**. Show that $I_p(x)$ is a solution of the differential equation
$$x^2 y'' + xy' - (x^2 + p^2)y = 0,$$

and obtain its series representation by the method of Frobenius.

29. Show that another solution of the differential equation in Problem 28 is the **modified Bessel function of the second kind of order** p:
$$K_p(x) = \frac{\pi}{2\sin p\pi}[I_{-p}(x) - I_p(x)].$$

30. Prove (see Problem 25) that
$$e^{(x/2)[t+(1/t)]} = I_0(x) + \sum_{n=1}^{\infty} I_n(x)[t^n + t^{-n}].$$

31. Prove the following identities:

   a. $\dfrac{d}{dx}[x^p I_p(x)] = x^p I_{p-1}(x)$

   b. $\dfrac{d}{dx}[x^{-p} I_p(x)] = x^{-p} I_{p+1}(x)$

   c. $x(I_{p-1} - I_{p+1}) = 2pI_p$

32. a. Prove for all integers $n > 0$ that
$$\int_0^\infty J_{n+1}(x)\,dx = \int_0^\infty J_{n-1}(x)\,dx$$
by means of equation (17), given the approximation
$$J_n(x) \approx \sqrt{\frac{2}{\pi x}} \cos\left(x - \frac{\pi}{4} - \frac{n\pi}{2}\right).$$

   b. Prove a similar fact for Y_n given the approximation
$$Y_n(x) \approx \sqrt{\frac{2}{\pi x}} \sin\left(x - \frac{\pi}{4} - \frac{n\pi}{2}\right).$$

   c. Show that $\int_0^\infty J_n(x)\,dx = 1$.

   d. Prove that $\int_0^\infty [J_n(x)/x]\,dx = 1/n$.

33. In Section 4.1 we considered the vibrations of mass-spring systems whose springs had a constant elastic force per unit elongation (the spring constant k). In practice, the elastic force per unit elongation $k(t)$ of a spring decays with time. Suppose $k(t) = k_0 e^{-at} + k_1$, with $a > 0$, so that equation (4.1.2) becomes
$$\frac{d^2 x}{dt^2} + \frac{k_0 e^{-at} + k_1}{m} x = 0,$$
$$x(0) = x_0, \quad x'(0) = v_0.$$

   a. Use the substitution $u = ce^{-at/2}$, $c = \dfrac{2}{a}\sqrt{\dfrac{k_0}{m}}$, to convert this differential equation into a Bessel equation.

   b. Solve the Bessel equation, and obtain a solution of the given initial-value problem.

5.5 Legendre Polynomials

Another very important differential equation that arises in many applications is the **Legendre differential equation**

$$(1 - x^2)y'' - 2xy' + p(p+1)y = 0, \tag{1}$$

where p is a given real number. Any solution of equation (1) is called a **Legendre function**.†

Dividing equation (1) by $(1 - x^2)$, we obtain the equation

$$y'' - \frac{2x}{1-x^2}y' + \frac{p(p+1)}{1-x^2}y = 0,$$

and we observe, using the geometric series

$$\frac{1}{1-x^2} = 1 + x^2 + x^4 + x^6 + \cdots,$$

that the coefficient functions

$$a(x) = -\frac{2x}{1-x^2} = -2x(1 + x^2 + x^4 + \cdots)$$

$$b(x) = \frac{p(p+1)}{1-x^2} = p(p+1)(1 + x^2 + x^4 + \cdots)$$

---
†Named after a famous French mathematician Adrien Marie Legendre (1752–1833). See the accompanying biographical sketch.

ADRIEN MARIE LEGENDRE (1752–1833)

Adrien Marie Legendre (The Bettmann Archive)

Adrien Marie Legendre is known in the history of elementary mathematics principally for his very popular *Eléments de géométrie*, in which he attempted a pedagogical improvement of Euclid's *Elements* by considerably rearranging and simplifying many of the propositions. This work was very favorably received in America and became the prototype of the geometry textbooks in this country. In fact, the first English translation of Legendre's geometry was made in 1819 by John Farrar of Harvard University. Three years later another English translation was made, by the famous Scottish litterateur Thomas Carlyle, who early in life was a teacher of mathematics. Carlyle's translation, as later revised by Charles Davies, and later still by J. H. Van Amringe, ran through thirty-three American editions. In later editions of his geometry, Legendre attempted to prove the parallel postulate (see his Section 13-6). Legendre's chief work in higher mathematics centered around number theory, elliptic functions, the method of least squares, and integrals. He was also an assiduous computer of mathematical tables. Legendre's name is today connected with the second-order differential equation

$$(1 - x^2) y'' - 2xy' + p(p+1) y = 0,$$

which is of considerable importance in applied mathematics. Functions satisfying this differential equation are called **Legendre functions** (of order p). When p is a nonnegative integer, the equation has polynomial solutions of special interest called **Legendre polynomials**. Legendre's name is also associated with the symbol $(c|p)$ of number theory. The **Legendre symbol** $(c|p)$ is equal to ± 1 according to whether the integer c, which is prime to p, is or is not a quadratic residue of the odd prime p. [For example, $(6|19) = 1$ since the congruence $x^2 \equiv 6 \pmod{19}$ has a solution, and $(39|47) = -1$ since the congruence $x^2 \equiv 39 \pmod{47}$ has no solution.]

In addition to his *Eléments de géométrie*, which appeared in 1794, Legendre published a two-volume 859-page work, *Essai sur la théorie des nombres* (1797–1798), which was the first treatise devoted exclusively to number theory. He later wrote a three-volume treatise, *Exercices du calcul intégral* (1811–1819), that, for comprehensiveness and authoritativeness, rivaled the similar work of Euler. Legendre later expanded parts of this work into another three-volume treatise, *Traité des fonctions elliptiques et des intégrals eulériennes* (1825–1832). In geodesy, Legendre achieved considerable fame for his triangulation of France.

---

have convergent power series representations in the interval $|x| < 1$. By Theorem 5.2.1, it follows that (1) must have a power series representation valid in the interval $|x| < 1$. Substituting $y = \sum_{n=0}^{\infty} c_n x^n$ and its derivatives into equation (1), we have

$$(1 - x^2) \sum_{n=2}^{\infty} c_n n(n-1) x^{n-2} - 2x \sum_{n=1}^{\infty} c_n n x^{n-1} + p(p+1) \sum_{n=0}^{\infty} c_n x^n = 0,$$

or
$$\sum_{n=0}^{\infty} \{(n+2)(n+1)c_{n+2} - c_n[n(n+1) - p(p+1)]\}x^n = 0. \quad (2)$$

Setting the coefficients of the sum (2) to zero, we obtain the recurrence relation
$$(n+2)(n+1)c_{n+2} = c_n(n^2 + n - p^2 - p) = c_n(n-p)(n+p+1).$$
Thus we have
$$c_{n+2} = -\frac{(p-n)(p+n+1)}{(n+2)(n+1)} c_n. \quad (3)$$

Therefore
$$c_2 = -\frac{p(p+1)}{2!} c_0, \quad c_3 = -\frac{(p-1)(p+2)}{3!} c_1,$$
$$c_4 = -\frac{(p-2)(p+3)}{4 \cdot 3} c_2 = \frac{(p-2)p(p+1)(p+3)}{4!} c_0, \ldots.$$

Inserting these values for the coefficients into the power series expansion for $y(x)$ yields
$$y(x) = c_0 y_1(x) + c_1 y_2(x), \quad (4)$$
where
$$y_1(x) = 1 - p(p+1)\frac{x^2}{2!} + (p-2)p(p+1)(p+3)\frac{x^4}{4!} - \cdots, \quad (5)$$
$$y_2(x) = x - (p-1)(p+2)\frac{x^3}{3!} + (p-3)(p-1)(p+2)(p+4)\frac{x^5}{5!} - \cdots. \quad (6)$$

Dividing equation (6) by equation (5), we have
$$\frac{y_2(x)}{y_1(x)} = x + \frac{(p^2+p+1)}{3} x^3 + \cdots,$$

which obviously is nonconstant, implying that y_1 and y_2 are linearly independent. Thus equation (4) is the general solution of the Legendre equation (1) for $|x| < 1$.

In many applications, the parameter p in the Legendre equation is a nonnegative integer. When this occurs, the right-hand side of equation (3) is zero for $n = p$, implying that $c_{p+2} = c_{p+4} = c_{p+6} = \cdots = 0$. Thus one of the equations (5) or (6) reduces to a polynomial of degree p (for even p, it is y_1; for odd p, it is y_2). These polynomials, multiplied by an appropriate constant, are called the **Legendre polynomials**. It is customary to set
$$c_p = \frac{(2p)!}{2^p (p!)^2}, \quad p = 0, 1, 2, \ldots, \quad (7)$$
so, by equation (3),
$$c_{p-2} = -\frac{p(p-1)}{2(2p-1)} c_p = -\frac{(2p-2)!}{2^p(p-1)!(p-2)!},$$
$$c_{p-4} = -\frac{(p-2)(p-3)}{4(2p-3)} c_{p-2} = \frac{(2p-4)!}{2^p 2!(p-2)!(p-4)!}, \ldots.$$

and in general

$$c_{p-2k} = \frac{(-1)^k(2p-2k)!}{2^p k!(p-k)!(p-2k)!}.$$

Then the **Legendre polynomials of degree p** are given by

$$P_p(x) = \sum_{k=0}^{M} \frac{(-1)^k(2p-2k)!}{2^p k!(p-k)!(p-2k)!} x^{p-2k}, \qquad p = 0, 1, 2, \ldots, \qquad (8)$$

where M is the largest integer not greater than $p/2$. In particular, we have

$$P_0(x) = 1, \qquad P_1(x) = x, \qquad P_2(x) = \tfrac{1}{2}(3x^2 - 1), \qquad P_3(x) = \tfrac{1}{2}(5x^3 - 3x),$$

$$P_4(x) = \tfrac{1}{8}(35x^4 - 30x^2 + 3), \ldots.$$

As these particular results illustrate, as a consequence of choice (7) of the value of c_p, we have $P_p(1) = 1$ and $P_p(-1) = (-1)^p$ for all integers $p \geq 0$.

To obtain an even more concise form than equation (8) for the Legendre polynomials, we observe that we can write

$$P_p(x) = \sum_{k=0}^{M} \frac{(-1)^k}{2^p k!(p-k)!} \frac{d^p}{dx^p}(x^{2p-2k}),$$

since

$$\frac{d^p}{dx^p}(x^{2p-2k}) = (2p-2k)\frac{d^{p-1}}{dx^{p-1}}(x^{2p-2k-1}) = \cdots$$

multiply and divide by $(p-2k)!$

$$= (2p-2k) \cdots (p-2k+1) x^{p-2k} = \frac{(2p-2k)!}{(p-2k)!} x^{p-2k}.$$

Hence

$$P_p(x) = \frac{1}{2^p p!} \frac{d^p}{dx^p} \sum_{k=0}^{M} (-1)^k \frac{p!}{k!(p-k)!} (x^2)^{p-k}.$$

We may now extend the range of this sum by letting k range from zero to p. This extension does not affect the result, since the added terms are a polynomial of degree $< p$, so the pth derivative is zero. Thus

$$P_p(x) = \frac{1}{2^p p!} \frac{d^p}{dx^p} \sum_{k=0}^{p} \frac{p!}{k!(p-k)!} (x^2)^{p-k} (-1)^k,$$

and by the binomial formula we have

$$P_p(x) = \frac{1}{2^p p!} \frac{d^p}{dx^p}(x^2 - 1)^p, \qquad p = 0, 1, 2, \ldots. \qquad (9)$$

This formula, called the **Rodrigues formula**,[†] provides an easy way of computing successive Legendre polynomials.

[†] Named after the French mathematician and banker Olinde Rodrigues (1794–1851).

Example 1 Show that $P_2(x) = \frac{1}{2}(3x^2 - 1)$.

Solution By the Rodrigues formula,

$$P_2(x) = \frac{1}{2^2 2!} \frac{d^2}{dx^2}(x^4 - 2x^2 + 1) = \frac{1}{8}(12x^2 - 4) = \frac{1}{2}(3x^2 - 1).$$

We can use the Rodrigues formula to obtain several useful recurrence relations. Observe that

$$P'_{p+1} = \frac{d}{dx}\left[\frac{1}{2^{p+1}(p+1)!} \frac{d^{p+1}}{dx^{p+1}}(x^2 - 1)^{p+1}\right]$$

$$= \frac{d}{dx}\left\{\frac{1}{2^p p!} \frac{d^p}{dx^p}\left[x(x^2-1)^p\right]\right\} = \frac{1}{2^p p!} \frac{d^{p+1}}{dx^{p+1}}\left[x(x^2-1)^p\right]. \quad (10)$$

Hence, taking the derivative of the term in brackets, we have

$$P'_{p+1} = \frac{1}{2^p p!} \frac{d^p}{dx^p}\left[(x^2-1)^p + 2px^2(x^2-1)^{p-1}\right]$$

$$= \frac{1}{2^p p!} \frac{d^p}{dx^p}\left[(2p+1)(x^2-1)^p + 2p(x^2-1)^{p-1}\right]$$

$$= (2p+1)P_p + P'_{p-1}, \quad p = 1, 2, 3, \ldots.$$

We can get another recurrence relation from equation (10) if we consider the effect of repeated differentiations on a product of the form $xf(x)$. Note that

$$\frac{d}{dx}[xf(x)] = x\frac{d}{dx}f(x) + f(x),$$

$$\frac{d^2}{dx^2}[xf(x)] = x\frac{d^2}{dx^2}f(x) + 2\frac{d}{dx}f(x),$$

and in general

$$\frac{d^{p+1}}{dx^{p+1}}[xf(x)] = x\frac{d^{p+1}}{dx^{p+1}}f(x) + (p+1)\frac{d^p}{dx^p}f(x). \quad (11)$$

Applying equation (11) to the expression in brackets in equation (10), we obtain

$$P'_{p+1} = \frac{1}{2^p p!}\left[x\frac{d^{p+1}}{dx^{p+1}}(x^2-1)^p + (p+1)\frac{d^p}{dx^p}(x^2-1)^p\right]$$

$$= xP'_p + (p+1)P_p, \quad p = 0, 1, 2, \ldots.$$

Thus we have proved the identities

$$(p+1)P_p = P'_{p+1} - xP'_p, \quad (2p+1)P_p = P'_{p+1} - P'_{p-1}. \quad (12)$$

Subtract the first identity in (12) from the second one, yielding

$$pP_p = xP'_p - P'_{p-1}, \quad p = 1, 2, \ldots. \quad (13)$$

Finally, we note that from (12) and (13) we can get

$$(p+1)P_{p+1} - (2p+1)xP_p + pP_{p-1} = (xP'_{p+1} - P'_p) - x(P'_{p+1} - P'_{p-1})$$
$$+ (P'_p - xP'_{p-1}) = 0,$$

so we can eliminate all derivatives and obtain the relation

$$\boxed{(p+1)P_{p+1} + pP_{p-1} = (2p+1)xP_p, \qquad p = 1, 2, \ldots.} \qquad (14)$$

Equation (14) can be used to generate all the Legendre polynomials. We illustrate this iterative technique in the next example.

Example 2 Starting with $P_0 = 1$ and $P_1 = x$, calculate the polynomials P_2, P_3, and P_4.

Solution By equation (14),

$$P_{p+1} = \frac{(2p+1)xP_p - pP_{p-1}}{p+1},$$

so

$$P_2 = \frac{3xP_1 - P_0}{2} = \frac{3x^2 - 1}{2},$$

$$P_3 = \frac{5xP_2 - 2P_1}{3} = \frac{15x^3 - 5x - 4x}{6} = \frac{5x^3 - 3x}{2},$$

$$P_4 = \frac{7xP_3 - 3P_2}{4} = \frac{35x^4 - 21x^2 - 9x^2 + 3}{8} = \frac{35x^4 - 30x^2 + 3}{8}.$$

PROBLEMS 5.5

1. Calculate P_5, P_6, P_7, and P_8 by means of equation (14).
2. Prove that series (5) has a radius of convergence $R = 1$.
3. Prove that series (6) has a radius of convergence $R = 1$.
4. Calculate P_4 by means of the Rodrigues formula.
5. Prove that $P_{2p+1}(0) = 0$, for all integers $p \geq 0$.
6. Prove that
$$P_{2p}(0) = \frac{(-1)^p (2p)!}{2^{2p}(p!)^2},$$
for all $p \geq 0$.
7. Prove that, for all integers $p \geq 0$,
   a. $P'_{2p}(0) = 0$;
   b. $P'_{2p+1}(0) = \frac{(-1)^p(2p+1)!}{2^{2p}(p!)^2}$.
8. Show that, for all integers $p > 0$,
   a. $\int_0^1 P_p(x)\,dx = \frac{1}{p+1}P_{p-1}(0)$;
   b. $\int_0^1 P_{2p}(x)\,dx = 0$;
   c. $\int_0^1 P_{2p+1}(x)\,dx = (-1)^p \frac{(2p)!}{2^{2p+1}p!(p+1)!}$.
   d. Compute these integrals for $p = 0$.
9. Consider the **Hermite equation**
$$y'' - 2xy' + 2py = 0. \qquad (i)$$

a. Use the method of Frobenius to show that all solutions of equation (i) are of the form

$$c_0\left[1 + \sum_{n=1}^{\infty} \frac{2^n(-p)(2-p)\cdots(2n-2-p)x^{2n}}{(2n)!}\right] + c_1\left[x + \sum_{n=1}^{\infty} \frac{2^n(1-p)(3-p)\cdots(2n-1-p)x^{2n+1}}{(2n+1)!}\right],$$

where c_0 and c_1 are arbitrary constants.

b. Show that equation (i) has a polynomial solution of degree p for a nonnegative integer p. These polynomials, denoted by $H_p(x)$, are called the **Hermite polynomials of degree p**.

c. Show that

$$H_p(x) = \sum_{n=0}^{M} \frac{(-1)^n p!(2x)^{p-2n}}{n!(p-2n)!},$$

where M is the greatest integer $\leq p/2$.

d. Calculate H_0, H_1, H_2, H_3, and H_4.

10. Consider the **Laguerre equation**[†]

$$xy'' + (1-x)y' + py = 0. \qquad \text{(ii)}$$

a. Show that if p is a nonnegative integer, there is a polynomial solution to equation (ii) of the form

$$L_p(x) = \sum_{n=0}^{p} \frac{(-1)^n p! x^n}{(p-n)!(n!)^2}.$$

The functions $L_p(x)$ are known as the **Laguerre polynomials**.

b. Calculate $L_0(x)$, $L_1(x)$, $L_2(x)$, $L_3(x)$, and $L_4(x)$.

11. Use the binomial theorem (see Problem 5.1.18) to prove that

$$\frac{1}{\sqrt{1-2xz+z^2}} = P_0(x) + P_1(x)z + P_2(x)z^2 + \cdots + P_n(x)z^n + \cdots,$$

where $P_n(x)$ is the nth Legendre polynomial. This identity is called the **generating function** for Legendre polynomials.

[†] Edmond Laguerre (1834–1886) was a French mathematician whose research was in geometry and infinite series.

5.6 An Excursion: Oscillatory Solutions of Linear Second-Order Differential Equations with Variable Coefficients

In this chapter we showed how some linear differential equations with variable coefficients could be solved by power series methods. In this excursion we show how interesting results about such equations can be found without solving the equations.

All solutions of the linear differential equation $y'' + \omega^2 y = 0$ are of the form $y = A\sin(\omega x + \delta)$ (see p. 140). These functions oscillate with period $2\pi/\omega$. In addition, each solution has an infinite number of zeros

$$x = -\frac{\delta}{\omega}, \frac{\pm\pi - \delta}{\omega}, \frac{\pm 2\pi - \delta}{\omega}, \ldots.$$

If a linear second-order differential equation has constant coefficients, then all its solutions can be found. The solutions oscillate if and only if the roots of the characteristic equation are complex conjugates. But what if we have variable coefficients? Can we determine when they have oscillatory solutions? It is this question that we explore here.

Consider the differential equation

$$y'' + a(x)y' + b(x)y = 0, \qquad (1)$$

where the function a has a continuous derivative and b is continuous. The

substitution
$$y(x) = z(x)e^{-(1/2)\int a(x)\,dx} \qquad (2)$$
yields
$$y'(x) = [z'(x) - \tfrac{1}{2}a(x)z(x)]e^{-(1/2)\int a(x)\,dx}$$
and
$$y''(x) = [z''(x) - a(x)z'(x) - \tfrac{1}{2}a'(x)z(x) + \tfrac{1}{4}a(x)^2 z(x)]e^{(-1/2)\int a(x)\,dx}.$$
Thus
$$\left[\left(z'' - az' - \frac{a'z}{2} + \frac{a^2 z}{4}\right) + a\left(z' - \frac{az}{2}\right) + bz\right] e^{-(1/2)\int a(x)\,dx}$$
$$= y'' + ay' + by = 0,$$
or
$$\left[z'' + \left(b - \frac{a'}{2} - \frac{a^2}{4}\right)z\right] e^{-(1/2)\int a(x)\,dx} = 0.$$
Since the exponential is nonzero, we obtain the equation
$$z'' + p(x)z = 0,$$
where
$$p(x) = b(x) - \frac{a'(x)}{2} - \frac{a^2(x)}{4}.$$
Thus equation (1) can be transformed into an equation of the form
$$y'' + p(x)y = 0 \qquad (3)$$
by a suitable change of variables. Since the exponential is nonzero, we need only discuss the oscillatory behavior of the simpler equation (3).

Oscillation

We say that $y(x)$ **oscillates** on the interval $[c, \infty)$ if y is nonconstant and has infinitely many zeros on $[c, \infty)$.

Example 1 $y = A\sin(\omega x + \delta)$ oscillates on $[c, \infty)$ for any real number c.

THEOREM 1[†]

Let y_1 and y_2 be linearly independent solutions of (3) and let x_1 and x_2 be consecutive zeros of y_1. Then y_2 has exactly one zero in the interval (x_1, x_2).

Proof Consider the Wronskian of y_1 and y_2 (see p. 90):
$$W(y_1, y_2)(x) = y_1(x)y_2'(x) - y_2(x)y_1'(x) \neq 0,$$
since y_1 and y_2 are independent. Now
$$W(y_1, y_2)(x_1) = y_1(x_1)y_2'(x_1) - y_2(x_1)y_1'(x_1) \stackrel{y_1(x_1)=0}{=} -y_2(x_1)y_1'(x_1) \qquad (4)$$

[†] This theorem was first proved by the Swiss mathematician Jacques Charles François Sturm (1803–1855).

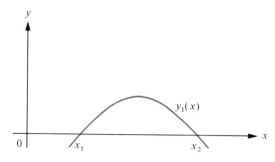

Figure 5.2

and
$$W(y_1, y_2)(x_2) = -y_2(x_2)y_1'(x_2). \tag{5}$$

Since $y_1(x) \neq 0$ in the interval (x_1, x_2) and y_1 is continuous, we conclude that y_1 is either always positive or always negative on that interval. Assume that y_1 is positive. The proof is similar in the other case. The situation is illustrated in Figure 5.2.

From (4) and (5), we know that $y_2(x_1) \neq 0$ and $y_2(x_2) \neq 0$. Assume that $y_2(x_1) > 0$. Figure 5.2 shows that $y_1'(x_1) \geq 0$. But $y_1'(x_1) = 0$ is impossible, because then y_1 would satisfy the initial-value problem

$$y'' + p(x)y = 0, \qquad y_1(x_1) = y_1'(x_1) = 0, \tag{6}$$

which (by the existence-uniqueness theorem, p. 84) would imply $y_1 \equiv 0$. Hence $y_1'(x_1) > 0$, so from (4)

$$W(y_1, y_2)(x_1) < 0. \tag{7}$$

Since W is continuous and nonzero (because y_1 and y_2 are linearly independent), we must have

$$W(y_1, y_2)(x_2) = -y_2(x_2)y_1'(x_2) < 0.$$

[Otherwise, if $W(y_1, y_2)(x_2) > 0$, then W would have a zero in (x_1, x_2), contradicting linear independence.] But, from Figure 5.2, $y_1'(x_2) < 0$, implying that $y_2(x_2) < 0$.

Since $y_2(x_1) > 0$ and $y_2(x_2) < 0$, we conclude from the intermediate-value theorem that y_2 has at least one zero in (x_1, x_2). However, y_2 cannot have two or more zeros in (x_1, x_2). If $x_3 < x_4$ were two such zeros, then reversing the role of y_1 and y_2 in the proof above shows that y_1 has a zero in (x_3, x_4); call it x_5. But then $x_1 < x_5 < x_2$ are three zeros of y_1, contradicting the hypothesis that x_1 and x_2 are consecutive zeros of y_1. Thus y_2 has exactly one zero in (x_1, x_2). ∎

Theorem 1 shows that if one solution of (3) oscillates, then every other nontrivial solution of (3) oscillates. Conversely, if one nontrivial solution of (3) does not oscillate, no solution oscillates. Thus we can talk about equation (3) being **oscillatory** or **nonoscillatory**.

Recall that all solutions of $y'' + \omega^2 y = 0$ are of the form $y = A\sin(\omega x + \delta)$ and oscillate with period $2\pi/\omega$. If we increase the size of ω, the oscillations become more frequent. Is this also true for second-order linear equations with variable coefficients? The following result, called the **Sturm comparison theorem**, compares the rates of oscillation of solutions of two equations

$$y'' + p(x)y = 0, \tag{8}$$
$$z'' + q(x)z = 0. \tag{9}$$

THEOREM 2: STURM COMPARISON THEOREM

Let y be a solution of (8) with consecutive zeros at x_1 and x_2. Suppose $q(x) \geq p(x)$, with p and q continuous and strict equality holding for at least one point of the interval $[x_1, x_2]$. Then any solution z of (9) that has a zero at x_1 has another zero in the interval (x_1, x_2). Roughly speaking, the larger q is, the more rapidly solutions oscillate.

Proof We can assume that $y(x) > 0$ on the interval (x_1, x_2); the proof is similar in the other case. Then, by the comments we made about the initial-value problem (6), it follows that $y'(x_1) > 0$ and $y'(x_2) < 0$ (see Figure 5.3).

Suppose $z'(x_1) > 0$; a similar proof holds in the other case. Multiply equation (8) by $-z$ and equation (9) by y to obtain

$$-zy'' - pyz = 0,$$
$$yz'' + qyz = 0.$$

Adding these two equations, we have

$$yz'' - zy'' + (q - p)yz = 0. \tag{10}$$

Observe that

$$(yz' - zy')' = yz'' + y'z' - y'z' - zy'' = yz'' - zy''.$$

Hence, if we integrate (10) from x_1 to x_2, we get

$$[y(x)z'(x) - z(x)y'(x)]\Big|_{x_1}^{x_2} + \int_{x_1}^{x_2} [q(x) - p(x)] y(x)z(x)\, dx = 0,$$

or, since $y(x_1) = y(x_2) = z(x_1) = 0$,

$$-z(x_2)y'(x_2) + \int_{x_1}^{x_2} [q(x) - p(x)] y(x)z(x)\, dx = 0.$$

Thus

$$\int_{x_1}^{x_2} [q(x) - p(x)] y(x)z(x)\, dx = z(x_2)y'(x_2). \tag{11}$$

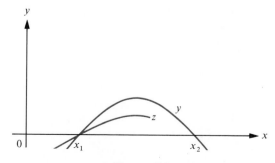

Figure 5.3

Now assume $z(x) > 0$ on (x_1, x_2). Then the integral on the left side of (11) is positive while the right side is not [since $y'(x_2) < 0$]. From this contradiction we can infer the truth of the theorem. ∎

COROLLARY

If $q(x) \geq \omega^2 > 0$ on $[c, \infty)$, then
$$y'' + q(x)y = 0$$
oscillates.

Proof In Example 1 we showed that $y'' + \omega^2 y = 0$ had the oscillatory solution $y = A\sin(\omega x + \delta)$. By Theorem 2, it follows that the solutions to $y'' + qy = 0$ oscillate as well. ∎

Example 2 Consider the Bessel equation of order zero:
$$x^2 y'' + xy' + x^2 y = 0. \tag{12}$$
Dividing both sides by x^2, we get
$$y'' + \frac{1}{x}y' + y = 0. \tag{13}$$
Substituting
$$y \stackrel{\text{equation (2)}}{=} ze^{-(1/2)\int dx/x} = z/\sqrt{x}$$
we have
$$y' = \frac{z'}{\sqrt{x}} - \frac{z}{2x^{3/2}} = \frac{1}{\sqrt{x}}\left(z' - \frac{z}{2x}\right),$$
$$y'' = \frac{z''}{\sqrt{x}} - \frac{z'}{x^{3/2}} + \frac{3z}{4x^{5/2}} = \frac{1}{\sqrt{x}}\left(z'' - \frac{z'}{x} + \frac{3z}{4x^2}\right),$$
and (13) becomes
$$0 = y'' + \frac{1}{x}y' + y = \frac{1}{\sqrt{x}}\left[\left(z'' - \frac{z'}{x} + \frac{3z}{4x^2}\right) + \frac{1}{x}\left(z' - \frac{z}{2x}\right) + z\right],$$
or
$$z'' + \left(1 + \frac{1}{4x^2}\right)z = 0. \tag{14}$$

But $1 + 1/4x^2 > 1$ on $[c, \infty)$ for $c > 0$, so all nontrivial solutions z to (14) oscillate by the corollary. Since $\sqrt{x} > 0$, $y = z/\sqrt{x}$ must also oscillate. Thus, without solving the Bessel equation of order zero, we can show that each of its nontrivial solutions oscillates. This result is not at all obvious even after the equation has been solved.

THEOREM 3

If $p(x)$ is continuous on $[a, \infty)$ and
$$\int_a^\infty p(x)\, dx = +\infty, \tag{15}$$

then every nontrivial solution of
$$y'' + p(x)y = 0 \qquad (16)$$
oscillates on $[a, \infty)$.

Proof[†] Suppose that some solution y of (16) has at most a finite number of zeros on $[a, \infty)$. Then there is a number $b \geq a$ such that $y(x) \neq 0$ on $[b, \infty)$. Let
$$z(x) = y'(x)/y(x).$$
Then
$$z' = \frac{yy'' - (y')^2}{y^2} = \frac{y''}{y} - z^2$$
and we obtain the Riccati equation
$$z'(x) + z^2(x) = \frac{y''(x)}{y(x)} = -p(x). \qquad (17)$$

If we integrate both ends of (17) from b to c ($> b$), we get
$$z(x)\Big|_b^c + \int_b^c z^2(x)\,dx = -\int_b^c p(x)\,dx,$$
or
$$z(c) + \int_b^c z^2(x)\,dx = z(b) - \int_b^c p(x)\,dx. \qquad (18)$$

Since $z(b)$ is finite and the integral on the right side of (18) tends to infinity as c approaches infinity, the left side of (18) is negative for all c no smaller than some number x_0. Hence, replacing c on the left side of (18) by $x \geq x_0$ and changing variables of integration to avoid confusion,
$$0 \leq \int_b^x z^2(t)\,dt \leq -z(x). \qquad (19)$$

Set $R(x) = \int_b^x z^2(t)\,dt$. Then, squaring both sides of (19), we get
$$R^2(x) = \left[\int_b^x z^2(t)\,dt\right]^2 \leq z^2(x). \qquad (20)$$

But $R'(x) = z^2(x)$, so inequality (20) can be rewritten as
$$R^2(x) \leq R'(x).$$
Separating variables, we have
$$\int_{x_0}^x dx \leq \int_{R(x_0)}^{R(x)} \frac{dR}{R^2},$$
or
$$x - x_0 \leq \frac{1}{R(x_0)} - \frac{1}{R(x)}.$$
Therefore
$$x \leq x_0 + \frac{1}{R(x_0)}, \qquad (21)$$

[†] This is a special case of the proof by W. J. Coles, "A Simple Proof of a Well-Known Oscillation Theorem," *Proceedings of the American Mathematical Society* 19 (1968): 507.

since R is positive. But the left side of (21) can be made arbitrarily large, while the right side is bounded, leading to a contradiction. Hence the assumption that y has only finitely many zeros is false. ∎

Example 3 The nontrivial solutions of

$$y'' + \frac{1}{x}y = 0, \qquad x \geq 1,$$

are oscillatory since

$$\int_1^\infty \frac{dx}{x} = +\infty.$$

This excursion presents only a brief glimpse of a well-developed theory on the oscillation or nonoscillation of linear second-order differential equations.

Review Exercises for Chapter 5

1. Show that when $-1 \leq x < 1$,
$$\ln(1-x) = -\left(x + \frac{x^2}{2} + \frac{x^3}{3} + \frac{x^4}{4} + \cdots\right).$$

2. Use Taylor's theorem to show that
$$\sin x = \sin a + \cos a (x - a)$$
$$- \frac{\sin a}{2!}(x-a)^2 - \frac{\cos a}{3!}(x-a)^3 + \cdots.$$

3. Verify that
$$\ln \frac{1+x}{1-x} = 2\left(x + \frac{x^3}{3} + \frac{x^5}{5} + \cdots \right.$$
$$\left. + \frac{x^{2n-1}}{2n-1} + \cdots\right), \quad |x| < 1.$$

4. Show that
$$\ln \frac{x^2}{x^2-1} = \frac{1}{x^2} + \frac{1}{2x^4} + \frac{1}{3x^6} + \cdots,$$
for $|x| > 1$. [*Hint:* Consider the series $\ln(1+x)$
$$= x - \frac{x^2}{2} + \frac{x^3}{3} - \frac{x^4}{4} + \cdots, \ |x| < 1.]$$

5. Use Taylor's theorem to show that
$$e^{\tan x} = 1 + x + \frac{x^2}{2!} + \frac{3x^3}{3!} + \frac{9x^4}{4!} + \frac{37x^5}{5!} + \cdots,$$
for $|x| < \pi/2$.

In Exercises 6–11 use power series to find the solution of the given initial-value problem.

6. $xy'' + xy' - y = -e^{-x}, \quad y(0) = 1, \quad y'(0) = 0$
7. $xy'' - y' + (1-x)y = x^2(1-x), \quad y(0) = y'(0) = 1$
8. $(1+x)y'' + (2-x-x^2)y' + y = 0, \quad y(0) = 0, \quad y'(0) = 1$
9. $xy'' - 2y' + xy = x^2 - 2 - 2\sin x, \quad y(0) = 0, \quad y'(0) = 1$
10. $y'' - 3x^2 y' - 6xy = 0, \quad y(0) = 1, \quad y'(0) = 0$
11. $y'' + 2xy' + 2y = 0, \quad y(0) = 1, \quad y'(0) = 0$

In Exercises 12–17 use the method of Frobenius to solve the given differential equations.

12. $4xy'' + 2y' - y = 0$
13. $2x(1-2x)y'' + (1+4x^2)y' - (1+2x)y = 0$
14. $2x(1-2x)y'' + (1+x)y' - y = 0$
15. $x(1-x)y'' + 2(1-2x)y' - 2y = 0$
16. $2x(1-x)y'' + (1-x)y' + 3y = 0$
17. $2x^2 y'' + x(1-x)y' - y = 0$
18. Gauss's[†] hypergeometric equation is given by
$$x(1-x)y'' + [c - (a+b+1)x]y' - aby = 0,$$
where a, b, and c are constants. Show that it has a solution of the form
$$y_1(x) = 1 + \frac{ab}{c}x + \frac{a(a+1)b(b+1)}{c(c+1)} \frac{x^2}{2!}$$
$$+ \frac{a(a+1)(a+2)b(b+1)(b+2)}{c(c+1)(c+2)} \frac{x^3}{3!} + \cdots.$$

[†]Carl Friedrich Gauss (1777–1855) is sometimes referred to as the greatest mathematician. He spent his entire prolific life in Germany. An infant prodigy, he made brilliant discoveries in geometry, algebra, number theory, statistics, and astronomy.

This series is called the **hypergeometric series** and is denoted by the symbol $F(a, b, c; x)$.

19. Prove, using the result of Exercise 18, that
    a. $F(-a, b, b; -x) = (1 + x)^a$;
    b. $xF(1, 1, 2; -x) = \ln(1 + x)$.

20. Show that the **Chebyshëv equation**†
$$(1 - x^2) y'' - xy' + p^2 y = 0$$
has a polynomial solution when p is an integer.

21. Prove the following identities if $J_0(a) = J_0(b) = 0$:
    a. $\int_0^1 x J_0(ax) J_0(bx) \, dx = 0, \quad a \neq b$
    b. $\int_0^1 x J_0^2(ax) \, dx = \frac{1}{2} J_1^2(ax)$
    c. $x^2 J_0'(x) = x[J_1(x) - J_0(x)] - (x^2 - 1) J_1(x)$
    d. $[x J_0(x) J_1(x)]' = x[J_0^2(x) - J_1^2(x)]$

22. Find solutions in terms of Bessel functions for the following differential equations:
    a. $x^2 y'' + (x^2 - 2) y = 0$
    b. $x^2 y'' - xy' + (1 + x^2 - k^2) y = 0$

23. Using the method of Frobenius, find a solution to
$$x^2 y'' - xy' + (x^2 + 1) y = 0.$$

24. Prove the following identities for integral values of n:
    a. $\int_{-1}^{1} x P_n(x) P_{n-1}(x) \, dx = \dfrac{2n}{4n^2 - 1}$
    b. $\int_{-1}^{1} (1 - x^2)[P_n'(x)]^2 \, dx = \dfrac{2n(n+1)}{2n+1}$

---

†Pafnuti Lvovich Chebyshëv (1821–1894) was the leading Russian mathematician of his time.

c. $(m + n + 1) \int_0^1 x^m P_n(x) \, dx = m \int_0^1 x^{m-1} P_{n-1}(x) \, dx$

25. Find a solution for the differential equation
$$y'' + 2(\cot x) y' + n(n + 1) y = 0.$$
[*Hint*: Make the trigonometric substitution $t = \cos x$.]

26. Power series methods can also be used to obtain approximate solutions of nonlinear differential equations. In Problem 4.1.11 we showed that an ideal pendulum satisfies the nonlinear initial-value problem
$$\frac{d^2 \theta}{dt^2} + \frac{g}{l} \sin \theta = 0, \quad \theta(0) = \theta_0, \quad \theta'(0) = 0.$$
a. Show that the substitutions $y = \theta$ and $x = \sqrt{g/l}\, t$ lead to the initial-value problem
$$\frac{d^2 y}{dx^2} + \sin y = 0, \quad y(0) = \theta_0, \quad y'(0) = 0.$$
*b. Replace $\sin y$ by its Maclaurin series, substitute $y = \sum_{n=0}^{\infty} c_n x^n$, and show that
$$1 \cdot 2 c_2 + \sin c_0 = 0$$
$$2 \cdot 3 c_3 + c_1 \cos c_0 = 0$$
$$3 \cdot 4 c_4 + c_2 \cos c_0 - \frac{c_1^2}{2} \sin c_0 = 0$$
$$4 \cdot 5 c_5 + c_3 \cos c_0 - c_1 c_2 \sin c_0 - \frac{c_1^3}{3} \cos c_0 = 0$$

c. Conclude that the Taylor series of $\theta(t)$ has the form
$$\theta(t) = \theta_0 - \frac{g}{l} (\sin \theta_0) \frac{t^2}{2!} + \frac{g^2}{l^2} (\sin \theta_0 \cos \theta_0) \frac{t^4}{4!}$$
$$+ (\text{terms in } t^6 \text{ and higher order}).$$

6

Laplace Transforms

6.1 Introduction: Definition and Basic Properties of the Laplace Transform

One of the most efficient methods of solving certain ordinary and partial differential equations is to use Laplace[†] transforms. The effectiveness of the Laplace transform is due to its ability to convert a differential equation into an algebraic equation whose solution yields the solution of the differential equation when the transformation is inverted.

Improper Integral

Recall from calculus that, if we integrate a function from a given real number t_0 to infinity, then the integral is called an **improper integral**. Formally, if t_0 is a given real number, then for any function $f(t)$, we define

$$\int_{t_0}^{\infty} f(t)\, dt = \lim_{A \to \infty} \int_{t_0}^{A} f(t)\, dt.$$

If this limit exists and is finite, we say that the improper integral **converges**; otherwise, it **diverges**.

Example 1 Let $f(t) = e^{at}$, where $a \neq 0$. Then

$$\int_0^{\infty} e^{at}\, dt = \lim_{A \to \infty} \int_0^A e^{at}\, dt = \lim_{A \to \infty} \left.\frac{1}{a} e^{at}\right|_0^A = \lim_{A \to \infty} \frac{1}{a}(e^{aA} - 1).$$

[†] Named after the French mathematician Pierre Simon de Laplace (1749–1827). See the accompanying biographical sketch.

PIERRE SIMON DE LAPLACE (1749–1827)

Pierre-Simon de Laplace was born of poor parents in 1749. His mathematical ability early won him good teaching posts, and as a political opportunist he ingratiated himself with whichever party happened to be in power during the uncertain days of the French Revolution. His most outstanding work was done in the fields of celestial mechanics, probability, differential equations, and geodesy. He published two monumental works, *Traité de mécanique céleste* (five volumes, 1799–1825) and *Théorie analytique des probabilités* (1812), each of which was preceded by an extensive nontechnical exposition. The five-volume *Traité de mécanique céleste*, which earned him the title of "the Newton of France," embraced all previous discoveries in celestial mechanics along with Laplace's own contributions and marked the author as the unrivaled master in the subject. It may be of interest to repeat a couple of anecdotes often told in connection with this work. When Napoleon teasingly remarked that God was not mentioned in his treatise, Laplace replied, "Sire, I did not need that hypothesis." And when he translated Laplace's treatise into English, the American astronomer Nathaniel Bowditch remarked, "I never come across one of Laplace's 'thus it plainly appears' without feeling sure that I have hours of hard work before me to fill up the chasm and find out and show how it plainly appears." Laplace's name is connected with the *nebular hypothesis* of cosmogony and with the so-called *Laplace equation* of potential theory (though neither of these contributions originated with Laplace), with the so-called *Laplace transform* that later became the key to the operational calculus of Heaviside, and with the *Laplace expansion* of a determinant. Laplace died in 1827, exactly one hundred years after the death of Isaac Newton. According to one report, his last words were "What we know is slight; what we don't know is immense."

Pierre-Simon de Laplace (Bettmann Archive)

The following story about Laplace offers a valuable suggestion to one applying for a position. When Laplace arrived as a young man in Paris seeking a professorship of mathematics, he submitted his recommendations by prominent people to d'Alembert but was not received. Returning to his lodgings, Laplace wrote d'Alembert a brilliant letter on the general principles of mechanics. This opened the door, and d'Alembert replied, "Sir, you notice that I paid little attention to your recommendations. You don't need any; you have introduced yourself better." A few days later Laplace was appointed professor of mathematics at the Military School of Paris.

Laplace was very generous to beginners in mathematical research. He called these beginners his stepchildren, and there are several instances in which he withheld publication of a discovery to allow a beginner the opportunity to publish first. Sadly, such generosity is rare in mathematics.

We close our brief account of Laplace with two quotations due to him: "All the effects of nature are only mathematical consequences of a small number of immutable laws." "In the final analysis, the theory of probability is only common sense expressed in numbers."

Clearly this limit is finite (and is equal to $-1/a$) if $a < 0$ and diverges to $+\infty$ if $a > 0$.

Example 2 Let $f(t) = \cos t$. Then

$$\int_0^\infty \cos t \, dt = \lim_{A \to \infty} \int_0^A \cos t \, dt = \lim_{A \to \infty} \sin t \Big|_0^A = \lim_{A \to \infty} \sin A.$$

But $\sin A$ has no limit as $A \to \infty$. Therefore, $\int_0^\infty \cos t \, dt$ diverges, even though $-1 \leq \int_0^A \cos t \, dt \leq 1$ for every $A \geq 0$.

In the rest of this chapter we do not rigorously calculate improper integrals. However, you should not forget that an improper integral is a special kind of limit.

LAPLACE TRANSFORM: DEFINITION

Let $f(t)$ be a real-valued function that is defined for $t \geq 0$. Suppose that $f(t)$ is multiplied by e^{-st} and the result is integrated with respect to t from zero to infinity. If the integral

$$F(s) = \mathscr{L}\{f(t)\} = \int_0^\infty e^{-st} f(t) \, dt \tag{1}$$

converges, the resulting function of s[†] is called the **Laplace transform** of the function f. We stress that the original function f is a function of the variable t, whereas its Laplace transform F is a function of the variable s.

Inverse Transform

It should now be apparent why the word *transform* is associated with this operation. The operation transforms the original function $f(t)$ into a new function $F(s) = \mathscr{L}\{f(t)\}$. Since we are interested in reversing the procedure, we call the original function f the **inverse transform** of $F = \mathscr{L}\{f\}$ and denote it by $f = \mathscr{L}^{-1}\{F\}$. Note that the last statement translates into the mathematical symbols $\mathscr{L}^{-1}\{F\} = \mathscr{L}^{-1}\{\mathscr{L}\{f\}\} = f$; that is,

$$\text{if } \mathscr{L}\{f(t)\} = F(s), \text{ then } \mathscr{L}^{-1}\{F(s)\} = f(t). \tag{2}$$

Example 3 Let $f(t) = e^{at}$, where a is constant. Then

$$\mathscr{L}\{e^{at}\} = \int_0^\infty e^{-st} e^{at} \, dt = \int_0^\infty e^{-(s-a)t} \, dt = \lim_{A \to \infty} \frac{e^{-(s-a)t}}{a-s} \Big|_0^A,$$

and the integral converges, if $s - a > 0$, to

$$\mathscr{L}\{e^{at}\} = \frac{1}{s-a}.$$

[†] In general, the variable s is complex, but in this book we will only consider real values of s.

Thus the Laplace transform of the function e^{at} is the function $F(s) = 1/(s-a)$ for $s > a$. Note that we can also conclude, from (2), that

$$\mathscr{L}^{-1}\{F(s)\} = \mathscr{L}^{-1}\left\{\frac{1}{s-a}\right\} = e^{at}.$$

The last example shows that $\mathscr{L}\{f(t)\}$ may not be defined for all values of s; but if it is defined, then it will exist for suitably large values of s. Indeed, $\mathscr{L}\{e^{at}\}$ is defined only for values $s > a$.

Example 4 Let $f(t) = 1/t$. Then

$$\int_0^\infty \frac{e^{-st}}{t}\, dt = \int_0^1 \frac{e^{-st}}{t}\, dt + \int_1^\infty \frac{e^{-st}}{t}\, dt. \tag{3}$$

But for t in the interval $0 \le t \le 1$, $e^{-st} \ge e^{-s}$ if $s > 0$. Thus,

$$\int_0^\infty \frac{e^{-st}}{t}\, dt \ge e^{-s}\int_0^1 \frac{dt}{t} + \int_1^\infty \frac{e^{-st}}{t}\, dt.$$

However,

$$\int_0^1 t^{-1}\, dt = \lim_{A \to 0}\int_A^1 t^{-1}\, dt = \lim_{A \to 0} \ln t\, \Big|_A^1$$

$$= \lim_{A \to 0} (\ln 1 - \ln A) = \lim_{A \to 0} (-\ln A) = +\infty.$$

Therefore the integral (3) diverges, so $f(t) = 1/t$ has no Laplace transform.

This example and the preceding discussion pinpoint the need for answers to the following questions:

a. Which functions f have Laplace transforms?

b. Can two functions f and g have the same Laplace transform?

In answering these questions, we will be satisfied if a class of functions, large enough to contain virtually all functions that arise in practice, can be found for which the Laplace transforms exist and whose inverses are unique. We will see in Section 6.2 that certain differential equations can be solved by first taking the Laplace transform of both sides of the equation, then finding the Laplace transform of the solution, and finally arriving at the answer by taking the unique inverse transform of the transform of the solution. The need for a unique inverse arises from the fact that we need to reverse the transformation to find the solution of a given problem, and this step is not possible if the inverses within the given class of functions are not unique.

In order to give simple conditions that guarantee the existence of a Laplace transform, we require the following definitions:

Jump Discontinuity

A function f has a **jump discontinuity** at a point t^* if the function has different (finite) limits as it approaches t^* from the left and from the right or if the two limits are equal but different from $f(t^*)$. Note that $f(t^*)$ may or may not be equal to either

$$\lim_{t \to t^{*+}} f(t) \quad \text{or} \quad \lim_{t \to t^{*-}} f(t).$$

Piecewise Continuous Function

In fact, $f(t^*)$ may not even be defined. A function f defined on $(0, \infty)$ is **piecewise continuous** if it is continuous on every finite interval $0 \leq t \leq b$, except possibly at finitely many points where it has jump discontinuities. Such a function is illustrated in Figure 6.1. The class of piecewise continuous functions includes every continuous function, as well as many important discontinuous functions such as the unit step function, square waves, and the staircase function, which we will encounter later in this chapter.

Figure 6.1

THEOREM 1: EXISTENCE THEOREM

Let f be piecewise continuous on $t \geq 0$ and satisfy the condition

$$|f(t)| \leq Me^{at}, \tag{4}$$

for $t \geq T$, where a, M, and T are fixed nonnegative constants. Then $\mathscr{L}\{f(t)\}$ exists for all $s > a$.

Remark Theorem 1 implies that all bounded continuous functions have Laplace transforms. It also implies that every polynomial $p(t)$ has a Laplace transform, because repeated application of L'Hôpital's rule yields

$$\lim_{t \to \infty} \frac{p(t)}{e^{at}} = 0, \quad \text{for } a > 0.$$

Thus a constant M exists for which $|p(t)| \leq Me^{at}$, for t sufficiently large.

Proof Since f is piecewise continuous, $e^{-st}f(t)$ has a finite integral over any finite interval on $t \geq 0$, and

$$|\mathscr{L}\{f(t)\}| = \left|\int_0^\infty e^{-st}f(t)\,dt\right| \leq \left|\int_0^T e^{-st}f(t)\,dt\right| + \int_T^\infty e^{-st}|f(t)|\,dt. \tag{5}$$

The first integral on the right-hand side of equation (5) exists and, by equation (4),

$$\int_T^\infty e^{-st}|f(t)|\,dt \leq M\int_T^\infty e^{-(s-a)t}\,dt = \left.\frac{Me^{-(s-a)t}}{a-s}\right|_T^\infty,$$

which converges to $Me^{-(s-a)T}/(s-a)$ as t approaches infinity if $s > a$. Thus it follows by the comparison theorem of integrals that $\mathscr{L}\{f(t)\}$ exists. ∎

The conditions in Theorem 1 are easy to test. For example, t^n has a Laplace transform (which we obtain in Example 5), because

$$t^n \leq n!e^t, \quad \text{for any } t > 0 \text{ and integer } n \geq 0.$$

To see this, observe that

$$e^t = 1 + t + \frac{t^2}{2!} + \cdots + \frac{t^n}{n!} + \cdots,$$

so $t^n/n! \leq e^t$, or $t^n \leq n!e^t$, for $t > 0$. On the other hand,

$$e^{t^2} > Me^{at},$$

since

$$t^2 > \ln M + at$$

for sufficiently large t, regardless of the choice of M and a. Exponentiating both sides, we get

$$e^{t^2} > e^{\ln M + at} = e^{\ln M} e^{at} = Me^{at}.$$

It is not hard to show that e^{t^2} has no Laplace transform.

It should also be noted that there are functions having Laplace transforms that do not satisfy the hypotheses of Theorem 1. For example, $f(t) = 1/\sqrt{t}$ is infinite at $t = 0$; but setting $x^2 = st$, we have

$$\mathscr{L}\{t^{-1/2}\} = \int_0^\infty e^{-st} t^{-1/2} \, dt = \frac{2}{\sqrt{s}} \int_0^\infty e^{-x^2} \, dx = \sqrt{\frac{\pi}{s}},$$

according to Formula 216 in Appendix 1.

The proof of uniqueness would take us too far afield at this point. However, it can be shown that two functions having the same Laplace transform cannot differ at every point in an interval of positive length,[†] although they may differ at several isolated points. Thus, two different piecewise continuous functions having the same Laplace transform can differ only at isolated points. Such differences are generally of no importance in applications. Hence the Laplace transform has an essentially unique inverse. In particular, different continuous functions have different Laplace transforms.

One of the most important properties of the Laplace transform is stated in the following theorem.

THEOREM 2: LINEARITY PROPERTY OF THE LAPLACE TRANSFORM

If $\mathscr{L}\{f(t)\}$ and $\mathscr{L}\{g(t)\}$ exist, then $\mathscr{L}\{af(t) + bg(t)\}$ exists and

$$\boxed{\mathscr{L}\{af(t) + bg(t)\} = a\mathscr{L}\{f(t)\} + b\mathscr{L}\{g(t)\}.}$$

[†]See, for example, I. S. Sokolnikoff and R. M. Redheffer, *Mathematics of Physics and Modern Engineering* (New York: McGraw-Hill, 1966), p. 217.

6.1 / INTRODUCTION: DEFINITION AND BASIC PROPERTIES OF THE LAPLACE TRANSFORM

Proof By definition

$$\mathscr{L}\{af(t) + bg(t)\} = \int_0^\infty e^{-st}[af(t) + bg(t)] \, dt$$

$$= a\int_0^\infty e^{-st}f(t) \, dt + b\int_0^\infty e^{-st}g(t) \, dt$$

$$= a\mathscr{L}\{f(t)\} + b\mathscr{L}\{g(t)\} = aF(s) + bG(s). \quad \blacksquare$$

COROLLARY: LINEARITY PROPERTY OF THE INVERSE LAPLACE TRANSFORM

If $F(s) = \mathscr{L}\{f(t)\}$ and $G(s) = \mathscr{L}\{g(t)\}$, then

$$\boxed{\begin{aligned}\mathscr{L}^{-1}\{aF(s) + bG(s)\} &= a\mathscr{L}^{-1}\{F(s)\} + b\mathscr{L}^{-1}\{G(s)\} \\ &= af(t) + bg(t).\end{aligned}}$$

The proof of the corollary is left as an exercise (Problem 79).

The linearity property allows us to deal with a linear equation term by term in order to obtain its Laplace transform.

It is essential that we begin to recognize the Laplace transforms of many different functions as quickly as possible. Table 6.1 gives the transforms of seven basic functions. A much more extensive list can be found in Appendix 2. Note that, by definition, $0! = 1$.

Equation (6) has already been obtained in Example 3. The derivations of the other six transforms are given below. For equation (7), we have

$$\mathscr{L}\{c\} = \int_0^\infty e^{-st}c \, dt = \left. \frac{ce^{-st}}{-s} \right|_0^\infty = \frac{c}{s}, \quad \text{for } s > 0.$$

TABLE 6.1

$f(t)$	$\mathscr{L}\{f(t)\}$	Domain of definition	
e^{at}	$\dfrac{1}{s-a}$	$s > a$ (real)	(6)
c (a constant)	$\dfrac{c}{s}$	$s > 0$	(7)
t^n	$\dfrac{n!}{s^{n+1}}$	$s > 0$	(8)
$\sin at$	$\dfrac{a}{s^2 + a^2}$	$s > 0$	(9)
$\cos at$	$\dfrac{s}{s^2 + a^2}$	$s > 0$	(10)
$\sinh at$	$\dfrac{a}{s^2 - a^2}$	$s > |a|$	(11)
$\cosh at$	$\dfrac{s}{s^2 - a^2}$	$s > |a|$	(12)

Example 5 Let n be a positive integer. Then equation (8) is obtained by integrating

$$\mathscr{L}\{t^n\} = \int_0^\infty e^{-st}t^n\,dt$$

by parts with $u = t^n$ and $dv = e^{-st}\,dt$. We have $du = nt^{n-1}\,dt$ and $v = -(1/s)e^{-st}$:

$$\mathscr{L}\{t^n\} = \left.\frac{-e^{-st}t^n}{s}\right|_0^\infty + \frac{n}{s}\int_0^\infty e^{-st}t^{n-1}\,dt$$

$$= \frac{n}{s}\mathscr{L}\{t^{n-1}\}, \qquad \text{for } s > 0,$$

because $\lim_{t\to\infty} e^{-st}t^n/s = 0$, if $s > 0$, by n applications of L'Hôpital's rule. Thus

$$\mathscr{L}\{t^n\} = \frac{n}{s}\mathscr{L}\{t^{n-1}\} = \frac{n(n-1)}{s^2}\mathscr{L}\{t^{n-2}\} = \cdots = \frac{n!}{s^n}\mathscr{L}\{t^0\}.$$

But $t^0 = 1$, so by equation (7) we have $\mathscr{L}\{1\} = 1/s$ and

$$\mathscr{L}\{t^n\} = \frac{n!}{s^{n+1}}.$$

Example 6 The Laplace transform of $\sin at$ can also be obtained by integrating by parts. Let $u = \sin at$ and $dv = e^{-st}\,dt$; then $du = a\cos at\,dt$ and $v = (-1/s)e^{-st}$, so

$$\mathscr{L}\{\sin at\} = \int_0^\infty e^{-st}\sin at\,dt = \left.\frac{-e^{-st}\sin at}{s}\right|_0^\infty + \frac{a}{s}\int_0^\infty e^{-st}\cos at\,dt,$$

or

$$\mathscr{L}\{\sin at\} = \frac{a}{s}\mathscr{L}\{\cos at\}. \tag{13}$$

We must integrate this last integral again by parts with $u = \cos at$ and $dv = e^{-st}\,dt$; we have $du = -a\sin at\,dt$ and $v = -(1/s)e^{-st}$, yielding

$$\mathscr{L}\{\sin at\} = \frac{a}{s}\int_0^\infty e^{-st}\cos at\,dt = \frac{a}{s}\left[\left.\frac{-e^{-st}\cos at}{s}\right|_0^\infty - \frac{a}{s}\int_0^\infty e^{-st}\sin at\,dt\right],$$

or

$$\mathscr{L}\{\sin at\} = \frac{a}{s^2} - \frac{a^2}{s^2}\mathscr{L}\{\sin at\}. \tag{14}$$

Moving the last term on the right-hand side of equation (14) to the left-hand side, we get

$$\left(1 + \frac{a^2}{s^2}\right)\mathscr{L}\{\sin at\} = \frac{a}{s^2},$$

or

$$\frac{s^2 + a^2}{s^2}\mathscr{L}\{\sin at\} = \frac{a}{s^2}.$$

Multiplying both sides by $s^2/(s^2+a^2)$, we obtain

$$\mathscr{L}\{\sin at\} = \frac{a}{s^2+a^2}.$$

Equation (10) follows at once from equations (9) and (13), since

$$\mathscr{L}\{\cos at\} = \frac{s}{a}\mathscr{L}\{\sin at\} = \frac{s}{s^2+a^2}.$$

(Explain why these computations are valid only if $s > 0$.)

Problem 41 illustrates another method for finding the Laplace transform of $\cos at$ and $\sin at$ by using the Euler formula (see Appendix 5)

$$e^{iat} = \cos(at) + i\sin(at).$$

Example 7 Since $\sinh at = (e^{at} - e^{-at})/2$, we can use Theorem 2 and equation (6) to obtain

$$\mathscr{L}\{\sinh at\} = \mathscr{L}\left\{\frac{e^{at}-e^{-at}}{2}\right\} = \frac{1}{2}\left(\mathscr{L}\{e^{at}\} - \mathscr{L}\{e^{-at}\}\right)$$

$$= \frac{1}{2}\left(\frac{1}{s-a} - \frac{1}{s+a}\right) = \frac{a}{s^2-a^2}.$$

Similarly,

$$\mathscr{L}\{\cosh at\} = \mathscr{L}\left\{\frac{e^{at}+e^{-at}}{2}\right\} = \frac{1}{2}\left(\mathscr{L}\{e^{at}\} + \mathscr{L}\{e^{-at}\}\right)$$

$$= \frac{1}{2}\left(\frac{1}{s-a} + \frac{1}{s+a}\right) = \frac{s}{s^2-a^2}.$$

Example 8 Compute $\mathscr{L}\{3t^5 - t^8 + 4 - 5e^{2t} + 6\cos 3t\}$.

Solution Using the linearity property and the results in Table 6.1, we have

$$\mathscr{L}\{3t^5 - t^8 + 4 - 5e^{2t} + 6\cos 3t\}$$
$$= 3\mathscr{L}\{t^5\} - \mathscr{L}\{t^8\} + \mathscr{L}\{4\} - 5\mathscr{L}\{e^{2t}\} + 6\mathscr{L}\{\cos 3t\}$$
$$= 3\frac{5!}{s^6} - \frac{8!}{s^9} + \frac{4}{s} - \frac{5}{s-2} + \frac{6s}{s^2+9}$$
$$= \frac{360}{s^6} - \frac{40{,}320}{s^9} + \frac{4}{s} - \frac{5}{s-2} + \frac{6s}{s^2+9}.$$

There are many other facts that can be used to facilitate the computation of Laplace transforms. One of these is given below. Other facts will be given in the exercises and in the next three sections.

The next theorem presents a quick way of computing the Laplace transform $\mathscr{L}\{e^{at}f(t)\}$ when $\mathscr{L}\{f(t)\}$ is known.

CHAPTER 6/LAPLACE TRANSFORMS

THEOREM 3: FIRST SHIFTING PROPERTY OF THE LAPLACE TRANSFORM

Suppose that $F(s) = \mathscr{L}\{f(t)\}$ exists for $s > b$. If a is a real number, then

$$\boxed{\mathscr{L}\{e^{at}f(t)\} = F(s-a), \quad \text{for } s > a + b.} \tag{15}$$

Rewriting (15) in terms of the inverse Laplace transform, we have

$$\boxed{\mathscr{L}^{-1}\{F(s-a)\} = e^{at}f(t).} \tag{16}$$

Proof By definition,

$$\mathscr{L}\{e^{at}f(t)\} = \int_0^\infty e^{-st} e^{at} f(t)\, dt = \int_0^\infty e^{-(s-a)t} f(t)\, dt$$

$$= \mathscr{L}\{f(t)\}\big|_{s-a} = F(s-a),$$

if $s - a > b$ or $s > a + b$. As the formula suggests, to find $\mathscr{L}\{e^{at}f(t)\}$ we simply replace each s in $\mathscr{L}\{f(t)\}$ by $s - a$. ∎

Example 9 Compute $\mathscr{L}\{e^{2t}\cos 3t\}$.

Solution Since

$$F(s) = \mathscr{L}\{\cos 3t\} = \frac{s}{s^2 + 9}$$

and $a = 2$, we have

$$\mathscr{L}\{e^{2t}\cos 3t\} = F(s-2) = \frac{s-2}{(s-2)^2 + 9}.$$

If we apply the first shifting property to the Laplace transforms in Table 6.1, we obtain the results shown in Table 6.2.

TABLE 6.2

$f(t)$	$\mathscr{L}\{f(t)\}$	Domain of definition
$e^{at}t^n$	$\dfrac{n!}{(s-a)^{n+1}}$	$s > a$
$e^{at}\sin bt$	$\dfrac{b}{(s-a)^2 + b^2}$	$s > a$
$e^{at}\cos bt$	$\dfrac{s-a}{(s-a)^2 + b^2}$	$s > a$
$e^{at}\sinh bt$	$\dfrac{b}{(s-a)^2 - b^2}$	$s > a + |b|$
$e^{at}\cosh bt$	$\dfrac{s-a}{(s-a)^2 - b^2}$	$s > a + |b|$

In Section 6.2 we will have frequent need to evaluate inverse Laplace transforms. The next example demonstrates the use of completing the square in finding inverse Laplace transforms.

Example 10 Compute
$$\mathscr{L}^{-1}\left\{\frac{s+9}{s^2+6s+13}\right\}.$$

Solution Example 9 provides a hint. We begin by completing the square in the denominator:
$$\frac{s+9}{s^2+6s+13} = \frac{s+9}{(s+3)^2+4}.$$

But $4 = 2^2$ and $s + 9 = (s + 3) + 6$, so we can write
$$\frac{s+9}{s^2+6s+13} = \frac{(s+3)+6}{(s+3)^2+2^2}.$$

By the linearity property of the inverse Laplace transform (see the corollary to Theorem 2), the inverse transform of a sum equals the sum of the inverse transforms. Hence

$$\mathscr{L}^{-1}\left\{\frac{(s+3)+6}{(s+3)^2+2^2}\right\} = \mathscr{L}^{-1}\left\{\frac{(s+3)}{(s+3)^2+2^2}\right\} + 3\mathscr{L}^{-1}\left\{\frac{2}{(s+3)^2+2^2}\right\},$$

and using Table 6.2 (or the first shifting property and Table 6.1), we have
$$\mathscr{L}^{-1}\left\{\frac{(s+3)+6}{(s+3)^2+2^2}\right\} = e^{-3t}\cos 2t + 3e^{-3t}\sin 2t.$$

PROBLEMS 6.1

Find the Laplace transforms of the functions in Problems 1–40, where a, b, and c are real constants. For what values of s are the transforms defined?

1. $5t + 2$
2. $7t - 8$
3. $9t^2 - 7$
4. $16t^2 - 4t$
5. $t^2 + 8t - 16$
6. $27t^3 - 9t + 4$
7. $\dfrac{t^3}{8} + \dfrac{t^2}{4} + \dfrac{t}{2} + 1$
8. $\dfrac{t^5}{120} + \dfrac{t^2}{6} + 1$
9. $at + b$
10. $at^2 + bt + c$
11. e^{5t+2}
12. e^{7t-8}
13. $e^{t/2}$
14. $e^{-t/3}$
15. $e^{-t-1/2}$
16. e^{at+b}
17. $\sin(3t)$
18. $\sin\left(\dfrac{t}{2}\right)$
19. $\cos(7t)$
20. $\cos(-t/3)$
21. $\sin(5t + 2)$
22. $\cos(7t - 8)$
23. $\cos(at + b)$
24. $\sin(at + b)$
25. $\cosh(t/2)$
26. $\sinh(-t/3)$
27. $\cosh(5t - 2)$
28. $\sinh(7t + 8)$
29. $\sinh(at + b)$
30. $\cosh(at + b)$
31. te^t
32. $t^2 e^{2t}$
33. $(t^3 - 1)e^{-t}$
34. $e^{3t}(t^2 + t)$
35. $e^t \sin t$
36. $e^{-t}\sin 2t$
37. $e^{4t}\cos 2t$
38. $e^{-t}\sinh 2t$
39. $e^{-t}(\sin t + \cos t)$
40. $e^{2t}(t + \cosh t)$
41. Recall that $e^{iat} = \cos at + i \sin at$ (see equation (14) in Appendix 5).

 a. Show that $\mathscr{L}\{e^{iat}\} = \dfrac{1}{s - ia}$; $s > 0$.

 b. Show that $\dfrac{1}{s - ia} = \dfrac{s + ia}{s^2 + a^2}$.

 c. Use parts (a) and (b) to derive (without integration by parts) the formulas for $\mathscr{L}\{\sin at\}$ and $\mathscr{L}\{\cos at\}$. [*Hint*: Equate real and imaginary parts.]

In Problems 42–55 find $f(t)$ where $F(s) = \mathcal{L}\{f(t)\}$ is given. If necessary, complete the square in the denominator.

42. $\dfrac{7}{s^2}$

43. $\dfrac{18}{s^3} + \dfrac{7}{s}$

44. $\dfrac{a_1}{s} + \dfrac{a_2}{s^2} + \dfrac{a_3}{s^3} + \cdots + \dfrac{a_{n+1}}{s^{n+1}}$

45. $\dfrac{s+1}{s^2+1}$

46. $\dfrac{7}{s-3}$

47. $\dfrac{s-2}{s^2-2}$

48. $\dfrac{s-2}{s^2+3}$

49. $\dfrac{1}{(s-1)^2}$

50. $\dfrac{3}{s^2+2s+2}$

51. $\dfrac{3}{s^2+4s+9}$

52. $\dfrac{s+12}{s^2+10s+35}$

53. $\dfrac{2s-1}{s^2+2s+8}$

54. $\dfrac{7s-8}{s^2+9s+25}$

55. $\dfrac{cs+d}{s^2+2as+b}$ $\quad b > a^2 > 0$; a, b, c, d are real

In Problems 56–61 express each given hyperbolic function in terms of exponentials and apply the first shifting theorem to prove the given equality.

56. $\mathcal{L}\{\cosh^2 at\} = \dfrac{s^2-2a^2}{s(s^2-4a^2)}$

57. $\mathcal{L}\{\sinh^2 at\} = \dfrac{2a^2}{s(s^2-4a^2)}$

58. $\mathcal{L}\{\cosh at \sin at\} = \dfrac{a(s^2+2a^2)}{s^4+4a^4}$

59. $\mathcal{L}\{\cosh at \cos at\} = \dfrac{s^3}{s^4+4a^4}$

60. $\mathcal{L}\{\sinh at \sin at\} = \dfrac{2a^2 s}{s^4+4a^4}$

61. $\mathcal{L}\{\sinh at \cos at\} = \dfrac{a(s^2-2a^2)}{s^4+4a^4}$

Using the method above, find the Laplace transforms in Problems 62–67.

62. $\mathcal{L}\{\cosh at \cosh bt\}$
63. $\mathcal{L}\{\sinh at \sinh bt\}$
64. $\mathcal{L}\{\cosh at \sin bt\}$
65. $\mathcal{L}\{\cosh at \cos bt\}$
66. $\mathcal{L}\{\sinh at \sin bt\}$
67. $\mathcal{L}\{\sinh at \cos bt\}$

68. Suppose that $F(s) = \mathcal{L}\{f(t)\}$ exists for $s > a$. Show that

$$\mathcal{L}\{tf(t)\} = -F'(s), \quad \text{for } s > a. \quad (i)$$

[*Hint:* Assume that you can interchange the derivative and integral on the right-hand side of equation (i).]

69. Use equation (i) to show that

$$\mathcal{L}\{t^n f(t)\} = (-1)^n \dfrac{d^n}{ds^n} F(s), \quad \text{for } s > a. \quad (ii)$$

Use equations (i) and (ii) to compute the Laplace transform of the functions given in Problems 70–78. Assume that a and b are real.

70. te^t
71. $t^3 e^{-t}$
72. $t \sin t$
73. $t^2 \cos 3t$
74. $te^t \sin t$
75. $te^{at} \cos bt$
76. $te^{at} \sin bt$
77. $3te^{-t} \cosh t$
78. $te^{-t} \sinh 2t$

79. Suppose that $f(t) = \mathcal{L}^{-1}\{F(s)\}$ and that $g(t) = \mathcal{L}^{-1}\{G(s)\}$. Prove the linearity property of the inverse Laplace transform:

$$af(t) + bg(t) = \mathcal{L}^{-1}\{aF(s) + bG(s)\},$$

where a and b are any real constants.

80. The **gamma function** is defined by

$$\Gamma(x) = \int_0^\infty e^{-u} u^{x-1} \, du, \quad x > 0.$$

 a. Show that $\Gamma(x+1) = \int_0^\infty e^{-u} u^x \, du$.
 b. By integrating by parts, show that $\Gamma(x+1) = x\Gamma(x)$.
 c. Show that $\Gamma(1) = 1$.
 d. Using the results of parts (b) and (c), show that if n is a positive integer, then $\Gamma(n+1) = n!$.
 e. By making the substitution $u = st$ in part (a), show that

$$\mathcal{L}\{t^x\} = \dfrac{\Gamma(x+1)}{s^{x+1}}, \quad s > 0, \ x > -1.$$

81. It can be shown that $\Gamma(1/2) = \sqrt{\pi}$. Use this fact and the results of Problem 80 to compute the following:

 a. $\mathcal{L}\left\{\dfrac{1}{\sqrt{t}}\right\}$
 b. $\mathcal{L}\{\sqrt{t}\}$
 c. $\mathcal{L}\{t^{5/2}\}$

6.2 Solving Initial-Value Problems by Laplace Transform Methods

In this section we show how the theory developed in Section 6.1 can be applied to solve linear initial-value problems. We see that the Laplace transform converts linear initial-value problems with constant coefficients into algebraic equations whose solutions are the Laplace transforms of the solutions to the initial-value problems.

The most important property of Laplace transforms for solving differential equations concerns the transform of the derivative of a function f. We prove below that differentiation of f roughly corresponds to multiplication of the transform by s.

THEOREM 1: DIFFERENTIATION PROPERTY

Let $f(t)$ satisfy the condition

$$|f(t)| \leq Me^{at} \tag{1}$$

for $t \geq T$, for fixed nonnegative constants a, M, and T, and suppose that $f'(t)$ is piecewise continuous for $t \geq 0$. Then the Laplace transform of $f'(t)$ exists for all $s > a$, and

$$\boxed{\mathscr{L}\{f'(t)\} = s\mathscr{L}\{f(t)\} - f(0).} \tag{2}$$

Proof Since f is differentiable, it is also continuous. Hence it satisfies the conditions of the existence theorem (Theorem 6.1.1) and has a Laplace transform. Suppose, first, that $f'(t)$ is continuous on $t \geq 0$. Then integrating $\mathscr{L}\{f'(t)\}$ by parts, we set $u = e^{-st}$ and $dv = f'(t)\,dt$; then $du = -se^{-st}\,dt$, $v = f(t)$, and

$$\mathscr{L}\{f'(t)\} = \int_0^\infty e^{-st} f'(t)\,dt = e^{-st} f(t)\Big|_0^\infty + s\int_0^\infty e^{-st} f(t)\,dt. \tag{3}$$

Since $f(t)$ satisfies equation (1), the first term on the right-hand side in equation (3) vanishes at the upper limit when $s > a$, and by definition we obtain $\mathscr{L}\{f'(t)\} = s\mathscr{L}\{f(t)\} - f(0)$. When $f'(t)$ is piecewise continuous, the proof is similar. We simply break up the range of integration into parts on each of which $f'(t)$ is continuous and integrate by parts as in equation (3). All first terms cancel out or vanish except $-f(0)$, and the second terms combine to yield $s\mathscr{L}\{f(t)\}$. ∎

Theorem 1 may be extended to apply to piecewise continuous functions $f(t)$ (see Problem 50).

Equation (2) may be applied repeatedly to obtain the Laplace transform of higher-order derivatives:

$$\mathscr{L}\{f''(t)\} = s\mathscr{L}\{f'(t)\} - f'(0) = s[s\mathscr{L}\{f(t)\} - f(0)] - f'(0),$$

or

$$\boxed{\mathscr{L}\{f''(t)\} = s^2\mathscr{L}\{f(t)\} - sf(0) - f'(0).} \tag{4}$$

Similarly,
$$\mathscr{L}\{f'''(t)\} = s^3\mathscr{L}\{f(t)\} - s^2 f(0) - sf'(0) - f''(0),$$

leading by induction to the following extension of Theorem 1.

THEOREM 2

Let $f^{(k)}(t)$ satisfy inequality (1) for $k = 0, 1, 2, \ldots, n-1$, and suppose that $f^{(n)}(t)$ is piecewise continuous on $t \geq 0$. Then $\mathscr{L}\{f^{(n)}(t)\}$ exists and is given by

$$\mathscr{L}\{f^{(n)}(t)\} = s^n\mathscr{L}\{f(t)\} - s^{n-1}f(0) - s^{n-2}f'(0) - \cdots - f^{(n-1)}(0),$$

or

$$\mathscr{L}\{f^{(n)}(t)\} = s^n\mathscr{L}\{f(t)\} - \sum_{j=0}^{n-1} s^{n-j-1}f^{(j)}(0). \qquad (5)$$

Theorems 1 and 2 are important, since they are used to reduce the Laplace transform of a differential equation into an equation involving only the transform of the solution. Several such applications are considered in this section. However, these theorems are also useful in determining the transforms of certain functions.

Example 1 Compute $\mathscr{L}\{\sin^2 at\}$.

Solution Let $f(t) = \sin^2 at$. Then
$$f'(t) = 2a \sin at \cos at = a \sin 2at,$$

so
$$\frac{2a^2}{s^2 + 4a^2} = \mathscr{L}\{f'\} = s\mathscr{L}\{f\} - f(0).$$

Since $f(0) = 0$, it follows that
$$\mathscr{L}\{\sin^2 at\} = \frac{2a^2}{s(s^2 + 4a^2)}, \quad s > 0.$$

Example 2 Compute $\mathscr{L}\{t \sin at\}$.

Solution Suppose that $f(t) = t \sin at$. Then
$$f'(t) = \sin at + at \cos at,$$
$$f''(t) = 2a \cos at - a^2 t \sin at.$$

Thus, since $f(0) = f'(0) = 0$,
$$2a\mathscr{L}\{\cos at\} - a^2\mathscr{L}\{f(t)\} = \mathscr{L}\{f''(t)\} = s^2\mathscr{L}\{f(t)\},$$

so
$$(s^2 + a^2)\mathscr{L}\{f(t)\} = 2a\mathscr{L}\{\cos at\} = \frac{2as}{s^2 + a^2},$$
or
$$\mathscr{L}\{f(t)\} = \frac{2as}{(s^2 + a^2)^2}, \qquad s > 0.$$

An alternate technique for computing this answer is provided later in this section (p. 228).

We now apply Theorems 1 and 2 to solve initial-value problems. In what follows, we take Laplace transforms without worrying about their existence. Any solution so obtained must be checked by substitution into the original equation.

Example 3   Find the solution of the initial-value problem
$$y'' - 4y = 0, \qquad y(0) = 1, \qquad y'(0) = 2. \tag{6}$$

Solution   Taking the Laplace transform of both sides of the differential equation in (6) and using the differentiation property, we transform equation (6) into the algebraic equation
$$\left[s^2\mathscr{L}\{y\} - sy(0) - y'(0)\right] - 4\mathscr{L}\{y\} = \left[s^2\mathscr{L}\{y\} - s - 2\right] - 4\mathscr{L}\{y\} = 0;$$
thus
$$\mathscr{L}\{y\} = \frac{s+2}{s^2 - 4} = \frac{1}{s - 2}.$$

By Table 6.1 (p. 215) we have
$$y(t) = e^{2t},$$
which satisfies all the conditions in (6).

Example 4   Solve the initial-value problem
$$y'' + 4y = 0, \qquad y(0) = 1, \qquad y'(0) = 2. \tag{7}$$

Solution   Using the differentiation property, we obtain
$$s^2\mathscr{L}\{y\} - sy(0) - y'(0) + 4\mathscr{L}\{y\} = s^2\mathscr{L}\{y\} - s - 2 + 4\mathscr{L}\{y\} = 0.$$
Solving for $\mathscr{L}\{y\}$, we have
$$\mathscr{L}\{y\} = \frac{s+2}{s^2 + 4} = \left(\frac{s}{s^2 + 4}\right) + \left(\frac{2}{s^2 + 4}\right).$$

By referring to Table 6.1 we find that
$$y(t) = \cos 2t + \sin 2t,$$
which can readily be verified to be the solution of (7). In calculating the inverse transform, we used the linearity of $\mathscr{L}^{-1}$.

The preceding example indicates the necessity of writing $\mathscr{L}\{y\}$ as a linear combination of terms for which the inverse Laplace transforms are known.

Example 5 Find the solution of the initial-value problem
$$y'' - 3y' + 2y = 4t - 6, \quad y(0) = 1, \quad y'(0) = 3. \tag{8}$$

Solution Taking the Laplace transform of both sides and using the differentiation property, we have, from Table 6.1,
$$[s^2 \mathscr{L}\{y\} - s - 3] - 3[s\mathscr{L}\{y\} - 1] + 2\mathscr{L}\{y\} = \frac{4}{s^2} - \frac{6}{s},$$

so
$$(s^2 - 3s + 2)\mathscr{L}\{y\} = s + \frac{4}{s^2} - \frac{6}{s} = \frac{s^3 - 6s + 4}{s^2}.$$

Hence, factoring the numerator and denominator, we obtain
$$\mathscr{L}\{y\} = \frac{s^3 - 6s + 4}{s^2(s^2 - 3s + 2)}$$
$$= \frac{(s-2)(s^2 + 2s - 2)}{s^2(s-2)(s-1)} = \frac{s^2 + 2s - 2}{s^2(s-1)}.$$

But
$$\frac{s^2 + 2s - 2}{s^2(s-1)} = \frac{s^2}{s^2(s-1)} + \frac{2s - 2}{s^2(s-1)} = \frac{1}{s-1} + \frac{2}{s^2},$$

so
$$\mathscr{L}\{y\} = \frac{1}{s-1} + \frac{2}{s^2}.$$

Using Table 6.1, we obtain the solution
$$y = e^t + 2t$$
to the initial-value problem in equation (8).

We now discuss the most general second-order linear initial-value problem with constant coefficients. Suppose that we wish to solve the nonhomogeneous differential equation with constant coefficients
$$y'' + ay' + by = f(t), \quad y(0) = y_0, \quad y'(0) = y_1. \tag{9}$$

The general existence-uniqueness theorem (Theorem 3.1.1 on p. 84) states that the initial-value problem (9) has a unique solution if $f(t)$ is continuous. Assuming that this is the case, and taking the Laplace transforms of both sides, we obtain
$$\mathscr{L}\{y''\} + a\mathscr{L}\{y'\} + b\mathscr{L}\{y\} = \mathscr{L}\{f\}.$$

Now by Theorems 1 and 2 (differentiation properties) we have
$$[s^2\mathscr{L}\{y\} - sy(0) - y'(0)] + a[s\mathscr{L}\{y\} - y(0)] + b\mathscr{L}\{y\} = \mathscr{L}\{f\}.$$

Then
$$[s^2 + as + b]\mathscr{L}\{y\} - [sy(0) + ay(0) + y'(0)] = \mathscr{L}\{f\},$$

so
$$\mathcal{L}\{y\} = \frac{(s+a)y(0) + y'(0) + \mathcal{L}\{f\}}{s^2 + as + b}. \tag{10}$$

Three facts are evident from equation (10):

a. Initial conditions must be given so that the first two terms in the numerator of (10) are determined.

b. The function f must have a Laplace transform so that the last term in the numerator is determined.

c. We must be able to find $\mathcal{L}^{-1}$ of the right-hand side of (10) in order to determine the solution y of the initial-value problem (9).

Thus Laplace transform methods are primarily intended for the solution of linear initial-value problems with constant coefficients.

It should be clear that the major difficulty in solving problem (9) lies in finding the inverse transform of the right-hand side of equation (10). There is a general formula that provides the solution as an integral, but some knowledge of complex variable theory is required to take full advantage of this formula. Fortunately, many of the transforms you will encounter in solving initial-value problems can be inverted using techniques from calculus. We illustrate with some examples.

Example 6 Solve the initial-value problem
$$y'' - 5y' + 4y = e^{2t}, \qquad y(0) = 1, \qquad y'(0) = 0.$$

Solution Making use of the differentiation property and Table 6.1, we have
$$[s^2\mathcal{L}\{y\} - sy(0) - y'(0)] - 5[s\mathcal{L}\{y\} - y(0)] + 4\mathcal{L}\{y\} = \mathcal{L}\{e^{2t}\},$$
or
$$[s^2\mathcal{L}\{y\} - s] - 5[s\mathcal{L}\{y\} - 1] + 4\mathcal{L}\{y\} = \frac{1}{s-2},$$
so
$$(s^2 - 5s + 4)\mathcal{L}\{y\} = s - 5 + \frac{1}{s-2} = \frac{s^2 - 7s + 11}{s-2}.$$

Then
$$\mathcal{L}\{y\} = \frac{s^2 - 7s + 11}{(s-2)(s^2 - 5s + 4)} = \frac{s^2 - 7s + 11}{(s-2)(s-1)(s-4)}. \tag{11}$$

Review of Partial Fractions

At this point we pause. Remember that when you studied techniques of integration in calculus, you integrated functions like the right-hand side of equation (11) by using the method of **partial fractions**. This method is useful here. We seek constants A, B, and C such that
$$\frac{A}{s-2} + \frac{B}{s-1} + \frac{C}{s-4} = \frac{s^2 - 7s + 11}{(s-2)(s-1)(s-4)}. \tag{12}$$

Why? Because we know that $\mathscr{L}^{-1}\{1/(s-2)\} = e^{2t}$, so $\mathscr{L}^{-1}\{A/(s-2)\} = Ae^{2t}$, and so on.

There is an easy method for finding these constants:

$$A = \left.\frac{s^2 - 7s + 11}{(s-1)(s-4)}\right|_{s=2} = -\frac{1}{2};$$

$$B = \left.\frac{s^2 - 7s + 11}{(s-2)(s-4)}\right|_{s=1} = \frac{5}{3};$$

$$C = \left.\frac{s^2 - 7s + 11}{(s-2)(s-1)}\right|_{s=4} = -\frac{1}{6}.$$

Observe that we eliminate the denominator $(s-a)$ of each term on the left-hand side of equation (12) from the right-hand side of equation (12) and evaluate the resulting equation at $s = a$ to obtain the desired constant.

To understand why this procedure works, let us multiply both sides of equation (12) by $(s-2)$. Then we have

$$A + (s-2)\left(\frac{B}{s-1} + \frac{C}{s-4}\right) = \frac{s^2 - 7s + 11}{(s-1)(s-4)}. \tag{13}$$

Setting $s = 2$ on both sides eliminates all but the constant A on the left-hand side of equation (13), and therefore

$$A = \left.\frac{s^2 - 7s + 11}{(s-1)(s-4)}\right|_{s=2}.$$

Returning to our problem, we see that

$$\mathscr{L}\{y\} = \frac{\left(-\frac{1}{2}\right)}{s-2} + \frac{\left(\frac{5}{3}\right)}{s-1} + \frac{\left(-\frac{1}{6}\right)}{s-4},$$

which implies, according to Table 6.1, that

$$y(t) = -\frac{e^{2t}}{2} + \frac{5e^t}{3} - \frac{e^{4t}}{6}.$$

Example 7   Solve
$$y'' + 2y' + 2y = t, \qquad y(0) = y'(0) = 1.$$

Solution   Using the differentiation property, we obtain

$$[s^2\mathscr{L}\{y\} - s - 1] + 2[s\mathscr{L}\{y\} - 1] + 2\mathscr{L}\{y\} = \frac{1}{s^2},$$

or

$$(s^2 + 2s + 2)\mathscr{L}\{y\} = \frac{1}{s^2} + s + 3 = \frac{s^3 + 3s^2 + 1}{s^2};$$

thus

$$\mathscr{L}\{y\} = \frac{s^3 + 3s^2 + 1}{s^2(s^2 + 2s + 2)}. \tag{14}$$

The expression $s^2 + 2s + 2$ does not have real roots, since $2^2 - 4(1)(2) = -4 < 0$, so we write the right-hand side of equation (14) as

$$\frac{s^3 + 3s^2 + 1}{s^2(s^2 + 2s + 2)} = \frac{As + B}{s^2 + 2s + 2} + \frac{C}{s} + \frac{D}{s^2}. \tag{15}$$

Why do we do this? Because

$$\frac{s+1}{s^2 + 2s + 2} = \frac{s+1}{(s+1)^2 + 1},$$

and so

$$\mathscr{L}^{-1}\left\{\frac{s+1}{s^2 + 2s + 1}\right\} = \mathscr{L}^{-1}\left\{\frac{s+1}{(s+1)^2 + 1}\right\} = e^{-t}\cos t$$

as a consequence of the first shifting theorem (see Table 6.2). Similarly,

$$\mathscr{L}^{-1}\left\{\frac{1}{(s+1)^2 + 1}\right\} = e^{-t}\sin t.$$

Also, by Table 6.1,

$$\mathscr{L}^{-1}\left\{\frac{1}{s}\right\} = 1 \quad \text{and} \quad \mathscr{L}^{-1}\left\{\frac{1}{s^2}\right\} = t.$$

Further Review of Partial Fractions

There are tricks for finding the constants A, B, C, and D in equation (15), but these are more complicated than the method we used in Example 6. We can find these constants directly by combining terms:

$$\frac{(As + B)s^2 + C(s^2 + 2s + 2)s + D(s^2 + 2s + 2)}{s^2(s^2 + 2s + 2)} = \frac{s^3 + 3s^2 + 1}{s^2(s^2 + 2s + 2)}.$$

Equating coefficients of like powers of s, we obtain

$$\begin{aligned} A + C &= 1 \quad \text{these are the coefficients of } s^3 \\ B + 2C + D &= 3 \quad \text{these are the coefficients of } s^2 \\ 2C + 2D &= 0 \quad \text{these are the coefficients of } s \\ 2D &= 1 \quad \text{these are the constant terms} \end{aligned}$$

From the last equation we see that $D = \frac{1}{2}$, so working backward, we obtain $C = -\frac{1}{2}$, $B = \frac{7}{2}$, and $A = \frac{3}{2}$. Then

$$y = \mathscr{L}^{-1}\left\{\frac{s^3 + 3s^2 + 1}{s^2(s^2 + 2s + 2)}\right\}$$

$$= \mathscr{L}^{-1}\left\{\frac{\frac{3}{2}s + \frac{7}{2}}{s^2 + 2s + 2} + \frac{(-\frac{1}{2})}{s} + \frac{(\frac{1}{2})}{s^2}\right\}$$

$$= \mathscr{L}^{-1}\left\{\frac{(\frac{3}{2})(s+1)}{(s+1)^2 + 1} + \frac{2}{(s+1)^2 + 1} + \frac{(-\frac{1}{2})}{s} + \frac{(\frac{1}{2})}{s^2}\right\},$$

or
$$y = \tfrac{3}{2}e^{-t}\cos t + 2e^{-t}\sin t - \tfrac{1}{2} + \tfrac{1}{2}t,$$
which is the solution to our differential equation.

The methods used in the last two examples apply to the problem of inverting a Laplace transform obtained in trying to solve an initial-value problem with constant coefficients. In some special cases, we can use these techniques to solve linear problems with variable coefficients. First, however, we need to prove the identities that were stated in Problems 6.1.68 and 6.1.69.

Consider the derivative
$$\frac{d}{ds}\mathscr{L}\{f(t)\} = \frac{d}{ds}\int_0^\infty e^{-st}f(t)\,dt. \tag{16}$$

If we reverse the order in which the operations of differentiation and integration are performed on the right-hand side of equation (16), we obtain
$$\frac{d}{ds}\mathscr{L}\{f(t)\} = \int_0^\infty \frac{d}{ds}e^{-st}f(t)\,dt = \int_0^\infty -te^{-st}f(t)\,dt = -\mathscr{L}\{tf(t)\}.$$

We have proved the following:

THEOREM 3

If $\mathscr{L}\{f(t)\}$ exists for $s > a$, then $\mathscr{L}\{tf(t)\}$ exists for $s > a$ and

$$\boxed{\mathscr{L}\{tf(t)\} = -\frac{d}{ds}\mathscr{L}\{f(t)\}.} \tag{17}$$

Using equation (17) repeatedly, we obtain
$$\mathscr{L}\{t^n f(t)\} = -\frac{d}{ds}\mathscr{L}\{t^{n-1}f(t)\} = (-1)^2 \frac{d^2}{ds^2}\mathscr{L}\{t^{n-2}f(t)\}$$
$$= \cdots = (-1)^n \frac{d^n}{ds^n}\mathscr{L}\{f(t)\}. \tag{18}$$

Of course it may not be legitimate to change the order of differentiation and integration in equation (16) (pathological examples do exist), but if the method succeeds in providing a correct solution to our problem, we need not be concerned. This lack of rigor reemphasizes the need for checking the final solution when solving problems by Laplace transform techniques.

Example 2 (revisited) Compute $\mathscr{L}\{t\sin at\}$.

Solution By (17),
$$\mathscr{L}\{t\sin at\} = -\frac{d}{ds}\mathscr{L}\{\sin at\} = -\frac{d}{ds}\left(\frac{a}{s^2+a^2}\right) = \frac{2as}{(s^2+a^2)^2}.$$

The following example illustrates how equations (17) and (18) can be used in conjunction with the differentiation property to solve some initial-value problems with variable coefficients.

Example 8 Solve the equation with variable coefficients
$$ty'' - ty' - y = 0, \qquad y(0) = 0, \qquad y'(0) = 3.$$

Solution If we let $Y(s) = \mathscr{L}\{y(t)\}$, then, by the differentiation property and equation (17),

$$\mathscr{L}\{ty''\} = -\frac{d}{ds}\mathscr{L}\{y''\} = -\frac{d}{ds}\{s^2 Y(s) - sy(0) - y'(0)\}$$
$$= -s^2 Y' - 2sY - y(0) = -s^2 Y' - 2sY$$

and

$$\mathscr{L}\{ty'\} = -\frac{d}{ds}\mathscr{L}\{y'\} = -\frac{d}{ds}\{sY - y(0)\} = -sY' - Y.$$

Substituting these expressions into the Laplace transform of the original equation yields

$$-s^2 Y' - 2sY + sY' + Y - Y = 0.$$

Rearranging and canceling terms, we have

$$(s^2 - s)Y' + 2sY = 0.$$

We now divide both sides by $s^2 - s = s(s-1)$ to obtain

$$Y' + \frac{2}{s-1} Y = 0.$$

Separating variables, we have

$$\frac{dY}{Y} = -\frac{2}{s-1} ds,$$

and an integration yields

$$\ln|Y| = -2\ln|s-1| + C,$$

or

$$Y(s) = \frac{c}{(s-1)^2}.$$

Thus, by Table 6.2,

$$y(t) = cte^t.$$

Note that $y(0) = 0$. To find c, we differentiate and use the second initial condition to obtain

$$3 = y'(0) = c(t+1)e^t\big|_{t=0} = c.$$

Thus the unique solution to the initial-value problem is given by

$$y(t) = 3te^t.$$

We caution the reader not to expect to be able to solve all variable coefficient equations by this method. It works only when

a. the coefficients $a_i(t)$ are polynomials in t;

b. the differential equation involving $Y(s)$ can be solved; and

c. the inverse transform of $Y(s)$ can be found.

It is rare that all these conditions can be met (see Problem 49).

Example 9 In Section 4.4 we applied Kirchhoff's law to obtain the following differential equation [equation (4.4.1)] relating the current (I), charge (Q), resistance (R), inductance (L), and capacitance (C) of the electric circuit shown in Figure 6.2:

$$L\frac{dI}{dt} + RI + \frac{Q}{C} = E. \tag{19}$$

Figure 6.2

We then differentiated equation (19) using the fact that

$$\frac{dQ}{dt} = I \tag{20}$$

and assuming that E was constant, and we obtained a homogeneous, second-order equation that we then solved. Now we make our model more realistic by assuming that the electromotive force (emf) $E = E(t)$ is a nonconstant function of time. Replacing each I in equation (19) by the identity in equation (20), we obtain

$$L\frac{dQ^2}{dt^2} + R\frac{dQ}{dt} + \frac{Q}{C} = E(t). \tag{21}$$

If $Q(0) = a$ and $Q'(0) = I(0) = b$, we can solve equation (21) by Laplace transform methods and use that solution to point out a feature common to many physical and biological systems.

We set $\hat{Q}(s) = \mathscr{L}\{Q(t)\}$, $\hat{E}(s) = \mathscr{L}\{E(t)\}$, and take the transform of both sides of equation (21), using the differentiation property, to obtain

$$L[s^2\hat{Q}(s) - as - b] + R[s\hat{Q}(s) - a] + \frac{1}{C}\hat{Q}(s) = \hat{E}(s),$$

or

$$\hat{Q}(s)\left[Ls^2 + Rs + \frac{1}{C}\right] = [Las + Lb + Ra] + \hat{E}(s). \tag{22}$$

Now let $U(s) = Ls^2 + Rs + 1/C$ and $V(s) = Las + Lb + Ra$ and rewrite equation (22) as

$$\hat{Q}(s) = \frac{\hat{E}(s) + V(s)}{U(s)} = \hat{E}(s)\left[\frac{1 + V(s)/\hat{E}(s)}{U(s)}\right]. \tag{23}$$

Defining $T(s) = (1 + V/\hat{E})/U$, we can rewrite equation (23) in the form

$$\hat{Q}(s) = \hat{E}(s)T(s). \tag{24}$$

Black Boxes and Transfer Functions

Engineers find the compartment or "black box" concept to be very useful in their work. In a **black box** something goes in (the input) and is transformed into something that comes out (the output). In this example, we have the black box setup shown in Figure 6.3. In equation (24), we have a relationship between the transforms of the input and the output. The function $T(s)$ is called a **transfer function** and describes, precisely, the inner workings of the black box subject to the driving function $E(t)$. Equation (24) tells us exactly what we get out in terms of what we put in and gives us a simple equation relating input to output.

Figure 6.3

In Example 10 and Sections 6.4 and 6.5, we will discuss several techniques that apply in solving black box problems.

Example 10 Suppose that the circuit in Figure 6.2 is connected at $t = 0$ to the emf $E(t) = \cos t$, so that $Q(0) = 0$ and $I(0) = 0$. Assume that $L = 1$ henry, $R = 6$ ohms, and $C = \frac{1}{9}$ farad. Then we have the initial-value problem

$$\frac{dI}{dt} + 6I + 9Q = \cos t, \qquad Q(0) = 0, \qquad I(0) = 0.$$

Using the identity (20), we obtain

$$\frac{d^2Q}{dt^2} + 6\frac{dQ}{dt} + 9Q = \cos t, \qquad Q(0) = Q'(0) = 0, \tag{25}$$

and if we use the differentiation property of Laplace transforms, equation (25) becomes

$$s^2 \hat{Q} + 6s\hat{Q} + 9\hat{Q} = \frac{s}{s^2 + 1},$$

or

$$\hat{Q}(s) \cdot (s^2 + 6s + 9) = \frac{s}{s^2 + 1}. \tag{26}$$

Comparing equations (22) and (26), we note that $U(s) = s^2 + 6s + 9 = (s + 3)^2$ and $V(s) = 0$, so the transfer function $T(s) = 1/U(s)$. Solving for $\hat{Q}$ we have

$$\hat{Q}(s) = \frac{s}{s^2 + 1} \cdot \frac{1}{(s + 3)^2},$$

which we wish to write in partial fraction form:

$$\frac{s}{(s^2 + 1)(s + 3)^2} = \frac{As + B}{s^2 + 1} + \frac{C}{s + 3} + \frac{D}{(s + 3)^2}. \tag{27}$$

Getting the common divisor on the right-hand side of equation (27), we have

$$s = (As + B)(s + 3)^2 + C(s + 3)(s^2 + 1) + D(s^2 + 1)$$
$$= (A + C)s^3 + (6A + B + 3C + D)s^2 + (9A + 6B + C)s + (9B + 3C + D).$$

Equating like powers of s on both sides of the equation, we obtain the system

$$\begin{aligned} A \quad + C \quad &= 0, \\ 6A + B + 3C + D &= 0, \\ 9A + 6B + C \quad &= 1, \\ 9B + 3C + D &= 0. \end{aligned} \quad (28)$$

Subtracting the first equation from the third and the fourth from the second, we get

$$\begin{aligned} 8A + 6B &= 1, \\ 6A - 8B &= 0, \end{aligned}$$

from which we obtain $A = 0.08$, $B = 0.06$, $C = -0.08$, and $D = -0.3$. Hence

$$\hat{Q}(s) = \frac{0.08s}{s^2 + 1} + \frac{0.06}{s^2 + 1} - \frac{0.08}{s + 3} - \frac{0.3}{(s + 3)^2},$$

which, by Tables 6.1.1 and 6.1.2, yields

$$Q(t) = 0.08 \cos t + 0.06 \sin t - 0.08 e^{-3t} - 0.3 t e^{-3t}. \quad (29)$$

Differentiating equation (29), we obtain the expression for the current:

$$I(t) = -0.08 \sin t + 0.06 \cos t - 0.06 e^{-3t} + 0.9 t e^{-3t}.$$

There are many other situations that can be modeled using the abstract of a black box and an appropriate transfer function relating output to input.

The following brief list of Laplace transforms will be useful in doing the exercises.

TABLE 6.3 Short Table of Laplace Transforms

$f(t)$	$\mathscr{L}\{f(t)\}$	$f(t)$	$\mathscr{L}\{f(t)\}$
c	$\dfrac{c}{s}$	e^{at}	$\dfrac{1}{s - a}$
t^n	$\dfrac{n!}{s^{n+1}}$	$e^{at} t^n$	$\dfrac{n!}{(s - a)^{n+1}}$
$\sin bt$	$\dfrac{b}{s^2 + b^2}$	$e^{at} \sin bt$	$\dfrac{b}{(s - a)^2 + b^2}$
$\cos bt$	$\dfrac{s}{s^2 + b^2}$	$e^{at} \cos bt$	$\dfrac{s - a}{(s - a)^2 + b^2}$
$\sinh bt$	$\dfrac{b}{s^2 - b^2}$	$e^{at} \sinh bt$	$\dfrac{b}{(s - a)^2 - b^2}$
$\cosh bt$	$\dfrac{s}{s^2 - b^2}$	$e^{at} \cosh bt$	$\dfrac{s - a}{(s - a)^2 - b^2}$

$$\mathscr{L}\{f'(t)\} = s\mathscr{L}\{f(t)\} - f(0), \quad \mathscr{L}\{f''(t)\} = s^2 \mathscr{L}\{f(t)\} - sf(0) - f'(0)$$

$$\mathscr{L}\{tf(t)\} = -\frac{d}{ds}\mathscr{L}\{f(t)\}, \quad \mathscr{L}\{t^n f(t)\} = (-1)^n \frac{d^n}{ds^n}\mathscr{L}\{f(t)\}.$$

PROBLEMS 6.2

In Problems 1–20 solve the given initial-value problems.

1. $y'' + y = 0$, $y(0) = 1$, $y'(0) = 0$
2. $y'' + y' = 0$, $y(0) = 0$, $y'(0) = 1$
3. $y'' - a^2 y = 0$, $y(0) = A$, $y'(0) = B$
4. $y'' - ay' = 0$, $y(0) = 1$, $y'(0) = a$
5. $y'' + 2y' + 5y = 0$, $y(0) = y'(0) = 1$
6. $y'' - y' + y = 0$, $y(0) = y'(0) = 1$
7. $y'' - 4y' + 3y = 1$, $y(0) = 1$, $y'(0) = 4$
8. $y'' - 2y' - 3y = 5$, $y(0) = 0$, $y'(0) = 1$
9. $y'' - 9y = t$, $y(0) = 1$, $y'(0) = 2$
10. $y'' - 3y' - 4y = t^2$, $y(0) = 2$, $y'(0) = 1$
11. $y''' + y = 0$, $y(0) = y''(0) = 1$, $y'(0) = -1$
12. $y^{(4)} - y = 0$, $y(0) = y''(0) = 1$, $y'(0) = y'''(0) = 0$
13. $y^{(4)} - y = 0$, $y(0) = y''(0) = 0$, $y'(0) = y'''(0) = 1$
14. $y''' - 3y' - 2y = e^{2t}$, $y(0) = y'(0) = 0$, $y''(0) = 1$
15. $y'' + k^2 y = \cos kt$, $y(0) = 0$, $y'(0) = k$ [*Hint:* Look at Example 2.]
16. $y'' + 4y = \cos t$, $y(0) = y'(0) = 0$
17. $y'' + a^2 y = \sin at$, $y(0) = a$, $y'(0) = a^2$
18. $y'' - y = te^t$, $y(0) = y'(0) = 1$
19. $y^{(4)} - y = \cos t$, $y(0) = y''(0) = 1$, $y'(0) = y'''(0) = 0$
20. $y^{(4)} - y = \sinh t$, $y(0) = y''(0) = 0$, $y'(0) = y'''(0) = 1$

In Problems 21–29 find the Laplace transform of each function by using the differentiation property or equation (18).

21. $\cos^2 at$
22. $t \cos at$
23. $t^2 \sin at$
24. $t^2 \cos at$
25. $t \sin^2 t$
26. $t \cos^2 t$
27. $t \sin^2 at$
28. $t^2 \cos^2 3t$
29. $t^2 \sin^2 2t$

30. By reversing the order of integration, show that, if $F(s) = \mathscr{L}\{f(t)\}$, then

$$\int_s^\infty F(s)\, ds = \mathscr{L}\left\{\frac{f(t)}{t}\right\}. \quad \text{(i)}$$

In particular, it follows that

$$\mathscr{L}^{-1}\left\{\int_s^\infty F(s)\, ds\right\} = \frac{f(t)}{t}.$$

31. Let $g(t) = \int_0^t f(u)\, du$. Using calculus and the differentiation property (Theorem 1), show that
    *a. $g'(t) = f(t)$ at all points of continuity of $f(t)$;
    b. $\mathscr{L}\{f(t)\} = s\mathscr{L}\{g(t)\} - g(0)$;
    c. $g(0) = 0$.
    Finally, using parts (b) and (c), conclude that

$$\mathscr{L}\left\{\int_0^t f(u)\, du\right\} = \frac{1}{s}\mathscr{L}\{f(t)\}. \quad \text{(ii)}$$

In Problems 32–44 use equations (i) and (ii) to compute the Laplace transform of the given function.

32. $\dfrac{\cos t - 1}{t}$
33. $\dfrac{\sin t}{t}$
34. $\dfrac{\sinh t}{t}$
35. $\dfrac{\sin 3t}{t}$
36. $\dfrac{\sinh kt}{t}$
37. $\dfrac{\sin kt}{t}$
38. $\dfrac{1 - \cos at}{t}$
39. $\dfrac{1 - \cosh at}{t}$
40. $\displaystyle\int_0^t \frac{\sin ku}{u}\, du$
41. $\displaystyle\int_0^t \frac{1 - \cosh au}{u}\, du$
42. $\displaystyle\int_0^t \frac{1 - \cos au}{u}\, du$
43. $\operatorname{erf}(t) = \dfrac{2}{\sqrt{\pi}} \displaystyle\int_0^t e^{-u^2}\, du$
44. $\dfrac{e^{-k^2/4t}}{\sqrt{\pi t}}$

Find the inverse Laplace transform of the functions in Problems 45–48. Use derivatives and integrals.

45. $\ln\left(1 + \dfrac{a^2}{s^2}\right)$
46. $\ln\dfrac{s-a}{s-b}$
47. $\arctan\dfrac{1}{s}$
*48. $\dfrac{1}{s}\arctan\dfrac{1}{s}$

49. Consider the equation

$$y'' + ty = 0, \quad y(0) = 0, \quad y'(0) = 1.$$

a. Obtain a differential equation for $Y(s) = \mathscr{L}\{y(t)\}$.
b. Solve the differential equation and find $Y(s)$. [Note that it is not possible to invert this transform by the methods we have discussed.]

*50. Let $f(t)$ be continuous, except for a jump discontinuity at $t = a\ (>0)$, and let it satisfy all other conditions of Theorem 1. Prove that

$$\mathscr{L}\{f'(t)\} = s\mathscr{L}\{f(t)\} - f(0)$$
$$- e^{-as}[f(a+0) - f(a-0)],$$

where

$$f(a+0) = \lim_{h \to 0^+} f(a+h)$$

and

$$f(a-0) = \lim_{h \to 0^-} f(a+h).$$

51. In Example 9, find $I(t)$ if $E(t) = \sin t$, $Q(0) = I(0) = 0$, $L = 2$ henrys, $R = 20$ ohms, and $C = 0.02$ farad.
52. Let $L = 1$ henry, $R = 100$ ohms, $C = 10^{-4}$ farad, and $E = 1000 \sin t$ volts in the circuit in Figure 6.2. Suppose that no charge and no current are initially present. Find the current and charge at all times t.
53. Let L, R, and C be as in Problem 51, but let $E(t) = 10 \sin 10t$. If $Q(0) = I(0) = 0$, find $Q(t)$.
54. A 50-kilogram mass is suspended from a spring with spring constant 20 newtons per meter. When the system is vibrating freely, the maximum displacement of each consecutive cycle decreases by 20 percent. Assume that a force equal to $10 \cos \omega t$ newtons acts on the system. Find the amplitude of the resultant steady-state motion if
    a. $\omega = 8$ radians per second;
    b. $\omega = 10$ radians per second;
    c. $\omega = 12$ radians per second;
    d. $\omega = 14$ radians per second;
    e. $\omega = 18$ radians per second.

6.3 Step Functions, Impulse Functions, and Periodic Functions

In Section 6.2 we saw how to solve linear differential equations in which the forcing function $f(t)$ was continuous. In a great number of applications, however, the forcing function is either a step function or an impulse function. We define these terms and show how to compute the Laplace transforms of these two important types of functions in this section. In Section 6.4 we will give some examples of differential equations with discontinuous forcing functions.

UNIT STEP FUNCTION

The following function, which is extremely important for practical applications, is known as the **unit step function** or **Heaviside function**[†] [see Figure 6.4(a)]:

$$H(t) = \begin{cases} 0, & t < 0, \\ 1, & t > 0. \end{cases} \tag{1}$$

In particular, if a is any fixed constant, we can shift the Heaviside function by a units [see Figure 6.4(b)] by defining

$$H(t-a) = \begin{cases} 0, & t < a, \\ 1, & t > a. \end{cases} \tag{2}$$

Then $H(t-a)$ has a jump discontinuity at $t = a$. Note that $H(t-a)$ is not defined at $t = a$.

A practical use of such a function might be to model a light switch that is switched on at time $t = a$.

[†] Named after the British physicist Oliver Heaviside (1850–1925). Heaviside had difficulty getting his research published because he made use of unusual methods in solving problems.

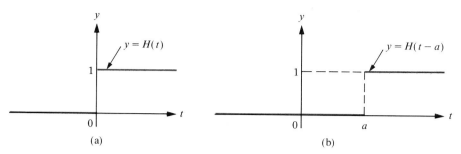

Figure 6.4

For $a \geq 0$ and $s > 0$, we obtain

$$\mathscr{L}\{H(t-a)\} = \int_0^\infty e^{-st}H(t-a)\,dt = \int_a^\infty e^{-st}\,dt = \frac{e^{-as}}{s}. \qquad (3)$$

The following theorem shows that multiplying a function by a unit step function has the effect of multiplying its transform by an exponential function.

THEOREM 1: THE SECOND SHIFTING PROPERTY OF LAPLACE TRANSFORMS

Let $a > 0$. Then, if $f(t)$ has a Laplace transform, so does $f(t-a)H(t-a)$ and

$$\mathscr{L}\{f(t-a)H(t-a)\} = e^{-as}\mathscr{L}\{f(t)\}, \qquad (4)$$

or, in terms of inverse transforms with $F(s) = \mathscr{L}\{f(t)\}$,

$$\mathscr{L}^{-1}\{e^{-as}F(s)\} = f(t-a)H(t-a). \qquad (5)$$

Proof A graph illustrating a function $f(t-a)H(t-a)$ is given in Figure 6.5.

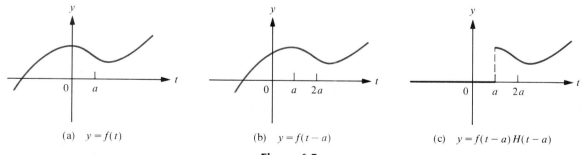

Figure 6.5
An arbitrary function f is graphed in (a). The graph of $y = f(t-a)$
in (b) is simply a shift to the right by a units of the original graph.
For the graph in (c), truncate the graph in (b) at $t = a$,
replacing it by $y = 0$ for $t < a$.

Using the definition and the substitution $x = t - a$, we find that

$$\mathcal{L}\{f(t-a)H(t-a)\} = \int_0^\infty e^{-st}f(t-a)H(t-a)\,dt = \int_a^\infty e^{-st}f(t-a)\,dt$$

let $x = t - a$
$$\downarrow$$
$$= \int_0^\infty e^{-s(x+a)}f(x)\,dx = e^{-as}\mathcal{L}\{f(t)\}. \quad \blacksquare$$

The next four examples illustrate the use of the second shifting property.

Example 1 $\mathcal{L}\{\sin a(t-b)H(t-b)\} = e^{-bs}\mathcal{L}\{\sin at\} = ae^{-bs}/(s^2 + a^2)$.

Example 2 Compute $\mathcal{L}\{f(t)\}$ when

$$f(t) = \begin{cases} e^t, & 0 \leq t < 2\pi, \\ e^t + \cos t, & t > 2\pi. \end{cases}$$

Solution The function $f(t)$ has a jump discontinuity at $t = 2\pi$. We may write

$$f(t) = e^t + H(t - 2\pi)\cos(t - 2\pi),$$

since

$$H(t-2\pi)\cos(t-2\pi) = \begin{cases} 0, & t < 2\pi, \\ \cos(t-2\pi) = \cos t, & t > 2\pi. \end{cases}$$

Thus, by the second shifting property (Theorem 1),

$$\mathcal{L}\{f(t)\} = \mathcal{L}\{e^t\} + e^{-2\pi s}\mathcal{L}\{\cos t\}$$
$$= \frac{1}{s-1} + \frac{se^{-2\pi s}}{1+s^2}.$$

Example 3 Compute

$$\mathcal{L}^{-1}\left\{\frac{1 - e^{-\pi s/2}}{1 + s^2}\right\}.$$

Solution Observe that

$$\mathcal{L}^{-1}\left\{\frac{1 - e^{-\pi s/2}}{1 + s^2}\right\} = \mathcal{L}^{-1}\left\{\frac{1}{1+s^2}\right\} - \mathcal{L}^{-1}\left\{\frac{e^{-\pi s/2}}{1+s^2}\right\}$$
$$= \sin t - H(t - \pi/2)\sin(t - \pi/2)$$
$$= \sin t + H(t - \pi/2)\cos t.$$

The unit step function can be used as a building block in the construction of other functions; for example,

$$f_1(t) = H(t-a) - H(t-b), \qquad a < b,$$

is a square wave between a and b [see Figure 6.6(a)], whereas

$$f_2(t) = H(t-a) + H(t-2a) + H(t-3a), \qquad a > 0,$$

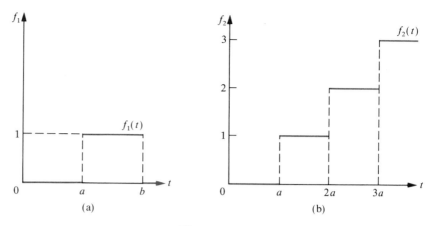

Figure 6.6

yields a three-step staircase [see Figure 6.6(b)]. By the linearity property of Laplace transforms (Theorem 6.1.2) we obtain

$$\mathscr{L}\{f_1(t)\} = \frac{1}{s}(e^{-as} - e^{-bs}),$$

and

$$\mathscr{L}\{f_2(t)\} = \frac{1}{s}(e^{-as} + e^{-2as} + e^{-3as}).$$

Example 4 Compute $\mathscr{L}\{t^2 H(t-3)\}$.

Solution Let $f(t-3) = t^2$; then, since $t = (t-3) + 3$, $f(t) = (t+3)^2 = t^2 + 6t + 9$,

$$\mathscr{L}\{t^2 H(t-3)\} = e^{-3s}\mathscr{L}\{t^2 + 6t + 9\} = e^{-3s}\left(\frac{2}{s^3} + \frac{6}{s^2} + \frac{9}{s}\right),$$

by Theorem 1.

Example 5 Compute the Laplace transform of the infinite staircase [obtained by continuing the staircase in Figure 6.6(b) forever]:

$$f(t) = H(t) + H(t-a) + H(t-2a) + H(t-3a) + \cdots, \qquad a > 0. \quad (6)$$

Solution Since $e^{-as} < 1$, if $as > 0$ we can use the formula for the sum of a geometric series:

$$\sum_{n=0}^{\infty} x^n = 1 + x + x^2 + \cdots = \frac{1}{1-x}, \qquad |x| < 1. \quad (7)$$

Then, for $s > 0$, by Table 6.1,

$$\mathscr{L}\{f(t)\} = \frac{1}{s}(1 + e^{-as} + e^{-2as} + e^{-3as} + \cdots) = \frac{1}{s(1 - e^{-as})}. \quad (8)$$

Example 6 Let $f(t)$ be the periodic square wave shown in Figure 6.7. We can write $f(t)$ in the form
$$f(t) = H(t) - 2H(t-a) + 2H(t-2a) - 2H(t-3a) + \cdots,$$

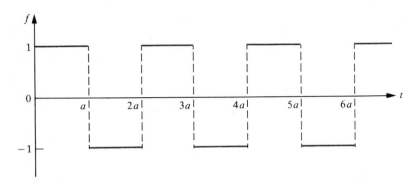

Figure 6.7

from which it follows that
$$\mathscr{L}\{f(t)\} = \frac{1}{s}(1 - 2e^{-as} + 2e^{-2as} - 2e^{-3as} + \cdots)$$

$$= \frac{1}{s}\left(\frac{2}{1+e^{-as}} - 1\right) = \frac{1-e^{-as}}{s(1+e^{-as})} \overset{\text{multiply and divide by } e^{as/2}}{=} \frac{1}{s}\left(\frac{e^{as/2} - e^{-as/2}}{e^{as/2} + e^{-as/2}}\right)$$

$$= \frac{1}{s}\tanh\left(\frac{1}{2}as\right).$$

UNIT IMPULSE FUNCTION

The **unit impulse function** (also called the **Dirac delta function**[†]) $\delta(t-a)$ is loosely described as a "function" that is zero everywhere except at $t = a$ and has the property that
$$\int_{-\infty}^{\infty} \delta(t-a)\, dt = 1. \tag{9}$$

As an illustration, we describe the Dirac delta function for $a = 0$. For any $\epsilon > 0$, consider the approximate delta functions (see Figure 6.8)
$$\delta_\epsilon(t) = \begin{cases} 1/2\epsilon, & -\epsilon < t < \epsilon, \\ 0, & |t| > \epsilon. \end{cases}$$

Clearly, $\delta_\epsilon(t)$ is piecewise continuous and
$$\int_{-\infty}^{\infty} \delta_\epsilon(t)\, dt = \int_{-\epsilon}^{\epsilon} \frac{1}{2\epsilon}\, dt = 1.$$

[†] This function was first discussed by the British physicist Paul A. M. Dirac (1902–1984) in 1932. In 1933 he received the Nobel Prize in Physics (jointly with E. Schrödinger) for his work in quantum theory.

6.3 / STEP FUNCTIONS, IMPULSE FUNCTIONS, AND PERIODIC FUNCTIONS 239

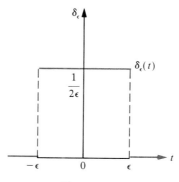

Figure 6.8

Then $\delta(t)$ may be defined by

$$\delta(t) = \lim_{\epsilon \to 0} \delta_\epsilon(t).$$

Of course, $\delta(t)$ is not a function. However, because it is the limit of piecewise continuous functions, we may treat $\delta(t)$ as if it were a legitimate function. We do not prove this fact here but we nevertheless make use of it in all further discussions of the delta function.

Since $\delta(t-a)$ concentrates all its "mass" at $t=a$, we see that

$$H(t-a) = \int_{-\infty}^{t} \delta(u-a)\, du, \tag{10}$$

since the integral in equation (10) is zero if $t < a$ and equals 1 when $t > a$. By the integration property (see Problem 6.2.31), if $a \geq 0$,

$$\frac{1}{s}\mathscr{L}\{\delta(t-a)\} = \mathscr{L}\left\{\int_0^t \delta(u-a)\, du\right\} = \mathscr{L}\{H(t-a)\} = \frac{e^{-as}}{s}.$$

Thus

$$\mathscr{L}\{\delta(t-a)\} = e^{-as}, \quad \text{for } a \geq 0 \quad \text{and} \quad s > 0. \tag{11}$$

The Heaviside function can be used in solving differential equations with discontinuous forcing functions:

Example 7 Solve the initial-value problem

$$y'' + y = f(t), \quad y(0) = y'(0) = 0,$$

where

$$f(t) = \begin{cases} 1, & 0 \leq t \leq 1, \\ 0, & t > 1. \end{cases}$$

Solution We can rewrite $f(t)$ using Heaviside functions as $f(t) = H(t) - H(t-1)$. Taking Laplace transforms, we get

$$s^2 \mathscr{L}\{y\} + \mathscr{L}\{y\} = \frac{1}{s} - \frac{e^{-s}}{s},$$

or

$$(s^2 + 1)\mathscr{L}\{y\} = \frac{1 - e^{-s}}{s}.$$

Hence

$$\mathscr{L}\{y\} = \frac{1 - e^{-s}}{s(s^2 + 1)},$$

and since

$$\frac{1}{s(s^2 + 1)} = \frac{1}{s} - \frac{s}{s^2 + 1},$$

we have

$$\mathscr{L}\{y\} = \frac{1}{s} - \frac{s}{s^2 + 1} - \frac{e^{-s}}{s} + \frac{se^{-s}}{s^2 + 1}.$$

Using the second shifting property (Theorem 1), we obtain

$$y(t) = 1 - \cos t - H(t-1) + H(t-1)\cos(t-1)$$
$$= 1 - \cos t - H(t-1)[1 - \cos(t-1)].$$

The unit impulse function is sometimes the forcing function in an initial-value problem

Example 8 Solve the differential equation

$$y'' + 2y' + y = \delta(t-1), \qquad y(0) = 2, \qquad y'(0) = 3.$$

Solution Taking Laplace transforms and using equation (11), we obtain

$$[s^2 \mathscr{L}\{y\} - 2s - 3] + 2[s\mathscr{L}\{y\} - 2] + \mathscr{L}\{y\} = e^{-s},$$

or

$$(s^2 + 2s + 1)\mathscr{L}\{y\} = 2s + 7 + e^{-s}.$$

Hence we get

$$\mathscr{L}\{y\} = \frac{2s + 7 + e^{-s}}{s^2 + 2s + 1} = \frac{2(s+1)}{(s+1)^2} + \frac{5}{(s+1)^2} + \frac{e^{-s}}{(s+1)^2}$$

$$= \frac{2}{(s+1)} + \frac{5}{(s+1)^2} + \frac{e^{-s}}{(s+1)^2}.$$

Since $\mathscr{L}\{te^{-t}\} = (s+1)^{-2}$, it follows from Theorem 1 that

$$\frac{e^{-s}}{(s+1)^2} = e^{-s}\mathscr{L}\{te^{-t}\} = \mathscr{L}\{(t-1)e^{-(t-1)}H(t-1)\}.$$

Finally, the solution is

$$y(t) = 2e^{-t} + 5te^{-t} + (t-1)e^{-(t-1)}H(t-1)$$
$$= e^{-t}[2 + 5t + e(t-1)H(t-1)].$$

LAPLACE TRANSFORMS OF PERIODIC FUNCTIONS

Before going further, we prove a result that is very useful for finding the Laplace transform of a wide variety of functions.

THEOREM 2: PERIODICITY PROPERTY OF THE LAPLACE TRANSFORM

Let $f(t)$ be continuous in $[0, \omega]$ and periodic with period ω ($\omega > 0$); that is, $f(t + \omega) = f(t)$, for each $t \geq 0$. Then $f(t)$ has the Laplace transform

$$F(s) = \mathscr{L}\{f(t)\} = \frac{\int_0^\omega e^{-st}f(t)\,dt}{1 - e^{-\omega s}}, \tag{12}$$

valid for every $s > 0$.

Proof By definition,

$$F(s) = \int_0^\infty e^{-st}f(t)\,dt$$
$$= \int_0^\omega e^{-st}f(t)\,dt + \int_\omega^{2\omega} e^{-st}f(t)\,dt + \cdots$$
$$= \sum_{k=0}^\infty \int_{k\omega}^{(k+1)\omega} e^{-st}f(t)\,dt. \tag{13}$$

Making the substitution $u = t - k\omega$, we obtain

$$\int_{k\omega}^{(k+1)\omega} e^{-st}f(t)\,dt = \int_0^\omega e^{-s(u+k\omega)}f(u+k\omega)\,du$$
$$= e^{-sk\omega}\int_0^\omega e^{-su}f(u)\,du, \tag{14}$$

because of the periodicity of f. Thus, substituting equation (14) into equation (13) and using equation (7) with $x = e^{-s\omega}$, we have

$$F(s) = \sum_{k=0}^\infty e^{-sk\omega}\int_0^\omega e^{-su}f(u)\,du$$
$$= \left[\int_0^\omega e^{-su}f(u)\,du\right]\sum_{k=0}^\infty (e^{-\omega s})^k$$
$$= \frac{\int_0^\omega e^{-su}f(u)\,du}{1 - e^{-\omega s}}.$$

Note that if $s > 0$, then $\omega s > 0$ and $e^{-\omega s} < 1$, so the use of formula (7) is valid. ∎

Example 9 Find the Laplace transform of the function
$$f(t) = |\sin at|, \quad a > 0.$$

Solution Note that $f(t)$ has period $\omega = \pi/a$. By Theorem 2 we have
$$\mathscr{L}\{|\sin at|\} = \frac{\int_0^{\pi/a} e^{-st}\sin at\, dt}{1 - e^{-\pi s/a}},$$
since $|\sin at| = \sin at$ in $[0, \pi/a]$. Now using Formula 168 in Appendix 1, we have
$$\int_0^{\pi/a} e^{-st}\sin at\, dt = \frac{e^{-st}}{s^2 + a^2}(-s\sin at - a\cos at)\Big|_0^{\pi/a}$$
$$= \frac{a(e^{-\pi s/a} + 1)}{s^2 + a^2},$$
so
$$\mathscr{L}\{|\sin at|\} = \frac{a}{s^2 + a^2}\frac{1 + e^{-\pi s/a}}{1 - e^{-\pi s/a}},$$
which can be simplified by using hyperbolic functions to
$$\mathscr{L}\{|\sin at|\} = \frac{a}{s^2 + a^2}\coth\left(\frac{\pi s}{2a}\right).$$

Example 10 Compute the Laplace transform of the periodic square wave
$$f(t) = H(t) - 2H(t - a) + 2H(t - 2a) - 2H(t - 3a) + \cdots, \quad a > 0.$$

Solution We solved this problem in Example 5. We now solve it using Theorem 2 by noting that f is periodic of period $2a$ (see Figure 6.6). We have, from (12),
$$F(s) = \frac{\int_0^{2a} e^{-st}f(t)\, dt}{1 - e^{-2as}}.$$
But $f(t) = 1$ for $0 < t < a$ and $f(t) = -1$ for $a < t < 2a$, so
$$\int_0^{2a} e^{-st}f(t) = \int_0^a e^{-st}\, dt - \int_a^{2a} e^{-st}\, dt = -\frac{1}{s}e^{-st}\Big|_0^a + \frac{1}{s}e^{-st}\Big|_a^{2a}$$
$$= \frac{1}{s}(-e^{-as} + 1 + e^{-2as} - e^{-as}) = \frac{1}{s}(1 - e^{-as})^2.$$
Since $1 - e^{-2as} = (1 - e^{-as})(1 + e^{-as})$, we obtain
$$F(s) = \frac{1}{s}\frac{(1 - e^{-as})(1 - e^{-as})}{(1 - e^{-as})(1 + e^{-as})} = \frac{1 - e^{-as}}{s(1 + e^{-as})}.$$
We can write this in another way as
$$\frac{1}{s}\frac{(1 - e^{-as})}{(1 + e^{-as})} \overset{\text{multiply top and bottom by } e^{as/2}}{=} \frac{1}{s}\left(\frac{e^{as/2} - e^{-as/2}}{e^{as/2} + e^{-as/2}}\right) = \frac{1}{s}\tanh\frac{as}{2}.$$

6.3 / STEP FUNCTIONS, IMPULSE FUNCTIONS, AND PERIODIC FUNCTIONS

Thus

$$\mathscr{L}\{f\} = \frac{1}{s}\tanh\frac{as}{2}.$$

Example 11 Compute the Laplace transform of the sawtooth function given in Figure 6.9.

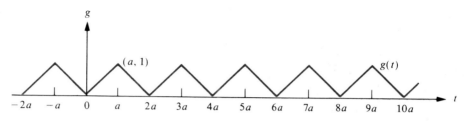

Figure 6.9

Solution Clearly g is periodic of period $2a$. For $0 \le t \le a$, g is the line segment that passes through $(0,0)$ and $(a,1)$. The slope of this line is $1/a$, so $g(t) = t/a$ for $0 \le t \le a$. For $a \le t \le 2a$, g is the line that passes through $(a,1)$ and $(2a,0)$. The slope of this line is $-1/a$ and the equation of the line is $g(t) = 2 - t/a$. Thus

$$g(t) = \begin{cases} t/a, & 0 \le t \le a, \\ 2 - t/a, & a \le t \le 2a, \end{cases}$$

and, from (12),

$$\mathscr{L}\{g\} = \frac{\int_0^{2a} e^{-st} g(t)\, dt}{1 - e^{-2as}}. \tag{15}$$

But using entry 164 in Appendix 1,

$$\int_0^{2a} e^{-st} g(t)\, dt = \frac{1}{a}\int_0^a t e^{-st}\, dt + \frac{1}{a}\int_a^{2a} (2a - t) e^{-st}\, dt$$

$$= \frac{1}{a}\left[\frac{e^{-st}}{-s}\left(t + \frac{1}{s}\right)\right]\bigg|_{t=0}^{t=a} + \frac{1}{a}\left[\frac{e^{-st}}{-s}\left(2a - t - \frac{1}{s}\right)\right]\bigg|_{t=a}^{t=2a}$$

$$= \frac{1}{as^2}(1 - 2e^{-as} + e^{-2as}) = \frac{(1 - e^{-as})^2}{as^2}.$$

We can complete the problem by substituting $e^x - e^{-x} = 2\sinh x$ and $e^x + e^{-x} = 2\cosh x$ for $x = as/2$, but there is another way to obtain the solution. Observe that

$$g'(t) = \begin{cases} 1/a, & 0 < t < a, \\ -1/a, & a < t < 2a; \end{cases}$$

that is, $g' = f/a$ where f is the periodic function of Example 10. Writing this another way, we see that

$$g(t) = \frac{1}{a}\int_0^t f(u)\, du.$$

Then, from the result (ii) of Problem 6.2.31 and Example 10, we see that

$$\mathscr{L}\{g\} = \frac{1}{s}\mathscr{L}\left\{\frac{f}{a}\right\} = \frac{1}{s}\cdot\frac{1}{as}\tanh\frac{as}{2} = \frac{1}{as^2}\tanh\frac{as}{2}.$$

Example 12 Solve the initial-value problem
$$y'' + 2y' + 5y = f(t), \qquad y(0) = y'(0) = 0,$$
where $f(t)$ is the periodic square wave in Figure 6.10.

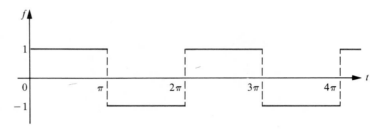

Figure 6.10

Solution Taking Laplace transforms and using the result in Example 5 (or Example 10), we have

$$(s^2 + 2s + 5)\mathscr{L}\{y\} = \frac{1 - e^{-\pi s}}{s(1 + e^{-\pi s})},$$

or

$$\mathscr{L}\{y\} = \frac{1}{s(s^2 + 2s + 5)} \cdot \frac{1 - e^{-\pi s}}{1 + e^{-\pi s}}.$$

But

$$\frac{1}{s(s^2 + 2s + 5)} = \frac{1}{5}\left[\frac{1}{s} - \frac{s+2}{s^2 + 2s + 5}\right]$$

$$= \frac{1}{5}\left[\frac{1}{s} - \frac{s+2}{(s+1)^2 + 2^2}\right]$$

and, by the geometric series in Example 4,

$$\frac{1 - e^{-\pi s}}{1 + e^{-\pi s}} = (1 - e^{-\pi s})(1 - e^{-\pi s} + e^{-2\pi s} - e^{-3\pi s} + \cdots)$$

$$= 1 - 2e^{-\pi s} + 2e^{-2\pi s} - 2e^{-3\pi s} + \cdots.$$

Thus
$$\mathscr{L}\{y\} = \frac{1}{5}\left[\frac{1}{s} - \frac{s+2}{(s+1)^2 + 2^2}\right](1 - 2e^{-\pi s} + 2e^{-2\pi s} - 2e^{-3\pi s} + \cdots).$$

By the first and second shifting theorems we have

$$\mathscr{L}^{-1}\left\{\frac{1}{5}\left[\frac{1}{s} - \frac{(s+1)+1}{(s+1)^2 + 2^2}\right]\right\} = \frac{1}{5}\left[1 - \overbrace{e^{-t}\left(\cos 2t + \frac{1}{2}\sin 2t\right)}^{\equiv g(t)}\right]$$

$$= \frac{1}{5}[1 - g(t)]$$

and
$$\mathcal{L}^{-1}\left\{\frac{2}{5}\left[\frac{1}{s}-\frac{(s+1)+1}{(s+1)^2+2^2}\right]e^{-k\pi s}\right\}=\frac{2}{5}[1-g(t-k\pi)]H(t-k\pi).$$

But
$$g(t-k\pi)=e^{-(t-k\pi)}\left[\cos 2(t-k\pi)+\tfrac{1}{2}\sin 2(t-k\pi)\right]$$
$$=e^{k\pi}g(t),$$

so
$$y(t)=\tfrac{1}{5}[1-g(t)]-\tfrac{2}{5}[1-e^{\pi}g(t)]H(t-\pi)+\tfrac{2}{5}[1-e^{2\pi}g(t)]H(t-2\pi)$$
$$-\tfrac{2}{5}[1-e^{3\pi}g(t)]H(t-3\pi)+\cdots$$
$$=\tfrac{1}{5}[1-2H(t-\pi)+2H(t-2\pi)-2H(t-3\pi)+\cdots]$$
$$-\frac{g(t)}{5}[1-2e^{\pi}H(t-\pi)+2e^{2\pi}H(t-2\pi)-2e^{3\pi}H(t-3\pi)+\cdots].$$

Then
$$y(t)=\tfrac{1}{5}(f(t)-g(t)[1-2e^{\pi}H(t-\pi)+2e^{2\pi}H(t-2\pi)-\cdots]).$$

Hence, if $n\pi < t < (n+1)\pi$,
$$y(t)=\frac{1}{5}\left[(-1)^n-g(t)(1-2e^{\pi}+\cdots+(-1)^n 2e^{n\pi})\right]$$
$$=\frac{1}{5}\left((-1)^n-g(t)\left[2\left(\frac{1+(-1)^n e^{(n+1)\pi}}{1+e^{\pi}}\right)-1\right]\right)$$
$$=\frac{1}{5}\left((-1)^n+g(t)-2g(t)\left(\frac{1+(-1)^n e^{(n+1)\pi}}{1+e^{\pi}}\right)\right).$$

PROBLEMS 6.3

In Problems 1–8 use the second shifting theorem to show the given equation.

1. $\mathcal{L}\{tH(t-1)\}=e^{-s}\left(\dfrac{1}{s^2}+\dfrac{1}{s}\right)$

2. $\mathcal{L}\{t^2 H(t-1)\}=e^{-s}\left(\dfrac{2}{s^3}+\dfrac{2}{s^2}+\dfrac{1}{s}\right)$

3. $\mathcal{L}\{e^t H(t-1)\}=\dfrac{e^{-(s-1)}}{s-1}$

4. $\mathcal{L}\{e^{at}H(t-b)\}=\dfrac{e^{-b(s-a)}}{s-a}$

5. $\mathcal{L}\left\{\sin t\cdot H\left(t-\dfrac{\pi}{2}\right)\right\}=\dfrac{se^{-\pi s/2}}{s^2+1}$

6. $\mathcal{L}\{\cos a(t-b)\cdot H(t-b)\}=\dfrac{e^{-bs}s}{s^2+a^2}$

7. $\mathcal{L}\{\sinh a(t-b)\cdot H(t-b)\}=\dfrac{e^{-bs}a}{s^2-a^2}$

8. $\mathcal{L}\{\cosh a(t-b)\cdot H(t-b)\}=\dfrac{e^{-bs}}{s^2-a^2}$

In Problems 9–11 represent the graphed functions in terms of unit step functions and find their respective Laplace transforms.

10.

11.

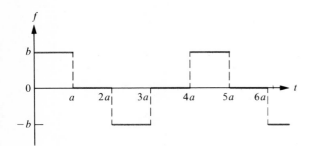

12. Let $g(t)$ be the function shown below.

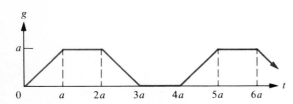

a. Show that $g'(t)$ is piecewise continuous and that $g'(t) = f(t)$, where f is the step function of Problem 11 with $b = 1$.

b. Use the differentiation property to compute $\mathcal{L}\{g\}$.

In Problems 13 and 14 use the method of Problem 12 to find the Laplace transform of the given functions.

13.

14.

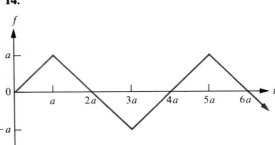

15. Compute the Laplace transform of the **half-wave rectifier** whose graph is given below.

16. Compute the Laplace transform of the sawtooth wave given below.

17. Let $f(t) = \begin{cases} \sin t, & t < 4\pi, \\ \sin t + \cos t, & t > 4\pi. \end{cases}$

Compute $\mathcal{L}\{f(t)\}$.

18. Let $f(t) = \begin{cases} \cos t, & t < 3\pi/2, \\ \cos t + \sin t, & t > 3\pi/2. \end{cases}$

Compute $\mathcal{L}\{f(t)\}$.

In Problems 19–24 compute the inverse of the given Laplace transform.

19. $\dfrac{e^{-\pi s}}{1 + s^2}$

20. $\dfrac{se^{-3\pi s}}{1 + s^2}$

21. $\dfrac{s - se^{-\pi s}}{1 + s^2}$

22. $\dfrac{1 - e^{-2s}}{s^2}$

23. $\dfrac{1 + e^{-4s}}{s^5}$

24. $\dfrac{e^{-2s}}{s^2 - 1}$ [*Hint:* Use partial fractions.]

25. Let
$$f(t) = \begin{cases} 1, & t < 1, \\ 3, & 1 < t < 7, \\ 5, & t > 7. \end{cases}$$
Write $f(t)$ as a step function and compute $\mathscr{L}\{f(t)\}$.

26. Let
$$f(t) = \begin{cases} -2, & 0 < t < 1, \\ 0, & 1 < 6 < 10, \\ 2, & t > 10. \end{cases}$$
Write $f(t)$ as a step function and compute $\mathscr{L}\{f(t)\}$.

In Problems 27–33 solve the given initial-value problem with discontinuous forcing function $f(t)$. Draw a graph of the solution.

27. $y'' + y = f(t)$, $y(0) = y'(0) = 0$, where
$$f(t) = \begin{cases} t, & 0 < t < \pi \\ 0, & t > \pi \end{cases}$$

28. $y'' - y = f(t)$, $y(0) = 1$, $y'(0) = 0$, where
$$f(t) = \begin{cases} 1, & 0 < t < 1 \\ 0, & t > 1 \end{cases}$$

*29. $y'' + 2y' + 10y = f(t)$, $y(0) = y'(0) = 0$, where $f(t)$ is the function in Problem 28

*30. $y'' + 2y' + 5y = f(t)$, $y(0) = 0$, $y'(0) = 1$, where $f(t)$ is defined in Problem 27

*31. $y'' + 2y' + 10y = f(t)$, $y(0) = 1$, $y'(0) = 0$, where $f(t)$ is defined in Problem 28

32. $y'' - y' + 6y = \delta(t - 2)$, $y(0) = 1$, $y'(0) = -2$

33. $y'' - 4y' + 13y = \delta(t - 1)$, $y(0) = 0$, $y'(0) = 3$

*34. A model rocket of 1-kilogram mass blasts off from a playground field with a propulsion force of 10 newtons. Assume that the motor runs out of propellant after 10 seconds, that the propellant weighs 500 grams, and that the propellant is consumed at a constant rate.

a. Write the equations of motion for this situation assuming a damping force (due to air resistance) proportional to the velocity of the rocket.

b. Solve the equation in part (a) using the Laplace transform methods of this section.

6.4 Some Differential Equations with Discontinuous Forcing Functions: Applications to Electric Circuits (Optional)

There are many physical models that give rise to differential equations with discontinuous forcing functions. In this section we discuss several models involving electric circuits. We begin with the *RLC* circuit discussed in Section 4.4.

Consider the differential equation

$$L\frac{dI}{dt} + RI + \frac{Q}{C} = E. \tag{1}$$

[See equation (4.4.1) p. 152.] This is the equation that describes the circuit shown (again) in Figure 6.11.

We assume that the voltage source, which may be a battery, is controlled by a switch that, initially, is turned off, and that $Q(0) = I(0) = 0$. At some later time t_1, the switch is turned on and the voltage is then equal to some constant

Figure 6.11

value E_0. In this situation it is easy to see that
$$E(t) = E_0 H(t - t_1). \qquad (2)$$
To find the current at all times $t > 0$, we use the technique of Example 6.2.9 and the fact that $s\hat{Q} = \hat{I}$ (since $dQ/dt = I$ and $Q(0) = 0$) to find that
$$\hat{I}(s) = T(s)\hat{E}(s), \qquad (3)$$
where the transfer function is given by
$$T(s) = \frac{s/L}{s^2 + (R/L)s + (1/LC)}. \qquad (4)$$
As before, the symbol $\wedge$ denotes the Laplace transform. From (2), we have
$$\hat{E}(s) = \mathcal{L}\{E_0 H(t - t_1)\}$$
$$= \frac{E_0}{s} e^{-t_1 s}.$$
We can use this to solve for $\hat{I}(s)$ in equation (3).

Remark $E(t)$ is not defined at $t = t_1$ and, evidently, is not differentiable at that point. Therefore the use of equation (3) is not strictly valid. However, we can get around this difficulty by observing that, according to equation (6.3.10), $H'(t - t_1) = \delta(t - t_1)$. Of course, $\delta(t - t_1)$ is not a function in the traditional sense but, since $\delta(t - t_1)$ has a Laplace transform [see equation (6.3.11)], we will not worry about this difficulty.

Example 1 An RLC circuit with $R = 10$ ohms, $L = 1$ henry, and $C = 0.01$ farad is hooked up to a battery that delivers a steady voltage of 20 volts when switched on. If the switch, initially off, is turned on after 10 seconds, find the current for all future values of t. Assume that Q and I are zero when the switch is turned on.

Solution Using equations (3) and (4) and the values given above, we have
$$\hat{I}(s) = \frac{s}{s^2 + 10s + 100}\left(\frac{E_0}{s}e^{-10s}\right)$$
$$= \frac{E_0 e^{-10s}}{s^2 + 10s + 100} = \frac{20 e^{-10s}}{(s+5)^2 + 75}.$$

From the second shifting theorem, we obtain
$$\frac{e^{-10s}}{(s+5)^2 + 75} = \mathcal{L}\{f(t-10)H(t-10)\},$$
where $f(t)$ is the function whose Laplace transform is $[(s+5)^2 + 75]^{-1}$. Thus
$$f(t) = \mathcal{L}^{-1}\left\{\frac{1}{(s+5)^2 + 75}\right\}$$
$$= \frac{1}{\sqrt{75}} \mathcal{L}^{-1}\left\{\frac{\sqrt{75}}{(s+5)^2 + 75}\right\} = \frac{1}{\sqrt{75}} e^{-5t} \sin\sqrt{75}\, t.$$

Next,
$$f(t-10) = \frac{1}{\sqrt{75}} e^{-5(t-10)} \sin\sqrt{75}\,(t-10),$$

so, finally, we have
$$I(t) = \frac{20}{\sqrt{75}} e^{-5(t-10)} \sin\sqrt{75}\,(t-10)\,H(t-10).$$

Note that there is no current if $t < 10$.

Example 2 Consider the LC circuit given in Figure 6.12 with $I(0) = I'(0) = 0$. The voltage is given by
$$E(t) = \begin{cases} 25t, & 0 \le t \le 4, \\ 100, & t > 4. \end{cases}$$

Find the current for all values of $t \ge 0$.

Figure 6.12

Solution Using equation (1) we have (since $1/0.04 = 25$)
$$\frac{dI}{dt} + 25Q = E$$

and, differentiating,
$$\frac{d^2I}{dt^2} + 25I = E'(t) = \begin{cases} 25, & 0 \le t \le 4, \\ 0, & t > 4, \end{cases}$$

or
$$\frac{d^2I}{dt^2} + 25I = 25 - 25H(t-4), \quad I(0) = I'(0) = 0.$$

Taking Laplace transforms and using $I(0) = I'(0) = 0$, we have
$$s^2\hat{I}(s) + 25\hat{I}(s) = \frac{25}{s} - \frac{25e^{-4s}}{s} = -\frac{25}{s}(e^{-4s} - 1),$$

so
$$\hat{I}(s) = -\frac{25(e^{-4s} - 1)}{s(s^2 + 25)}.$$

Note that
$$\frac{25}{s(s^2 + 25)} = \frac{1}{s} - \frac{s}{s^2 + 25}$$

and
$$\hat{I}(s) = (e^{-4s} - 1)\left(\frac{-1}{s} + \frac{s}{s^2 + 25}\right)$$
$$= -\frac{e^{-4s}}{s} + e^{-4s}\frac{s}{s^2 + 25} + \frac{1}{s} - \frac{s}{s^2 + 25}.$$

Thus
$$I(t) = 1 - \cos 5t + H(t-4)[\cos 5(t-4) - 1].$$

Example 3 Consider the parallel electric circuit shown in Figure 6.13, where the arrows denote the direction of current flow over each component of the circuit. We assume that $I(t) = CE_0 \delta(t - t_1)$. Clearly there is no current except at $t = t_1$. Show that if $E(0) = 0$, then $I(t)$ yields enough current to charge the capacitor to the voltage E_0 immediately.

Figure 6.13

Solution We refer to Kirchhoff's laws given in Section 2.5. We have the system
$$E = RI_R, \tag{5}$$
$$E = \frac{1}{C}Q_C,$$

or, differentiating,
$$E' = \frac{1}{C}\frac{dQ_C}{dt} = \frac{1}{C}I_C(t). \tag{6}$$

The total current is given by
$$I = I_R + I_C. \tag{7}$$

Thus
$$I(t) = I_R(t) + I_C(t) = \frac{E}{R} + CE',$$

or
$$E'(t) + \frac{1}{RC}E(t) = \frac{1}{C}I(t) = E_0\delta(t - t_1).$$

Taking transforms and using the fact that $E(0) = 0$, we have
$$s\hat{E}(s) + \frac{1}{RC}\hat{E}(s) = E_0 e^{-st_1},$$

or
$$\hat{E}(s) = \frac{E_0 e^{-st_1}}{s + 1/RC} = \mathscr{L}\{H(t - t_1)f(t - t_1)\},$$

where
$$\mathcal{L}\{f(t)\} = \frac{E_0}{s + 1/RC}.$$

Thus
$$f(t) = E_0 e^{-(1/RC)t}$$

and
$$E(t) = E_0 e^{-(1/RC)(t-t_1)} H(t - t_1).$$

Note that the voltage on the capacitor is zero before time $t = t_1$. At that time it jumps to the value E_0.

Example 4 In Example 3, what happens to the current after time t_1?

Solution For $t > t_1$, $I(t) = 0$. Thus for $t > t_1$ the only current is the current through the resistor and capacitor. From equation (5), we have, for $t > t_1$,

$$I_R(t) = \frac{1}{R} E(t) = \frac{1}{R} E_0 e^{-(1/RC)(t-t_1)} H(t - t_1)$$
$$= \frac{E_0}{R} e^{-(1/RC)(t-t_1)}.$$

Thus the current through the resistor decreases exponentially after the time $t = t_1$; that is, the current "bleeds off" through the resistor after the instant at which the capacitor is charged.

PROBLEMS 6.4

1. Find the current for all t in the RLC circuit of Example 1, if $R = 20$ ohms, $L = \frac{1}{2}$ henry, and $C = 0.002$ farad, and a steady voltage 50-volt battery, which is initially off, is turned on 30 seconds later.

2. Answer the question in Problem 1 if $R = 15$ ohms, $L = 2$ henrys, $C = 0.04$ farad, and E is a steady-voltage battery of 25 volts that is turned on 1 minute later.

In Problems 3–5 find the current for all values of t in the LC circuit of Example 2 using the given data. Graph the solution.

3. $L = 1$ henry, $C = \frac{1}{16}$ farad,
$$E(t) = \begin{cases} 16t, & 0 \le t \le 5, \\ 80, & t > 5. \end{cases} \text{ volts}$$

4. $L = 1$ henry, $C = 0.04$ farad,
$$E(t) = \begin{cases} 0, & 0 \le t < 2, \\ 20t, & 2 \le t \le 4, \\ 80, & t > 4. \end{cases} \text{ volts}$$

5. $L = 1$ henry, $C = 0.1$ farad,
$$E(t) = \begin{cases} 10t, & 0 \le t \le 2, \\ 20, & 2 \le t \le 4, \\ 20t, & t > 4. \end{cases} \text{ volts}$$

6. In Example 3 find the voltage across the resistor at $t = 20$ seconds if $E_0 = 10$ volts, $t_1 = 10$ seconds, $R = 10$ ohms, and $C = 0.1$ farad.

7. An undamped spring (see p. 147) supports an object of 1-kilogram mass. The spring constant of the spring is 4 newtons per meter. Suppose that a force $f(t)$ is applied to the object, where
$$f(t) = \begin{cases} 2t, & 0 \le t < \pi/2 \\ 0, & t \ge \pi/2 \end{cases} \text{ newtons.}$$
   a. Find the equation of motion of the mass.
   b. Assuming that, initially, the mass is displaced downward 1 meter before $f(t)$ is applied, find the position of the object for all $t > 0$.

8. Answer the questions in Problem 7 if
$$f(t) = \begin{cases} 0, & t < \pi/2 \\ 4, & t > \pi/2 \end{cases} \text{ newtons.}$$

6.5 The Transform of Convolution Integrals

It often occurs that in the process of solving a linear differential equation by transforms, we end up with a transform that is the product of two other transforms. Although we proved in Problem 6.1.79 that

$$\mathscr{L}^{-1}\{F+G\} = \mathscr{L}^{-1}\{F\} + \mathscr{L}^{-1}\{G\},$$

it is not true that $\mathscr{L}^{-1}\{FG\} = \mathscr{L}^{-1}\{F\}\mathscr{L}^{-1}\{G\}$; for example, if $F(s) = 1/s$ and $G(s) = 1/s^2$, then $F(s)G(s) = 1/s^3$, but

$$\mathscr{L}^{-1}\{FG\} = \mathscr{L}^{-1}\left\{\frac{1}{s^3}\right\} = \frac{t^2}{2}, \qquad \mathscr{L}^{-1}\{F\} = 1, \qquad \mathscr{L}^{-1}\{G\} = t$$

and, clearly, $\mathscr{L}^{-1}\{FG\} \neq \mathscr{L}^{-1}\{F\}\mathscr{L}^{-1}\{G\}$.

In this section we define the convolution of two functions f and g and show that $\mathscr{L}^{-1}\{FG\}$ is equal to the convolution of $\mathscr{L}^{-1}\{F\}$ and $\mathscr{L}^{-1}\{G\}$. We then apply this fact in a variety of ways.

Convolution

If f and g are piecewise continuous functions, then the **convolution** of f and g, written $(f*g)$, is defined by

$$\boxed{(f*g)(t) = \int_0^t f(t-u)g(u)\,du.} \tag{1}$$

The notation $(f*g)(t)$ indicates that the convolution $f*g$ is a function of the independent variable t.

Using the change of variables $v = t - u$, we see that

$$(f*g)(t) = -\int_t^0 f(v)g(t-v)\,dv = \int_0^t g(t-v)f(v)\,dv$$
$$= (g*f)(t). \tag{2}$$

Hence $(f*g)(t) = (g*f)(t)$, and we can take the convolution in either order without altering the result. We may now state the main result of this section.

THEOREM 1: CONVOLUTION THEOREM FOR LAPLACE TRANSFORMS

If $F(s) = \mathscr{L}\{f(t)\}$ and $G(s) = \mathscr{L}\{g(t)\}$ exist, then $\mathscr{L}\{(f*g)(t)\}$ exists and

$$\boxed{\mathscr{L}\{(f*g)(t)\} = F(s)G(s).}$$

Proof By definition,

$$F(s)G(s) = \left[\int_0^\infty e^{-su}f(u)\,du\right]\left[\int_0^\infty e^{-sv}g(v)\,dv\right]$$
$$= \int_0^\infty \int_0^\infty e^{-s(u+v)}f(u)g(v)\,dv\,du. \tag{3}$$

If we make the change of variables $t = u + v$, then $dt = dv$ and the integral (3)

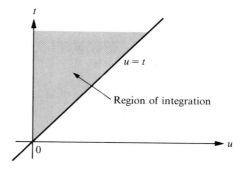

Figure 6.14

is equal to

$$F(s)G(s) = \int_0^\infty \int_u^\infty e^{-st} f(u) g(t-u) \, dt \, du. \tag{4}$$

Changing the order of integration† and noting that

$$\int_0^\infty \int_u^\infty dt \, du = \int_0^\infty \int_0^t du \, dt$$

(see Figure 6.14), we have the integral in (4) equal to

$$F(s)G(s) = \int_0^\infty \int_0^t e^{-st} f(u) g(t-u) \, du \, dt$$

$$= \int_0^\infty e^{-st} \left[\int_0^t g(t-u) f(u) \, du \right] dt$$

$$= \int_0^\infty e^{-st} (g*f)(t) \, dt = \int_0^\infty e^{-st} (f*g)(t) \, dt$$

$$= \mathscr{L}\{f*g\}. \quad \blacksquare$$

COROLLARY

If $F(s) = \mathscr{L}\{f(t)\}$ and $G(s) = \mathscr{L}\{g(t)\}$, then

$$\boxed{\mathscr{L}^{-1}\{F(s)G(s)\} = (f*g)(t).} \tag{5}$$

Our first applications of this theorem and corollary are in the computation of inverse transforms.

Example 1 Compute $\mathscr{L}^{-1}\{s/(s^2+1)^2\}$.

Solution Since

$$\mathscr{L}\{\cos t\} = \frac{s}{s^2+1} \quad \text{and} \quad \mathscr{L}\{\sin t\} = \frac{1}{s^2+1},$$

---

† This may not always be possible. Conditions under which reversing the order of integration is permissible are found in most advanced calculus texts.

we have

$$\mathscr{L}^{-1}\left\{\frac{s}{(s^2+1)^2}\right\} = \mathscr{L}^{-1}\left\{\frac{s}{s^2+1} \cdot \frac{1}{s^2+1}\right\}.$$

By Theorem 1, this is the convolution of $\cos t$ and $\sin t$. But

$$\sin t * \cos t = \int_0^t \sin(t-u)\cos u\,du$$

$$= \int_0^t (\sin t \cos u - \cos t \sin u)\cos u\,du$$

$$= \sin t \int_0^t \cos^2 u\,du - \cos t \int_0^t \sin u \cos u\,du,$$

$$= \left[\sin t\left(\frac{\sin u \cos u + u}{2}\right) - \cos t \frac{\sin^2 u}{2}\right]\Bigg|_{u=0}^{u=t}$$

$$= \frac{t \sin t}{2}.$$

Therefore

$$\mathscr{L}^{-1}\left\{\frac{s}{(s^2+1)^2}\right\} = \frac{t \sin t}{2}.$$

Example 2 Compute $\mathscr{L}^{-1}\{e^{-as}/s^{n+1}\}$, where $n \geq 1$ is an integer and a is a real number.

Solution Note that

$$\frac{1}{s^{n+1}} e^{-as} = \frac{1}{s^n} \frac{e^{-as}}{s} = \mathscr{L}\left\{\frac{t^{n-1}}{(n-1)!}\right\} \mathscr{L}\{H(t-a)\}.$$

Thus

$$\mathscr{L}^{-1}\left\{\frac{1}{s^{n+1}} e^{-as}\right\} = \int_0^t \frac{(t-u)^{n-1}}{(n-1)!} H(u-a)\,du.$$

If $t < a$, then $H(u-a) = 0$, but if $t > a$, then

$$\int_0^t \frac{(t-u)^{n-1}}{(n-1)!} H(u-a)\,du = \int_a^t \frac{(t-u)^{n-1}}{(n-1)!}\,du$$

$$= -\frac{(t-u)^n}{n!}\bigg|_a^t = \frac{(t-a)^n}{n!}.$$

Thus

$$\mathscr{L}^{-1}\left\{\frac{1}{s^{n+1}} e^{-as}\right\} = \begin{cases} 0, & t < a, \\ \dfrac{(t-a)^n}{n!}, & t > a, \end{cases}$$

or

$$\mathscr{L}^{-1}\left\{\frac{1}{s^{n+1}}e^{-as}\right\} = \frac{(t-a)^n}{n!}H(t-a). \qquad (6)$$

Convolutions are very useful in solving differential equations with discontinuous right-hand sides.

Note This problem can also be solved using the second shifting theorem.

Example 3 Solve the initial-value problem
$$y'' + y = f(t), \qquad y(0) = 0, \qquad y'(0) = 1,$$
where
$$f(t) = \begin{cases} 1, & 0 < t < 1, \\ 0, & t > 1. \end{cases}$$

Solution Since $f(t) = H(t) - H(t-1)$, taking Laplace transforms of the differential equation we have
$$s^2 \mathscr{L}\{y\} - 1 + \mathscr{L}\{y\} = \frac{1-e^{-s}}{s},$$
or
$$\mathscr{L}\{y\} = \frac{1+s-e^{-s}}{s(s^2+1)} = \frac{1}{s} - \frac{s-1}{s^2+1} - \frac{e^{-s}}{s} \cdot \frac{1}{s^2+1}.$$

Using the convolution theorem for Laplace transforms (Theorem 1), we get
$$y(t) = 1 - \cos t + \sin t - \sin t * H(t-1).$$

But
$$\sin t * H(t-1) = \int_0^t \sin(t-u) H(u-1) \, du$$

and, by the definition of the Heaviside function,
$$\sin t * H(t-1) = H(t-1) \int_1^t \sin(t-u) \, du$$
$$= H(t-1) \cos(t-u)\big|_1^t$$
$$= H(t-1)[1 - \cos(t-1)],$$
so
$$y(t) = 1 - \cos t + \sin t - H(t-1)[1 - \cos(t-1)].$$

VOLTERRA INTEGRAL EQUATIONS

Although the convolution theorem is obviously very useful in calculating inverse transforms, it also has important applications in a very different area. In 1931 the Italian mathematician Vito Volterra[†] published a book that

---

[†] V. Volterra, *Leçons sur la théorie mathématique de la lutte pour la vie* (Paris: Gauthier-Villars, 1931).

contained a fairly sophisticated model of population growth. It would be beyond the scope of this book to go into a derivation of Volterra's model. However, a central equation in this model is of the form

$$\boxed{x(t) = f(t) + \int_0^t a(t-u)x(u)\,du.} \tag{7}$$

An equation of this type, where $f(t)$ and $a(t)$ can be assumed to be continuous, is called a **Volterra integral equation**. Since the publication of Volterra's papers, many diverse phenomena in thermodynamics, electric systems theory, nuclear reactor theory, and chemotherapy have been modeled with Volterra integral equations.

It is quite easy to see how Laplace transforms can be used to solve an equation in the form of equation (7). Taking transforms on both sides of equation (7), using the convolution theorem, and denoting transforms by the appropriate capital letters, we obtain

$$X(s) = F(s) + A(s)X(s),$$

or

$$X(s)[1 - A(s)] = X(s) - A(s)X(s) = F(s),$$

so

$$X(s) = \frac{F(s)}{1 - A(s)}. \tag{8}$$

Looking at equation (8), we immediately see that if $F(s)$ and $A(s)$ are defined for $s \geq s_0$, then $X(s)$ is similarly defined as long as $A(s) \neq 1$. Once $X(s)$ is known, we may (if possible) calculate the solution $x(t) = \mathscr{L}^{-1}\{X(s)\}$.

Example 4 Consider the integral equation

$$x(t) = t^2 + \int_0^t \sin(t-u)x(u)\,du. \tag{9}$$

Taking transforms, we have

$$X(s) = \frac{2}{s^3} + \frac{1}{s^2+1}X(s),$$

or

$$X(s) = \frac{2/s^3}{1 - 1/(s^2+1)}$$

$$= \frac{2(s^2+1)}{s^5}$$

$$= \frac{2}{s^3} + \frac{2}{s^5}.$$

Hence the solution to equation (9) is given by

$$x(t) = t^2 + \frac{1}{12}t^4.$$

There are other applications of the very useful convolution theorem given in the exercises. The student, however, should always keep in mind that the greatest difficulty in using any of these methods is that it is frequently difficult to calculate inverse transforms. Unfortunately, most problems that arise lead to inverting transforms that do not fit into familiar patterns. For this reason, methods have been devised for estimating such inverses. The interested reader should consult a more advanced book on Laplace transforms, such as the excellent book by Widder.[†]

[†] D. V. Widder, *The Laplace Transformation* (Princeton, N.J.: Princeton University Press, 1941).

PROBLEMS 6.5

In Problems 1–5 find the Laplace transform of each given convolution integral.

1. $f(t) = \int_0^t (t-u)^3 \sin u \, du$

2. $f(t) = \int_0^t e^{-(t-u)} \cos 2u \, du$

3. $f(t) = \int_0^t (t-u)^3 u^5 \, du$

4. $f(t) = \int_0^t \sinh 4(t-u) \cosh 5u \, du$

5. $f(t) = \int_0^t e^{17(t-u)} u^{19} \, du$

In Problems 6–12 use the convolution theorem to calculate the inverse Laplace transforms of the given functions.

6. $F(s) = \dfrac{1}{s^2(s^2 + a^2)}$

7. $F(s) = \dfrac{3}{s^4(s^2 + 1)}$

8. $F(s) = \dfrac{1}{(s^2 + 1)^2}$

9. $F(s) = \dfrac{1}{s(s^2 + a^2)}$

10. $F(s) = \dfrac{1}{(s^2 + 1)^3}$

11. $F(s) = \dfrac{e^{-3s}}{s^3}$

12. $F(s) = \dfrac{e^{-10s}}{s^5}$

13. Solve the Volterra integral equation

$$x(t) = e^{-t} - 2\int_0^t \cos(t-u) x(u) \, du.$$

14. Solve the Volterra integral equation

$$x(t) = t + \frac{1}{6}\int_0^t (t-u)^3 x(u) \, du.$$

15. Find the solution of the initial-value problem

$$y'' + 4y' + 13y = f(t), \quad y(0) = y'(0) = 0,$$

where

$$f(t) = \begin{cases} 1, & t < \pi, \\ 0, & t > \pi. \end{cases}$$

16. Find the solution of the initial-value problem

$$y'' - 2y' + 2y = f(t), \quad y(0) = y'(0) = 0,$$

where

$$f(t) = \begin{cases} 0, & t < \pi/2, \\ 1, & \pi/2 < t < 3\pi/2, \\ 2, & t > 3\pi/2. \end{cases}$$

17. Use the convolution theorem to show that

$$\mathscr{L}\left\{\int_0^t f(u) \, du\right\} = \frac{F(s)}{s}, \quad \text{where } F(s) = \mathscr{L}\{f(t)\}.$$

[*Hint:* $\int_0^t f(u)\,du = (1*f)(t)$.]

18. Compute $\mathscr{L}\{\int_0^{t-a} f(u)\,du\}$. [*Hint:* $\int_0^{t-a} f(u)\,du = \int_0^t f(u) H(t-a-u)\,du$.]

19. If $f(0) = g(0) = 0$, show that
    a. $f' * g = f * g'$;
    b. $(f*g)' = \frac{1}{2}[f' * g + f * g']$.

Review Exercises for Chapter 6

In Exercises 1–17 find the Laplace transform of the given function.

1. $3t - 2$
2. $t^3 + 4t^2 - 2t + 1$
3. e^{2t-1}
4. $\cos(2t + 1)$
5. te^{-t}
6. $e^{-t}\cos 2t$
7. $\sinh(3t - 4)$
8. $t^3 e^{2t}$
9. $te^t \cos t$
10. $\cosh^2 2t$
11. $\cos^2 t$
12. $\int_0^t u \cos u\, du$
13. $t^2 \cos 2t$
14. $\cos t \cdot H(t - 2\pi)$
15. $\delta(t - 3)$
16. the function shown in the figure below:

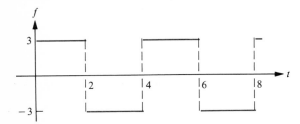

17. $f(t) = \begin{cases} \sin t, & t < 2\pi \\ \sin t + \cos t, & t > 2\pi \end{cases}$

In Exercises 18–32 find the inverse of the given Laplace transform.

18. $\dfrac{3}{s^2}$
19. $\dfrac{-14}{s}$
20. $\dfrac{s + 2}{s^2 + 4}$
21. $\dfrac{1}{(s - 2)^2}$
22. $\dfrac{s - 3}{s^2 - 3}$
23. $\dfrac{1}{s^2 + 4s + 5}$
24. $\dfrac{s + 3}{s^2 + 8s + 17}$
25. $\dfrac{s}{s^2 - 3s + 2}$
26. $\dfrac{s + 2}{s(s^2 + 4)}$
27. $\dfrac{s^2 + 8s - 3}{(s^2 + 2s + 1)(s^2 + 1)}$
28. $\ln \dfrac{s - 3}{s - 2}$
29. $\ln\left(1 + \dfrac{4}{s^2}\right)$
30. $\dfrac{se^{-(\pi/2)s}}{1 + s^2}$
31. $\dfrac{s - se^{-(\pi/2)s}}{1 + s^2}$
32. $\dfrac{e^{-3s}}{s^2 - 1}$

In Exercises 33–41 solve the given initial-value problem by Laplace transform methods.

33. $y'' + y = 0$, $y(0) = 1$, $y'(0) = 3$
34. $y'' - 4y = 0$, $y(0) = 2$, $y'(0) = -5$
35. $y'' - 5y' + 6y = 0$, $y(0) = 2$, $y'(0) = 1$
36. $y'' + 4y' + 4y = 0$, $y(0) = -1$, $y'(0) = 2$
37. $y'' - y = te^{2t}$, $y(0) = 0$, $y'(0) = 1$
38. $y'' + 9y = \sin t$, $y(0) = y'(0) = 0$
39. $y'' + 2y' + 2y = H(t - 3)$, $y(0) = y'(0) = 0$
40. $y' - 6y = H(t - 2)$, $y(0) = 0$
41. $y'' - 4y' + 4y = \delta(t - 1)$, $y(0) = 0$, $y'(0) = 1$

In Exercises 42–44 find the Laplace transform of the given convolution integral.

42. $\int_0^t (t - u)^2 \sin u\, du$
43. $\int_0^t (t - u)^4 u^7\, du$
44. $\int_0^t e^{15(t - u)} u^{25}\, du$

45. Use the convolution theorem to compute the inverse Laplace transform of

  a. $F(s) = \dfrac{2}{s^3(s^2 + 1)}$

  b. $F(s) = \dfrac{1}{s(s^2 + 4)}$

46. Solve the initial-value problem
$$y'' + 6y' + 18y = f(t), \quad y(0) = y'(0) = 0,$$
where
$$f(t) = \begin{cases} 2, & t < \pi, \\ 0, & t > \pi. \end{cases}$$

7

Systems of Linear Differential Equations

This chapter presents three different approaches to the solution of systems of linear differential equations. *It is not our intention that you study all three techniques; instead, concentrate on the approach that best meets the requirements of your plan of study.* Each technique presented in this chapter has advantages and disadvantages:

1. The **method of elimination** and its allied **method of determinants** (Section 7.3) provide an efficient and elementary procedure for solving systems of two first-order linear differential equations in two unknown functions. As the name suggests, the technique involves eliminating one of the unknown functions by converting the system of two first-order equations into a single linear second-order equation in the remaining dependent variable. This second-order equation may then be solved using the methods in Chapter 3. Although the method is easily extendable to systems of n linear first-order equations in n unknown variables, the task of eliminating $(n-1)$ dependent variables becomes prohibitive for large n.

2. The **method of Laplace transforms** (Section 7.4) converts a system of n linear first-order differential equations into a system of n linear equations, which may then be solved by the techniques of linear algebra. Once the transform of each unknown function has been determined, we may use the procedures in Chapter 6 to obtain a solution to the system. This last part may be the most difficult step in this procedure for large n, as the partial-fraction decomposition may be extremely tedious to obtain.

3. The rest of this chapter (Sections 7.5–7.11) provides a discussion of **matrix methods** for the solution of systems of linear differential equations.

Although a brief review of matrices is given in Section 7.5, additional familiarity with linear algebra is recommended. Two methods for obtaining the principal matrix solution for a homogeneous system are given in Sections 7.9 and 7.11; nonhomogeneous systems are treated in Section 7.10. This method provides a powerful but sophisticated technique for solving large systems with constant coefficients.

Each of the approaches makes use of the introductory material in Section 7.1 and the theoretical material in Section 7.2. The procedures in approach (1) are extremely elementary. The Laplace transform approach requires familiarity with the concepts of Sections 6.1–6.3. The third approach, via matrix methods, requires substantial class time.

Study Guide

Approach	Required	Option	Remarks
1	Sections 7.1, 7.2	Section 7.3	Most elementary approach
2	Sections 7.1, 7.2	Section 7.4	For students who covered Sections 6.1–6.3
3	Sections 7.1, 7.2	Sections 7.5–7.11	For students with a background in linear algebra

7.1 Systems of First-Order Equations

In the differential equations we have discussed so far in this book, there was exactly one unknown function. A typical problem was to find the solution of, say, $x'' + a(t)x' + b(t)x = f(t)$, where a, b, and f are given. Here the solution consists of the single function $x(t)$.

In many applications, however, there is more than one unknown function. In these cases we are often led to a number of interrelated differential equations. These equations together form what we call a **system** of differential equations. One typical system is given below:

$$\begin{aligned} x_1' &= f_1(t, x_1, x_2, \ldots, x_n), \\ x_2' &= f_2(t, x_1, x_2, \ldots, x_n), \\ &\vdots \qquad \vdots \\ x_n' &= f_n(t, x_1, x_2, \ldots, x_n), \end{aligned} \qquad (1)$$

where each f_i, $i = 1, 2, \ldots, n$, is a function of $n+1$ variables. The problem is to find n differentiable functions $x_1(t), x_2(t), \ldots, x_n(t)$ that satisfy (1) for all t in some interval $\alpha < t < \beta$.

Solution of a System of Differential Equations

Any collection of n differentiable functions $x_1(t), x_2(t), \ldots, x_n(t)$ that satisfy the system (1) for all values t in some interval $\alpha < t < \beta$ is called a **solution** of the system (1) on that interval.

First-Order System

System (1) is called a **first-order** system because it involves only first derivatives of the functions $x_1, x_2, \ldots, x_n$.

We see in this section that certain higher-order differential equations or systems of higher-order differential equations can be written as first-order systems of form (1). Therefore we limit our considerations to such first-order systems, that is, to systems that can be written in form (1).

In much the same way that differential equations are classified as linear or nonlinear, systems are either linear or nonlinear.

Linear System

A first-order system is said to be **linear** if it can be written in the form

$$\begin{aligned}
\frac{dx_1}{dt} &= a_{11}(t)x_1 + a_{12}(t)x_2 + \cdots + a_{1n}(t)x_n + f_1(t), \\
\frac{dx_2}{dt} &= a_{21}(t)x_1 + a_{22}(t)x_2 + \cdots + a_{2n}(t)x_n + f_2(t), \\
&\vdots \\
\frac{dx_n}{dt} &= a_{n1}(t)x_1 + a_{n2}(t)x_2 + \cdots + a_{nn}(t)x_n + f_n(t),
\end{aligned} \qquad (2)$$

where a_{ij} and f_i are functions that are continuous on some common interval $\alpha < t < \beta$. Any system for which this cannot be done is said to be **nonlinear**. The functions a_{ij} are called the **coefficients** of the linear system (2). If every a_{ij} is constant, system (2) is said to have **constant coefficients**. If every function f_i satisfies $f_i(t) \equiv 0$ for $\alpha < t < \beta$, $i = 1, 2, \ldots, n$, then system (2) is said to be **homogeneous**; otherwise it is called a **nonhomogeneous** system.

Example 1 Suppose that a chemical solution flows from one container, at a rate proportional to its volume, into a second container. It flows out from the second container at a constant rate. Let $x_1(t)$ and $x_2(t)$ denote the volumes of solution in the first and second containers, respectively, at time t. (The containers may be, for example, cells, in which case we are describing a diffusion process across a cell wall.) To establish the necessary equations, we note that the change in volume equals the difference between input and output in each container. The change in volume is the derivative of volume with respect to time. Since no chemical is flowing into the first container, the change in its volume equals the output:

$$\frac{dx_1}{dt} = -c_1 x_1,$$

where c_1 is a positive constant of proportionality. The amount of solution $c_1 x_1$ flowing out of the first container is the input of the second container. Let c_2 be the constant output of the second container. Then the change in volume in the second container equals the difference between its input and output:

$$\frac{dx_2}{dt} = c_1 x_1 - c_2.$$

Thus we can describe the flow of solution by means of two differential equations. Since more than one differential equation is involved, we have obtained the linear nonhomogeneous system with constant coefficients

$$\frac{dx_1}{dt} = -c_1 x_1,$$
$$\frac{dx_2}{dt} = c_1 x_1 - c_2, \tag{3}$$

(c_1 and c_2 are positive constants). By a solution of system (3) we mean a pair of functions $(x_1(t), x_2(t))$ that simultaneously satisfy the two equations in (3).

Initial-Value Problem

In addition to the given system of differential equations (1) or (2), initial conditions of the form

$$x_1(t_0) = x_{10}, x_2(t_0) = x_{20}, \ldots, x_n(t_0) = x_{n0} \tag{4}$$

may also be given for some t_0 satisfying $\alpha < t_0 < \beta$. We call system (1) or (2) together with the initial condition (4) an **initial-value problem**.

Example 1 (continued) Recall system (3) of Example 1:

$$\frac{dx_1}{dt} = -c_1 x_1,$$
$$\frac{dx_2}{dt} = c_1 x_1 - c_2.$$

It is easy to solve this system by solving the two equations successively (this is not usually possible). If we denote the initial volumes in the two containers by $x_1(0)$ and $x_2(0)$, respectively (these are the initial conditions), we see that the first equation has the solution

$$x_1(t) = x_1(0) e^{-c_1 t}. \tag{5}$$

Substituting (5) into the second equation of (3), we obtain the equation

$$\frac{dx_2}{dt} = c_1 x_1(0) e^{-c_1 t} - c_2,$$

which, upon integration from 0 to t, yields the solution

$$x_2(t) = x_2(0) + x_1(0)(1 - e^{-c_1 t}) - c_2 t. \tag{6}$$

Equations (5) and (6) together constitute the unique solution of system (3) that satisfies the given initial conditions.

As the definitions above indicate, many of the terms we have used in studying a single differential equation are also used to describe systems of differential equations. This is deliberate, because the system reduces to a single differential equation if we set $n = 1$ in (1) or (2).

As in the case with initial-value problems involving a single differential equation (see Theorem 1.2.1), it is necessary to impose conditions on the functions $f_1, f_2, \ldots, f_n$ in order to guarantee that the initial-value problem (1) and (4) has a unique solution. The following result, which is proved in Appendix 3 (Theorem 7), provides sufficient conditions for a local solution to the initial-value problem (1) and (4).

THEOREM 1 LOCAL EXISTENCE–UNIQUENESS THEOREM

Let $f_1, f_2, \ldots, f_n$ and each of the first partials $\partial f_1/\partial x_1, \ldots, \partial f_1/\partial x_n$, $\partial f_2/\partial x_1, \ldots, \partial f_n/\partial x_n$ be continuous in an $(n+1)$-dimensional rectangular region R given by $\alpha < t < \beta$, $c_1 < x_1 < d_1, \ldots, c_n < x_n < d_n$ containing the point $(t_0, x_{10}, \ldots, x_{n0})$. Then, in some interval $|t - t_0| < h$ contained in (α, β), there is a unique solution of system (1) that also satisfies the initial conditions (4).

For linear systems the existence-uniqueness theorem is simpler and provides a global solution. We prove the following result in Appendix 3 (Theorem 8).

THEOREM 2 GLOBAL EXISTENCE–UNIQUENESS THEOREM

If $a_{11}, a_{12}, \ldots, a_{nn}, f_1, \ldots, f_n$ are continuous functions on the interval $\alpha < t < \beta$ containing the point t_0, then system (2) has a unique solution that satisfies the initial conditions (4). Furthermore, this solution is valid throughout the interval $\alpha < t < \beta$.

Observe that in Theorem 1 we were only guaranteed a solution in a subinterval of $\alpha < t < \beta$, whereas in Theorem 2 the solution is valid throughout the interval in which the hypotheses hold. In particular, if $a_{11}, a_{12}, \ldots, a_{nn}, f_1, \ldots, f_n$ are continuous functions on the real numbers $\mathbb{R}$, the unique solution is valid on all of $\mathbb{R}$.

Warning Without stating it, we will always assume that a_{ij} and f_i are continuous functions of t on some common interval containing t_0. If they are all constants, the solution is unique for all real numbers $\mathbb{R}$. For example, the solution we obtained in Example 1 is valid for all t in the real numbers $\mathbb{R}$, since the functions $a_{11} = -c$, $a_{12} = 0$, $f_1 = 0$, $a_{21} = c_1$, $a_{22} = 0$, $f_2 = -c_2$ are constant and therefore continuous on $\mathbb{R}$.

We mentioned earlier that certain higher-order differential equations, such as

$$y^{(n)} = F(t, y, y', y'', \ldots, y^{(n-1)}), \tag{7}$$

could be written as a system of first-order equations of form (1). The procedure is very easy: we define n new variables $x_1, x_2, \ldots, x_n$ by setting

$$x_1 = y, \; x_2 = y', \; x_3 = y'', \ldots, x_n = y^{(n-1)}. \tag{8}$$

Then, since $y' = x_1' = x_2$, $y'' = x_2' = x_3, \ldots, y^{(n-1)} = x_{n-1}' = x_n$ and $y^{(n)} = x_n'$, we have the following first-order system

$$\begin{aligned} x_1' &= x_2, \\ x_2' &= x_3, \\ &\vdots \\ x_{n-1}' &= x_n, \\ x_n' &= F(t, x_1, x_2, \ldots, x_n), \end{aligned} \tag{9}$$

which is a special case of system (1). The technique we have developed above can easily be applied to certain higher-order systems, as the following example illustrates.

Example 2 Consider the mass-spring system of Figure 7.1, which is a direct generalization of the system described in Section 4.1 (see p. 138). In this example we have two objects of mass m_1 and m_2 suspended by springs in series with spring constants k_1 and k_2. If the vertical displacements from equilibrium of the two point masses are denoted by $x_1(t)$ and $x_2(t)$, respectively, then using assumptions (a) and (b) (Hooke's law, p. 138), we find that the net forces acting on the two masses are given by

$$F_1 = -k_1 x_1 + k_2(x_2 - x_1),$$
$$F_2 = -k_2(x_2 - x_1).$$

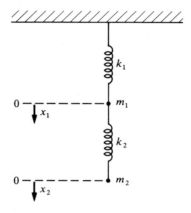

Figure 7.1

Here the positive direction is downward. Note that the first spring is compressed when $x_1 < 0$ and the second spring is compressed when $x_1 > x_2$. The equations of motion are a system of two linear second-order differential equations:

$$m_1 \frac{d^2 x_1}{dt^2} = -k_1 x_1 + k_2(x_2 - x_1) = -(k_1 + k_2)x_1 + k_2 x_2,$$
$$m_2 \frac{d^2 x_2}{dt^2} = -k_2(x_2 - x_1) = k_2 x_1 - k_2 x_2. \tag{10}$$

To rewrite system (10) as a first-order system, we define the new variables $x_3 = x_1'$ and $x_4 = x_2'$. Then $x_3' = x_1''$, $x_4' = x_2''$ and (10) can be expressed as the system of four first-order equations

$$x_1' = x_3,$$
$$x_2' = x_4,$$
$$x_1'' = x_3' = -\left(\frac{k_1 + k_2}{m_1}\right) x_1 + \left(\frac{k_2}{m_1}\right) x_2, \tag{11}$$
$$x_2'' = x_4' = \left(\frac{k_2}{m_2}\right) x_1 - \left(\frac{k_2}{m_2}\right) x_2.$$

This is a linear homogeneous system with constant coefficients.

We now show that linear differential equations (and systems) of *any* order can be converted, by the introduction of new variables, into a linear system of first-order differential equations. This concept is very important, since it means that the study of first-order linear systems provides a unified theory for all linear differential equations and systems. From a practical point of view, it means, for example, that once we know how to solve first-order linear systems with constant coefficients, we will be able to solve any constant-coefficient linear differential equation or system.

THEOREM 3

The linear nth-order differential equation
$$x^{(n)} + a_1(t)x^{(n-1)} + a_2(t)x^{(n-2)} + \cdots + a_{n-1}(t)x' + a_n(t)x = f(t) \quad (12)$$
can be rewritten as a system of n first-order linear equations.

Proof Define $x_1 = x$, $x_2 = x'$, $x_3 = x''$, ..., $x_n = x^{(n-1)}$. Then we have the linear nonhomogeneous system

$$\begin{aligned}
x_1' &= x_2, \\
x_2' &= x_3, \\
&\vdots \\
x_{n-1}' &= x_n, \\
x_n' &= -a_n x_1 - a_{n-1} x_2 - \cdots - a_1 x_n + f.
\end{aligned} \quad (13)$$
∎

Suppose that n initial conditions are specified for the nth-order equation (12):
$$x(t_0) = c_1,\ x'(t_0) = c_2, \ldots, x^{(n-1)}(t_0) = c_n.$$

These initial conditions can be immediately transformed into an initial condition for system (13):
$$x_1(t_0) = c_1,\ x_2(t_0) = c_2, \ldots, x_n(t_0) = c_n. \quad (14)$$

Example 3 Consider the initial-value problem
$$t^3 x''' + 4t^2 x'' - 8tx' + 8x = 0, \qquad x(2) = 3,\ x'(2) = -6,\ x''(2) = 14.$$
Defining $x_1 = x$, $x_2 = x'$, $x_3 = x''$, we obtain the system (since $x_3' = x'''$)
$$\begin{aligned}
x_1' &= x_2, \\
x_2' &= x_3, \\
x_3' &= -\frac{8}{t^3} x_1 + \frac{8}{t^2} x_2 - \frac{4}{t} x_3,
\end{aligned}$$
with the initial condition $x_1(2) = 3$, $x_2(2) = -6$, $x_3(2) = 14$.

We will see, in the sections that follow, that almost every concept that we are already familiar with for differential equations extends to systems. As in the case of differential equations, nonlinear systems are extremely difficult to solve, so virtually all our discussion will concern linear first-order systems.

PROBLEMS 7.1

In Problems 1–7 transform each given equation into a system of first-order equations.
1. $x'' + 2x' + 3x = 0$
2. $x'' - 6tx' + 3t^3 x = \cos t$
3. $x''' - x'' + (x')^2 - x^3 = t$
4. $x^{(4)} - \cos x(t) = t$
5. $x''' + xx'' - x'x^4 = \sin t$
6. $xx'x''x''' = t^5$
7. $x''' - 3x'' + 4x' - x = 0$
8. A mass m moves in xyz-space according to the following equations of motion:
$$mx'' = f(t, x, y, z),$$
$$my'' = g(t, x, y, z),$$
$$mz'' = h(t, x, y, z).$$
Transform these equations into a system of six first-order equations.
9. Consider the uncoupled system
$$x_1' = x_1, \quad x_2' = x_2.$$
   a. What is the general solution of this system?
   *b. Show that there is no second-order equation equivalent to this system. [*Hint*: Show that any second-order equation has solutions that are not solutions of this system.] This shows that first-order systems are more general than higher-order equations in the sense that any of the latter can be written as a first-order system, but not vice versa.
10. Let $x = x_1(t)$, $y = y_1(t)$ and $x = x_2(t)$, $y = y_2(t)$ be two solutions of the linear homogeneous system
$$\begin{aligned} x' &= a_{11}(t)x + a_{12}(t)y, \\ y' &= a_{21}(t)x + a_{22}(t)y. \end{aligned} \quad (i)$$
Show that
$$x = c_1 x_1(t) + c_2 x_2(t), \quad y = c_1 y_1(t) + c_2 y_2(t),$$
is also a solution of (i), for any constants c_1 and c_2. This is the **principle of superposition for systems**.
11. Consider the nonhomogeneous system
$$\begin{aligned} x' &= a_{11}(t)x + a_{12}(t)y + f_1(t), \\ y' &= a_{21}(t)x + a_{22}(t)y + f_2(t). \end{aligned} \quad (ii)$$
Show that if $x = x_1(t)$, $y = y_1(t)$ and $x = x_2(t)$, $y = y_2(t)$ are two solutions of the nonhomogeneous system (ii), then $x = x_1(t) - x_2(t)$, $y = y_1(t) - y_2(t)$ is a solution of the homogeneous system (i).

7.2 Linear Systems: Theory

In this section we consider the linear system of two first-order equations
$$\begin{aligned} x' &= a_{11}(t)x + a_{12}(t)y + f_1(t), \\ y' &= a_{21}(t)x + a_{22}(t)y + f_2(t), \end{aligned} \quad (1)$$
and the associated homogeneous system (i.e., $f_1 = f_2 = 0$)
$$\begin{aligned} x' &= a_{11}(t)x + a_{12}(t)y, \\ y' &= a_{21}(t)x + a_{22}(t)y. \end{aligned} \quad (2)$$

The point of view here emphasizes the similarities between such systems and the linear second-order equations discussed in Section 3.1. That there is a parallel between the two theories should not be surprising, since we have already shown in Section 7.1 that any linear second-order equation can always be transformed into a system of form (1).

Throughout our discussion we assume that the functions a_{ij} and f_i are all continuous on a common interval I. As we stated in Section 7.1, this guarantees the following result, provided t_0 is a point in that interval (see Appendix 3):

THEOREM 1 EXISTENCE–UNIQUENESS THEOREM

If the functions $a_{11}(t)$, $a_{12}(t)$, $a_{21}(t)$, $a_{22}(t)$, $f_1(t)$, and $f_2(t)$ are continuous on I, then given any numbers t_0, x_0, and y_0, with t_0 in I, there exists exactly one solution $(x(t), y(t))$ of system (1) that satisfies $x(t_0) = x_0$ and $y(t_0) = y_0$.

7.2/LINEAR SYSTEMS: THEORY

By a **solution** of system (1) [or (2)] we mean a *pair* of functions $(x(t), y(t))$ that possess first derivatives and that satisfy the given equations.

Warning Whenever we compare two or more solutions or pairs of functions, we assume that all the functions are continuous on a common interval I.

Linear Combination

The pair of functions $(x_3(t), y_3(t))$ is a **linear combination** of the pairs $(x_1(t), y_1(t))$ and $(x_2(t), y_2(t))$ if there exist constants c_1 and c_2 such that the following two equations hold:

$$x_3(t) = c_1 x_1(t) + c_2 x_2(t),$$
$$y_3(t) = c_1 y_1(t) + c_2 y_2(t). \tag{3}$$

The next theorem is the systems analogue of Theorem 3.1.3 (p. 89). Its proof is left as an exercise.

THEOREM 2

If the pairs $(x_1(t), y_1(t))$ and $(x_2(t), y_2(t))$ are solutions of the homogeneous system (2), then any linear combination of them is also a solution of system (2).

Example 1 Consider the system

$$x' = -x + 6y,$$
$$y' = x - 2y. \tag{4}$$

It is easy to verify that the pairs $(-2e^{-4t}, e^{-4t})$ and $(3e^t, e^t)$ are solutions of system (4). Hence, by Theorem 2, the pair $(-2c_1 e^{-4t} + 3c_2 e^t, c_1 e^{-4t} + c_2 e^t)$ is a solution of (4) for any constants c_1 and c_2.

Linear Independence

We define two pairs of functions $(x_1(t), y_1(t))$ and $(x_2(t), y_2(t))$ to be **linearly independent** on an interval I if, whenever the equations

$$c_1 x_1(t) + c_2 x_2(t) = 0,$$
$$c_1 y_1(t) + c_2 y_2(t) = 0 \tag{5}$$

hold for all values of t in I, then $c_1 = c_2 = 0$. In Example 1, the two given pairs of solutions are linearly independent since $c_1 e^{-4t} + c_2 e^t$ vanishes for all t only when $c_1 = c_2 = 0$.

Wronskian

Given two solutions $(x_1(t), y_1(t))$ and $(x_2(t), y_2(t))$ of system (2), we define the **Wronskian** of the two solutions by the following determinant:

$$W(t) = \begin{vmatrix} x_1(t) & y_1(t) \\ x_2(t) & y_2(t) \end{vmatrix} = x_1(t) y_2(t) - x_2(t) y_1(t). \tag{6}$$

We can then prove the next theorem.

THEOREM 3

Let $(x_1(t), y_1(t))$ and $(x_2(t), y_2(t))$ be solutions of the homogeneous system (2) satisfying $W(t) \neq 0$ for every t in some interval I. If $(x^*(t), y^*(t))$ is any

other solution of system (2), there exist constants c_1 and c_2 such that
$$x^* = c_1 x_1 + c_2 x_2,$$
$$y^* = c_1 y_1 + c_2 y_2. \tag{7}$$

Thus $(c_1 x_1(t) + c_2 x_2(t), c_1 y_1(t) + c_2 y_2(t))$ is the **general solution** of system (2).

Proof Let t_0 in I be given and consider the linear system of two equations in the unknown quantities c_1 and c_2:
$$c_1 x_1(t_0) + c_2 x_2(t_0) = x^*(t_0),$$
$$c_1 y_1(t_0) + c_2 y_2(t_0) = y^*(t_0). \tag{8}$$

The determinant of this system is $W(t_0)$, which is nonzero by assumption. Thus there is a unique pair of constants (c_1, c_2) satisfying (8). By Theorem 2,
$$(c_1 x_1(t) + c_2 x_2(t), c_1 y_1(t) + c_2 y_2(t))$$
is a solution of (2). But, by (8), this solution satisfies the same initial conditions at t_0 as the solution $(x^*(t), y^*(t))$. By the uniqueness part of Theorem 1, these solutions must be identical for all t in I. ∎

Example 2 In Example 1 the Wronskian $W(t)$ is
$$W(t) = \begin{vmatrix} -2e^{-4t} & e^{-4t} \\ 3e^t & e^t \end{vmatrix} = -2e^{-3t} - 3e^{-3t} = -5e^{-3t} \neq 0.$$

Hence we need look no further for the general solution of system (4). The general solution is given by
$$(x(t), y(t)) = (-2c_1 e^{-4t} + 3c_2 e^t, c_1 e^{-4t} + c_2 e^t).$$

A particular solution can be obtained, for example, by setting $c_1 = 2$ and $c_2 = -1$, to obtain the solution
$$(-4e^{-4t} - 3e^t, 2e^{-4t} - e^t).$$

This should be checked by differentiating and inserting the results into system (4).

In view of the condition required in Theorem 3 that the Wronskian $W(t)$ never vanish on I, we must consider the properties of the Wronskian more carefully. Let (x_1, y_1) and (x_2, y_2) be two solutions of the homogeneous system (2). Since $W(t) = x_1 y_2 - x_2 y_1$, we have
$$W'(t) = x_1 y_2' + x_1' y_2 - x_2 y_1' - x_2' y_1$$
$$= x_1(a_{21} x_2 + a_{22} y_2) + y_2(a_{11} x_1 + a_{12} y_1) - x_2(a_{21} x_1 + a_{22} y_1)$$
$$- y_1(a_{11} x_2 + a_{12} y_2).$$

Multiplying these expressions through and canceling like terms, we obtain
$$W' = a_{11} x_1 y_2 + a_{22} x_1 y_2 - a_{11} x_2 y_1 - a_{22} x_2 y_1$$
$$= (a_{11} + a_{22})(x_1 y_2 - x_2 y_1) = (a_{11} + a_{22}) W.$$

Thus

$$W(t) = W(t_0) \exp\left(\int_{t_0}^{t}[a_{11}(u) + a_{22}(u)]\, du\right). \tag{9}$$

We have shown the following theorem to be true.

THEOREM 4

Let (x_1, y_1) and (x_2, y_2) be two solutions of the homogeneous system (2). Then the Wronskian $W(t)$ is either always zero or never zero in the interval I (since $e^x \neq 0$ for any x).

We are now ready to state the theorem that links linear independence with a nonvanishing Wronskian (see Theorem 3.1.5).

THEOREM 5

Two solutions $(x_1(t), y_1(t))$ and $(x_2(t), y_2(t))$ of the homogeneous system (2) are linearly independent on an interval I if and only if $W(t) \neq 0$ in I.

Proof Let the solutions be linearly independent and suppose that $W(t) = 0$ in I. Then $x_1 y_2 = x_2 y_1$ or $x_1/x_2 = y_1/y_2 = f(t)$, for some function $f(t)$. Now

$$f' = \left(\frac{x_1}{x_2}\right)' = \frac{x_2 x_1' - x_1 x_2'}{x_2^2} = \frac{x_2(a_{11}x_1 + a_{12}y_1) - x_1(a_{11}x_2 + a_{12}y_2)}{x_2^2}$$

$$= \frac{a_{12}(x_2 y_1 - x_1 y_2)}{x_2^2} = \frac{a_{12} W}{x_2^2} = 0,$$

since $W = 0$. Thus $f'(t) = 0$ so $f(t) = c$, a constant. Then $x_1 = cx_2$ and $y_1 = cy_2$, so the solutions are dependent, which is a contradiction. Hence $W(t) \neq 0$. Conversely, let $W(t) \neq 0$ in I. If the solutions were dependent, then there would exist constants c_1 and c_2, not both zero, such that

$$c_1 x_1 + c_2 x_2 = 0,$$
$$c_1 y_1 + c_2 y_2 = 0.$$

Assuming that $c_1 \neq 0$, we then have $x_1 = cx_2$, $y_1 = cy_2$, where $c = -c_2/c_1$. But then

$$W(t) = x_1 y_2 - x_2 y_1 = cx_2 y_2 - cx_2 y_2 = 0,$$

which is again a contradiction. Therefore the solutions are linearly independent. ∎

General Solution

We may redefine what we mean by a **general solution** in terms of these concepts: Let (x_1, y_1) and (x_2, y_2) be solutions of the homogeneous linear system

$$\begin{aligned} x' &= a_{11}x + a_{12}y, \\ y' &= a_{21}x + a_{22}y. \end{aligned} \tag{10}$$

Then $(c_1x_1 + c_2x_2, c_1y_1 + c_2y_2)$ is the general solution of system (10) provided that $W(t) \neq 0$; that is, provided that the solutions (x_1, y_1) and (x_2, y_2) are linearly independent.

Finally, let us consider the nonhomogeneous system (1). The following theorem is the direct analogue of Theorem 3.1.6. Its proof is left as an exercise.

THEOREM 6

Let (x^*, y^*) be the general solution of system (1), and let (x_p, y_p) be any solution of (1). Then $(x^* - x_p, y^* - y_p)$ is the general solution of the homogeneous equation (10). In other words, the general solution of system (1) can be written as the sum of the general solution of the homogeneous system (10) and any particular solution of the nonhomogeneous system (1).

Example 3 Consider the system

$$x' = 3x + 3y + t,$$
$$y' = -x - y + 1. \tag{11}$$

Note first that $(1, -1)$ and $(-3e^{2t}, e^{2t})$ are solutions to the homogeneous system

$$x' = 3x + 3y,$$
$$y' = -x - y.$$

A particular solution to system (11) is $(-\frac{1}{4}(t^2 + 9t + 3), \frac{1}{4}(t^2 + 7t))$. The general solution to (11) is, therefore,

$$(x(t), y(t)) = \left(c_1 - 3c_2e^{2t} - \tfrac{1}{4}(t^2 + 9t + 3), -c_1 + c_2e^{2t} + \tfrac{1}{4}(t^2 + 7t)\right).$$

We close this section by noting that the theorems in this section can easily be generalized to apply to systems of three or more equations.

PROBLEMS 7.2

1. a. Show that $(e^{-3t}, -e^{-3t})$ and $((1-t)e^{-3t}, te^{-3t})$ are solutions to

$$x' = -4x - y,$$
$$y' = x - 2y.$$

b. Calculate the Wronskian of these solutions and verify that the solutions are linearly independent on $\mathbb{R}$.

c. Write the general solution to the system.

2. a. Show that $(e^{2t}\cos 2t, -2e^{2t}\sin 2t)$ and $(e^{2t}\sin 2t, 2e^{2t}\cos 2t)$ are solutions of the system

$$x' = 2x + y,$$
$$y' = -4x + 2y.$$

b. Calculate the Wronskian of these solutions and show that they are linearly independent on $\mathbb{R}$.

c. Show that $(\tfrac{1}{4}te^{2t}, -\tfrac{11}{4}e^{2t})$ is a solution of the nonhomogeneous system

$$x' = 2x + y + 3e^{2t},$$
$$y' = -4x + 2y + te^{2t}.$$

d. Combining (a) and (c), write the general solution of the nonhomogeneous equation in (c).

3. a. Show that $(\sin t^2, 2t\cos t^2)$ and $(\cos t^2, -2t\sin t^2)$ are solutions of the system

$$x' = y,$$
$$y' = -4t^2x + \frac{1}{t}y.$$

b. Show where the solutions are linearly independent.

c. Show that $W(0) = 0$.

d. Explain the apparent contradiction of Theorem 5.

4. **a.** Show that $(\sin \ln t^2, (2/t)\cos \ln t^2)$ and $(\cos \ln t^2, -(2/t)\sin \ln t^2)$ are linearly independent solutions of the system
$$x' = y,$$
$$y' = -\frac{4}{t^2}x - \frac{1}{t}y.$$

b. Calculate the Wronskian $W(t)$.

5. Prove Theorem 2.

6. Prove Theorem 6. [*Hint*: See the proof of Theorem 3.1.6.]

Nonmatrix Methods

7.3 The Method of Elimination and the Method of Determinants for Linear Systems with Constant Coefficients

In this section we discuss an elementary method for solving a system of simultaneous first-order linear differential equations by converting the system into a single higher-order linear differential equation that may then be solved by the methods we have seen in previous chapters. It is easiest to motivate this method with an example.

Example 1 Let tank X contain 100 gallons of brine in which 100 pounds of salt is dissolved and tank Y contain 100 gallons of water. Suppose that water flows into tank X at the rate of 2 gallons per minute, and that the mixture flows from tank X into tank Y at 3 gallons per minute. Each minute, 1 gallon is pumped from Y back to X (establishing **feedback**) while 2 gallons are flushed away. We wish to find the amount of salt in both tanks at all time t (see Figure 7.2).

Solution If we let $x(t)$ and $y(t)$ represent the number of pounds of salt in tanks X and Y at time t and note that the change in weight equals the difference between input and output, we can again derive a system of linear first-order equations. Tanks X and Y initially contain $x(0) = 100$ and $y(0) = 0$

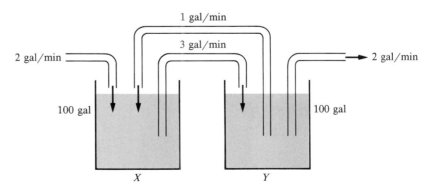

Figure 7.2

pounds of salt, respectively, at time $t = 0$. The quantities $x/100$ and $y/100$ are, respectively, the amounts of salt contained in each gallon of water taken from tanks X and Y at time t. Three gallons are being removed from tank X and added to tank Y, while only 1 of the 3 gallons removed from tank Y is put in tank X. Thus we have the system

$$\frac{dx}{dt} = -3\frac{x}{100} + \frac{y}{100}, \qquad x(0) = 100,$$
$$\frac{dy}{dt} = 3\frac{x}{100} - 3\frac{y}{100}, \qquad y(0) = 0. \tag{1}$$

Since both equations in system (1) involve *both* dependent variables, we cannot immediately solve for one of the variables, as we did in Example 7.1.1. Instead, we use the operation of differentiation to eliminate one of the dependent variables. Suppose we begin by solving the second equation for $x(t)$ in terms of the dependent variable y and its derivative:

$$\frac{3}{100} x = \frac{3}{100} y + \frac{dy}{dt},$$

or

$$x = y + \frac{100}{3} \frac{dy}{dt}. \tag{2}$$

We now differentiate equation (2) and equate the result to the first equation in (1):

$$\frac{dx}{dt} = \frac{dy}{dt} + \frac{100}{3} \frac{d^2y}{dt^2} = \frac{-3}{100} x + \frac{y}{100}. \tag{3}$$

But, from the second equation of (1),

$$\frac{3}{100} x = \frac{dy}{dt} + \frac{3}{100} y,$$

so, inserting this into the last expression in (3), we obtain

$$\frac{100}{3} \frac{d^2y}{dt^2} + \frac{dy}{dt} = \overbrace{-\frac{dy}{dt} - \frac{3}{100} y}^{-\frac{3}{100} x} + \frac{1}{100} y.$$

This yields the second-order equation

$$\frac{100}{3} \frac{d^2y}{dt^2} + 2\frac{dy}{dt} + \frac{2y}{100} = 0. \tag{4}$$

The initial conditions for (4) are obtained directly from system (1), since $y(0) = 0$ and

$$y'(0) = 3\frac{x(0)}{100} - 3\frac{y(0)}{100} = 3.$$

Multiplying both sides of (4) by $\frac{3}{100}$, we have the initial-value problem

$$y'' + \frac{6}{100} y' + \frac{6}{(100)^2} y = 0, \qquad y(0) = 0, \quad y'(0) = 3. \tag{5}$$

The characteristic equation for (5) is

$$\lambda^2 + \frac{6}{100}\lambda + \frac{6}{(100)^2} = 0, \tag{6}$$

which has the roots

$$\left.\begin{array}{c}\lambda_1 \\ \lambda_2\end{array}\right\} = \frac{\frac{-6}{100} \pm \sqrt{\frac{36}{(100)^2} - \frac{24}{(100)^2}}}{2}$$

$$= \frac{-6 \pm \sqrt{12}}{200},$$

or

$$\lambda_1 = \frac{-3 + \sqrt{3}}{100}, \qquad \lambda_2 = \frac{-3 - \sqrt{3}}{100}.$$

Thus the general solution is

$$y(t) = c_1 e^{[(-3+\sqrt{3})t]/100} + c_2 e^{[(-3-\sqrt{3})t]/100}.$$

Using the initial conditions, we obtain the simultaneous equations

$$c_1 + c_2 = 0,$$

$$\frac{-3+\sqrt{3}}{100}c_1 - \frac{3+\sqrt{3}}{100}c_2 = 3.$$

These have the unique solution $c_1 = -c_2 = 50\sqrt{3}$. Hence

$$y(t) = 50\sqrt{3}\left[e^{[(-3+\sqrt{3})t]/100} - e^{[(-3-\sqrt{3})t]/100}\right],$$

and substituting this function into the right-hand side of equation (2), we obtain, after some algebra,

$$x(t) = 50\left[e^{[(-3+\sqrt{3})t]/100} + e^{[(-3-\sqrt{3})t]/100}\right].$$

Note that, as is evident from the problem, the amounts of salt in the two tanks approach zero as time tends to infinity.

The technique we have used in solving Example 1 is called the **method of elimination**, since all but one dependent variable is eliminated by repeated differentiation. The method is quite elementary but requires many calculations. We will introduce a direct way of obtaining the characteristic equation without doing the elimination procedure. Nevertheless, because of its simplicity, elimination can be a very useful tool.

Example 2 Consider the system

$$\begin{aligned} x' &= x + y, & x(0) &= 1, \\ y' &= -3x - y, & y(0) &= 0. \end{aligned} \tag{7}$$

Differentiating the first equation and substituting the second equation for y',

we have

$$x'' = x' + \overbrace{(-3x - y)}^{y'}$$

y from the first equation of (7)

$$= x' - 3x - \overbrace{(x' - x)},$$

or

$$x'' + 2x = 0.$$

Thus

$$x(t) = c_1 \cos \sqrt{2}\, t + c_2 \sin \sqrt{2}\, t.$$

Then, according to the first equation of system (7),

$$y = x' - x = -\sqrt{2}\, c_1 \sin \sqrt{2}\, t + \sqrt{2}\, c_2 \cos \sqrt{2}\, t - c_1 \cos \sqrt{2}\, t - c_2 \sin \sqrt{2}\, t$$
$$= (c_2\sqrt{2} - c_1) \cos \sqrt{2}\, t - (\sqrt{2}\, c_1 + c_2) \sin \sqrt{2}\, t.$$

Using the initial conditions, we find that

$$x(0) = c_1 = 1, \quad y(0) = c_2\sqrt{2} - c_1 = 0, \quad \text{or} \quad c_2 = 1/\sqrt{2}.$$

Therefore the unique solution of system (7) is given by the pair of functions

$$x(t) = \cos \sqrt{2}\, t + \frac{1}{\sqrt{2}} \sin \sqrt{2}\, t,$$

$$y(t) = -\left(\sqrt{2} + \frac{1}{\sqrt{2}}\right) \sin \sqrt{2}\, t = -\frac{3}{\sqrt{2}} \sin \sqrt{2}\, t.$$

Example 3 Consider the system

$$x' = 2x + y + t,$$
$$y' = x + 2y + t^2.$$

Proceeding as before, we obtain

$$x'' = 2x' + y' + 1 = 2x' + (x + 2y + t^2) + 1$$
$$= 2x' + x + (2x' - 4x - 2t) + t^2 + 1,$$

or

$$x'' - 4x' + 3x = t^2 - 2t + 1 = (t - 1)^2. \tag{8}$$

The solution to the homogeneous part of equation (8) is $x(t) = c_1 e^t + c_2 e^{3t}$. A particular solution of (8) is easily found to be $\frac{1}{3}t^2 + \frac{2}{9}t + \frac{11}{27}$, so the general solution of (8) is

$$x(t) = c_1 e^t + c_2 e^{3t} + \tfrac{1}{3}t^2 + \tfrac{2}{9}t + \tfrac{11}{27}.$$

As before, since $y = x' - 2x - t$, we obtain

$$y(t) = c_1 e^t + 3c_2 e^{3t} + \tfrac{2}{3}t + \tfrac{2}{9} - 2c_1 e^t - 2c_2 e^{3t} - \tfrac{2}{3}t^2 - \tfrac{4}{9}t - \tfrac{22}{27} - t,$$

or

$$y(t) = -c_1 e^t + c_2 e^{3t} - \tfrac{2}{3}t^2 - \tfrac{7}{9}t - \tfrac{16}{27}.$$

7.3/THE METHOD OF ELIMINATION AND THE METHOD OF DETERMINANTS

The method illustrated in the last three examples can easily be generalized to apply to linear systems with three or more equations. A liner system of n first-order equations usually reduces to an nth-order linear differential equation, because it generally requires one differentiation to eliminate each variable $x_2, \ldots, x_n$ from the system.

THE METHOD OF DETERMINANTS

We now develop a more efficient way of obtaining the characteristic equation of the differential equation that is obtained by eliminating all but one dependent variable of a homogeneous system (nonhomogeneous systems are discussed in Problems 18–23).

Consider the homogeneous system

$$x' = a_{11}x + a_{12}y,$$
$$y' = a_{21}x + a_{22}y, \qquad (9)$$

where the a_{ij} are constants. Our main tool for solving second-order linear homogeneous equations with constant coefficients involved obtaining a characteristic equation by "guessing" that the solution had the form $y = e^{\lambda x}$.

Parallel to the method of Section 3.3, we guess that there is a solution to system (9) of the form $(\alpha e^{\lambda t}, \beta e^{\lambda t})$, where α, β, and λ are constants yet to be determined. Substituting $x(t) = \alpha e^{\lambda t}$ and $y(t) = \beta e^{\lambda t}$ into system (9), we obtain

$$x' = \alpha \lambda e^{\lambda t} = a_{11}\alpha e^{\lambda t} + a_{12}\beta e^{\lambda t},$$
$$y' = \beta \lambda e^{\lambda t} = a_{21}\alpha e^{\lambda t} + a_{22}\beta e^{\lambda t}.$$

After dividing by $e^{\lambda t}$, we obtain the linear system

$$(a_{11} - \lambda)\alpha + a_{12}\beta = 0,$$
$$a_{21}\alpha + (a_{22} - \lambda)\beta = 0. \qquad (10)$$

We would like to find values for λ such that the system of equations (10) has a solution (α, β) where α and β are not both zero. According to the theory of determinants (see Appendix 4), such a solution occurs whenever the determinant of the system

$$\boxed{\begin{aligned} D &= \begin{vmatrix} a_{11} - \lambda & a_{12} \\ a_{21} & a_{22} - \lambda \end{vmatrix} \\ &= (a_{11} - \lambda)(a_{22} - \lambda) - a_{21}a_{12} \end{aligned}} \qquad (11)$$

equals zero. Solving the equation $D = 0$, we obtain the quadratic equation

Characteristic Equation

$$\boxed{\lambda^2 - (a_{11} + a_{22})\lambda + (a_{11}a_{22} - a_{21}a_{12}) = 0.} \qquad (12)$$

We define this to be the **characteristic equation** of system (9). That we are using the same term again is no accident, as we now demonstrate. Suppose we differentiate the first equation in system (9) and eliminate the function $y(t)$:

$$x'' = a_{11}x' + a_{12}\overbrace{(a_{21}x + a_{22}y)}^{y'}.$$

Then

$$x'' - a_{11}x' - a_{12}a_{21}x = a_{22}a_{12}y \overbrace{= a_{22}(x' - a_{11}x)}^{a_{12}y = x' - a_{11}x},$$

and gathering like terms, we obtain the homogeneous equation

$$x'' - (a_{11} + a_{22})x' + (a_{11}a_{22} - a_{12}a_{21})x = 0. \qquad (13)$$

The characteristic equation for (13) is exactly the same as (12). Hence the algebraic steps needed to obtain (13) can be avoided by setting the determinant $D = 0$.

As in Sections 3.3 and 3.4, there are three cases to consider, depending on whether the two roots λ_1 and λ_2 of the characteristic equation are real and distinct, real and equal, or complex conjugates. We deal with the three cases separately. Rather than discuss the theory, we demonstrate each case with an example.

Case 1: Distinct Real Roots

Example 4 Consider the system

$$\begin{aligned} x' &= -x + 6y, \\ y' &= x - 2y. \end{aligned} \qquad (14)$$

Here $a_{11} = -1$, $a_{12} = 6$, $a_{21} = 1$, $a_{22} = -2$, and equation (12) becomes

$$D = \begin{vmatrix} -1 - \lambda & 6 \\ 1 & -2 - \lambda \end{vmatrix} = (\lambda + 2)(\lambda + 1) - 6 = \lambda^2 + 3\lambda - 4 = 0,$$

which has the roots $\lambda_1 = -4$, $\lambda_2 = 1$.

At this point we can write down the general solution of *either* dependent variable:

$$x(t) = c_1 e^{-4t} + c_2 e^t, \qquad (15)$$

or

$$y(t) = k_1 e^{-4t} + k_2 e^t. \qquad (16)$$

It does not matter which expression, (15) or (16), we choose. However, once we pick one, we must determine the other dependent variable in terms of the first. For example, if we choose (15), then using the first equation in (14), we have

$$y = \tfrac{1}{6}(x' + x) = \tfrac{1}{6}(-4c_1 e^{-4t} + c_2 e^t + c_1 e^{-4t} + c_2 e^t)$$
$$= \tfrac{1}{6}(-3c_1 e^{-4t} + 2c_2 e^t) = -\tfrac{1}{2}c_1 e^{-4t} + \tfrac{1}{3}c_2 e^t.$$

Thus the general solution is

$$x(t) = c_1 e^{-4t} + c_2 e^t,$$
$$y(t) = -\tfrac{1}{2}c_1 e^{-4t} + \tfrac{1}{3}c_2 e^t,$$

which can be written as
$$(x(t), y(t)) = \left(c_1 e^{-4t} + c_2 e^t, -\tfrac{1}{2} c_1 e^{-4t} + \tfrac{1}{3} c_2 e^t\right).$$

Case 2: One Real Root

Example 5 Consider the system
$$\begin{aligned} x' &= -4x - y, \\ y' &= x - 2y. \end{aligned} \tag{17}$$

Equation (12) is
$$D = \begin{vmatrix} -4 - \lambda & -1 \\ 1 & -2 - \lambda \end{vmatrix} = (\lambda + 4)(\lambda + 2) + 1 = \lambda^2 + 6\lambda + 9 = 0,$$

which has the double root $\lambda_1 = \lambda_2 = -3$. Hence the general solution to (17) for x is given by
$$x(t) = c_1 e^{-3t} + c_2 t e^{-3t} = (c_1 + c_2 t) e^{-3t}.$$

From the first equation in (17),
$$\begin{aligned} y &= -x' - 4x = -e^{-3t}(-3c_1 - 3c_2 t + c_2) - e^{-3t}(4c_1 + 4c_2 t) \\ &= e^{-3t}(-c_1 - c_2 - c_2 t). \end{aligned}$$

Thus the general solution to (17) is
$$(x(t), y(t)) = \left(e^{-3t}(c_1 + c_2 t), e^{-3t}(-c_1 - c_2 - c_2 t)\right).$$

Case 3: Complex Conjugate Roots

Example 6 Consider the system
$$\begin{aligned} x' &= 4x + y, \\ y' &= -8x + 8y, \end{aligned} \tag{18}$$

with
$$D = \begin{vmatrix} 4 - \lambda & 1 \\ -8 & 8 - \lambda \end{vmatrix} = \lambda^2 - 12\lambda + 40 = 0.$$

The roots of the characteristic equation are $\lambda_1 = 6 + 2i$ and $\lambda_2 = 6 - 2i$. Then the general solution to (18) for y is
$$y(t) = e^{6t}(c_1 \cos 2t + c_2 \sin 2t).$$

Using the second equation of (18), we have
$$8x = 8y - y' \quad \text{or} \quad x = y - \tfrac{1}{8} y',$$

so
$$\begin{aligned} x(t) &= e^{6t}(c_1 \cos 2t + c_2 \sin 2t) \\ &\quad - \tfrac{1}{8}\underbrace{\left[6 e^{6t}(c_1 \cos 2t + c_2 \sin 2t) + e^{6t}(-2c_1 \sin 2t + 2c_2 \cos 2t)\right]}_{y'} \\ &= e^{6t}\left[\left(c_1 - \tfrac{6}{8} c_1 - \tfrac{2}{8} c_2\right) \cos 2t + \left(c_2 - \tfrac{6}{8} c_2 + \tfrac{2}{8} c_1\right) \sin 2t\right] \\ &= e^{6t}\left[\tfrac{1}{4}(c_1 - c_2) \cos 2t + \tfrac{1}{4}(c_1 + c_2) \sin 2t\right]. \end{aligned}$$

Thus the general solution to (18) is

$$(x(t), y(t))$$
$$= \left(\frac{e^{6t}}{4}[(c_1 - c_2)\cos 2t + (c_1 + c_2)\sin 2t], e^{6t}(c_1 \cos 2t + c_2 \sin 2t)\right).$$

The method of elimination, particularly when we use determinants to obtain the characteristic equation of the system, provides a quick way to find the general solution to a homogeneous system of two equations in two unknown functions. The procedure can be extended to systems of more than two equations, but the algebra involved is often extremely tedious. Rather than discussing this modification, we will use matrix methods (in Sections 7.5–7.11) to obtain general solutions to both homogeneous and nonhomogeneous linear systems of first-order equations. We summarize the results of Examples 4, 5, and 6.

To find the general solution of the system
$$x' = a_{11}x + a_{12}y,$$
$$y' = a_{21}x + a_{22}y,$$
in the case $a_{12} \neq 0$, first find two numbers λ_1 and λ_2 that satisfy the characteristic equation

$$\begin{vmatrix} a_{11} - \lambda & a_{12} \\ a_{21} & a_{22} - \lambda \end{vmatrix} = 0.$$

Case 1 If $\lambda_1 \neq \lambda_2$ are real, then
$$x(t) = c_1 e^{\lambda_1 t} + c_2 e^{\lambda_2 t}, \tag{19}$$
and $y(t)$ can be obtained from the equation
$$y = \frac{1}{a_{12}}(x' - a_{11}x). \tag{20}$$

Case 2 If $\lambda_1 = \lambda_2$, then
$$x(t) = e^{\lambda_1 t}(c_1 + c_2 t), \tag{21}$$
and $y(t)$ can be obtained from (20).

Case 3 If $\lambda_1 = a + ib$ and $\lambda_2 = a - ib$, then
$$x(t) = e^{at}(c_1 \cos bt + c_2 \sin bt), \tag{22}$$
and again $y(t)$ can be obtained from (20).

Remark 1 If $a_{12} = 0$ and $a_{21} \neq 0$, then $y(t)$ can be written in form (19), (21), or (22) and $x(t)$ can then be obtained from the equation

$$x = \frac{1}{a_{21}}(y' - a_{22}y). \tag{23}$$

7.3 / THE METHOD OF ELIMINATION AND THE METHOD OF DETERMINANTS

Remark 2 If $a_{12} = a_{21} = 0$, then the system is **uncoupled** and $x(t)$ and $y(t)$ can be obtained immediately:

$$x(t) = c_1 e^{a_{11}t} \quad \text{and} \quad y(t) = c_2 e^{a_{22}t}.$$

PROBLEMS 7.3

In Problems 1–8 find the general solution of each system of equations. When initial conditions are given, find the unique solution.

1. $x' = x + 2y,$
   $y' = 3x + 2y$
2. $x' = x + 2y + t - 1, \quad x(0) = 0,$
   $y' = 3x + 2y - 5t - 2, \quad y(0) = 4$
3. $x' = -4x - y,$
   $y' = x - 2y$
4. $x' = x + y, \quad x(0) = 1,$
   $y' = y, \quad y(0) = 0$
5. $x' = 8x - y,$
   $y' = 4x + 12y$
6. $x' = 2x + y + 3e^{2t},$
   $y' = -4x + 2y + te^{2t}$
7. $x' = 3x + 3y + t,$
   $y' = -x - y + 1$
8. $x' = 4x + y, \quad x(\pi/4) = 0,$
   $y' = -8x + 8y, \quad y(\pi/4) = 1$
9. By elimination, find a solution to the following nonlinear system:

   $x' = x + \sin x \cos x + 2y,$
   $y' = (x + \sin x \cos x + 2y)\sin^2 x + x.$

Use the method of elimination to solve the systems in Problems 10 and 11.

10. $x_1' = x_1,$
    $x_2' = 2x_1 + x_2 - 2x_3,$
    $x_3' = 3x_1 + 2x_2 + x_3$
11. $x_1' = x_1 + x_2 + x_3,$
    $x_2' = 2x_1 + x_2 - x_3,$
    $x_3' = -8x_1 - 5x_2 - 3x_3$

In Problems 12–17 use the method of determinants to find two linearly independent solutions for each given system.

12. $x' = 4x - 3y,$
    $y' = 5x - 4y$
13. $x' = 7x + 6y,$
    $y' = 2x + 6y$
14. $x' = -x + y,$
    $y' = -5x + 3y$
15. $x' = x + y,$
    $y' = -x + 3y$
16. $x' = -4x - y,$
    $y' = x - 2y$
17. $x' = 4x - 2y,$
    $y' = 5x + 2y$
18. Consider the nonhomogeneous equations

    $x' = a_{11}x + a_{12}y + f_1,$
    $y' = a_{21}x + a_{22}y + f_2.$ \quad (i)

Let (x_1, y_1) and (x_2, y_2) be two linearly independent solution pairs of the homogeneous system (9). Show that

$$x_p(t) = v_1(t)x_1(t) + v_2(t)x_2(t),$$
$$y_p(t) = v_1(t)y_1(t) + v_2(t)y_2(t),$$

is a particular solution of system (i) if v_1 and v_2 satisfy the equations

$$v_1'x_1 + v_2'x_2 = f_1,$$
$$v_1'y_1 + v_2'y_2 = f_2.$$

This process for finding a particular solution of the nonhomogeneous system (i) is called the **method of variation of parameters for systems**. Note the close parallel between this method and the method given in Section 3.6.

In Problems 19–23 use the method of variation of parameters to find a particular solution for each given nonhomogeneous system.

19. $x' = 2x + y + 3e^{2t},$
    $y' = -4x + 2y + te^{2t}$
20. $x' = 3x + 3y + t,$
    $y' = -x - y + 1$
21. $x' = -2x + y,$
    $y' = -3x + 2y + 2\sin t$
22. $x' = -x + y + \cos t,$
    $y' = -5x + 3y$
23. $x' = 3x - 2y + t,$
    $y' = 2x - 2y + 3e^t$
24. In Example 1, when does tank Y contain a maximum amount of salt? How much salt is in tank Y at that time?
25. Suppose that in Example 1 the rate of flow from tank Y to tank X is 2 gallons per minute (instead of 1) and all other facts are unchanged. Find the differential equations for the amount of salt in each tank at all times t.
26. Tank X contains 500 gallons of brine in which 500 pounds of salt are dissolved. Tank Y contains 500 gallons of water. Water flows into tank X at the rate of 30 gallons per minute, and the mixture flows into Y at the rate of 40 gallons per minute. From Y the solution is pumped back into X at the rate of 10 gallons per minute and into a third tank at the rate of 30 gallons per minute. Find the maximum amount of salt in Y. When does this concentration occur?

27. Suppose that in Problem 26 tank X contains 1000 gallons of brine. Solve the problem, given that all other conditions are unchanged.

28. Consider the mass-spring system illustrated below. Here three objects of mass m_1, m_2, and m_3 are suspended in series by three springs with spring constants k_1, k_2, and k_3, respectively. Formulate a system of second-order differential equations that describes this system.

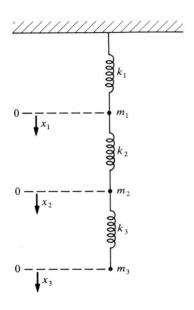

29. Find a single fourth-order linear differential equation in terms of the dependent variable x_1 for system (7.1.11) on p. 264. Find a solution to the system if $m_1 = 1$ kg, $m_2 = 2$ kg, $k_1 = 5$ N/m, $k_2 = 4$ N/m.

30. In a study concerning the distribution of radioactive potassium ^{42}K between red blood cells and the plasma of the human blood, C. W. Sheppard and W. R. Martin added ^{42}K to freshly drawn blood.[†] They discovered that although the total amount of potassium (stable and radioactive) in the red cells and in the plasma remained practically constant during the experiment, the radioactivity was gradually transmitted from the plasma to the red cells. Thus the behavior of the radioactivity is that of a linear closed two-compartment system. If the fractional transfer coefficient from the plasma to the cells is $k_{12} = 30.1$ percent per hour, while $k_{21} = 1.7$ percent per hour, and the initial radioactivity was 800 counts per minute in the plasma and 25 counts per minute in the red cells, what is the number of counts per minute in the red cells after 300 minutes?

31. Temperature inversions and low wind speeds often trap air pollutants in mountain valleys for an extended period of time. Gaseous sulfur compounds are often a significant air pollution problem, and their study is complicated by their rapid oxidation. Hydrogen sulfide (H_2S) oxidizes into sulfur dioxide (SO_2), which in turn oxidizes into a sulphate. The following model has been proposed for determining the concentrations $x(t)$ and $y(t)$ of H_2S and SO_2, respectively, in a fixed airshed.[‡] Let

$$\frac{dx}{dt} = -\alpha x + \gamma, \quad \frac{dy}{dt} = \alpha x - \beta y + \delta,$$

where the constants α and β are the conversion rates of H_2S into SO_2 and SO_2 into sulphate, respectively, and γ and δ are the production rates of H_2S and SO_2, respectively. Solve the equations sequentially and estimate the concentration levels that could be reached under a prolonged air pollution episode.

[†] *Journal of General Physiology* 33 (1950): 703–722.

[‡] R. L. Bohac, "A Mathematical Model for the Conversion of Sulphur Compounds in the Missoula Valley Airshed," *Proceedings of the Montana Academy of Science* (1974).

7.4 Laplace Transform Methods for Systems

Laplace transform techniques are very useful for solving systems of differential equations with given initial conditions. Consider the system

$$\begin{aligned}\frac{dx}{dt} &= a_{11}x + a_{12}y + f(t), \\ \frac{dy}{dt} &= a_{21}x + a_{22}y + g(t),\end{aligned} \quad (1)$$

with initial conditions $x(0) = x_0$, $y(0) = y_0$. Taking the Laplace transform of both equations in system (1) and letting the corresponding capital letters represent the Laplace transforms of the functions x, y, f, and g (that is, we let $X = L\{x\}$, $Y = L\{y\}$, and so on), we obtain

$$sX - x_0 = a_{11}X + a_{12}Y + F,$$
$$sY - y_0 = a_{21}X + a_{22}Y + G. \quad (2)$$

Gathering all the terms involving X and Y on the left-hand side, we obtain from (2) the system of simultaneous equations

$$(s - a_{11})X - a_{12}Y = F + x_0,$$
$$-a_{21}X + (s - a_{22})Y = G + y_0. \quad (3)$$

If $F + x_0$ and $G + y_0$ are not both identically zero, we may solve equations (3) simultaneously, obtaining

$$X = \frac{(s - a_{22})(F + x_0) + a_{12}(G + y_0)}{s^2 - (a_{11} + a_{22})s + (a_{11}a_{22} - a_{12}a_{21})},$$
$$Y = \frac{(s - a_{11})(G + y_0) + a_{21}(F + x_0)}{s^2 - (a_{11} + a_{22})s + (a_{11}a_{22} - a_{12}a_{21})}. \quad (4)$$

Note that the denominators of the fractions in equations (4) are identical to the characteristic polynomial for system (1) that we obtained in equation (7.3.12.) Thus we may rewrite equations (4) in the form

$$X = \frac{(s - a_{22})(F + x_0) + a_{12}(G + y_0)}{(s - \lambda_1)(s - \lambda_2)},$$
$$Y = \frac{(s - a_{11})(G + y_0) + a_{21}(F + x_0)}{(s - \lambda_1)(s - \lambda_2)}, \quad (5)$$

where λ_1 and λ_2 are the solutions of the characteristic equation

$$\lambda^2 - (a_{11} + a_{22})\lambda + (a_{11}a_{22} - a_{12}a_{21}) = 0.$$

We can now find the solution (x, y) of system (1) with given initial conditions by inverting (if possible) equations (5).

It should be apparent from the discussion above that the Laplace transform allows us to convert a system of differential equations with given initial conditions into a system of simultaneous algebraic equations. This method clearly can be generalized to apply to systems of n linear first-order differential equations with constant coefficients, thus yielding a corresponding system of n simultaneous linear algebraic equations. As in the case of a single differential equation (see Section 6.2), Laplace transform methods are primarily intended for the solution of systems of linear differential equations with constant coefficients and given initial conditions. Any deviation from these conditions can so complicate the problem as to make no solution obtainable.

Example 1 Consider the initial-value problem

$$x' = x - y - e^{-t}, \quad x(0) = 1,$$
$$y' = 2x + 3y + e^{-t}, \quad y(0) = 0. \quad (6)$$

Using the differentiation property of Laplace transforms, we have
$$sX - 1 = X - Y - 1/(s+1),$$
$$sY = 2X + 3Y + 1/(s+1). \tag{7}$$

System (7) may be rewritten as
$$(s-1)X + Y = s/(s+1),$$
$$-2X + (s-3)Y = 1/(s+1), \tag{8}$$

from which we find that

$$X = \frac{s^2 - 3s - 1}{(s+1)[(s-2)^2 + 1]} = \frac{A_1}{s+1} + \frac{A_2(s-2) + A_3}{(s-2)^2 + 1},$$

$$Y = \frac{3s - 1}{(s+1)[(s-2)^2 + 1]} = \frac{B_1}{s+1} + \frac{B_2(s-2) + B_3}{(s-2)^2 + 1}. \tag{9}$$

Using the methods developed in Section 6.2, we obtain
$$A_1 = \tfrac{3}{10}, \qquad A_2 = \tfrac{7}{10}, \qquad A_3 = -\tfrac{11}{10},$$
$$B_1 = -\tfrac{2}{5}, \qquad B_2 = \tfrac{2}{5}, \quad \text{and} \quad B_3 = \tfrac{9}{5}.$$

Therefore we have the solution
$$(x, y) = \left(\frac{3}{10} e^{-t} + \frac{e^{2t}}{10}(7 \cos t - 11 \sin t), \; -\frac{2}{5} e^{-t} + \frac{e^{2t}}{5}(2 \cos t + 9 \sin t) \right).$$

Example 2 Here we give a far more complicated model—a description of the motion of a particle projected from the earth where we do not ignore the effects of the earth's rotation on the particle.[†] To begin, we set up the needed three-dimensional coordinate system.

Assume that the positive x-axis points south, the positive y-axis points east, and the positive z-axis points in the direction opposite the direction of acceleration g due to gravity.

This model was first studied by the German physicist B. M. Planck.[‡] Planck took the origin to have latitude β and the angular velocity of the earth to be ω. Assuming the mass of the particle to be negligible, Planck derived the following system of second-order equations:

$$\frac{d^2 x}{dt^2} = 2\omega \sin \beta \frac{dy}{dt}, \tag{10}$$

$$\frac{d^2 y}{dt^2} = -2\omega \left(\sin \beta \frac{dx}{dt} + \cos \beta \frac{dz}{dt} \right), \tag{11}$$

$$\frac{d^2 z}{dt^2} = 2\omega \cos \beta \frac{dy}{dt} - g, \tag{12}$$

[†] This problem is very difficult computationally. Skip it if you balk at lengthy calculations. Unfortunately, real problems rarely have easy solutions. This problem is typical of those encountered beyond textbook examples.

[‡] *Einführung in die Allgemeine Mechanik*, 4th ed. (Leipzig: S. Hirzel, 1928), p. 81.

7.4/LAPLACE TRANSFORM METHODS FOR SYSTEMS

with
$$0 = x(0) = y(0) = z(0), \quad x'(0) = u, \quad y'(0) = v, \quad z'(0) = w. \quad (13)$$

Taking transforms of both sides of equation (10), we obtain
$$s^2 X - sx(0) - x'(0) = 2\omega \sin\beta [sY - y(0)],$$

and, using the conditions in equation (13),
$$s^2 X - u = 2\omega (\sin\beta) sY,$$

or
$$s^2 X - 2\omega (\sin\beta) sY = u. \quad (14)$$

Similarly (see Problem 13), we obtain
$$2\omega (\sin\beta) sX + s^2 Y + 2\omega (\cos\beta) sZ = v \quad (15)$$

and
$$-2\omega (\cos\beta) sY + s^2 Z = w - \frac{g}{s}. \quad (16)$$

Equations (14), (15), and (16) constitute a system of three equations in the three unknowns X, Y, and Z and have a unique solution if and only if the determinant of the system if nonzero. This determinant is

$$D = \begin{vmatrix} s^2 & -2\omega(\sin\beta)s & 0 \\ 2\omega(\sin\beta)s & s^2 & 2\omega(\cos\beta)s \\ 0 & -2\omega(\cos\beta)s & s^2 \end{vmatrix}$$
$$= s^6 + 4\omega^2(\sin^2\beta)s^4 + 4\omega^2(\cos^2\beta)s^4$$
$$= s^6 + 4\omega^2 s^4 = s^4(s^2 + 4\omega^2).$$

Thus for $s > 0$ the determinant is nonzero and the system has a unique solution. Using Cramer's rule (see Appendix 4), we obtain

$$X(s) = \frac{\begin{vmatrix} u & -2\omega(\sin\beta)s & 0 \\ v & s^2 & 2\omega(\cos\beta)s \\ w - \dfrac{g}{s} & -2\omega(\cos\beta)s & s^2 \end{vmatrix}}{D}$$

$$= \frac{us^4 - 4\omega^2 \sin\beta(\cos\beta)s^2\left(w - \dfrac{g}{s}\right) + 2\omega(\sin\beta)vs^3 + 4\omega^2(\cos^2\beta)s^2 u}{s^4(s^2 + 4\omega^2)}$$

$$= \frac{u}{s^2 + 4\omega^2} + v\left[\frac{2\omega \sin\beta}{s(s^2 + 4\omega^2)}\right] + u\left[\frac{4\omega^2 \cos^2\beta}{s^2(s^2 + 4\omega^2)}\right]$$

$$- w\left[\frac{4\omega^2 \sin\beta \cos\beta}{s^2(s^2 + 4\omega^2)}\right] + g\left[\frac{4\omega^2 \sin\beta \cos\beta}{s^3(s^2 + 4\omega^2)}\right]. \quad (17)$$

Similarly (see Problem 14), we obtain

$$Y(s) = -u\left[\frac{2\omega \sin \beta}{s(s^2 + 4\omega^2)}\right] + v\left[\frac{1}{s^2 + 4\omega^2}\right]$$
$$-w\left[\frac{2\omega \cos \beta}{s(s^2 + 4\omega^2)}\right] + g\left[\frac{2\omega \cos \beta}{s^2(s^2 + 4\omega^2)}\right] \quad (18)$$

and

$$Z(s) = -u\left[\frac{4\omega^2 \sin \beta \cos \beta}{s^2(s^2 + 4\omega^2)}\right] + v\left[\frac{2\omega \cos \beta}{s(s^2 + 4\omega^2)}\right]$$
$$+ w\left[\frac{1}{s^2 + 4\omega^2} + \frac{4\omega^2 \sin^2\beta}{s^2(s^2 + 4\omega^2)}\right] - g\left[\frac{1}{s(s^2 + 4\omega^2)} + \frac{4\omega^2 \sin^2\beta}{s^3(s^2 + 4\omega^2)}\right]. \quad (19)$$

Next (see Problem 15), equations (17), (18), and (19) can be inverted, and using the trigonometric identity $2\sin^2 A = 1 - \cos 2A$ we obtain

$$x(t) = \frac{u}{2\omega}(2\omega t \cos^2\beta + \sin^2\beta \sin 2\omega t) + \frac{v}{\omega}\sin \beta \sin^2\omega t$$
$$- \frac{w}{2\omega}\sin \beta \cos \beta (2\omega t - \sin 2\omega t) + \frac{g}{2\omega^2}\sin \beta \cos \beta(\omega^2 t^2 - \sin^2\omega t); \quad (20)$$

$$y(t) = -\frac{u}{\omega}\sin \beta \sin^2\omega t + \frac{v}{2\omega}\sin 2\omega t$$
$$- \frac{w}{\omega}\cos \beta \sin^2\omega t + \frac{g}{4\omega^2}\cos \beta(2\omega t - \sin 2\omega t); \quad (21)$$

$$z(t) = -\frac{u}{2\omega}\sin \beta \cos \beta (2\omega t - \sin 2\omega t) + \frac{v}{\omega}\cos \beta \sin^2\omega t$$
$$+ \frac{w}{2\omega}(2\omega t \sin^2\beta + \sin 2\omega t \cos^2\beta)$$
$$- \frac{g}{2\omega^2}(\omega^2 t^2 \sin^2\beta + \cos^2\beta \sin^2\omega t). \quad (22)$$

These are the equations of motion of the projectile when the earth's rotation is taken into account. Finally, we note by L'Hôpital's rule that

$$\lim_{\omega \to 0}\frac{\sin 2\omega t}{2\omega} = \lim_{\omega \to 0}\frac{2t \cos \omega t}{2} = t$$

and

$$\lim_{\omega \to 0}\frac{\sin^2\omega t}{\omega} = \lim_{\omega \to 0}\frac{2t \sin \omega t \cos \omega t}{1} = 0.$$

Thus

$$\lim_{\omega \to 0} x(t) = ut, \quad \lim_{\omega \to 0} y(t) = vt, \quad \text{and} \quad \lim_{\omega \to 0} z(t) = wt - \tfrac{1}{2}gt^2, \quad (23)$$

showing that equations (20), (21), and (22) reduce to the usual equations of motion when the earth's rotation is ignored.

PROBLEMS 7.4

In Problems 1–12 solve the initial-value problems involving systems of differential equations by the Laplace transform method.

1. $x' = y$, $x(0) = 1$
   $y' = x$, $y(0) = 0$
2. $x' = -3x + 4y$, $x(0) = 3$
   $y' = -2x + 3y$, $y(0) = 2$
3. $x' = 4x - 2y$, $x(0) = 2$
   $y' = 5x + 2y$, $y(0) = -2$
4. $x' = x - 2y$, $x(0) = 1$
   $y' = 4x + 5y$, $y(0) = -2$
5. $x' = -3x + 4y + \cos t$, $x(0) = 0$
   $y' = -2x + 3y + t$, $y(0) = 1$
6. $x' = 4x - 2y + e^t$, $x(0) = 1$
   $y' = 5x + 2y - t$, $y(0) = 0$
7. $x' = x - 2y + t^2$, $x(0) = 1$
   $y' = 4x + 5y - e^t$, $y(0) = -1$
8. $x'' + x + y = 0$, $x(0) = x'(0) = 0$
   $x' + y' = 0$, $y(0) = 1$
9. $x'' = y + \sin t$, $x(0) = 1$, $x'(0) = 0$
   $y'' = -x' + \cos t$, $y(0) = -1$, $y'(0) = -1$
10. $x'' = 2y + 2$, $x(0) = 2$, $x'(0) = 2$
    $y' = -x + 5e^{2t} + 1$, $y(0) = 1$
11. $x = z'$, $x(0) = y(0) = z(0) = 1$
    $x' + y + z = 1$
    $-x + y' + z = 2 \sin t$
12. $x'' = y - z$, $x(0) = x'(0) = 0$
    $y'' = x' + z'$, $y(0) = -1$, $y'(0) = 1$
    $z'' = -(1 + x + y)$, $z(0) = 0$, $z'(0) = 1$
13. Beginning with equations (11) and (12), use the differentiation property to obtain equations (15) and (16).
14. Use Cramer's rule in systems (14), (15), and (16) to obtain expressions (18) and (19) for Y and Z.
*15. Invert the transforms in equations (17), (18), and (19) to solve for $x(t)$, $y(t)$, and $z(t)$.

Matrix Methods

7.5 Review of Matrices

Matrix

An $m \times n$ **matrix**† A is a rectangular array of mn numbers arranged in m rows and n columns:

$$A = \begin{pmatrix} a_{11} & a_{12} & \cdots & a_{1j} & \cdots & a_{1n} \\ a_{21} & a_{22} & \cdots & a_{2j} & \cdots & a_{2n} \\ \vdots & \vdots & & \vdots & & \vdots \\ a_{i1} & a_{i2} & \cdots & a_{ij} & \cdots & a_{in} \\ \vdots & \vdots & & \vdots & & \vdots \\ a_{m1} & a_{m2} & \cdots & a_{mj} & \cdots & a_{mn} \end{pmatrix}. \quad (1)$$

The ijth component of A, denoted a_{ij}, is the number appearing in the ith row and jth column of A. We will sometimes write the matrix A as $A = (a_{ij})$. Usually matrices will be denoted by capital letters.

---

† The term "matrix" was first used in 1850 by the British mathematician James Joseph Sylvester (1814–1897) to distinguish matrices from determinants. In fact, the term "matrix" was intended to mean "mother of determinants."

CHAPTER 7/SYSTEMS OF LINEAR DIFFERENTIAL EQUATIONS

If A is an $m \times n$ matrix with $m = n$, then A is called a **square matrix**. An $m \times n$ matrix with all components equal to zero is called the $m \times n$ **zero matrix**.

Equality

An $m \times n$ matrix is said to have the **size** $m \times n$. Two matrices $A = (a_{ij})$ and $B = (b_{ij})$ are **equal** if (a) they have the same size and (b) corresponding components are equal.

Example 1 Five matrices of different sizes are given below:

$$\begin{pmatrix} 1 & 3 \\ 4 & 2 \end{pmatrix}, 2 \times 2 \text{ (square)} \qquad \begin{pmatrix} -1 & 3 \\ 4 & 0 \\ 1 & -2 \end{pmatrix}, 3 \times 2$$

$$\begin{pmatrix} -1 & 4 & 1 \\ 3 & 0 & 2 \end{pmatrix}, 2 \times 3 \qquad \begin{pmatrix} 1 & 6 & -2 \\ 3 & 1 & 4 \\ 2 & -6 & 5 \end{pmatrix}, 3 \times 3 \text{ (square)}$$

$$\begin{pmatrix} 0 & 0 & 0 & 0 \\ 0 & 0 & 0 & 0 \end{pmatrix}, 2 \times 4 \text{ zero matrix}$$

Vectors

A **row vector** is a matrix with one row. A **column vector** is a matrix with one column. Vectors are usually denoted by lowercase boldface letters.

Example 2 The following are vectors:

a. $\mathbf{v} = (1, 2)$ row vector

b. $\mathbf{v} = \begin{pmatrix} 1 \\ 3 \\ -2 \end{pmatrix}$ column vector

c. $\mathbf{0} = \begin{pmatrix} 0 \\ 0 \\ 0 \end{pmatrix}$ zero vector

d. $\mathbf{v} = (2, -1, 5, 7, 4)$ row vector

e. $\mathbf{v} = \begin{pmatrix} 1 \\ -1 \\ 2 \\ 3 \\ 5 \end{pmatrix}$ column vector

---

Note Unless stated otherwise, we will always assume that the numbers in a matrix or vector are real.

Matrices and vectors can be added together and multiplied by scalars (real numbers).

DEFINITION 1: ADDITION OF MATRICES†

Let $A = (a_{ij})$ and $B = (b_{ij})$ be two $m \times n$ matrices. Then the sum of A and B is the $m \times n$ matrix $A + B$ given by

$$A + B = (a_{ij} + b_{ij}) = \begin{pmatrix} a_{11} + b_{11} & a_{12} + b_{12} & \cdots & a_{1n} + b_{1n} \\ a_{21} + b_{21} & a_{22} + b_{22} & \cdots & a_{2n} + b_{2n} \\ \vdots & \vdots & & \vdots \\ a_{m1} + b_{m1} & a_{m2} + b_{m2} & \cdots & a_{mn} + b_{mn} \end{pmatrix}. \quad (2)$$

In other words, $A + B$ is the $m \times n$ matrix obtained by adding the corresponding components of A and B.

Warning The sum of two matrices is defined only when both matrices have the same size. Thus, for example, it is not possible to add together the matrices

$$\begin{pmatrix} 1 & 2 & 3 \\ 4 & 5 & 6 \end{pmatrix} \quad \text{and} \quad \begin{pmatrix} -1 & 0 \\ 2 & -5 \\ 4 & 7 \end{pmatrix}.$$

Example 3

$$\begin{pmatrix} 2 & 4 & -6 & 7 \\ 1 & 3 & 2 & 1 \\ -4 & 3 & -5 & 5 \end{pmatrix} + \begin{pmatrix} 0 & 1 & 6 & -2 \\ 2 & 3 & 4 & 3 \\ -2 & 1 & 4 & 4 \end{pmatrix} = \begin{pmatrix} 2 & 5 & 0 & 5 \\ 3 & 6 & 6 & 4 \\ -6 & 4 & -1 & 9 \end{pmatrix}.$$

DEFINITION 2: MULTIPLICATION OF A MATRIX BY A SCALAR

If $A = (a_{ij})$ is an $m \times n$ matrix and if α is a scalar, then the $m \times n$ matrix αA is given by

$$\alpha A = (\alpha a_{ij}) = \begin{pmatrix} \alpha a_{11} & \alpha a_{12} & \cdots & \alpha a_{1n} \\ \alpha a_{21} & \alpha a_{22} & \cdots & \alpha a_{2n} \\ \vdots & \vdots & & \vdots \\ \alpha a_{m1} & \alpha a_{m2} & \cdots & \alpha a_{mn} \end{pmatrix}. \quad (3)$$

---

† The algebra of matrices, that is, the rules by which matrices can be added and multiplied, was developed by the English mathematician Arthur Cayley (1821–1895) in 1857. Matrices arose with Cayley in connection with linear transformations of the type

$$x^* = ax + by,$$
$$y^* = cx + dy,$$

where a, b, c, d are real numbers, and which may be thought of as mapping the point (x, y) into the point (x^*, y^*). Clearly, the transformation above is completely determined by the four coefficients a, b, c, d, and so the transformation can be symbolized by the square array

$$\begin{pmatrix} a & b \\ c & d \end{pmatrix},$$

which we have called a (**square**) **matrix**.

In other words, $\alpha A = (\alpha a_{ij})$ is the matrix obtained by multiplying each component of A by α.

Example 4 Let
$$A = \begin{pmatrix} 1 & -3 & 4 & 2 \\ 3 & 1 & 4 & 6 \\ -2 & 3 & 5 & 7 \end{pmatrix}.$$

Then
$$2A = \begin{pmatrix} 2 & -6 & 8 & 4 \\ 6 & 2 & 8 & 12 \\ -4 & 6 & 10 & 14 \end{pmatrix},$$

$$-3A = \begin{pmatrix} -3 & 9 & -12 & -6 \\ -9 & -3 & -12 & -18 \\ 6 & -9 & -15 & -21 \end{pmatrix}, \quad \text{and} \quad 0A = \begin{pmatrix} 0 & 0 & 0 & 0 \\ 0 & 0 & 0 & 0 \\ 0 & 0 & 0 & 0 \end{pmatrix}.$$

We now define the multiplication of two matrices. Quite obviously, we could define the product of two $m \times n$ matrices $A = (a_{ij})$ and $B = (b_{ij})$ to be the $m \times n$ matrix whose ijth component is $a_{ij}b_{ij}$. However, for just about all the important applications involving matrices, another kind of product is needed.

DEFINITION 3: PRODUCT OF TWO MATRICES

Let $A = (a_{ij})$ be an $m \times n$ matrix whose ith row is denoted $\mathbf{a}_i$. Let $B = (b_{ij})$ be an $n \times p$ matrix whose jth column is denoted $\mathbf{b}_j$. Then the product of A and B is an $m \times p$ matrix $C = (c_{ij})$, where

$$\boxed{c_{ij} = a_{i1}b_{1j} + a_{i2}b_{2j} + \cdots + a_{in}b_{nj}.} \tag{4}$$

The product in (4) is called the **dot product** of the vectors in $\mathbf{a}_i$ and $\mathbf{b}_j$ and is denoted $\mathbf{a}_i \cdot \mathbf{b}_j$. Thus (4) can be written as

$$\boxed{c_{ij} = \mathbf{a}_i \cdot \mathbf{b}_j.} \tag{5}$$

Warning Two matrices can be multiplied together only if the number of columns of the first is equal to the number of rows of the second. Otherwise the vectors $\mathbf{a}_i$ and $\mathbf{b}_j$ have different numbers of components and the dot product in (4) and (5) is not defined.

Example 5 If
$$A = \begin{pmatrix} 1 & 3 \\ -2 & 4 \end{pmatrix} \quad \text{and} \quad B = \begin{pmatrix} 3 & -2 \\ 5 & 6 \end{pmatrix},$$

calculate AB and BA.

Solution Let $C = (c_{ij}) = AB$. Then

$$c_{11} = \mathbf{a}_1 \cdot \mathbf{b}_1 = (1 \quad 3) \cdot \begin{pmatrix} 3 \\ 5 \end{pmatrix} = 3 + 15 = 18;$$

$$c_{12} = \mathbf{a}_1 \cdot \mathbf{b}_2 = (1 \quad 3) \cdot \begin{pmatrix} -2 \\ 6 \end{pmatrix} = -2 + 18 = 16;$$

$$c_{21} = (-2 \quad 4) \cdot \begin{pmatrix} 3 \\ 5 \end{pmatrix} = -6 + 20 = 14;$$

$$c_{22} = (-2 \quad 4) \cdot \begin{pmatrix} -2 \\ 6 \end{pmatrix} = 4 + 24 = 28.$$

Thus,

$$C = AB = \begin{pmatrix} 18 & 16 \\ 14 & 28 \end{pmatrix}.$$

Similarly, leaving out the intermediate steps, we see that

$$C' = BA = \begin{pmatrix} 3 & -2 \\ 5 & 6 \end{pmatrix} \begin{pmatrix} 1 & 3 \\ -2 & 4 \end{pmatrix} = \begin{pmatrix} 3+4 & 9-8 \\ 5-12 & 15+24 \end{pmatrix} = \begin{pmatrix} 7 & 1 \\ -7 & 39 \end{pmatrix}.$$

Note that $AB \neq BA$; in fact none of the corresponding entries are equal. Thus matrix multiplication is **not commutative**.

Remark Observe that if A is an $m \times n$ matrix and B is an $n \times m$ matrix, then the product AB is an $m \times m$ matrix and BA is an $n \times n$ matrix. Thus, not only are the entries different, but the sizes of the two products are unequal if $m \neq n$.

Although matrix multiplication is not commutative, it is associative.

THEOREM 1 ASSOCIATIVE LAW OF MATRIX MULTIPLICATION

Let A be an $m \times n$ matrix, B an $n \times p$ matrix, and C a $p \times q$ matrix. Then $(AB)C$ and $A(BC)$ are $m \times q$ matrices and

$$\boxed{(AB)C = A(BC)}$$

The proof of this theorem is left as an exercise (see Problem 61).

DEFINITION 4: IDENTITY MATRIX

The $n \times n$ **identity matrix** is the $n \times n$ matrix with 1's down the **main diagonal**† and 0's everywhere else; that is,

$$\boxed{I_n = (\delta_{ij}), \quad \text{where } \delta_{ij} = \begin{cases} 1, & \text{if } i = j, \\ 0, & \text{if } i \neq j. \end{cases}} \tag{6}$$

---

† The main diagonal of $A = (a_{ij})$ consists of the components a_{11}, a_{22}, a_{33}, and so on. Unless otherwise stated, we refer to the main diagonal simply as the **diagonal**.

Example 6

$$I_3 = \begin{pmatrix} 1 & 0 & 0 \\ 0 & 1 & 0 \\ 0 & 0 & 1 \end{pmatrix} \quad \text{and} \quad I_5 = \begin{pmatrix} 1 & 0 & 0 & 0 & 0 \\ 0 & 1 & 0 & 0 & 0 \\ 0 & 0 & 1 & 0 & 0 \\ 0 & 0 & 0 & 1 & 0 \\ 0 & 0 & 0 & 0 & 1 \end{pmatrix}$$

THEOREM 2

Let A be a square $n \times n$ matrix. Then

$$AI_n = I_n A = A.$$

In other words, I_n commutes with every $n \times n$ matrix and leaves it unchanged after multiplication on the left or right.

Note that I_n behaves for $n \times n$ matrices in the way the number 1 behaves for real numbers (since $1 \cdot a = a \cdot 1 = a$ for every real number a).

Proof Let c_{ij} be the ijth element of AI_n. Then

$$c_{ij} = a_{i1}\delta_{1j} + a_{i2}\delta_{2j} + \cdots + a_{ij}\delta_{jj} + \cdots + a_{in}\delta_{nj}.$$

But, from (6), this sum is equal to a_{ij}. Thus $AI_n = A$. In a similar fashion we can show that $I_n A = A$, and this proves the theorem. ∎

Notation From now on we will write the identity matrix simply as I, since if A is $n \times n$, the products IA and AI are defined only if I is also $n \times n$.

DEFINITION 5: THE INVERSE OF A MATRIX

Let A and B be $n \times n$ matrices. Suppose that

$$AB = BA = I.$$

Then B is called the **inverse** of A and is written as A^{-1}. We then have

$$AA^{-1} = A^{-1}A = I.$$

If A has an inverse, then A is said to be **invertible**.

Remark From this definition it immediately follows that $(A^{-1})^{-1} = A$ if A is invertible.

Remark This definition does *not* state that every square matrix has an inverse. In fact, there are many square matrices that have no inverse.

In Definition 5 we defined *the* inverse of a matrix. This statement suggests that inverses are unique. This is indeed the case, as the following theorem shows.

THEOREM 3

If a square matrix A is invertible, then its inverse is unique.

Proof Suppose that B and C are two inverses for A. We can show that $B = C$. By definition, we have $AB = BA = I$ and $AC = CA = I$. Then $B(AC) = BI = B$ and $(BA)C = IC = C$. But $B(AC) = (BA)C$, since matrix multiplication is associative, by the associative law of matrix multiplication. Hence $B = C$ and the theorem is proved. ∎

Another important fact about inverses is given below.

THEOREM 4

Let A and B be invertible $n \times n$ matrices. Then AB is invertible and

$$\boxed{(AB)^{-1} = B^{-1}A^{-1}.}$$

Proof To prove this result, we refer to Definition 5; that is, $B^{-1}A^{-1} = (AB)^{-1}$ if and only if $B^{-1}A^{-1}(AB) = (AB)(B^{-1}A^{-1}) = I$. But this follows, since

$$(B^{-1}A^{-1})(AB) = B^{-1}(A^{-1}A)B = B^{-1}IB = B^{-1}B = I$$

and

$$(AB)(B^{-1}A^{-1}) = A(BB^{-1})A^{-1} = AIA^{-1} = AA^{-1} = I. \quad \blacksquare$$

There are two basic questions that come to mind once we have defined the inverse of a matrix:

a. What matrices do have inverses?

b. If a matrix has an inverse, how can we compute it?

There are two essentially different ways to compute the inverse of a square matrix.† The method we describe below involves less work when dealing with large matrices ($n > 3$).

COMPUTATION OF INVERSES BY GAUSS-JORDAN ELIMINATION

Consider the system

$$\begin{aligned} 2x_1 + 4x_2 &= 6 \\ 3x_1 + 5x_2 &= 2. \end{aligned} \quad (7)$$

We solve it as follows:

Step 1 Divide the first row by 2:

$$\begin{aligned} x_1 + 2x_2 &= 3 \\ 3x_1 + 5x_2 &= 2. \end{aligned}$$

---

† The other method involves the computation of det A and n^2 $(n-1) \times (n-1)$ determinants.

Step 2 Multiply the first row by -3, and add it to the second row:
$$x_1 + 2x_2 = 3$$
$$-x_2 = -7.$$

Step 3 Multiply the second row by -1:
$$x_1 + 2x_2 = 3$$
$$x_2 = 7.$$

Step 4 Multiply the second row by -2, and add it to the first row:
$$x_1 = -11$$
$$x_2 = 7.$$

The solution is now displayed.

Check
$$2(-11) + 4(7) = 6$$
$$3(-11) + 5(7) = 2$$

We see that performing certain **elementary row operations** on the original systems leads to an equivalent system in which the solution is displayed.

ELEMENTARY ROW OPERATIONS	
Operation	Symbol
(i) Multiply the ith row by a nonzero number c	$M_i(c)$
(ii) Multiply the ith row by c and add it to the jth row	$A_{ij}(c)$
(iii) Permute (interchange) the ith and jth rows	P_{ij}

We can write system (7) as an **augmented matrix**:
$$\begin{pmatrix} 2 & 4 & | & 6 \\ 3 & 5 & | & 2 \end{pmatrix}$$

The first line reads $2x_1 + 4x_2 = 6$, and so on. We repeat the solution using augmented matrices and our elementary row operations notation:

$$\begin{pmatrix} 2 & 4 & | & 6 \\ 3 & 5 & | & 2 \end{pmatrix} \xrightarrow{M_1(\frac{1}{2})} \begin{pmatrix} 1 & 2 & | & 3 \\ 3 & 5 & | & 2 \end{pmatrix} \xrightarrow{A_{1,2}(-3)} \begin{pmatrix} 1 & 2 & | & 3 \\ 0 & -1 & | & -7 \end{pmatrix}$$
$$\xrightarrow{M_2(-1)} \begin{pmatrix} 1 & 2 & | & 3 \\ 0 & 1 & | & 7 \end{pmatrix} \xrightarrow{A_{2,1}(-2)} \begin{pmatrix} 1 & 0 & | & -11 \\ 0 & 1 & | & 7 \end{pmatrix}$$

Again, we "see" the solution $x_1 = -11$, $x_2 = 7$. The process we have carried out is called **row reduction**.

System (7) can also be written in the matrix form
$$A\mathbf{x} = \mathbf{b} \qquad (8)$$
where
$$A = \begin{pmatrix} 2 & 4 \\ 3 & 5 \end{pmatrix}, \quad \mathbf{x} = \begin{pmatrix} x_1 \\ x_2 \end{pmatrix}, \quad \text{and} \quad \mathbf{b} = \begin{pmatrix} 6 \\ 2 \end{pmatrix}.$$

Suppose A is invertible. Then we can multiply both sides of (8) on the left by A^{-1} to obtain

$$A^{-1}A\mathbf{x} = A^{-1}\mathbf{b}$$
$$I\mathbf{x} = A^{-1}\mathbf{b}$$
$$\mathbf{x} = A^{-1}\mathbf{b}$$

That is, in the row reduction process

$$A\mathbf{x} = \mathbf{b} = I\mathbf{b}$$

has been transformed into

$$I\mathbf{x} = A^{-1}\mathbf{b}$$

and $A^{-1}\mathbf{b}$ is the unique solution to the system. This suggests the following procedure for finding the inverse of a matrix. More details and proofs can be found in any standard linear algebra text.[†] The following procedure is called **Gauss-Jordan elimination**.

PROCEDURE FOR COMPUTING THE INVERSE OF A SQUARE MATRIX A

Step 1 Write the augmented matrix $(A|I)$.

Step 2 Use row reduction to reduce A to either

(a) the identity matrix or

(b) a matrix with a row of zeros.

Step 3 Decide if A is invertible.

(a) If A can be reduced to the identity matrix I, then A^{-1} is the matrix to the right of the vertical bar.

(b) If the row reduction leads to a row of zeros to the left of the vertical bar, then A is not invertible.

Example 7 Let

$$A = \begin{pmatrix} 2 & 4 & 6 \\ 4 & 5 & 6 \\ 3 & 1 & -2 \end{pmatrix}.$$

Calculate A^{-1} if it exists.

Solution

Step 1

$$\begin{pmatrix} \overbrace{\begin{matrix}2 & 4 & 6\\4 & 5 & 6\\3 & 1 & -2\end{matrix}}^{A} & \bigg| & \overbrace{\begin{matrix}1 & 0 & 0\\0 & 1 & 0\\0 & 0 & 1\end{matrix}}^{I} \end{pmatrix}$$

[†] See, for example, S. I. Grossman, *Elementary Linear Algebra*, 3rd ed. (Belmont, Calif.: Wadsworth 1987), Chapter 1.

Step 2

$$\xrightarrow{M_1(\frac{1}{2})} \begin{pmatrix} 1 & 2 & 3 & | & \frac{1}{2} & 0 & 0 \\ 4 & 5 & 6 & | & 0 & 1 & 0 \\ 3 & 1 & -2 & | & 0 & 0 & 1 \end{pmatrix} \xrightarrow[A_{1,3}(-3)]{A_{1,2}(-4)} \begin{pmatrix} 1 & 2 & 3 & | & \frac{1}{2} & 0 & 0 \\ 0 & -3 & -6 & | & -2 & 1 & 0 \\ 0 & -5 & -11 & | & -\frac{3}{2} & 0 & 1 \end{pmatrix}$$

$$\xrightarrow{M_2(-\frac{1}{3})} \begin{pmatrix} 1 & 2 & 3 & | & \frac{1}{2} & 0 & 0 \\ 0 & 1 & 2 & | & \frac{2}{3} & -\frac{1}{3} & 0 \\ 0 & -5 & -11 & | & -\frac{3}{2} & 0 & 1 \end{pmatrix}$$

$$\xrightarrow[A_{2,3}(5)]{A_{2,1}(-2)} \begin{pmatrix} 1 & 0 & -1 & | & -\frac{5}{6} & \frac{2}{3} & 0 \\ 0 & 1 & 2 & | & \frac{2}{3} & -\frac{1}{3} & 0 \\ 0 & 0 & -1 & | & \frac{11}{6} & -\frac{5}{3} & 1 \end{pmatrix}$$

$$\xrightarrow{M_3(-1)} \begin{pmatrix} 1 & 0 & -1 & | & -\frac{5}{6} & \frac{2}{3} & 0 \\ 0 & 1 & 2 & | & \frac{2}{3} & -\frac{1}{3} & 0 \\ 0 & 0 & 1 & | & -\frac{11}{6} & \frac{5}{3} & -1 \end{pmatrix}$$

$$\xrightarrow[A_{3,2}(-2)]{A_{3,1}(1)} \overbrace{\begin{pmatrix} 1 & 0 & 0 \\ 0 & 1 & 0 \\ 0 & 0 & 1 \end{pmatrix}}^{I} \; \overbrace{\begin{pmatrix} -\frac{8}{3} & \frac{7}{3} & -1 \\ \frac{13}{3} & -\frac{11}{3} & 2 \\ -\frac{11}{6} & \frac{5}{3} & -1 \end{pmatrix}}^{A^{-1}}$$

Step 3 Since A has been reduced to I, we conclude that

$$A^{-1} = \begin{pmatrix} -\frac{8}{3} & \frac{7}{3} & -1 \\ \frac{13}{3} & -\frac{11}{3} & 2 \\ -\frac{11}{6} & \frac{5}{3} & -1 \end{pmatrix}.$$

After doing all these computations, it is always a good idea to check your answer.

Check

$$A^{-1}A = \begin{pmatrix} -\frac{8}{3} & \frac{7}{3} & -1 \\ \frac{13}{3} & -\frac{11}{3} & 2 \\ -\frac{11}{6} & \frac{5}{3} & -1 \end{pmatrix} \begin{pmatrix} 2 & 4 & 6 \\ 4 & 5 & 6 \\ 3 & 1 & -2 \end{pmatrix} = \begin{pmatrix} 1 & 0 & 0 \\ 0 & 1 & 0 \\ 0 & 0 & 1 \end{pmatrix}$$

Similarly, $AA^{-1} = I$.

Example 8 Let

$$A = \begin{pmatrix} 1 & -3 & 4 \\ 2 & -5 & 7 \\ 0 & -1 & 1 \end{pmatrix}.$$

Calculate A^{-1} if it exists.

Solution We proceed as before:

$$\begin{pmatrix} 1 & -3 & 4 & | & 1 & 0 & 0 \\ 2 & -5 & 7 & | & 0 & 1 & 0 \\ 0 & -1 & 1 & | & 0 & 0 & 1 \end{pmatrix} \xrightarrow{A_{1,2}(-2)} \begin{pmatrix} 1 & -3 & 4 & | & 1 & 0 & 0 \\ 0 & 1 & -1 & | & -2 & 1 & 0 \\ 0 & -1 & 1 & | & 0 & 0 & 1 \end{pmatrix}$$

$$\xrightarrow[A_{2,3}(1)]{A_{2,1}(3)} \begin{pmatrix} 1 & 0 & 1 & | & -5 & 3 & 0 \\ 0 & 1 & -1 & | & -2 & 1 & 0 \\ 0 & 0 & 0 & | & -2 & 1 & 1 \end{pmatrix}.$$

There is now a row of zeros to the left of the vertical bar. We conclude that A is not invertible.

It is possible to determine whether a matrix is invertible by computing its determinant.

THEOREM 5

A square matrix A is invertible if and only if $\det A \neq 0$.

Example 9 In Examples 7 and 8 we have

$$\det \begin{pmatrix} 2 & 4 & 6 \\ 4 & 5 & 6 \\ 3 & 1 & -2 \end{pmatrix} = 6 \neq 0 \quad \text{and} \quad \det \begin{pmatrix} 1 & -3 & 4 \\ 2 & -5 & 7 \\ 0 & -1 & 1 \end{pmatrix} = 0.$$

Thus the first matrix is invertible while the second one is not.

It is especially easy to compute the inverse of a 2×2 matrix. The following theorem can be proved by matrix multiplication.

THEOREM 6

Let

$$A = \begin{pmatrix} a_{11} & a_{12} \\ a_{21} & a_{22} \end{pmatrix}$$

be a 2×2 matrix. Then A is invertible if and only if $\det A = a_{11}a_{22} - a_{12}a_{21} \neq 0$. If $\det A \neq 0$, then

$$A^{-1} = \frac{1}{\det A} \begin{pmatrix} a_{22} & -a_{12} \\ -a_{21} & a_{11} \end{pmatrix} = \begin{pmatrix} \dfrac{a_{22}}{a_{11}a_{22} - a_{12}a_{21}} & \dfrac{-a_{12}}{a_{11}a_{22} - a_{12}a_{21}} \\ \dfrac{-a_{21}}{a_{11}a_{22} - a_{12}a_{21}} & \dfrac{a_{11}}{a_{11}a_{22} - a_{12}a_{21}} \end{pmatrix}$$

(9)

Remark Theorem 6 shows us that finding the inverse of an invertible 2×2 matrix is very simple. We just reverse the diagonal components, multiply each off-diagonal component by -1, and divide the result by $\det A$.

Example 10 Compute the inverse (if it exists) of
$$A = \begin{pmatrix} 2 & 5 \\ -7 & 4 \end{pmatrix}.$$

Solution Since $\det A = 43$, A is invertible. Then, from (9),
$$A^{-1} = \tfrac{1}{43}\begin{pmatrix} 4 & -5 \\ 7 & 2 \end{pmatrix} = \begin{pmatrix} \tfrac{4}{43} & -\tfrac{5}{43} \\ \tfrac{7}{43} & \tfrac{2}{43} \end{pmatrix}.$$

Check $A^{-1}A = \tfrac{1}{43}\begin{pmatrix} 4 & -5 \\ 7 & 2 \end{pmatrix}\begin{pmatrix} 2 & 5 \\ -7 & 4 \end{pmatrix} = \tfrac{1}{43}\begin{pmatrix} 43 & 0 \\ 0 & 43 \end{pmatrix} = \begin{pmatrix} 1 & 0 \\ 0 & 1 \end{pmatrix} = I.$

Similarly, $AA^{-1} = I$; this shows that our answer is correct.

MATRIX FUNCTIONS, DERIVATIVES, AND INTEGRALS

We have discussed matrix theory here in order to apply it to the solution of systems of differential equations. In such a context, the matrices we encounter often have components that are functions rather than real numbers. For that reason, we provide some elementary definitions.

Vector Function

An n-component **vector function**
$$\mathbf{v}(t) = \begin{pmatrix} v_1(t) \\ v_2(t) \\ \vdots \\ v_n(t) \end{pmatrix} \tag{10}$$

is an n-vector each of whose components is a function (usually assumed to be continuous).

Matrix Function

An $n \times n$ **matrix function** $A(t)$ is an $n \times n$ matrix
$$A(t) = \begin{pmatrix} a_{11}(t) & a_{12}(t) & \cdots & a_{1n}(t) \\ a_{21}(t) & a_{22}(t) & \cdots & a_{2n}(t) \\ \vdots & \vdots & & \vdots \\ a_{n1}(t) & a_{n2}(t) & \cdots & a_{nn}(t) \end{pmatrix} \tag{11}$$

each of whose n^2 components is a function. We may add and multiply vector and matrix functions in the same way that we add and multiply constant vectors and matrices. Thus if, for example,
$$A(t) = \begin{pmatrix} a_{11}(t) & a_{12}(t) \\ a_{21}(t) & a_{22}(t) \end{pmatrix} \quad \text{and} \quad B(t) = \begin{pmatrix} b_{11}(t) & b_{12}(t) \\ b_{21}(t) & b_{22}(t) \end{pmatrix},$$
then
$$A(t)B(t) = \begin{pmatrix} a_{11}(t)b_{11}(t) + a_{12}(t)b_{21}(t) & a_{11}(t)b_{12}(t) + a_{12}(t)b_{22}(t) \\ a_{21}(t)b_{11}(t) + a_{22}(t)b_{21}(t) & a_{21}(t)b_{12}(t) + a_{22}(t)b_{22}(t) \end{pmatrix}.$$

Derivatives and Integrals

We may also differentiate and integrate vector and matrix functions, componentwise. Thus if $\mathbf{v}(t)$ is given by (10), then

$$\mathbf{v}'(t) = \begin{pmatrix} v_1'(t) \\ v_2'(t) \\ \vdots \\ v_n'(t) \end{pmatrix} \quad \text{and} \quad \int_{t_0}^{t} \mathbf{v}(s)\, ds = \begin{pmatrix} \int_{t_0}^{t} v_1(s)\, ds \\ \int_{t_0}^{t} v_2(s)\, ds \\ \vdots \\ \int_{t_0}^{t} v_n(s)\, ds \end{pmatrix}.$$

Similarly, if $A(t)$ is given by (11), then

$$A'(t) = \begin{pmatrix} a_{11}'(t) & a_{12}'(t) & \cdots & a_{1n}'(t) \\ a_{21}'(t) & a_{22}'(t) & \cdots & a_{2n}'(t) \\ \vdots & \vdots & & \vdots \\ a_{n1}'(t) & a_{n2}'(t) & \cdots & a_{nn}'(t) \end{pmatrix}$$

and

$$\int_{t_0}^{t} A(s)\, ds = \begin{pmatrix} \int_{t_0}^{t} a_{11}(s)\, ds & \int_{t_0}^{t} a_{12}(s)\, ds & \cdots & \int_{t_0}^{t} a_{1n}(s)\, ds \\ \int_{t_0}^{t} a_{21}(s)\, ds & \int_{t_0}^{t} a_{22}(s)\, ds & \cdots & \int_{t_0}^{t} a_{2n}(s)\, ds \\ \vdots & \vdots & & \vdots \\ \int_{t_0}^{t} a_{n1}(s)\, ds & \int_{t_0}^{t} a_{n2}(s)\, ds & \cdots & \int_{t_0}^{t} a_{nn}(s)\, ds \end{pmatrix}.$$

Example 11 Let

$$\mathbf{v}(t) = \begin{pmatrix} t \\ t^2 \\ \sin t \\ e^t \end{pmatrix}.$$

Then

$$\mathbf{v}'(t) = \begin{pmatrix} 1 \\ 2t \\ \cos t \\ e^t \end{pmatrix} \quad \text{and} \quad \int_0^t \mathbf{v}(s)\, ds = \begin{pmatrix} t^2/2 \\ t^3/3 \\ 1 - \cos t \\ e^t - 1 \end{pmatrix}.$$

PROBLEMS 7.5

In Problems 1–12 perform the indicated computation with

$$A = \begin{pmatrix} 1 & 3 \\ 2 & 5 \\ -1 & 2 \end{pmatrix},$$

$$B = \begin{pmatrix} -2 & 0 \\ 1 & 4 \\ -7 & 5 \end{pmatrix}, \quad \text{and} \quad C = \begin{pmatrix} -1 & 1 \\ 4 & 6 \\ -7 & 3 \end{pmatrix}.$$

1. $3A$
2. $A + B$
3. $A - C$
4. $2C - 5A$
5. $0B$ (0 is the scalar zero)
6. $-7A + 3B$
7. $A + B + C$
8. $C - A - B$
9. $2A - 3B + 4C$
10. $7C - B + 2A$
11. Find a matrix D such that $2A + B - D$ is the 3×2 zero matrix.

12. Find a matrix E such that $A + 2B - 3C + E$ is the 3×2 zero matrix.

In Problems 13–20 perform the indicated computation with

$$A = \begin{pmatrix} 1 & -1 & 2 \\ 3 & 4 & 5 \\ 0 & 1 & -1 \end{pmatrix},$$

$$B = \begin{pmatrix} 0 & 2 & 1 \\ 3 & 0 & 5 \\ 7 & -6 & 0 \end{pmatrix}, \text{ and } C = \begin{pmatrix} 0 & 0 & 2 \\ 3 & 1 & 0 \\ 0 & -2 & 4 \end{pmatrix}.$$

13. $A - 2B$
14. $3A - C$
15. $A + B + C$
16. $2A - B + 2C$
17. $C - A - B$
18. $4C - 2B + 3A$
19. Find a matrix D such that $A + B + C + D$ is the 3×3 zero matrix.
20. Find a matrix E such that $3C - 2B + 8A - 4E$ is the 3×3 zero matrix.

In Problems 21–34 perform the indicated computation.

21. $\begin{pmatrix} 2 & 3 \\ -1 & 2 \end{pmatrix}\begin{pmatrix} 4 & 1 \\ 0 & 6 \end{pmatrix}$

22. $\begin{pmatrix} 3 & -2 \\ 1 & 4 \end{pmatrix}\begin{pmatrix} -5 & 6 \\ 1 & 3 \end{pmatrix}$

23. $\begin{pmatrix} 1 & -1 \\ 1 & 1 \end{pmatrix}\begin{pmatrix} -1 & 0 \\ 2 & 3 \end{pmatrix}$

24. $\begin{pmatrix} -5 & 6 \\ 1 & 3 \end{pmatrix}\begin{pmatrix} 3 & -2 \\ 1 & 4 \end{pmatrix}$

25. $\begin{pmatrix} -4 & 5 & 1 \\ 0 & 4 & 2 \end{pmatrix}\begin{pmatrix} 3 & -1 & 1 \\ 5 & 6 & 4 \\ 0 & 1 & 2 \end{pmatrix}$

26. $\begin{pmatrix} 7 & 1 & 4 \\ 2 & -3 & 5 \end{pmatrix}\begin{pmatrix} 1 & 6 \\ 0 & 4 \\ -2 & 3 \end{pmatrix}$

27. $\begin{pmatrix} 1 & 6 \\ 0 & 4 \\ -2 & 3 \end{pmatrix}\begin{pmatrix} 7 & 1 & 4 \\ 2 & -3 & 5 \end{pmatrix}$

28. $\begin{pmatrix} 1 & 4 & -2 \\ 3 & 0 & 4 \end{pmatrix}\begin{pmatrix} 0 & 1 \\ 2 & 3 \end{pmatrix}$

29. $\begin{pmatrix} 1 & 4 & 6 \\ -2 & 3 & 5 \\ 1 & 0 & 4 \end{pmatrix}\begin{pmatrix} 2 & -3 & 5 \\ 1 & 0 & 6 \\ 2 & 3 & 1 \end{pmatrix}$

30. $\begin{pmatrix} 2 & -3 & 5 \\ 1 & 0 & 6 \\ 2 & 3 & 1 \end{pmatrix}\begin{pmatrix} 1 & 4 & 6 \\ -2 & 3 & 5 \\ 1 & 0 & 4 \end{pmatrix}$

31. $(1 \ 4 \ 0 \ 2)\begin{pmatrix} 3 & -6 \\ 2 & 4 \\ 1 & 0 \\ -2 & 3 \end{pmatrix}$

32. $\begin{pmatrix} 3 & 2 & 1 & -2 \\ -6 & 4 & 0 & 3 \end{pmatrix}\begin{pmatrix} 1 \\ 4 \\ 0 \\ 2 \end{pmatrix}$

33. $\begin{pmatrix} 3 & -2 & 1 \\ 4 & 0 & 6 \\ 5 & 1 & 9 \end{pmatrix}\begin{pmatrix} 1 & 0 & 0 \\ 0 & 1 & 0 \\ 0 & 0 & 1 \end{pmatrix}$

34. $\begin{pmatrix} 1 & 0 & 0 \\ 0 & 1 & 0 \\ 0 & 0 & 1 \end{pmatrix}\begin{pmatrix} 3 & -2 & 1 \\ 4 & 0 & 6 \\ 5 & 1 & 9 \end{pmatrix}$

35. Let A be a square matrix. Then A^2 is defined simply as AA. Calculate

$$\begin{pmatrix} 2 & -1 \\ 4 & 6 \end{pmatrix}^2.$$

36. Calculate A^2, where

$$A = \begin{pmatrix} 1 & -2 & 4 \\ 2 & 0 & 3 \\ 1 & 1 & 5 \end{pmatrix}.$$

37. Calculate A^3, where

$$A = \begin{pmatrix} -1 & 2 \\ 3 & 4 \end{pmatrix}.$$

38. Calculate A^2, A^3, A^4, and A^5, where

$$A = \begin{pmatrix} 0 & 1 & 0 & 0 \\ 0 & 0 & 1 & 0 \\ 0 & 0 & 0 & 1 \\ 0 & 0 & 0 & 0 \end{pmatrix}.$$

In Problems 39–48 use the methods of this section to determine whether the given matrix is invertible. If it is, compute the inverse.

39. $\begin{pmatrix} 3 & 2 \\ 1 & 2 \end{pmatrix}$

40. $\begin{pmatrix} 3 & 6 \\ -4 & -8 \end{pmatrix}$

41. $\begin{pmatrix} 0 & 1 \\ 1 & 0 \end{pmatrix}$

42. $\begin{pmatrix} 1 & 1 & 1 \\ 0 & 2 & 3 \\ 5 & 5 & 1 \end{pmatrix}$

43. $\begin{pmatrix} 3 & 2 & 1 \\ 0 & 2 & 2 \\ 0 & 1 & -1 \end{pmatrix}$

44. $\begin{pmatrix} 1 & 1 & 1 \\ 0 & 1 & 1 \\ 0 & 0 & 1 \end{pmatrix}$

45. $\begin{pmatrix} 1 & 2 & 3 \\ 1 & 1 & 2 \\ 0 & 1 & 2 \end{pmatrix}$

46. $\begin{pmatrix} 3 & 1 & 0 \\ 1 & -1 & 2 \\ 1 & 1 & 1 \end{pmatrix}$

47. $\begin{pmatrix} 2 & -1 & 4 \\ -1 & 0 & 5 \\ 19 & -7 & 3 \end{pmatrix}$

48. $\begin{pmatrix} 1 & 6 & 2 \\ -2 & 3 & 5 \\ 7 & 12 & -4 \end{pmatrix}$

49. Use the methods of this section to find the inverses of the following matrices with complex entries:

a. $\begin{pmatrix} i & 2 \\ 1 & -i \end{pmatrix}$

b. $\begin{pmatrix} 1-i & 0 \\ 0 & 1+i \end{pmatrix}$

c. $\begin{pmatrix} 1 & i & 0 \\ -i & 0 & 1 \\ 0 & 1+i & 1-i \end{pmatrix}$

50. Show that for every real number θ the matrix

$$\begin{pmatrix} \sin\theta & \cos\theta & 0 \\ \cos\theta & -\sin\theta & 0 \\ 0 & 0 & 1 \end{pmatrix}$$

is invertible and find its inverse.

51. Calculate the inverse of

$$A = \begin{pmatrix} 2 & 0 & 0 \\ 0 & 3 & 0 \\ 0 & 0 & 4 \end{pmatrix}.$$

52. A square matrix $A = (a_{ij})$ is called **diagonal** if all its elements off the main diagonal are zero; that is, $a_{ij} = 0$ if $i \neq j$. [The matrix of Problem 51 is diagonal.] Show that a diagonal matrix is invertible if and only if each of its diagonal components is nonzero.

53. Let

$$A = \begin{pmatrix} a_{11} & 0 & \cdots & 0 \\ 0 & a_{22} & \cdots & 0 \\ & & \ddots & \\ 0 & 0 & \cdots & a_{nn} \end{pmatrix}$$

be a diagonal matrix such that each of its diagonal components is nonzero. Calculate A^{-1}.

54. Calculate the inverse of

$$A = \begin{pmatrix} 2 & 1 & -1 \\ 0 & 3 & 4 \\ 0 & 0 & 5 \end{pmatrix}.$$

55. Show that the matrix

$$A = \begin{pmatrix} 1 & 0 & 0 \\ -2 & 0 & 0 \\ 4 & 6 & 1 \end{pmatrix}$$

is not invertible.

*56. A square matrix is called **upper (lower) triangular** if all its elements below (above) the main diagonal are zero. [The matrix of Problem 54 is upper triangular and the matrix of Problem 55 is lower triangular.] Show that an upper or lower triangular matrix is invertible if and only if each of its diagonal elements is nonzero.

In Problems 57–60 find the derivative and integral of each vector or matrix function.

57. $\mathbf{x}(t) = (t, \sin t)$

58. $A(t) = \begin{pmatrix} \sqrt{t} & t^2 \\ e^{2t} & \sin 2t \end{pmatrix}$

59. $\mathbf{y}(t) = \begin{pmatrix} e^t \\ \cos t \\ \tan t \end{pmatrix}$

60. $B(t) = \begin{pmatrix} \ln t & e^t \sin t & e^t \cos t \\ t^{5/2} & -\cos t & -\sin t \\ 1/t & te^{t^2} & t^2 e^{t^3} \end{pmatrix}$

*61. Prove Theorem 1. [*Hint:* Use a double sum to show that the ijth component of $(AB)C$ is equal to the ijth component of $A(BC)$.]

7.6 Matrices and Linear Systems of Differential Equations

Consider the following linear system of n first-order equations

$$\begin{aligned} x_1' &= a_{11}(t)x_1 + a_{12}(t)x_2 + \cdots + a_{1n}(t)x_n + f_1(t), \\ x_2' &= a_{21}(t)x_1 + a_{22}(t)x_2 + \cdots + a_{2n}(t)x_n + f_2(t), \\ &\vdots \\ x_n' &= a_{n1}(t)x_1 + a_{n2}(t)x_2 + \cdots + a_{nn}(t)x_n + f_n(t), \end{aligned} \quad (1)$$

which is **nonhomogeneous** if at least one of the functions $f_i(t)$, $i = 1, 2, \ldots, n$ is not the zero function, and the **associated homogeneous system**

$$\begin{aligned} x_1' &= a_{11}(t)x_1 + a_{12}(t)x_2 + \cdots + a_{1n}(t)x_n, \\ x_2' &= a_{21}(t)x_1 + a_{22}(t)x_2 + \cdots + a_{2n}(t)x_n, \\ &\vdots \\ x_n' &= a_{n1}(t)x_1 + a_{n2}(t)x_2 + \cdots + a_{nn}(t)x_n. \end{aligned} \quad (2)$$

We assume that all the coefficient functions a_{ij} and the functions f_i are continuous on some common interval containing the point t_0.

We now define the vector function $\mathbf{x}(t)$, the matrix function $A(t)$, and the vector function $\mathbf{f}(t)$ as follows:

$$\mathbf{x}(t) = \begin{pmatrix} x_1(t) \\ x_2(t) \\ \vdots \\ x_n(t) \end{pmatrix}, \quad A(t) = \begin{pmatrix} a_{11}(t) & a_{12}(t) & \cdots & a_{1n}(t) \\ a_{21}(t) & a_{22}(t) & \cdots & a_{2n}(t) \\ \vdots & \vdots & & \vdots \\ a_{n1}(t) & a_{n2}(t) & \cdots & a_{nn}(t) \end{pmatrix},$$

$$\mathbf{f}(t) = \begin{pmatrix} f_1(t) \\ f_2(t) \\ \vdots \\ f_n(t) \end{pmatrix}. \tag{3}$$

Vector Differential Equation

Then, using (3), we can rewrite system (1) as the **vector differential equation**

$$\boxed{\mathbf{x}'(t) = A(t)\mathbf{x}(t) + \mathbf{f}(t).} \tag{4}$$

System (2) becomes

$$\boxed{\mathbf{x}'(t) = A(t)\mathbf{x}(t).} \tag{5}$$

It is clear from what has already been said that any linear differential equation or system can be written in form (5) if it is homogeneous, and in form (4) if it is nonhomogeneous. The reason for writing a system in these forms is that, besides the obvious advantage of compactness of notation, equations (4) and (5) behave very much like first-order linear differential equations, as we will see. It will be very easy to work with systems of equations in this way once we get used to the notation.

Example 1 Consider the system

$$x_1' = (2t)x_1 + (\sin t)x_2 - e^t x_3 - e^t,$$
$$x_2' = -t^3 x_1 + e^{\sin t} x_2 - (\ln t)x_3 + \cos t,$$
$$x_3' = 2x_1 - 5tx_2 + 2tx_3 + \tan t.$$

It can be rewritten as

$$\begin{pmatrix} x_1 \\ x_2 \\ x_3 \end{pmatrix}' = \begin{pmatrix} 2t & \sin t & -e^t \\ -t^3 & e^{\sin t} & -\ln t \\ 2 & -5t & 2t \end{pmatrix} \begin{pmatrix} x_1 \\ x_2 \\ x_3 \end{pmatrix} + \begin{pmatrix} -e^t \\ \cos t \\ \tan t \end{pmatrix}.$$

Example 2 Consider the general second-order linear differential equation

$$x'' + a(t)x' + b(t)x = f(t).$$

Defining $x_1 = x$ and $x_2 = x'$, we obtain the equivalent system
$$x_1' = x_2,$$
$$x_2' = -b(t)x_1 - a(t)x_2 + f(t),$$
which can be rewritten as
$$\mathbf{x}' = A(t)\mathbf{x} + \mathbf{f}(t),$$
where
$$\mathbf{x} = \begin{pmatrix} x_1 \\ x_2 \end{pmatrix}, \quad A(t) = \begin{pmatrix} 0 & 1 \\ -b(t) & -a(t) \end{pmatrix}, \quad \mathbf{f}(t) = \begin{pmatrix} 0 \\ f(t) \end{pmatrix}.$$

Example 3 Consider the third-order equation with constant coefficients
$$x''' - 6x'' + 11x' - 6x = 0. \tag{6}$$
Defining $x_1 = x$, $x_2 = x'$, and $x_3 = x''$, we obtain the system
$$x_1' = x_2,$$
$$x_2' = x_3,$$
$$x_3' = 6x_1 - 11x_2 + 6x_3,$$
or
$$\mathbf{x}' = A\mathbf{x}, \tag{7}$$
where
$$\mathbf{x} = \begin{pmatrix} x_1 \\ x_2 \\ x_3 \end{pmatrix} \quad \text{and} \quad A = \begin{pmatrix} 0 & 1 & 0 \\ 0 & 0 & 1 \\ 6 & -11 & 6 \end{pmatrix}.$$

Let us now consider the initial-value problem
$$\boxed{\mathbf{x}'(t) = A(t)\mathbf{x}(t) + \mathbf{f}(t), \qquad \mathbf{x}(t_0) = \mathbf{x}_0,} \tag{8}$$
where
$$\mathbf{x}(t) = \begin{pmatrix} x_1(t) \\ x_2(t) \\ \vdots \\ x_n(t) \end{pmatrix} \quad \text{and} \quad \mathbf{x}_0 = \begin{pmatrix} x_{10} \\ x_{20} \\ \vdots \\ x_{n0} \end{pmatrix}.$$

Solution to a Vector Differential Equation

We say that a vector function
$$\boldsymbol{\phi}(t) = \begin{pmatrix} \phi_1(t) \\ \vdots \\ \phi_n(t) \end{pmatrix}$$
is a **solution** to equation (8) if $\boldsymbol{\phi}$ is differentiable and satisfies the differential equation and the given initial condition in some interval containing t_0. The following theorem is proved in Appendix 3.

THEOREM 1   EXISTENCE–UNIQUENESS THEOREM

Let $A(t)$ and $\mathbf{f}(t)$ be continuous matrix and vector functions, respectively, on some interval $[a, b]$ containing t_0 (that is, the component functions of both

$A(t)$ and $\mathbf{f}(t)$ are continuous). Then there exists a unique vector function $\boldsymbol{\phi}(t)$ that is a solution to the initial-value problem (8) on the entire interval $[a, b]$.

Example 4 Referring to Example 3, we find by the methods of Chapter 3 that e^t, e^{2t}, and e^{3t} are solutions of equation (6). Since a solution vector for this problem is

$$\boldsymbol{\phi}(t) = \begin{pmatrix} x(t) \\ x'(t) \\ x''(t) \end{pmatrix},$$

we see that three vector solutions of equation (7) are

$$\boldsymbol{\phi}_1(t) = \begin{pmatrix} e^t \\ e^t \\ e^t \end{pmatrix} = e^t \begin{pmatrix} 1 \\ 1 \\ 1 \end{pmatrix}, \qquad \boldsymbol{\phi}_2(t) = \begin{pmatrix} e^{2t} \\ 2e^{2t} \\ 4e^{2t} \end{pmatrix} = e^{2t} \begin{pmatrix} 1 \\ 2 \\ 4 \end{pmatrix},$$

and

$$\boldsymbol{\phi}_3(t) = \begin{pmatrix} e^{3t} \\ 3e^{3t} \\ 9e^{3t} \end{pmatrix} = e^{3t} \begin{pmatrix} 1 \\ 3 \\ 9 \end{pmatrix}. \quad \text{(Check!)}$$

Thus the general solution is

$$\boldsymbol{\phi}(t) = c_1 \boldsymbol{\phi}_1(t) + c_2 \boldsymbol{\phi}_2(t) + c_3 \boldsymbol{\phi}_3(t)$$

$$= \begin{pmatrix} c_1 e^t + c_2 e^{2t} + c_3 e^{3t} \\ c_1 e^t + 2c_2 e^{2t} + 3c_3 e^{3t} \\ c_1 e^t + 4c_2 e^{2t} + 9c_3 e^{3t} \end{pmatrix}.$$

If we specify the initial condition

$$\mathbf{x}(0) = \begin{pmatrix} 2 \\ -3 \\ 5 \end{pmatrix},$$

then it is easily verified that $c_1 = 16$, $c_2 = -23$, and $c_3 = 9$, so that the unique solution vector is

$$\boldsymbol{\phi} = 16\boldsymbol{\phi}_1 - 23\boldsymbol{\phi}_2 + 9\boldsymbol{\phi}_3,$$

or

$$\begin{pmatrix} 16e^t - 23e^{2t} + 9e^{3t} \\ 16e^t - 46e^{2t} + 27e^{3t} \\ 16e^t - 92e^{2t} + 81e^{3t} \end{pmatrix}.$$

Example 5 The system

$$x_1' = -4x_1 - x_2, \qquad x_1(0) = 1,$$
$$x_2' = x_1 - 2x_2, \qquad x_2(0) = 2,$$

can be written as

$$\begin{pmatrix} x_1 \\ x_2 \end{pmatrix}' = \begin{pmatrix} -4 & -1 \\ 1 & -2 \end{pmatrix} \begin{pmatrix} x_1 \\ x_2 \end{pmatrix}, \qquad \begin{pmatrix} x_1(0) \\ x_2(0) \end{pmatrix} = \begin{pmatrix} 1 \\ 2 \end{pmatrix}.$$

It can easily be verified that

$$\phi_1(t) = \begin{pmatrix} e^{-3t} \\ -e^{-3t} \end{pmatrix} \quad \text{and} \quad \phi_2(t) = \begin{pmatrix} (1-t)e^{-3t} \\ te^{-3t} \end{pmatrix}$$

are solution vectors. It can also be immediately verified that the unique solution vector that satisfies the given initial conditions is

$$\phi(t) = \begin{pmatrix} (1-3t)e^{-3t} \\ (2+3t)e^{-3t} \end{pmatrix}.$$

The central problem of the remainder of the chapter is to derive properties of vector solutions and, where possible, to calculate them. In the next section we will show how all solutions to the homogeneous system $\mathbf{x}' = A\mathbf{x}$ can be represented in a convenient form, and in Section 7.10 we will show how information about the solutions to this homogeneous system can be used to find a particular solution to the nonhomogeneous system $\mathbf{x}' = A\mathbf{x} + \mathbf{f}$.

PROBLEMS 7.6

In Problems 1–6 write each given equation or system in the matrix vector form (4), (5), or (8).

1. $x_1' = 2x_1 + 3x_2,$
   $x_2' = 4x_1 - 6x_2$

2. $x_1' = (\cos t)x_1 - (\sin t)x_2 + e^{t^2},$
   $x_2' = e^t x_1 + 2tx_2 - \ln t,$
   $x_1(2) = 3, \quad x_2(2) = 7$

3. $x''' - 2x'' + 4tx' - x = \sin t$

4. $x^{(4)} + 2x''' - 3x'' + 4x' - 7x = 0,$
   $x(0) = 1, \quad x'(0) = 2, \quad x''(0) = 3, \quad x'''(0) = 4$

5. $x_1' = 2tx_1 - 3t^2 x_2 + (\sin t)x_3,$
   $x_2' = 2x_1 - 4x_3 - \sin t,$
   $x_3' = 17x_2 + 4tx_3 + e^t$

6. $x''' + a(t)x'' + b(t)x' + c(t)x = f(t),$
   $x(t_0) = d_1, \quad x'(t_0) = d_2, \quad x''(t_0) = d_3$

In Problems 7–14 verify that each given vector function is a solution to the given system.

7. $\mathbf{x}' = \begin{pmatrix} 1 & 1 \\ -3 & -1 \end{pmatrix} \mathbf{x},$
   $\phi(t) = -\begin{pmatrix} \cos\sqrt{2}\,t \\ \sqrt{2}\sin\sqrt{2}\,t - \cos\sqrt{2}\,t \end{pmatrix}$

8. $\mathbf{x}' = \begin{pmatrix} 2 & 1 \\ 1 & 2 \end{pmatrix}\mathbf{x} + \begin{pmatrix} t \\ t^2 \end{pmatrix},$
   $\phi(t) = \begin{pmatrix} e^t + \frac{1}{3}t^2 + \frac{2}{9}t + \frac{11}{27} \\ -e^t - \frac{2}{3}t^2 - \frac{7}{9}t - \frac{16}{27} \end{pmatrix}$

9. $\mathbf{x}' = \begin{pmatrix} -1 & 6 \\ 1 & -2 \end{pmatrix}\mathbf{x}, \quad \phi(t) = \begin{pmatrix} 3e^t \\ e^t \end{pmatrix}$

10. $\mathbf{x}' = \begin{pmatrix} -4 & -1 \\ 1 & -2 \end{pmatrix}\mathbf{x}, \quad \phi(t) = \begin{pmatrix} (1+t)e^{-3t} \\ (-2-t)e^{-3t} \end{pmatrix}$

11. $\mathbf{x}' = \begin{pmatrix} 4 & 1 \\ -8 & 8 \end{pmatrix}\mathbf{x},$
    $\phi(t) = \begin{pmatrix} e^{6t}(\cos 2t - \frac{1}{2}\sin 2t) \\ e^{6t}(\cos 2t - 3\sin 2t) \end{pmatrix}$

12. $\mathbf{x}' = \begin{pmatrix} 1 & 1 & 1 \\ -1 & -1 & 0 \\ -1 & 0 & 1 \end{pmatrix}\mathbf{x},$
    $\phi(t) = \begin{pmatrix} \sin t - \cos t \\ \cos t \\ \cos t \end{pmatrix}$

13. $\mathbf{x}' = \begin{pmatrix} 1 & -1 & 1 & -1 \\ 0 & -1 & 2 & -2 \\ 0 & 0 & 2 & -3 \\ 0 & 0 & 0 & -2 \end{pmatrix}\mathbf{x},$
    $\phi(t) = \begin{pmatrix} e^{-2t} \\ 2e^{-2t} \\ 3e^{-2t} \\ 4e^{-2t} \end{pmatrix}$

14. $\mathbf{x}' = \begin{pmatrix} 3 & 2 & 1 \\ -1 & 0 & -1 \\ 1 & 1 & 2 \end{pmatrix}\mathbf{x},$
    $\phi(t) = \begin{pmatrix} e^{2t} + te^{2t} \\ -te^{2t} \\ te^{2t} \end{pmatrix}$

15. Let $\phi_1(t)$ and $\phi_2(t)$ be any two vector solutions of the homogeneous system $\mathbf{x}' = A(t)\mathbf{x}$. Show that $\phi(t) = c_1\phi_1(t) + c_2\phi_2(t)$ is also a solution.

16. Let $\phi_1(t)$ and $\phi_2(t)$ be vector solutions of the nonhomogeneous system (4). Show that their difference,

$$\phi(t) = \phi_1(t) - \phi_2(t),$$

is a solution to the homogeneous system (5).

7.7 Fundamental and Principal Matrix Solutions of a Homogeneous System of Differential Equations

In this section we discuss properties of the homogeneous system

$$\mathbf{x}' = A(t)\mathbf{x}, \tag{1}$$

where $\mathbf{x}(t)$ is an n-vector and $A(t)$ is an $n \times n$ matrix of functions. Throughout this section we assume that the components of $A(t)$ are functions continuous on some common interval containing the point t_0.

Linearly Independent Solutions

Let $\phi_1(t), \phi_2(t), \ldots, \phi_m(t)$ be m vector solutions of system (1). We say that they are **linearly independent** on an interval if the equation (for all t in the common interval)

$$c_1\phi_1(t) + c_2\phi_2(t) + \cdots + c_m\phi_m(t) = \mathbf{0}$$

holds only for $c_1 = c_2 = \cdots = c_m = 0$. Since system (1) is equivalent to an nth-order equation, it is natural for us to seek n linearly independent solutions to the system. Any set of n linearly independent solutions of (1) is called a **fundamental set of solutions**.

Fundamental Set of Solutions

In Section 7.3 we used the method of determinants to obtain the general solution for linear homogeneous systems with constant coefficients. If a general solution is known, it is an easy task to obtain a fundamental set of solutions: simply set *one* of the arbitrary coefficients c_j equal to 1 and set all the rest of them equal to zero for $j = 1, 2, \ldots, n$.

Example 1 Let $\mathbf{x}' = A\mathbf{x}$, where

$$A = \begin{pmatrix} -1 & 6 \\ 1 & -2 \end{pmatrix}.$$

In Example 7.3.4 we verified that

$$x_1(t) = c_1 e^{-4t} + c_2 e^t,$$
$$x_2(t) = -\tfrac{1}{2}c_1 e^{-4t} + \tfrac{1}{3}c_2 e^t,$$

is the general solution to this system. We can arrange this general solution in vector form as

$$\mathbf{x}(t) = c_1 \begin{pmatrix} e^{-4t} \\ -\tfrac{1}{2}e^{-4t} \end{pmatrix} + c_2 \begin{pmatrix} e^t \\ \tfrac{1}{3}e^t \end{pmatrix}.$$

Thus, setting $c_1 = 1$ and $c_2 = 0$, we obtain the solution vector

$$\phi_1(t) = \begin{pmatrix} e^{-4t} \\ -\tfrac{1}{2}e^{-4t} \end{pmatrix},$$

while $c_1 = 0$ and $c_2 = 1$ yields

$$\phi_2 = \begin{pmatrix} e^t \\ \frac{1}{3}e^t \end{pmatrix}.$$

The vectors ϕ_1 and ϕ_2 form a fundamental set of solutions since

$$W(t) = \begin{vmatrix} e^{-4t} & e^t \\ -\frac{1}{2}e^{-4t} & \frac{1}{3}e^t \end{vmatrix} = \left(\frac{1}{3} + \frac{1}{2}\right)e^{-3t} \neq 0.$$

Example 2 Let $\mathbf{x}' = A\mathbf{x}$, where

$$A = \begin{pmatrix} -4 & -1 \\ 1 & -2 \end{pmatrix}.$$

In Example 7.3.5 we obtained the general solution (which we now write in vector form)

$$\mathbf{x}(t) = \begin{pmatrix} c_1 e^{-3t} + c_2 t e^{-3t} \\ (-c_1 - c_2)e^{-3t} - c_2 t e^{-3t} \end{pmatrix},$$

or

$$\mathbf{x}(t) = c_1 \begin{pmatrix} e^{-3t} \\ -e^{-3t} \end{pmatrix} + c_2 \begin{pmatrix} te^{-3t} \\ (-1-t)e^{-3t} \end{pmatrix} = c_1 \phi_1(t) + c_2 \phi_2(t).$$

The vectors ϕ_1 and ϕ_2 form a fundamental set of solutions since

$$W(t) = \begin{vmatrix} e^{-3t} & te^{-3t} \\ -e^{-3t} & (-1-t)e^{-3t} \end{vmatrix} = -e^{-6t} \neq 0.$$

Example 3 Consider the system $\mathbf{x}' = A\mathbf{x}$, where

$$A = \begin{pmatrix} 4 & 1 \\ -8 & 8 \end{pmatrix}.$$

In Example 7.3.6 we obtained the general solution (written in vector form)

$$\mathbf{x}(t) = \begin{pmatrix} \frac{1}{4}e^{6t}[(c_1 - c_2)\cos 2t + (c_1 + c_2)\sin 2t] \\ e^{6t}(c_1 \cos 2t + c_2 \sin 2t) \end{pmatrix}$$

$$= c_1 \begin{pmatrix} \frac{1}{4}e^{6t}(\cos 2t + \sin 2t) \\ e^{6t}\cos 2t \end{pmatrix} + c_2 \begin{pmatrix} \frac{1}{4}e^{6t}(-\cos 2t + \sin 2t) \\ e^{6t}\sin 2t \end{pmatrix}.$$

Thus, by first setting $c_1 = 1$ and $c_2 = 0$ and then setting $c_1 = 0$ and $c_2 = 1$, we get the solutions

$$\phi_1(t) = \begin{pmatrix} \frac{1}{4}e^{6t}(\cos 2t + \sin 2t) \\ e^{6t}\cos 2t \end{pmatrix}, \quad \phi_2(t) = \begin{pmatrix} \frac{1}{4}e^{6t}(-\cos 2t + \sin 2t) \\ e^{6t}\sin 2t \end{pmatrix},$$

and the Wronskian of these solutions is

$$W(t) = \begin{vmatrix} \frac{1}{4}e^{6t}(\cos 2t + \sin 2t) & \frac{1}{4}e^{6t}(-\cos 2t + \sin 2t) \\ e^{6t}\cos 2t & e^{6t}\sin 2t \end{vmatrix} = \frac{1}{4}e^{12t} \neq 0.$$

Thus ϕ_1 and ϕ_2 form a fundamental set of solutions.

Example 4 Let $\mathbf{x}' = A\mathbf{x}$, where

$$A = \begin{pmatrix} 1 & 1 & -2 \\ -1 & 2 & 1 \\ 0 & 1 & -1 \end{pmatrix}.$$

Then it is not difficult to show that

$$\boldsymbol{\phi}_1(t) = \begin{pmatrix} e^{-t} \\ 0 \\ e^{-t} \end{pmatrix} = e^{-t} \begin{pmatrix} 1 \\ 0 \\ 1 \end{pmatrix}, \quad \boldsymbol{\phi}_2(t) = \begin{pmatrix} 3e^t \\ 2e^t \\ e^t \end{pmatrix} = e^t \begin{pmatrix} 3 \\ 2 \\ 1 \end{pmatrix},$$

$$\boldsymbol{\phi}_3(t) = \begin{pmatrix} e^{2t} \\ 3e^{2t} \\ e^{2t} \end{pmatrix} = e^{2t} \begin{pmatrix} 1 \\ 3 \\ 1 \end{pmatrix}$$

are solutions of this system. To show that they are a fundamental set of solutions, we consider their Wronskian:

$$W(t) = \begin{vmatrix} e^{-t} & 3e^t & e^{2t} \\ 0 & 2e^t & 3e^{2t} \\ e^{-t} & e^t & e^{2t} \end{vmatrix}.$$

By Theorem 7.2.4 (which extends to larger systems), the Wronskian is either always zero or never zero. Thus, we need only check the value of $W(t)$ at a single point, say $t = 0$:

$$W(0) = \begin{vmatrix} 1 & 3 & 1 \\ 0 & 2 & 3 \\ 1 & 1 & 1 \end{vmatrix} = 6 \neq 0.$$

Hence $\boldsymbol{\phi}_1$, $\boldsymbol{\phi}_2$, and $\boldsymbol{\phi}_3$ are linearly independent, so they form a fundamental set of solutions.

Matrix Solution

Let $\boldsymbol{\phi}_1, \boldsymbol{\phi}_2, \ldots, \boldsymbol{\phi}_n$ be n-vector solutions of $\mathbf{x}' = A(t)\mathbf{x}$. Let $\Phi(t)$ be the matrix whose columns are the vectors $\boldsymbol{\phi}_1, \boldsymbol{\phi}_2, \ldots, \boldsymbol{\phi}_n$; that is,

$$\Phi(t) = (\boldsymbol{\phi}_1(t), \ldots, \boldsymbol{\phi}_n(t)) = \begin{pmatrix} \phi_{11}(t) & \phi_{12}(t) & \cdots & \phi_{1n}(t) \\ \phi_{21}(t) & \phi_{22}(t) & \cdots & \phi_{2n}(t) \\ \vdots & \vdots & & \vdots \\ \phi_{n1}(t) & \phi_{n2}(t) & \cdots & \phi_{nn}(t) \end{pmatrix}. \quad (2)$$

Such a matrix is called a **matrix solution** of the system $\mathbf{x}' = A\mathbf{x}$. Equivalently, an $n \times n$ matrix function $\Phi(t)$ is a matrix solution of $\mathbf{x}' = A\mathbf{x}$ if and only if each of its columns is a solution vector of $\mathbf{x}' = A\mathbf{x}$.

Wronskian of a Matrix Solution

We define the **Wronskian** of a matrix solution $\Phi(t)$ of the system $\mathbf{x}' = A(t)\mathbf{x}$ by

$$\boxed{W(t) = \det \Phi(t).} \quad (3)$$

This is precisely the same definition of the Wronskian that we have used in Examples 1–4.

Fundamental Matrix Solution

If $W(t) \neq 0$, then the vectors $\phi_1, \phi_2, \ldots, \phi_n$ form a fundamental set of solutions (that is, they are linearly independent), and $\Phi(t)$ is called a **fundamental matrix solution**. In what follows, we show that fundamental matrix solutions play a central role in the theory of linear systems of differential equations.

Example 5 In the four previous examples of this section, fundamental matrix solutions are, respectively,

1. $\begin{pmatrix} e^{-4t} & e^t \\ -\frac{1}{2}e^{-4t} & \frac{1}{3}e^t \end{pmatrix}$

2. $\begin{pmatrix} e^{-3t} & te^{-3t} \\ -e^{-3t} & (-1-t)e^{-3t} \end{pmatrix}$

3. $\begin{pmatrix} \frac{1}{4}e^{6t}(\cos 2t + \sin 2t) & \frac{1}{4}e^{6t}(-\cos 2t + \sin 2t) \\ e^{6t}\cos 2t & e^{6t}\sin 2t \end{pmatrix}$

4. $\begin{pmatrix} e^{-t} & 3e^t & e^{2t} \\ 0 & 2e^t & 3e^{2t} \\ e^{-t} & e^t & e^{2t} \end{pmatrix}$

Fundamental matrix solutions are not unique, due to the fact that a solution vector may be multiplied by any constant and still remain a solution. In addition, any linear combination of solutions is still a solution (see Problem 6). However, there is one particular fundamental matrix solution that is unique:

Principal Matrix Solution

> The fundamental matrix solution $\Psi(t)$ for the system
> $$\mathbf{x}' = A(t)\mathbf{x}$$
> that satisfies $\Psi(t_0) = I$, where I is the identity matrix, is called the **principal matrix solution**.[†]

Remark In the definition above we are assuming that the point t_0 is contained within the interval of continuity that each of the coefficient functions $a_{ij}(t)$ have in common.

The next result proves that principal matrix solutions (and hence also fundamental matrix solutions) always exist for a linear homogeneous system.

THEOREM 1 **EXISTENCE–UNIQUENESS THEOREM**

Let $A(t)$ be continuous on some interval $[a, b]$. Then for any t_0, $a \leq t_0 \leq b$, there exists a unique fundamental matrix solution $\Psi(t)$ of the system $\mathbf{x}' = A(t)\mathbf{x}$ satisfying the condition $\Psi(t_0) = I$.

[†] In some books this is called a **state transition matrix**.

Proof Let δ_i, $i = 1, 2, \ldots, n$, denote the n-column vector that has a one in the ith position (row) and a zero everywhere else:

$$\delta_1 = \begin{pmatrix} 1 \\ 0 \\ 0 \\ \vdots \\ 0 \end{pmatrix}, \quad \delta_2 = \begin{pmatrix} 0 \\ 1 \\ 0 \\ \vdots \\ 0 \end{pmatrix}, \quad \ldots, \quad \delta_n = \begin{pmatrix} 0 \\ 0 \\ \vdots \\ 0 \\ 1 \end{pmatrix}.$$

By the basic existence-uniqueness theorem (Theorem 7.6.1) with $\mathbf{f}(t) \equiv \mathbf{0}$, there exists a unique vector solution $\boldsymbol{\phi}_i(t)$ of $\mathbf{x}' = A\mathbf{x}$ that satisfies $\boldsymbol{\phi}_i(t_0) = \delta_i$, $i = 1, 2, \ldots, n$. Define the matrix function

$$\Psi(t) = (\boldsymbol{\phi}_1(t), \boldsymbol{\phi}_2(t), \ldots, \boldsymbol{\phi}_n(t)). \tag{4}$$

Then $\Psi(t)$ is the matrix whose columns are the vector solutions $\boldsymbol{\phi}_i$, $i = 1, 2, \ldots, n$. Clearly $\Psi(t)$ is a matrix solution and $\Psi(t_0) = (\boldsymbol{\phi}_1(t_0), \ldots, \boldsymbol{\phi}_n(t_0)) = (\delta_1, \delta_2, \ldots, \delta_n) = I$. It remains to be shown that $\Psi(t)$ is a fundamental matrix solution. This is easy to do. We simply note that $\det \Psi(t_0) = \det I = 1 \neq 0$. ■

The calculation of a fundamental or principal matrix solution is generally impossible if $A(t)$ is nonconstant. If $A(t)$ is a constant matrix, then, as we will see in Section 7.9, the principal matrix solution can always, in principle, be obtained.

What does it mean to say that $\Phi(t)$ is a matrix solution of $\mathbf{x}' = A(t)\mathbf{x}$? By definition, we know that each column of $\boldsymbol{\phi}_i(t)$ of $\Phi(t)$ is a solution to $\mathbf{x}' = A(t)\mathbf{x}$; that is,

$$\boldsymbol{\phi}_i' = A(t)\boldsymbol{\phi}_i.$$

Hence, by matrix multiplication it follows that

$$\boxed{\Phi' = A(t)\Phi.} \tag{5}$$

Example 6 We have seen in Example 1 that

$$\boldsymbol{\phi}_1(t) = \begin{pmatrix} e^{-4t} \\ -\tfrac{1}{2}e^{-4t} \end{pmatrix} \quad \text{and} \quad \boldsymbol{\phi}_2(t) = \begin{pmatrix} e^t \\ \tfrac{1}{3}e^t \end{pmatrix}$$

are solutions to $\mathbf{x}' = A\mathbf{x}$, where

$$A = \begin{pmatrix} -1 & 6 \\ 1 & -2 \end{pmatrix}.$$

Let

$$\Phi(t) = (\boldsymbol{\phi}_1(t), \boldsymbol{\phi}_2(t)) = \begin{pmatrix} e^{-4t} & e^t \\ -\tfrac{1}{2}e^{-4t} & \tfrac{1}{3}e^t \end{pmatrix}.$$

Then, since matrix differentiation involves differentiation of each component,

$$\Phi'(t) = \begin{pmatrix} -4e^{-4t} & e^t \\ 2e^{-4t} & \tfrac{1}{3}e^t \end{pmatrix},$$

and
$$A(t)\Phi(t) = \begin{pmatrix} -1 & 6 \\ 1 & -2 \end{pmatrix}\begin{pmatrix} e^{-4t} & e^t \\ -\frac{1}{2}e^{-4t} & \frac{1}{3}e^t \end{pmatrix} = \begin{pmatrix} -4e^{-4t} & e^t \\ 2e^{-4t} & \frac{1}{3}e^t \end{pmatrix}.$$
Hence $\Phi'(t) = A(t)\Phi(t)$.

THEOREM 2

Let Φ be a matrix solution of $\mathbf{x}' = A(t)\mathbf{x}$ and let C be any constant square matrix. Then $\Phi_1 = \Phi C$ is also a matrix solution of $\mathbf{x}' = A(t)\mathbf{x}$.

Proof Since a matrix solution of $\mathbf{x}' = A(t)\mathbf{x}$ is also a solution of equation (5), we must show that Φ_1 is also a solution of (5). But $\Phi_1' = (\Phi C)' = \Phi'C + \Phi C' = \Phi'C$, since $C' = 0$, C being constant. Finally, since Φ is a solution,
$$\Phi_1' = \Phi'C = A\Phi C = A\Phi_1. \quad \blacksquare$$

Example 7 We have seen in Example 6 that
$$\Phi(t) = \begin{pmatrix} e^{-4t} & e^t \\ -\frac{1}{2}e^{-4t} & \frac{1}{3}e^t \end{pmatrix}$$
is a matrix solution of $\mathbf{x}' = A(t)\mathbf{x}$, with
$$A = \begin{pmatrix} -1 & 6 \\ 1 & -2 \end{pmatrix}.$$
Let
$$C_1 = \begin{pmatrix} 1 & 2 \\ 3 & 4 \end{pmatrix} \quad \text{and} \quad C_2 = \begin{pmatrix} 1 & 2 \\ 2 & 4 \end{pmatrix}.$$
Then
$$\Phi_1 = \Phi C_1 = \begin{pmatrix} e^{-4t} + 3e^t & 2e^{-4t} + 4e^t \\ -\frac{1}{2}e^{-4t} + e^t & -e^{-4t} + \frac{4}{3}e^t \end{pmatrix}$$
and
$$\Phi_2 = \Phi C_2 = \begin{pmatrix} e^{-4t} + 2e^t & 2e^{-4t} + 4e^t \\ -\frac{1}{2}e^{-4t} + \frac{2}{3}e^t & -e^{-4t} + \frac{4}{3}e^t \end{pmatrix}$$
are also matrix solutions (verify this!). Note that although Φ and Φ_1 are fundamental matrix solutions, Φ_2 is not since
$$\det \Phi_2(0) = \begin{vmatrix} 3 & 6 \\ \frac{1}{6} & \frac{1}{3} \end{vmatrix} = 0.$$

OBTAINING PRINCIPAL MATRIX SOLUTIONS FROM FUNDAMENTAL MATRIX SOLUTIONS

Theorem 2 gives us an easy way of finding the principal matrix solution when a fundamental matrix solution is known. To see this, let $\Phi(t)$ be a fundamental matrix solution. Since $\det \Phi(t_0) \neq 0$, $\Phi(t_0)$ is invertible, and we define $C = \Phi^{-1}(t_0)$ and $\Psi(t) = \Phi(t)C$. By Theorem 2, $\Psi(t)$ is a matrix solution and $\Psi(t_0) = \Phi(t_0)C = \Phi(t_0)\Phi^{-1}(t_0) = I$, so $\Psi(t)$ is the principal matrix solution.

Thus

> if $\Phi(t)$ is a fundamental matrix solution for $\mathbf{x}' = A(t)\mathbf{x}$, then
> $$\Psi(t) = \Phi(t)\Phi^{-1}(t_0)$$
> is the principal matrix solution (associated with the point t_0).

Example 8 In Example 7 we showed that

$$\Phi(t) = \begin{pmatrix} e^{-4t} & e^t \\ -\frac{1}{2}e^{-4t} & \frac{1}{3}e^t \end{pmatrix}$$

is a fundamental matrix solution for the system

$$\mathbf{x}' = \begin{pmatrix} -1 & 6 \\ 1 & -2 \end{pmatrix} \mathbf{x}.$$

Assume $t_0 = 0$, so that

$$\Phi(t_0) = \Phi(0) = \begin{pmatrix} 1 & 1 \\ -\frac{1}{2} & \frac{1}{3} \end{pmatrix} \quad \text{and} \quad \Phi^{-1}(0) = \begin{pmatrix} \frac{2}{5} & -\frac{6}{5} \\ \frac{3}{5} & \frac{6}{5} \end{pmatrix}.$$

Then

$$\Psi(t) = \Phi(t)\Phi^{-1}(0) = \frac{1}{5}\begin{pmatrix} 2e^{-4t} + 3e^t & -6e^{-4t} + 6e^t \\ -e^{-4t} + e^t & 3e^{-4t} + 2e^t \end{pmatrix}$$

is the principal matrix solution. Note that

$$\Psi(0) = \frac{1}{5}\begin{pmatrix} 5 & 0 \\ 0 & 5 \end{pmatrix} = I.$$

THEOREM 3

Let $\Phi(t)$ be a fundamental matrix solution and let $X(t)$ be any other matrix solution of the system $\mathbf{x}' = A(t)\mathbf{x}$. Then there exists a constant matrix C such that $X(t) = \Phi(t)C$; that is, *any solution vector of $\mathbf{x}' = A(t)\mathbf{x}$ can be written as a linear combination of vectors in a fundamental set* (compare this theorem with Theorem 7.2.3, p. 267).

Before giving the proof, we should warn the reader that it is important to state on which side we are multiplying the matrix Φ by C, since matrix multiplication is not, in general, commutative.

Proof Since $\Phi(t)$ is a fundamental matrix solution, $\det \Phi(t) \neq 0$ and $\Phi^{-1}(t)$ exists for every t. We will show that

$$\frac{d}{dt}[\Phi^{-1}(t)X(t)] = 0.$$

This will imply that $\Phi^{-1}(t)X(t)$ is a constant matrix C, proving the theorem.

7.7 / FUNDAMENTAL AND PRINCIPAL MATRIX SOLUTIONS

First, we calculate
$$\frac{d}{dt}[\Phi^{-1}(t)].$$

Using the product rule of differentiation (which can easily be shown to apply to matrix products), we have

$$0 = \frac{dI}{dt} = \frac{d}{dt}(\Phi\Phi^{-1}) = \frac{d\Phi}{dt}\Phi^{-1} + \Phi\frac{d\Phi^{-1}}{dt}, \qquad (6)$$

or, after multiplying both sides of (6) on the left by Φ^{-1} and solving for $d\Phi^{-1}/dt$,

$$\frac{d\Phi^{-1}}{dt} = -\Phi^{-1}\frac{d\Phi}{dt}\Phi^{-1}. \qquad (7)$$

Note the analogy between (7) and the identity

$$\frac{d}{dt}\left(\frac{1}{f(t)}\right) = -\frac{f'(t)}{[f(t)]^2}.$$

Now, by the product formula of derivatives,

$$\frac{d}{dt}(\Phi^{-1}X) = \left(\frac{d}{dt}\Phi^{-1}\right)X + \Phi^{-1}\frac{dX}{dt}, \qquad (8)$$

and since both Φ and X are solutions of $\mathbf{x}' = A(t)\mathbf{x}$, equation (8) becomes

$$\frac{d}{dt}(\Phi^{-1}X) = \left(-\Phi^{-1}\frac{d\Phi}{dt}\Phi^{-1}\right)X + \Phi^{-1}(AX)$$
$$= -\Phi^{-1}A\Phi\Phi^{-1}X + \Phi^{-1}AX$$
$$= -\Phi^{-1}AX + \Phi^{-1}AX = 0. \qquad \blacksquare$$

Example 9 Consider the system $\mathbf{x}' = A\mathbf{x}$, where

$$A = \begin{pmatrix} 1 & -2 \\ 2 & -3 \end{pmatrix}.$$

It is easy to verify that

$$\Phi_1(t) = \begin{pmatrix} e^{-t} & (2t+2)e^{-t} \\ e^{-t} & (2t+1)e^{-t} \end{pmatrix} = e^{-t}\begin{pmatrix} 1 & 2t+2 \\ 1 & 2t+1 \end{pmatrix}$$

is a fundamental matrix solution. Another matrix solution is

$$\Phi_2(t) = e^{-t}\begin{pmatrix} 4t+7 & 8t+1 \\ 4t+5 & 8t-3 \end{pmatrix}.$$

There is a matrix C such that $\Phi_2 = \Phi_1 C$. But $\Phi_1^{-1}(t)\Phi_2(t) = C$ holds for every value of t, in particular for $t = 0$. Thus

$$C = \Phi_1^{-1}(0)\Phi_2(0) = \begin{pmatrix} -1 & 2 \\ 1 & -1 \end{pmatrix}\begin{pmatrix} 7 & 1 \\ 5 & -3 \end{pmatrix} = \begin{pmatrix} 3 & -7 \\ 2 & 4 \end{pmatrix}.$$

Example 10 Consider the system $\mathbf{x}' = A\mathbf{x}$, where

$$A = \begin{pmatrix} 3 & -1 & 1 \\ -1 & 5 & -1 \\ 1 & -1 & 3 \end{pmatrix}.$$

A fundamental matrix solution is

$$\Phi(t) = \begin{pmatrix} e^{2t} & e^{3t} & e^{6t} \\ 0 & e^{3t} & -2e^{6t} \\ -e^{2t} & e^{3t} & e^{6t} \end{pmatrix}.$$

Another matrix solution is

$$X(t) = \begin{pmatrix} e^{2t} + 2e^{3t} + 3e^{6t} & e^{2t} - 3e^{3t} - 2e^{6t} & 2e^{2t} + 5e^{3t} + 7e^{6t} \\ 2e^{3t} - 6e^{6t} & -3e^{3t} + 4e^{6t} & 5e^{3t} - 14e^{6t} \\ -e^{2t} + 2e^{2t} + 3e^{6t} & -e^{2t} - 3e^{3t} - 2e^{6t} & -2e^{2t} + 5e^{3t} + 7e^{6t} \end{pmatrix}.$$

As in the previous example, a matrix C such that $X(t) = \Phi(t)C$ is given by

$$C = \Phi^{-1}(0)X(0) = \frac{1}{6}\begin{pmatrix} 3 & 0 & -3 \\ 2 & 2 & 2 \\ 1 & -2 & 1 \end{pmatrix}\begin{pmatrix} 6 & -4 & 14 \\ -4 & 1 & -9 \\ 4 & -6 & 10 \end{pmatrix}$$

$$= \frac{1}{6}\begin{pmatrix} 6 & 6 & 12 \\ 12 & -18 & 30 \\ 18 & -12 & 42 \end{pmatrix} = \begin{pmatrix} 1 & 1 & 2 \\ 2 & -3 & 5 \\ 3 & -2 & 7 \end{pmatrix}.$$

THEOREM 4

Let $\Phi(t)$ be a fundamental matrix solution and let $\mathbf{x}(t)$ be any solution of $\mathbf{x}' = A(t)\mathbf{x}$. Then there exists a constant vector $\mathbf{c}$ such that

$$\mathbf{x}(t) = \Phi(t)\mathbf{c}. \tag{9}$$

Proof This theorem is an immediate consequence of Theorem 3 if we form the matrix solution $X(t) = (\mathbf{x}(t), \mathbf{x}(t), \ldots, \mathbf{x}(t))$ whose n columns are each the vector solution $\mathbf{x}(t)$. Then a matrix C exists such that $X(t) = \Phi(t)C$. Since each column of X is identical, every column of C is the same vector $\mathbf{c}$. ∎

General Solution

Theorem 4 asserts that the **general solution** of the system $\mathbf{x}' = A(t)\mathbf{x}$ is obtained by multiplying any fundamental matrix solution $\Phi(t)$ by an arbitrary vector of constants $\mathbf{c}$; that is,

$$\Phi(t)\mathbf{c} = (\phi_1(t), \phi_2(t), \ldots, \phi_n(t))\begin{pmatrix} c_1 \\ c_2 \\ \vdots \\ c_n \end{pmatrix} = c_1\phi_1(t) + c_2\phi_2(t) + \cdots + c_n\phi_n(t).$$

This completely justifies our previous use of the term "general solution."

In this section we defined both fundamental and principal matrix solutions (relative to a point t_0) to $\mathbf{x}'(t) = A(t)\mathbf{x}(t)$. Why go to the extra trouble of computing the principal matrix solution? The answer is that in Section 7.10, when we discuss nonhomogeneous systems, we will need to compute the inverse of a fundamental matrix solution. In general, this involves a great deal of algebra. However, the computation of the inverse of a principal matrix

7.7 / FUNDAMENTAL AND PRINCIPAL MATRIX SOLUTIONS

solution is sometimes very easy. The following theorem is proved in Section 7.11 (Theorem 7.11.3).

THEOREM 5

Let $\Psi(t)$ be the principal matrix solution for the system $\mathbf{x}'(t) = A\mathbf{x}(t)$, relative to the point t_0, where A is a constant matrix. Then

$$\boxed{\Psi^{-1}(t) = \Psi(-t).} \tag{10}$$

Example 11 In Example 8 we found that

$$\Psi(t) = \frac{1}{5}\begin{pmatrix} 2e^{-4t} + 3e^t & -6e^{-4t} + 6e^t \\ -e^{-4t} + e^t & 3e^{-4t} + 2e^t \end{pmatrix} \tag{11}$$

is the principal matrix solution relative to $t_0 = 0$ for the system

$$\mathbf{x}' = \begin{pmatrix} -1 & 6 \\ 1 & -2 \end{pmatrix}\mathbf{x}.$$

Then, from (10), we obtain

Replace t by $-t$ in (11)
$$\downarrow$$
$$\Psi^{-1}(t) = \frac{1}{5}\begin{pmatrix} 2e^{4t} + 3e^{-t} & -6e^{4t} + 6e^{-t} \\ -e^{4t} + e^{-t} & 3e^{4t} + 2e^{-t} \end{pmatrix}. \tag{12}$$

It is not difficult to verify that (12) is the inverse matrix for (11).

PROBLEMS 7.7

In Problems 1–5 decide whether each given set of solution vectors constitutes a fundamental set of the given system by (a) determining whether the vectors are linearly independent, and (b) using the method of Wronskians to determine whether or not $W(t)$ is zero.

1. $\mathbf{x}' = \begin{pmatrix} 2 & 5 \\ 0 & 2 \end{pmatrix}\mathbf{x}, \quad \boldsymbol{\phi}_1(t) = \begin{pmatrix} e^{2t}(1+10t) \\ 2e^{2t} \end{pmatrix},$

$\boldsymbol{\phi}_2(t) = \begin{pmatrix} e^{2t}(-3+20t) \\ 4e^{2t} \end{pmatrix}$

2. $\mathbf{x}' = \begin{pmatrix} 4 & -13 \\ 2 & -6 \end{pmatrix}\mathbf{x},$

$\boldsymbol{\phi}_1(t) = \begin{pmatrix} e^{-t}(13\cos t - 26\sin t) \\ e^{-t}(7\cos t - 9\sin t) \end{pmatrix},$

$\boldsymbol{\phi}_2(t) = \begin{pmatrix} e^{-t}(26\cos t - 52\sin t) \\ e^{-t}(14\cos t - 18\sin t) \end{pmatrix}$

3. $\mathbf{x}' = \begin{pmatrix} 1 & 1 \\ 4 & 1 \end{pmatrix}\mathbf{x}, \quad \boldsymbol{\phi}_1(t) = \begin{pmatrix} e^{3t} - e^{-t} \\ 2e^{3t} + 2e^{-t} \end{pmatrix},$

$\boldsymbol{\phi}_2(t) = \begin{pmatrix} 2e^{3t} \\ 4e^{3t} \end{pmatrix}$

4. $\mathbf{x}' = \begin{pmatrix} 1 & -1 & 4 \\ 3 & 2 & -1 \\ 2 & 1 & -1 \end{pmatrix}\mathbf{x},$

$\boldsymbol{\phi}_1(t) = \begin{pmatrix} e^t + 2e^{-2t} + 3e^{3t} \\ -4e^t - 2e^{-2t} + 6e^{3t} \\ -e^t - 2e^{-2t} + 3e^{3t} \end{pmatrix},$

$\boldsymbol{\phi}_2(t) = \begin{pmatrix} -2e^t + 2e^{-2t} \\ 8e^t - 2e^{-2t} \\ 2e^t - 2e^{-2t} \end{pmatrix},$

$\boldsymbol{\phi}_3(t) = \begin{pmatrix} 3e^t - 6e^{-2t} + 3e^{3t} \\ -12e^t + 6e^{-2t} + 6e^{3t} \\ -3e^t + 6e^{-2t} + 3e^{3t} \end{pmatrix}$

5. $\mathbf{x}' = \begin{pmatrix} 3 & 2 & 1 \\ -1 & 0 & -1 \\ 1 & 1 & 2 \end{pmatrix} \mathbf{x}$,

$$\phi_1(t) = \begin{pmatrix} -e^t + te^{2t} + e^{2t} \\ e^t - te^{2t} \\ te^{2t} \end{pmatrix},$$

$$\phi_2(t) = \begin{pmatrix} 2e^{2t} + te^{2t} \\ -e^{2t} - te^{2t} \\ e^{2t} + te^{2t} \end{pmatrix},$$

$$\phi_3(t) = \begin{pmatrix} -e^t + 3e^{2t} + 2te^{2t} \\ e^t - e^{2t} - 2te^{2t} \\ e^{2t} + 2te^{2t} \end{pmatrix}$$

6. Let $\phi_1(t), \phi_2(t), \ldots, \phi_m(t)$ be m solutions of the homogeneous system (1). Show that

$$\phi(t) = c_1\phi_1(t) + c_2\phi_2(t) + \cdots + c_m\phi_m(t)$$

is also a solution.

In each of Problems 7–9, two matrix functions Φ_1 and Φ_2 are given. Find a matrix C such that $\Phi_2(t) = \Phi_1(t)C$.

7. $\Phi_1(t) = e^{6t} \begin{pmatrix} \cos 2t - \frac{1}{2}\sin 2t & \sin 2t + \frac{1}{2}\cos 2t \\ \cos 2t - 3\sin 2t & \sin 2t + 3\cos 2t \end{pmatrix}$,

$\Phi_2(t)$

$= e^{6t} \begin{pmatrix} \frac{1}{2}\cos 2t - \frac{3}{2}\sin 2t & -\frac{3}{2}\cos 2t + 2\sin 2t \\ -2\cos 2t - 4\sin 2t & \cos 2t + 7\sin 2t \end{pmatrix}$

8. $\Phi_1(t) = \begin{pmatrix} \sin e^t & \cos e^t \\ e^t\cos e^t & -e^t\sin e^t \end{pmatrix}$,

$\Phi_2(t) = \begin{pmatrix} \cos e^t & 3\sin e^t + 2\cos e^t \\ -e^t\sin e^t & 3e^t\cos e^t - 2e^t\sin e^t \end{pmatrix}$

9. $\Phi_1(t) = \begin{pmatrix} e^{-t} & 3e^t & e^{2t} \\ 0 & 2e^t & 3e^{2t} \\ e^{-t} & e^t & e^{2t} \end{pmatrix}$,

$\Phi_2(t) = \begin{pmatrix} e^{-t} + e^{2t} & -e^{-t} + 3e^t & e^{-t} + 3e^t \\ 3e^{2t} & 2e^t & 2e^t \\ e^{-t} + e^{2t} & -e^{-t} + e^t & e^{-t} + e^t \end{pmatrix}$

10. Let $\Phi_1(t)$ be a fundamental matrix solution of $\mathbf{x}' = A\mathbf{x}$. Then $\Phi_2 = \Phi_1 C$ is a matrix solution for any constant matrix C. Show that Φ_2 is a fundamental matrix solution if and only if C is nonsingular.

11. Let $\Phi(t)$ be a matrix solution of the system $\mathbf{x}' = A(t)\mathbf{x}$ where $A(t)$ is a 3×3 matrix. Prove that

$W(t)$

$= W(t_0)\exp\left(\int_{t_0}^t [a_{11}(s) + a_{22}(s) + a_{33}(s)]\, ds\right).$

*12. Using the result of Problem 11, prove that

$$W(t) = W(t_0)\exp\left(\int_{t_0}^t \text{tr}\, A(s)\, ds\right),$$

where the trace of A, written $\text{tr}\, A(t)$, is the sum of the diagonal elements of the matrix $A(t)$:

$$\text{tr}\, A(t) = a_{11}(t) + a_{22}(t) + \cdots + a_{nn}(t),$$

for the case of $A(t)$ being an $n \times n$ matrix.

In Problems 13–16 (a) find the principal matrix solution $\Psi(t)$ for each given fundamental matrix solution $\Phi(t)$, assuming that $t_0 = 0$; (b) compute $\Psi^{-1}(t)$.

13. $\Phi(t) = e^{6t}\begin{pmatrix} \cos 2t - \frac{1}{2}\sin 2t & \sin 2t + \frac{1}{2}\cos 2t \\ \cos 2t - 3\sin 2t & \sin 2t + 3\cos 2t \end{pmatrix}$

14. $\Phi(t) = \begin{pmatrix} \sin e^t & \cos e^t \\ e^t\cos e^t & -e^t\sin e^t \end{pmatrix}$

15. $\Phi(t) = \begin{pmatrix} e^{-t} & 3e^t & e^{2t} \\ 0 & 2e^t & 3e^{2t} \\ e^{-t} & e^t & e^{2t} \end{pmatrix}$

16. $\Phi(t) = \begin{pmatrix} 2e^t + te^t & te^t + e^t & e^t \\ 0 & e^t & e^t \\ -3e^t - te^t & -te^t - 2e^t & -e^t \end{pmatrix}$

17. Consider the system

$$\mathbf{x}' = \begin{pmatrix} 3 & -2 \\ 2 & -1 \end{pmatrix}\mathbf{x}.$$

It is easy to verify that

$$\Phi(t) = \begin{pmatrix} 2te^t + e^t & 2te^t \\ 2te^t & -e^t + 2te^t \end{pmatrix}$$

is a fundamental matrix solution. Find a solution that satisfies each of the following initial conditions:

a. $\mathbf{x}(0) = \begin{pmatrix} 1 \\ 2 \end{pmatrix}$ b. $\mathbf{x}(0) = \begin{pmatrix} -2 \\ 3 \end{pmatrix}$

c. $\mathbf{x}(1) = \begin{pmatrix} 0 \\ 1 \end{pmatrix}$ d. $\mathbf{x}(-1) = \begin{pmatrix} 2 \\ 1 \end{pmatrix}$

e. $\mathbf{x}(3) = \begin{pmatrix} 3 \\ 3 \end{pmatrix}$ f. $\mathbf{x}(a) = \begin{pmatrix} b \\ c \end{pmatrix}$

18. In Example 4 we saw that a fundamental matrix solution to the system

$$\mathbf{x}' = \begin{pmatrix} 1 & 1 & -2 \\ -1 & 2 & 1 \\ 0 & 1 & -1 \end{pmatrix}\mathbf{x}$$

is

$$\Phi(t) = \begin{pmatrix} e^{-t} & 3e^t & e^{2t} \\ 0 & 2e^t & 3e^{2t} \\ e^{-t} & e^t & e^{2t} \end{pmatrix}.$$

Find the particular solutions that satisfy the following conditions:

a. $\mathbf{x}(0) = \begin{pmatrix} 1 \\ -1 \\ 2 \end{pmatrix}$ **b.** $\mathbf{x}(0) = \begin{pmatrix} 3 \\ 1 \\ 2 \end{pmatrix}$

c. $\mathbf{x}(1) = \begin{pmatrix} 1 \\ 0 \\ 1 \end{pmatrix}$ **d.** $\mathbf{x}(-1) = \begin{pmatrix} 2 \\ -3 \\ 5 \end{pmatrix}$

19. Consider the second-order equation
$$x'' + a(t)x' + b(t)x = 0. \quad \text{(i)}$$

a. Write (i) in the form $\mathbf{x}' = A(t)\mathbf{x}$.

b. Given that
$$\Phi(t) = \begin{pmatrix} \phi_1 & \phi_2 \\ \phi_1' & \phi_2' \end{pmatrix}$$
is a fundamental matrix solution, show that
$$\det \Phi(t) = \det \Phi(t_0) \exp\left(-\int_{t_0}^{t} a(s)\, ds \right).$$

c. Show that the formula in part (b) can be rearranged as
$$\phi_2' - \frac{\phi_1'}{\phi_1}\phi_2 = \frac{\det \Phi(t_0)}{\phi_1} \exp\left(-\int_{t_0}^{t} a(s)\, ds \right). \quad \text{(ii)}$$
Therefore, if one solution $\phi_1(t)$ of equation (i) is known, then another solution can be calculated by solving this equation with $\det \Phi(t_0) = 1$.

20. Given that $\phi_1(t) = \sin(\ln t)$ is a solution, find a second linearly independent solution of
$$x'' + \frac{1}{t}x' + \frac{1}{t^2}x = 0.$$

21. Given that $\phi_1(t) = e^{t^2}$ is a solution of
$$x'' - 2tx' - 2x = 0,$$
find a second linearly independent solution.

22. Given that $\phi_1(t) = \sin t^2$ is a solution of
$$tx'' - x' + 4t^3x = 0,$$
find a second linearly independent solution.

7.8 Eigenvalues and Eigenvectors

In the next section we will present a method for computing the principal matrix solution of the system

$$\mathbf{x}' = A\mathbf{x}, \quad (1)$$

where the matrix A is constant. The method involves the use of the eigenvalues and corresponding eigenvectors of the matrix A.

Eigenvalue

Let A be an $n \times n$ matrix with real[†] components. The number λ (real or complex) is called an **eigenvalue** of A if there is a *nonzero* vector $\mathbf{v}$ with real or complex entries such that

$$\boxed{A\mathbf{v} = \lambda \mathbf{v}.} \quad (2)$$

Eigenvector

The vector $\mathbf{v} \neq 0$ is called an **eigenvector of A corresponding to the eigenvalue** λ.

Note *Eigen* is the German word for "own" or "proper." Eigenvalues are also called **proper values** or **characteristic values** and eigenvectors are called **proper vectors** or **characteristic vectors**.

Remark As we will see (for example, in Example 6), a matrix with real components can have complex eigenvalues and eigenvectors. That is why, in

---
[†] This definition is also valid if A has complex components, but since the matrices we deal with have real components, the definition is sufficient for our purposes.

the definition, we have asserted that λ and the components of $\mathbf{v}$ may be complex.

Example 1 Let

$$A = \begin{pmatrix} 10 & -18 \\ 6 & -11 \end{pmatrix}.$$

Then

$$A\begin{pmatrix} 2 \\ 1 \end{pmatrix} = \begin{pmatrix} 10 & -18 \\ 6 & -11 \end{pmatrix}\begin{pmatrix} 2 \\ 1 \end{pmatrix} = \begin{pmatrix} 2 \\ 1 \end{pmatrix}.$$

Thus $\lambda_1 = 1$ is an eigenvalue of A with corresponding eigenvector $\mathbf{v}_1 = \begin{pmatrix} 2 \\ 1 \end{pmatrix}$. Similarly,

$$A\begin{pmatrix} 3 \\ 2 \end{pmatrix} = \begin{pmatrix} 10 & -18 \\ 6 & -11 \end{pmatrix}\begin{pmatrix} 3 \\ 2 \end{pmatrix} = \begin{pmatrix} -6 \\ -4 \end{pmatrix} = -2\begin{pmatrix} 3 \\ 2 \end{pmatrix},$$

so $\lambda_2 = -2$ is an eigenvalue of A with corresponding eigenvector $\mathbf{v}_2 = \begin{pmatrix} 3 \\ 2 \end{pmatrix}$. As we soon see, these are the only eigenvalues of A.

Example 2 Let $A = I$. Then for any $\mathbf{v}$, $A\mathbf{v} = I\mathbf{v} = \mathbf{v}$. Thus 1 is the only eigenvalue of A and every nonzero $\mathbf{v}$ is an eigenvector of I.

We compute the eigenvalues and eigenvectors of many matrices in this section. But first we need to prove some facts that can simplify our computations.

Suppose that λ is an eigenvalue of A. Then there exists a nonzero vector

$$\mathbf{v} = \begin{pmatrix} x_1 \\ x_2 \\ \vdots \\ x_n \end{pmatrix} \neq \mathbf{0}$$

such that $A\mathbf{v} = \lambda\mathbf{v} = \lambda I\mathbf{v}$.

Rewriting this, we have

$$(\lambda I - A)\mathbf{v} = \mathbf{0}. \tag{3}$$

If A is an $n \times n$ matrix, equation (3) is a homogeneous system of n equations in the unknowns $x_1, x_2, \ldots, x_n$. Since, by assumption, the system has nontrivial solutions, we conclude that $\det(\lambda I - A) = 0$ (see Theorem 2 in Appendix 4). Conversely, if $\det(\lambda I - A) = 0$, then (3) has nontrivial solutions and λ is an eigenvalue of A. On the other hand, if $\det(\lambda I - A) \neq 0$, then (3) has only the solution $\mathbf{v} = \mathbf{0}$, so λ is *not* an eigenvalue of A. Summing up these facts, we have the following.

THEOREM 1

Let A be an $n \times n$ matrix. Then λ is an eigenvalue of A if and only if

$$\boxed{p(\lambda) = \det(\lambda I - A) = 0.} \tag{4}$$

Characteristic Equation
Characteristic Polynomial

Equation (4) is called the **characteristic equation** of A and $p(\lambda)$ is called the **characteristic polynomial** of A. Compare this with equation (7.3.12) on p. 275.

As becomes apparent in the examples, $p(\lambda)$ is a polynomial of degree n in λ; for example, if

$$A = \begin{pmatrix} a & b \\ c & d \end{pmatrix},$$

then

$$\lambda I - A = \begin{pmatrix} \lambda & 0 \\ 0 & \lambda \end{pmatrix} - \begin{pmatrix} a & b \\ c & d \end{pmatrix} = \begin{pmatrix} \lambda - a & -b \\ -c & \lambda - d \end{pmatrix}$$

and

$$p(\lambda) = \det(\lambda I - A) = (\lambda - a)(\lambda - d) - bc$$
$$= \lambda^2 - (a+d)\lambda + (ad - bc).$$

Similarly, if

$$A = \begin{pmatrix} a_{11} & a_{12} & \cdots & a_{1n} \\ a_{21} & a_{22} & \cdots & a_{2n} \\ \vdots & \vdots & & \vdots \\ a_{n1} & a_{n2} & \cdots & a_{nn} \end{pmatrix},$$

then

$$p(\lambda) = \det(\lambda I - A) = \begin{vmatrix} \lambda - a_{11} & -a_{12} & \cdots & -a_{1n} \\ -a_{21} & \lambda - a_{22} & \cdots & -a_{2n} \\ \vdots & \vdots & & \vdots \\ -a_{n1} & -a_{n2} & \cdots & \lambda - a_{nn} \end{vmatrix},$$

and $p(\lambda)$ can be written in the form

$$p(\lambda) = \lambda^n + b_{n-1}\lambda^{n-1} + \cdots + b_1\lambda + b_0 = 0. \tag{5}$$

By the fundamental theorem of algebra, any polynomial of degree n with real or complex coefficients has exactly n roots (counting multiplicities). By this we mean, for example, that the polynomial $(\lambda - 1)^5$ has five roots, all equal to the number 1. Since any eigenvalue of A is a root of the characteristic equation of A, an $n \times n$ matrix has n eigenvalues, some of which may be repeated. If $\lambda_1, \lambda_2, \ldots, \lambda_m$ are the distinct roots of equation (5) with multiplicities $r_1, r_2, \ldots, r_m$, respectively, then (5) may be factored to obtain

$$p(\lambda) = (\lambda - \lambda_1)^{r_1}(\lambda - \lambda_2)^{r_2} \cdots (\lambda - \lambda_m)^{r_m} = 0. \tag{6}$$

The numbers $r_1, r_2, \ldots, r_m$ are called the **algebraic multiplicities** of the eigenvalues $\lambda_1, \lambda_2, \ldots, \lambda_m$, respectively.

THEOREM 2

Let $\mathbf{v}_1$ and $\mathbf{v}_2$ be eigenvectors of A corresponding to the same eigenvalue λ. Then $\mathbf{v} = c_1\mathbf{v}_1 + c_2\mathbf{v}_2$ is an eigenvector of A corresponding to λ; that is,

> any linear combination of eigenvectors corresponding to the same eigenvalue is an eigenvector corresponding to that eigenvalue.

Proof
$$A\mathbf{v} = A(c_1\mathbf{v}_1 + c_2\mathbf{v}_2) = c_1 A\mathbf{v}_1 + c_2 A\mathbf{v}_2 = c_1\lambda\mathbf{v}_1 + c_2\lambda\mathbf{v}_2$$
$$= \lambda(c_1\mathbf{v}_1 + c_2\mathbf{v}_2) = \lambda\mathbf{v}. \blacksquare$$

According to Theorem 2, corresponding to each eigenvalue of a square matrix A there are an infinite number of eigenvectors. Clearly, then, listing every eigenvector is infeasible. Rather, our aim is to find as many linearly independent eigenvectors as possible corresponding to each eigenvalue. The following result is very useful and is given without proof.

THEOREM 3

If λ is a simple eigenvalue of A (that is, the algebraic multiplicity of λ is 1), then there is only one linearly independent eigenvector corresponding to λ.

We now prove another useful result.

THEOREM 4

Let A be an $n \times n$ matrix and let $\lambda_1, \lambda_2, \ldots, \lambda_m$ be distinct eigenvalues of A with corresponding eigenvectors $\mathbf{v}_1, \mathbf{v}_2, \ldots, \mathbf{v}_m$. Then $\mathbf{v}_1, \mathbf{v}_2, \ldots, \mathbf{v}_m$ are linearly independent; that is, *eigenvectors corresponding to distinct eigenvalues are linearly independent.*[†]

Proof We prove this by mathematical induction. We start with $m = 2$. Suppose that
$$c_1\mathbf{v}_1 + c_2\mathbf{v}_2 = \mathbf{0}. \tag{7}$$
Then, multiplying both sides of equation (7) by A, we have
$$\mathbf{0} = A(c_1\mathbf{v}_1 + c_2\mathbf{v}_2) = c_1 A\mathbf{v}_1 + c_2 A\mathbf{v}_2,$$
or
$$c_1\lambda_1\mathbf{v}_1 + c_2\lambda_2\mathbf{v}_2 = \mathbf{0}. \tag{8}$$
We then multiply (7) by λ_1 and subtract it from (8) to obtain
$$(c_1\lambda_1\mathbf{v}_1 + c_2\lambda_2\mathbf{v}_2) - (c_1\lambda_1\mathbf{v}_1 + c_2\lambda_1\mathbf{v}_2) = \mathbf{0},$$

[†] The definition of linear independence of vectors is the same as the definition of linear independence of functions (see p. 87).

or
$$c_2(\lambda_2 - \lambda_1)\mathbf{v}_2 = \mathbf{0}.$$

Since $\mathbf{v}_2 \neq \mathbf{0}$ (by the definition of an eigenvector) and since $\lambda_2 \neq \lambda_1$, we conclude that $c_2 = 0$. Then inserting $c_2 = 0$ in (7), we obtain $c_1 = 0$, which proves the theorem in the case $m = 2$. Now suppose that the theorem is true for $m = k$; that is, assume that any k eigenvectors corresponding to distinct eigenvalues are linearly independent. We prove the theorem for $m = k + 1$. So we assume that

$$c_1\mathbf{v}_1 + c_2\mathbf{v}_2 + \cdots + c_k\mathbf{v}_k + c_{k+1}\mathbf{v}_{k+1} = \mathbf{0}. \tag{9}$$

Then, multiplying both sides of equation (9) by A and using the fact that $A\mathbf{v}_i = \lambda_i \mathbf{v}_i$, we obtain

$$c_1\lambda_1\mathbf{v}_1 + c_2\lambda_2\mathbf{v}_2 + \cdots + c_k\lambda_k\mathbf{v}_k + c_{k+1}\lambda_{k+1}\mathbf{v}_{k+1} = \mathbf{0}. \tag{10}$$

We multiply both sides of (9) by λ_{k+1} and subtract it from (10):

$$c_1(\lambda_1 - \lambda_{k+1})\mathbf{v}_1 + c_2(\lambda_2 - \lambda_{k+1})\mathbf{v}_2 + \cdots + c_k(\lambda_k - \lambda_{k+1})\mathbf{v}_k = \mathbf{0}.$$

But, by the induction assumption, $\mathbf{v}_1, \mathbf{v}_2, \ldots, \mathbf{v}_k$ are linearly independent. Thus

$$c_1(\lambda_1 - \lambda_{k+1}) = c_2(\lambda_2 - \lambda_{k+1}) = \cdots = c_k(\lambda_k - \lambda_{k+1}) = 0,$$

and, since $\lambda_i \neq \lambda_{k+1}$ for $i = 1, 2, \ldots, k$, we conclude that $c_1 = c_2 = \cdots = c_k = 0$. But, from (9), this means that $c_{k+1} = 0$. Thus the theorem is true for $m = k + 1$ and the proof is complete. ∎

We now proceed to calculate eigenvalues and corresponding eigenvectors. We do this using a three-step procedure:

TO COMPUTE EIGENVALUES AND EIGENVECTORS

a. Find $p(\lambda) = \det(\lambda I - A)$. (11)

b. Find the roots $\lambda_1, \lambda_2, \ldots, \lambda_m$ of $p(\lambda) = 0$. (12)

c. Corresponding to each eigenvalue λ_i, solve the homogeneous system $(\lambda_i I - A)\mathbf{v} = \mathbf{0}$ or $(A - \lambda_i I)\mathbf{v} = \mathbf{0}$. (13)

Remark Step (b) is often the hardest one to carry out.

Example 3 Let

$$A = \begin{pmatrix} 4 & 2 \\ 3 & 3 \end{pmatrix}.$$

Then

$$\det(\lambda I - A) = \begin{vmatrix} \lambda - 4 & -2 \\ -3 & \lambda - 3 \end{vmatrix} = (\lambda - 4)(\lambda - 3) - 6$$
$$= \lambda^2 - 7\lambda + 6 = (\lambda - 1)(\lambda - 6) = 0.$$

Thus the eigenvalues of A are $\lambda_1 = 1$ and $\lambda_2 = 6$. For $\lambda_1 = 1$, we solve

$(A - I)\mathbf{v} = \mathbf{0}$ or
$$\begin{pmatrix} 3 & 2 \\ 3 & 2 \end{pmatrix} \begin{pmatrix} x_1 \\ x_2 \end{pmatrix} = \begin{pmatrix} 0 \\ 0 \end{pmatrix}.$$

Clearly, any eigenvector corresponding to $\lambda_1 = 1$ satisfies $3x_1 + 2x_2 = 0$, or $x_1 = -2x_2/3$. One such eigenvector is
$$\mathbf{v}_1 = \begin{pmatrix} 2 \\ -3 \end{pmatrix}.$$

According to Theorem 3, we cannot find any other linearly independent eigenvector corresponding to $\lambda = 1$.

Similarly, the equation $(A - 6I)\mathbf{v} = \mathbf{0}$ means that
$$\begin{pmatrix} -2 & 2 \\ 3 & -3 \end{pmatrix} \begin{pmatrix} x_1 \\ x_2 \end{pmatrix} = \begin{pmatrix} 0 \\ 0 \end{pmatrix},$$

or $x_1 = x_2$. Thus $\mathbf{v}_2 = \begin{pmatrix} 1 \\ 1 \end{pmatrix}$ is an eigenvector corresponding to $\lambda_2 = 6$. Note that $\mathbf{v}_1$ and $\mathbf{v}_2$ are linearly independent by Theorem 4.

Example 4 Let
$$A = \begin{pmatrix} 1 & -1 & 4 \\ 3 & 2 & -1 \\ 2 & 1 & -1 \end{pmatrix}.$$

Then
$$\det(\lambda I - A) = \begin{vmatrix} \lambda - 1 & 1 & -4 \\ -3 & \lambda - 2 & 1 \\ -2 & -1 & \lambda + 1 \end{vmatrix} = \lambda^3 - 2\lambda^2 - 5\lambda + 6$$
$$= (\lambda - 1)(\lambda + 2)(\lambda - 3) = 0.$$

Thus the eigenvalues of A are $\lambda_1 = 1$, $\lambda_2 = -2$, and $\lambda_3 = 3$. Since each eigenvalue is simple, we need only find one eigenvector corresponding to each eigenvalue. Corresponding to $\lambda = 1$, we have
$$(A - I)\mathbf{v} = \begin{pmatrix} 0 & -1 & 4 \\ 3 & 1 & -1 \\ 2 & 1 & -2 \end{pmatrix} \begin{pmatrix} x_1 \\ x_2 \\ x_3 \end{pmatrix} = \begin{pmatrix} 0 \\ 0 \\ 0 \end{pmatrix}. \tag{14}$$

We write out the equations in (14):
$$\begin{aligned} 0x_1 - x_2 + 4x_3 &= 0, \\ 3x_1 + x_2 - x_3 &= 0, \\ 2x_1 + x_2 - 2x_3 &= 0. \end{aligned} \tag{15}$$

We solve this system by Gauss-Jordan elimination. The process is to write system (15) as an augmented matrix and then row reduce:

$$\begin{pmatrix} 0 & -1 & 4 & | & 0 \\ 3 & 1 & -1 & | & 0 \\ 2 & 1 & -2 & | & 0 \end{pmatrix} \xrightarrow{\substack{A_{1,2}(1) \\ A_{1,3}(1)}} \begin{pmatrix} 0 & -1 & 4 & | & 0 \\ 3 & 0 & 3 & | & 0 \\ 2 & 0 & 2 & | & 0 \end{pmatrix}$$

$$\xrightarrow{M_2(\frac{1}{3})} \begin{pmatrix} 0 & -1 & 4 & | & 0 \\ 1 & 0 & 1 & | & 0 \\ 2 & 0 & 2 & | & 0 \end{pmatrix} \xrightarrow{A_{2,3}(-2)} \begin{pmatrix} 0 & -1 & 4 & | & 0 \\ 1 & 0 & 1 & | & 0 \\ 0 & 0 & 0 & | & 0 \end{pmatrix}.$$

This denotes the system of equations
$$-x_2 + 4x_3 = 0,$$
$$x_1 \;\;\;\;\;\; + x_3 = 0$$

Thus $x_1 = -x_3$, $x_2 = 4x_3$, and an eigenvector corresponding to $\lambda = 1$ is

$$\mathbf{v}_1 = \begin{pmatrix} -1 \\ 4 \\ 1 \end{pmatrix}.$$

Because of Theorem 3, we know that any other eigenvector of A corresponding to $\lambda = 1$ is a multiple of $\mathbf{v}_1$.

For $\lambda_2 = -2$, we have
$$[A - (-2I)]\mathbf{v} = (A + 2I)\mathbf{v} = \mathbf{0},$$
or
$$\begin{pmatrix} 3 & -1 & 4 \\ 3 & 4 & -1 \\ 2 & 1 & 1 \end{pmatrix} \begin{pmatrix} x_1 \\ x_2 \\ x_3 \end{pmatrix} = \begin{pmatrix} 0 \\ 0 \\ 0 \end{pmatrix}.$$

This leads to

$$\begin{pmatrix} 3 & -1 & 4 & | & 0 \\ 3 & 4 & -1 & | & 0 \\ 2 & 1 & 1 & | & 0 \end{pmatrix} \xrightarrow{\substack{A_{1,2}(4) \\ A_{1,3}(1)}} \begin{pmatrix} 3 & -1 & 4 & | & 0 \\ 15 & 0 & 15 & | & 0 \\ 5 & 0 & 5 & | & 0 \end{pmatrix}$$

$$\xrightarrow{M_2(\frac{1}{15})} \begin{pmatrix} 3 & -1 & 4 & | & 0 \\ 1 & 0 & 1 & | & 0 \\ 5 & 0 & 5 & | & 0 \end{pmatrix} \xrightarrow{\substack{A_{2,1}(-4) \\ A_{2,3}(-5)}} \begin{pmatrix} -1 & -1 & 0 & | & 0 \\ 1 & 0 & 1 & | & 0 \\ 0 & 0 & 0 & | & 0 \end{pmatrix}.$$

Thus $x_2 = -x_1$, $x_3 = -x_1$, and an eigenvector corresponding to $\lambda = -2$ is

$$\mathbf{v}_2 = \begin{pmatrix} 1 \\ -1 \\ -1 \end{pmatrix}.$$

Finally, for $\lambda_3 = 3$, we have

$$(A - 3I)\mathbf{v} = \begin{pmatrix} -2 & -1 & 4 \\ 3 & -1 & -1 \\ 2 & 1 & -4 \end{pmatrix} \begin{pmatrix} x_1 \\ x_2 \\ x_3 \end{pmatrix} = \begin{pmatrix} 0 \\ 0 \\ 0 \end{pmatrix}$$

and

$$\begin{pmatrix} -2 & -1 & 4 & | & 0 \\ 3 & -1 & -1 & | & 0 \\ 2 & 1 & -4 & | & 0 \end{pmatrix} \xrightarrow{\substack{A_{3,1}(1) \\ A_{3,2}(1)}} \begin{pmatrix} 0 & 0 & 0 & | & 0 \\ 5 & 0 & -5 & | & 0 \\ 2 & 1 & -4 & | & 0 \end{pmatrix}$$

$$\xrightarrow{M_2(\frac{1}{5})} \begin{pmatrix} 0 & 0 & 0 & | & 0 \\ 1 & 0 & -1 & | & 0 \\ 2 & 1 & -4 & | & 0 \end{pmatrix} \xrightarrow{A_{2,3}(-4)} \begin{pmatrix} 0 & 0 & 0 & | & 0 \\ 1 & 0 & -1 & | & 0 \\ -2 & 1 & 0 & | & 0 \end{pmatrix}.$$

Thus $x_3 = x_1$, $x_2 = 2x_1$, and an eigenvector corresponding to $\lambda = 3$ is

$$\mathbf{v}_3 = \begin{pmatrix} 1 \\ 2 \\ 1 \end{pmatrix}.$$

Remark In this and every other example, there is always an infinite number of choices for each eigenvector. We arbitrarily choose a simple one by setting one or more of the x_i's equal to 1. Other normalizations are sometimes useful.

In the last example we found three eigenvectors that, according to Theorem 4, must be linearly independent. There is an easy way to verify this. The following result is standard in the theory of determinants.

THEOREM 5

The $n \times 1$ vectors $\mathbf{v}_1, \mathbf{v}_2, \ldots, \mathbf{v}_n$ are linearly independent if and only if the determinant of the matrix whose columns are $\mathbf{v}_1, \mathbf{v}_2, \ldots, \mathbf{v}_n$ is nonzero.

Example 5 In Example 4 we can verify that the vectors

$$\mathbf{v}_1 = \begin{pmatrix} -1 \\ 4 \\ 1 \end{pmatrix}, \quad \mathbf{v}_2 = \begin{pmatrix} 1 \\ -1 \\ -1 \end{pmatrix}, \quad \text{and} \quad \mathbf{v}_3 = \begin{pmatrix} 1 \\ 2 \\ 1 \end{pmatrix}$$

are linearly independent because

$$\begin{vmatrix} -1 & 1 & 1 \\ 4 & -1 & 2 \\ 1 & -1 & 1 \end{vmatrix} = -6 \neq 0.$$

Example 6 Let

$$A = \begin{pmatrix} 3 & -5 \\ 1 & -1 \end{pmatrix}.$$

Then

$$\det(\lambda I - A) = \begin{vmatrix} \lambda - 3 & 5 \\ -1 & \lambda + 1 \end{vmatrix} = \lambda^2 - 2\lambda + 2 = 0.$$

Then

$$\begin{matrix} \lambda_1 \\ \lambda_2 \end{matrix} \bigg\} = \frac{-(-2) \pm \sqrt{4 - 4(1)(2)}}{2} = \frac{2 \pm \sqrt{-4}}{2} = \frac{2 \pm 2i}{2} = 1 \pm i.$$

Thus $\lambda_1 = 1 + i$ and $\lambda_2 = 1 - i$. Then

$$[A - (1+i)I]\mathbf{v} = \begin{pmatrix} 2-i & -5 \\ 1 & -2-i \end{pmatrix} \begin{pmatrix} x_1 \\ x_2 \end{pmatrix} = \begin{pmatrix} 0 \\ 0 \end{pmatrix}$$

and we obtain $(2-i)x_1 - 5x_2 = 0$ and $x_1 + (-2-i)x_2 = 0.^\dagger$ Thus $x_1 = (2+i)x_2$, which yields the eigenvector (corresponding to $\lambda_1 = 1+i$)

$$\mathbf{v}_1 = \begin{pmatrix} 2+i \\ 1 \end{pmatrix}.$$

Similarly,

$$[A - (1-i)I]\mathbf{v} = \begin{pmatrix} 2+i & -5 \\ 1 & -2+i \end{pmatrix} \begin{pmatrix} x_1 \\ x_2 \end{pmatrix} = \begin{pmatrix} 0 \\ 0 \end{pmatrix},$$

---

† Note that

$$(-2-i)\begin{pmatrix} 2-i \\ 1 \end{pmatrix} = \begin{pmatrix} -5 \\ -2-i \end{pmatrix}.$$

or $x_1 + (-2 + i)x_2 = 0$, which yields $x_1 = (2 - i)x_2$, and we obtain the eigenvector (corresponding to $\lambda_2 = 1 - i$)

$$\mathbf{v}_2 = \begin{pmatrix} 2 - i \\ 1 \end{pmatrix}.$$

Remark This example illustrates that a real matrix may have complex eigenvalues and eigenvectors. It should be pointed out that some texts define eigenvalues of real matrices to be the *real* roots of the characteristic equation. With this definition, the matrix of the last example has *no* eigenvalues. This might make the computations simpler, but it also significantly reduces the usefulness of the theory of eigenvalues and eigenvectors. We will see an important illustration of the use of complex eigenvalues in Section 7.9 in our computation of principal matrix solutions.

Example 7 Let

$$A = \begin{pmatrix} 4 & 1 \\ 0 & 4 \end{pmatrix}.$$

Then

$$\det(\lambda I - A) = \begin{vmatrix} \lambda - 4 & -1 \\ 0 & \lambda - 4 \end{vmatrix} = (\lambda - 4)^2 = 0,$$

so $\lambda = 4$ is an eigenvalue of algebraic multiplicity 2 and we have

$$(A - 4I)\mathbf{v} = \begin{pmatrix} 0 & 1 \\ 0 & 0 \end{pmatrix}\begin{pmatrix} x_1 \\ x_2 \end{pmatrix} = \begin{pmatrix} x_2 \\ 0 \end{pmatrix}.$$

Thus $x_2 = 0$ and x_1 is arbitrary. Therefore the only linearly independent eigenvector is

$$\mathbf{v} = \begin{pmatrix} 1 \\ 0 \end{pmatrix}.$$

Here is an instance where a 2×2 matrix has only one linearly independent eigenvector.

Example 8 Let

$$A = \begin{pmatrix} 3 & 2 & 4 \\ 2 & 0 & 2 \\ 4 & 2 & 3 \end{pmatrix}.$$

Then

$$\det(\lambda I - A) = \begin{vmatrix} \lambda - 3 & -2 & -4 \\ -2 & \lambda & -2 \\ -4 & -2 & \lambda - 3 \end{vmatrix}$$

$$= \lambda^3 - 6\lambda^2 - 15\lambda - 8 = (\lambda + 1)^2(\lambda - 8) = 0,$$

so the eigenvalues are $\lambda_1 = 8$ and $\lambda_2 = -1$ (with algebraic multiplicity 2). For $\lambda_1 = 8$, we obtain

$$(A - 8I)\mathbf{v} = \begin{pmatrix} -5 & 2 & 4 \\ 2 & -8 & 2 \\ 4 & 2 & -5 \end{pmatrix}\begin{pmatrix} x_1 \\ x_2 \\ x_3 \end{pmatrix} = \begin{pmatrix} 0 \\ 0 \\ 0 \end{pmatrix},$$

or, row reducing,

$$\begin{pmatrix} -5 & 2 & 4 & | & 0 \\ 2 & -8 & 2 & | & 0 \\ 4 & 2 & -5 & | & 0 \end{pmatrix} \xrightarrow{\substack{A_{1,2}(4) \\ A_{1,3}(-1)}} \begin{pmatrix} -5 & 2 & 4 & | & 0 \\ -18 & 0 & 18 & | & 0 \\ 9 & 0 & -9 & | & 0 \end{pmatrix}$$

$$\xrightarrow{M_2(\frac{1}{18})} \begin{pmatrix} -5 & 2 & 4 & | & 0 \\ -1 & 0 & 1 & | & 0 \\ 9 & 0 & -9 & | & 0 \end{pmatrix} \xrightarrow{A_{2,3}(9)} \begin{pmatrix} 0 & 2 & -1 & | & 0 \\ -1 & 0 & 1 & | & 0 \\ 0 & 0 & 0 & | & 0 \end{pmatrix}.$$

Hence $x_3 = 2x_2$, $x_1 = x_3$, and we obtain an eigenvector

$$\mathbf{v}_1 = \begin{pmatrix} 2 \\ 1 \\ 2 \end{pmatrix}.$$

For $\lambda_2 = -1$, we have

$$(A + I)\mathbf{v} = \begin{pmatrix} 4 & 2 & 4 \\ 2 & 1 & 2 \\ 4 & 2 & 4 \end{pmatrix} \begin{pmatrix} x_1 \\ x_2 \\ x_3 \end{pmatrix} = \begin{pmatrix} 0 \\ 0 \\ 0 \end{pmatrix},$$

which gives us the single equation $2x_1 + x_2 + 2x_3 = 0$ or $x_2 = -2x_1 - 2x_3$. If $x_1 = 1$ and $x_3 = 0$, we obtain

$$\mathbf{v}_2 = \begin{pmatrix} 1 \\ -2 \\ 0 \end{pmatrix}.$$

If $x_1 = 0$ and $x_3 = 1$, we obtain

$$\mathbf{v}_3 = \begin{pmatrix} 0 \\ -2 \\ 1 \end{pmatrix}.$$

Here there are two linearly independent eigenvectors corresponding to an eigenvalue of multiplicity 2.

Example 9 Let

$$A = \begin{pmatrix} -5 & -5 & -9 \\ 8 & 9 & 18 \\ -2 & -3 & -7 \end{pmatrix}.$$

Then

$$\det(\lambda I - A) = \begin{vmatrix} \lambda + 5 & 5 & 9 \\ -8 & \lambda - 9 & -18 \\ 2 & 3 & \lambda + 7 \end{vmatrix} = \lambda^3 + 3\lambda^2 + 3\lambda + 1 = (\lambda + 1)^3 = 0.$$

Thus $\lambda = -1$ is an eigenvalue of algebraic multiplicity 3. To find eigenvectors, we set

$$(A + I)\mathbf{v} = \begin{pmatrix} -4 & -5 & -9 \\ 8 & 10 & 18 \\ -2 & -3 & -6 \end{pmatrix} \begin{pmatrix} x_1 \\ x_2 \\ x_3 \end{pmatrix} = \begin{pmatrix} 0 \\ 0 \\ 0 \end{pmatrix}$$

and row reduce to obtain, successively,

$$\begin{pmatrix} -4 & -5 & -9 & | & 0 \\ 8 & 10 & 18 & | & 0 \\ -2 & -3 & -6 & | & 0 \end{pmatrix} \xrightarrow{\substack{A_{3,1}(-2) \\ A_{3,2}(4)}} \begin{pmatrix} 0 & 1 & 3 & | & 0 \\ 0 & -2 & -6 & | & 0 \\ -2 & -3 & -6 & | & 0 \end{pmatrix}$$

$$\xrightarrow{\substack{A_{1,2}(2) \\ A_{1,3}(3)}} \begin{pmatrix} 0 & 1 & 3 & | & 0 \\ 0 & 0 & 0 & | & 0 \\ -2 & 0 & 3 & | & 0 \end{pmatrix}.$$

This yields $x_2 = -3x_3$ and $2x_1 = 3x_3$. Setting $x_3 = 2$, we obtain only one linearly independent eigenvector:

$$\mathbf{v}_1 = \begin{pmatrix} 3 \\ -6 \\ 2 \end{pmatrix}.$$

Example 10   Let

$$A = \begin{pmatrix} -1 & -3 & -9 \\ 0 & 5 & 18 \\ 0 & -2 & -7 \end{pmatrix}.$$

Then

$$\det(\lambda I - A) = \begin{vmatrix} \lambda + 1 & 3 & 9 \\ 0 & \lambda - 5 & -18 \\ 0 & 2 & \lambda + 7 \end{vmatrix} = (\lambda + 1)^3 = 0.$$

Thus, as in Example 9, $\lambda = -1$ is an eigenvalue of algebraic multiplicity 3. To find eigenvectors, we compute

$$(A + I)\mathbf{v} = \begin{pmatrix} 0 & -3 & -9 \\ 0 & 6 & 18 \\ 0 & -2 & -6 \end{pmatrix} \begin{pmatrix} x_1 \\ x_2 \\ x_3 \end{pmatrix} = \begin{pmatrix} 0 \\ 0 \\ 0 \end{pmatrix}.$$

Thus $-2x_2 - 6x_3 = 0$ or $x_2 = -3x_3$, and x_1 is arbitrary. Setting $x_1 = 0$ and $x_3 = 1$, we obtain

$$\mathbf{v}_1 = \begin{pmatrix} 0 \\ -3 \\ 1 \end{pmatrix}.$$

Setting $x_1 = 1$ and $x_3 = 1$ yields

$$\mathbf{v}_2 = \begin{pmatrix} 1 \\ -3 \\ 1 \end{pmatrix}.$$

Here there are two linearly independent eigenvectors corresponding to an eigenvalue of multiplicity 3.

Example 11   Let

$$A = \begin{pmatrix} -1 & 0 & 0 \\ 0 & -1 & 0 \\ 0 & 0 & -1 \end{pmatrix} = -I.$$

The characteristic equation is $(\lambda + 1)^3 = 0$, and $\lambda = -1$ is an eigenvalue of

algebraic multiplicity 3. Clearly, for any 3×1 vector $\mathbf{v}$,
$$A\mathbf{v} = -I\mathbf{v} = -\mathbf{v} = (-1)\mathbf{v}$$
and $\mathbf{v}(\neq \mathbf{0})$ is an eigenvector. Three linearly independent eigenvectors are
$$\mathbf{v}_1 = \begin{pmatrix} 1 \\ 0 \\ 0 \end{pmatrix}, \quad \mathbf{v}_2 = \begin{pmatrix} 0 \\ 1 \\ 0 \end{pmatrix}, \quad \text{and} \quad \mathbf{v}_3 = \begin{pmatrix} 0 \\ 0 \\ 1 \end{pmatrix}.$$

Remark Examples 9, 10, and 11 illustrate the fact that if λ is an eigenvalue of algebraic multiplicity 3, then there can be one, two, or three linearly independent eigenvectors corresponding to λ. A similar result is true if the multiplicity is n, where $n > 1$.

PROBLEMS 7.8

In Problems 1–20 calculate the eigenvalues and eigenvectors of the given matrix. If the algebraic multiplicity of an eigenvalue is greater than 1, determine the number of linearly independent eigenvectors that correspond to it.

1. $\begin{pmatrix} -2 & -2 \\ -5 & 1 \end{pmatrix}$

2. $\begin{pmatrix} -12 & 7 \\ -7 & 2 \end{pmatrix}$

3. $\begin{pmatrix} 2 & -1 \\ 5 & -2 \end{pmatrix}$

4. $\begin{pmatrix} -3 & 0 \\ 0 & -3 \end{pmatrix}$

5. $\begin{pmatrix} -3 & 2 \\ 0 & -3 \end{pmatrix}$

6. $\begin{pmatrix} 3 & 2 \\ -5 & 1 \end{pmatrix}$

7. $\begin{pmatrix} 1 & -1 & 0 \\ -1 & 2 & -1 \\ 0 & -1 & 1 \end{pmatrix}$

8. $\begin{pmatrix} 1 & 1 & -2 \\ -1 & 2 & 1 \\ 0 & 1 & -1 \end{pmatrix}$

9. $\begin{pmatrix} 5 & 4 & 2 \\ 4 & 5 & 2 \\ 2 & 2 & 2 \end{pmatrix}$

10. $\begin{pmatrix} 1 & 2 & 2 \\ 0 & 2 & 1 \\ -1 & 2 & 2 \end{pmatrix}$

11. $\begin{pmatrix} 0 & 1 & 0 \\ 0 & 0 & 1 \\ 1 & -3 & 3 \end{pmatrix}$

12. $\begin{pmatrix} -3 & -7 & -5 \\ 2 & 4 & 3 \\ 1 & 2 & 2 \end{pmatrix}$

13. $\begin{pmatrix} 1 & -1 & -1 \\ 1 & -1 & 0 \\ 1 & 0 & -1 \end{pmatrix}$

14. $\begin{pmatrix} 7 & -2 & -4 \\ 3 & 0 & -2 \\ 6 & -2 & -3 \end{pmatrix}$

15. $\begin{pmatrix} 4 & 6 & 6 \\ 1 & 3 & 2 \\ -1 & -5 & -2 \end{pmatrix}$

16. $\begin{pmatrix} 4 & 1 & 0 & 1 \\ 2 & 3 & 0 & 1 \\ -2 & 1 & 2 & -3 \\ 2 & -1 & 0 & 5 \end{pmatrix}$

17. $\begin{pmatrix} a & 0 & 0 & 0 \\ 0 & a & 0 & 0 \\ 0 & 0 & a & 0 \\ 0 & 0 & 0 & a \end{pmatrix}$

18. $\begin{pmatrix} a & b & 0 & 0 \\ 0 & a & 0 & 0 \\ 0 & 0 & a & 0 \\ 0 & 0 & 0 & a \end{pmatrix}, \quad b \neq 0$

19. $\begin{pmatrix} a & b & 0 & 0 \\ 0 & a & c & 0 \\ 0 & 0 & a & 0 \\ 0 & 0 & 0 & a \end{pmatrix}, \quad bc \neq 0$

20. $\begin{pmatrix} a & b & 0 & 0 \\ 0 & a & c & 0 \\ 0 & 0 & a & d \\ 0 & 0 & 0 & a \end{pmatrix}, \quad bcd \neq 0$

21. Show that for any real numbers a and b, the matrix
$$A = \begin{pmatrix} a & b \\ -b & a \end{pmatrix}$$
has the eigenvectors
$$\begin{pmatrix} 1 \\ i \end{pmatrix} \quad \text{and} \quad \begin{pmatrix} 1 \\ -i \end{pmatrix}.$$

In Problems 22–28 assume that the matrix A has the eigenvalues $\lambda_1, \lambda_2, \ldots, \lambda_k$.

*22. Show that the eigenvalues of A^t are $\lambda_1, \lambda_2, \ldots, \lambda_k$.†

23. Show that the eigenvalues of αA are $\alpha\lambda_1, \alpha\lambda_2, \ldots, \alpha\lambda_k$.

24. Show that A^{-1} exists if and only if $\lambda_1 \lambda_2 \cdots \lambda_k \neq 0$.

*25. If A^{-1} exists, show that the eigenvalues of A^{-1} are $1/\lambda_1, 1/\lambda_2, \ldots, 1/\lambda_k$.

---

† $A^t =$ the **transpose** of $A =$ the matrix obtained by interchanging the rows and columns of A.

26. Show that the matrix $A - \alpha I$ has the eigenvalues $\lambda_1 - \alpha, \lambda_2 - \alpha, \ldots, \lambda_k - \alpha$.

*27. Show that the eigenvalues of A^2 are $\lambda_1^2, \lambda_2^2, \ldots, \lambda_k^2$.

*28. Show that the eigenvalues of A^m are $\lambda_1^m, \lambda_2^m, \ldots, \lambda_k^m$ for $m = 1, 2, 3, \ldots$.

29. Let λ be an eigenvalue of A with corresponding eigenvector $\mathbf{v}$. Let $p(\lambda) = a_0 + a_1\lambda + a_2\lambda^2 + \cdots + a_n\lambda^n$. Define the matrix $p(A)$ by $p(A) = a_0 I + a_1 A + a_2 A^2 + \cdots + a_n A^n$. Show that $p(A)\mathbf{v} = p(\lambda)\mathbf{v}$.

30. Using the result of Problem 29, show that if $\lambda_1, \lambda_2, \ldots, \lambda_k$ are eigenvalues of A, then $p(\lambda_1), p(\lambda_2), \ldots, p(\lambda_k)$ are eigenvalues of $p(A)$.

31. Show that if A is an upper triangular matrix, then the eigenvalues of A are the diagonal components of A.

32. Let
$$A_1 = \begin{pmatrix} 2 & 0 & 0 & 0 \\ 0 & 2 & 0 & 0 \\ 0 & 0 & 2 & 0 \\ 0 & 0 & 0 & 2 \end{pmatrix}, \quad A_2 = \begin{pmatrix} 2 & 1 & 0 & 0 \\ 0 & 2 & 0 & 0 \\ 0 & 0 & 2 & 0 \\ 0 & 0 & 0 & 2 \end{pmatrix},$$
$$A_3 = \begin{pmatrix} 2 & 1 & 0 & 0 \\ 0 & 2 & 1 & 0 \\ 0 & 0 & 2 & 0 \\ 0 & 0 & 0 & 2 \end{pmatrix}, \text{ and } A_4 = \begin{pmatrix} 2 & 1 & 0 & 0 \\ 0 & 2 & 1 & 0 \\ 0 & 0 & 2 & 1 \\ 0 & 0 & 0 & 2 \end{pmatrix}.$$

Show that, for each matrix, $\lambda = 2$ is an eigenvalue of algebraic multiplicity 4. In each case determine the number of linearly independent eigenvectors that correspond to $\lambda = 2$.

*33. Let A be a real $n \times n$ matrix. Show that if λ_1 is a complex eigenvalue of A with corresponding eigenvector $\mathbf{v}_1$, then $\bar{\lambda}_1$ is an eigenvalue of A with corresponding eigenvector $\bar{\mathbf{v}}_1$. [$\bar{\lambda}_1$ is the complex conjugate of λ_1.]

7.9 Computation of the Principal Matrix Solution

In this section we provide a method for finding the principal matrix solution $\Psi(t)$ for the homogeneous initial-value problem

$$\mathbf{x}'(t) = A\mathbf{x}(t), \quad \mathbf{x}(t_0) = \mathbf{x}_0, \tag{1}$$

where A is a constant matrix. There are many procedures for computing $\Psi(t)$; another method will be given in Section 7.11. Our procedure in this section is to compute $\Psi(t)$ in two steps:

PROCEDURE FOR FINDING THE PRINCIPAL MATRIX SOLUTION

i. Find a fundamental matrix solution $\Phi(t)$.

ii. Find the principal matrix solution $\Psi(t)$ by computing (see p. 310)

$$\Psi(t) = \Phi(t)\Phi^{-1}(t_0). \tag{2}$$

Usually step (i) is the most difficult. However, if the $n \times n$ matrix A has n linearly independent eigenvectors, then the task is no more difficult than computing the eigenvalues and eigenvectors of A, as the following theorem suggests.

THEOREM 1

Suppose that the constant $n \times n$ matrix A has n linearly independent eigenvectors $\mathbf{v}_1, \mathbf{v}_2, \ldots, \mathbf{v}_n$ corresponding to the eigenvalues $\lambda_1, \lambda_2, \ldots, \lambda_n$ (not necessarily distinct), respectively. Then there exists a fundamental set of solutions to system (1) of the form

$$\phi_1(t) = \mathbf{v}_1 e^{\lambda_1 t}, \quad \phi_2(t) = \mathbf{v}_2 e^{\lambda_2 t}, \ldots, \phi_n = \mathbf{v}_n e^{\lambda_n t}, \qquad (3)$$

so the matrix $\Phi(t)$ whose columns are $\phi_1(t), \ldots, \phi_n(t)$ is a fundamental matrix solution for equation (1).

Proof We first show that $\phi_i(t)$ is a solution of equation (1) for $i = 1, 2, \ldots, n$. We have

$$\phi_i'(t) = (\mathbf{v}_i e^{\lambda_i t})' = \lambda_i \mathbf{v}_i e^{\lambda_i t}$$
$$= A \mathbf{v}_i e^{\lambda_i t} = A \phi_i(t),$$

since $\mathbf{v}_i$ is an eigenvector of A corresponding to the eigenvalue λ_i. To show that $\Phi(t)$ is a fundamental matrix solution, we simply note that

$$\det \Phi(0) = \det(\mathbf{v}_1, \mathbf{v}_2, \ldots, \mathbf{v}_n) \neq 0,$$

(by Theorem 7.8.5), since the eigenvectors $\mathbf{v}_i$ are linearly independent. ∎

Using Theorem 1 we can compute a fundamental matrix solution and, therefore, the principal matrix solution for any t_0 when A has n linearly independent eigenvectors. This is always the case when A has n distinct eigenvalues since, according to Theorem 7.8.4, the eigenvectors corresponding to these eigenvalues are linearly independent. The only problem occurs when A has eigenvalues of algebraic multiplicity greater than 1 *and* not as many linearly independent eigenvectors as the multiplicity. We deal with this problem later in this section.

We divide our procedure into three cases. *In all cases we choose $t_0 = 0$.*

Case 1 A has n linearly independent eigenvectors and the eigenvalues of A are all real. Here we simply use Theorem 1.

Example 1 Find the principal matrix solution for the system

$$\begin{pmatrix} x_1 \\ x_2 \end{pmatrix}' = \begin{pmatrix} 4 & 2 \\ 3 & 3 \end{pmatrix} \begin{pmatrix} x_1 \\ x_2 \end{pmatrix}. \qquad (4)$$

Solution In Example 7.8.3 we found that the eigenvalues of this system are $\lambda_1 = 1$ and $\lambda_2 = 6$ with corresponding linearly independent eigenvectors

$$\mathbf{v}_1 = \begin{pmatrix} 2 \\ -3 \end{pmatrix} \quad \text{and} \quad \mathbf{v}_2 = \begin{pmatrix} 1 \\ 1 \end{pmatrix}.$$

Therefore a fundamental set of solutions of equation (4) is

$$\phi_1(t) = \begin{pmatrix} 2 \\ -3 \end{pmatrix} e^t = \begin{pmatrix} 2e^t \\ -3e^t \end{pmatrix} \quad \text{and} \quad \phi_2(t) = \begin{pmatrix} 1 \\ 1 \end{pmatrix} e^{6t} = \begin{pmatrix} e^{6t} \\ e^{6t} \end{pmatrix},$$

so a fundamental matrix solution is

$$\Phi(t) = \begin{pmatrix} 2e^t & e^{6t} \\ -3e^t & e^{6t} \end{pmatrix}.$$

Here

$$\Phi(0) = \begin{pmatrix} 2 & 1 \\ -3 & 1 \end{pmatrix}, \qquad \Phi^{-1}(0) = \frac{1}{5}\begin{pmatrix} 1 & -1 \\ 3 & 2 \end{pmatrix},$$

and, from (2), the principal matrix solution is given by

$$\Psi(t) = \frac{1}{5}\begin{pmatrix} 2e^t & e^{6t} \\ -3e^t & e^{6t} \end{pmatrix}\begin{pmatrix} 1 & -1 \\ 3 & 2 \end{pmatrix} = \frac{1}{5}\begin{pmatrix} 2e^t + 3e^{6t} & -2e^t + 2e^{6t} \\ -3e^t + 3e^{6t} & 3e^t + 2e^{6t} \end{pmatrix}.$$

Example 2 Find the principal matrix solution for the system

$$\begin{pmatrix} x_1 \\ x_2 \\ x_3 \end{pmatrix}' = \begin{pmatrix} 1 & -1 & 4 \\ 3 & 2 & -1 \\ 2 & 1 & -1 \end{pmatrix}\begin{pmatrix} x_1 \\ x_2 \\ x_3 \end{pmatrix}. \tag{5}$$

Solution   As was shown in Example 7.8.4, the eigenvalues of A are $\lambda_1 = 1$, $\lambda_2 = -2$, and $\lambda_3 = 3$ with corresponding eigenvectors

$$\mathbf{v}_1 = \begin{pmatrix} -1 \\ 4 \\ 1 \end{pmatrix}, \qquad \mathbf{v}_2 = \begin{pmatrix} 1 \\ -1 \\ -1 \end{pmatrix}, \qquad \mathbf{v}_3 = \begin{pmatrix} 1 \\ 2 \\ 1 \end{pmatrix}.$$

Therefore a fundamental set of solutions to system (5) is

$$\boldsymbol{\phi}_1(t) = \begin{pmatrix} -1 \\ 4 \\ 1 \end{pmatrix}e^t = \begin{pmatrix} -e^t \\ 4e^t \\ e^t \end{pmatrix}, \qquad \boldsymbol{\phi}_2(t) = \begin{pmatrix} 1 \\ -1 \\ -1 \end{pmatrix}e^{-2t} = \begin{pmatrix} e^{-2t} \\ -e^{-2t} \\ -e^{-2t} \end{pmatrix},$$

and

$$\boldsymbol{\phi}_3(t) = \begin{pmatrix} 1 \\ 2 \\ 1 \end{pmatrix}e^{3t} = \begin{pmatrix} e^{3t} \\ 2e^{3t} \\ e^{3t} \end{pmatrix},$$

yielding the fundamental matrix solution

$$\Phi(t) = \begin{pmatrix} -e^t & e^{-2t} & e^{3t} \\ 4e^t & -e^{-2t} & 2e^{3t} \\ e^t & -e^{-2t} & e^{3t} \end{pmatrix}.$$

Here

$$\Phi(0) = \begin{pmatrix} -1 & 1 & 1 \\ 4 & -1 & 2 \\ 1 & -1 & 1 \end{pmatrix}, \qquad \Phi^{-1}(0) = -\frac{1}{6}\begin{pmatrix} 1 & -2 & 3 \\ -2 & -2 & 6 \\ -3 & 0 & -3 \end{pmatrix}, \qquad \text{and}$$

$$\Psi(t) = -\frac{1}{6}\begin{pmatrix} -e^t & e^{-2t} & e^{3t} \\ 4e^t & -e^{-2t} & 2e^{3t} \\ e^t & -e^{-2t} & e^{3t} \end{pmatrix}\begin{pmatrix} 1 & -2 & 3 \\ -2 & -2 & 6 \\ -3 & 0 & -3 \end{pmatrix}$$

$$= -\frac{1}{6}\begin{pmatrix} -e^t - 2e^{-2t} - 3e^{3t} & 2e^t - 2e^{-2t} & -3e^t + 6e^{-2t} - 3e^{3t} \\ 4e^t + 2e^{-2t} - 6e^{3t} & -8e^t + 2e^{-2t} & 12e^t - 6e^{-2t} - 6e^{3t} \\ e^t + 2e^{-2t} - 3e^{3t} & -2e^t + 2e^{-2t} & 3e^t - 6e^{-2t} - 3e^{3t} \end{pmatrix}$$

Example 3 **a.** Find the principal matrix solution for the system

$$\begin{pmatrix} x_1 \\ x_2 \\ x_3 \end{pmatrix}' = \begin{pmatrix} 3 & 2 & 4 \\ 2 & 0 & 2 \\ 4 & 2 & 3 \end{pmatrix} \begin{pmatrix} x_1 \\ x_2 \\ x_3 \end{pmatrix}. \tag{6}$$

b. Find the particular solution that satisfies

$$\mathbf{x}(0) = \begin{pmatrix} 2 \\ -1 \\ 3 \end{pmatrix}.$$

Solution **a.** In Example 7.8.8 we found the simple eigenvalue $\lambda_1 = 8$ with eigenvector

$$\mathbf{v}_1 = \begin{pmatrix} 2 \\ 1 \\ 2 \end{pmatrix}$$

and the eigenvalue $\lambda_2 = -1$ of multiplicity 2 with the eigenvectors

$$\mathbf{v}_2 = \begin{pmatrix} 1 \\ -2 \\ 0 \end{pmatrix} \quad \text{and} \quad \mathbf{v}_3 = \begin{pmatrix} 0 \\ -2 \\ 1 \end{pmatrix}.$$

Therefore a fundamental set of solutions to system (6) is

$$\boldsymbol{\phi}_1(t) = e^{8t} \begin{pmatrix} 2 \\ 1 \\ 2 \end{pmatrix} = \begin{pmatrix} 2e^{8t} \\ e^{8t} \\ 2e^{8t} \end{pmatrix}, \quad \boldsymbol{\phi}_2(t) = e^{-t} \begin{pmatrix} 1 \\ -2 \\ 0 \end{pmatrix} = \begin{pmatrix} e^{-t} \\ -2e^{-t} \\ 0 \end{pmatrix}, \quad \text{and}$$

$$\boldsymbol{\phi}_3(t) = e^{-t} \begin{pmatrix} 0 \\ -2 \\ 1 \end{pmatrix} = \begin{pmatrix} 0 \\ -2e^{-t} \\ e^{-t} \end{pmatrix}.$$

A fundamental matrix solution is

$$\Phi(t) = \begin{pmatrix} 2e^{8t} & e^{-t} & 0 \\ e^{8t} & -2e^{-t} & -2e^{-t} \\ 2e^{8t} & 0 & e^{-t} \end{pmatrix}.$$

Here

$$\Phi(0) = \begin{pmatrix} 2 & 1 & 0 \\ 1 & -2 & -2 \\ 2 & 0 & 1 \end{pmatrix}, \quad \Phi^{-1}(0) = -\frac{1}{9} \begin{pmatrix} -2 & -1 & -2 \\ -5 & 2 & 4 \\ 4 & 2 & -5 \end{pmatrix}, \quad \text{and}$$

$$\Psi(t) = -\frac{1}{9} \begin{pmatrix} 2e^{8t} & e^{-t} & 0 \\ e^{8t} & -2e^{-t} & -2e^{-t} \\ 2e^{8t} & 0 & e^{-t} \end{pmatrix} \begin{pmatrix} -2 & -1 & -2 \\ -5 & 2 & 4 \\ 4 & 2 & -5 \end{pmatrix}$$

$$= -\frac{1}{9} \begin{pmatrix} -4e^{8t} - 5e^{-t} & -2e^{8t} + 2e^{-t} & -4e^{8t} + 4e^{-t} \\ -2e^{8t} + 2e^{-t} & -e^{8t} - 8e^{-t} & -2e^{8t} + 2e^{-t} \\ -4e^{8t} + 4e^{-t} & -4e^{8t} + 4e^{-t} & -4e^{8t} - 5e^{-t} \end{pmatrix}.$$

Remark This example illustrates the fact that the technique suggested in Theorem 1 works even when A has fewer than n distinct eigenvalues (here $n = 3$), provided that there are n linearly independent eigenvectors.

b. Let $\phi(t) = \Psi(t)\mathbf{x}_0$. Then $\phi(0) = \Psi(0)\mathbf{x}_0 = I\mathbf{x}_0 = \mathbf{x}_0$, since $\Psi(t)$ is the principal matrix solution. Thus the solution to the initial-value problem is

$$\phi(t) = \Psi(t)\mathbf{x}_0$$

$$= -\frac{1}{9}\begin{pmatrix} -4e^{8t} - 5e^{-t} & -2e^{8t} + 2e^{-t} & -4e^{8t} + 4e^{-t} \\ -2e^{8t} + 2e^{-t} & -e^{8t} - 8e^{-t} & -2e^{8t} + 2e^{-t} \\ -4e^{8t} + 4e^{-t} & -4e^{8t} + 4e^{-t} & -4e^{8t} - 5e^{-t} \end{pmatrix}\begin{pmatrix} 2 \\ -1 \\ 3 \end{pmatrix},$$

or

$$\phi(t) = -\frac{1}{9}\begin{pmatrix} -18e^{8t} \\ -9e^{8t} + 18e^{-t} \\ -16e^{8t} - 11e^{-t} \end{pmatrix} = \begin{pmatrix} 2e^{8t} \\ e^{8t} - 2e^{-t} \\ \frac{16}{9}e^{8t} + \frac{11}{9}e^{-t} \end{pmatrix}.$$

Remark We proved the following fact in the last example:

> The unique solution to the initial-value problem
> $$\mathbf{x}' = A\mathbf{x}, \qquad \mathbf{x}(t_0) = \mathbf{x}_0$$
> is given by
> $$\phi(t) = \Psi(t)\mathbf{x}_0, \tag{7}$$
> where $\Psi(t)$ is the principal matrix solution relative to the initial point t_0.

Case 2 A has n linearly independent eigenvectors and some of the eigenvalues are complex conjugates. This case is not very different from case 1, except that now we have to deal with complex numbers. Several facts simplify our answers. From the Euler formula (see equation (14) in Appendix 5),

$$e^{(\alpha + i\beta)t} = e^{\alpha t}e^{i\beta t} = e^{\alpha t}(\cos \beta t + i \sin \beta t) \tag{8}$$

and

$$e^{(\alpha - i\beta)t} = e^{\alpha t}e^{-i\beta t} = e^{\alpha t}(\cos \beta t - i \sin \beta t). \tag{9}$$

Adding (8) and (9) and dividing by 2 yields

$$e^{\alpha t}\cos \beta t = \tfrac{1}{2}[e^{(\alpha + i\beta)t} + e^{(\alpha - i\beta)t}]. \tag{10}$$

Subtracting (9) from (8) and dividing by $2i$ results in

$$e^{\alpha t}\sin \beta t = \frac{1}{2i}[e^{(\alpha + i\beta)t} - e^{(\alpha - i\beta)t}]. \tag{11}$$

If A has real entries, then the eigenvalues of A occur in complex conjugate pairs. If $\alpha \pm i\beta$ is such a pair, then the corresponding columns of the fundamental matrix solution have entries of the form $ce^{(\alpha+i\beta)t}$ and $de^{(\alpha-i\beta)t}$. When we multiply the fundamental matrix solution on the right by $\Phi^{-1}(0)$ it would be nice if, in the resulting principal matrix solution, only terms involving $e^{\alpha t}\cos\beta t$ and $e^{\alpha t}\sin\beta t$ occurred. In fact, this always happens! The following remarkable fact will be proved in Section 7.11.

THEOREM 2

Let the real matrix A have the complex conjugate eigenvalues $\alpha \pm i\beta$. Then the principal matrix solution of $\mathbf{x}' = A\mathbf{x}$ contains no complex terms; that is, the terms $e^{(\alpha+i\beta)t}$ and $e^{(\alpha-i\beta)t}$ always appear in pairs in such a way that they can be expressed as a *real* linear combination of the terms $e^{\alpha t}\cos\beta t$ and $e^{\alpha t}\sin\beta t$.

We illustrate this with an example.

Example 4 Find the principal matrix solution of the system

$$\begin{pmatrix} x_1 \\ x_2 \end{pmatrix}' = \begin{pmatrix} 3 & -5 \\ 1 & -1 \end{pmatrix} \begin{pmatrix} x_1 \\ x_2 \end{pmatrix}. \tag{12}$$

Solution As was shown in Example 7.8.6, the eigenvalues of this system are $\lambda_1 = 1 + i$ and $\lambda_2 = 1 - i$, with corresponding eigenvectors

$$\mathbf{v}_1 = \begin{pmatrix} 2+i \\ 1 \end{pmatrix} \quad \text{and} \quad \mathbf{v}_2 = \begin{pmatrix} 2-i \\ 1 \end{pmatrix}. \quad \text{Note that } \mathbf{v}_1 = \bar{\mathbf{v}}_2$$

A fundamental set of solutions to system (12) is

$$\boldsymbol{\phi}_1(t) = \begin{pmatrix} 2+i \\ 1 \end{pmatrix} e^{(1+i)t} = \begin{pmatrix} (2+i)e^{(1+i)t} \\ e^{(1+i)t} \end{pmatrix},$$

and

$$\boldsymbol{\phi}_2(t) = \begin{pmatrix} 2-i \\ 1 \end{pmatrix} e^{(1-i)t} = \begin{pmatrix} (2-i)e^{(1-i)t} \\ e^{(1-i)t} \end{pmatrix},$$

with fundamental matrix solution

$$\Phi(t) = \begin{pmatrix} (2+i)e^{(1+i)t} & (2-i)e^{(1-i)t} \\ e^{(1+i)t} & e^{(1-i)t} \end{pmatrix}.$$

Here

$$\Phi(0) = \begin{pmatrix} 2+i & 2-i \\ 1 & 1 \end{pmatrix} \quad \text{and} \quad \Phi^{-1}(0) = \frac{1}{2i}\begin{pmatrix} 1 & -2+i \\ -1 & 2+i \end{pmatrix},$$

so

$$\Psi(t) = \frac{1}{2i}\begin{pmatrix} (2+i)e^{(1+i)t} & (2-i)e^{(1-i)t} \\ e^{(1+i)t} & e^{(1-i)t} \end{pmatrix} \begin{pmatrix} 1 & -2+i \\ -1 & 2+i \end{pmatrix}.$$

Before continuing, we note that

$$(2+i)(-2+i) = -4 + i^2 = -5 \quad \text{and} \quad (2-i)(2+i) = 4 - i^2 = 5.$$

Then

$$\Psi(t) = \frac{1}{2i}\begin{pmatrix} (2+i)e^{(1+i)t} - (2-i)e^{(1-i)t} & -5e^{(1+i)t} + 5e^{(1-i)t} \\ e^{(1+i)t} - e^{(1-i)t} & (-2+i)e^{(1+i)t} + (2+i)e^{(1-i)t} \end{pmatrix}$$

$$= \begin{pmatrix} 2\left(\dfrac{e^{(1+i)t} - e^{(1-i)t}}{2i}\right) + \left(\dfrac{e^{(1+i)t} + e^{(1-i)t}}{2}\right) & -5\left(\dfrac{e^{(1+i)t} - e^{(1-i)t}}{2i}\right) \\ \left(\dfrac{e^{(1+i)t} - e^{(1-i)t}}{2i}\right) & -2\left(\dfrac{e^{(1+i)t} - e^{(1-i)t}}{2i}\right) + \left(\dfrac{e^{(1+i)t} + e^{(1-i)t}}{2}\right) \end{pmatrix}$$

from (10) and (11)

$$\downarrow$$

$$= \begin{pmatrix} 2e^t \sin t + e^t \cos t & -5e^t \sin t \\ e^t \sin t & -2e^t \sin t + e^t \cos t \end{pmatrix}$$

$$= e^t \begin{pmatrix} 2\sin t + \cos t & -5\sin t \\ \sin t & \cos t - 2\sin t \end{pmatrix}.$$

Case 3: The matrix A has fewer than n linearly independent eigenvectors
This situation is more complicated than the preceding one. To motivate our procedure, we begin with an example.

Example 5 Find the principal matrix solution for the system

$$\begin{pmatrix} x_1 \\ x_2 \end{pmatrix}' = \begin{pmatrix} 7 & -1 \\ 9 & 1 \end{pmatrix}\begin{pmatrix} x_1 \\ x_2 \end{pmatrix}. \tag{13}$$

Solution

$$\det(\lambda I - A) = \begin{vmatrix} \lambda - 7 & 1 \\ -9 & \lambda - 1 \end{vmatrix} = (\lambda - 7)(\lambda - 1) + 9 = \lambda^2 - 8\lambda + 16 = (\lambda - 4)^2.$$

Thus $\lambda = 4$ is the only eigenvalue of A. It has algebraic multiplicity 2. Then

$$(A - \lambda I)\begin{pmatrix} x_1 \\ x_2 \end{pmatrix} = \begin{pmatrix} 3 & -1 \\ 9 & -3 \end{pmatrix}\begin{pmatrix} x_1 \\ x_2 \end{pmatrix} = \begin{pmatrix} 0 \\ 0 \end{pmatrix},$$

which implies that $x_2 = 3x_1$ and that $\mathbf{v}_1 = \begin{pmatrix} 1 \\ 3 \end{pmatrix}$ is an eigenvector. There is no other linearly independent eigenvector. One solution to (13) is, therefore,

$$\boldsymbol{\phi}_1(t) = e^{4t}\begin{pmatrix} 1 \\ 3 \end{pmatrix} = \begin{pmatrix} e^{4t} \\ 3e^{4t} \end{pmatrix}.$$

Recall that when the characteristic equation of a second-order differential equation has a double root λ, then the general solution to the equation takes the form

$$y(t) = (c_1 + c_2 t)e^{\lambda t}.$$

Hence, in order to find a second linearly independent solution, it is reasonable to seek two *vectors*, $\mathbf{u}$ and $\mathbf{w}$, such that

$$\boldsymbol{\phi}_2 = e^{4t}\mathbf{u} + te^{4t}\mathbf{w} \tag{14}$$

is a solution to (13). (It might be tempting to seek a solution having the form $te^{4t}\mathbf{w}$, but this does not work. Try it.) Inserting (14) into the equation $\mathbf{x}' = A\mathbf{x}$

yields
$$\phi'_2 = 4e^{4t}\mathbf{u} + (1+4t)e^{4t}\mathbf{w} = A\phi_2 = e^{4t}A\mathbf{u} + te^{4t}A\mathbf{w}.$$
We equate coefficients of e^{4t} and te^{4t} to obtain
$$4\mathbf{u} + \mathbf{w} = A\mathbf{u} \tag{15}$$
and
$$4\mathbf{w} = A\mathbf{w}. \tag{16}$$
Rewriting (15) and (16) leads to the equations
$$(A - 4I)\mathbf{w} = \mathbf{0}, \tag{17}$$
$$(A - 4I)\mathbf{u} = \mathbf{w}. \tag{18}$$

From (17), we see that $\mathbf{w}$ is an eigenvector of A. Any vector $\mathbf{u}$ that satisfies (18) is called a **generalized eigenvector** of A. In this case, if
$$\mathbf{u} = \begin{pmatrix} u_1 \\ u_2 \end{pmatrix} \quad \text{and} \quad \mathbf{w} = \begin{pmatrix} 1 \\ 3 \end{pmatrix},$$
we obtain, from (18),
$$\begin{pmatrix} 3 & -1 \\ 9 & -3 \end{pmatrix} \begin{pmatrix} u_1 \\ u_2 \end{pmatrix} = \begin{pmatrix} 1 \\ 3 \end{pmatrix}.$$
Thus
$$3u_1 - u_2 = 1$$
$$9u_1 - 3u_2 = 3.$$
Then $u_1 = (1 + u_2)/3$, and (setting $u_2 = 2$) a generalized eigenvector of A is $\mathbf{u} = \begin{pmatrix} 1 \\ 2 \end{pmatrix}$. Thus a second linearly independent solution to (13) is given by
$$\phi_2 = e^{4t}(\mathbf{u} + t\mathbf{w}) = e^{4t}\left[\begin{pmatrix} 1 \\ 2 \end{pmatrix} + t\begin{pmatrix} 1 \\ 3 \end{pmatrix}\right] = e^{4t}\begin{pmatrix} 1+t \\ 2+3t \end{pmatrix}.$$
A fundamental matrix solution is given by
$$\Phi(t) = e^{4t}\begin{pmatrix} 1 & 1+t \\ 3 & 2+3t \end{pmatrix}.$$
Finally,
$$\Phi(0) = \begin{pmatrix} 1 & 1 \\ 3 & 2 \end{pmatrix} \quad \text{and} \quad \Phi^{-1}(0) = \begin{pmatrix} -2 & 1 \\ 3 & -1 \end{pmatrix},$$
so
$$\Psi(t) = e^{4t}\begin{pmatrix} 1 & 1+t \\ 3 & 2+3t \end{pmatrix}\begin{pmatrix} -2 & 1 \\ 3 & -1 \end{pmatrix} = e^{4t}\begin{pmatrix} 1+3t & -t \\ 9t & 1-3t \end{pmatrix}.$$

In the last example we sought a solution to the equation $(A - \lambda I)\mathbf{u} = \mathbf{w}$, where $\mathbf{w}$ was an eigenvector of A. The following theorem asserts that, under certain conditions, this equation always has a solution. The proof is difficult and is omitted.

THEOREM 3

Let λ be an eigenvalue of A of algebraic multiplicity 2. Suppose that $\mathbf{w}$ is an eigenvector of A corresponding to λ and that there is no other linearly independent eigenvector of A corresponding to λ. Then the equation $(A - \lambda I)\mathbf{u} = \mathbf{w}$ has a solution.

The technique of Example 5 can be generalized. We indicate how in the next example.

Example 6 Find the principal matrix solution for the system

$$\begin{pmatrix} x_1 \\ x_2 \\ x_3 \end{pmatrix}' = \begin{pmatrix} -5 & -5 & -9 \\ 8 & 9 & 18 \\ -2 & -3 & -7 \end{pmatrix} \begin{pmatrix} x_1 \\ x_2 \\ x_3 \end{pmatrix}. \tag{19}$$

Solution In Example 7.8.9 we saw that A had the single eigenvalue $\lambda = -1$ of algebraic multiplicity 3 and the single (independent) eigenvector

$$\mathbf{v}_1 = \begin{pmatrix} 3 \\ -6 \\ 2 \end{pmatrix}.$$

Thus one solution is

$$\boldsymbol{\phi}_1 = e^{-t} \begin{pmatrix} 3 \\ -6 \\ 2 \end{pmatrix}.$$

Using the multiple root theory of Section 3.8 as a model, we seek two other solutions having the forms

$$\boldsymbol{\phi}_2 = e^{-t}(\mathbf{u}_1 + t\mathbf{u}_2) \tag{20}$$

and

$$\boldsymbol{\phi}_3 = e^{-t}\left(\mathbf{w}_1 + t\mathbf{w}_2 + \frac{t^2}{2}\mathbf{w}_3\right). \tag{21}$$

Substituting (20) into $\mathbf{x}' = A\mathbf{x}$, we obtain

$$(A + I)\mathbf{u}_2 = 0,$$
$$(A + I)\mathbf{u}_1 = \mathbf{u}_2.$$

Then

$$\mathbf{u}_2 = \begin{pmatrix} 3 \\ -6 \\ 2 \end{pmatrix}$$

(because it is an eigenvector), and

$$\mathbf{u}_1 = \begin{pmatrix} x_1 \\ x_2 \\ x_3 \end{pmatrix}$$

satisfies

$$\begin{pmatrix} -4 & -5 & -9 \\ 8 & 10 & 18 \\ -2 & -3 & -6 \end{pmatrix} \begin{pmatrix} x_1 \\ x_2 \\ x_3 \end{pmatrix} = \begin{pmatrix} 3 \\ -6 \\ 2 \end{pmatrix}. \tag{22}$$

One solution to (22) is found by row reduction:

$$\begin{pmatrix} -4 & -5 & -9 & | & 3 \\ 8 & 10 & 18 & | & -6 \\ -2 & -3 & -6 & | & 2 \end{pmatrix} \xrightarrow{\substack{A_{3,1}(-2) \\ A_{3,2}(4)}} \begin{pmatrix} 0 & 1 & 3 & | & -1 \\ 0 & -2 & -6 & | & 2 \\ -2 & -3 & -6 & | & 2 \end{pmatrix}$$

$$\xrightarrow{\substack{A_{1,2}(2) \\ A_{1,3}(3)}} \begin{pmatrix} 0 & 1 & 3 & | & -1 \\ 0 & 0 & 0 & | & 0 \\ -2 & 0 & 3 & | & -1 \end{pmatrix}.$$

Hence $x_2 = -1 - 3x_3$ and $-2x_1 = -1 - 3x_3$. One solution (obtained by setting $x_3 = 1$) is

$$\mathbf{u}_1 = \begin{pmatrix} 2 \\ -4 \\ 1 \end{pmatrix}.$$

Then

$$\boldsymbol{\phi}_2 = e^{-t}\left[\begin{pmatrix} 2 \\ -4 \\ 1 \end{pmatrix} + t\begin{pmatrix} 3 \\ -6 \\ 2 \end{pmatrix}\right] = e^{-t}\begin{pmatrix} 2 + 3t \\ -4 - 6t \\ 1 + 2t \end{pmatrix}.$$

Inserting (21) into $\mathbf{x}' = A\mathbf{x}$ yields, after simplification,

$$(A + I)\mathbf{w}_3 = \mathbf{0},$$
$$(A + I)\mathbf{w}_2 = \mathbf{w}_3,$$
$$(A + I)\mathbf{w}_1 = \mathbf{w}_2.$$

Leaving out the details, we obtain (since $\mathbf{w}_3$ is an eigenvector)

$$\mathbf{w}_3 = \begin{pmatrix} 3 \\ -6 \\ 2 \end{pmatrix}, \quad \mathbf{w}_2 = \begin{pmatrix} 2 \\ -4 \\ 1 \end{pmatrix}, \quad \text{and} \quad \mathbf{w}_1 = \begin{pmatrix} 1 \\ -3 \\ 1 \end{pmatrix},$$

so

$$\boldsymbol{\phi}_3(t) = e^{-t}\left(\mathbf{w}_1 + t\mathbf{w}_2 + \frac{t^2}{2}\mathbf{w}_3\right) = e^{-t}\begin{pmatrix} 1 + 2t + \frac{3}{2}t^2 \\ -3 - 4t - 3t^2 \\ 1 + t + t^2 \end{pmatrix}.$$

Thus a fundamental matrix solution is

$$\Phi(t) = e^{-t}\begin{pmatrix} 3 & 2 + 3t & 1 + 2t + \frac{3}{2}t^2 \\ -6 & -4 - 6t & -3 - 4t - 3t^2 \\ 2 & 1 + 2t & 1 + t + t^2 \end{pmatrix}.$$

Finally,

$$\Phi(0) = \begin{pmatrix} 3 & 2 & 1 \\ -6 & -4 & -3 \\ 2 & 1 & 1 \end{pmatrix} \quad \text{and} \quad \Phi^{-1}(0) = \begin{pmatrix} 1 & 1 & 2 \\ 0 & -1 & -3 \\ -2 & -1 & 0 \end{pmatrix},$$

so

$$\Psi(t) = e^{-t}\begin{pmatrix} 3 & 2 + 3t & 1 + 2t + \frac{3}{2}t^2 \\ -6 & -4 - 6t & -3 - 4t - 3t^2 \\ 2 & 1 + 2t & 1 + t + t^2 \end{pmatrix}\begin{pmatrix} 1 & 1 & 2 \\ 0 & -1 & -3 \\ -2 & -1 & 0 \end{pmatrix}$$

$$= e^{-t}\begin{pmatrix} 1 - 4t - 3t^2 & -5t - \frac{3}{2}t^2 & -9t \\ 8t + 6t^2 & 1 + 10t + 3t^2 & 18t \\ -2t - 2t^2 & -3t - t^2 & 1 - 6t \end{pmatrix}.$$

Remark The technique in the last two examples works (although it is tedious) whenever A has a multiple eigenvalue and one eigenvector. If it has a multiple root and more than one eigenvector, and the number of linearly independent eigenvectors is less than the algebraic multiplicity of the eigenvalue (as in Example 7.8.10), then some other technique must be used.

There are some very elegant techniques that work in all cases. Two of these involve the Jordan canonical form of a matrix and the Cayley-Hamilton theorem. We do not discuss Jordan canonical form here, primarily because it involves more advanced matrix theory and because our methods work in all but the most exceptional cases. We will discuss the Cayley-Hamilton theorem in Section 7.11.

PROBLEMS 7.9

Use the methods of this section to calculate the principal matrix solution of the system $\mathbf{x}' = A\mathbf{x}$, where A is the given constant matrix. If initial conditions are given, find the solution of the initial-value problem.

1. $\begin{pmatrix} -2 & -2 \\ -5 & 1 \end{pmatrix}$

2. $\begin{pmatrix} -12 & 7 \\ -7 & 2 \end{pmatrix}$

3. $\begin{pmatrix} 2 & -1 \\ 5 & -2 \end{pmatrix}$

4. $\begin{pmatrix} 3 & 2 \\ -5 & 1 \end{pmatrix}$

5. $\begin{pmatrix} 3 & -2 \\ 8 & -5 \end{pmatrix}$

6. $\begin{pmatrix} 1 & 1 \\ 1 & -1 \end{pmatrix}$

7. $\begin{pmatrix} 3 & -2 \\ 8 & -5 \end{pmatrix}$, $\mathbf{x}(0) = \begin{pmatrix} \frac{3}{4} \\ 1 \end{pmatrix}$

8. $\begin{pmatrix} 3 & -2 \\ 8 & -5 \end{pmatrix}$, $\mathbf{x}(0) = \begin{pmatrix} 1 \\ 2 \end{pmatrix}$

9. $\begin{pmatrix} 3 & -2 \\ 8 & -5 \end{pmatrix}$, $\mathbf{x}(0) = \begin{pmatrix} 2 \\ 5 \end{pmatrix}$

10. $\begin{pmatrix} 4 & 1 \\ -8 & 8 \end{pmatrix}$, $\mathbf{x}(0) = \begin{pmatrix} 1 \\ 0 \end{pmatrix}$

11. $\begin{pmatrix} 1 & -1 & 0 \\ -1 & 2 & -1 \\ 0 & -1 & 1 \end{pmatrix}$, $\mathbf{x}(0) = \begin{pmatrix} 1 \\ -2 \\ 3 \end{pmatrix}$

12. $\begin{pmatrix} 1 & 1 & -2 \\ -1 & 2 & 1 \\ 0 & 1 & -1 \end{pmatrix}$, $\mathbf{x}(0) = \begin{pmatrix} 2 \\ 5 \\ 7 \end{pmatrix}$

13. $\begin{pmatrix} 4 & 6 & 6 \\ 1 & 3 & 2 \\ -1 & -5 & -2 \end{pmatrix}$

14. $\begin{pmatrix} 7 & -2 & -4 \\ 3 & 0 & -2 \\ 6 & -2 & -3 \end{pmatrix}$

15. $\begin{pmatrix} 5 & 4 & 2 \\ 4 & 5 & 2 \\ 2 & 2 & 2 \end{pmatrix}$

16. $\begin{pmatrix} -3 & 0 & 2 \\ 1 & -1 & 0 \\ -2 & -1 & 0 \end{pmatrix}$

17. $\begin{pmatrix} 4 & 1 & 0 & 1 \\ 2 & 3 & 0 & 1 \\ -2 & 1 & 2 & -3 \\ 2 & -1 & 0 & 5 \end{pmatrix}$

18. $\begin{pmatrix} 0 & -1 & -2 \\ 1 & 0 & 1 \\ 2 & -1 & 0 \end{pmatrix}$

19. $\begin{pmatrix} 1 & -1 & 0 \\ -1 & 2 & -1 \\ 0 & -1 & 1 \end{pmatrix}$, $\mathbf{x}(0) = \begin{pmatrix} 1 \\ 0 \\ 1 \end{pmatrix}$

20. $\begin{pmatrix} 5 & 4 & 2 \\ 4 & 5 & 2 \\ 2 & 2 & 2 \end{pmatrix}$, $\mathbf{x}(0) = \begin{pmatrix} 1 \\ 2 \\ 3 \end{pmatrix}$

21. $\begin{pmatrix} 4 & 6 & 6 \\ 1 & 3 & 2 \\ -1 & -5 & -2 \end{pmatrix}$, $\mathbf{x}(0) = \begin{pmatrix} -1 \\ 0 \\ 2 \end{pmatrix}$

22. $\begin{pmatrix} 4 & 1 & 0 & 1 \\ 2 & 3 & 0 & 1 \\ -2 & 1 & 2 & -3 \\ 2 & -1 & 0 & 5 \end{pmatrix}$, $\mathbf{x}(0) = \begin{pmatrix} 4 \\ 5 \\ 6 \\ 2 \end{pmatrix}$

7.10 Nonhomogeneous Systems

We now present a method for solving the nonhomogeneous system

$$\mathbf{x}' = A(t)\mathbf{x} + \mathbf{f}(t), \tag{1}$$

given that a fundamental matrix solution $\Phi(t)$ for the homogeneous system

$$\mathbf{x}' = A(t)\mathbf{x} \tag{2}$$

is known. Such a solution can be found if $A(t)$ is a constant matrix (by the methods of Section 7.9).

Throughout this section we assume that the components of $A(t)$ and $\mathbf{f}(t)$ are functions continuous on some common interval containing the point t_0.

THEOREM 1

Let $\boldsymbol{\phi}_p(t)$ and $\boldsymbol{\phi}_q(t)$ be two solutions of system (1). Then their difference,

$$\boldsymbol{\phi}(t) = \boldsymbol{\phi}_p(t) - \boldsymbol{\phi}_q(t),$$

is a solution of equation (2).

Proof $\boldsymbol{\phi}' = (\boldsymbol{\phi}_p - \boldsymbol{\phi}_q)' = (A\boldsymbol{\phi}_p + \mathbf{f}) - (A\boldsymbol{\phi}_q + \mathbf{f}) = A(\boldsymbol{\phi}_p - \boldsymbol{\phi}_q) = A\boldsymbol{\phi}$. Thus, as in the case of linear scalar equations (Chapter 3), it is necessary to find only one particular solution of equation (1). ∎

If $\boldsymbol{\phi}_p(t)$ is such a solution, then *the general solution of the nonhomogeneous system* (1) *is of the form*

$$\boldsymbol{\phi}(t) = \Phi(t)\mathbf{c} + \boldsymbol{\phi}_p(t), \tag{3}$$

where $\mathbf{c}$ *is a vector of arbitrary constants and* $\Phi(t)$ *is a fundamental matrix solution of the homogeneous equation* (2). That equation (3) is a solution can be verified as follows:

$$\boldsymbol{\phi}'(t) = \Phi'(t)\mathbf{c} + \boldsymbol{\phi}'_p(t)$$
$$= [A(t)\Phi(t)\mathbf{c}] + [A(t)\boldsymbol{\phi}_p(t) + \mathbf{f}(t)],$$

since Φ is a solution of the associated matrix equation and $\boldsymbol{\phi}_p$ is a particular solution of (2). Combining terms, we have

$$\boldsymbol{\phi}'(t) = A(t)[\Phi(t)\mathbf{c} + \boldsymbol{\phi}_p(t)] + \mathbf{f}(t) = A(t)\boldsymbol{\phi}(t) + \mathbf{f}(t),$$

and $\boldsymbol{\phi}$ is a solution of (2).

Variation of Parameters

We now derive a *variation-of-parameters* formula for the nonhomogeneous system

$$\mathbf{x}' = A(t)\mathbf{x} + \mathbf{f}(t). \tag{4}$$

All variation-of-parameters formulas begin by assuming that a solution to the homogeneous equation $\mathbf{x}' = A(t)\mathbf{x}$ is known. Assuming that Φ is a fundamental matrix solution of the homogeneous equation, we seek a particular solution to (4) of the form

$$\boldsymbol{\phi}_p(t) = \Phi(t)\mathbf{c}(t), \tag{5}$$

where $\mathbf{c}(t)$ is a vector function in t. Differentiating both sides of (5) with respect to t, we have

$$\phi_p' = \Phi'c + \Phi c' = A\Phi c + \Phi c' = A\phi_p + \Phi c'.$$

Since ϕ_p is a particular solution of (4), it follows that $\Phi c' = f$. But every fundamental matrix solution has an inverse, so we can integrate $c' = \Phi^{-1}f$, obtaining

$$\phi_p(t) = \Phi(t)c(t) = \Phi(t)\int \Phi^{-1}(t)f(t)\,dt. \tag{6}$$

This is the **variation-of-parameters formula** for a particular solution to the nonhomogeneous system (4). Thus, the general solution to system (4) has the form

$$\boxed{\phi(t) = \Phi(t)c + \Phi(t)\int \Phi^{-1}(t)f(t)\,dt,} \tag{7}$$

where c is an arbitrary *constant* vector.

For the initial-value problem

$$x' = A(t)x + f(t), \qquad x(t_0) = x_0, \tag{8}$$

it is convenient to choose a particular solution $\phi_p(t)$ that vanishes at t_0. This can be done by selecting the limits of integration in (6) to be from t_0 to t; that is,

$$\phi(t) = \Phi(t)c + \Phi(t)\int_{t_0}^{t} \Phi^{-1}(s)f(s)\,ds.$$

Substituting $t = t_0$ in this equation, we obtain

$$x_0 = \phi(t_0) = \Phi(t_0)c,$$

which implies that $c = \Phi^{-1}(t_0)x_0$. Hence the solution of the initial-value problem (8) is

$$\boxed{\begin{aligned}\phi(t) &= \Phi(t)\Phi^{-1}(t_0)x_0 + \Phi(t)\int_{t_0}^{t}\Phi^{-1}(s)f(s)\,ds \\ &= \phi_h(t) + \phi_p(t),\end{aligned}} \tag{9}$$

where ϕ_h and ϕ_p are the homogeneous and particular solutions, respectively. Note that if $\Psi(t)$ is the principal matrix solution of $x' = Ax$, then $\Psi(t_0) = \Psi^{-1}(t_0) = I$. Thus (9) takes the simpler form

$$\boxed{\phi(t) = \Psi(t)x_0 + \Psi(t)\int_{t_0}^{t}\Psi^{-1}(s)f(s)\,ds} \tag{10}$$

when the principal matrix solution $\Psi(t)$ of $x' = A(t)x$ is used.

340 CHAPTER 7/SYSTEMS OF LINEAR DIFFERENTIAL EQUATIONS

If $A(t) = A$, a constant matrix, things become simpler. Computing the inverse of a matrix function is generally quite tedious. However, if $A(t)$ is constant, then, according to Theorem 7.7.5,

$$\Psi^{-1}(t) = \Psi(-t). \tag{11}$$

Inserting (11) into (10), we obtain the form of the variation-of-parameters formula that we will use most frequently:

$$\boxed{\phi(t) = \Psi(t)\mathbf{x}_0 + \Psi(t)\int_{t_0}^{t}\Psi(-s)\mathbf{f}(s)\,ds.} \tag{12}$$

We summarize these results in the theorem below.

THEOREM 2

Let $\Psi(t)$ be the principal matrix solution of the homogeneous system

$$\mathbf{x}' = A(t)\mathbf{x}, \qquad \mathbf{x}(t_0) = \mathbf{x}_0. \tag{13}$$

Then the solution to the initial-value problem

$$\mathbf{x}' = A(t)\mathbf{x} + \mathbf{f}(t), \qquad \mathbf{x}(t_0) = \mathbf{x}_0, \tag{14}$$

is given by

$$\phi(t) = \Psi(t)\mathbf{x}_0 + \Psi(t)\int_{t_0}^{t}\Psi^{-1}(s)\mathbf{f}(s)\,ds.$$

If $A(t)$ is constant, then

$$\phi(t) = \Psi(t)\mathbf{x}_0 + \Psi(t)\int_{t_0}^{t}\Psi(-s)\mathbf{f}(s)\,ds. \tag{15}$$

Example 1 Find the unique solution to the system

$$\mathbf{x}' = \begin{pmatrix} x_1 \\ x_2 \end{pmatrix}' = \begin{pmatrix} 4 & 2 \\ 3 & 3 \end{pmatrix}\begin{pmatrix} x_1 \\ x_2 \end{pmatrix} + \begin{pmatrix} e^t \\ e^{2t} \end{pmatrix} = A\mathbf{x} + \mathbf{f}(t), \qquad \mathbf{x}(0) = \begin{pmatrix} 1 \\ 2 \end{pmatrix}.$$

Solution The principal matrix solution for the homogeneous system is (see Example 7.9.1)

$$\Psi(t) = \frac{1}{5}\begin{pmatrix} 2e^t + 3e^{6t} & -2e^t + 2e^{6t} \\ -3e^t + 3e^{6t} & 3e^t + 2e^{6t} \end{pmatrix}.$$

Then

$$\Psi^{-1}(s) = \Psi(-s) = \frac{1}{5}\begin{pmatrix} 2e^{-s} + 3e^{-6s} & -2e^{-s} + 2e^{-6s} \\ -3e^{-s} + 3e^{-6s} & 3e^{-s} + 2e^{-6s} \end{pmatrix}.$$

By (12), the solution is given by

$$\phi(t) = \Psi(t)\mathbf{x}_0 + \Psi(t)\int_0^t \Psi(-s)\mathbf{f}(s)\,ds$$

$$= \frac{1}{5}\begin{pmatrix} 2e^t + 3e^{6t} & -2e^t + 2e^{6t} \\ -3e^t + 3e^{6t} & 3e^t + 2e^{6t} \end{pmatrix}\begin{pmatrix} 1 \\ 2 \end{pmatrix}$$

$$+ \frac{1}{5}\Psi(t)\int_0^t \begin{pmatrix} 2e^{-s} + 3e^{-6s} & -2e^{-s} + 2e^{-6s} \\ -3e^{-s} + 3e^{-6s} & 3e^{-s} + 2e^{-6s} \end{pmatrix}\begin{pmatrix} e^s \\ e^{2s} \end{pmatrix} ds$$

$$= \frac{1}{5}\begin{pmatrix} -2e^t + 7e^{6t} \\ 3e^t + 7e^{6t} \end{pmatrix} + \frac{1}{5}\Psi(t)\int_0^t \begin{pmatrix} 2 + 3e^{-5s} - 2e^s + 2e^{-4s} \\ -3 + 3e^{-5s} + 3e^s + 2e^{-4s} \end{pmatrix} ds$$

$$= \frac{1}{5}\begin{pmatrix} -2e^t + 7e^{6t} \\ 3e^t + 7e^{6t} \end{pmatrix} + \frac{1}{5}\Psi(t)\begin{pmatrix} \left(2s - \frac{3}{5}e^{-5s} - 2e^s - \frac{1}{2}e^{-4s}\right)\big|_0^t \\ \left(-3s - \frac{3}{5}e^{-5s} + 3e^s - \frac{1}{2}e^{-4s}\right)\big|_0^t \end{pmatrix}$$

$$= \frac{1}{5}\begin{pmatrix} -2e^t + 7e^{6t} \\ 3e^t + 7e^{6t} \end{pmatrix}$$

$$+ \frac{1}{25}\begin{pmatrix} 2e^t + 3e^{6t} & -2e^t + 2e^{6t} \\ -3e^t + 3e^{6t} & 3e^t + 2e^{6t} \end{pmatrix}\begin{pmatrix} 2t - \frac{3}{5}e^{-5t} - 2e^t - \frac{1}{2}e^{-4t} + \frac{31}{10} \\ -3t - \frac{3}{5}e^{-5t} + 3e^t - \frac{1}{2}e^{-4t} - \frac{19}{10} \end{pmatrix}$$

$$= \begin{pmatrix} -\frac{2}{5}e^t + \frac{7}{5}e^{6t} \\ \frac{3}{5}e^t + \frac{7}{5}e^{6t} \end{pmatrix} + \begin{pmatrix} \frac{2}{5}te^t - \frac{1}{2}e^{2t} + \frac{7}{25}e^t + \frac{11}{50}e^{6t} \\ \frac{-3}{5}te^t + \frac{1}{2}e^{2t} - \frac{18}{25}e^t + \frac{11}{50}e^{6t} \end{pmatrix}$$

$$= \begin{pmatrix} \frac{2}{5}te^t - \frac{1}{2}e^{2t} - \frac{3}{25}e^t + \frac{81}{50}e^{6t} \\ \frac{-3}{5}te^t + \frac{1}{2}e^{2t} - \frac{3}{25}e^t + \frac{81}{50}e^{6t} \end{pmatrix}.$$

Note, as a check, that $\phi(0) = \begin{pmatrix} 1 \\ 2 \end{pmatrix}$.

Example 2 Consider the initial-value problem

$$x'' + x = 2\cos t, \qquad x(0) = 5, \qquad x'(0) = 2.$$

Using the substitution $x_1 = x$, $x_2 = x'$, we can write this in matrix form

$$\begin{pmatrix} x_1 \\ x_2 \end{pmatrix}' = \begin{pmatrix} 0 & 1 \\ -1 & 0 \end{pmatrix}\begin{pmatrix} x_1 \\ x_2 \end{pmatrix} + \begin{pmatrix} 0 \\ 2\cos t \end{pmatrix},$$

where the homogeneous system has the principal matrix solution

$$\Psi(t) = \begin{pmatrix} \cos t & \sin t \\ -\sin t & \cos t \end{pmatrix}.$$

Then

$$\Psi^{-1}(t) = \Psi(-t) = \begin{pmatrix} \cos t & -\sin t \\ \sin t & \cos t \end{pmatrix} \qquad \cos(-t) = \cos t \text{ and } \sin(-t) = -\sin t$$

and

$$\int_0^t \Psi^{-1}(s)\mathbf{f}(s)\,ds = \int_0^t \begin{pmatrix} \cos s & -\sin s \\ \sin s & \cos s \end{pmatrix}\begin{pmatrix} 0 \\ 2\cos s \end{pmatrix} ds$$

$$= \int_0^t \begin{pmatrix} -2\sin s \cos s \\ 2\cos^2 s \end{pmatrix} ds = \begin{pmatrix} \cos^2 t - 1 \\ t + \sin t \cos t \end{pmatrix}.$$

Thus, since the solution $\boldsymbol{\phi}(t)$ satisfies the initial conditions

$$\boldsymbol{\phi}(0) = \begin{pmatrix} 5 \\ 2 \end{pmatrix},$$

we obtain, from equation (12),

$$\boldsymbol{\phi}(t) = \begin{pmatrix} \cos t & \sin t \\ -\sin t & \cos t \end{pmatrix}\begin{pmatrix} 5 \\ 2 \end{pmatrix} + \begin{pmatrix} \cos t & \sin t \\ -\sin t & \cos t \end{pmatrix}\begin{pmatrix} \cos^2 t - 1 \\ t + \sin t \cos t \end{pmatrix}$$

$$= \begin{pmatrix} 5\cos t + 2\sin t + t\sin t \\ -4\sin t + 2\cos t + t\cos t \end{pmatrix},$$

where we have used the fact that

$$\cos^3 t + \sin^2 t \cos t = \cos t(\cos^2 t + \sin^2 t) = \cos t.$$

Example 3 Consider the system

$$\begin{pmatrix} x_1 \\ x_2 \\ x_3 \end{pmatrix}' = \begin{pmatrix} 0 & 1 & 0 \\ 0 & 0 & 1 \\ -2/t^3 & 2/t^2 & 1/t \end{pmatrix}\begin{pmatrix} x_1 \\ x_2 \\ x_3 \end{pmatrix} + \begin{pmatrix} 2t^2 \\ -t^3 \\ t^5 \end{pmatrix}, \qquad (16)$$

with initial conditions $x_1(1) = 2$, $x_2(1) = 0$, and $x_3(1) = -1$. A fundamental matrix solution (check!) is

$$\Phi(t) = \begin{pmatrix} t & 1/t & t^2 \\ 1 & -1/t^2 & 2t \\ 0 & 2/t^3 & 2 \end{pmatrix}.$$

Note that this solution is valid only for $t > 0$, since $\Phi(t)$ is not defined at $t = 0$. Then

$$\Phi^{-1}(t) = \frac{1}{6}\begin{pmatrix} 6/t & 0 & -3t \\ 2t & -2t^2 & t^3 \\ -2/t^2 & 2/t & 2 \end{pmatrix}.$$

Hence, by (9), the solution to the initial-value problem (16) is

$$\boldsymbol{\phi}(t) = \Phi(t)\Phi^{-1}(1)\begin{pmatrix} 2 \\ 0 \\ -1 \end{pmatrix} + \Phi(t)\int_1^t \Phi^{-1}(s)\mathbf{f}(s)\,ds = \Phi(t)\mathbf{c} + \boldsymbol{\phi}_p(t).$$

Setting $t = 1$, we have

$$\Phi^{-1}(1) = \frac{1}{6}\begin{pmatrix} 6 & 0 & -3 \\ 2 & -2 & 1 \\ -2 & 2 & 2 \end{pmatrix},$$

so

$$\Phi(t)\Phi^{-1}(1)\begin{pmatrix} 2 \\ 0 \\ 1 \end{pmatrix} = \frac{1}{6}\begin{pmatrix} 15t + 3/t - 6t^2 \\ 15 - 3/t^2 - 12t \\ 6/t^3 - 12 \end{pmatrix}.$$

After a great deal of arithmetic, we arrive at

$$\phi(t) = \frac{1}{6}\begin{pmatrix} 15t + 3/t - 6t^2 \\ 15 - 3/t^2 - 12t \\ 6/t^3 - 12 \end{pmatrix} + \frac{1}{6}\begin{pmatrix} t & 1/t & t^2 \\ 1 & -1/t^2 & 2t \\ 0 & 2/t^3 & 2 \end{pmatrix} \int_1^t \begin{pmatrix} 12s - 3s^6 \\ 4s^3 + 2s^5 + s^8 \\ -4 - 2s^2 + 2s^5 \end{pmatrix} ds$$

$$= \begin{pmatrix} \frac{2}{27}t^8 - \frac{1}{14}t^7 - \frac{1}{18}t^5 - \frac{1}{2}t^3 + \frac{13}{18}t^2 + \frac{5}{2}t - \frac{13}{14} + \frac{7}{27t} \\ -\frac{17}{189}t^7 + \frac{1}{9}t^6 - \frac{5}{18}t^4 - \frac{1}{2}t^2 - \frac{5}{9}t^2 + \frac{11}{7} - \frac{7}{27t^2} \\ \frac{4}{27}t^6 - \frac{1}{9}t^3 - t - \frac{5}{9} + \frac{14}{27t^3} \end{pmatrix}$$

An answer this complex is difficult to check. However, you should at least verify that the initial conditions are satisfied.

PROBLEMS 7.10

In each of Problems 1–9 calculate the principal matrix solution for the associated homogeneous system and then use the variation-of-parameters formula (12) to obtain a particular solution to the given nonhomogeneous system. Where initial conditions are given, find the unique solution that satisfies them.

1. $\mathbf{x}' = \begin{pmatrix} -2 & -2 \\ -5 & 1 \end{pmatrix}\mathbf{x} + \begin{pmatrix} e^t \\ e^{2t} \end{pmatrix}$

2. $\mathbf{x}' = \begin{pmatrix} -12 & 7 \\ -7 & 2 \end{pmatrix}\mathbf{x} + \begin{pmatrix} t \\ 2 \end{pmatrix}$, $x_1(0) = 1$, $x_2(0) = 0$

3. $\mathbf{x}' = \begin{pmatrix} 2 & -1 \\ 5 & -2 \end{pmatrix}\mathbf{x} + \begin{pmatrix} \sin t \\ \cos t \end{pmatrix}$, $x_1(0) = 0$, $x_2(0) = 1$

4. $\mathbf{x}' = \begin{pmatrix} 3 & 2 \\ -5 & 1 \end{pmatrix}\mathbf{x} + \begin{pmatrix} 2\sin 3t \\ \cos 3t \end{pmatrix}$

5. $\mathbf{x}' = \begin{pmatrix} 1 & 1 & -2 \\ -1 & 2 & 1 \\ 0 & 1 & -1 \end{pmatrix}\mathbf{x} + \begin{pmatrix} e^t \\ e^{2t} \\ e^{3t} \end{pmatrix}$,
$x_1(0) = 0$, $x_2(0) = 1$, $x_3(0) = -1$

6. $\mathbf{x}' = \begin{pmatrix} 0 & 1 & 0 \\ 0 & 0 & 1 \\ 1 & -3 & 3 \end{pmatrix}\mathbf{x} + \begin{pmatrix} t^2 \\ 0 \\ 1 \end{pmatrix}$

7. $\mathbf{x}' = \begin{pmatrix} 4 & 6 & 6 \\ 1 & 3 & 2 \\ -1 & -5 & -2 \end{pmatrix}\mathbf{x}$
$+ \begin{pmatrix} 1 - 4t - 6t^2 - 6t^4 \\ t - 3t^2 - 2t^4 \\ t + 5t^2 + 4t^3 + 2t^4 \end{pmatrix}$

8. $\mathbf{x}' = \begin{pmatrix} 1 & -1 & -1 \\ 1 & -1 & 0 \\ 1 & 0 & -1 \end{pmatrix}\mathbf{x} + \begin{pmatrix} 2 \\ e^{-t} \\ e^{-t} \end{pmatrix}$,
$x_1(0) = 1$, $x_2(0) = -1$, $x_3(0) = 0$

9. $\mathbf{x}' = \begin{pmatrix} 2 & -5 \\ 1 & -2 \end{pmatrix}\mathbf{x} + \begin{pmatrix} 0 \\ \cot t \end{pmatrix}$, $0 < t < \pi$

In each of Problems 10–12, one homogeneous solution to a given system is given. Use the method of Problem 7.7.19 to obtain a fundamental matrix solution. Then use this solution to find the general solution of the nonhomogeneous system.

10. $\mathbf{x}' = \begin{pmatrix} 0 & 1 \\ -1/4t^2 & 0 \end{pmatrix}\mathbf{x} + \begin{pmatrix} \sqrt{t} \\ 2 \end{pmatrix}$,

$\phi_1(t) = \begin{pmatrix} \sqrt{t} \\ 1/2\sqrt{t} \end{pmatrix}$, $t > 0$.

11. $\mathbf{x}' = \begin{pmatrix} 0 & 1 \\ -1/t^2 & -3/t \end{pmatrix}\mathbf{x} + \begin{pmatrix} t \\ e^t \end{pmatrix}$,

$\phi_1(t) = \begin{pmatrix} 1/t \\ -1/t^2 \end{pmatrix}$, $t > 0$.

12. $\mathbf{x}' = \begin{pmatrix} \dfrac{-t}{1-t^2} & \dfrac{1}{1-t^2} \\ \dfrac{1}{1-t^2} & \dfrac{-t}{1-t^2} \end{pmatrix}\mathbf{x} + \begin{pmatrix} t^2 \\ t^3 \end{pmatrix}$,

$\phi_1(t) = \begin{pmatrix} 1 \\ t \end{pmatrix}$, $-1 < t < 1$

13. Let $\phi_1(t)$ be a solution to $\mathbf{x}'(t) = A\mathbf{x}(t) + \mathbf{b}_1(t)$, $\phi_2(t)$ a solution to $\mathbf{x}'(t) = A\mathbf{x}(t) + \mathbf{b}_2(t), \ldots$, and $\phi_n(t)$ a solution to $\mathbf{x}'(t) = A\mathbf{x}(t) + \mathbf{b}_n(t)$. Prove that $\phi_1(t) + \phi_2(t) + \cdots + \phi_n(t)$ is a solution to

$$\mathbf{x}'(t) = A\mathbf{x}(t) + \mathbf{b}_1(t) + \mathbf{b}_2(t) + \cdots + \mathbf{b}_n(t).$$

This again is called the **principle of superposition**.

7.11 An Excursion: The Matrix Exponential

In this section we show how the principal matrix solution to $\mathbf{x}' = A\mathbf{x}$, where A is a constant matrix, can be represented in an exponential form. This form allows us to prove two important results stated earlier.

Consider the system

$$\mathbf{x}'(t) = A\mathbf{x}(t), \tag{1}$$

where A is a constant $n \times n$ matrix, and the related scalar equation

$$x'(t) = ax(t). \tag{2}$$

Equation (2) is equation (1) in the case $n = 1$. Equation (2) has a solution

$$\psi(t) = e^{at}. \tag{3}$$

The function $\psi(t)$ is the principal 1×1 matrix solution to (2) since $\psi(0) = 1$. By analogy, we define the matrix function e^{At} in the case where A is an $n \times n$ matrix.

Recall that the exponential function can be defined as the power series

$$e^{at} = 1 + at + \frac{(at)^2}{2!} + \cdots + \frac{(at)^m}{m!} + \cdots. \tag{4}$$

We use this expansion to define the matrix function:

$$\boxed{e^{At} = I + At + \frac{(At)^2}{2!} + \frac{(At)^3}{3!} + \cdots + \frac{(At)^m}{m!} + \cdots.} \tag{5}$$

Note that since powers of the matrix A are $n \times n$ matrices, the right side of (5) is an $n \times n$ matrix if the series converges.

The question of convergence for the series in (5) for an arbitrary $n \times n$ matrix A is settled in the following theorem, which we give without proof.

THEOREM 1

The series

$$e^{At} = I + At + \frac{(At)^2}{2!} + \frac{(At)^3}{3!} + \cdots \tag{6}$$

converges for all t, can be differentiated term by term, and is the principal matrix solution of the system $\mathbf{x}' = A\mathbf{x}$.

Note We are assuming in this section that $t_0 = 0$ so that the principal matrix solution $\Psi(t) = e^{At}$ satisfies $\Psi(0) = I$. If $t_0 \neq 0$, then the principal matrix solution is given by $\Psi(t) = e^{A(t-t_0)}$. There are many techniques for computing

e^{At}. For example

If
$$A = \begin{pmatrix} \lambda_1 & 0 & \cdots & 0 \\ 0 & \lambda_2 & \cdots & 0 \\ \vdots & \vdots & & \vdots \\ 0 & 0 & \cdots & \lambda_n \end{pmatrix},$$
then
$$e^{At} = \begin{pmatrix} e^{\lambda_1 t} & 0 & \cdots & 0 \\ 0 & e^{\lambda_2 t} & \cdots & 0 \\ \vdots & \vdots & & \vdots \\ 0 & 0 & \cdots & e^{\lambda_n t} \end{pmatrix}.$$

(7)

Example 1 Let
$$A = \begin{pmatrix} 1 & 0 & 0 \\ 0 & 2 & 0 \\ 0 & 0 & 3 \end{pmatrix}.$$

Then
$$A^2 = \begin{pmatrix} 1 & 0 & 0 \\ 0 & 2^2 & 0 \\ 0 & 0 & 3^2 \end{pmatrix}, \quad A^3 = \begin{pmatrix} 1 & 0 & 0 \\ 0 & 2^3 & 0 \\ 0 & 0 & 3^3 \end{pmatrix}, \ldots, \quad A^m = \begin{pmatrix} 1 & 0 & 0 \\ 0 & 2^m & 0 \\ 0 & 0 & 3^m \end{pmatrix},$$

so

$$e^{At} = I + At + \frac{A^2 t^2}{2} + \cdots + \frac{A^m t^m}{m} + \cdots = \begin{pmatrix} 1 & 0 & 0 \\ 0 & 1 & 0 \\ 0 & 0 & 1 \end{pmatrix} + \begin{pmatrix} t & 0 & 0 \\ 0 & 2t & 0 \\ 0 & 0 & 3t \end{pmatrix}$$

$$+ \cdots + \begin{pmatrix} \frac{t^m}{m!} & 0 & 0 \\ 0 & \frac{(2t)^m}{m!} & 0 \\ 0 & 0 & \frac{(3t)^m}{m!} \end{pmatrix} + \cdots = \begin{pmatrix} e^t & 0 & 0 \\ 0 & e^{2t} & 0 \\ 0 & 0 & e^{3t} \end{pmatrix}.$$

The "brute force" method of calculating e^{At} by using definition (5) is useful only for particularly simple matrices A, such as the diagonal matrix in

Example 1. Indeed, the procedure we used in Example 1 can be easily extended to obtain our assertion in (7).

When the $n \times n$ matrix A has n linearly independent eigenvectors $\phi_1, \phi_2, \ldots, \phi_n$, we know that $A\phi_i = \lambda_i \phi_i$, where λ_i is the eigenvalue corresponding to the eigenvector ϕ_i. Hence, if we form the matrix $\Phi = (\phi_1, \phi_2, \ldots, \phi_n)$ whose columns are the n eigenvectors of A, we have

$$A\Phi = A(\phi_1, \phi_2, \ldots, \phi_n) = (A\phi_1, A\phi_2, \ldots, A\phi_n)$$

$$= (\lambda_1 \phi_1, \lambda_2 \phi_2, \ldots, \lambda_n \phi_n)$$

$$= \Phi \begin{pmatrix} \lambda_1 & 0 & \cdots & 0 \\ 0 & \lambda_2 & \cdots & 0 \\ \vdots & \vdots & \ddots & \vdots \\ 0 & 0 & \cdots & \lambda_n \end{pmatrix} = \Phi D, \qquad (8)$$

where D is a diagonal matrix. Since the columns of Φ are linearly independent, Φ is invertible, so

$$A = \Phi D \Phi^{-1}. \qquad (9)$$

Then, using (5), we have

$$e^{At} = I + At + A^2 \frac{t^2}{2} + A^3 \frac{t^3}{3!} + \cdots$$

$$= \Phi \Phi^{-1} + \Phi D \Phi^{-1} t + \Phi D \underbrace{\Phi^{-1} \Phi}_{=I} D \Phi^{-1} \frac{t^2}{2!} + \cdots$$

$$= \Phi I \Phi^{-1} + \Phi D \Phi^{-1} t + \Phi D^2 \Phi^{-1} \frac{t^2}{2!} + \Phi D^3 \Phi^{-1} \frac{t^3}{3!} + \cdots$$

$$= \Phi \left(I + Dt + D^2 \frac{t^2}{2!} + D^3 \frac{t^3}{3!} + \cdots \right) \Phi^{-1} = \Phi e^{Dt} \Phi^{-1}.$$

Thus the matrix exponential for A is obtained by multiplying the matrix exponential for D (which is easily obtained from [7]) by Φ on the left and Φ^{-1} on the right:

$$\boxed{e^{At} = \Phi e^{Dt} \Phi^{-1}.} \qquad (10)$$

Example 2   In Example 7.8.3 we found that for the system

$$\mathbf{x}' = \begin{pmatrix} 4 & 2 \\ 3 & 3 \end{pmatrix} \mathbf{x} = A\mathbf{x}, \qquad (11)$$

the matrix A has the eigenvalues $\lambda_1 = 1$ and $\lambda_2 = 6$ with corresponding eigenvectors

$$\phi_1 = \begin{pmatrix} 2 \\ -3 \end{pmatrix} \quad \text{and} \quad \phi_2 = \begin{pmatrix} 1 \\ 1 \end{pmatrix}. \qquad (12)$$

Setting $\Phi = (\phi_1, \phi_2)$, we see that

$$A\Phi = \begin{pmatrix} 4 & 2 \\ 3 & 3 \end{pmatrix}\begin{pmatrix} 2 & 1 \\ -3 & 1 \end{pmatrix} = \begin{pmatrix} 2 & 6 \\ -3 & 6 \end{pmatrix},$$

$$\Phi D = \begin{pmatrix} 2 & 1 \\ -3 & 1 \end{pmatrix}\begin{pmatrix} 1 & 0 \\ 0 & 6 \end{pmatrix} = \begin{pmatrix} 2 & 6 \\ -3 & 6 \end{pmatrix},$$

so $A\Phi = \Phi D$. Since

$$\Phi^{-1} = \frac{1}{5}\begin{pmatrix} 1 & -1 \\ 3 & 2 \end{pmatrix},$$

it follows by (10) that

$$e^{At} = \Phi e^{Dt} \Phi^{-1} \qquad e^{Dt} = \begin{pmatrix} e^{\lambda_1 t} & 0 \\ 0 & e^{\lambda_2 t} \end{pmatrix}$$

$$= \frac{1}{5}\begin{pmatrix} 2 & 1 \\ -3 & 1 \end{pmatrix}\begin{pmatrix} e^t & 0 \\ 0 & e^{6t} \end{pmatrix}\begin{pmatrix} 1 & -1 \\ 3 & 2 \end{pmatrix}$$

$$= \frac{1}{5}\begin{pmatrix} 2 & 1 \\ -3 & 1 \end{pmatrix}\begin{pmatrix} e^t & -e^t \\ 3e^{6t} & 2e^{6t} \end{pmatrix} = \frac{1}{5}\begin{pmatrix} 2e^t + 3e^{6t} & -2e^t + 2e^{6t} \\ -3e^t + 3e^{6t} & 3e^t + 2e^{6t} \end{pmatrix}. \quad (13)$$

This is exactly the principal matrix solution for (11) that we determined in Example 7.9.1.

When the $n \times n$ matrix A does not have n linearly independent eigenvectors, the procedure in (10) cannot be used. We now give a procedure, using the Cayley-Hamilton theorem, that can be used in this case.

THEOREM 2: THE CAYLEY-HAMILTON THEOREM[†]

Every square matrix satisfies its own characteristic equation; that is, if $p(\lambda) = 0$ is the characteristic equation of A, then $p(A) = 0$.

We are now ready to calculate e^{At} for any square matrix A. Although the procedure may appear to be complicated, it is very easy to use.

Before explaining the procedure, we define the notation we plan to use. Let $p(\lambda) = \det(\lambda I - A)$ be the characteristic polynomial of the $n \times n$ matrix A, and suppose that

$$p(\lambda) = (\lambda - \lambda_1)^{r_1}(\lambda - \lambda_2)^{r_2}\ldots(\lambda - \lambda_k)^{r_k}, \quad (14)$$

where $\lambda_1, \lambda_2, \ldots, \lambda_k$ are the eigenvalues of A of multiplicities $r_1, r_2, \ldots, r_k$, respectively. Using partial fractions, we can write

$$\frac{1}{p(\lambda)} = \frac{a_1(\lambda)}{(\lambda - \lambda_1)^{r_1}} + \frac{a_2(\lambda)}{(\lambda - \lambda_2)^{r_2}} + \cdots + \frac{a_k(\lambda)}{(\lambda - \lambda_k)^{r_k}}, \quad (15)$$

where, for each polynomial $a_i(\lambda)$,

$$\deg a_i(\lambda) \le r_i - 1. \quad (16)$$

[†] Named after William Rowan Hamilton (1805–1865) and Arthur Cayley (1821–1895). Cayley published the first discussion of this famous theorem in 1858. Independently, Hamilton discovered the result in his work on quaternions. A proof of the Cayley-Hamilton theorem can be found in S. I. Grossman, *Elementary Linear Algebra*, 3d ed. (Belmont, Calif.: Wadsworth, 1987), Section 6.8.

Multiplying both sides of (15) by $p(\lambda)$, we get

$$\boxed{1 = a_1(\lambda)q_1(\lambda) + a_2(\lambda)q_2(\lambda) + \cdots + a_k(\lambda)q_k(\lambda),} \qquad (17)$$

where $q_i(\lambda)$ is the polynomial consisting of all but the $(\lambda - \lambda_i)^{r_i}$ term of $p(\lambda)$:

$$\boxed{p(\lambda) = q_i(\lambda)(\lambda - \lambda_i)^{r_i}.} \qquad (18)$$

We are now ready to begin the procedure for calculating e^{At} for any $n \times n$ matrix A. We illustrate each step in the procedure by applying it to the matrix

$$A = \begin{pmatrix} 1 & 2 & 3 \\ 0 & 0 & 4 \\ 0 & 0 & 0 \end{pmatrix}.$$

Step 1 *Find the characteristic polynomial $p(\lambda)$ and use it to determine the polynomials $a_i(\lambda)$ and $q_i(\lambda)$ in equations (17) and (18).*

In this case we have

$$p(\lambda) = \det(\lambda I - A) = \begin{vmatrix} \lambda - 1 & -2 & -3 \\ 0 & \lambda & -4 \\ 0 & 0 & \lambda \end{vmatrix} = \lambda^2(\lambda - 1),$$

so $\lambda_1 = 0$ and $\lambda_2 = 1$, with $r_1 = 2$ and $r_2 = 1$, respectively. By (18) we easily obtain

$$q_1(\lambda) = \lambda - 1 \quad \text{and} \quad q_2(\lambda) = \lambda^2. \qquad (19)$$

Using (16), we note that $\deg a_1(\lambda) \leq 1$ and $\deg a_2(\lambda) \leq 0$, so the general forms of the polynomials a_1 and a_2 are

$$a_1(\lambda) = a\lambda + b \quad \text{and} \quad a_2(\lambda) = c.$$

Substituting these polynomials into (17) and equating like powers of λ, we obtain

$$a_1(\lambda) = -\lambda - 1 \quad \text{and} \quad a_2(\lambda) = 1. \qquad (20)$$

At this point we need to develop some additional facts. Recall the Cayley-Hamilton theorem. Since A satisfies its characteristic equation, (18) becomes

$$0 = p(A) = q_i(A)(A - \lambda_i I)^{r_i}. \qquad (21)$$

Equation (17) is also satisfied by the matrix A, since it was derived from the characteristic equation. Thus

$$I = a_1(A)q_1(A) + a_2(A)q_2(A) + \cdots + a_k(A)q_k(A). \qquad (22)$$

We can easily check (22) for our particular example:

$$I = (-A - I)(A - I) + A^2.$$

Using (7), we observe that $e^{\lambda_i tI} = e^{\lambda_i t}I$, so we have

$$e^{At} = e^{\lambda_i tI} e^{(A-\lambda_i I)t} = e^{\lambda_i t} \sum_{j=0}^{\infty} \frac{(A-\lambda_i I)^j t^j}{j!}. \tag{23}$$

Multiplying both ends of (23) on the left by $q_i(A)$, we obtain

$$q_i(A) e^{At} = e^{\lambda_i t} \sum_{j=0}^{\infty} \frac{q_i(A)(A-\lambda_i I)^j t^j}{j!}. \tag{24}$$

By (21), all the terms in the series for $j \geq r_i$ are equal to the zero matrix, so (24) reduces to

$$q_i(A) e^{At} = e^{\lambda_i t} \sum_{j=0}^{r_i-1} \frac{q_i(A)(A-\lambda_i I)^j t^j}{j!}. \tag{25}$$

Multiplying each side of (25) on the left by $a_i(A)$ yields

$$a_i(A) q_i(A) e^{At} = e^{\lambda_i t} \sum_{j=0}^{r_i-1} \frac{a_i(A) q_i(A)(A-\lambda_i I)^j t^j}{j!}. \tag{26}$$

Finally, summing (26) over all the indices i and using (22), we conclude that

$$e^{At} = I e^{At} = \sum_{i=1}^{k} a_i(A) q_i(A) e^{At},$$

or

$$e^{At} = \sum_{i=1}^{k} \left\{ e^{\lambda_i t} a_i(A) q_i(A) \sum_{j=0}^{r_i-1} \frac{(A-\lambda_i I)^j t^j}{j!} \right\}.$$

Although this equation looks formidable, for most applications it is very easy to use.

Step 2 Using the polynomials $a_i(\lambda)$, $q_i(\lambda)$, and the eigenvalues λ_i with multiplicities r_i that were obtained in step 1, compute

$$\boxed{e^{At} = \sum_{i=1}^{k} \left\{ e^{\lambda_i t} a_i(A) q_i(A) \sum_{j=0}^{r_i-1} \frac{(A-\lambda_i I)^j t^j}{j!} \right\}.} \tag{27}$$

For the particular matrix A we are considering, we have

$$\lambda_1 = 0, \quad r_1 = 2, \quad \lambda_2 = 1, \quad r_2 = 1,$$

and

$$q_1(A) = A - I, \quad q_2(A) = A^2, \quad a_1(A) = -A - I, \quad a_2(A) = I.$$

Substituting these values in (27), we get

$$e^{At} = \left\{ e^{0t}(-A-I)(A-I) \sum_{j=0}^{1} \frac{A^j t^j}{j!} \right\} + \left\{ e^t I A^2 \sum_{j=0}^{0} \frac{(A-I)^j t^j}{j!} \right\}$$

$$= (A+I)(I-A)(I+At) + e^t A^2$$

$$= (I-A^2)(I+At) + e^t A^2.$$

Since

$$A^2 = \begin{pmatrix} 1 & 2 & 3 \\ 0 & 0 & 4 \\ 0 & 0 & 0 \end{pmatrix} \begin{pmatrix} 1 & 2 & 3 \\ 0 & 0 & 4 \\ 0 & 0 & 0 \end{pmatrix} = \begin{pmatrix} 1 & 2 & 11 \\ 0 & 0 & 0 \\ 0 & 0 & 0 \end{pmatrix},$$

we have

$$e^{At} = \begin{pmatrix} 0 & -2 & -11 \\ 0 & 1 & 0 \\ 0 & 0 & 1 \end{pmatrix} \begin{pmatrix} 1+t & 2t & 3t \\ 0 & 1 & 4t \\ 0 & 0 & 1 \end{pmatrix} + e^t \begin{pmatrix} 1 & 2 & 11 \\ 0 & 0 & 0 \\ 0 & 0 & 0 \end{pmatrix}$$

$$= \begin{pmatrix} 0 & -2 & -8t-11 \\ 0 & 1 & 4t \\ 0 & 0 & 1 \end{pmatrix} + \begin{pmatrix} e^t & 2e^t & 11e^t \\ 0 & 0 & 0 \\ 0 & 0 & 0 \end{pmatrix}$$

$$= \begin{pmatrix} e^t & 2(e^t-1) & 11e^t - 8t - 11 \\ 0 & 1 & 4t \\ 0 & 0 & 1 \end{pmatrix}.$$

CALCULATION OF e^{At} FOR ANY $n \times n$ MATRIX A

1. Find the characteristic polynomial for the matrix A,

$$p(\lambda) = \det(\lambda I - A)$$
$$= (\lambda - \lambda_1)^{r_1}(\lambda - \lambda_2)^{r_2} \cdots (\lambda - \lambda_k)^{r_k};$$

use it first to determine the polynomials

$$q_i(\lambda) = \frac{p(\lambda)}{(\lambda - \lambda_i)^{r_i}}$$

and then the polynomials $a_i(\lambda)$ that satisfy (28)

$$a_1(\lambda)q_1(\lambda) + a_2(\lambda)q_2(\lambda) + \cdots + a_k(\lambda)q_k(\lambda) = 1.$$

2. Replace each λ^m by A^m in the expressions for $a_i(\lambda)$ and $q_i(\lambda)$ and compute

$$e^{At} = \sum_{i=1}^{k} \left\{ e^{\lambda_i t} a_i(A) q_i(A) \sum_{j=0}^{r_i - 1} \frac{(A - \lambda_i I)^j t^j}{j!} \right\}.$$

7.11/AN EXCURSION: THE MATRIX EXPONENTIAL

Two final facts about the principal matrix solution for a system $\mathbf{x}' = A\mathbf{x}$, where A is a constant matrix, can easily be obtained using the notion of a matrix exponential:

THEOREM 3 (THEOREM 7.7.5)

Let $\Psi(t)$ be the principal matrix solution for the system $\mathbf{x}'(t) = A\mathbf{x}(t), \mathbf{x}(t_0) = \mathbf{x}_0$, where A is a constant matrix. Then

$$\Psi^{-1}(t) = \Psi(-t).$$

Proof If $t_0 = 0$, we need to show that

$$(e^{At})^{-1} = e^{-At}.$$

But

$$e^{At}e^{-At} = \left(I + At + \frac{A^2 t^2}{2!} + \frac{A^3 t^3}{3!} + \cdots\right)\left(I - At + \frac{A^2 t^2}{2!} - \frac{A^3 t^3}{3!} + \cdots\right)$$

$$= \left[I + (A - A)t + \left(\frac{A^2}{2!} + \frac{A^2}{2!} - A^2\right)t^2 + \cdots\right] = I,$$

since all terms except I cancel. Similarly, $e^{-At}e^{At} = I$, which shows that $e^{-At} = (e^{At})^{-1}$. See the note on p. 344 for the case $t_0 \neq 0$. ∎

THEOREM 4 (THEOREM 7.9.2)

Let the real matrix A have the complex eigenvalues $\alpha \pm i\beta$. Then the principal solution of $\mathbf{x}' = A\mathbf{x}$ contains no complex terms.

Proof This is easy. We have

$$\Psi(t) = e^{At} = I + At + \frac{A^2 t^2}{2!} + \frac{A^3 t^3}{3!} + \cdots.$$

Since A is real, A^k is real for every k and the sum of real matrices is a real matrix. ∎

PROBLEMS 7.11

In Problems 1–9 use the method in Example 1 or Example 2 to compute e^{At}.

1. $A = \begin{pmatrix} 1 & 0 \\ 0 & 3 \end{pmatrix}$

2. $A = \begin{pmatrix} 1 & -1 \\ 2 & 3 \end{pmatrix}$

3. $A = \begin{pmatrix} -2 & -2 \\ -5 & 1 \end{pmatrix}$

4. $A = \begin{pmatrix} -12 & 7 \\ -7 & 2 \end{pmatrix}$

5. $A = \begin{pmatrix} 2 & -1 \\ 5 & -2 \end{pmatrix}$

6. $A = \begin{pmatrix} 0 & 1 & 0 \\ 0 & 0 & 1 \\ 1 & -3 & 3 \end{pmatrix}$

7. $A = \begin{pmatrix} 1 & -1 & 0 \\ -1 & 2 & -1 \\ 0 & -1 & 1 \end{pmatrix}$

8. $A = \begin{pmatrix} 4 & 6 & 6 \\ 1 & 3 & 2 \\ -1 & -5 & -2 \end{pmatrix}$

9. $A = \begin{pmatrix} 5 & 4 & 2 \\ 4 & 5 & 2 \\ 2 & 2 & 2 \end{pmatrix}$

10. The Cayley–Hamilton theorem can be used to calculate the inverse of a matrix if A^{-1} exists. If

$$p(\lambda) = \lambda^n + a_{n-1}\lambda^{n-1} + \cdots + a_1\lambda + a_0,$$

then

$$p(A) = A^n + a_{n-1}A^{n-1} + \cdots + a_1 A + a_0 I = 0,$$

and

$$A^{-1}p(A) = A^{n-1} + a_{n-1}A^{n-2} + \cdots + a_2 A$$
$$+ a_1 I + a_0 A^{-1} = 0.$$

Thus

$$A^{-1} = \frac{1}{a_0}\left(-A^{n-1} - a_{n-1}A^{n-2} - \cdots - a_2 A - a_1 I\right).$$

Use this procedure to find the inverse of

$$A = \begin{pmatrix} 1 & -1 & 4 \\ 3 & 2 & -1 \\ 2 & 1 & -1 \end{pmatrix}.$$

In Problems 11–20 use the two-step procedure (28) to find e^{At} for each given matrix A.

11. $A = \begin{pmatrix} 4 & 1 \\ 0 & 4 \end{pmatrix}$ 12. $A = \begin{pmatrix} 2 & -1 \\ -4 & 2 \end{pmatrix}$

13. $A = \begin{pmatrix} 1 & 2 \\ 0 & 1 \end{pmatrix}$ 14. $A = \begin{pmatrix} 3 & 2 \\ -5 & 1 \end{pmatrix}$

15. $A = \begin{pmatrix} 3 & -2 \\ 8 & -5 \end{pmatrix}$ 16. $A = \begin{pmatrix} 1 & 1 \\ 1 & -1 \end{pmatrix}$

17. $A = \begin{pmatrix} 1 & -1 & 0 \\ -1 & 2 & -1 \\ 0 & -1 & 1 \end{pmatrix}$

18. $A = \begin{pmatrix} 1 & 1 & -2 \\ -1 & 2 & 1 \\ 0 & 1 & -1 \end{pmatrix}$

19. $A = \begin{pmatrix} 4 & 6 & 6 \\ 1 & 3 & 2 \\ -1 & -5 & -2 \end{pmatrix}$

20. $A = \begin{pmatrix} 7 & -2 & -4 \\ 3 & 0 & -2 \\ 6 & -2 & -3 \end{pmatrix}$

Review Exercises for Chapter 7

In Exercises 1–3 transform the equation into a first-order system.

1. $x''' - 6x'' + 2x' - 5x = 0$
2. $x'' - 3x' + 4t^2 x = \sin t$
3. $xx'' + x'x''' = \ln t$

Find the general solution (or particular solution when initial conditions are given) by the method of elimination for Exercises 4–10.

4. $x' = x + y,$
   $y' = 9x + y$

5. $x' = 4x - y,$
   $y' = x + 2y$

6. $x' = x - 4y,$
   $y' = x + y$

7. $x' = x + 2y,$
   $y' = 4x + 3y$

8. $x' = 3x + 2y,$
   $y' = -5x + y$

9. $x' = -x - 3e^{-2t},$
   $y' = -2x - y - 6e^{-2t}$

10. $x' = -4x - 6y + 9e^{-3t},$ $x(0) = -9,$
    $y' = x + y - 5e^{-3t},$ $y(0) = 4$

In Exercises 11–14 write the given system of equations in vector matrix form.

11. $x_1' = 3x_1 - 4x_2,$
    $x_2' = -2x_1 + 7x_2$

12. $x_1' = (\sin t)x_1 + e^t x_2,$
    $x_2' = -x_1 + (\tan t)x_2$

13. $x_1' = x_1 + x_2 + e^t,$
    $x_2' = -3x_1 + 2x_2 + e^{2t}$

14. $x_1' = -tx_1 + t^2 x_2 + t^3,$
    $x_2' = -\sqrt{t}\, x_1 + \sqrt[3]{t}\, x_2 + t^{3/5}$

15. Consider the system

$$\mathbf{x}' = \begin{pmatrix} 4 & 2 \\ 3 & 3 \end{pmatrix}\mathbf{x}.$$

A fundamental matrix solution is

$$\Phi(t) = \begin{pmatrix} 2e^t & e^{6t} \\ -3e^t & e^{6t} \end{pmatrix}.$$

Find a solution that satisfies each of the following initial conditions.

a. $\mathbf{x}(0) = \begin{pmatrix} 2 \\ 3 \end{pmatrix}$

b. $\mathbf{x}(0) = \begin{pmatrix} -1 \\ 0 \end{pmatrix}$

c. $\mathbf{x}(0) = \begin{pmatrix} 0 \\ 0 \end{pmatrix}$

d. $\mathbf{x}(0) = \begin{pmatrix} 7 \\ -2 \end{pmatrix}$

e. $\mathbf{x}(0) = \begin{pmatrix} a \\ b \end{pmatrix}$

In Exercises 16–21 calculate the eigenvalues and eigenvectors of the given matrix.

16. $\begin{pmatrix} 5 & -1 \\ 8 & 1 \end{pmatrix}$ 17. $\begin{pmatrix} 2 & 5 \\ 0 & 2 \end{pmatrix}$

18. $\begin{pmatrix} 1 & 0 & 0 \\ 3 & 7 & 0 \\ -2 & 4 & -5 \end{pmatrix}$

19. $\begin{pmatrix} 1 & -1 & 0 \\ 1 & 2 & 1 \\ -2 & 1 & -1 \end{pmatrix}$

20. $\begin{pmatrix} 5 & -2 & 0 & 0 \\ 4 & -1 & 0 & 0 \\ 0 & 0 & 3 & -1 \\ 0 & 0 & 2 & 3 \end{pmatrix}$

21. $\begin{pmatrix} -2 & 1 & 0 \\ 0 & -2 & 1 \\ 0 & 0 & -2 \end{pmatrix}$

In Exercises 22–26 find the principal matrix solution at $t = 0$ of the given system.

22. $\mathbf{x}' = \begin{pmatrix} -3 & 4 \\ -2 & 3 \end{pmatrix} \mathbf{x}$

23. $\mathbf{x}' = \begin{pmatrix} 3 & -1 \\ -2 & 4 \end{pmatrix} \mathbf{x}$

24. $\mathbf{x}' = \begin{pmatrix} -3 & -4 \\ -2 & 1 \end{pmatrix} \mathbf{x}$

25. $\mathbf{x}' = \begin{pmatrix} -1 & -18 & -7 \\ 1 & -13 & -4 \\ -1 & 25 & 8 \end{pmatrix} \mathbf{x}$

26. $\mathbf{x}' = \begin{pmatrix} 2 & 1 & 0 \\ -2 & -1 & 2 \\ 1 & 1 & 1 \end{pmatrix} \mathbf{x}$

27. Solve the system

$$\mathbf{x}' = \begin{pmatrix} 2 & 1 \\ -4 & 2 \end{pmatrix} \mathbf{x} + \begin{pmatrix} 3 \\ t \end{pmatrix} e^{2t}, \quad \mathbf{x}(0) = \begin{pmatrix} 3 \\ 2 \end{pmatrix}$$

28. Solve the system

$$\mathbf{x}' = \begin{pmatrix} 2 & 1 & 0 \\ -2 & -1 & 2 \\ 1 & 1 & 1 \end{pmatrix} \mathbf{x} + \begin{pmatrix} 0 \\ 1 \\ e^t \end{pmatrix}, \quad \mathbf{x}(0) = \begin{pmatrix} 1 \\ 2 \\ 3 \end{pmatrix}$$

8

Applications of Systems of Differential Equations

In many problems of engineering, physics, chemistry, biology, and economics, there are many dependent variables, each a function of a single independent variable (usually time). When these problems involve the rates of change of the dependent variables, their solutions are usually obtained by solving systems of differential equations.

In Chapter 7 we saw how to solve a variety of systems of linear differential equations. In this chapter we discuss a number of applications that give rise to such systems. Three principal methods of solution were presented in Chapter 7. However, because you may not have covered matrix methods, all problems in this chapter are solved using the method of elimination or determinants.

8.1 Electric Circuits with Several Loops

Here we make use of the concepts developed in Sections 2.5 and 4.4 to study electric networks with two or more coupled closed circuits. The two fundamental principles governing such networks are the two laws of Kirchhoff:

> i. The algebraic sum of all voltage drops around any closed circuit is zero.
> ii. The algebraic sum of the currents flowing into any junction in the network is zero.

Example 1   Consider the two-loop electric circuit in Figure 8.1. Suppose we wish to find the current in each loop as a function of time, given that all currents are zero when the switch is closed at $t = 0$. Let I be the current

8.1 / ELECTRIC CIRCUITS WITH SEVERAL LOOPS

Figure 8.1

flowing through the inductor L_1 and let I_R and I_L denote the current flowing through the resistor and the inductor L_2, respectively. By Kirchhoff's current law (ii), $I = I_R + I_L$, so when we apply Kirchhoff's voltage law (i) to each loop, we obtain the system

$$L_1 \frac{dI}{dt} + RI_R = E, \qquad (1)$$

$$L_2 \frac{dI_L}{dt} - RI_R = 0. \qquad (2)$$

Replacing I by $I_R + I_L$ in (1), we have the system of simultaneous linear differential equations

$$L_1 \frac{dI_R}{dt} + L_1 \frac{dI_L}{dt} + RI_R = E, \qquad (3)$$

$$L_2 \frac{dI_L}{dt} - RI_R = 0. \qquad (4)$$

If we multiply (4) by L_1/L_2, we obtain

$$L_1 \frac{dI_L}{dt} = \frac{L_1 R}{L_2} I_R,$$

which we can substitute into (3) to eliminate the I_L variable, yielding

$$L_1 \frac{dI_R}{dt} + \left(\frac{L_1}{L_2} + 1 \right) RI_R = E. \qquad (5)$$

If L_1, L_2, and R are constant, we can solve the linear first-order differential equation by the method in Section 2.3. Assume that $L_1 = 1$ henry, $L_2 = \frac{1}{2}$ henry, $R = 20$ ohms, and $E = 50$ volts. Then

$$\frac{dI_R}{dt} + 60 I_R = 50.$$

Multiplying both sides by the integrating factor e^{60t}, we have

$$\frac{d}{dt}\left(e^{60t} I_R \right) = 50 e^{60t},$$

and an integration yields

$$e^{60t} I_R = \tfrac{5}{6} e^{60t} + k_1.$$

Since $I_R(0) = 0$, it follows that $k_1 = -\tfrac{5}{6}$ and

$$I_R(t) = \tfrac{5}{6}(1 - e^{-60t}). \qquad (6)$$

We can find I_L by substituting (6) into (4) to get

$$\frac{1}{2}\frac{dI_L}{dt} = \frac{100}{6}(1 - e^{-60t}),$$

from which it follows that

$$I_L(t) = \frac{100}{3}\left(t + \frac{e^{-60t}}{60}\right) + k_2.$$

Setting $t = 0$, so that $I_L(0) = 0$, we obtain $k_2 = -\frac{5}{9}$ and

$$I_L(t) = \tfrac{100}{3}t + \tfrac{5}{9}(e^{-60t} - 1). \tag{7}$$

Example 2 Consider the circuit in Figure 8.2. There are two loops. By Kirchhoff's voltage law, we obtain

$$L\frac{dI_L}{dt} + RI_R = E, \tag{8}$$

$$\frac{Q_C}{C} - RI_R = 0. \tag{9}$$

Since $I = dQ/dt$, the second equation may be rewritten as

$$\frac{I_C}{C} - R\frac{dI_R}{dt} = 0. \tag{10}$$

By Kirchhoff's current law, we have

$$I_L = I_C + I_R,$$

which, if substituted into (10), yields, together with (8), the nonhomogeneous system of linear first-order differential equations

$$\begin{aligned}\frac{dI_L}{dt} &= -\frac{R}{L}I_R + \frac{E}{L}, \\ \frac{dI_R}{dt} &= \frac{I_L}{RC} - \frac{I_R}{RC}.\end{aligned} \tag{11}$$

The characteristic equation of this system is

$$D = \begin{vmatrix} -\lambda & -R/L \\ 1/RC & -\lambda - 1/RC \end{vmatrix} = \lambda(\lambda + 1/RC) + 1/LC \tag{12}$$

$$= \lambda^2 + \lambda/RC + 1/LC = 0.$$

The roots of (12) are $(-L \pm \sqrt{L^2 - 4R^2LC})/2RLC$.

Figure 8.2

The value of the discriminant
$$L^2 - 4R^2LC = L(L - 4R^2C)$$
is now important. For simplicity, assume that $R = 100$ ohms, $C = 1.5 \times 10^{-4}$ farad, $E = 100$ volts, and $L = 8$ henrys. Then
$$L - 4R^2C = 8 - 4(100)^2 \cdot 1.5(10^{-4}) = 2,$$
so the roots of (12) are $\lambda_1 = -50$ and $\lambda_2 = -50/3$.

Consider the homogeneous system

$$\frac{dI_L}{dt} = -\frac{R}{L}I_R, \qquad (13)$$
$$\frac{dI_R}{dt} = \frac{I_L}{RC} - \frac{I_R}{RC}.$$

From equations (7.3.19) and (7.3.20), the general solution to (13) is given by
$$I_L(t) = c_1 e^{-50t} + c_2 e^{-(50/3)t}$$
and
$$I_R(t) = -\frac{L}{R}\frac{dI_L}{dt} = -\frac{8}{100}\left[-50c_1 e^{-50t} - \frac{50}{3}c_2 e^{-(50/3)t}\right]$$
$$= 4c_1 e^{-50t} + \frac{4}{3}c_2 e^{-(50/3)t}. \qquad (14)$$

Since $E = 100$ volts is constant, when we use the method of undetermined coefficients we assume a particular solution of the form
$$(I_L, I_R)_p = (A, B), \qquad (15)$$
where A and B are constants we must determine. Substituting (15) into the nonhomogeneous system (11), we get
$$0 = -\frac{100}{8}B + \frac{100}{8},$$
$$0 = \frac{A - B}{(100)1.5 \times 10^{-4}},$$
so $B = 1 = A$. Finally, we find the constants c_1 and c_2 by using the initial conditions $I_L(0) = 0 = I_R(0)$ in the equations
$$I_L = c_1 e^{-50t} + c_2 e^{-(50/3)t} + 1,$$
$$I_R = 4c_1 e^{-50t} + \tfrac{4}{3}c_2 e^{-(50/3)t} + 1.$$

We obtain
$$I_L(0) = c_1 + c_2 + 1 = 0,$$
$$I_R(0) = 4c_1 + \tfrac{4}{3}c_2 + 1 = 0,$$

with solution $c_1 = \tfrac{1}{8}$ and $c_2 = -\tfrac{9}{8}$. Thus the unique solution to the initial-value problem is
$$(I_L, I_R) = \left(\tfrac{1}{8}e^{-50t} - \tfrac{9}{8}e^{(-50/3)t} + 1, \tfrac{1}{2}e^{-50t} - \tfrac{3}{2}e^{(-50/3)t} + 1\right).$$

Example 3 If $L = 6$ henrys in Example 2 and all other facts remain the same, then $L - 4R^2C = 0$ and (12) has the double root $\lambda_1 = \lambda_2 = -1/2RC = -100/3$. Hence, from equation (7.3.21), the general solution to the homogeneous system (13) is

$$I_L(t) = (c_1 + c_2 t)e^{-(100/3)t}$$

and

$$I_R(t) = -\frac{L}{R}\frac{dI_L}{dt} = -\frac{6}{100}e^{(-100/3)t}\left[\frac{-100}{3}(c_1 + c_2 t) + c_2\right]$$

$$= \left[\left(2c_1 - \frac{3}{50}c_2\right) + 2c_2 t\right]e^{-(100/3)t}.$$

As in Example 2, we find the particular solutions to (11) to be

$$(I_L, I_R) = (1, 1).$$

Hence the general solution to (11) is

$$I_L(t) = (c_1 + c_2 t)e^{(-100/3)t} + 1,$$
$$I_R(t) = \left[(2c_1 - \tfrac{3}{50}c_2) + 2c_2 t\right]e^{(-100/3)t} + 1.$$

Using the initial conditions, we have

$$I_L(0) = c_1 + 1 = 0,$$
$$I_R(0) = 2c_1 - \tfrac{3}{50}c_2 + 1 = 0,$$

or

$$c_1 = -1 \quad \text{and} \quad c_2 = -\tfrac{50}{3}.$$

Thus the unique solution to the initial-value problem is

$$(I_L(t), I_R(t)) = \left(\left(-1 - \tfrac{50}{3}t\right)e^{(-100/3)t} + 1, \left(-1 - \tfrac{100}{3}t\right)e^{(-100/3)t} + 1\right).$$

Example 4 Suppose, in Example 2, that $L = 3$ henrys and all other facts remain unchanged. Then $L - 4R^2C = -3$, so the characteristic equation (12) has the roots $100(-1 \pm i)/3$. Then, from equation (7.3.22), solutions to the homogeneous system (13) are

$$I_L(t) = e^{-(100/3)t}\left(c_1 \cos\tfrac{100}{3}t + c_2 \sin\tfrac{100}{3}t\right)$$

and

$$I_R(t) = \frac{L}{R}\left(-\frac{dI_L}{dt}\right) = \frac{3}{100}\left(-\frac{dI_L}{dt}\right)$$

$$= \frac{3}{100}e^{(-100/3)t}\left[\frac{100}{3}\left(c_1\cos\frac{100}{3}t + c_2\sin\frac{100}{3}t\right) + c_1\sin\frac{100}{3}t - c_2\cos\frac{100}{3}t\right]$$

$$= e^{(-100/3)t}\left[(c_1 - c_2)\cos\frac{100}{3}t + (c_1 + c_2)\sin\frac{100}{3}t\right].$$

The nonhomogeneous term in system (11) is E/L, a constant. Thus it is reasonable to find a particular solution to (11) of the form $I_L = A$ and $I_R = B$, where A and B are constants. Inserting these values into (11), we obtain, as in

Example 3,
$$I_L = I_R = 1.$$
The general solution to (11) is, therefore,
$$I_L(t) = e^{(-100/3)t}\left(c_1\cos\tfrac{100}{3}t + c_2\sin\tfrac{100}{3}t\right) + 1$$
$$I_R(t) = e^{(-100/3)t}\left[(c_1 - c_2)\cos\tfrac{100}{3}t + (c_1 + c_2)\sin\tfrac{100}{3}t\right] + 1.$$
Finally, setting $I_R(0) = I_L(0) = 0$, we obtain
$$c_1 + 1 = 0,$$
$$c_1 - c_2 + 1 = 0,$$
with solution $c_1 = -1$, $c_2 = 0$.

Thus the unique solution to our initial-value problem is
$$I_L(t) = 1 - e^{(-100/3)t}\cos\tfrac{100}{3}t$$
$$I_R(t) = 1 - e^{(-100/3)t}\left(\cos\tfrac{100}{3}t + \sin\tfrac{100}{3}t\right).$$

PROBLEMS 8.1

1. Let $R = 100$ ohms, $L = 4$ henrys, $C = 10^{-4}$ farad, and $E = 100$ volts in the network in Figure 8.1. Assume that the currents I_R and I_L are both zero at time $t = 0$. Find the currents when $t = 0.001$ second.

2. Let $L = 1$ henry in Problem 1 and assume that all the other facts are unchanged. Find the currents when $t = 0.001$ second and 0.1 second.

3. Let $L = 8$ henrys in Problem 1 and assume that all the other facts are unchanged. Find the currents when $t = 0.001$ second and 0.1 second.

4. Assume that $E = 100 e^{-1000t}$ volts and that all the other values are unchanged in Problem 1. Do
   a. Problem 1.
   b. Problem 2.
   c. Problem 3.

5. Repeat Problem 4 for $E = 100\sin 60\pi t$ volts.

6. Find the current at time t in each loop of the network shown below, given that $E = 100$ volts, $R = 10$ ohms, and $L = 10$ henrys.

7. Repeat Problem 6 for $E = 10\sin t$ volts.

8. Consider the air-core transformer network shown below with $E = 10\cos t$ volts, $R = 1$ ohm, $L = 2$ henrys, and mutual inductance $L_* = -1$ (which depends on the relative modes of winding of the two coils involved). Treating the mutual inductance as an inductance for each circuit, find the two circuit currents at all times t assuming they are zero at $t = 0$.

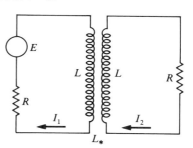

9. Consider the following circuit. Find the current in the top loop if all the resistances are $R = 100$ ohms and the inductances are $L = 1$ henry. Assume that $E = 100$ volts and that there is no current flowing when the switch is closed at time $t = 0$.

10. Repeat Problem 9 for the circuit shown below, where $C = 10^{-4}$ farad.

*11. Consider the **string insulator** illustrated here. Assume that the voltage at $E_0 = A \cos \omega t$. Find the voltage at the kth junction E_k for each k. [*Hint*: Express E_k as a function of the voltages at previous junctions.]

*12. Consider the following **low-pass filter**, where we *assume* that the voltage in the line is alternating and that the voltage at each junction is given by $E_k = A_k \cos \omega t$ for fixed ω. Show that the only nontrivial solutions satisfy

$$\omega_N = \frac{2}{\sqrt{CL}} \sin \frac{N\pi}{2(n+1)}, \quad N = 1, 2, \ldots, n.$$

[Thus waves of frequency higher than $\omega_n < 2/\sqrt{CL}$ are damped out as t increases.]

*13. Consider the network shown below, called a **high-pass filter**. Show that it damps out all waves below a certain **cut-off** frequency.

*14. The illustrated network is called a **band-pass filter**, since frequencies outside a certain band are damped out. Find the two cut-off frequencies.

*15. R. FitzHugh[†] proposed the electric circuit shown below as a model for the transmission of current in a myelinated axon. Find the voltage at all junctions x_k in the network. Assume that $x_0 = A_0 \cos \omega t$.

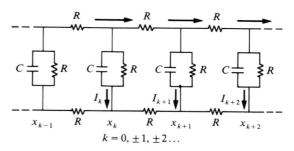

$k = 0, \pm 1, \pm 2 \ldots$

[†] "Computation of Impulse Initiation and Saltatory Conduction in Myelinated Fibres: Theoretical Basis of the Velocity-Diameter Relation," *Biophysics Journal* 2 (1962):11–21.

8.2 Chemical Mixture and Population Biology Problems

In Section 7.3 we presented an example of a chemical mixture problem involving feedback. The problem set of that section included similar situations, as well as applications of systems to medicine and air pollution. In this section we present further examples related to these topics and to population biology. We solve the first example by the method of determinants.

Example 1 Let tank X contain 10 gallons of brine in which 10 pounds of salt is dissolved and tank Y contain 20 gallons of water. Suppose water flows into

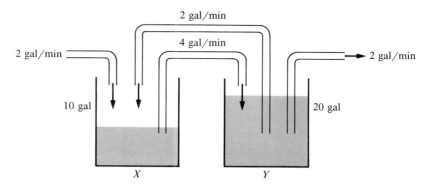

Figure 8.3

tank X at the rate of 2 gallons per minute, and that the mixture flows from tank X into tank Y at 4 gallons per minute. From Y, 2 gallons are pumped back to X each minute (establishing **feedback**) while 2 gallons are flushed away. (a) Find the amount of salt in both tanks at all time t (see Figure 8.3). (b) At what time is the amount of salt in tank Y a maximum?

Solution

a. If we let $x(t)$ and $y(t)$ represent the number of pounds of salt in tanks X and Y at time t and note that the change in weight equals the difference between input and output, we can again derive a system of linear first-order equations. Tanks X and Y initially contain $x(0) = 10$ and $y(0) = 0$ pounds of salt, respectively, at time $t = 0$. The quantities

$$\frac{x}{10} \quad \text{and} \quad \frac{y}{20}$$

are, respectively, the amounts of salt contained in each gallon of water taken from tanks X and Y at time t. Thus we have the system

$$\frac{dx}{dt} = -4\frac{x}{10} + 2\frac{y}{20}, \quad x(0) = 10,$$
$$\frac{dy}{dt} = 4\frac{x}{10} - 4\frac{y}{20}, \quad y(0) = 0. \tag{1}$$

From (1) we obtain the determinant

$$D = \begin{vmatrix} -\frac{2}{5} - \lambda & \frac{1}{10} \\ \frac{2}{5} & -\frac{1}{5} - \lambda \end{vmatrix} = \left(\lambda + \frac{1}{5}\right)\left(\lambda + \frac{2}{5}\right) - \frac{1}{25} = 0,$$

or

$$\lambda^2 + \tfrac{3}{5}\lambda + \tfrac{1}{25} = 0. \tag{2}$$

The characteristic equation (2) has the roots

$$\lambda_1 = \frac{-3 + \sqrt{5}}{10} \quad \text{and} \quad \lambda_2 = \frac{-3 - \sqrt{5}}{10}.$$

From equation (7.3.19),

$$x(t) = c_1 e^{(-3+\sqrt{5})t/10} + c_2 e^{(-3-\sqrt{5})t/10}.$$

From the first equation of (1),

$$y(t) = 10\frac{dx}{dt} + 4x$$

after some algebra
↓

$$= (1+\sqrt{5})c_1 e^{(-3+\sqrt{5})t/10} + (1-\sqrt{5})c_2 e^{(-3-\sqrt{5})t/10}.$$

Using the initial conditions, we obtain

$$x(0) = c_1 + c_2 = 10,$$
$$y(0) = (1+\sqrt{5})c_1 + (1-\sqrt{5})c_2 = 0.$$

Then $c_1 = 5 - \sqrt{5}$, $c_2 = 5 + \sqrt{5}$, and the unique solution to the initial-value problem is

$$(x(t), y(t)) = \left((5-\sqrt{5})e^{(-3+\sqrt{5})t/10} + (5+\sqrt{5})e^{(-3-\sqrt{5})t/10},\right.$$
$$\left. 4\sqrt{5}\,e^{(-3+\sqrt{5})t/10} - 4\sqrt{5}\,e^{(-3-\sqrt{5})t/10}\right).$$

b. $y(t)$ is a maximum when $y'(t) = 0$. But

$$y'(t) = 4\sqrt{5}\left[\frac{-3+\sqrt{5}}{10}e^{(-3+\sqrt{5})t/10} + \frac{3+\sqrt{5}}{10}e^{(-3-\sqrt{5})t/10}\right] = 0.$$

Multiplying through by $(10/4\sqrt{5})e^{3t/10}$ and simplifying, we obtain

$$(-3+\sqrt{5})e^{\sqrt{5}\,t/10} = (-3-\sqrt{5})e^{-\sqrt{5}\,t/10},$$

so

$$e^{\sqrt{5}\,t/5} = \frac{-3-\sqrt{5}}{-3+\sqrt{5}} = \frac{7+3\sqrt{5}}{2}.$$

Hence

$$t = \sqrt{5}\,\ln\left(\frac{7+3\sqrt{5}}{2}\right) \approx 4.3 \text{ minutes.}$$

Example 2 Consider two species that inhabit the same ecosystem. Denote one of the species by x and the other by y and assume that $x(t)$ and $y(t)$ are the sizes of the respective populations at time t.

If the two populations were isolated from each other, and resources were plentiful, it would be reasonable to assume that the growth rate of their populations would be proportional to their number;

$$\frac{dx}{dt} = ax \quad \text{and} \quad \frac{dy}{dt} = by,$$

where a and b are positive constants (the average growth rate per individual of each population).

However, if the two populations *compete* for the same resource (say they both consume the same forage), then the growth rate of each population is influenced by the size of the other population. In such a case we might arrive at the linear **competition system**

$$\frac{dx}{dt} = ax - Ay,$$
$$\frac{dy}{dt} = by - Bx, \tag{3}$$

where A and B are positive constants (the average rate of competition per individual of the other population). The characteristic equation for system (3) is

$$\lambda^2 - (a+b)\lambda + (ab - AB) = 0. \tag{4}$$

If (4) has the roots $\lambda_1 \neq \lambda_2$, then

$$x(t) = c_1 e^{\lambda_1 t} + c_2 e^{\lambda_2 t}$$

and

$$y(t) = \frac{1}{A}\left(ax - \frac{dx}{dt}\right) = \frac{1}{A}\left[(a - \lambda_1)c_1 e^{\lambda_1 t} + (a - \lambda_2)c_2 e^{\lambda_2 t}\right].$$

The constants c_1 and c_2 can be determined from the initial populations $x(0)$ and $y(0)$.

Example 3 Many biological systems are controlled by the production of enzymes or hormones that stimulate or inhibit the secretion of some compound. For example, the pancreatic hormone glucagon stimulates the release of glucose from the liver to the plasma. A rise in blood glucose inhibits the secretion of glucagon but causes an increase in the production of the hormone insulin. Insulin, in turn, aids in the removal of glucose from the blood and in its conversion to glycogen in the muscle tissue. Let G and I be the deviations of plasma glucose and plasma insulin from the normal (fasting) level, respectively. We then have the system

$$\frac{dG}{dt} = -k_{11}G - k_{12}I,$$
$$\frac{dI}{dt} = k_{21}G - k_{22}I, \tag{5}$$

where the positive constants k_{ij} are model parameters, some of which may be determined experimentally. This system is known to exhibit a strongly damped oscillatory behavior; direct injection of glucose into the blood produces a fall of blood glucose to a level below fasting in about one and a half hours followed by a rise slightly above the fasting level in about three hours. Hence the characteristic equation of system (5),

$$D = \begin{vmatrix} -k_{11} - \lambda & -k_{12} \\ k_{21} & -k_{22} - \lambda \end{vmatrix} = (k_{11} + \lambda)(k_{22} + \lambda) + k_{12}k_{21}$$
$$= \lambda^2 + (k_{11} + k_{22})\lambda + (k_{11}k_{22} + k_{12}k_{21}) = 0,$$

must have complex conjugate roots $-a \pm ib$, with $a = (k_{11} + k_{22})/2$ and $b = \sqrt{k_{12}k_{21} - (k_{11} - k_{22})^2/4}$, since only complex roots can lead to oscillatory behavior. Thus our solution has the form

$$(G(t), I(t)) = \left(e^{-at}(c_1\cos bt + c_2\sin bt),\ e^{-at}(d_1\cos bt + d_2\sin bt)\right),$$

where c_1, c_2, d_1, and d_2 are determined by the initial conditions and one of the equations in (5); and $b = 2\pi/3$ if we measure time in hours.[†]

Assume now that a glucose injection is administered at a time when plasma insulin and glucose are at fasting levels and that the glucose is diffused completely in the blood before the insulin level begins to increase ($t = 0$). Then $G(0) = G_0$ equals the ratio of the volume of glucose administered to blood volume, and $I(0) = 0$. Since $G(t)$ is at a maximum when $t = 0$, it follows that $c_1 = G_0$ and $c_2 = 0$. Hence

$$G(t) = G_0 e^{-at}\cos bt. \tag{6}$$

But

$$d_1 = I(0) = 0, \tag{7}$$

and from the first equation in (5)

$$dG/dt = G_0 e^{-at}(-a\cos bt - b\sin bt)$$
$$= -k_{11}G_0 e^{-at}\cos bt - k_{12}e^{at}(d_2\sin bt),$$

or, equating sine and cosine terms,

$$aG_0 = k_{11}G_0 \quad \text{and} \quad bG_0 = k_{12}d_2. \tag{8}$$

Thus

$$I(t) = \frac{bG_0}{k_{12}}e^{-at}\sin bt$$

and $k_{11} = a = (k_{11} + k_{22})/2$, so $k_{11} = k_{22}$.

If the minimum level $G(3/2)$ (< 0) is known, then by equation (6),

$$e^{-3a/2} = \frac{|G(\tfrac{3}{2})|}{G_0},$$

so

$$k_{11} = a = -\frac{2}{3}\ln\frac{|G(\tfrac{3}{2})|}{G_0}.$$

If we determine the plasma insulin at any given time $t_0 > 0$, we can then evaluate the parameters k_{12} and k_{21}.

[†] The function $A\cos bt$ has period $2\pi/b$. If the period is 3 hours, then $(2\pi)/b = 3$ and $b = (2\pi)/3$.

PROBLEMS 8.2

1. Suppose that in Example 1 only $1\tfrac{1}{2}$ gallons per minute are pumped back from tank Y to tank X, while $2\tfrac{1}{2}$ gallons per minute are flushed away. If $2\tfrac{1}{2}$ gallons per minute of water flows into X, when does Y contain the maximum amount of salt? How much salt does it contain?

2. Tank X contains 500 gallons of brine in which 500 pounds of salt are dissolved. Tank Y contains 1500 gallons of water. Water flows into tank X at the rate of 30 gallons per minute, and the mixture flows into Y at the rate of 40 gallons per minute. From Y, the solution is pumped back into X at the rate of 10 gallons per minute and into a third tank at the rate of 30 gallons per minute. Find the maximum amount of salt in Y. When does this concentration occur?

3. Suppose that in Problem 2 tank X contains 2000 gallons of brine. Solve the problem, given that all other conditions are unchanged.

4. In some instances, populations in an ecosystem interact in such a way that each species, by its presence, benefits the other. For example, a pollinating insect may feed on the pollen of a plant species. Such an interaction is often called **mutualism**.

   a. Obtain a system of equations similar to those in Example 2 for a linear mutualism system, with all constants positive.

   b. Analyze the system in (a) and find its solution assuming given initial conditions.

5. In some biological populations, one species preys exclusively on another. If $x(t)$ is the population of the prey at time t and $y(t)$ is the population of the predator, the system

$$\frac{dx}{dt} = ax - Ay,$$
$$\frac{dy}{dt} = Bx - by,$$
(i)

   describes a linear **predator-prey** model. Here a is the growth rate per individual of the prey, A is the predation rate per predator, b is the death rate per individual of the predators, and Bx provides a measurement of the food supply that regulates the growth rate of the predators. Solve (i) for arbitrary initial values of x and y.

6. Linear nonhomogeneous systems of differential equations

$$x' = a_{11}x + a_{12}y + b_1,$$
$$y' = a_{21}x + a_{22}y + b_2,$$
(ii)

   have been used to describe periodic fluctuations of the concentration of hormones in the blood.[†] If $a_{11}, a_{12}, a_{21}, a_{22}, b_1,$ and b_2 are constants with $b_2 = 0$, find the general solution to (ii).

7. Repeat Problem 6 with $b_1, b_2 \neq 0$.[‡]

8. Lewis F. Richardson describes a model of an arms race between two nations X and Y.[§] Only one weapon (missile or bomb) is involved; $x(t)$ denotes the number of weapons nation X has at time t, and $y(t)$ denotes those of Y at that time. Then

$$\frac{dx}{dt} = -a_1 x + a_2 y + a_3,$$
$$\frac{dy}{dt} = b_1 x - b_2 y + b_3,$$
(iii)

   where a_1 and b_2 are "expense and fatigue" coefficients, b_1 and a_2 are "defense" coefficients, and a_3 and b_3 involve the effects of all other factors on the production of weapons. If all the coefficients are positive constants, obtain a general solution to (iii).

9. Repeat Problem 8 assuming that $b_3 = 0$ and $a_3 = a(1 - \cos(\pi t/4))$.

10. In an experiment of cholesterol turnover in humans, radioactive cholesterol $-4-{}^{14}C$ was injected intravenously and the total plasma cholesterol and radioactivity were measured. It was discovered that the turnover of cholesterol behaves like a two-compartment system.[‖] The compartment consisting of the organs and blood has a rapid turnover, while the turnover in the other compartment is much slower. Assume that the body intakes and excretes all cholesterol through the first compartment. Let $x(t)$ and $y(t)$ denote the deviations from normal cholesterol levels in each compartment. Suppose that the daily fractional transfer coefficient from compartment x is 0.134, of which 0.036 is the input to compartment y, and that the transfer coefficient from compartment y is 0.02.

    a. Describe the problem discussed above as a system of homogeneous linear differential equations.

    b. Obtain the general solution of the system.

[†] L. Danziger and G. L. Elmergreen, *Bulletin of Mathematical Biophysics* 19 (1957):9–18.

[‡] Such a model has been used by E. Ackerman, L. C. Gatewood, J. W. Rosevear, and G. D. Molnar, "Blood Glucose Regulation and Diabetes," in *Concepts and Models of Biomathematics*, ed. F. Heinmets (New York: Marcel Dekker, 1969).

[§] *Generalized Foreign Politics*. British Journal of Psychology Monograph Supplement 23 (1939).

[‖] D. S. Goodman and R. P. Noble, "Turnover of Plasma Cholesterol in Man," *Journal of Clinical Investigations* 47 (1968):231–241.

8.3 Mechanical Systems

In Section 7.1 we discussed a mass-spring system consisting of two masses and two springs, with one mass and spring pair suspended from another and the latter suspended from a fixed object. In this section we present some additional examples and problems of this type.

Example 1 Consider the mass-spring system in Figure 8.4, where the point masses m_1 and m_2 are connected in series by springs with spring constants k_1 and k_2. If we denote the vertical displacements from equilibrium of the two masses by $x_1(t)$ and $x_2(t)$, respectively, then using the assumptions of Hooke's law (p. 138), we find that the net forces acting on the two masses are given by

$$F_1 = -k_1 x_1 + k_2(x_2 - x_1),$$
$$F_2 = -k_2(x_2 - x_1).$$

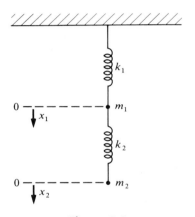

Figure 8.4

Here the positive direction is downward. Note that the first spring is compressed when $x_1 < 0$ and the second spring is compressed when $x_1 > x_2$. The equations of motion are

$$m_1 \frac{d^2 x_1}{dt^2} = -k_1 x_1 + k_2(x_2 - x_1) = -(k_1 + k_2) x_1 + k_2 x_2,$$
$$m_2 \frac{d^2 x_2}{dt^2} = -k_2(x_2 - x_1) = k_2 x_1 - k_2 x_2, \tag{1}$$

a system of two second-order linear differential equations with constant coefficients.

To rewrite (1) as a system of first-order linear equations we define the variables $x_3 = x_1'$ and $x_4 = x_2'$. Then $x_3' = x_1''$, $x_4' = x_2''$, and (1) can be ex-

pressed as the system of four first-order equations
$$x_1' = x_3,$$
$$x_2' = x_4,$$
$$x_1'' = x_3' = -\left(\frac{k_1+k_2}{m_1}\right)x_1 + \left(\frac{k_2}{m_1}\right)x_2, \quad (2)$$
$$x_2'' = x_4' = \left(\frac{k_2}{m_2}\right)x_1 - \left(\frac{k_2}{m_2}\right)x_2.$$

The characteristic equation for this system is given by

$$D = \begin{vmatrix} -\lambda & 0 & 1 & 0 \\ 0 & -\lambda & 0 & 1 \\ -\left(\dfrac{k_1+k_2}{m_1}\right) & \dfrac{k_2}{m_1} & -\lambda & 0 \\ \dfrac{k_2}{m_2} & -\dfrac{k_2}{m_2} & 0 & -\lambda \end{vmatrix} = 0. \quad (3)$$

For simplicity, we assume that $k_1 = 9m_1/2$, $k_2 = 2m_2$, and $7m_1 = 4m_2$. Then (3) becomes

See Appendix 4
↓

$$0 = \begin{vmatrix} -\lambda & 0 & 1 & 0 \\ 0 & -\lambda & 0 & 1 \\ -8 & \tfrac{7}{2} & -\lambda & 0 \\ 2 & -2 & 0 & -\lambda \end{vmatrix} = \lambda^4 + 10\lambda^2 + 9 = (\lambda^2+9)(\lambda^2+1). \quad (4)$$

The four roots are $\lambda_1 = 3i$, $\lambda_2 = -3i$, $\lambda_3 = i$, $\lambda_4 = -i$. Hence, from the theory in Section 7.3, we know that there are constants $a_1, a_2, a_3, a_4, b_1, b_2, b_3,$ and b_4 such that

$$x_1 = a_1\cos 3t + a_2\sin 3t + a_3\cos t + a_4\sin t,$$
$$x_2 = b_1\cos 3t + b_2\sin 3t + b_3\cos t + b_4\sin t. \quad (5)$$

Substituting (5) into (1), we get

$$a_1 = -\tfrac{7}{2}b_1, \qquad a_2 = -\tfrac{7}{2}b_2, \qquad a_3 = \tfrac{1}{2}b_3, \qquad a_4 = \tfrac{1}{2}b_4. \quad (6)$$

If initial conditions are given
$$x_1(0) = x_{10}, \qquad x_1'(0) = x_3(0) = x_{30},$$
$$x_2(0) = x_{20}, \qquad x_2'(0) = x_4(0) = x_{40},$$
we obtain the system of equations
$$a_1 + a_3 = x_{10},$$
$$3a_2 + a_4 = x_{30},$$
$$b_1 + b_3 = x_{20},$$
$$3b_2 + b_4 = x_{40}.$$

Together with (6), this system determines all the coefficients in (5):

$$b_1 = \frac{x_{20} - 2x_{10}}{8}, \quad b_2 = \frac{x_{40} - 2x_{30}}{24},$$

$$b_3 = \frac{7x_{20} + 2x_{10}}{8}, \quad b_4 = \frac{7x_{40} + 2x_{30}}{8}.$$

PROBLEMS 8.3

1. Show that the oscillations exhibited by the masses m_1 and m_2 in Example 1 can be represented by the superposition of *two* cosines.

2. Solve Example 1 with $k_1 = 5m_1/2$, $k_2 = 2m_2$, and $3m_1 = 4m_2$.

3. Solve Example 1 with $k_1 = 3m_1/2$, $k_2 = 2m_2$, and $m_1 = 4m_2$.

4. When a water wave travels past any point, the water rises as the wave approaches the point and recedes as it moves past the point. If a ship is in the trough of a wave, its angular acceleration from the vertical is given by

$$\frac{d^2\psi}{dt^2} = -b^2\psi + A\sin\omega t. \quad\text{(i)}$$

   a. Rewrite (i) as a system of first-order differential equations.

   b. Solve the resulting system by the method of determinants.

   c. Solve the system in part (a) by finding its principal solution.

*5. A rotating, straight slender shaft may be dynamically unstable at high speeds. When the period ω of rotation is nearly equal to one of its periods of lateral vibration, the shaft is then said to **whirl**. A whirling shaft satisfies the differential equation

$$EI\frac{d^4y}{dx^4} = m\omega^2 y, \quad\text{(ii)}$$

where $y = y(x)$ is the distance of the shaft from its geometric axis, E denotes Young's modulus of elasticity, I is the moment of inertia of the shaft, m is the mass. Assume that the shaft is hinged at both ends, has length L, and satisfies

$$y(0) = y(L) = y''(0) = y''(L) = 0.$$

   a. Express (ii) as a system of first-order differential equations.

   b. Show that nontrivial solutions exist if and only if $\omega = n^2\pi^2\sqrt{EI/m}/L^2$.

   c. Find the rotational speed necessary to produce whirling of a steel shaft $1\frac{1}{2}$ inches in diameter and 8 feet long. [*Hint:* $E = 4.32 \times 10^9$, $I = \pi r^4/4$, $m = 6/32.2$.]

6. Consider the coupled mechanical system shown below, where two masses m_1 and m_2 rest on a frictionless plane and are attached to fixed walls by two springs with spring constants k_1 and k_3. The masses m_1 and m_2 are connected by a spring with spring constant k_2.

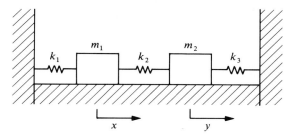

   a. Obtain the system of differential equations describing the motion of this mechanical system.

   b. Solve the system in part (a), and show that the motion of each mass is a superposition of two simple harmonic motions.

*7. Determine the equations of motion of a double pendulum consisting of two simple pendulums of masses m_1 and m_2 and lengths l_1 and l_2, respectively, shown below. You may assume that the angular displacements are so small that $\sin\theta \approx \theta$ and $\sin\phi \approx \phi$.

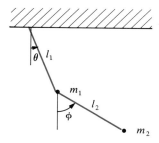

8. Using the system in Problem 7, neglecting the terms containing $(\theta')^2$ and $(\phi')^2$, and replacing $\cos(\theta - \phi)$ by 1, show that one obtains the system

$$l_1(m_1 + m_2)\theta'' + m_2 l_2 \phi'' + g(m_1 + m_2)\theta = 0,$$
$$l_1\theta'' \qquad\qquad + l_2\phi'' + g\phi \qquad = 0.$$

9. Find the general solution of the system in Problem 8.

8.4 An Excursion: A Model for Epidemics

In recent years many mathematicians and biologists have attempted to find reasonable mathematical models to describe the growth of an epidemic in a population. One such model has been used by the Center for Disease Control in Atlanta, Georgia, to help form a public testing policy to limit the spread of gonorrhea.[†] In this section we provide a simple model to describe what may happen in an epidemic.

An **epidemic** is the spread of an infectious disease through a community that affects a significant proportion of the population. The epidemic may begin when a certain number of infected individuals enter the community. This could result, for example, from new people moving to the community or old residents returning from a trip. We make the following assumptions:

a. Everyone in the community is initially susceptible to the disease; that is, no one is immune.

b. The disease is spread only by direct contact between a susceptible person (hereafter called a **susceptible**) and an infected person (called an **infective**).

c. Everyone who has had the disease and has **recovered** is immune. The people who die from the disease are also considered to be in the recovered class in this model.

d. After the disease has been introduced into the community, the total population N of the community remains fixed.

e. The infectives are introduced into the community (that is, the epidemic "starts") at time $t = 0$.

In order to model the spread of the disease, we define three variables:

$x(t)$ is the number of susceptibles at time t;

$y(t)$ is the number of infectives at time t;

$z(t)$ is the number of recovered persons at time t.

Then, by assumption (d), we have

$$x(t) + y(t) + z(t) = N. \tag{1}$$

Also, we have

$$y(0) = \text{number of initial infectives}, \tag{2}$$
$$x(0) = N - y(0), \tag{3}$$
$$z(0) = 0. \tag{4}$$

Equation (4) states the obvious fact that no one has yet recovered or died from the disease at the time the epidemic begins.

To get a better idea of what is going on, look at Figure 8.5. Here we see that a susceptible can become infective and an infective can recover. These are the

---

[†] J. A. Yorke, H. W. Hethcote, and A. Nold, "Dynamics and Control of the Transmission of Gonorrhea," *Sexually Transmitted Diseases* 5(2) (1978):51–56.

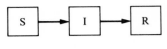

Figure 8.5

only possibilities. For reasons that are obvious, the model we are describing is, in the literature, usually referred to as an **SIR model**.

What mechanism regulates the rate at which the disease spreads? Most likely the disease spreads more rapidly (that is, susceptibles become infectives) if the number of infectives or the number of susceptibles increases, because chances of contact between infectives and susceptibles increase. Thus it is reasonable to assume that the rate of change of the number of susceptibles is proportional both to the number of susceptibles and to the number of infectives. In mathematical terms, we have

$$x'(t) = -\alpha x(t) y(t), \tag{5}$$

where α is a constant of proportionality and the minus sign indicates that the number of susceptibles is decreasing. Equation (5) is called the **law of mass action**.

On the other hand, it is reasonable to assume that the rate at which people recover or die from the disease is proportional to the number of infectives. This gives us the equation

$$z'(t) = \beta y(t). \tag{6}$$

Equations (5) and (6) constitute the system of equations defining our epidemic model. The constant α is often called the **infection rate** and β is called the **removal rate**.

We begin our analysis by dividing equation (5) by equation (6):

$$\frac{x'(t)}{z'(t)} = -\frac{\alpha x(t) y(t)}{\beta y(t)} = -\frac{\alpha}{\beta} x(t). \tag{7}$$

Rearranging the terms in (7) yields

$$\frac{x'(t)}{x(t)} = -\frac{\alpha}{\beta} z'(t),$$

and after integrating both sides we have

$$\int \frac{x'(t)}{x(t)} dt = -\frac{\alpha}{\beta} \int z'(t) dt + C,$$

or

$$\ln x(t) = -\frac{\alpha}{\beta} z(t) + C. \tag{8}$$

Setting $t = 0$ in (8) and noting that $z(0) = 0$, we have

$$\ln x(0) = 0 + C \quad \text{or} \quad C = \ln x(0).$$

Thus

$$\ln x(t) = -\frac{\alpha}{\beta} z(t) + \ln x(0),$$

so

$$\ln \frac{x(t)}{x(0)} = -\frac{\alpha}{\beta}z(t) \quad \text{or} \quad \frac{x(t)}{x(0)} = e^{(-\alpha/\beta)z(t)},$$

and, finally,

$$x(t) = x(0)e^{(-\alpha/\beta)z(t)}. \tag{9}$$

A major question about any epidemic is, Can it be controlled? It is clear that the quantity $y'(t)$ is a measure of how bad the epidemic is. The bigger $y'(t)$ is, the greater the number of people becoming infected. We can say that the epidemic has been **controlled** if, at some point, $y'(t) \leq 0$. Certainly, at some point $y'(t)$ must become negative. This follows from the fact that the population size N is fixed. Then, if everyone becomes infected, $y'(t)$ will be negative as infectives die or recover, and there cannot possibly be any new infectives (there is no one left to become infected). Thus the epidemic is controlled if $y'(t) \leq 0$ for every $t > t_0$.

To analyze this situation, from (1) we have

$$y(t) = N - x(t) - z(t),$$

so

$$y'(t) = -x'(t) - z'(t) = \alpha x(t)y(t) - \beta y(t), \tag{10}$$

or

$$y'(t) = \alpha y(t)\left[x(t) - \frac{\beta}{\alpha}\right]. \tag{11}$$

Thus $y'(t) \leq 0$ whenever $x(t) \leq \beta/\alpha$. The term β/α is called the **relative removal rate**. We can now answer our question. Since $x' \leq 0$ by equation (5), if $x(0) \leq \beta/\alpha$, then $x(t) \leq \beta/\alpha$ for all t; that is

> the epidemic will be controlled at the start if the condition $x(0) \leq \beta/\alpha$ holds.

Biologically, this means that

> an epidemic will not ensue if the initial number of susceptibles (N minus the number of initial infectives) does not exceed the relative removal rate.

Example 1 Let $N = 1000$, $y(0) = 50$, $\alpha = 0.0001$, and $\beta = 0.01$. Then $x(0) = 950$ and $\beta/\alpha = 100$, so $x(0) > \beta/\alpha$ and an epidemic will ensue.

Example 2 Let $N = 1000$, $y(0) = 50$, $\alpha = 0.0001$, and $\beta = 0.1$. Then $x(0) = 950$, $\beta/\alpha = 1000$, and an epidemic will not ensue.

We now ask another question. Suppose that $x(0) > \beta/\alpha$ and an epidemic does ensue. How many people eventually become infected? Equivalently, after the epidemic has run its course, how many susceptibles remain? Since $z(t) \leq N$, equation (9) provides a *positive* lower bound for the number of susceptibles at every time t:

$$x(t) = x(0)e^{(-\alpha/\beta)z(t)} \geq x(0)e^{-(\alpha/\beta)N} > 0. \qquad (12)$$

Denote the number of susceptibles that never get infected by x^*. The fact that x^* is positive [and greater than or equal to the right side of (12)] has some interesting consequences. By (5) note that $x'(t) < 0$ whenever $x(t)$ and $y(t)$ are positive. Since $x(t)$ does not decrease any lower than x^*, this implies that $x' \to 0$ as $t \to \infty$. But $x \geq x^*$, hence $y \to 0$ as $t \to \infty$. Call $W = N - x^*$ the **extent** of the epidemic: it is the total number of individuals infected. We can obtain a formula for W as follows: Divide equation (10) by equation (5) to obtain

$$\frac{y'(t)}{x'(t)} = \frac{\alpha x(t) y(t) - \beta y(t)}{-\alpha x(t) y(t)} = \frac{\beta}{\alpha x(t)} - 1.$$

Rearranging and integrating both sides, we have

$$\int y'(t)\, dt = \frac{\beta}{\alpha} \int \frac{x'(t)}{x(t)}\, dx - \int x'(t)\, dt + C$$

or

$$y(t) = \frac{\beta}{\alpha} \ln x(t) - x(t) + C,$$

so that

$$x(t) + y(t) - \frac{\beta}{\alpha} \ln x(t) = C.$$

Setting $t = 0$ and recalling that $N = x(0) + y(0)$, we get

$$C = N - \frac{\beta}{\alpha} \ln x(0),$$

or

$$\boxed{x(t) + y(t) - N = \frac{\beta}{\alpha} \ln \frac{x(t)}{x(0)}.} \qquad (13)$$

Letting $t \to \infty$, we have $x(t) \to x^*$ and $y(t) \to 0$, so equation (13) becomes

$$x^* - N = \frac{\beta}{\alpha} \ln \frac{x^*}{x(0)},$$

or

$$-W = \frac{\beta}{\alpha} \ln \frac{N - W}{x(0)}. \qquad (14)$$

Multiplying both sides of (14) by α/β and exponentiating, we have

$$e^{(-\alpha/\beta)W} = \frac{N-W}{x(0)},$$

from which we obtain

$$\boxed{N - x(0)e^{(-\alpha/\beta)W} - W = 0.} \tag{15}$$

If α and β are > 0, it is impossible to find an explicit solution for W in terms of α, β, and N. The best we can do is solve it numerically. This is a reasonable thing to do using Newton's method from calculus, as we show in the next example.

Example 3 Let $N = 1000$, $y(0) = 50$, $\alpha = 0.0001$, and $\beta = 0.09$. Compute the extent of the epidemic.

Solution We first note that $x(0) = 950$ and $\beta/\alpha = 900$, so $x(0) > \beta/\alpha$ and an epidemic will ensue. Since $\alpha/\beta = 1/900$, we must find a root of the equation

$$f(W) = 1000 - 950e^{-W/900} - W = 0. \tag{16}$$

Then

$$f'(W) = \tfrac{950}{900}e^{-W/900} - 1,$$

and by Newton's method we obtain the iterates

$$W_{n+1} = W_n - \frac{f(W_n)}{f'(W_n)},$$

or

$$W_{n+1} = W_n - \frac{1000 - 950e^{-W_n/900} - W_n}{(950/900)e^{-W_n/900} - 1}. \tag{17}$$

We have no idea, initially, what W is (although $W > y(0) = 50$), so we start with a guess: $W_0 = 100$. We then obtain the iterates given in Table 8.1. After

TABLE 8.1

n	W_n	$e^{-W_n/900}$	(a) $1000 - 950e^{-W_n/900} - W_n$	(b) $\frac{950}{900}e^{-W_n/900} - 1$	$\frac{(a)}{(b)}$	$W_{n+1} = W_n - \frac{(a)}{(b)}$
0	100.0000000	0.8948393168	-49.9026490400	-0.0554473878	-900.0000000000	1000.0000000
1	1000.0000000	0.3291929878	-312.7333384000	-0.6525185129	479.2712119000	520.7287882
2	520.7287882	0.5606897580	-53.3840582900	-0.4081608110	130.7917293000	389.9370589
3	389.9370589	0.6483896843	-5.9072589340	-0.3155886666	18.7182226700	371.2188362
4	371.2188362	0.6620161196	-0.1341497754	-0.3012052071	0.4453766808	370.7734595
5	370.7734595	0.6623438079	-0.0000770206	-0.3008593139	0.0002560020	370.7732035
6	370.7732035	0.6623439963	-0.0000000004	-0.3008591150	-0.0000000013	370.7732035

six iterations we obtain the value $W_6 = 370.7732035$, which is correct to ten significant figures, as can be verified by substituting it into equation (16). Thus $W \approx 371$, which means that by the time the epidemic has run its course, 371 individuals have been infected. Of course, we cannot say how many of these remain infected, recover, or die.

Before leaving this section, we mention some of the limitations of this model. The constants α and β often vary with time. Also, the recovered individuals may, after a time, lose their immunity (or never acquire it) and reenter the susceptible state. This could give rise to periodic epidemics in which individuals get the disease many times. This lack of immunity holds for many diseases, notably gonorrhea and certain types of influenza. Finally, there may be many factors other than relative population sizes that control the sizes of the three classes. Such factors might include weather, the available food supply, living conditions, and the presence of other diseases in the community. However, even a simple model like this one can give us the kind of insight needed to study more complicated situations. In a few cases this has been done with great success.[†]

[†] If you are interested in learning more about epidemic models, consult the following: Paul Waltman, *Deterministic Threshold Models in the Theory of Epidemics. Lecture Notes in Biomathematics*, vol. 1. (New York: Springer-Verlag, 1974); Klaus Dietz, "Epidemics and Rumours: A Survey," *Journal of the Royal Statistical Society* Series A., 130 (1967):505–528.

PROBLEMS 8.4

In Problems 1–5 values for N, $y(0)$, α, and β are given. Determine whether an epidemic will occur and, if so, find the extent of the epidemic.

1. $N = 1000$, $y(0) = 100$, $\alpha = 0.001$, $\beta = 0.4$.
2. $N = 1000$, $y(0) = 100$, $\alpha = 0.0001$, $\beta = 0.1$.
3. $N = 10{,}000$, $y(0) = 1500$, $\alpha = 10^{-5}$, $\beta = 0.2$.
4. $N = 10{,}000$, $y(0) = 1500$, $\alpha = 10^{-5}$, $\beta = 0.02$.
5. $N = 25{,}000$, $y(0) = 5000$, $\alpha = 10^{-5}$, $\beta = 0.1$.
6. Let $f(W)$ be given by (15). Show that there is exactly one positive value of W for which $f(W) = 0$. [*Hint*: Show that $f(0) > 0$ and that $f''(W) < 0$ for all W and that $f(W) < 0$ if W is sufficiently large.]
7. Prove that the maximum number of infectives, $y_{\max}$, is given by

$$y_{\max} = N + \frac{\beta}{\alpha}\left[\ln\left(\frac{\beta}{\alpha x(0)}\right) - 1\right]$$

$$\geq N - \frac{\beta}{\alpha}\left[1 + \ln\left(\frac{\alpha N}{\beta}\right)\right] > 0,$$

whenever $x(0) > \beta/\alpha$.

Review Exercises for Chapter 8

1. A direct-current transmission line of length L connecting a power source to a distant receiver is subject to (a) drops in voltage due to the resistance of the line, and (b) leakage of current along the line due to imperfect insulation. If $E(x)$ and $I(x)$ are the voltage and current at a distance x from the power source, R is the resistance (ohms) per unit length of line, and G is the leakance (conductance) (mhos) per unit length, find

   a. a system of differential equations describing the voltage and current in the transmission line;
   b. the solution to part (a);
   c. the current and voltage at the end of the line.

2. J. P. Brady and C. Marmasse obtained the initial-value problem

   $$ay'' + y' + by = \beta, \quad y(0) = \alpha, \quad y'(0) = \beta,$$

to describe avoidance learning in rats.[†] Here $y(t)$ is the value of the learning curve of the rat at time t, α and β are the initial values, and a and b are constants.

a. Obtain a system of first-order equations equivalent to the given initial-value problem.

b. Solve the system in part (a).

3. Tank X contains 150 gallons of pure water and tank Y 150 gallons of brine in which 60 pounds of salt has been dissolved. Liquid circulates from each tank to the other at the rate of 9 gallons per minute, with the mixture kept uniform in each tank by stirring.

a. How much salt is in each tank at any time t?

b. How much salt is in each tank as $t \to \infty$?

c. When does tank X contain the maximum amount of salt?

4. Find the current I_2 in the indicated loop for the illustrated circuit, where $R = 10$ ohms and $L = 1$ henry, assuming $I_1 = I_2 = 0$ when the switch is closed at time $t = 0$. Assume $E = 10$ volts.

5. S. Grossberg obtained the initial-value problem

$$\frac{dx}{dt} = a(b-x) - cy, \qquad x(0) = b,$$

$$\frac{dy}{dt} = d(x-y) - (a+c)y, \quad y(0) = \frac{bd}{a+d},$$

where a, b, c, d are positive constants.[‡] Show that y decays monotonically to a positive minimum.

6. A transformer having a spark gap in the primary circuit (see illustration) is called an **oscillation transformer**. Suppose that the capacitor C_1 is initially charged to e_0 volts, and that this is the only source of energy. If C_1 is allowed to discharge through the primary circuit, the current jumps the spark gap G and continues around the circuit through the coil L_1. Find the currents in the circuits at all subsequent times. Let M denote the mutual inductance.

[†] *Psychological Record* 12 (1962):361–368.

[‡] *Journal of Theoretical Biology* (1969):325–364.

9

Numerical Methods

9.1 Error Analysis

In every chapter of this book we have performed numerical computations: solving differential equations, applying the methods of power series and Frobenius, and using Laplace transforms. With few exceptions, we have limited our examples to problems involving a small number of variables—not because most applications have only two or three variables but because the computations would have been too tedious otherwise.

With the recent and widespread use of calculators and computers, the situation has been altered. The remarkable strides made in the last few years in the theory of numerical methods for solving certain computational problems have made it possible to perform, quickly and accurately, the calculations mentioned in the first paragraph.

The use of the computer presents new difficulties, however. Computers do not store numbers such as $\frac{2}{3}$, $7\frac{3}{8}$, $\sqrt{2}$, and π. Rather, every computer uses what is called **floating-point arithmetic**. In this system every number is represented in the form

$$x = \pm 0.d_1 d_2 \cdots d_k \times 10^n, \qquad d_1 \neq 0, \tag{1}$$

where $d_1, d_2, \ldots, d_k$ are single-digit integers and n is an integer. Any number written in this form is called a **floating-point number**. In equation (1), the number $\pm 0.d_1 d_2 \cdots d_k$ is called the **mantissa** and the number n is called the **exponent**. The number k is called the **number of significant digits** in the expression.

Example 1 The following numbers are expressed in floating-point form:

a. $\frac{1}{4} = 0.25$

b. $2378 = 0.2378 \times 10^4$

c. $-0.000816 = -0.816 \times 10^{-3}$

d. $83.27 = 0.8327 \times 10^2$

If the number of significant digits were unlimited, we would have no problem. Almost every time numbers are introduced into a computer, however, errors begin to accumulate. This can happen in one of two ways:

a. **Truncation:** All significant digits after k digits are simply "cut off." For example, if truncation is used, $\frac{2}{3} = 0.666666\ldots$ is stored (with $k = 8$) as $\frac{2}{3} = 0.66666666 \times 10^0$.

b. **Rounding:** If $d_{k+1} \geq 5$, then 1 is added to d_k and the resulting number is truncated. Otherwise, the number is simply truncated. For example, with rounding (and $k = 8$), $\frac{2}{3}$ is stored as $\frac{2}{3} = 0.66666667 \times 10^0$.

We can illustrate how some numbers are stored with truncation and rounding by using eight significant digits:

Number	Truncated number	Rounded number
$\frac{8}{3}$	0.26666666×10^1	0.26666667×10^1
π	0.31415926×10^1	0.31415927×10^1
$-\frac{1}{57}$	$-0.17543859 \times 10^{-1}$	$-0.17543860 \times 10^{-1}$

Individual round-off or truncation errors do not seem very significant. When thousands of computational steps are involved, however, the *accumulated* round-off error can be devastating. Thus, in discussing any numerical scheme, it is necessary to know not only whether you will get the right answer, theoretically, but also how badly the round-off errors will accumulate. To keep track of things, we define two types of error. If x is the actual value of a number and x^* is the number that appears in the computer, then the **absolute error** ε_a is defined by

$$\boxed{\varepsilon_a = |x^* - x|.} \qquad (2)$$

More interesting in most situations (as we see in Example 2) is the **relative error** ε_r, defined by

$$\boxed{\varepsilon_r = \left| \frac{x^* - x}{x} \right|.} \qquad (3)$$

Example 2 Let $x = 2$ and $x^* = 2.1$. Then $\varepsilon_a = 0.1$ and $\varepsilon_r = 0.1/2 = 0.05$. If $x_1 = 2000$ and $x_1^* = 2000.1$, then, again, $\varepsilon_a = 0.1$. But now $\varepsilon_r = 0.1/2000 = 0.00005$. Most people would agree that the 0.1 error in the first case is more significant than the 0.1 error in the second.

Much of numerical analysis is concerned with questions of **convergence** and **stability**. If x is the answer to a problem and our computational method gives us approximating values x_n, then the method converges if, theoretically, x_n approaches x as n gets large. If, moreover, it can be shown that the round-off errors do not accumulate in such a way as to make the answer unreliable, then the method is stable.

It is easy to give an example of a procedure in which round-off error can be quite large. Suppose we wish to compute $y = 1/(x - 0.66666665)$. For $x = \frac{2}{3}$, if the computer truncates, then $x = 0.66666666$ and $y = 1/0.00000001 = 10^8 = 10 \times 10^7$. If the computer rounds, then $x = 0.66666667$ and $y = 1/0.00000002 = 5 \times 10^7$. The difference here is enormous. The correct answer is $1/\left(\frac{2}{3} - \frac{66666665}{100000000}\right) = 60{,}000{,}000 = 6 \times 10^7$.

In this chapter we examine several numerical procedures for approximating the solution of a differential equation. In certain instances we also discuss convergence and stability. This is but a very superficial view of numerical methods, however; entire books and courses are devoted to the subject. For a more exhaustive view, you are encouraged to consult the following references:

1. Blum, E. K. *Numerical Analysis and Computation: Theory and Practice*. Reading, Mass.: Addison-Wesley, 1972.

2. Burden, R. L., J. D. Faires, and A. C. Reynolds. *Numerical Analysis*. Boston: Prindle, Weber and Schmidt, 1978.

3. Conte, S. D. *Elementary Numerical Analysis*, 2nd Ed. New York: McGraw-Hill, 1972.

4. Faddeev, D. K., and V. N. Faddeeva. *Computational Methods of Linear Algebra*. San Francisco: Freeman, 1963.

PROBLEMS 9.1

In Problems 1–13 convert the number to a floating-point number with eight decimal places of accuracy. Either truncate (T) or round off (R) as indicated.

1. $\frac{1}{3}$ (T)
2. $\frac{7}{8}$
3. -0.000035
4. $\frac{7}{9}$ (R)
5. $\frac{7}{9}$ (T)
6. $\frac{33}{7}$ (T)
7. $\frac{85}{11}$ (R)
8. $-18\frac{5}{6}$ (T)
9. $-18\frac{5}{6}$ (R)
10. $237{,}059{,}628$ (T)
11. $237{,}059{,}628$ (R)
12. -23.7×10^{15}
13. 8374.2×10^{-24}

In Problems 14–21 the number x and an approximation x^* are given. Find the absolute and relative errors ε_a and ε_r.

14. $x = 5,\ x^* = 0.49 \times 10^1$
15. $x = 500,\ x^* = 0.4999 \times 10^3$
16. $x = 3720,\ x^* = 0.3704 \times 10^4$
17. $x = \frac{1}{8},\ x^* = 0.12 \times 10^0$
18. $x = \frac{1}{800},\ x^* = 0.12 \times 10^{-2}$
19. $x = -5\frac{5}{6},\ x^* = -0.583 \times 10^1$
20. $x = 0.70465,\ x^* = 0.70466 \times 10^0$
21. $x = 70465,\ x^* = 0.70466 \times 10^5$

9.2 Euler Methods for First-Order Differential Equations

In Section 1.3 we described the Euler method of approximating the solution of the first-order initial-value problem

$$\frac{dy}{dx} = f(x, y), \quad y(x_0) = y_0, \tag{1}$$

at the points

$$x_0, \ x_1 = x_0 + h, \ x_2 = x_0 + 2h, \ldots, \ x_n = x_0 + nh$$

for h, some nonzero real number. The method, defined by the iterative formula

$$y_{n+1} = y_n + hf(x_n, y_n), \tag{2}$$

Euler Method

with $y_0 = y(x_0)$, approximates the solution by following the tangent to the solution curve passing through (x_n, y_n) for a small horizontal distance. Note that if we replace $f(x_n, y_n)$ by the derivative $y'_n = y'(x_n)$, then equation (2) resembles the first two terms of the Taylor series for $y(x)$ at the value x_n. The following example illustrates how the work can be organized in tabular form.

Example 1 Find an approximate value for the solution of the initial-value problem

$$y' = x + y^2, \quad y(1) = 0,$$

at $x = 1, 1.1, 1.2, 1.3, 1.4, 1.5$.

Solution We arrange our work in columns: the first column contains the values of x at $1, 1.1, \ldots, 1.5$; the second contains the values of y beginning with the initial condition; the third, the computed value of $f(x_n, y_n) = x_n + y_n^2$ for the given x and y on that row; and the last the computed value of y_{n+1} according to equation (2). The value in the last column is then transferred to the y entry in the next row to compute the entries in the third and fourth columns (see Table 9.1).

TABLE 9.1 Euler Method with $h = 0.1$, for $y' = x + y^2$, $y(1) = 0$

x_n	y_n	$f(x_n, y_n)$	$y_n + hf(x_n, y_n)$
1.00	0.00	1.00	0.10
1.10	0.10	1.11	0.21
1.20	0.21	1.24	0.34
1.30	0.34	1.41	0.48
1.40	0.48	1.63	0.64
1.50	0.64		

As we saw in Section 1.3 (see p. 24), we can usually reduce the discretization error by reducing the step size. For example, if instead of a step size of $h = 0.1$ we use $h = 0.05$, our solution would be that shown in Table 9.2, while

TABLE 9.2 Euler Method with $h = 0.5$, for $y' = x + y^2$, $y(1) = 0$

x_n	y_n	$f(x_n, y_n)$	$y_n + hf(x_n, y_n)$
1.00	0.00000	1.0000	0.05000
1.05	0.05000	1.0525	0.10263
1.10	0.10263	1.1105	0.15815
1.15	0.15815	1.1750	0.21690
1.20	0.21690	1.2470	0.27925
1.25	0.27925	1.3280	0.34565
1.30	0.34565	1.4195	0.41663
1.35	0.41663	1.5236	0.49281
1.40	0.49281	1.6429	0.57495
1.45	0.57495	1.7806	0.66398
1.50	0.66398		

TABLE 9.3 Euler Method with $h = 0.025$, for $y' = x + y^2$, $y(1) = 0$

x_n	y_n	$f(x_n, y_n)$	$y_n + hf(x_n, y_n)$
1.000	0.00000	1.0000	0.02500
1.025	0.02500	1.0256	0.05064
1.050	0.05064	1.0526	0.07695
1.075	0.07695	1.0809	0.10398
1.100	0.10398	1.1108	0.13175
1.125	0.13175	1.1424	0.16031
1.150	0.16031	1.1757	0.18970
1.175	0.18970	1.2110	0.21997
1.200	0.21997	1.2484	0.25118
1.225	0.25118	1.2881	0.28339
1.250	0.28339	1.3303	0.31664
1.275	0.31664	1.3753	0.35103
1.300	0.35103	1.4232	0.38661
1.325	0.38661	1.4745	0.42347
1.350	0.42347	1.5293	0.46170
1.375	0.46170	1.5882	0.50140
1.400	0.50140	1.6514	0.54269
1.425	0.54269	1.7195	0.58568
1.450	0.58568	1.7930	0.63050
1.475	0.63050	1.8725	0.67732
1.500	0.67732		

$h = 0.025$ would yield Table 9.3. Observe that in each case the approximate value of $y(1.5)$ increases. This is the behavior that one would anticipate if the solution curves in the direction field are all increasing in the interval $1 \leq x \leq 1.5$. Figure 9.1 illustrates this assertion: the approximation with step size $h/2$ lies above the one of step size h, because the slope of the solution curves, near the true solution at $x_0 + h/2$, exceeds the slope at x_0. Thus two steps of size $h/2$ incur less error, in this situation, than one step of size h.

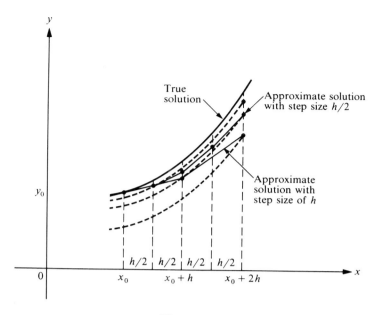

Figure 9.1

IMPROVED EULER METHOD

The improved Euler method was developed to reduce the discretization error that occurs in the Euler method. The procedure *averages* the slopes at the left and right endpoints of each step and uses this average slope to compute the next y-value. To see why this is a reasonable procedure, integrate both sides of equation (1) between x_0 and x_1:

$$y(x_1) - y(x_0) = \int_{x_0}^{x_1} \frac{dy}{dx}\, dx = \int_{x_0}^{x_1} f(x, y(x))\, dx,$$

or

$$y(x_1) = y(x_0) + \int_{x_0}^{x_1} f(x, y(x))\, dx. \qquad (3)$$

If we approximate the integral in (3) by the **trapezoidal rule** of calculus, we obtain

$$y(x_1) \approx y(x_0) + \frac{h}{2}[f(x_0, y(x_0)) + f(x_1, y(x_1))]. \qquad (4)$$

Since $y(x_1)$ is not known, we replace it by the value found by the Euler method, which we call z_1; then (4) can be replaced by the system of equations

$$z_1 = y_0 + hf(x_0, y_0),$$

$$y_1 = y_0 + \frac{h}{2}[f(x_0, y_0) + f(x_1, z_1)].$$

Figure 9.2 illustrates the effect of averaging the slopes at the endpoint of the interval and provides a geometric interpretation of each step of the general procedure.

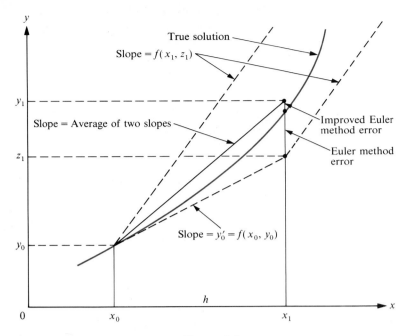

Figure 9.2
Improved Euler Method

IMPROVED EULER METHOD
$$z_{n+1} = y_n + hf(x_n, y_n),$$
$$y_{n+1} = y_n + \frac{h}{2}[f(x_n, y_n) + f(x_{n+1}, z_{n+1})]. \tag{5}$$

Example 2   Consider the initial-value problem that we solved using the Euler method in Example 1.3.2 (p. 24):
$$\frac{dy}{dx} = y + x^2, \qquad y(0) = 1. \tag{6}$$

We wish to determine $y(1)$. We saw in Section 1.3 that with $h = 0.2$, the Euler method yields the approximation $y(1) \approx 2.77$, which is off by about 12 percent from the exact solution $y(1) = 3e - 5 \approx 3.1548$. Using $x_0 = 0$, $y_0 = 1$, and $h = 0.2$, equation (5) yields

$$z_1 = y_0 + hf(x_0, y_0) = 1 + 0.2(1 + 0^2) = 1.2,$$
$$y_1 = y_0 + \frac{h}{2}[f(x_0, y_0) + f(x_1, z_1)] = 1 + 0.1[(1 + 0^2) + 1.2 + 0.2^2]$$
$$= 1 + 0.1(2.24) = 1.224 \approx 1.22,$$
$$z_2 = y_1 + hf(x_1, y_1) = 1.22 + 0.2[1.22 + (0.2)^2] = 1.472 \approx 1.47,$$
$$y_2 = y_1 + \frac{h}{2}[f(x_1, y_1) + f(x_2, z_2)]$$
$$= 1.22 + 0.1[1.22 + (0.2)^2 + 1.47 + (0.4)^2] = 1.509 \approx 1.51,$$

9.2/EULER METHODS FOR FIRST-ORDER DIFFERENTIAL EQUATIONS

TABLE 9.4 Improved Euler Method with $h = 0.2$, for $y' = y + x^2$, $y(0) = 1$

x_n	y_n	$f(x_n, y_n) = y_n + x_n^2$	z_{n+1}	$f(x_{n+1}, z_{n+1}) = z_{n+1} + x_{n+1}^2$	y_{n+1}
0.0	1.00	1.00	1.20	1.24	1.22
0.2	1.22	1.26	1.47	1.63	1.51
0.4	1.51	1.67	1.84	2.20	1.90
0.6	1.90	2.26	2.35	2.99	2.43
0.8	2.43	3.07	3.04	4.04	3.14
1.0	3.14				

TABLE 9.5 Improved Euler Method with $h = 0.1$, for $y' = y + x^2$, $y(0) = 1$

x_n	y_n	$f(x_n, y_n)$	z_{n+1}	$f(x_{n+1}, z_{n+1})$	y_{n+1}
0.00	1.00	1.00	1.10	1.11	1.11
0.10	1.11	1.12	1.22	1.26	1.22
0.20	1.22	1.26	1.35	1.44	1.36
0.30	1.36	1.45	1.50	1.66	1.52
0.40	1.52	1.68	1.68	1.93	1.70
0.50	1.70	1.95	1.89	2.25	1.91
0.60	1.91	2.27	2.13	2.62	2.15
0.70	2.15	2.64	2.41	3.05	2.43
0.80	2.43	3.07	2.74	3.55	2.77
0.90	2.77	3.58	3.12	4.12	3.15
1.00	3.15				

TABLE 9.6 A Comparison of Results Using Euler and Improved Euler Methods with $h = 0.2$ and $h = 0.1$, for the Initial-Value Problem $y' = y + x^2$, $y(0) = 1$

x	Euler	Improved Euler	Exact
$h = 0.2$			
0.0	1.000000	1.00000000	1.00000000
0.2	1.200000	1.22400000	1.22420828
0.4	1.448000	1.51408000	1.51547409
0.6	1.769600	1.90237760	1.90635640
0.8	2.195520	2.42810067	2.43662279
1.0	2.762624	3.13908282	3.15484548
$h = 0.1$			
0.0	1.00000000	1.00000000	1.00000000
0.1	1.10000000	1.10550000	1.10551275
0.2	1.21100000	1.22412750	1.22420828
0.3	1.33610000	1.35936089	1.35957642
0.4	1.47871000	1.51504378	1.51547409
0.5	1.64258100	1.69542338	1.69616381
0.6	1.83183910	1.90519283	1.90635640
0.7	2.05102301	2.14953808	2.15125812
0.8	2.30512531	2.43418958	2.43662279
0.9	2.59963784	2.76547948	2.76880934
1.0	2.94060163	3.15040483	3.15484548

and so on. Table 9.4 shows the approximate values of the solution of (6); the absolute error this time is less than 1 percent.

If we reduce the step size to $h = 0.1$, we obtain the results in Table 9.5. The actual value obtained using the improved Euler method to five significant figures is $y(1) \approx 3.1504$, an absolute error of 0.14 percent. A comparison of both methods and step sizes is shown in Table 9.6.

Example 3 If we apply the improved Euler method to the initial-value problem in Example 1,

$$y' = x + y^2, \qquad y(1) = 0,$$

with $h = 0.1$, we obtain the values in Table 9.7. Here $y(1) \approx 0.6918$ to four

TABLE 9.7 Improved Euler Method with $h = 0.1$, for $y' = x + y^2$, $y(1) = 0$

x_n	y_n	$f(x_n, y_n)$	z_{n+1}	$f(x_{n+1}, z_{n+1})$	y_{n+1}
1.00	0.00	1.00	0.10	1.11	0.11
1.10	0.11	1.11	0.22	1.25	0.22
1.20	0.22	1.25	0.35	1.42	0.36
1.30	0.36	1.43	0.50	1.65	0.51
1.40	0.51	1.66	0.68	1.96	0.69
1.50	0.69				

TABLE 9.8 A Comparison of Results Using Euler and Improved Euler Methods with $h = 0.1$ and $h = 0.05$, for the Initial-Value Problem $y' = x + y^2$, $y(1) = 0$

x	Euler	Improved Euler
$h = 0.1$		
1.0	0.000000000	0.000000000
1.1	0.100000000	0.105500000
1.2	0.211000000	0.223402573
1.3	0.335452100	0.356966908
1.4	0.476704911	0.510823653
1.5	0.639429668	0.691781574
$h = 0.05$		
1.00	0.000000000	0.000000000
1.05	0.050000000	0.051312500
1.10	0.102625000	0.105398434
1.15	0.158151595	0.162573808
1.20	0.216902191	0.223209962
1.25	0.279254519	0.287746157
1.30	0.345653673	0.356705840
1.35	0.416627496	0.430717867
1.40	0.492806420	0.510544562
1.45	0.574949328	0.597119307
1.50	0.663977664	0.691597674

decimal places. If we change the step size to $h = 0.05$, we obtain $y(1) \approx 0.6916$, suggesting that the calculation in Table 9.7 probably has an absolute error of no more than 0.05 percent. A comparison of the improved Euler and Euler methods for both step sizes is given in Table 9.8.

An exact solution of this initial-value problem can be obtained: use the substitution $y = -z'/z$ to transform the given Riccati equation into the second-order equation $z'' + xz = 0$, which can then be solved in terms of Bessel functions by the methods of Section 5.4. However, the resulting quotient of Bessel functions is difficult to evaluate. Numerical methods provide information when either no exact solution is known or the solution is difficult to evaluate. Comparisons between various methods, for varying step sizes, may be used to justify the selection of a given value.

In Section 9.5, we will discuss the discretization error of the Euler method. A similar study could be done for the improved Euler method, but we do not do so in this text.

PROBLEMS 9.2

In Problems 1 through 10 solve each problem exactly using the methods of Chapter 2. Then

a. Use the Euler method with the indicated value of h;

b. Use the improved Euler method with the given value of h.

Compare the accuracy of the two methods with the exact answer. What is the relative error?

1. $\dfrac{dy}{dx} = x + y$, $y(0) = 1$. Find $y(1)$ with $h = 0.2$.

2. $\dfrac{dy}{dx} = x - y$, $y(1) = 2$. Find $y(3)$ with $h = 0.4$.

3. $\dfrac{dy}{dx} = \dfrac{x - y}{x + y}$, $y(2) = 1$. Find $y(1)$ with $h = -0.2$.

4. $\dfrac{dy}{dx} = \dfrac{y}{x} + \left(\dfrac{y}{x}\right)^2$, $y(1) = 1$. Find $y(2)$ with $h = 0.2$.

5. $\dfrac{dy}{dx} = x\sqrt{1 + y^2}$, $y(1) = 0$. Find $y(3)$ with $h = 0.4$.

6. $\dfrac{dy}{dx} = x\sqrt{1 - y^2}$, $y(1) = 0$. Find $y(2)$ with $h = 0.125$.

7. $\dfrac{dy}{dx} = \dfrac{y}{x} - \dfrac{5}{2}x^2 y^3$, $y(1) = \dfrac{1}{\sqrt{2}}$. Find $y(2)$ with $h = 0.125$.

8. $\dfrac{dy}{dx} = \dfrac{-y}{x} + x^2 y^2$, $y(1) = \dfrac{2}{9}$. Find $y(3)$ with $h = \dfrac{1}{3}$. [*Hint:* This is a Bernoulli equation.]

9. $\dfrac{dy}{dx} = ye^x$, $y(0) = 2$. Find $y(2)$ with $h = 0.2$.

10. $\dfrac{dy}{dx} = xe^y$, $y(0) = 0$. Find $y(1)$ with $h = 0.1$.

In Problems 11 through 20 use (a) the Euler method, or (b) the improved Euler method to graph approximately the solution of the given initial-value problem by plotting the points (x_k, y_k) over the indicated range, where $x_k = x_0 + kh$.

11. $y' = xy^2 + y^3$, $y(0) = 1$, $h = 0.02$, $0 \le x \le 0.1$
12. $y' = x + \sin(\pi y)$, $y(1) = 0$, $h = 0.2$, $1 \le x \le 2$
13. $y' = x + \cos(\pi y)$, $y(0) = 0$, $h = 0.4$, $0 \le x \le 2$
14. $y' = \cos(xy)$, $y(0) = 0$, $h = \pi/4$, $0 \le x \le \pi$
15. $y' = \sin(xy)$, $y(0) = 1$, $h = \pi/4$, $0 \le x \le 2\pi$
16. $y' = \sqrt{x^2 + y^2}$, $y(0) = 1$, $h = 0.5$, $0 \le x \le 5$
17. $y' = \sqrt{y^2 - x^2}$, $y(0) = 1$, $h = 0.1$, $0 \le x \le 1$
18. $y' = \sqrt{x + y^2}$, $y(0) = 1$, $h = 0.2$, $0 \le x \le 1$
19. $y' = \sqrt{x + y^2}$, $y(1) = 2$, $h = -0.2$, $0 \le x \le 1$
20. $y' = \sqrt{x^2 + y^2}$, $y(1) = 5$, $h = -0.2$, $0 \le x \le 1$
21. Let

$$\dfrac{dy}{dx} = \dfrac{x - y}{x + y}, \quad y(2) = 0.$$

Use the improved Euler method to approximate $y(1)$ with $h = -0.2$. Compare your answer with the exact value and explain why the method failed.

22. Use the improved Euler method to approximate $y(3)$ with $h = 0.4$ for the initial-value problem

$$\frac{dy}{dx} + \frac{3}{x}y = x^2y^2, \quad y(1) = 2.$$

Compare your answer to the exact answer and explain why the numerical technique failed.

23. Repeat Problem 22 for the initial-value problem

$$\frac{dy}{dx} + \frac{y}{x} = x^3y^3, \quad y(1) = 1.$$

*24. Let $y' = e^{xy}$, $y(0) = 1$, and obtain a value for $y(4)$ with $h = 0.5$ and $h = 0.1$ using the improved Euler method. What difficulties are encountered?

How much confidence do you have in your answer?

*25. The **three-term Taylor series method** is given by

$$y_{n+1} = y_n + hy'_n + \frac{h^2}{2}y''_n,$$

where $y'_n = f(x_n, y_n)$, $x_n = x_0 + nh$, and

$$y''_n = \frac{d}{dx}[f(x, y)]\bigg|_{(x_n, y_n)}$$

$$= \frac{\partial f}{\partial x}(x_n, y_n) + \frac{\partial f}{\partial y}(x_n, y_n)y'_n.$$

Use the three-term Taylor series method to find $y(1)$ for the initial-value problem

$$y' = y + x^2, \quad y(0) = 1,$$

with $h = 0.2$ and $h = 0.1$. Compare your results with those in Table 9.6.

9.3 The Runge-Kutta Method

This powerful method gives accurate results without a large number of steps (that is, without the need to make the step size too small). The efficiency is obtained by using a version of Simpson's rule (a frequently used method of numerical integration) in evaluating the integral in equation (9.2.3):

$$y(x_1) = y(x_0) + \int_{x_0}^{x_1} f(x, y(x))\, dx, \quad (1)$$

which yields the solution at x_1 of the initial-value problem

$$y' = f(x, y), \quad y(x_0) = y_0. \quad (2)$$

If $f(x, y(x))$ is a function of x alone, then **Simpson's rule** of calculus states that

$$\int_{x_0}^{x_0+h} f(x)\, dx \approx \frac{(h/2)}{3}\left[f(x_0) + 4f\left(x_0 + \frac{h}{2}\right) + f(x_0 + h)\right]$$

$$\approx \frac{1}{6}\left[hf(x_0) + 4hf\left(x_0 + \frac{h}{2}\right) + hf(x_0 + h)\right]. \quad (3)$$

We modify this rule (a justification is provided later in this section) to adjust for the fact that f also depends on y with the following approximation:

$$\int_{x_0}^{x_0+h} f(x, y(x))\, dx \approx \tfrac{1}{6}(m_1 + 2m_2 + 2m_3 + m_4), \quad (4)$$

where

$$\begin{aligned} m_1 &= hf(x_0, y_0), \\ m_2 &= hf\left(x_0 + \frac{h}{2}, y_0 + \frac{m_1}{2}\right), \\ m_3 &= hf\left(x_0 + \frac{h}{2}, y_0 + \frac{m_2}{2}\right), \\ m_4 &= hf(x_0 + h, y_0 + m_3). \end{aligned} \quad (5)$$

Note that if f depends only on x, then $m_2 = m_3$ and (4) becomes Simpson's rule. Setting $x_1 = x_0 + h$, $x_2 = x_0 + 2h, \ldots, x_n = x_0 + nh$, we obtain the following recursive equation, called the **Runge-Kutta method**.

RUNGE-KUTTA METHOD

$$y_{n+1} = y_n + \tfrac{1}{6}(m_1 + 2m_2 + 2m_3 + m_4), \qquad (6)$$

where,

$$\begin{aligned}
m_1 &= hf(x_n, y_n), \\
m_2 &= hf\left(x_n + \frac{h}{2}, y_n + \frac{m_1}{2}\right), \\
m_3 &= hf\left(x_n + \frac{h}{2}, y_n + \frac{m_2}{2}\right), \\
m_4 &= hf(x_n + h, y_n + m_3).
\end{aligned} \qquad (7)$$

Note that the values m_1/h, m_2/h, m_3/h, and m_4/h are four slopes between $x_n \le x \le x_{n+1}$, so that (6) is a weighted average of these slopes—a procedure similar to the one we used in the improved Euler method.

The following two examples illustrate the use of (6) and (7).

Example 1 Consider the initial-value problem in Example 9.2.2 (and Example 1.3.2) in which we wish to evaluate $y(1)$ given

$$\frac{dy}{dx} = y + x^2, \quad y(0) = 1.$$

If we apply the Runge-Kutta method with $h = 1$, then $y(1) = y_1$, which is obtained by first calculating the expressions in (7):

$$\begin{aligned}
m_1 &= f(0, 1) = 1, \\
m_2 &= f\left(\tfrac{1}{2}, \tfrac{3}{2}\right) = \tfrac{7}{4}, \\
m_3 &= f\left(\tfrac{1}{2}, \tfrac{15}{8}\right) = \tfrac{17}{8}, \\
m_4 &= f\left(1, \tfrac{25}{8}\right) = \tfrac{33}{8}.
\end{aligned}$$

Thus, using (6),

$$y_1 = 1 + \tfrac{1}{6}\left(1 + \tfrac{7}{2} + \tfrac{17}{4} + \tfrac{33}{8}\right) = \tfrac{151}{48} \approx 3.146.$$

In one step this method got us even closer to the correct value than the improved Euler method did in five steps.

If we use five steps ($h = 0.2$) to calculate $y(1)$, we can arrange our work in tabular form as shown in Table 9.9. A comparison of the Runge-Kutta method for step sizes of $h = 0.2$ and $h = 0.1$ with the Euler and improved Euler methods is given in Table 9.10. The $y(1)$ entry for both step sizes, when displayed to four significant figures, is $y(1) \approx 3.1548$, which is the exact answer to that many significant figures (see Table 9.10).

TABLE 9.9 Runge-Kutta Method with $h = 0.2$, for $y' = y + x^2$, $y(0) = 1$

x_n	y_n	m_1	m_2	m_3	m_4
0.00	1.00	0.20	0.22	0.22	0.25
0.20	1.22	0.25	0.29	0.29	0.34
0.40	1.52	0.34	0.39	0.39	0.45
0.60	1.91	0.45	0.52	0.53	0.62
0.80	2.44	0.62	0.71	0.72	0.83
1.00	3.15				

Note: These values were obtained with much higher accuracy on a computer and then rounded to two decimal places.

TABLE 9.10 A Comparison of Results Using the Euler, Improved Euler, and Runge-Kutta Methods with $h = 0.2$ and $h = 0.1$, for the Initial-Value Problem $y' = y + x^2$, $y(0) = 1$

x	Euler	Improved Euler	Runge-Kutta	Exact
$h = 0.2$				
0.0	1.000000	1.00000000	1.00000000	1.00000000
0.2	1.200000	1.22400000	1.22420667	1.22420828
0.4	1.448000	1.51408000	1.51546869	1.51547409
0.6	1.769600	1.90237760	1.90634413	1.90635640
0.8	2.195520	2.42810067	2.43659938	2.43662279
1.0	2.762624	3.13908282	3.15480515	3.15484548
$h = 0.1$				
0.0	1.00000000	1.00000000	1.00000000	1.00000000
0.1	1.10000000	1.10550000	1.10551271	1.10551275
0.2	1.21100000	1.22412750	1.22420815	1.22420828
0.3	1.33610000	1.35936089	1.35957618	1.35957642
0.4	1.47871000	1.51504378	1.51547370	1.51547409
0.5	1.64258100	1.69542338	1.69616320	1.69616381
0.6	1.83183910	1.90519283	1.90635551	1.90635640
0.7	2.05102301	2.14953808	2.15125689	2.15125812
0.8	2.30512531	2.43418958	2.43662112	2.43662279
0.9	2.59963784	2.76547948	2.76880714	2.76880934
1.0	2.94060163	3.15040483	3.15484264	3.15484548

TABLE 9.11 Runge-Kutta Method with $h = 0.1$, for $y' = x + y^2$, $y(1) = 0$

x_n	y_n	m_1	m_2	m_3	m_4
1.00	0.00	0.10	0.11	0.11	0.11
1.10	0.11	0.11	0.12	0.12	0.12
1.20	0.22	0.12	0.13	0.13	0.14
1.30	0.36	0.14	0.15	0.15	0.17
1.40	0.51	0.17	0.18	0.18	0.20
1.50	0.69				

TABLE 9.12 A Comparison of Results Using the Euler, Improved Euler, and Runge-Kutta Methods with $h = 0.1$ and $h = 0.05$, for the Initial-Value Problem $y' = x + y^2$, $y(1) = 0$

x	Euler	Improved Euler	Runge-Kutta
$h = 0.1$			
1.0	0.000000000	0.000000000	0.000000000
1.1	0.100000000	0.105500000	0.105360367
1.2	0.211000000	0.223402573	0.223135961
1.3	0.335452100	0.356966908	0.356601567
1.4	0.476704911	0.510823653	0.510424419
1.5	0.639429668	0.691781574	0.691496736
$h = 0.05$			
1.00	0.000000000	0.000000000	0.000000000
1.05	0.050000000	0.051312500	0.051293290
1.10	0.102625000	0.105398434	0.105360321
1.15	0.158151595	0.162573808	0.162517274
1.20	0.216902191	0.223209962	0.223135802
1.25	0.279254519	0.287746157	0.287655681
1.30	0.345653673	0.356705840	0.356601169
1.35	0.416627496	0.430717867	0.430602377
1.40	0.492806420	0.510544562	0.510423556
1.45	0.574949328	0.597119307	0.597001050
1.50	0.663977664	0.691597674	0.691494999

Example 2 Consider the initial-value problem

$$y' = x + y^2, \qquad y(1) = 0$$

(see Examples 9.2.1 and 9.2.3). We wish to find approximate values of the solution at $x = 1, 1.1, 1.2, 1.3, 1.4, 1.5$. Setting $h = 0.1$ and applying the Runge-Kutta method, equations (6) and (7), we obtain the values in Table 9.11.

A comparison of the Runge-Kutta method with the Euler and improved Euler methods is made in Table 9.12.

In the Euler method we used one value of the derivative $y' = f(x, y)$ for each iteration. In the improved Euler method we used two values. In the Runge-Kutta formulas, we made use of four values of the derivative for each iteration. We now show how to find the four "best" values in a sense to be made precise later.

To begin, we need to recall the Taylor series expansions of functions of one variable:

$$y(x_0 + h) = y(x_0) + hy'(x_0) + \frac{h^2}{2!}y''(x_0) + \frac{h^3}{3!}y'''(x_0) + \cdots, \qquad (8)$$

and of two variables:
$$f(x_0 + mh, y_0 + nh) = f(x_0, y_0) + h(mf_x + nf_y)$$
$$+ \frac{h^2}{2!}(m^2 f_{xx} + 2mn f_{xy} + n^2 f_{yy})$$
$$+ \frac{h^3}{3!}(m^3 f_{xxx} + 3m^2 n f_{xxy} + 3mn^2 f_{xyy} + n^3 f_{yyy})$$
$$+ \cdots, \tag{9}$$

where the partials are all evaluated at the point (x_0, y_0). Since
$$y' = f(x, y), \tag{10}$$
we find that
$$y'' = f_x + (f_y) y' = f_x + ff_y,$$
$$y''' = f_{xx} + 2ff_{xy} + f^2 f_{yy} + f_y(f_x + ff_y),$$

and so on. (Note that $f_{xy} = f_{yx}$ when these derivatives are continuous.) Thus (8) can be written in the form
$$y_1 - y_0 = hf + \frac{h^2}{2}(f_x + ff_y) + \frac{h^3}{6}\left[f_{xx} + 2ff_{xy} + f^2 f_{yy} + f_y(f_x + ff_y)\right] + \cdots. \tag{11}$$

The main idea is somehow to select several points (x, y) so that the Taylor series expansions (9) of the corresponding $f(x, y)$ terms coincide with the terms on the right-hand side of (11). Suppose we let
$$m_1 = hf(x_0, y_0),$$
$$m_2 = hf(x_0 + nh, y_0 + nm_1),$$
$$m_3 = hf(x_0 + ph, y_0 + pm_2),$$
$$m_4 = hf(x_0 + qh, y_0 + qm_3).$$

(The reason for doing this is made clear shortly.) Using (9), we may write these values as
$$m_1 = hf,$$
$$m_2 = h\left[f + nh(f_x + ff_y) + \frac{(nh)^2}{2}(f_{xx} + 2ff_{xy} + f^2 f_{yy}) + \cdots\right],$$
$$m_3 = h\left\{f + ph(f_x + ff_y)\right.$$
$$\left. + \frac{h^2}{2}\left[p^2(f_{xx} + 2ff_{xy} + f^2 f_{yy}) + 2npf_y(f_x + ff_y)\right] + \cdots\right\},$$
$$m_4 = h\left\{f + qh(f_x + ff_y)\right.$$
$$\left. + \frac{h^2}{2}\left[q^2(f_{xx} + 2ff_{xy} + f^2 f_{yy}) + 2pqf_y(f_x + ff_y)\right] + \cdots\right\},$$

where all functions are evaluated at the point (x_0, y_0).

We now consider an expression of the form

$$am_1 + bm_2 + cm_3 + dm_4$$

and try to equate it to the right-hand side of (11). This has the effect of giving us a numerical scheme that agrees with the solution to (10) up to and including third-order terms. Then the error is no greater than terms like kh^4, and so on. Matching like expressions, we find that

$$\begin{aligned}
\text{coefficient of } hf & & a + b + c + d &= 1, \\
\text{coefficient of } h^2(f_x + ff_y) & & bn + cp + dq &= \tfrac{1}{2}, \\
\text{coefficient of } \tfrac{h^3}{2}(f_{xx} + 2ff_{xy} + f^2 f_{yy}) & & bn^2 + cp^2 + dq^2 &= \tfrac{1}{3}, \\
\text{coefficient of } h^3[f_y(f_x + ff_y)] & & cnp + dpq &= \tfrac{1}{6}.
\end{aligned} \quad (12)$$

Any solution of these equations produces a method in which there is no error up to the third-order terms. Suppose we take $n = p = \tfrac{1}{2}$ and $q = 1$. Then (12) reduces to the system of equations

$$\begin{aligned}
a + b + c + d &= 1, \\
b + c + 2d &= 1, \\
3b + 3c + 12d &= 4, \\
3c + 6d &= 2,
\end{aligned}$$

which has the solution $a = d = \tfrac{1}{6}$, $b = c = \tfrac{1}{3}$. Thus the Runge-Kutta formula [equations (6) and (7)] agrees with $y_1 - y_0$ for all terms up to and including the terms in h^3. Actually, with quite a bit more work, one can show that they agree in the h^4 terms, too. Thus the error (if any) involves only terms in h^5 and higher. Hence, for small h, we should expect to get very good results.

Other formulas are also readily derivable. Suppose we choose $n = \tfrac{1}{3}$, $p = \tfrac{2}{3}$, $q = 1$. Then (12) yields

$$\begin{aligned}
a + b + c + d &= 1, \\
2b + 4c + 6d &= 3, \\
b + 4c + 9d &= 3, \\
4c + 12d &= 3,
\end{aligned}$$

which has a solution $a = d = \tfrac{1}{8}$, $b = c = \tfrac{3}{8}$. The formula, known as the **Kutta-Simpson $\tfrac{3}{8}$-rule**, may be written

$$y_1 = y_0 + \tfrac{1}{8}(m_1 + 3m_2 + 3m_3 + m_4),$$

where

$$\begin{aligned}
m_1 &= hf(x_0, y_0), \\
m_2 &= hf(x_0 + h/3, y_0 + m_1/3), \\
m_3 &= hf(x_0 + 2h/3, y_0 + 2m_2/3), \\
m_4 &= hf(x_0 + h, y_0 + m_3).
\end{aligned} \quad (13)$$

Similarly, the choice $n = \tfrac{1}{3}$, $p = \tfrac{2}{3}$, $q = 1$ also yields a solution $a = c = 0$, $b = \tfrac{3}{4}$, $d = \tfrac{1}{4}$; hence

$$y_1 = y_0 + \tfrac{1}{4}(3m_2 + m_4),$$

where m_2 and m_4 are defined as in (13). Since the number of possible choices

PROBLEMS 9.3

In Problems 1 through 10 solve each problem exactly using the methods of Chapter 2. Then use the Runge-Kutta method with the given h. Compare the accuracy of this method with the exact answer.

1. $\dfrac{dy}{dx} = x + y$, $y(0) = 1$. Find $y(1)$ with $h = 0.2$.

2. $\dfrac{dy}{dx} = x - y$, $y(1) = 2$. Find $y(3)$ with $h = 0.4$.

3. $\dfrac{dy}{dx} = \dfrac{x - y}{x + y}$, $y(2) = 1$. Find $y(1)$ with $h = -0.2$.

4. $\dfrac{dy}{dx} = \dfrac{y}{x} + \left(\dfrac{y}{x}\right)^2$, $y(1) = 1$. Find $y(2)$ with $h = 0.2$.

5. $\dfrac{dy}{dx} = x\sqrt{1 + y^2}$, $y(1) = 0$. Find $y(3)$ with $h = 0.4$.

6. $\dfrac{dy}{dx} = x\sqrt{1 - y^2}$, $y(1) = 0$. Find $y(2)$ with $h = 0.125$.

7. $\dfrac{dy}{dx} = \dfrac{y}{x} - \dfrac{5}{2}x^2y^3$, $y(1) = \dfrac{1}{\sqrt{2}}$. Find $y(2)$ with $h = 0.125$.

8. $\dfrac{dy}{dx} = \dfrac{-y}{x} + x^2y^2$, $y(1) = \dfrac{2}{9}$. Find $y(3)$ with $h = \dfrac{1}{3}$.

9. $\dfrac{dy}{dx} = ye^x$, $y(0) = 2$. Find $y(2)$ with $h = 0.2$.

10. $\dfrac{dy}{dx} = xe^y$, $y(0) = 0$. Find $y(1)$ with $h = 0.1$.

In Problems 11 through 20 use the Runge-Kutta method to graph approximately the solution of the given initial-value problem by plotting the points (x_k, y_k) over the indicated range, where $x_k = x_0 + kh$.

11. $y' = xy^2 + y^3$, $y(0) = 1$, $h = 0.02$, $0 \le x \le 0.1$

12. $y' = x + \sin(\pi y)$, $y(1) = 0$, $h = 0.2$, $1 \le x \le 2$

13. $y' = x + \cos(\pi y)$, $y(0) = 0$, $h = 0.4$, $0 \le x \le 2$

14. $y' = \cos(xy)$, $y(0) = 0$, $h = \pi/4$, $0 \le x \le \pi$

15. $y' = \sin(xy)$, $y(0) = 1$, $h = \pi/4$, $0 \le x \le 2\pi$

16. $y' = \sqrt{x^2 + y^2}$, $y(0) = 1$, $h = 0.5$, $0 \le x \le 5$

17. $y' = \sqrt{y^2 - x^2}$, $y(0) = 1$, $h = 0.1$, $0 \le x \le 1$

18. $y' = \sqrt{x + y^2}$, $y(0) = 1$, $h = 0.2$, $0 \le x \le 1$

19. $y' = \sqrt{x + y^2}$, $y(1) = 2$, $h = -0.2$, $0 \le x \le 1$

20. $y' = \sqrt{x^2 + y^2}$, $y(1) = 5$, $h = -0.2$, $0 \le x \le 1$

21. Let
$$\dfrac{dy}{dx} = \dfrac{x - y}{x + y}, \quad y(2) = 0.$$
Use the Runge-Kutta method to approximate $y(1)$ with $h = -0.2$. Compare your answer with the exact value and explain why the method fails.

22. Use the Runge-Kutta method to approximate $y(3)$ with $h = 0.4$ for the initial-value problem
$$\dfrac{dy}{dx} + \dfrac{3}{x}y = x^2y^2, \quad y(1) = 2.$$
Compare your answer to the exact answer and explain why the numerical technique fails.

23. Repeat Problem 22 for the initial-value problem
$$\dfrac{dy}{dx} + \dfrac{y}{x} = x^3y^3, \quad y(1) = 1.$$

*24. Let $y' = e^{xy}$, $y(0) = 1$, and obtain a value for $y(4)$ with $h = 0.5$ using the Runge-Kutta method. What difficulties are encountered? How much confidence do you have in your answer?

25. Set $d = 0$ in equation (12) and let $n = \frac{1}{3}$, $p = \frac{2}{3}$. Then find the coefficients a, b, and c for these choices. The resulting formula is called **Heun's formula**.

26. Set $a = d = 0$, $b = \frac{3}{4}$, $c = \frac{1}{4}$ in equation (12) and find n and p.

27. Set $a = \frac{2}{8}$, $b = c = \frac{3}{8}$, $d = 0$ in equation (12) and find n and p.

*28. Prove that the Runge-Kutta formula agrees with equation (11) up to and including the h^4 terms.

29. Explain why the choice $n = \frac{1}{3}$, $p = \frac{2}{3}$, $q = 1$ in equation (12) yielded two sets of solutions a, b, c, and d. Are other solutions possible? Does the choice $n = p = \frac{1}{2}$, $q = 1$ allow for multiple solutions?

9.4 Predictor-Corrector Formulas

The simplest type of predictor-corrector formula is one that we have already used, namely, the improved Euler method. Recall that with this method we first obtain a value for y_n by the Euler method (the *predicting* part of the process) and then improve on the accuracy of the method by applying a trapezoidal rule (the *correcting* phase of the process). In this section we indicate how such methods are derived and provide several procedures that perform their tasks with a high degree of accuracy.

We begin by developing a procedure for obtaining **quadrature formulas**. These formulas are designed to obtain the approximate value of a definite integral by using equally spaced values of the integrand. Two well-known quadrature formulas are the trapezoidal rule and Simpson's rule,[†] which can be written in the forms

Trapezoidal rule:
$$y_1 - y_0 = \frac{h}{2}(y_0' + y_1'), \tag{1}$$

Simpson's rule:
$$y_2 - y_0 = \frac{h}{3}(y_0' + 4y_1' + y_2'), \tag{2}$$

respectively, where h is the step size, $x_n = x_0 + nh$, $y_n = y(x_n)$, and $y_n' = f(x_n, y_n)$. Note that, unlike the previously discussed methods, here we have slightly shifted the point of view by representing the right-hand side as a sum of derivatives instead of values of the function f.

In equation (1) we are using only the points y_0' and y_1', whereas in equation (2) we also use y_2'. For this reason, (1) is called a **2-point quadrature formula**, whereas (2) is a **3-point quadrature formula**. In general, the more points involved, the higher the accuracy. In the present case, (1) provides the exact integral only for straight lines, whereas (2) is also exact for quadratic polynomials (see Problems 15 and 16).

We now develop a method for obtaining 3-point quadrature formulas. The method is easily adapted to n-point quadrature formulas.

3-POINT QUADRATURE FORMULAS

Assume that the function $y(x)$ is a quadratic polynomial

$$y(x) = a_0 + a_1 x + a_2 x^2. \tag{3}$$

Let h and x_0 be fixed real numbers and define $x_k = x_0 + kh$, for all integers k. Let $y_k = y(x_k)$. We wish to find coefficients A_0, A_1, A_2 such that

$$y_j - y_i = h(A_0 y_0' + A_1 y_1' + A_2 y_2'), \tag{4}$$

where $i, j = 0, 1, 2$ and $i \neq j$. This leads to an integration scheme that is exact for quadratic polynomials. For example, if $i = 0$, then the left-hand side of (4) becomes

$$y_j - y_0 = a_1(x_j - x_0) + a_2(x_j^2 - x_0^2)$$
$$= a_1 jh + a_2 \left[2x_0 jh + (jh)^2\right] = j(a_1 h + 2a_2 x_0 h) + j^2(a_2 h^2). \tag{5}$$

[†] These rules can be found in most calculus books.

On the other hand, using the facts that $y' = a_1 + 2a_2 x$ and $x_k = x_0 + kh$, we find that the right-hand side of (4) becomes

$$h[A_0 y_0' + A_1 y_1' + A_2 y_2']$$
$$= h[A_0(a_1 + 2a_2 x_0) + A_1(a_1 + 2a_2 x_1) + A_2(a_1 + 2a_2 x_2)]$$
$$= (A_0 + A_1 + A_2)(a_1 h + 2a_2 x_0 h) + (2A_1 + 4A_2)(a_2 h^2). \quad (6)$$

Equating equations (5) and (6), we obtain the simultaneous equations

$$A_0 + A_1 + A_2 = j,$$
$$2A_1 + 4A_2 = j^2. \quad (7)$$

Since this system is underdetermined, it has an infinite number of solutions A_0, A_1, A_2, each leading to a different quadrature formula. For example, if $j = 1$, we can pick $A_0 = \frac{5}{6}, A_1 = -\frac{1}{6}, A_2 = \frac{1}{3}$ and arrive at the formula

$$y_1 - y_0 = \frac{h}{6}(5y_0' - y_1' + 2y_2'). \quad (8)$$

If $j = 2$, then selecting $A_0 = \frac{1}{3}, A_1 = \frac{4}{3}, A_2 = \frac{1}{3}$ yields Simpson's rule (2). In addition, quadrature formulas may be added or subtracted to yield new formulas. For example, subtracting (8) from (2) yields the **Adams-Bashforth formula**,

$$y_2 - y_1 = \frac{h}{2}(-y_0' + 3y_1'). \quad (9)$$

PREDICTOR-CORRECTOR METHODS

Let us now illustrate how predictor-corrector formulas are used. Suppose we are given the initial-value problem

$$\frac{dy}{dx} = f(x, y), \quad y(x_0) = y_0. \quad (10)$$

The value y_0 is given, and y_0' can be obtained by evaluating equation (10) at $x = x_0, y = y_0$. We next perform an improved Euler method computation to obtain y_1 [and y_1' by means of (10)]. At this point it is often desirable to apply repeatedly the trapezoidal rule to the process until the value of y_1 stabilizes (that is, remains unchanged to a given number of decimal places; see Example 1). Then since y_0, y_0', y_1, and y_1' are all known, we may use (9) to *predict* the value of y_2. This value is then used in (10) to obtain y_2', and the trapezoidal rule, (1), is used to *correct* the y_2 value previously determined. The process is now repeated to predict and correct y_3 and y_3' in terms of the known values y_1, y_1', y_2, y_2'. We use the Adams-Bashforth formula

$$y_{n+1} = y_n + \frac{h}{2}(-y_{n-1}' + 3y_n') \quad (11)$$

to *predict* the value of y_{n+1}, equation (10) to obtain y_{n+1}', and the trapezoidal rule

$$y_{n+1} = y_n + \frac{h}{2}(y_{n+1}' + y_n') \quad (12)$$

to *correct* the value of y_{n+1}, and equation (10) again to obtain y_{n+1}'.

9.4/PREDICTOR-CORRECTOR FORMULAS

Example 1 Let $dy/dx = y + x^2$, $y(0) = 1$, and suppose we wish to find $y(1)$, with $h = 0.2$, by using the predictor-corrector formulas (11) and (12).

The predictor formula (11) requires that we know the values of y_0', y_1, and y_1' before it can be used to generate y_2. At this point, the only one of these three values that we know is

$$y_0' = f(x_0, y_0) = y_0 + x_0^2 = 1,$$

since $x_0 = 0$, $y_0 = y(x_0) = y(0) = 1$. Therefore, before we can begin using predictor formula (11) we must determine y_1. [Once we know y_1 we can compute $y_1' = f(x_1, y_1)$.] The usual method for obtaining y_1 is to use the improved Euler method *repeatedly*. We illustrate this procedure below:

x_0	y_0	y_0'	$\bar{y}_1$	$\bar{y}_1'$	y_1
0.0	1.0	1.0	1.20	1.24	1.22
			1.22	1.26	1.23
			1.23	1.27	1.23

The first entry in the $\bar{y}_1$ column is obtained by the Euler formula: $y_1 = y_0 + hy_0'$. The entries in the $\bar{y}_1'$ column are obtained from the differential equation $\bar{y}_1' = f(x_1, \bar{y}_1) = \bar{y}_1 + x_1^2$. The last column is determined by the improved Euler formula

$$y_1 = y_0 + \frac{h}{2}(y_0' + \bar{y}_1'),$$

and the resulting value for y_1 is transferred to the next row in the $\bar{y}_1$ column. We then repeat the steps in the $\bar{y}_1'$ and y_1 columns. We see that the process stabilizes with $y_1 = 1.23$ and $y_1' = 1.27$ (provided that we are using two-decimal-place accuracy).

We now apply (11) and (12) where we have arranged the calculations in Table 9.13. The result is better than the answer obtained by the improved Euler method (with slightly more work). A comparison with the other methods we have discussed in this chapter for this initial-value problem is given in Table 9.14. A similar comparison for Example 9.3.2 is given in Table 9.15.

TABLE 9.13 Predictor-Corrector Method with $h = 0.2$, for $y' = y + x^2$, $y(0) = 1$

x_n	y_n	$y_n' = f(x_n, y_n)$	Predictor $y_{n+1} = y_n + \frac{h}{2}(-y_{n-1}' + 3y_n')$	y_{n+1}'	Corrector $y_{n+1} = y_n + \frac{h}{2}(y_{n+1}' + y_n')$
0.0	1.00	1.00			
0.2	1.23	1.27	$1.23 + (0.1)(-1.00 + 3.81) = 1.51$	1.67	$1.23 + (0.1)(1.67 + 1.27) = 1.52$
0.4	1.52	1.68	$1.52 + (0.1)(-1.27 + 5.04) = 1.90$	2.26	$1.52 + (0.1)(2.26 + 1.68) = 1.91$
0.6	1.91	2.27	$1.91 + (0.1)(-1.68 + 6.81) = 2.42$	3.06	$1.91 + (0.1)(3.06 + 2.27) = 2.44$
0.8	2.44	3.08	$2.44 + (0.1)(-2.27 + 9.24) = 3.14$	4.14	$2.44 + (0.1)(4.14 + 3.08) = 3.16$
1.0	3.16				

TABLE 9.14 A Comparison of Results Using the Euler, Improved Euler, Runge-Kutta, and Predictor-Corrector Methods with $h = 0.2$ and $h = 0.1$, for the Initial-Value Problem $y' = y + x^2$, $y(0) = 1$

x	Euler	Improved Euler	Runge-Kutta	Predictor-Corrector	Exact
$h = 0.2$					
0.0	1.000000	1.00000000	1.00000000	1.00000000	1.00000000
0.2	1.200000	1.22400000	1.22420667	1.22666667	1.22420828
0.4	1.448000	1.51408000	1.51546869	1.52000000	1.51547409
0.6	1.769600	1.90237760	1.90634413	1.91373333	1.90635640
0.8	2.195520	2.42810067	2.43659938	2.44789200	2.43662279
1.0	2.762624	3.13908282	3.15480515	3.17136982	3.15484548
$h = 0.1$					
0.0	1.00000000	1.00000000	1.00000000	1.00000000	1.00000000
0.1	1.10000000	1.10550000	1.10551271	1.10578950	1.10551275
0.2	1.21100000	1.22412750	1.22420815	1.22473687	1.22420828
0.3	1.33610000	1.35936089	1.35957618	1.36040661	1.35957642
0.4	1.47871000	1.51504378	1.51547370	1.51666348	1.51547409
0.5	1.64258100	1.69542338	1.69616320	1.69777879	1.69616381
0.6	1.83183910	1.90519283	1.90635551	1.90847335	1.90635640
0.7	2.05102301	2.14953808	2.15125689	2.15396479	2.15125812
0.8	2.30512531	2.43418958	2.43662112	2.44001982	2.43662279
0.9	2.59963784	2.76547948	2.76880714	2.77301204	2.76880934
1.0	2.94060163	3.15040483	3.15484264	3.15998578	3.15484548

TABLE 9.15 A Comparison of Results Using the Euler, Improved Euler, Runge-Kutta, and Predictor-Corrector Methods with $h = 0.1$ and $h = 0.05$, for the Initial-Value Problem $y' = x + y^2$, $y(1) = 0$

x	Euler	Improved Euler	Runge-Kutta	Predictor-Corrector
$h = 0.1$				
1.0	0.000000000	0.000000000	0.000000000	0.000000000
1.1	0.100000000	0.105500000	0.105360367	0.100505000
1.2	0.211000000	0.223402573	0.223135961	0.218364951
1.3	0.335452100	0.356966908	0.356601567	0.351874547
1.4	0.476704911	0.510823653	0.510424419	0.505718937
1.5	0.639429668	0.691781574	0.691496736	0.686823519
$h = 0.05$				
1.00	0.000000000	0.000000000	0.000000000	0.000000000
1.05	0.050000000	0.051312500	0.051293290	0.050062600
1.10	0.102625000	0.105398434	0.105360321	0.104145660
1.15	0.158151595	0.162573808	0.162517274	0.161316022
1.20	0.216902191	0.223209962	0.223135802	0.221945882
1.25	0.279254519	0.287746157	0.287655681	0.286475712
1.30	0.345653673	0.356705840	0.356601169	0.355430697
1.35	0.416627496	0.430717867	0.430602377	0.429442224
1.40	0.492806420	0.510544562	0.510423556	0.509276341
1.45	0.574949328	0.597119307	0.597001050	0.595871978
1.50	0.663977664	0.691597674	0.691494999	0.690393049

A very accurate predictor-corrector method due to Milne uses a 4-point quadrature formula as a predictor and Simpson's rule as a corrector:

$$\text{Predictor:} \quad y_{n+4} - y_n = \frac{4h}{3}(2y'_{n+1} - y'_{n+2} + 2y'_{n+3}),$$

$$\text{Corrector:} \quad y_{n+4} - y_{n+2} = \frac{h}{3}(y'_{n+2} + 4y'_{n+3} + y'_{n+4}). \tag{13}$$

To apply this method, it is necessary to have good values for y_0, y'_0, y_1, y'_1, y_2, y'_2, y_3, and y'_3.

Predictor-corrector methods have many advantages from the point of view of accuracy and the amount of work involved. The extra step involved in "correcting" dramatically improves the accuracy, without requiring an inordinate amount of extra work.

PROBLEMS 9.4

In Problems 1 through 10 solve each problem exactly using the methods of Chapter 2. Then

a. use the predictor-corrector method of Example 1 [formulas (11) and (12)] and the indicated value of h to find an approximate solution to the given value of y;

b. use the predictor-corrector method in equation (13) with the given h to find the indicated value of y. Use the values generated in part (a) to initialize Milne's method.

Compare the accuracy of these methods with the exact answer.

1. $y' = x + y$, $y(0) = 1$. Find $y(1)$ with $h = 0.2$.
2. $y' = x - y$, $y(1) = 2$. Find $y(3)$ with $h = 0.4$.
3. $y' = \dfrac{x-y}{x+y}$, $y(2) = 1$. Find $y(1)$ with $h = -0.2$.
4. $y' = (y/x) + (y/x)^2$, $y(1) = 1$. Find $y(2)$ with $h = 0.2$.
5. $y' = x\sqrt{1 + y^2}$, $y(1) = 0$. Find $y(3)$ with $h = 0.4$.
6. $y' = x\sqrt{1 - y^2}$, $y(1) = 0$. Find $y(2)$ with $h = \frac{1}{8}$.
7. $y' = (y/x) - (5x^2 y^3/2)$, $y(1) = 1/\sqrt{2}$. Find $y(2)$ with $h = \frac{1}{8}$.
8. $y' = (-y/x) + x^2 y^2$, $y(1) = \frac{2}{9}$. Find $y(3)$ with $h = \frac{1}{3}$.
9. $y' = ye^x$, $y(0) = 2$. Find $y(2)$ with $h = 0.2$.
10. $y' = xe^y$, $y(0) = 0$. Find $y(1)$ with $h = 0.1$.
11. Obtain the trapezoidal rule (1) by using the 3-point quadrature formulas (4) and (7).

12. Obtain these 3-point quadrature formulas:

    a. $y_1 - y_0 = \dfrac{h}{12}(5y'_0 + 8y'_1 - y'_2)$;

    b. $y_2 - y_0 = \dfrac{h}{8}(y'_0 + 14y'_1 + y'_2)$;

    c. $y_2 - y_0 = \dfrac{h}{4}(y'_0 + 6y'_1 + y'_2)$.

13. Obtain the 4-point quadrature formula

    $$y_3 - y_0 = \frac{h}{2}(-y'_0 + 5y'_1 + 2y'_2).$$

14. Use the formula obtained in Problem 13 and Simpson's rule as a predictor-corrector to solve the initial-value problem

    $$y' = x + y, \quad y(0) = 1,$$

    for $y(1)$ with $h = 0.2$. Use the improved Euler method to find y_1 and y_2.

15. Let $y(x) = ax + b$. Show that the trapezoidal rule (1) provides the exact value for $\int_0^1 y(x)\,dx$ for step sizes $h = \frac{1}{2}$ and $\frac{1}{5}$.

16. Let $y(x) = ax^2 + bx + c$. Show that Simpson's rule (2) provides the exact value for $\int_0^1 y(x)\,dx$ for step sizes $h = \frac{1}{4}$ and $\frac{1}{8}$.

17. Show that the equations for 4-point quadrature formulas analogous to (7) are

    $$A_0 + A_1 + A_2 + A_3 = j,$$
    $$2A_1 + 4A_2 + 6A_3 = j^2,$$
    $$3A_1 + 12A_2 + 27A_3 = j^3.$$

Use these equations to derive Milne's equation (13).

*18. Obtain the underdetermined system of equations for 5-point quadrature formulas analogous to those in Problem 17.

19. Use the three-term Taylor series method (see Problem 9.2.25) as a predictor and Simpson's rule as a corrector to find an approximate solution of $y(1)$ for the initial-value problem

$$y' = x + y, \quad y(0) = 1,$$

with $h = 0.2$. [Compare this result with Problems 1 and 14.]

9.5 Error Analysis for the Euler Method: If Time Permits

In this section we discuss only the discretization errors encountered in the use of the Euler method. Round-off errors depend not only on the method and the number of steps in the calculation but also on the type of instrument (hand-calculator, computer, pencil and paper, etc.) used for computing the answer. Round-off error is not discussed in this section (it will be discussed in Section 9.6), although it should never be ignored.

Let us again consider the first-order initial-value problem

$$y' = f(x, y), \quad y(x_0) = y_0, \tag{1}$$

and use the iteration scheme

$$y_{n+1} = y_n + hf(x_n, y_n), \tag{2}$$

where h is a fixed step size.

We assume for the remainder of this section that $f(x, y)$ possesses continuous first partial derivatives. Then, on any finite interval, $\partial f(x, y)/\partial y$ is bounded by some constant that we denote by L (a continuous function is always bounded on a closed, bounded interval). Since $y'(x) = f(x, y)$, we obtain, by the chain rule,

$$y''(x) = \frac{\partial f}{\partial x}(x, y) + \frac{\partial f}{\partial y}(x, y) y'(x),$$

which must be continuous since it is the sum of continuous functions. Hence $y''(x)$ must be bounded on the interval $x_0 \le x \le a$. So we assume that $|y''(x)| < M$ for some positive constant M.

We now wish to estimate the error e_n at the nth step of the iteration defined by equation (2). Since $y(x_n)$ is the exact value of the solution $y(x)$ at the point $x_n = x_0 + nh$, and y_n is the approximate value at that point, the error at the nth step is given by

$$e_n = y_n - y(x_n). \tag{3}$$

Note that $y_0 = y(x_0)$, so $e_0 = 0$.

Now $y(x_{n+1}) = y(x_n + h)$ and $y''(x)$ is continuous. So we may use Taylor's theorem with remainder to obtain

$$y(x_{n+1}) = y(x_n + h) = y(x_n) + hy'(x_n) + \frac{h^2}{2} y''(\xi_n), \tag{4}$$

where $x_n \le \xi_n \le x_{n+1}$. We may now state the main result of this section.

THEOREM 1

Let $f(x, y)$ have continuous first partial derivatives and let y_n be the approximate solution of equation (1) generated by the Euler method (2). Suppose that $y(x)$ is defined and the inequalities

$$\left|\frac{\partial f}{\partial y}(x, y)\right| < L \quad \text{and} \quad |y''(x)| < M$$

hold on the bounded interval $x_0 \leq x \leq a$. Then the error $e_n = y_n - y(x_n)$ satisfies the inequality

$$|e_n| \leq \frac{hM}{2L}\left(e^{(x_n - x_0)L} - 1\right) = \frac{hM}{2L}\left(e^{nhL} - 1\right). \tag{5}$$

In particular, since $x_n - x_0 \leq a - x_0$ (which is finite), $|e_n|$ tends to zero as h tends to zero.

Proof A subtraction of (4) from (2) yields

$$y_{n+1} - y(x_{n+1}) = y_n - y(x_n) + h[f(x_n, y_n) - y'(x_n)] - \frac{h^2}{2}y''(\xi_n),$$

or

$$e_{n+1} = e_n + h[f(x_n, y_n) - f(x_n, y(x_n))] - \frac{h^2}{2}y''(\xi_n). \tag{6}$$

By the mean value theorem of differential calculus,

$$f(x_n, y_n) - f(x_n, y(x_n)) = \frac{\partial f}{\partial y}(x_n, \hat{y}_n)[y_n - y(x_n)]$$

$$= \frac{\partial f}{\partial y}(x_n, \hat{y}_n)e_n, \tag{7}$$

where $\hat{y}_n$ is between y_n and $y(x_n)$. We substitute (7) into (6) to obtain

$$e_{n+1} = e_n + h\frac{\partial f}{\partial y}(x_n, \hat{y}_n)e_n - \frac{h^2}{2}y''(\xi_n). \tag{8}$$

But $|\partial f/\partial y| \leq L$ and $|y''| \leq M$, so taking the absolute value of both sides of (8) and using the triangle inequality, we obtain

$$|e_{n+1}| \leq |e_n| + hL|e_n| + \frac{h^2}{2}M = (1 + hL)|e_n| + \frac{h^2}{2}M. \tag{9}$$

We now consider the difference equation (see Section 2.8)

$$r_{n+1} = (1 + hL)r_n + \frac{h^2}{2}M, \quad r_0 = 0, \tag{10}$$

and claim that if r_n is the solution to (10), then $|e_n| \leq r_n$. We show this by induction. It is true for $n = 0$, since $e_0 = r_0 = 0$. We assume it is true for $n = k$ and prove it for $n = k + 1$; that is, we assume that $|e_m| \leq r_m$, for $m = 0, 1, \ldots, k$.

Then

$$r_{k+1} = (1+hL)r_k + \frac{h^2}{2}M \geq (1+hL)|e_k| + \frac{h^2}{2}M \geq |e_{k+1}|,$$

and the claim is proved [the last step follows from (9)]. We can solve equation (10) by proceeding inductively: $r_1 = h^2M/2 = [(1+hL) - 1]hM/2L$, $r_2 = [(1+hL) + 1]h^2M/2 = [(1+hL)^2 - 1]hM/2L$. We find that

$$r_n = \frac{hM}{2L}(1+hL)^n - \frac{hM}{2L}.$$

Now $e^{hL} = 1 + hL + h^2L^2/2! + \cdots$, so

$$1 + hL \leq e^{hL} \quad \text{and} \quad (1+hL)^n \leq (e^{hL})^n = e^{nhL}.$$

Thus

$$|e_n| \leq r_n \leq \frac{hM}{2L}e^{nhL} - \frac{hM}{2L} = \frac{hM}{2L}(e^{nhL} - 1). \tag{11}$$

But $x_n = x_0 + nh$, so $x_n - x_0 = nh$ and (11) becomes

$$|e_n| \leq \frac{hM}{2L}(e^{(x_n-x_0)L} - 1).$$

Thus the theorem is proved. ■

Theorem 1 not only shows that the errors get small as h tends to zero; it also tells us *how fast* the errors decrease. If we define the constant k by

$$k = \frac{M}{2L}|e^{(a-x_0)L} - 1|, \tag{12}$$

we have

$$|e_n| \leq kh. \tag{13}$$

Thus the error is bounded by a *linear* function of h. (Note that $|e_n|$ is bounded by a term that depends only on h, not on n.) Roughly speaking, this implies that the error decreases at a rate proportional to the decrease in the step size. If, for example, we halve the step size, then we can expect at least to halve the error. Actually, since the estimates used in arriving at (13) were very crude, we can often do better, as in Example 1.3.2, where we halved the step size and decreased the error by a factor of four. Nevertheless, it is useful to have an upper bound for the error. It should be noted, however, that this bound may be difficult to obtain, since it is frequently difficult to find a bound for $y''(x)$.

Example 1 Consider the equation $y' = y$, $y(0) = 1$. We have

$$f(x, y) = y \quad \text{and} \quad \left|\frac{\partial f}{\partial y}(x, y)\right| = 1 = L.$$

Since the solution of the problem is $y(x) = e^x$, we have $|y''| \leq e^1 = M$ on the interval $0 \leq x \leq 1$. Then equation (12) becomes

$$k = \frac{e}{2}|e - 1| = \frac{e^2 - e}{2} \approx 2.34,$$

TABLE 9.16

x_n	$y_n' = f(x_n, y_n)$	$y_{n+1} = y_n + hy_n'$	$y(x_n) = e^{x_n}$	$e_n = y_n - y(x_n)$
0.0	1.00	1.10	1.00	0.00
0.1	1.10	1.21	1.11	−0.01
0.2	1.21	1.33	1.22	−0.01
0.3	1.33	1.46	1.35	−0.02
0.4	1.46	1.61	1.49	−0.03
0.5	1.61	1.77	1.65	−0.04
0.6	1.77	1.95	1.82	−0.05
0.7	1.95	2.15	2.01	−0.06
0.8	2.15	2.37	2.23	−0.08
0.9	2.37	2.61	2.46	−0.09
1.0	2.61		2.72	−0.11

so that the error is proportional to the step size:

$$|e_n| \leq 2.34h.$$

Therefore, using a step size of, say, $h = 0.1$, we can expect to have an error at each step of less than 0.234 (see Table 9.16). We note that the greatest actual error is about half of the maximum possible error according to equation (13).

It turns out that it is possible to derive error estimates like (11) or (13) for every method we discuss in this chapter for solving differential equations numerically. Actually, to derive these estimates would take us beyond the scope of this book,[†] but we should mention that for the Runge-Kutta method, which was discussed in Section 9.3, the discretization error e_n is of the form

$$|e_n| \leq kh^4,$$

for some appropriate constant k. Thus, halving the step size, for example, has the effect of decreasing the bound on the error by a factor of $2^4 = 16$. However, the price for this greater accuracy is to have to calculate $f(x, y)$ at four points [see equation (9.3.7)] for each step in the iteration.

[†] For a more detailed analysis, see for example, C. W. Gear, *Numerical Initial Value Problems in Ordinary Differential Equations* (Englewood Cliffs, N. J.: Prentice-Hall, 1971).

PROBLEMS 9.5

1. Consider the differential equation $y' = -y$, $y(0) = 1$. We wish to find $y(1)$.

    a. Calculate an upper bound on the error of the Euler method as a function of h.

    b. Calculate this bound for $h = 0.1$ and $h = 0.2$.

    c. Perform the iterations for $h = 0.2$ and $h = 0.1$ and compare the actual error with the maximum error.

2. Consider the equation of Problem 1. If we ignore round-off error, how many iterations have to be performed in order to guarantee that the calculation of $y(1)$ obtained by the Euler method is correct to (a) five decimal places? (b) six decimal places?

3. Answer the questions in Problem 2 for the equation

$$y' = 3y - x^2, \quad y(1) = 2$$

if we wish to find $y(1.5)$.

9.6 An Excursion: Numerical Instability Caused by Propagation of Round-Off Error

In this section we show how a theoretically very accurate method can produce results that are useless. A *multistep method* is one that involves information about the solution at more than one point. Consider the multistep method given by the equation

$$y_{n+1} = y_{n-1} + 2hf(x_n, y_n). \tag{1}$$

Here it is necessary to use both the nth and the $(n-1)$st iterate to obtain the $(n+1)$st iterate. It can be shown[†] that this method has the following error estimate:

$$|e_n| = |y_n - y(x_n)| \leq kh^2.$$

Since the error for the Euler method is $|e_n| \leq kh$, we would theoretically expect more accuracy in solving our initial-value problem by using equation (1) than by using the Euler method. However, this does not always turn out to be the case.

Example 1 Consider the initial-value problem

$$y' = -y + 2, \qquad y(0) = 1.$$

The solution to this equation is easily obtained: $y(x) = 2 - e^{-x}$. Let us obtain $y(5)$ by the Euler method and the method of (1). To use the latter, we need two initial values y_0 and y_1. Since we know the solution, we use the exact value $y_1 = y(x_1) = 2 - e^{-x_1}$. Table 9.17 illustrates the computation with a step size $h = 0.25$. The second column is the correct value of $y(x_n)$ to four decimal places. Column three gives the Euler iterates, and column four gives the iterates obtained by the two-step method (1). Column five is the Euler error, $e_n^{(\text{Euler})} = y_n^{(\text{Euler})} - y(x_n)$, and column six is the error of the two-step method, $e_n^{(2s)} = y_n^{(2s)} - y(x_n)$.

It is evident that the two-step method (1) produces a smaller error for small values of x_n than the Euler method. However, as x_n increases, the error in the Euler method decreases, whereas the error in the two-step method not only increases but does so with oscillating sign. This phenomenon is called **numerical instability**. As we see below, it is due to a propagation of round-off errors.

Let us now explain what leads to this instability. In the example, $f(x_n, y_n) = -y_n + 2$, so (1) is

$$y_{n+1} = y_{n-1} + 2h(2 - y_n),$$

or

$$y_{n+1} + 2hy_n - y_{n-1} = 4h, \qquad y_0 = 1. \tag{2}$$

This is a linear second-order **nonhomogeneous difference equation** (see Section 3.9). Its solution is obtained in a manner analogous to that used in solving linear nonhomogeneous differential equations. First we find the general solution of the **homogeneous** difference equation

$$y_{n+1} + 2hy_n - y_{n-1} = 0. \tag{3}$$

[†] See S. D. Conte, *Elements of Numerical Analysis* (New York: McGraw-Hill, 1965), section 6.6.

9.6/AN EXCURSION: NUMERICAL INSTABILITY CAUSED BY THE PROPAGATION OF ROUND-OFF ERROR

TABLE 9.17 A Comparison of the Error Using the Euler Method and That of Equation (1)

x_n	$y(x_n) = 2 - e^{-x_n}$ (exact)	$y_n^{(E)} = y_{n-1}^{(E)} + h(2 - y_{n-1}^{(E)})$	$y_n^{(2s)} = y_{n-2}^{(2s)} + 2h(2 - y_{n-1}^{(2s)})$	$e_n^{(E)}$	$e_n^{(2s)}$
0.00	1.0000	1.0000	1.0000	0.0000	0.0000
0.25	1.2212	1.2500	1.2212	0.0288	0.0000
0.50	1.3935	1.4375	1.3894	0.0440	−0.0041
0.75	1.5276	1.5781	1.5265	0.0505	−0.0011
1.00	1.6321	1.6836	1.6262	0.0515	−0.0059
1.25	1.7135	1.7627	1.7134	0.0492	−0.0001
1.50	1.7769	1.8220	1.7695	0.0453	−0.0074
1.75	1.8262	1.8665	1.8287	0.0403	+0.0025
2.00	1.8647	1.8999	1.8552	0.0352	−0.0095
2.25	1.8946	1.9249	1.9011	0.0303	+0.0065
2.50	1.9179	1.9437	1.9047	0.0258	−0.0132
2.75	1.9361	1.9578	1.9488	0.0217	+0.0127
3.00	1.9502	1.9684	1.9303	0.0182	−0.0199
3.25	1.9612	1.9763	1.9837	0.0151	+0.0225
3.50	1.9698	1.9822	1.9385	0.0124	−0.0313
3.75	1.9765	1.9867	2.0145	0.0102	+0.0380
4.00	1.9817	1.9900	1.9313	0.0083	−0.0504
4.25	1.9857	1.9925	2.0489	0.0068	+0.0632
4.50	1.9889	1.9944	1.9069	0.0055	−0.0820
4.75	1.9913	1.9958	2.0955	0.0045	+0.1042
5.00	1.9933	1.9969	1.8952	0.0036	−0.0981

To do this, we substitute $y_n = \lambda^n$ in (3), obtaining

$$\lambda^{n-1}(\lambda^2 + 2h\lambda - 1) = 0.$$

Since we are not interested in trivial solutions, assume $\lambda \neq 0$. Thus the two roots of the **characteristic equation**

$$\lambda^2 + 2h\lambda - 1 = 0 \tag{4}$$

given by

$$\lambda_1 = \frac{-2h + \sqrt{4h^2 + 4}}{2} = -h + \sqrt{1 + h^2} \quad \text{and} \quad \lambda_2 = -h - \sqrt{1 + h^2}$$

yield the solutions $y_{n,1} = (\lambda_1)^n$ and $y_{n,2} = (\lambda_2)^n$ to (3). Hence (3) has the general solution

$$y_n = c_1 \lambda_1^n + c_2 \lambda_2^n.$$

Finally, we need to find *any* particular solution to the nonhomogeneous equation (2). Since $4h$ is constant, using the method of undetermined coefficients, we set $y_{n,p} = A$. Then (2) becomes

$$A + 2hA - A = 4h,$$

so that $A = 2$. Hence the general solution of (2) is given by

$$y_n = c_1 \lambda_1^n + c_2 \lambda_2^n + 2. \tag{5}$$

By the binomial theorem (see Problem 5.1.18),
$$(1+h^2)^{1/2} = 1 + \tfrac{1}{2}h^2 - \tfrac{1}{8}h^4 + \tfrac{1}{16}h^6 - \cdots,$$
where the omitted terms are higher powers of h. Hence the roots of the characteristic equation (4) can be written as
$$\lambda_1 = 1 - h + \alpha(h) \quad \text{and} \quad \lambda_2 = -1 - h - \alpha(h), \tag{6}$$
where
$$\alpha(h) = \frac{h^2}{2} - \frac{h^4}{8} + \frac{h^6}{16} - \cdots.$$
Substituting (6) into (5) yields
$$y_n = c_1[1 - h + \alpha(h)]^n + c_2(-1)^n[1 + h + \alpha(h)]^n + 2. \tag{7}$$
From calculus we know that
$$\lim_{k \to \infty} \left(1 + \frac{1}{k}\right)^k = \lim_{h \to 0} (1+h)^{1/h} = e.$$
Therefore, since $x_n = 0 + nh = nh$, we have
$$\lim_{h \to 0} (1-h)^n = \lim_{h \to 0} (1-h)^{x_n/h} = e^{-x_n} \quad \text{and} \quad \lim_{h \to 0} (1+h)^n = e^{x_n}.$$
Hence as $h \to 0$ we may ignore the higher-order terms $\alpha(h)$ in (7) to obtain
$$y_n = c_1 e^{-x_n} + 2 + c_2(-1)^n e^{x_n}. \tag{8}$$
Here lies the problem. The exact solution of the problem requires that $c_1 = -1$ and $c_2 = 0$. However, even a small round-off error may cause c_2 to be nonzero, and this error grows exponentially while the real solution is approaching the constant two. This is the phenomenon we observed in Table 9.17. Note that the $(-1)^n$ in (8) causes the errors to oscillate (as we also observed).

The problem arose because we approximated a *first*-order differential equation by a *second*-order difference equation. Such approximations do not always lead to this kind of instability, but it is a possibility that cannot be ignored. In general, to analyze the effectiveness of a given method, we must not only estimate the discretization error but also show that the method is not numerically unstable (that is, *that it is numerically stable*), for the given problem.

9.7 Numerical Solution of Systems and Boundary Value Problems

The methods developed in Sections 9.2, 9.3, and 9.4 can be extended very easily to apply to higher-order equations and systems of equations.

EULER METHODS

For the Euler and improved Euler methods, it is necessary only to reinterpret the formulas
$$y_{n+1} = y_n + hf(x_n, y_n) \tag{1}$$

and

$$y_{n+1} = y_n + \frac{h}{2}[f(x_n, y_n) + f(x_{n+1}, z_{n+1})], \quad \text{with} \quad z_{n+1} = y_n + hf(x_n, y_n). \tag{2}$$

Here y_n is a vector with as many entries as there are dependent variables. In this case, the function f consists of a vector of functions also. The methods are best illustrated by examples.

Example 1   Consider the initial-value problem

$$\frac{dx}{dt} = -3x + 4y, \quad x(0) = 1,$$
$$\frac{dy}{dt} = -2x + 3y, \quad y(0) = 2. \tag{3}$$

Suppose we are seeking the values $x(1)$ and $y(1)$. In this problem, t is the independent variable, and x and y are the dependent variables. If we wish to use the Euler method, formula (1) translates into the equations

$$x_{n+1} = x_n + hx'_n = x_n + h(-3x_n + 4y_n),$$
$$y_{n+1} = y_n + hy'_n = y_n + h(-2x_n + 3y_n).$$

The initial values are $x_0 = 1$, $y_0 = 2$, and, for $h = 0.2$, the procedure is essentially the same as before (see Table 9.18).

TABLE 9.18

t_n	x_n	y_n	x'_n	y'_n	$x_{n+1} = x_n + hx'_n$	$y_{n+1} = y_n + hy'_n$
0.0	1.00	2.00	5.00	4.00	2.00	2.80
0.2	2.00	2.80	5.20	4.40	3.04	3.68
0.4	3.04	3.68	5.60	4.96	4.16	4.67
0.6	4.16	4.67	6.20	5.69	5.40	5.81
0.8	5.40	5.81	7.04	6.63	6.81	7.14
1.0	6.81	7.14				

The solution of (3) is given by

$$x(t) = 3e^t - 2e^{-t}, \quad y(t) = 3e^t - e^{-t},$$

so $x(1) = 3e - 2e^{-1} \approx 7.419$ and $y(1) = 3e - e^{-1} \approx 7.787$, implying that our method has an error of about 10 percent. The accuracy may be improved by selecting smaller values of h.

No additional difficulty is caused by having a nonhomogeneous or nonlinear system of equations.

PREDICTOR-CORRECTOR METHODS

Example 2   Adapt predictor-corrector formulas (11) and (12) in Section 9.4 to the system in Example 1.

TABLE 9.19

			Predictor						Corrector	
t_{n+1}	x_{n+1}	y_{n+1}	x'_{n+1}	y'_{n+1}	x_{n+2}	y_{n+2}	x'_{n+2}	y'_{n+2}	x_{n+2}	y_{n+2}
0.0	1.00	2.00	5.00	4.00						
0.2	2.00	2.80	5.20	4.40	3.06	3.72	5.70	5.04	3.09	3.74
0.4	3.09	3.74	5.69	5.04	4.28	4.81	6.40	5.87	4.30	4.83
0.6	4.30	4.83	6.42	5.89	5.66	6.09	7.38	6.95	5.68	6.11
0.8	5.68	6.11	7.40	6.97	7.26	7.61	8.66	8.31	7.29	7.64
1.0	7.29	7.64								

We assume from the calculations in Table 9.18 that $x_0 = 1$, $x'_0 = 5$, $x_1 = 2$, $x'_1 = 5.2$, $y_0 = 2$, $y'_0 = 4$, $y_1 = 2.8$, $y'_1 = 4.4$, and $h = 0.2$. With these values we first predict

$$x_{n+2} = x_{n+1} + \frac{h}{2}(-x'_n + 3x'_{n+1}),$$
$$y_{n+2} = y_{n+1} + \frac{h}{2}(-y'_n + 3y'_{n+1}),$$
(4)

and then use these values to compute x'_{n+2}, y'_{n+2}, using the system of differential equations (3). Then, to correct these values, we use the trapezoidal rules:

$$x_{n+2} = x_{n+1} + \frac{h}{2}(x'_{n+2} + x'_{n+1}),$$
$$y_{n+2} = y_{n+1} + \frac{h}{2}(y'_{n+2} + y'_{n+1}),$$
(5)

and recalculate x'_{n+2}, y'_{n+2} with (3).

It is clear that such methods are laborious, but they are easily carried out on a computer. The calculations shown in Table 9.19 agree far better with the exact values, $x(1) \approx 7.419$ and $y(1) \approx 7.787$, than those of Example 1.

RUNGE-KUTTA METHOD

The Runge-Kutta formula for a system of differential equations is a direct generalization of equations (9.3.6) and (9.3.7.) Suppose that we are given the system

$$\frac{dx}{dt} = f(t, x, y), \quad x(t_0) = x_0,$$
$$\frac{dy}{dt} = g(t, x, y), \quad y(t_0) = y_0.$$
(6)

The rule then becomes

$$x_1 = x_0 + \tfrac{1}{6}(m_1 + 2m_2 + 2m_3 + m_4),$$
$$y_1 = y_0 + \tfrac{1}{6}(n_1 + 2n_2 + 2n_3 + n_4),$$
(7)

where
$$m_1 = hf(t_0, x_0, y_0), \qquad n_1 = hg(t_0, x_0, y_0),$$
$$m_2 = hf\left(t_0 + \frac{h}{2}, x_0 + \frac{m_1}{2}, y_0 + \frac{n_1}{2}\right), \qquad n_2 = hg\left(t_0 + \frac{h}{2}, x_0 + \frac{m_1}{2}, y_0 + \frac{n_1}{2}\right),$$
$$m_3 = hf\left(t_0 + \frac{h}{2}, x_0 + \frac{m_2}{2}, y_0 + \frac{n_2}{2}\right), \qquad n_3 = hg\left(t_0 + \frac{h}{2}, x_0 + \frac{m_2}{2}, y_0 + \frac{n_2}{2}\right),$$
$$m_4 = hf(t_0 + h, x_0 + m_3, y_0 + n_3), \qquad n_4 = hg(t_0 + h, x_0 + m_3, y_0 + n_3).$$
(8)

It should now be apparent how this procedure is generalized for systems involving more dependent variables. We apply the formulas above to system (3) in Example 1 with $h = 1$:

$$m_1 = 5, \qquad n_1 = 4,$$
$$m_2 = \frac{11}{2}, \qquad n_2 = 5,$$
$$m_3 = \frac{27}{4}, \qquad n_3 = 6,$$
$$m_4 = \frac{35}{4}, \qquad n_4 = \frac{17}{2}.$$

We obtain
$$x_1 \approx 7.375, \qquad y_1 \approx 7.750.$$

Even though this process involves more complicated computations at each step, it involves less work than predictor-corrector methods. Since it is also quite accurate, it is thus the preferred method for hand calculations.

All of these methods can be applied to higher-order differential equations by converting each higher-order equation into a system of first-order equations using the procedure outlined in Section 7.1.

BOUNDARY VALUE PROBLEMS

Consider the differential equation
$$y'' = f(x, y, y'), \tag{9}$$
with boundary conditions $y(a) = y_a$ and $y(b) = y_b$. We now describe a procedure, sometimes called the *shooting method*, that is often used to solve such problems. The idea is to convert (9) into an initial-value problem with initial conditions $y(a) = y_a$ and $y'(a) = M_0$, where the number M_0 is arbitrarily selected. Using one of the previously described techniques (Euler method, Runge-Kutta, or predictor-corrector method), we now calculate the value $y(b) = N_0$ for the initial-value problem. This number will undoubtedly be different from the required value y_b, so we again solve (9) as an initial-value problem with $y(a) = y_a$ and $y'(a) = M_1$, obtaining $y(b) = N_1$. The assumption is now made that $y(b)$ varies linearly with the values $y'(a)$, so the next value we choose for $y'(a)(= M_2)$ is selected by solving the equation
$$\frac{M_2 - M_1}{M_0 - M_1} = \frac{y_b - N_1}{N_0 - N_1}, \tag{10}$$

or
$$M_2 = M_1 + \frac{y_b - N_1}{N_0 - N_1}(M_0 - M_1). \tag{11}$$

Equation (11) yields very accurate information if the user is prepared to repeat the process several times. Again, it is clear that since this process is very laborious, it is best performed by using a computer.

Finally, it should be noted that although shooting methods are useful for finding approximate solutions for many boundary value problems, they do not work, for a variety of reasons, for all such problems. Other methods, some very complex, are available to deal with these situations.

A Perspective

STIFF EQUATIONS

As soon as we begin to deal with a system of two or more first-order equations (or a higher-order equation), we encounter the phenomenon of **stiffness**. Stiffness arises in any problem in which the dependent variable can be expressed in terms of two or more very different scales in the independent variable. For example, consider the differential equation

$$y'' - 15y' - 100y = 0. \tag{12}$$

This second-order equation is equivalent to the system
$$y' = z,$$
$$z' = 100y + 15z.$$

The general solution has the form
$$y = c_1 e^{20x} + c_2 e^{-5x}, \tag{13}$$

since the characteristic equation is $\lambda^2 - 15\lambda - 100 = (\lambda - 20)(\lambda + 5) = 0$. Observe that the general solution depends (for increasing x) on a rapidly growing term and a rapidly decaying term.

Suppose we are interested in solving the initial-value problem consisting of equation (12) and the initial conditions
$$y(0) = 1, \quad y'(0) = -5.$$

In this case the exact solution is $y = e^{-5x}$. However, as Table 9.20 shows, each method begins with a numerical solution that decays as e^{-5x} but then rapidly deviates from the exact solution. The reason for this difficulty is simple: each step in every numerical procedure introduces a small error ϵ, so the *numerical solution* is now a linear combination of both e^{-5x} and e^{20x}:

$$y_{\text{numerical}} \approx e^{-5x} + \epsilon e^{20x}.$$

No matter how small ϵ is, the ϵe^{20x} term quickly dominates the other term (compare this situation to that in Section 9.6). Stiff equations can be handled if precautions are taken to deal with them[†]; modern computer codes include such precautions.

[†] For further information about this topic, see C. W. Gear, *Numerical Initial Value Problems in Ordinary Differential Equations* (Englewood Cliffs, N. J.: Prentice-Hall, 1971).

TABLE 9.20 A Comparison of the Runge-Kutta Method and the Exact Solution $y = e^{-5x}$, with $h = 0.02$ and $h = 0.2$, for the Initial-Value Problem $y'' - 15y' - 100y = 0$, $y(0) = 1$, $y'(0) = -5$

x	y, Runge-Kutta Method	y, Exact
$h = 0.02$		
0.00	1.000000000	1.000000000
0.02	0.862070833	0.904837418
0.04	0.635804065	0.818730753
0.06	0.300813510	0.740818221
0.08	−0.165364430	0.670320046
0.10	−0.786953158	0.606530660
0.12	−1.589425340	0.548811636
0.14	−2.598755640	0.496585304
0.16	−3.840388700	0.449328964
0.18	−5.337870700	0.406569660
0.20	−7.111090350	0.367879441
$h = 0.2$		
0.0	1.00000000	1.000000000
0.2	−7.29166667	0.367879441
0.4	−42.5121528	0.135335283
0.6	50.8496093	0.0497870684
0.8	1105.45961	0.0183156389
1.0	1836.41827	6.737947×10^{-3}
1.2	−20190.9270	$2.47875217 \times 10^{-3}$
1.4	−96487.7822	$9.11881964 \times 10^{-4}$
1.6	198302.943	$3.35462627 \times 10^{-4}$
1.8	2733507.62	$1.23409804 \times 10^{-4}$
2.0	2889103.65	$4.53999298 \times 10^{-5}$

PROBLEMS 9.7

In Problems 1–8 find $x(1)$ and $y(1)$ with $h = 0.2$ using (a) the Euler method, (b) the improved Euler method, and (c) the Runge-Kutta method for each initial-value system. Check your accuracy by calculating the exact value.

1. $x' = 4x - 2y$, $x(0) = 1$
   $y' = 5x + 2y$, $y(0) = 2$
2. $x' = x + y$, $x(0) = 1$
   $y' = x - y$, $y(0) = 0$
3. $x' = x + 2y$, $x(0) = 0$
   $y' = 3x + 2y$, $y(0) = 1$
4. $x' = -4x - y$, $x(0) = 0$
   $y' = x - 2y$, $y(0) = 1$
5. $x' = 2x + y + t$, $x(0) = 1$
   $y' = x + 2y + t^2$, $y(0) = 0$
6. $x' = x + 2y + t - 1$, $x(0) = 0$
   $y' = 3x + 2y - 5t - 2$, $y(0) = 4$
7. $x' = 3x + 3y + t$, $x(0) = 0$
   $y' = -x - y + 1$, $y(0) = 2$
8. $x' = 4x - 3y + t$, $x(0) = 1$
   $y' = 5x - 4y - 1$, $y(0) = -1$
9. Solve the initial-value problem
   $$y'' = y' + xy^2, \; y(0) = 1, \; y'(0) = 0,$$
   for $y(1)$ with $h = 0.2$ by the Euler method.
10. Use the "shooting method" to determine the value of $y'(0)$ for the boundary value problem
    $$y'' = -y^2, \; y(0) = y(1) = 0,$$

in such a way that y is positive over the interval $0 < x < 1$. Use the Runge-Kutta method [equations (7) and (8)] with $h = 1$.

11. Find the maximum value of y over the interval $0 \le x \le 1$ for the boundary value problem
$$y'' + yy' + 1 = 0, \quad y(0) = y(1) = 0.$$

12. Develop a BASIC (or FORTRAN) program to perform the "shooting method" using the improved Euler method for an equation of the form
$$y'' = f(x, y, y'), \quad y(a) = y_a, \quad y(b) = y_b.$$
Use it to find the maximum value of y over the interval $0 \le x \le 1$ for the boundary value problem
$$y'' + \sin y = 0, \quad y(0) = y(1) = 0.$$

Review Exercises for Chapter 9

In Exercises 1–6 solve the given initial-value problem using the methods of Chapter 2. Then use the (a) improved Euler method or (b) the Runge-Kutta method and the given value of h to obtain an approximate solution at the indicated value of x. Compare the numerical answer with the exact answer.

1. $\dfrac{dy}{dx} = \dfrac{e^x}{y}$, $y(0) = 2$. Find $y(3)$ with $h = \tfrac{1}{2}$.

2. $\dfrac{dy}{dx} = \dfrac{e^y}{x}$, $y(1) = 0$. Find $y(\tfrac{1}{2})$ with $h = -0.1$.

3. $\dfrac{dy}{dx} = \dfrac{y}{\sqrt{1+x^2}}$, $y(0) = 1$. Find $y(3)$ with $h = \tfrac{1}{2}$.

4. $xy\dfrac{dy}{dx} = y^2 - x^2$, $y(1) = 2$. Find $y(3)$ with $h = \tfrac{1}{2}$.

5. $\dfrac{dy}{dx} = y - xy^3$, $y(0) = 1$. Find $y(3)$ with $h = \tfrac{1}{2}$.

6. $\dfrac{dy}{dx} = \dfrac{2xy}{3x^2 - y^2}$, $y(-\tfrac{3}{8}) = -\tfrac{3}{4}$. Find $y(6)$ with $h = \tfrac{3}{8}$.

7. Consider the differential equation in Exercise 6 with the initial condition $y(0) = -1$. Use the Runge-Kutta method or the improved Euler method to calculate $y(6)$ with $h = 1$. Why does the numerical solution differ from the exact answer?

8. Suppose the initial condition in Exercise 4 is $y(1) = 1$. Can any of the methods of Section 9.2 or 9.4 provide the correct answer for $y(3)$?

9. Consider the initial-value problem
$$y' = 1 + y^2, \quad y(0) = 0.$$
Can any of the methods in Section 9.2 be used to obtain $y(2)$?

In Exercises 10–13 find the value of the dependent function(s) at 2 with $h = 0.2$ using (a) the Euler method, (b) the Runge-Kutta method, and (c) the improved Euler method as predictor, and the trapezoidal rule as corrector to solve numerically the following initial-value problems. Compare the numerical answer to the exact solution.

10. $x^2 y'' + xy' + y = 0$,
$y(1) = 1$, $y'(1) = 0$

11. $x^2 y'' + xy' + y = 0$,
$y(1) = 1$, $y'(1) = 1$

12. $x' = x - 2y$, $x(0) = 1$
$y' = 2x + 5y$, $y(0) = 0$

13. $x' = x + 2y$, $x(0) = 1$
$y' = 2x + 5y$, $y(0) = 0$

10

Fourier Series and Partial Differential Equations

Until this chapter we have considered differential equations whose solutions are determined by initial conditions. Here we discuss another equally important way to specify a particular solution of a differential equation; namely, specifying certain values of the function and its derivatives at two or more points. The differential equation together with the given conditions at two or more points is called a **boundary value problem** (this was defined earlier in Section 1.2). Boundary value problems arise in a number of ways; in particular they arise when we apply the method of separation of variables (see Section 10.7) to a linear **partial differential equation** (also defined in Section 1.2).

We present a brief introduction to partial differential equations, but we make no attempt to study the general theory. Instead, we examine certain simple examples that have a number of practical applications.

10.1 Introduction to Trigonometric Series

In this section we present some examples of boundary value problems, together with a discussion of some of the powerful methods that are used to solve them. We see how the concepts of eigenvalues and eigenfunctions arise naturally in these situations and discuss the importance of Fourier series techniques in solving nonhomogeneous boundary value problems.

Since a boundary value problem requires that conditions be given at two or more points, we can assume the differential equation involved to be at least of second order, because a first-order equation is usually completely specified by a single condition (this follows from the existence-uniqueness result that was cited in Section 1.2).

Example 1 Consider the simple harmonic motion of a point mass m attached to a coiled spring with spring constant k. Applying Newton's law of motion and Hooke's law to this situation, we obtained in Section 4.1 [see equation (4.1.2), p. 139], the differential equation

$$x'' + p^2 x = 0, \qquad p^2 = \frac{k}{m}, \tag{1}$$

where $x = x(t)$ denotes the displacement at time t of the mass from its equilibrium position. Suppose the mass is initially ($t = 0$) at its equilibrium position when it is given an unknown initial velocity $v_0 > 0$. How can we guarantee that the mass will again be at its equilibrium position after precisely one second? Obviously, such a mechanism provides a simple, though crude, timepiece (see Figure 10.1).

Figure 10.1

Our purpose now is to discover some of the properties of the unknown solution $x(t)$. If the time is measured in seconds, then we have the boundary conditions

$$x(0) = x(1) = 0. \tag{2}$$

The boundary conditions together with the differential equation (1) specify the boundary value problem we must solve.

Recall from Chapter 3 that to find the general solution of the differential equation (1) we need only solve the characteristic equation

$$\lambda^2 + p^2 = 0.$$

Since the roots of the characteristic equation are $\lambda = \pm ip$, the general solution of (1) is given by

$$x(t) = A \cos pt + B \sin pt. \tag{3}$$

Setting $t = 0$ in (3) and using the boundary condition $x(0) = 0$, we observe that $A = 0$, so the solution must consist only of multiples of $\sin pt$. For the boundary condition $x(1) = 0$ to hold, we must also have

$$B \sin p = x(1) = 0.$$

There are two cases to consider: either $B = 0$ or $\sin p = 0$. If $B = 0$, then solution (3) is constantly zero, which means that the mass remains at its equilibrium position at all times t. This is impossible, since the mass was given an initial velocity $v_0 > 0$. Therefore $\sin p = 0$, which is possible only if p is a nonzero multiple of π:

$$p = \pm\pi, \pm 2\pi, \pm 3\pi, \ldots.$$

[If $p = 0$, we see that (3) is again constantly zero.] Thus we have arrived at a surprising conclusion: There are infinitely many solutions of the form

$$x(t) = B \sin(n\pi)t, \tag{4}$$

where n is a nonzero integer, for the boundary value problem (1), (2). Furthermore, the constant B cannot be specified without additional information. Indeed, B depends on v_0 and n, since

$$v_0 = x'(0) = n\pi B,$$

or $B = v_0/n\pi$.

Actually, in doing the above calculations, we used one fact that was not part of the boundary value problem (1), (2). This was the assumption, given in the statement of the problem, that $v_0 > 0$. This assumption was used to disallow the *trivial solutions* $x(t) \equiv 0$ for all t. However, the results we have obtained illustrate one of the basic facts about boundary value problems: *Nontrivial solutions exist only for certain values of the parameter p.* All such nontrivial solutions (4) are called **eigenfunctions** of the problem, and the corresponding values of p^2 that yield these eigenfunctions are called **eigenvalues** of the problem. Rephrasing our work in this context, we see that

$$x(t) = \left(\frac{v_0}{n\pi}\right) \sin n\pi t \tag{5}$$

is an eigenfunction of this problem corresponding to the eigenvalue

$$\frac{k}{m} = p^2 = (n\pi)^2. \tag{6}$$

It is interesting to discover the physical significance of this solution. Suppose we are given a spring with (fixed) spring constant k. Solving (6) for m, we obtain the masses

$$m_n = \frac{k}{n^2\pi^2}, \quad n = 1, 2, 3, \ldots, \tag{7}$$

each corresponding to a different value of n. Since k and π are fixed quantities, it is *only* for these masses that a solution of the boundary value problem exists. For any other choice, no matter what initial velocity is given, the mass will not be back at its equilibrium position after exactly one second. Thus the construction of our clock depends on the selection of one of the masses in (7). Furthermore, (5) indicates that if mass m_n is chosen, it will be at its equilibrium position n times during the time period $0 < t \leq 1$. Since increasing n decreases both the weight m_n and the amplitude ($v_0/n\pi$) of the oscillations, we arrive at the conclusion that *the lighter the mass, the more rapid the oscillation*. This fact is easy to observe in practice.

Example 2 A population is subject to an influenza epidemic. Let $s(t)$ and $i(t)$ be, respectively, the numbers of susceptible and infected individuals at time t. Suppose that the rate of change of the number of susceptible individuals is proportional to the number of infected individuals, whereas the rate of change of the number of infected individuals is proportional to the number of susceptible individuals. (The difference between the change in the numbers of infected and susceptible individuals consists of those who have gained immunity or died.) The epidemic lasts a certain period of time, say, two months. [With this assumption, we can set $i(0) = i(2) = 0$.] We wish to discover the number of infected individuals at all time t (in months).

From the hypotheses above, we obtain the system of equations

$$\frac{ds}{dt} = -pi,$$
$$\frac{di}{dt} = qs, \tag{8}$$

where p and q are unknown positive constants of proportionality. Differentiating the second of the equations in system (8) with respect to t, and substituting the first equation for ds/dt, we obtain the second-order equation

$$\frac{d^2 i}{dt^2} = q\frac{ds}{dt} = -qpi,$$

or

$$i'' + k^2 i = 0, \qquad k^2 = pq, \tag{9}$$

with the boundary conditions

$$i(0) = i(2) = 0. \tag{10}$$

This boundary value problem is very similar to that of Example 1. We again find the general solution

$$i(t) = A \cos kt + B \sin kt.$$

Setting $t = 0$ and using the first boundary condition, we find that $0 = i(0) = A$. The second boundary condition $t = 2$ implies that

$$B \sin 2k = i(2) = 0.$$

Since we are not interested in the trivial solution $i(t) \equiv 0$, it follows that $B \neq 0$, so $2k$ must be a nonzero multiple of π. Hence the eigenvalues of this problem are

$$k^2 = \left(\frac{\pi}{2}\right)^2, (\pi)^2, \left(\frac{3\pi}{2}\right)^2, \ldots, \left(\frac{n\pi}{2}\right)^2, \ldots,$$

and the corresponding eigenfunctions $i_k(t)$ are given by

$$i_k(t) = B \sin \frac{n\pi t}{2}, \qquad k^2 = \left(\frac{n\pi}{2}\right)^2. \tag{11}$$

If the number of infected individuals is assumed to be positive during the entire epidemic $0 < t < 2$, then $n = 1$. This is a reasonable limitation since negative numbers of infected individuals do not have physical meaning.

The boundary value problems in these two examples will also arise in several different problems later in this chapter (see Examples 10.9.1 and 10.10.1).

Example 3 We return to Example 1, but we assume that the spring-mass system is also subject to a periodic external force $K \sin \omega t$, where $K > 0$ and $\omega > 0$. Using the theory of forced vibrations developed in Section 4.3, we obtain the nonhomogeneous second-order equation

$$m \frac{d^2 x}{dt^2} = -kx + K \sin \omega t.$$

Using the same boundary conditions as in Example 1, we are led to the nonhomogeneous boundary value problem

$$x'' + \frac{k}{m} x = \frac{K}{m} \sin \omega t, \qquad x(0) = x(1) = 0. \tag{12}$$

There are two cases to consider: either $\omega = n\pi$ for some integer n or $\omega \neq n\pi$ for every integer n.

First, we use the method of undetermined coefficients (Section 3.5) to find a particular solution. We suppose that

$$x(t) = A \sin \omega t + B \cos \omega t. \tag{13}$$

Since $x(0) = 0$, it follows that $B = 0$. Differentiating equation (13) twice and substituting into equation (12), we have

$$-A\omega^2 \sin \omega t + \frac{k}{m} A \sin \omega t = \frac{K}{m} \sin \omega t.$$

Canceling the $\sin \omega t$ on both sides of this equation and solving for A, we obtain

$$A = \frac{K/m}{(k/m) - \omega^2} = \frac{K}{k - m\omega^2}.$$

Case 1 If $\omega = n_0 \pi$, then

$$x_p(t) = \frac{K}{k - m n_0^2 \pi^2} \sin n_0 \pi t \tag{14}$$

is a particular solution of (12) provided that

$$\omega^2 = (n_0 \pi)^2 \neq k/m. \tag{15}$$

From Example 1 we know that the homogeneous boundary value problem

$$x'' + \frac{k}{m} x = 0, \qquad x(0) = x(1) = 0, \tag{16}$$

has the eigenfunction

$$x_n(t) = c \sin n\pi t$$

whenever

$$\frac{k}{m} = n^2 \pi^2, \qquad n = 1, 2, 3, \ldots,$$

and the trivial solution for all other values. Hence

$$x(t) = \begin{cases} c \sin n\pi t + \dfrac{K}{m\pi^2 (n^2 - n_0^2)} \sin n_0 \pi t, & \dfrac{k}{m} = n^2 \pi^2, \quad n \neq n_0, \\ \dfrac{K}{k - m\pi^2 n_0^2} \sin n_0 \pi t, & \dfrac{k}{m} \neq n^2 \pi^2. \end{cases}$$

If $k/m = n_0^2\pi^2 = \omega^2$, the method of undetermined coefficient requires that we multiply equation (13) by t to find a particular solution. In this case $A = 0$ and $B = -K/2\omega m$, so

$$x_p(t) = -\frac{K}{2\omega m}t\cos\omega t = \frac{-Kt}{2\pi n_0 m}\cos n_0\pi t.$$

But then $x_p(1) = -K\cos n_0\pi/2\pi n_0 m \neq 0$, so the boundary conditions cannot be satisfied. Thus there is *no* solution if $k/m = n_0^2\pi^2$.

Case 2 If $\omega \neq n\pi$, we see that the function we have obtained by the method of undetermined coefficients,

$$x_p(t) = \frac{K}{k - m\omega^2}\sin\omega t, \tag{17}$$

does not satisfy the boundary condition $x_p(1) = 0$. Since only linear combinations of periodic functions of the form

$$x_n(t) = \sin n\pi t \tag{18}$$

satisfy the boundary conditions $x(0) = x(1) = 0$, we try to approximate (17) as closely as possible by a sum of such functions. We can show, using the methods of Section 10.3, that the trigonometric series

$$2\pi\sin\omega\sum_{n=1}^{\infty}\frac{(-1)^n n}{\omega^2 - (n\pi)^2}\sin n\pi t, \tag{19}$$

called a **Fourier series**, converges to the function $\sin\omega t$ at all points in $0 \leq t < 1$, and to zero at $t = 1$. Making use of this fact, we seek a solution of the form

$$x(t) = 2\pi\sin\omega\sum_{n=1}^{\infty}\frac{(-1)^n nc_n}{\omega^2 - (n\pi)^2}\sin n\pi t. \tag{20}$$

Substituting (20) into (12) and using the Fourier series, (19), we can easily show that $c_n = K/(k - mn^2\pi^2)$. Using these values in (20), we obtain the general solution of the boundary value problem (12).

At this point one may wonder whether very many functions can be represented as a Fourier series. The question is significant, since forcing functions can be quite arbitrary. Actually, the class of functions that have Fourier series representations includes almost every practical forcing function. Although the Fourier series in (19) may appear complicated, it is really quite easy to work with once one becomes more familiar with it. We will see that the nth term in the series is independent of the other terms and behaves essentially like the nth coordinate of a certain vector. This phenomenon, called **orthogonality**, is one of the central themes of this chapter.

Example 4 Consider the ideal pendulum consisting of a weightless rod of length l supported at one end and attached to a particle of positive weight at the other (Figure 10.2). In Problem 4.1.11 we found that the angular motion

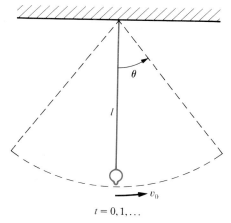

Figure 10.2

$\theta(t)$ of the pendulum is given by the nonlinear differential equation

$$\frac{d^2\theta}{dt^2} + \frac{g}{l}\sin\theta(t) = 0.$$

Suppose that initially $\theta(0) = 0$ (the rod is vertical) and that we give the pendulum an unknown initial velocity $\theta'(0) = v_0 > 0$. How can we guarantee that the mass will return to its initial position after precisely one second, that is, at $\theta(1) = 0$? The situation, which is essentially that of constructing a pendulum clock, reduces to the solution of the nonlinear boundary value problem

$$\theta'' + \frac{g}{l}\sin\theta = 0, \qquad \theta(0) = \theta(1) = 0.$$

In practice, the situation is much more complicated, since it involves damping effects due to friction and a forcing term (such as tension on a spring) to prevent the pendulum from stopping. We do not investigate nonlinear boundary value problems in this book; instead, we refer the interested reader to more advanced books on the subject.[†]

Finally, it should be noted that boundary value problems can involve differential equations of order greater than two. For example, in studying the deflection of a beam, one arrives at the fourth-order boundary value problem[‡]

$$\frac{d^4y}{dx^4} = k^2 y,$$

with specified boundary condition $y(0)$, $y'(0)$, $y(L)$, and $y'(L)$, where L is the length of the beam.

[†] See, for example, P. B. Bailey, L. F. Shampine, and P. E. Waltman, *Nonlinear Two-Point Boundary Value Problems* (New York: Academic Press, 1968).

[‡] This equation is derived in C. R. Wylie, *Advanced Engineering Mathematics* (New York: McGraw-Hill, 1966), p. 323.

PROBLEMS 10.1

1. What eigenfunctions and eigenvalues result in Example 1 if the boundary condition $x(1) = 0$ is replaced by $x(T) = 0$, where T is some fixed constant?

2. Consider the boundary value problem
$$\frac{d^2x}{dt^2} - kx(t) = 0, \quad x(0) = x(1) = 0,$$
where k is any real number. What are the eigenvalues and eigenfunctions of this problem?

3. Answer Problem 2 given that the boundary condition $x(1) = 0$ is replaced by $x(T) = 0$ for some fixed constant T.

4. Answer Problem 2 given that the boundary condition $x(0) = 0$ is replaced by $x(-1) = 0$, all other conditions remaining the same.

5. Find the eigenfunctions and eigenvalues of the boundary value problem
$$x''(t) + k^2x(t) = 0, \quad x(-T) = x(T) = 0,$$
where T is a fixed constant.

6. Find the eigenvalues and eigenfunctions of
$$x''(t) + k^2x(t) = 0, \quad x(0) = x'(1) = 0.$$

7. Answer Problem 6 if the condition $x'(1) = 0$ is replaced by $x'(T) = 0$ for some fixed constant T.

8. Answer Problem 6 if the conditions $x(0) = x'(1) = 0$ are replaced by $x'(0) = x(1) = 0$.

9. Answer Problem 6 if the conditions $x(0) = x'(1) = 0$ are replaced by $x'(0) = x'(1) = 0$.

*10. Find the eigenvalues and eigenvectors of the boundary value problem
$$(x^2 y')' + \lambda y = 0, \quad y(1) = y'(e) = 0.$$

*11. Answer Problem 10 if the boundary conditions are replaced by $y'(1) = y(e) = 0$.

10.2 Orthogonal Sets of Functions

Let $f_1(x)$ and $f_2(x)$ be two real-valued functions defined on an interval $a \leq x \leq b$, and suppose that the **inner product** of f_1 and f_2, given by

$$(f_1, f_2) = \int_a^b f_1(x) f_2(x)\, dx, \tag{1}$$

exists.† We say that the functions f_1 and f_2 are **orthogonal on the interval** $a \leq x \leq b$ if integral (1) vanishes; that is, if $(f_1, f_2) = 0$.‡ A set of real-valued functions $f_1(x), f_2(x), f_3(x), \ldots$ defined on $a \leq x \leq b$ is called an **orthogonal set of functions on the interval** $a \leq x \leq b$ if each integral

$$(f_n, f_m) = \int_a^b f_n(x) f_m(x)\, dx$$

exists and $(f_n, f_m) = 0$ whenever $n \neq m$. The nonnegative square root of the

---

† For the benefit of those readers familiar with the concept of the orthogonality of vectors in Euclidean n-space, we note that equation (1) is a generalization of the usual inner product

$$(\mathbf{f}_1, \mathbf{f}_2) = f_{11}f_{21} + f_{12}f_{22} + \cdots + f_{1n}f_{2n} = \sum_{j=1}^{n} f_{1j}f_{2j}.$$

‡ Two vectors $\mathbf{f}_1$ and $\mathbf{f}_2$ are orthogonal in Euclidean n-space if $(\mathbf{f}_1, \mathbf{f}_2) = 0$. Also the **norm** of a vector is its Euclidean length

$$\|\mathbf{f}_1\| = \sqrt{(\mathbf{f}_1, \mathbf{f}_1)} = \left(\sum_{j=1}^{n} f_{1j}^2 \right)^{1/2}$$

integral

$$(f_n, f_n) = \int_a^b f_n^2(x)\, dx$$

is called the **norm** of the function $f_n(x)$ and is often denoted by $\|f_n\|$ (see Problem 10):

$$\|f_n\| = \sqrt{(f_n, f_n)} = \left[\int_a^b f_n^2(x)\, dx\right]^{1/2}. \qquad (2)$$

An orthogonal set of functions $\{f_n(x)\}$ is called an **orthonormal set of functions on the interval** $a \le x \le b$ if $\|f_n\| = 1$ for all n.[†] Thus, if $\{f_n(x)\}$ is an orthonormal set of functions,

$$(f_n, f_m) = \int_a^b f_n(x) f_m(x)\, dx = \begin{cases} 0, & \text{whenever } n \ne m, \\ 1, & \text{whenever } n = m. \end{cases} \qquad (3)$$

Given any orthogonal set of functions $\{f_n(x)\}$ whose norms are nonzero, it is always possible to construct an orthonormal set of functions $\{g_n(x)\}$ by defining

$$g_n(x) = \frac{f_n(x)}{\|f_n\|}, \qquad a \le x \le b.$$

Then

$$(g_n, g_m) = \int_a^b \frac{f_n(x)}{\|f_n\|} \frac{f_m(x)}{\|f_m\|}\, dx = \frac{1}{\|f_n\|\,\|f_m\|}(f_n, f_m), \qquad (4)$$

which vanishes whenever $m \ne n$ since $\{f_n\}$ is an orthogonal set of functions. If $m = n$, then (4) yields

$$\|g_n\|^2 = (g_n, g_n)$$
$$= \frac{1}{\|f_n\|^2}(f_n, f_n) = 1,$$

so that $\|g_n\| = 1$ for all n.

Example 1   The functions $f_n(x) = \sin nx$, $n = 1, 2, 3, \ldots$, are an orthogonal set of functions on the interval $-\pi \le x \le \pi$. To see this, we use the trigonometric identities

$$\cos(m - n)x = \cos mx \cos nx + \sin mx \sin nx,$$
$$\cos(m + n)x = \cos mx \cos nx - \sin mx \sin nx,$$

which imply (after subtracting and dividing by 2) that

$$\sin nx \sin mx = \tfrac{1}{2}[\cos(m - n)x - \cos(m + n)x].$$

[†] This corresponds to taking an orthonormal basis in Euclidean n-space, such as the **unit vectors** $\mathbf{e}_1 = (1, 0, \ldots, 0)$, $\mathbf{e}_2 = (0, 1, 0, \ldots, 0)$, $\ldots$, $\mathbf{e}_n = (0, \ldots, 0, 1)$.

Then

$$(f_n, f_m) = \int_{-\pi}^{\pi} \sin nx \sin mx\, dx$$

$$= \frac{1}{2} \int_{-\pi}^{\pi} [\cos(m-n)x - \cos(m+n)x]\, dx$$

$$= \frac{1}{2} \left[\frac{\sin(m-n)x}{m-n} - \frac{\sin(m+n)x}{m+n} \right]_{-\pi}^{\pi} = 0, \qquad (5)$$

whenever $m \neq n$. If $m = n$, then $\cos(m-n)x = 1$ in the second integral of (5), so

$$(f_n, f_n) = \frac{1}{2} \int_{-\pi}^{\pi} (1 - \cos 2nx)\, dx$$

$$= \frac{1}{2} \left(x - \frac{\sin 2nx}{2n} \right)\bigg|_{-\pi}^{\pi} = \pi.$$

Hence $\|f_n\| = \sqrt{\pi}$, and we may construct the orthonormal set of functions

$$g_n(x) = \frac{\sin nx}{\sqrt{\pi}}, \qquad n = 1, 2, \ldots.$$

Example 2 The functions

$$\frac{1}{\sqrt{2\pi}}, \quad \frac{\cos x}{\sqrt{\pi}}, \quad \frac{\sin x}{\sqrt{\pi}}, \quad \frac{\cos 2x}{\sqrt{\pi}}, \quad \frac{\sin 2x}{\sqrt{\pi}}, \ldots$$

form an orthonormal set of functions on the interval $-\pi \leq x \leq \pi$. We have already demonstrated the orthonormality of the $\sin nx/\sqrt{\pi}$ terms, and clearly

$$\int_{-\pi}^{\pi} \frac{\cos nx}{\sqrt{\pi}} \frac{\cos mx}{\sqrt{\pi}}\, dx = \frac{1}{2\pi} \int_{-\pi}^{\pi} [\cos(m-n)x + \cos(m+n)x]\, dx = 0$$

whenever $m \neq n$, and

$$\int_{-\pi}^{\pi} \frac{\cos nx}{\sqrt{\pi}} \frac{\cos nx}{\sqrt{\pi}}\, dx = \frac{1}{2\pi} \int_{-\pi}^{\pi} (1 + \cos 2nx)\, dx = \frac{1}{2\pi} \left(x + \frac{\sin 2nx}{2n} \right)\bigg|_{-\pi}^{\pi} = 1.$$

The identities

$$\sin(m+n)x = \sin mx \cos nx + \cos mx \sin nx,$$
$$\sin(m-n)x = \sin mx \cos nx - \cos mx \sin nx$$

imply (after addition and division by 2) that

$$\int_{-\pi}^{\pi} \frac{\sin mx}{\sqrt{\pi}} \frac{\cos nx}{\sqrt{\pi}}\, dx = \frac{1}{2\pi} \int_{-\pi}^{\pi} [\sin(m-n)x + \sin(m+n)x]\, dx$$

$$= \frac{-1}{2\pi} \left[\frac{\cos(m-n)x}{m-n} + \frac{\cos(m+n)x}{m+n} \right]_{-\pi}^{\pi} = 0$$

for all m and n, since $\sin(m-n)x = 0$ if $m = n$, and $\cos(-\theta) = \cos \theta$. Finally, the integrals

$$\int_{-\pi}^{\pi} \frac{\sin mx}{\sqrt{\pi}} \frac{1}{\sqrt{2\pi}} dx = \frac{-1}{\pi\sqrt{2}} \left.\frac{\cos mx}{m}\right|_{-\pi}^{\pi} = 0,$$

$$\int_{-\pi}^{\pi} \frac{\cos mx}{\sqrt{\pi}} \frac{1}{\sqrt{2\pi}} dx = \frac{1}{\pi\sqrt{2}} \left.\frac{\sin mx}{m}\right|_{-\pi}^{\pi} = 0,$$

$$\int_{-\pi}^{\pi} \left(\frac{1}{\sqrt{2\pi}}\right)^2 dx = \left.\frac{1}{2\pi} x \right|_{-\pi}^{\pi} = 1$$

complete the verification of the orthonormality.

Some important sets of functions $\{f_n(x)\}$ that occur in physical applications are not orthogonal but have the property that for some nonnegative function $w(x)$ the integral

$$(f_n, f_m)_w = \int_a^b w(x) f_n(x) f_m(x) \, dx = 0 \tag{6}$$

whenever $n \neq m$. If this happens, the set $\{f_n(x)\}$ is said to be **orthogonal with respect to the weight function** $w(x)$ **on the interval** $a \leq x \leq b$. The **weighted norm** of f_n is defined as

$$\|f_n\|_w = \left[\int_a^b w(x) f_n(x)^2 \, dx\right]^{1/2}, \tag{7}$$

and $\{f_n(x)\}$ is said to be **orthonormal with respect to the weight function** $w(x)$, if, in addition to being orthogonal with respect to $w(x)$, it satisfies the condition $\|f_n\|_w = 1$ for all n.

Example 3   Let $y_n(x)$ be a polynomial solution of the **Laguerre equation**,[†]

$$(xe^{-x} y')' + ne^{-x} y = 0. \tag{8}$$

[Equation (8) is obtained by multiplying equation (ii), p. 201, by e^{-x}.] Then y_n is called a **Laguerre polynomial**. Consider the set of Laguerre polynomials $\{y_n(x)\}$, $n = 1, 2, 3, \ldots$, and suppose $m \neq n$. Then we have the equations

$$(xe^{-x} y_n')' + ne^{-x} y_n = 0,$$
$$(xe^{-x} y_m')' + me^{-x} y_m = 0.$$

We multiply the first equation by y_m and the second by $-y_n$ and add them to obtain

$$y_m (xe^{-x} y_n')' - y_n (xe^{-x} y_m')' = e^{-x}(m - n) y_n y_m. \tag{9}$$

[†] Edmond Laguerre (1834–1886), a French mathematician, made many contributions to geometry and the theory of infinite series.

Integrating both sides of (9) on the interval $0 \leq x < \infty$, we have

$$\int_0^\infty \left[y_m (xe^{-x} y_n')' - y_n (xe^{-x} y_m')' \right] dx = (m - n) \int_0^\infty e^{-x} y_n y_m \, dx. \quad (10)$$

But by the product formula for differentiation,

$$\left[xe^{-x} (y_m y_n' - y_n y_m') \right]' = (xe^{-x} y_n')' y_m + xe^{-x} y_n' y_m' - (xe^{-x} y_m')' y_n - xe^{-x} y_m' y_n'$$
$$= y_m (xe^{-x} y_n')' - y_n (xe^{-x} y_m')'.$$

Thus (10) becomes

$$(m - n) \int_0^\infty e^{-x} y_n y_m \, dx = xe^{-x} (y_m y_n' - y_n y_m') \Big|_0^\infty = 0, \quad m \neq n,$$

because $f(x) = x(y_m y_n' - y_n y_m')$ is a polynomial (since each term is a polynomial) and the quotient $f(x)/e^x$ tends to zero as $x \to \infty$. This may be verified by applying L'Hôpital's rule as many times as the degree of f to the quotient $f(x)/e^x$:

$$\lim_{x \to \infty} \frac{f(x)}{e^x} = \lim_{x \to \infty} \frac{f'(x)}{e^x} = \cdots = \lim_{x \to \infty} \frac{c}{e^x} = 0.$$

Therefore the set of functions $\{y_n\}$, $n = 1, 2, 3, \ldots$ is orthogonal with respect to the weight function e^{-x} on $0 \leq x < \infty$.

PROBLEMS 10.2

In Problems 1–8 show that each given set of functions is orthogonal on the given interval and determine the corresponding orthonormal set.

1. $\{\cos nx\}$, $n = 0, 1, 2, \ldots$, $0 \leq x \leq 2\pi$
2. $\left\{ \sin \dfrac{n\pi x}{T} \right\}$, $n = 1, 2, 3, \ldots$, $-T \leq x \leq T$
3. $\left\{ \cos \dfrac{2n\pi x}{T} \right\}$, $n = 0, 1, 2, \ldots$, $0 \leq x \leq T$
4. $\{\sin 2nx\}$, $n = 1, 2, 3, \ldots$, $0 \leq x \leq \pi$
5. $\{\cos 2nx\}$, $n = 0, 1, 2, \ldots$, $0 \leq x \leq \pi$
6. $\{\sin 3nx\}$, $n = 1, 2, 3, \ldots$, $-\pi \leq x \leq \pi$
7. $\{\cos 3nx\}$, $n = 0, 1, 2, \ldots$, $|x| \leq \pi$
8. $\{\sin 2nx, \cos 2nx\}$, $n = 1, 2, 3, \ldots$, $|x| \leq \pi$
9. **Hermite polynomials.**[†] The functions

$$H_0 = 1, \quad H_n(x) = (-1)^n e^{x^2} \frac{d^n}{dx^n} e^{-x^2},$$

$n = 1, 2, 3, \ldots$,

are called Hermite polynomials (see Problem 5.5.9). Prove that

a. $H_1(x) = 2x$, $H_2(x) = 4x^2 - 2$, $H_3(x) = 8x^3 - 12x$;

b. the Hermite polynomials satisfy the relation

$$H_{n+1}(x) = 2xH_n(x) - H_n'(x);$$

c. $H_n(x)$ is a solution of the **Hermite equation**:

$$y'' - 2xy' + 2ny = 0;$$

d. the set of functions $\{H_n(x)\}$ is orthogonal with respect to the weight function $\exp(-x^2)$ on the interval $-\infty < x < \infty$.

*10. The norm of a function $f(x)$ on an interval $a \leq x \leq b$ is defined in equation (2) by the integral

$$\|f\| = \left(\int_a^b f^2(x) \, dx \right)^{1/2}.$$

This concept is a generalization of the notion of the distance of a point (x, y) in the plane from the origin:

$$d(x, y) = \sqrt{x^2 + y^2}.$$

Assume that f and g are continuous functions on $a \leq x \leq b$. Prove that the norm satisfies the following three properties:

a. $\|f\| \geq 0$, where the equality holds if and only if $f(x) \equiv 0$;

b. $\|af\| = |a| \cdot \|f\|$ for any constant a;

c. $\|f + g\| \leq \|f\| + \|g\|$ (triangle inequality).

[†] Charles Hermite (1822–1901), a French mathematician, was known for his work in number theory.

[*Hint*: (a) Use the fact that if $f(x_0) \neq 0$, then by continuity $f^2(x) \geq \epsilon > 0$ for all x in some interval $x_0 - \delta \leq x \leq x_0 + \delta$, where $\delta > 0$. (c) Show that for any two real numbers α and β
$$0 \leq \|\alpha f - \beta g\|^2 = \alpha^2 \|f\|^2 - 2\alpha\beta(f, g) + \beta^2 \|g\|^2.$$

Then let $\alpha = \|g\|$ and $\beta = \|f\|$, showing that
$$(f, g) \leq \|f\| \cdot \|g\|.$$
Finally, use the last inequality to show that
$$\|f + g\|^2 \leq (\|f\| + \|g\|)^2.]$$

10.3 Fourier Series†

An infinite series of the form

$$\frac{a_0}{2} + \sum_{n=1}^{\infty} (a_n \cos nx + b_n \sin nx), \tag{1}$$

where a_n and b_n are constants, is generally referred to as a **trigonometric** or **Fourier series**. Note that equation (10.1.19) is a special case of this expression with $a_n = 0$ for all n.

If series (1) converges to a periodic function $F(x)$ on the interval $[-\pi, \pi]$, then it converges for all real x, since

$$F(x) = \frac{a_0}{2} + \sum_{n=1}^{\infty} (a_n \cos nx + b_n \sin nx)$$

is a **periodic function of period 2π**:

$$F(x + 2\pi) = F(x) \qquad \text{for all } x. \tag{2}$$

This observation is obvious, since the functions $\cos nx$ and $\sin nx$ are periodic of period 2π. The constants a_n and b_n are called the **Fourier coefficients** of $F(x)$ and are given by the **Euler formulas**:

$$\boxed{a_n = \frac{1}{\pi} \int_{-\pi}^{\pi} F(x) \cos nx \, dx, \qquad b_n = \frac{1}{\pi} \int_{-\pi}^{\pi} F(x) \sin nx \, dx.} \tag{3}$$

To see why (3) is true, observe that

$$\frac{1}{\pi} \int_{-\pi}^{\pi} F(x) \cos kx \, dx = \frac{1}{\pi} \int_{-\pi}^{\pi} \left(\frac{a_0}{2} + \sum_{n=1}^{\infty} a_n \cos nx + b_n \sin nx \right) \cos kx \, dx;$$

if term-by-term integration is permissible,‡ we have

$$\frac{1}{\pi} \int_{-\pi}^{\pi} F(x) \cos kx \, dx = \frac{a_0}{2\pi} \int_{-\pi}^{\pi} \cos kx \, dx + \sum_{n=1}^{\infty} \left(\frac{a_n}{\pi} \int_{-\pi}^{\pi} \cos nx \cos kx \, dx \right.$$
$$\left. + \frac{b_n}{\pi} \int_{-\pi}^{\pi} \sin nx \cos kx \, dx \right). \tag{4}$$

However, as we saw in Example 10.2.2, the functions

$$\frac{1}{\sqrt{2\pi}}, \quad \frac{\cos x}{\sqrt{\pi}}, \quad \frac{\sin x}{\sqrt{\pi}}, \quad \frac{\cos 2x}{\sqrt{\pi}}, \quad \frac{\sin 2x}{\sqrt{\pi}}, \ldots$$

---

† Named after the French physicist Jean Baptiste Joseph Fourier (1768–1830).
‡ Most advanced calculus textbooks contain proofs that term-by-term integration is permissible if series (1) converges *uniformly* on $a \leq x \leq b$. Uniform convergence is discussed in Appendix 3.

form an orthonormal set of functions on the interval $-\pi \le x \le \pi$. Hence all of the integrals in (4) are zero except

$$\frac{a_k}{\pi} \int_{-\pi}^{\pi} \cos^2 kx\, dx = a_k, \quad \text{if } k \ne 0, \quad \text{or} \quad \frac{a_0}{2\pi} \int_{-\pi}^{\pi} 1\, dx = a_0, \quad \text{if } k = 0.$$

Thus the first of the Euler formulas has been verified. The second Euler formula is proved similarly.

Remark There is nothing special about the period of $F(x)$. Suppose that instead of having period 2π, the function F has period $2T$:

$$F(x + 2T) = F(x).$$

Then define the function

$$G(x) = F\left(\frac{xT}{\pi}\right) = F(y),$$

where $y = xT/\pi$, and note that

$$G(x + 2\pi) = F\left(\frac{xT}{\pi} + 2T\right) = F\left(\frac{xT}{\pi}\right) = G(x).$$

If there is a Fourier series (1) that converges to $G(x)$, then

$$G(x) = \frac{a_0}{2} + \sum_{n=1}^{\infty} (a_n \cos nx + b_n \sin nx) \tag{5}$$

on the interval $-\pi \le x \le \pi$; the substitution $x = \pi y/T$ can be used in (5) to obtain a Fourier series for F:

$$F(y) = G(\pi y/T) = \frac{a_0}{2} + \sum_{n=1}^{\infty} \left(a_n \cos \frac{n\pi y}{T} + b_n \sin \frac{n\pi y}{T}\right). \tag{6}$$

The change of variables above can also be used to show the orthogonality of the functions

$$1, \quad \cos\frac{\pi y}{T}, \quad \sin\frac{\pi y}{T}, \quad \cos\frac{2\pi y}{T}, \quad \sin\frac{2\pi y}{T}, \ldots$$

over the interval $|x| \le T$.

So far we have concentrated only on properties of the Fourier series (1). We have not discussed how to go about determining the Fourier series for a given periodic function $F(x)$. The task is not difficult: we merely compute the coefficients a_n and b_n using the Euler formulas (3) and substitute these values in (1). The result is called the **Fourier series corresponding to** $F(x)$. The remaining question is whether the Fourier series so obtained actually converges to $F(x)$. If the Fourier series does converge to $F(x)$, we call it a **representation** of $F(x)$.

The class of periodic functions $F(x)$ that can be represented by Fourier series is very large, so large in fact, that it originally aroused a big controversy. The following theorem gives sufficient conditions for almost all conceivable practical applications.

THEOREM 1 FOURIER CONVERGENCE THEOREM

Let $F(x)$ be a periodic function with period $2T$ and such that $F(x)$ and $F'(x)$ are piecewise continuous[†] on the interval $-T \leq x \leq T$. Then $F(x)$ has a Fourier series

$$F(x) = \frac{a_0}{2} + \sum_{n=1}^{\infty}\left(a_n \cos\frac{n\pi x}{T} + b_n \sin\frac{n\pi x}{T}\right), \qquad (7)$$

whose coefficients are given by the **Euler formulas**

$$a_n = \frac{1}{T}\int_{-T}^{T} F(x)\cos\frac{n\pi x}{T}\,dx, \qquad n = 0,1,2,\ldots, \qquad (8)$$

$$b_n = \frac{1}{T}\int_{-T}^{T} F(x)\sin\frac{n\pi x}{T}\,dx, \qquad n = 1,2,\ldots. \qquad (9)$$

The Fourier series (7) converges to $F(x)$ at all points where F is continuous, and to $[F(x+0)+F(x-0)]/2$ at all points of jump discontinuity of F.[‡]

Remark Note that $[F(x+0)+F(x-0)]/2$ is the average of the right- and left-hand limits at the point x. At any point of continuity, both of these values coincide, so (7) converges to $[F(x+0)+F(x-0)]/2$ for all x in the interval $-T \leq x \leq T$.

Although the conditions given in this theorem guarantee the existence and convergence of a Fourier series for the function $F(x)$, they are *not* the most general of such conditions. Moreover, the convergence of a Fourier series to a function $F(x)$ does not imply that F satisfies the conditions given in this theorem. In summary, the conditions given in this theorem are neither necessary nor the most general sufficient conditions. Furthermore, even with these limitations, the proof of the theorem as stated is too complicated to be presented here.[§]

Although the theory of Fourier series is complicated, the application of these series is very easy. It should be clear from Theorem 1 that Fourier series apply to a much wider class of functions than Taylor series do (even under the assumption we have used in the proof), since discontinuous functions cannot have a Taylor series representation. It is also useful to consider some functions for which Theorem 1 *does not* apply.

[†] See Section 6.1.
[‡] $F(x+0) = \lim_{h \to 0+} F(x+h)$, $F(x-0) = \lim_{h \to 0+} F(x-h)$.
[§] Proofs of this theorem can be found in most books on advanced calculus or complex variables. See, for example, W. Kaplan, *Advanced Calculus* (Reading, Mass.: Addison-Wesley, 1973), p. 484; or W. R. Derrick, *Complex Analysis and Applications*, 2d ed. (Belmont, Calif.: Wadsworth, 1984), p. 267.

Example 1 The functions $F_1(x) = 1/x$ and

$$F_2(x) = (-1)^n, \qquad \frac{1}{n+1} < |x| \leq \frac{1}{n}, \qquad n = 1, 2, 3, \ldots,$$

do not satisfy the hypotheses of Theorem 1 in the interval $-1 \leq x \leq 1$, since neither function is piecewise continuous. The function F_1 has an infinite jump discontinuity at $x = 0$, whereas F_2 has an infinite number of discontinuities in $-1 \leq x \leq 1$ (see Figure 10.3).

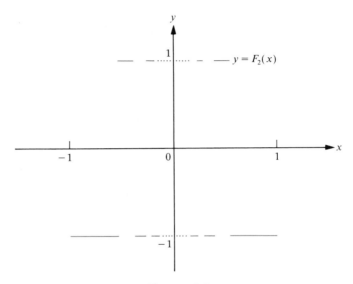

Figure 10.3

Let us now illustrate the procedure involved in obtaining the Fourier series of a function by some examples.

Example 2 Find the Fourier series of the periodic function (see Figure 10.4)

$$f(x) = (-1)^n k, \qquad n < x < n + 1.$$

Functions of this type occur as off-on controls in mechanical systems. Observe

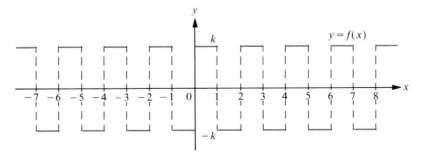

Figure 10.4

that $f(x)$ has period $2T = 2$; by equation (8) with $T = 1$,

$$a_n = \int_{-1}^{1} f(x) \cos n\pi x \, dx = -k \int_{-1}^{0} \cos n\pi x \, dx + k \int_{0}^{1} \cos n\pi x \, dx.$$

If $n = 0$, we obtain $a_0 = -k + k = 0$; and if $n \neq 0$,

$$a_n = \frac{-k}{n\pi} \sin n\pi x \bigg|_{-1}^{0} + \frac{k}{n\pi} \sin n\pi x \bigg|_{0}^{1} = 0.$$

Similarly, since $\cos(-t) = \cos t$,

$$b_n = -k \int_{-1}^{0} \sin n\pi x \, dx + k \int_{0}^{1} \sin n\pi x \, dx$$

$$= \frac{k}{n\pi} \cos n\pi x \bigg|_{-1}^{0} - \frac{k}{n\pi} \cos n\pi x \bigg|_{0}^{1} = \frac{2k}{n\pi} (1 - \cos n\pi),$$

which is zero for even n and equals $4k/n\pi$ for odd n. Thus

$$f(x) = \frac{4k}{\pi} \left(\sin \pi x + \frac{1}{3} \sin 3\pi x + \frac{1}{5} \sin 5\pi x + \cdots \right). \tag{10}$$

The graphs of $f(x)$ and the first three partial sums of equation (10) are shown in Figure 10.5.

As a bonus for our work above, note that if we set $x = \frac{1}{2}$ in (10), we obtain the series

$$k = \frac{4k}{\pi} \left(1 - \frac{1}{3} + \frac{1}{5} - \frac{1}{7} + \cdots \right),$$

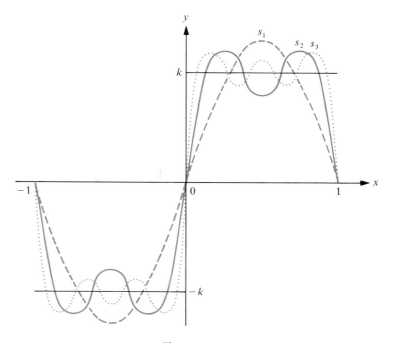

Figure 10.5

or

$$\frac{\pi}{4} = 1 - \frac{1}{3} + \frac{1}{5} - \frac{1}{7} + \cdots,$$

since $\sin \pi/2 = 1$, $\sin 3\pi/2 = -1$, and so on. This is a famous result that Leibniz obtained by means of a complicated geometrical construction.

As a final note, observe that the function $f(x)$ in this example satisfies the condition $f(-x) = -f(x)$. Functions having this property are said to be **odd** functions. In particular, observe that all functions of the form $\sin n\pi x$ are odd functions.

Example 3 Consider the sawtooth function

$$f(x) = \begin{cases} x + 1, & -1 \leq x \leq 0, \\ -x + 1, & 0 \leq x \leq 1, \end{cases} \qquad f(x+2) = f(x),$$

shown in Figure 10.6. Again $T = 1$ and

$$a_n = \int_{-1}^{0} (x+1)\cos n\pi x \, dx + \int_{0}^{1} (-x+1)\cos n\pi x \, dx.$$

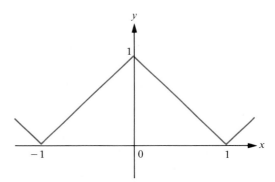

Figure 10.6

If $n = 0$, we have

$$a_0 = \left(\frac{x^2}{2} + x\right)\bigg|_{-1}^{0} + \left(x - \frac{x^2}{2}\right)\bigg|_{0}^{1} = 1,$$

and integrating by parts for $n \neq 0$, we obtain

$$a_n = \frac{(x+1)}{n\pi} \sin n\pi x \bigg|_{-1}^{0} - \frac{1}{n\pi} \int_{-1}^{0} \sin n\pi x \, dx$$

$$+ \frac{(1-x)}{n\pi} \sin n\pi x \bigg|_{0}^{1} + \frac{1}{n\pi} \int_{0}^{1} \sin n\pi x \, dx$$

$$= \frac{1}{(n\pi)^2} \cos n\pi x \bigg|_{-1}^{0} - \frac{1}{(n\pi)^2} \cos n\pi x \bigg|_{0}^{1} = \frac{2}{(n\pi)^2}(1 - \cos n\pi).$$

Therefore, a_n is zero for even $n \neq 0$ and equals $4/(n\pi)^2$ for odd n. Similarly,

$$b_n = \left.\frac{-(x+1)}{n\pi}\cos n\pi x\right|_{-1}^0 + \frac{1}{n\pi}\int_{-1}^0 \cos n\pi x\, dx$$

$$+ \left.\frac{(x-1)}{n\pi}\cos n\pi x\right|_0^1 - \frac{1}{n\pi}\int_0^1 \cos n\pi x\, dx$$

$$= \frac{-1}{n\pi} + \left.\frac{1}{(n\pi)^2}\sin n\pi x\right|_{-1}^0 + \frac{1}{n\pi} - \left.\frac{1}{(n\pi)^2}\sin n\pi x\right|_0^1 = 0.$$

Thus

$$f(x) = \frac{1}{2} + \frac{4}{\pi^2}\left(\cos \pi x + \frac{1}{3^2}\cos 3\pi x + \frac{1}{5^2}\cos 5\pi x + \frac{1}{7^2}\cos 7\pi x + \cdots\right). \quad (11)$$

Incidentally, note that if we set $x = 1$, then

$$0 = \frac{1}{2} - \frac{4}{\pi^2}\left(1 + \frac{1}{3^2} + \frac{1}{5^2} + \frac{1}{7^2} + \cdots\right),$$

or

$$\frac{\pi^2}{8} = 1 + \frac{1}{3^2} + \frac{1}{5^2} + \frac{1}{7^2} + \cdots.$$

We observe that the term-by-term derivative of (11) yields

$$f'(x) = \frac{-4}{\pi}\left(\sin \pi x + \frac{1}{3}\sin 3\pi x + \frac{1}{5}\sin 5\pi x + \cdots\right),$$

which is the same series as (10) for $k = -1$. An explanation for this fact is found by noting that $f'(x) = (-1)^n(-1)$ for $n < x < n+1$.

Finally, note that this function satisfies the condition $f(-x) = f(x)$. All such functions are called **even** functions. In particular, $\cos n\pi x$ is an even function for every integer n.

Example 4   Consider the clipped sine wave

$$f(x) = \begin{cases} k\sin Tx, & 0 < x < \pi/T, \\ 0, & -\pi/T < x < 0, \end{cases} \qquad f\left(x + \frac{2\pi}{T}\right) = f(x),$$

which is obtained by passing a sinusoidal voltage $k \sin Tx$ through a half-wave rectifier (see Figure 10.7). Here the period is $2\pi/T$, so

$$a_n = \frac{T}{\pi}\int_0^{\pi/T} k\sin Tx \cos nTx\, dx, \qquad a_0 = \frac{2k}{\pi}.$$

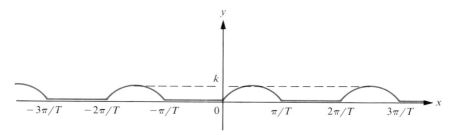

Figure 10.7

Using the identity
$$2\sin Tx \cos nTx = \sin(n+1)Tx - \sin(n-1)Tx$$
as in Example 10.2.2, we have

$$a_n = \frac{kT}{2\pi}\left[\frac{\cos(n-1)Tx}{(n-1)T} - \frac{\cos(n+1)Tx}{(n+1)T}\right]\Bigg|_0^{\pi/T}$$

$$= \frac{k}{2\pi}\left[\frac{\cos(n-1)\pi - 1}{n-1} - \frac{\cos(n+1)\pi - 1}{n+1}\right], \quad \text{for } n \neq 1,$$

and

$$a_1 = \frac{kT}{2\pi}\int_0^{\pi/T}\sin 2Tx\, dx = \frac{kT}{2\pi}\left(\frac{1-\cos 2\pi}{2T}\right) = 0.$$

When n is odd, $a_n = 0$, and when n is even,

$$a_n = \frac{k}{\pi}\left(\frac{1}{n+1} - \frac{1}{n-1}\right) = \frac{-2k}{\pi(n^2-1)}.$$

Similarly, for $n \neq 1$,

$$b_n = \frac{T}{\pi}\int_0^{\pi/T} k\sin Tx \sin nTx\, dx$$

$$= \frac{kT}{2\pi}\int_0^{\pi/T}\left[\cos(n-1)Tx - \cos(n+1)Tx\right]dx$$

$$= \frac{k}{2\pi}\left[\frac{\sin(n-1)Tx}{n-1} - \frac{\sin(n+1)Tx}{n+1}\right]\Bigg|_0^{\pi/T} = 0;$$

and if $n = 1$, by Formula 100 of Appendix 1,

$$b_1 = \frac{k}{2\pi}(Tx - \sin Tx \cos Tx)\Bigg|_0^{\pi/T} = \frac{k}{2}.$$

Thus, since only b_1 and the evenly subscripted terms $a_{2n} = -2k/\pi(4n^2 - 1)$ are nonzero,

$$f(x) = \frac{k}{\pi} + \frac{k}{2}\sin Tx - \frac{2k}{\pi}\sum_{n=1}^{\infty}\frac{\cos 2nTx}{4n^2 - 1}. \tag{12}$$

Incidentally, if we set $x = \pi/2T$, another intriguing identity results:

$$k = \frac{k}{\pi} + \frac{k}{2} - \frac{2k}{\pi}\sum_{n=1}^{\infty}\frac{\cos n\pi}{4n^2 - 1},$$

or

$$\frac{\pi}{2} - 1 = 2\left(\frac{1}{1\cdot 3} - \frac{1}{3\cdot 5} + \frac{1}{5\cdot 7} - \frac{1}{7\cdot 9} + \cdots\right).$$

Note that in this example the Fourier series (12) contains both sine and cosine terms.

Although we have concentrated in this section on trigonometric series, it is important to notice that other orthogonal sets of functions often provide useful series expansions. Let $\{f_n(x)\}$ be an orthogonal set of functions on an interval $a \le x \le b$ and suppose that the function $F(x)$ can be represented in terms of the functions $f_n(x)$ by a convergent series

$$F(x) = \sum_{n=1}^{\infty} c_n f_n(x) = c_1 f_1(x) + c_2 f_2(x) + \cdots. \qquad (13)$$

Series (13) is called a **generalized Fourier series of** $F(x)$ and the coefficients $c_1, c_2, \ldots$ are called the **generalized Fourier coefficients of** $F(x)$ **with respect to the orthogonal set of functions** $\{f_n(x)\}$. It is easy to determine the constants $c_1, c_2, \ldots$ by the following procedure. Fixing m, we multiply both sides of (13) by $f_m(x)$ and integrate the result over the interval $a \le x \le b$:

$$(F, f_m) = \int_a^b F(x) f_m(x)\, dx = \sum_{n=1}^{\infty} c_n \int_a^b f_n(x) f_m(x)\, dx = \sum_{n=1}^{\infty} c_n (f_n, f_m).$$

Note that we have again assumed term-by-term integration to be permissible—an assumption that sometimes is not valid. Since $(f_n, f_m) = 0$ whenever $n \ne m$, we have

$$(F, f_m) = c_m \|f_m\|^2,$$

so that

$$\boxed{c_m = \frac{(F, f_m)}{\|f_m\|^2} = \frac{1}{\|f_m\|^2} \int_a^b F(x) f_m(x)\, dx, \qquad m = 1, 2, \ldots.} \qquad (14)$$

Moreover, we can also use orthogonal sets of functions $\{f_n(x)\}$ **with respect to a weight function** $w(x)$ on an interval $a \le x \le b$. If $F(x)$ can be represented by a generalized Fourier series in these weighted orthogonal functions,

$$F(x) = \sum_{n=1}^{\infty} c_n f_n(x), \qquad (15)$$

then the general Fourier coefficients c_k can be determined by multiplying both sides of (15) by $w(x) f_k(x)$ and integrating over the interval $a \le x \le b$:

$$\int_a^b w(x) F(x) f_k(x)\, dx = \sum_{n=1}^{\infty} c_n \int_a^b w(x) f_n(x) f_k(x)\, dx = c_k \|f_k\|_w^2,$$

or

$$\boxed{c_k = \frac{(F, f_k)_w}{\|f_k\|_w^2}.} \qquad (16)$$

PROBLEMS 10.3

1. Find the smallest positive period of the functions $\cos 2x, \sin \pi x, \cos(2\pi n x/T), \sin 2k\pi x$.

2. Show that a constant function is periodic with any period $2T > 0$.

3. Suppose that $f(x)$ has period $2T$. What is the period of $f(ax/b)$?

4. Prove that a convergent infinite series of functions of period T is periodic of period T.

In Problems 5–13 find the Fourier series of each function $f(x)$ of period 2π, where one period is defined, and accurately plot the first three partial sums

$$\frac{a_0}{2} + \sum_{n=1}^{k} (a_n \cos nx + b_n \sin nx), \quad k = 1, 2, 3.$$

5. $f(x) = x, \quad |x| < \pi$

6. $f(x) = \begin{cases} 0, & -\pi < x < 0 \\ 1, & 0 < x < \pi \end{cases}$

7. $f(x) = x^2, \quad |x| < \pi$

8. $f(x) = \begin{cases} 0, & -\pi < x < 0 \\ x, & 0 < x < \pi \end{cases}$

9. $f(x) = |x|, \quad |x| < \pi$

10. $f(x) = \begin{cases} x, & -\pi < x < 0 \\ x - \pi, & 0 < x < \pi \end{cases}$

11. $f(x) = \begin{cases} -1, & -\pi < x < -1 \\ x, & -1 < x < 1 \\ 1, & 1 < x < \pi \end{cases}$

12. $f(x) = \begin{cases} \pi + x, & -\pi < x < 0 \\ \pi - x, & 0 < x < \pi \end{cases}$

13. $f(x) = e^x, \quad |x| < \pi$

In Problems 14–21 find the Fourier series of each function $f(x)$ of period T, where one of the periods is defined.

14. $f(x) = x, \quad |x| < 1, \quad T = 2$

15. $f(x) = x, \quad 0 < x < 2, \quad T = 2$

16. $f(x) = x, \quad 0 < x < 3, \quad T = 3$

17. $f(x) = x^2, \quad |x| < 1, \quad T = 2$

18. $f(x) = x^2, \quad 0 < x < 2, \quad T = 2$

19. $f(x) = \begin{cases} 0, & 0 < x < 1, \\ 1, & 1 < x < 2, \quad T = 2 \end{cases}$

20. $f(x) = \begin{cases} 0, & 0 < x < 1, \\ x - 1, & 1 < x < 2, \quad T = 2 \end{cases}$

21. $f(x) = \begin{cases} x, & 0 < x < 1, \\ 1, & 1 < x < 2, \quad T = 2 \end{cases}$

22. Find the Fourier series of the periodic function of period 2π

$$f(x) = \frac{x^2}{4}, \quad |x| < \pi,$$

and use this series to verify the identities:

$$\frac{\pi^2}{6} = 1 + \frac{1}{2^2} + \frac{1}{3^2} + \frac{1}{4^2} + \frac{1}{5^2} + \cdots,$$

$$\frac{\pi^2}{12} = 1 - \frac{1}{2^2} + \frac{1}{3^2} - \frac{1}{4^2} + \frac{1}{5^2} - \cdots,$$

$$\frac{\pi^2}{8} = 1 + \frac{1}{3^2} + \frac{1}{5^2} + \frac{1}{7^2} + \frac{1}{9^2} + \cdots.$$

10.4 Half-Range Expansions

In Section 10.3 we defined the concepts of odd and even functions. A function $f(x)$ is **odd** if it satisfies the condition

$$f(-x) = -f(x) \tag{1}$$

and **even** if

$$f(-x) = f(x). \tag{2}$$

As we saw, the function $\sin n\pi x$ is odd, whereas $\cos n\pi x$ is even for $n = 1, 2, 3, \ldots$.

Knowing that a function $f(x)$ is even or odd can help us avoid unnecessary work in computing the Fourier coefficients of $f(x)$. This claim is based on the following facts:

THEOREM 1

a. The product of two even or two odd functions is even.

b. The product of an even and an odd function is odd.

c. If $g(x)$ is an odd function, then

$$\int_{-T}^{0} g(x)\, dx = -\int_{0}^{T} g(x)\, dx \tag{3}$$

and, for every $T > 0$,
$$\int_{-T}^{T} g(x)\, dx = 0. \tag{4}$$

d. If $g(x)$ is an even function, then
$$\int_{-T}^{0} g(x)\, dx = \int_{0}^{T} g(x)\, dx$$
and
$$\int_{-T}^{T} g(x)\, dx = 2\int_{0}^{T} g(x)\, dx. \tag{5}$$

Proof

a. Let $f(x)$ and $g(x)$ be even functions. By (2),
$$f(-x)g(-x) = f(x)g(x),$$
indicating that their product is even. Now suppose that they are both odd functions. By (1),
$$f(-x)g(-x) = [-f(x)][-g(x)] = f(x)g(x),$$
so again their product is even.

b. If $f(x)$ is even and $g(x)$ is odd, then
$$f(-x)g(-x) = f(x)[-g(x)] = -f(x)g(x),$$
so the product is odd.

c. Let $g(x)$ be an odd function and let $x = -t$, so that $dx = -dt$. Then
$$\int_{-T}^{0} g(x)\, dx = -\int_{T}^{0} g(-t)\, dt = \int_{0}^{T} g(-t)\, dt = -\int_{0}^{T} g(t)\, dt,$$
so we have
$$\int_{-T}^{T} g(x)\, dx = \int_{-T}^{0} g(x)\, dx + \int_{0}^{T} g(x)\, dx = 0.$$

d.
$$\int_{-T}^{0} g(x)\, dx = -\int_{T}^{0} g(-t)\, dt = \int_{0}^{T} g(t)\, dt. \quad \blacksquare$$

What precisely is the effect of (4) and (5) in computing the Fourier coefficients of an even or odd periodic function $F(x)$? If $F(x)$ is even and of period $2T$, then $F(x)\sin(n\pi x/T)$ is odd and of period $2T$ by Theorem 1(b). Hence, by (4),
$$\int_{-T}^{T} F(x)\sin\frac{n\pi x}{T}\, dx = 0,$$
and all the Fourier coefficients b_n are zero when $F(x)$ is even. Similarly, if $F(x)$ is odd and of period $2T$, then $F(x)\cos(n\pi x/T)$ is odd and, by (5),
$$\int_{-T}^{T} F(x)\cos\frac{n\pi x}{T}\, dx = 0,$$
indicating that all the Fourier coefficients a_n are zero. Thus we have the following result.

THEOREM 2

Let $F(x)$ be a periodic function of period $2T$ that has a Fourier series. If $F(x)$ is an even function, then all the Fourier coefficients b_n are zero, whereas if $F(x)$ is an odd function, then the Fourier coefficients a_n are zero.

Remark The Fourier series of an even function,

$$F(x) = \frac{a_0}{2} + \sum_{n=1}^{\infty} a_n \cos \frac{n\pi x}{T}, \qquad (6)$$

is called a **Fourier cosine series**, whereas that of an odd function,

$$F(x) = \sum_{n=1}^{\infty} b_n \sin \frac{n\pi x}{T}, \qquad (7)$$

is said to be a **Fourier sine series**. Observe that Theorem 2 implies that for even and odd functions, we need to calculate only half as many coefficients as are generally required.

Example 1 Find the Fourier series of the periodic function of period 2π given by

$$F(x) = |x|, \qquad |x| \leq \pi.$$

Solution Since $F(-x) = |-x| = |x| = F(x)$ is even, we need only calculate a Fourier cosine series. By (5),

$$a_n = \frac{1}{\pi} \int_{-\pi}^{\pi} |x| \cos nx \, dx = \frac{2}{\pi} \int_{0}^{\pi} x \cos nx \, dx.$$

Integrating by parts, we have

$$a_n = \frac{2}{\pi} \left(\frac{x \sin nx}{n} \Big|_0^\pi - \frac{1}{n} \int_0^\pi \sin nx \, dx \right)$$

$$= \frac{2 \cos nx}{\pi n^2} \Big|_0^\pi$$

$$= \frac{2}{\pi n^2} [(-1)^n - 1], \qquad \text{for } n \geq 1.$$

Also, by (5),

$$a_0 = \frac{1}{\pi} \int_{-\pi}^{\pi} |x| \, dx = \frac{2}{\pi} \int_0^\pi x \, dx = \frac{x^2}{\pi} \Big|_0^\pi = \pi,$$

so

$$F(x) = \frac{\pi}{2} + \sum_{n=1}^{\infty} \frac{2}{\pi n^2} [(-1)^n - 1] \cos nx,$$

or

$$|x| = \frac{\pi}{2} - \frac{4}{\pi} \sum_{k=0}^{\infty} \frac{\cos(2k+1)x}{(2k+1)^2}, \qquad |x| \le \pi. \tag{8}$$

In particular, setting $x = 0$ in (8) we have

$$0 = \frac{\pi}{2} - \frac{4}{\pi} \sum_{k=0}^{\infty} \frac{1}{(2k+1)^2},$$

$$\frac{\pi^2}{8} = \sum_{k=0}^{\infty} \frac{1}{(2k+1)^2} = 1 + \frac{1}{3^2} + \frac{1}{5^2} + \cdots. \tag{9}$$

Example 2 What is the Fourier series of the periodic function

$$F(x) = x, \qquad |x| \le T?$$

Solution Since $F(-x) = -x = -F(x)$ is odd, all the coefficients $a_n = 0$. By Theorem 1(a) and (5),

$$b_n = \frac{1}{T} \int_{-T}^{T} x \sin \frac{n\pi x}{T} \, dx = \frac{2}{T} \int_{0}^{T} x \sin \frac{n\pi x}{T} \, dx.$$

Integrating by parts, we have

$$b_n = \frac{2}{T} \left[\frac{-Tx \cos(n\pi x/T)}{n\pi} \bigg|_0^T + \frac{T}{n\pi} \int_0^T \cos \frac{n\pi x}{T} \, dx \right]$$

$$= \frac{2}{T} \left[\frac{-(-1)^n T^2}{n\pi} + \left(\frac{T}{n\pi} \right)^2 \sin \frac{n\pi x}{T} \bigg|_0^T \right]$$

$$= \frac{-2(-1)^n T}{n\pi}.$$

Hence

$$x = \frac{-2T}{\pi} \sum_{n=1}^{\infty} \frac{(-1)^n}{n} \sin \frac{n\pi x}{T}, \qquad |x| < T. \tag{10}$$

In particular, setting $x = T/2$ in (10) we have

$$\frac{T}{2} = \frac{-2T}{\pi} \sum_{n=1}^{\infty} \frac{(-1)^n}{n} \sin \frac{n\pi}{2},$$

or

$$\frac{\pi}{4} = 1 - \frac{1}{3} + \frac{1}{5} - \frac{1}{7} + \cdots. \tag{11}$$

In many problems of physics and engineering there is a practical need to apply a Fourier series to a nonperiodic function $F(x)$ that is defined only on the interval $0 < x < T$. Because of physical or mathematical considerations, it may be permissible to *extend* $F(x)$ over the interval $-T < x < T$, making it periodic of period $2T$. Figure 10.8 illustrates two such extensions. The **odd extension** in Figure 10.8(b) has a Fourier *sine* series; the **even extension** in Figure 10.8(c) has a Fourier *cosine* series.

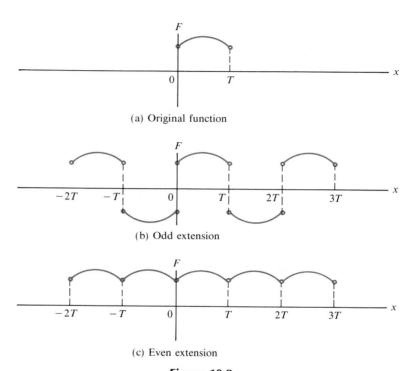

Figure 10.8

Example 3 Let $F(x) = x$ on $0 \leq x \leq \pi$. If we extend F as an even function of period 2π, we obtain the graph in Figure 10.9(c). Hence, for the even extension, $F_e(x) = |x|$ on $|x| \leq \pi$, so the result of Example 1 holds and

$$F_e(x) = |x| = \frac{\pi}{2} - \frac{4}{\pi} \sum_{k=0}^{\infty} \frac{\cos(2k+1)x}{(2k+1)^2}, \qquad |x| \leq \pi.$$

If we extend F as an odd function of period 2π [Figure 10.9(b)], then $F_0(x) = x$ on $|x| < \pi$ and the result (with $T = \pi$) of Example 2 applies:

$$F_0(x) = x = -2 \sum_{n=1}^{\infty} \frac{(-1)^n}{n} \sin nx, \qquad |x| < \pi.$$

Since the Fourier series is based on a function $F(x)$ defined on only half of the interval of periodicity, the Fourier series for either an even or an odd extension is called a **half-range expansion**. Using (4) we obtain the following result.

(a) Original function

(b) Odd extension

(c) Even extension

Figure 10.9

THEOREM 3

Let $F(x)$ be a function defined on the interval $0 \leq x \leq T$. Then an **even half-range expansion** of $F(x)$ is given by

$$F(x) = \frac{a_0}{2} + \sum_{n=1}^{\infty} a_n \cos \frac{n\pi x}{T},$$

where

$$a_n = \frac{2}{T} \int_0^T F(x) \cos \frac{n\pi x}{T} \, dx, \quad n = 0, 1, 2, \ldots,$$

and an **odd half-range expansion** is given by

$$F(x) = \sum_{n=1}^{\infty} b_n \sin \frac{n\pi x}{T},$$

where

$$b_n = \frac{2}{T} \int_0^T F(x) \sin \frac{n\pi x}{T} \, dx, \quad n = 1, 2, \ldots.$$

Examples of practical situations in which half-range expansions arise will be encountered in Sections 10.7–10.10.

PROBLEMS 10.4

In Problems 1–10 determine whether each function is even, odd, or neither.

1. x^2
2. x^3
3. $x \sin x$
4. e^x
5. $x + x^3$
6. $x + x^2$
7. $\ln x$
8. $\sin x + \cos x$

9. $F(x) = \begin{cases} x, & -T < x \leq 0, \\ 0, & 0 < x < T, \end{cases}$
   $F(x + 2T) = F(x)$

10. $F(x) = \begin{cases} x^2, & -T < x < 0, \\ -x^2, & 0 < x < T, \end{cases}$
    $F(x + 2T) = F(x)$

In Problems 11–20 find the Fourier series of the given functions, assuming that they are periodic. Pay particular attention to whether each function is even or odd.

11. $F(x) = x^2$, $|x| \leq \pi$
12. $F(x) = x^3$, $|x| < \pi$
13. $F(x) = |x^3|$, $|x| \leq \pi$
14. $F(x) = |\sin x|$, $|x| \leq \pi$

15. $F(x) = \begin{cases} x + T, & -T < x < 0 \\ -x + T, & 0 < x < T \end{cases}$

16. $F(x) = \begin{cases} -x - T, & -T < x < 0 \\ x - T, & 0 < x < T \end{cases}$

17. $F(x) = \begin{cases} -x - T, & -T < x < 0 \\ -x + T, & 0 < x < T \end{cases}$

18. $F(x) = \begin{cases} x + T, & -T < x < 0 \\ x - T, & 0 < x < T \end{cases}$

19. $F(x) = \begin{cases} (x+1)^2, & -1 < x < 0 \\ (x-1)^2, & 0 < x < 1 \end{cases}$

20. $F(x) = \begin{cases} (x+1)^2, & -1 < x < 0 \\ -(x-1)^2, & 0 < x < 1 \end{cases}$

Extend each of the functions in Problems 21–30 as both an even and an odd periodic function. Sketch the resulting graphs and compute their corresponding Fourier cosine or sine series.

21. $F(x) = k$, $0 < x < 1$
22. $F(x) = x^2$, $0 < x < 1$
23. $F(x) = x^3$, $0 < x < 1$
24. $F(x) = e^x$, $0 < x < 1$

25. $F(x) = \begin{cases} 1, & 0 < x < 1 \\ 0, & 1 < x < 2 \end{cases}$

26. $F(x) = \begin{cases} 1, & 0 < x < 1 \\ \frac{1}{2}, & 1 < x < 2 \end{cases}$

27. $F(x) = \begin{cases} 1, & 0 < x < 1 \\ -x + 2, & 1 < x < 2 \end{cases}$

28. $F(x) = \begin{cases} x, & 0 < x < 1 \\ 1, & 1 < x < 2 \end{cases}$

29. $F(x) = \begin{cases} x^2, & 0 < x < 1 \\ 1, & 1 < x < 2 \end{cases}$

30. $F(x) = \begin{cases} x^2, & 0 < x < \frac{1}{2} \\ 1, & \frac{1}{2} < x < 1 \end{cases}$

31. Using the formula $e^{ix} = \cos x + i \sin x$, show that

   a. $\cos nx = \frac{1}{2}(e^{inx} + e^{-inx})$,

   $\sin nx = \frac{1}{2i}(e^{inx} - e^{-inx})$;

   b. the Fourier series

   $$F(x) = \frac{a_0}{2} + \sum_{n=1}^{\infty}(a_n \cos nx + b_n \sin nx)$$

   can be written in the **complex form**

   $$F(x) = \sum_{n=-\infty}^{\infty} c_n e^{inx};$$

   c. the **complex Fourier coefficients** c_n in (b) are given by

   $$c_n = \frac{1}{2\pi} \int_{-\pi}^{\pi} F(x) e^{-inx} dx, \quad n = 0, \pm 1, \pm 2, \ldots.$$

32. Show that the complex Fourier coefficients of an even function are real.

33. Show that the complex Fourier coefficients of an odd function are purely imaginary.

10.5 Classification of Linear Second-Order Equations

In this section we are concerned only with second-order linear partial differential equations of two independent variables with constant coefficients, that is, equations of the form

$$a\frac{\partial^2 w}{\partial x^2}(x, y) + 2b\frac{\partial^2 w}{\partial x \partial y}(x, y) + c\frac{\partial^2 w}{\partial y^2}(x, y) + k\frac{\partial w}{\partial x}(x, y)$$

$$+ m\frac{\partial w}{\partial y}(x, y) + nw(x, y) = f(x, y), \quad (1)$$

where a, b, c, k, m, and n are constants (the number 2 in front of the coefficient b simplifies later computations). Since we want to consider second-order equations, we assume that at least one of the coefficients a, b, or c is nonzero. If so, equation (1) is called a **second-order** equation. The general second-order equation (1) can be transformed into one of four standard simpler forms:

$$\frac{\partial^2 z}{\partial x^2} + \frac{\partial^2 z}{\partial y^2} + \lambda z = G(x, y) \quad \text{(elliptic)}$$

$$\frac{\partial^2 z}{\partial x^2} - \frac{\partial^2 z}{\partial y^2} + \lambda z = G(x, y) \quad \text{(hyperbolic)}$$

$$\frac{\partial^2 z}{\partial x^2} + \lambda \frac{\partial z}{\partial y} = G(x, y) \quad \text{(parabolic)}$$

$$\frac{\partial^2 z}{\partial x^2} + \lambda z = G(x, y) \quad \text{(degenerate)}$$

λ constant.

In addition to being easier to solve, these standard forms arise naturally in a number of important applications. In Sections 10.6 and 10.7 we will study a particular hyperbolic equation, called the **wave equation**, that arises in modeling small transverse vibrations in a tightly stretched string. In Sections 10.8 and 10.9 we will consider a parabolic equation, called the **heat equation**, that is used to analyze heat conduction in thin round bars. In Sections 10.9 and 10.10 we will study a particular case of an elliptic equation, called **Laplace's equation**, that is used in determining steady-state flows (of fluids, heat, etc.). The degenerate equation has already been considered in Chapter 4, when we looked at spring-mass vibrations.

Initially, it is convenient to simplify equation (1) by a rotation of axes. Using the substitution

$$\begin{aligned} u &= x \cos \alpha + y \sin \alpha, \\ v &= -x \sin \alpha + y \cos \alpha, \end{aligned} \quad (2)$$

with $\tan 2\alpha = 2b/(a - c)$, we can eliminate the mixed second partial $\partial^2 w/\partial u \, \partial v$ and obtain an equation of the form

$$A \frac{\partial^2 w}{\partial u^2} + C \frac{\partial^2 w}{\partial v^2} + K \frac{\partial w}{\partial u} + M \frac{\partial w}{\partial v} + nw$$

$$= f(u \cos \alpha - v \sin \alpha, u \sin \alpha + v \cos \alpha) \quad (3)$$

$$= F(u, v),$$

where

$$\begin{aligned} A &= a \cos^2 \alpha + 2b \sin \alpha \cos \alpha + c \sin^2 \alpha, \\ C &= a \sin^2 \alpha - 2b \sin \alpha \cos \alpha + c \cos^2 \alpha, \\ K &= k \cos \alpha + m \sin \alpha, \\ M &= m \cos \alpha - k \sin \alpha. \end{aligned}$$

To see why this happens, observe that

$$\frac{\partial w}{\partial x} = \frac{\partial w}{\partial u}\frac{\partial u}{\partial x} + \frac{\partial w}{\partial v}\frac{\partial v}{\partial x} = \cos\alpha\frac{\partial w}{\partial u} - \sin\alpha\frac{\partial w}{\partial v},$$

$$\frac{\partial w}{\partial y} = \frac{\partial w}{\partial u}\frac{\partial u}{\partial y} + \frac{\partial w}{\partial v}\frac{\partial v}{\partial y} = \sin\alpha\frac{\partial w}{\partial u} + \cos\alpha\frac{\partial w}{\partial v},$$

and

$$\frac{\partial^2 w}{\partial x^2} = \frac{\partial}{\partial u}\left(\frac{\partial w}{\partial x}\right)\frac{\partial u}{\partial x} + \frac{\partial}{\partial v}\left(\frac{\partial w}{\partial x}\right)\frac{\partial v}{\partial x}$$

$$= \cos\alpha\left(\cos\alpha\frac{\partial^2 w}{\partial u^2} - \sin\alpha\frac{\partial^2 w}{\partial u\,\partial v}\right) - \sin\alpha\left(\cos\alpha\frac{\partial^2 w}{\partial u\,\partial v} - \sin\alpha\frac{\partial^2 w}{\partial v^2}\right)$$

$$= \cos^2\alpha\frac{\partial^2 w}{\partial u^2} - 2\sin\alpha\cos\alpha\frac{\partial^2 w}{\partial u\,\partial v} + \sin^2\alpha\frac{\partial^2 w}{\partial v^2}.$$

Similarly,

$$\frac{\partial^2 w}{\partial x\,\partial y} = \sin\alpha\cos\alpha\frac{\partial^2 w}{\partial u^2} + (\cos^2\alpha - \sin^2\alpha)\frac{\partial^2 w}{\partial u\,\partial v} - \sin\alpha\cos\alpha\frac{\partial^2 w}{\partial v^2},$$

and

$$\frac{\partial^2 w}{\partial y^2} = \sin^2\alpha\frac{\partial^2 w}{\partial u^2} + 2\sin\alpha\cos\alpha\frac{\partial^2 w}{\partial u\,\partial v} + \cos^2\alpha\frac{\partial^2 w}{\partial v^2}.$$

Then

$$a\frac{\partial^2 w}{\partial x^2} + 2b\frac{\partial^2 w}{\partial x\,\partial y} + c\frac{\partial^2 w}{\partial y^2} = A\frac{\partial^2 w}{\partial u^2} + 2B\frac{\partial^2 w}{\partial u\,\partial v} + C\frac{\partial^2 w}{\partial v^2},$$

where

$$B = (c-a)\sin\alpha\cos\alpha + b(\cos^2\alpha - \sin^2\alpha).$$

By the double angle formulas of trigonometry,

$$B = \frac{(c-a)}{2}\sin 2\alpha + b\cos 2\alpha = b\cos 2\alpha\left[\frac{(c-a)}{2b}\tan 2\alpha + 1\right].$$

Since

$$\tan 2\alpha = \frac{2b}{a-c},$$

it follows that $B = 0$. An easy but messy calculation (see Problem 21) shows that

$$b^2 - ac = B^2 - AC = -AC. \qquad (4)$$

Elliptic, Hyperbolic, and Parabolic Equations

By analogy with analytic geometry, we call the partial differential equation (1) an **elliptic** equation if $b^2 - ac < 0$, that is, if A and C have the same sign. If A and C have opposite signs or, equivalently, if $b^2 - ac > 0$, we say that (1) is a **hyperbolic** equation. Finally, if $b^2 - ac = 0$, (1) is called a **parabolic** equation.

In summary, (1) is called

$$\left.\begin{array}{r}\text{hyperbolic}\\\text{parabolic}\\\text{elliptic}\end{array}\right\} \text{ if } b^2 - ac \begin{cases} > 0, \\ = 0, \\ < 0. \end{cases} \qquad (5)$$

Example 1 Classify the second-order linear partial differential equation

$$\frac{\partial^2 w}{\partial x^2} + 2\frac{\partial^2 w}{\partial x \, \partial y} + 4\frac{\partial^2 w}{\partial y^2} + 7\frac{\partial w}{\partial x} + w = 0. \qquad (6)$$

Solution Here $a = b = 1$ and $c = 4$, so $b^2 - ac = -3$. Thus the partial differential equation (6) is elliptic.

Once we have eliminated the mixed second partial term, we can, for suitable values of α and β, use the substitution

$$w = e^{\alpha u + \beta v} z$$

to eliminate *at least one* of the first partials in (3). Note that by the product rule of differentiation,

$$\frac{\partial w}{\partial u} = \alpha w + e^{\alpha u + \beta v}\frac{\partial z}{\partial u}$$

and

$$\frac{\partial w}{\partial v} = \beta w + e^{\alpha u + \beta v}\frac{\partial z}{\partial v},$$

so that

$$\frac{\partial^2 w}{\partial u^2} = \alpha^2 w + 2\alpha e^{\alpha u + \beta v}\frac{\partial z}{\partial u} + e^{\alpha u + \beta v}\frac{\partial^2 z}{\partial u^2}$$

and

$$\frac{\partial^2 w}{\partial v^2} = \beta^2 w + 2\beta e^{\alpha u + \beta v}\frac{\partial z}{\partial v} + e^{\alpha u + \beta v}\frac{\partial^2 z}{\partial v^2}.$$

Equation (3) becomes

$$A\frac{\partial^2 z}{\partial u^2} + C\frac{\partial^2 z}{\partial v^2} + (2\alpha A + K)\frac{\partial z}{\partial u} + (2\beta C + M)\frac{\partial z}{\partial v}$$
$$+ (\alpha^2 A + \beta^2 C + \alpha K + \beta M + n)z = e^{-(\alpha u + \beta v)}F(u, v). \qquad (7)$$

If A and $C \neq 0$, then choosing $\alpha = -K/2A$ and $\beta = -M/2C$ allows us to eliminate the first partial terms, yielding the partial differential equation

$$A\frac{\partial^2 z}{\partial u^2} + C\frac{\partial^2 z}{\partial v^2} + \left(n - \frac{K^2}{4A} - \frac{M^2}{4C}\right)z = e^{-(\alpha u + \beta v)}F(u, v). \qquad (8)$$

If either A or C is zero, we cannot eliminate one of the first partial terms.

A change of scale in the independent variables provides a final simplification: If A and $C \neq 0$, set $u = \sqrt{|A|}\, x$ and $v = \sqrt{|C|}\, y$, so that (8) becomes

$$\frac{A}{|A|}\frac{\partial^2 z}{\partial x^2} + \frac{C}{|C|}\frac{\partial^2 z}{\partial y^2} + \left(n - \frac{K^2}{4A} - \frac{M^2}{4C}\right)z = G(x,y). \qquad (9)$$

From the calculations above, it follows that (1) can be transformed into one of the following four standard forms:

STANDARD FORMS FOR A SECOND-ORDER, LINEAR PARTIAL DIFFERENTIAL EQUATION WITH CONSTANT COEFFICIENTS

$$\frac{\partial^2 z}{\partial x^2} + \frac{\partial^2 z}{\partial y^2} + \lambda z = G(x,y) \quad \text{(elliptic)} \qquad (10)$$

$$\frac{\partial^2 z}{\partial x^2} - \frac{\partial^2 z}{\partial y^2} + \lambda z = G(x,y) \quad \text{(hyperbolic)} \qquad (11)$$

$$\frac{\partial^2 z}{\partial x^2} + \lambda \frac{\partial z}{\partial y} = G(x,y) \quad \text{(parabolic)} \qquad (12)$$

$$\frac{\partial^2 z}{\partial x^2} + \lambda z = G(x,y) \quad \text{(degenerate)} \qquad (13)$$

λ constant.

Remark Equation (10) occurs when A and C in (9) are nonzero and have the same sign. Equation (11) occurs when A and C are nonzero and have opposite signs. Equation (12) arises when $C = 0$ and $AM \neq 0$, by choosing $\alpha = -K/2A$ and $\beta = (K^2/4A - n)/M$; here $\lambda = M/A$. Equation (13) occurs when $C = M = 0$; here $\lambda = n - K^2/4A$.

The degenerate case may be treated as an ordinary differential equation with parameter v, in a manner very similar to equation (4.3.1). For this reason, we do not consider this case any further. In the next three sections we consider examples of elliptic, hyperbolic, and parabolic equations, showing how each arises in practice and indicating what steps can be taken to obtain a solution.

Example 2   To transform the equation

$$\frac{\partial^2 w}{\partial x^2} + 4\frac{\partial^2 w}{\partial x\, \partial y} + \frac{\partial^2 w}{\partial y^2} + \frac{\partial w}{\partial x} = 0$$

into standard form, observe that $b^2 - ac = 3$, since $a = c = 1$ and $b = 2$, and therefore that the equation is hyperbolic. Since $\tan 2\alpha$ is infinite, we have $\alpha = 45° = \pi/4$. Substituting this value into (3), we obtain

$$3\frac{\partial^2 w}{\partial u^2} - \frac{\partial^2 w}{\partial v^2} + \frac{1}{\sqrt{2}}\frac{\partial w}{\partial u} - \frac{1}{\sqrt{2}}\frac{\partial w}{\partial v} = 0.$$

Setting
$$w = z\exp\left[-\frac{\sqrt{2}}{12}(u+3v)\right]$$
(or $\alpha = -K/2A = -1/6\sqrt{2} = -\sqrt{2}/12$ and $\beta = -M/2C = -1/2\sqrt{2} = -\sqrt{2}/4$), we obtain from (8)
$$3\frac{\partial^2 z}{\partial u^2} - \frac{\partial^2 z}{\partial v^2} + \frac{z}{12} = 0.$$
Finally, letting $u = \sqrt{3}\,x$ and $v = y$, we get the hyperbolic equation
$$\frac{\partial^2 z}{\partial x^2} - \frac{\partial^2 z}{\partial y^2} + \frac{z}{12} = 0.$$

PROBLEMS 10.5

In Problems 1–10 classify the given equation as elliptic, parabolic, hyperbolic, or degenerate second-order partial differential equations.

1. $\dfrac{\partial^2 w}{\partial x^2} - \dfrac{\partial^2 w}{\partial y^2} = 0$
2. $\dfrac{\partial^2 w}{\partial x^2} + \dfrac{\partial^2 w}{\partial y^2} + w = 0$
3. $w_{xx} + 2w_{xy} + w_{yy} + w_x + w_y = 0$
4. $w_{xx} + w_{xy} + 4w_{yy} + 5w_x = 2$
5. $w_{xx} + 2w_{xy} + 2w_{yy} = 0$
6. $w_{xx} + 2w_{xy} + w_{yy} + w = 0$
7. $w_{xx} + 2w_{xy} + 4w_{yy} + 5w = 0$
8. $w_{xx} + 2w_{xy} - w_{yy} + w_y = 0$
9. $3w_{xx} - 5w_{xy} + 2w_{yy} = \sin(xy)$
10. $7w_{xy} + 5w_{yy} + 8w = x/y$

In Problems 11–20 transform the given equation into standard form.

11. $4\dfrac{\partial^2 w}{\partial x^2} + 3\dfrac{\partial^2 w}{\partial y^2} - w = 0$
12. $7\dfrac{\partial^2 w}{\partial x^2} + 4\dfrac{\partial^2 w}{\partial y^2} + 3w = 2$
13. $\dfrac{\partial^2 w}{\partial x^2} - \dfrac{\partial^2 w}{\partial y^2} + 2\dfrac{\partial w}{\partial x} - 5\dfrac{\partial w}{\partial y} + 7w = 0$
14. $\dfrac{\partial^2 w}{\partial x^2} + 2\dfrac{\partial^2 w}{\partial x\,\partial y} + 2\dfrac{\partial^2 w}{\partial y^2} = 0$
15. $\dfrac{\partial^2 w}{\partial x^2} + 2\dfrac{\partial^2 w}{\partial x\,\partial y} + \dfrac{\partial^2 w}{\partial y^2} + w = 0$
16. $\dfrac{\partial^2 w}{\partial x^2} + 2\dfrac{\partial^2 w}{\partial x\,\partial y} - \dfrac{\partial^2 w}{\partial y^2} = 0$
17. $\dfrac{\partial^2 w}{\partial x^2} + 2\dfrac{\partial^2 w}{\partial x\,\partial y} + 4\dfrac{\partial^2 w}{\partial y^2} + 5w = 0$
18. $\dfrac{\partial^2 w}{\partial x^2} + 2\dfrac{\partial^2 w}{\partial x\,\partial y} + \dfrac{\partial^2 w}{\partial y^2} + \dfrac{\partial w}{\partial y} = 0$
19. $\dfrac{\partial^2 w}{\partial x^2} - 4\dfrac{\partial^2 w}{\partial x\,\partial y} + \dfrac{\partial^2 w}{\partial y^2} + \dfrac{\partial w}{\partial y} = 0$
20. $w_{xx} - 4w_{xy} + 2w_{yy} + w_x - 3w_y + w = 0$
21. Obtain equation (4) by a direct calculation.

The classification of a partial differential equation can be done even if the coefficients in (1) are not constants. In such a case we divide the plane into the regions where the **discriminant**
$$\Delta = b^2 - ac$$
is everywhere positive, everywhere negative, or everywhere zero. For example
$$w_{xx} + xw_{yy} = 0$$
has discriminant $\Delta = -x$, so the partial differential equation is hyperbolic in the region $\{(x, y): x < 0\}$, elliptic in $\{(x, y): x > 0\}$, and parabolic in $\{(x, y): x = 0\}$. In Problems 22–27 describe the regions in the plane where the given equation is elliptic, parabolic, or hyperbolic.

22. $w_{xx} - xw_{yy} = 0$
23. $w_{xx} - xyw_{yy} = 0$
24. $x^2 w_{xx} + w_{yy} = xy$
25. $w_{xx} + (x^2 - y^2)w_{yy} = 0$
26. $w_{xx} + 2xw_{xy} + w_{yy} + 5w = 0$
27. $y\dfrac{\partial^2 w}{\partial x^2} - 2\dfrac{\partial^2 w}{\partial x\,\partial y} + e^x\dfrac{\partial^2 w}{\partial y^2} + y^2\dfrac{\partial w}{\partial y} - w = 0$

10.6 The Vibrating String: d'Alembert's Method

As our first physical application, consider a string that is tightly stretched between two fixed points 0 and L on the x-axis (see Figure 10.10). Suppose that the string is pulled back vertically a distance that is very small compared to the length L and released at time $t = 0$, causing it to vibrate. Our problem is to determine the displacement $y(x, t)$ of the point on the string that is x units away from the end 0, at any time t.

Figure 10.10

To avoid making our equation too complicated, we make two simplifying assumptions:

a. Our "ideal" string has uniform mass m per unit length and offers no resistance to bending.

b. The tension T in the string is so large that the gravitational force on the string may be neglected.

Consider a small segment of length Δx. By Newton's second law of motion, the total force acting on this piece of string is equal to the mass of the string multiplied by its acceleration:

$$F = ma = (m \, \Delta x) \frac{\partial^2 y}{\partial t^2}. \tag{1}$$

We assume in this equation that the string is moving only in the xy-plane and that each particle in the string moves only vertically.

Let T_x and $T_{x+\Delta x}$ be the tension vectors at the endpoints of the given segment. These forces are applied tangentially (see Figure 10.10) since the string offers no resistance to bending. Since there is no motion in the x-direction, the x-components of the tension vectors must coincide:

$$T_{x+\Delta x} \cos \beta = T_x \cos \alpha \equiv T. \tag{2}$$

Thus T is constant, since x and Δx are arbitrary. Similarly, the difference in the y-components of the tension vectors must equal the total force acting on the string. Then by equation (1),

$$T_{x+\Delta x} \sin \beta - T_x \sin \alpha = m \, \Delta x \frac{\partial^2 y}{\partial t^2}. \tag{3}$$

Dividing each term in (3) by the corresponding term in (2), we have

$$\frac{T_{x+\Delta x}\sin\beta}{T_{x+\Delta x}\cos\beta} - \frac{T_x\sin\alpha}{T_x\cos\alpha} = \frac{m}{T}\Delta x\frac{\partial^2 y}{\partial t^2},$$

or

$$\tan\beta - \tan\alpha = \frac{m}{T}\Delta x\frac{\partial^2 y}{\partial t^2}. \tag{4}$$

Since

$$\tan\alpha = \frac{\partial y}{\partial x}\bigg|_x \quad \text{and} \quad \tan\beta = \frac{\partial y}{\partial x}\bigg|_{x+\Delta x},$$

we may rewrite (4) in the form

$$\frac{1}{\Delta x}\left[\frac{\partial y}{\partial x}\bigg|_{x+\Delta x} - \frac{\partial y}{\partial x}\bigg|_x\right] = \frac{m}{T}\frac{\partial^2 y}{\partial t^2}.$$

Letting Δx tend to zero, we obtain in the limit the equation

$$\frac{\partial^2 y}{\partial x^2} = \frac{m}{T}\frac{\partial^2 y}{\partial t^2}.$$

Since m/T is positive, it is clear that this is a hyperbolic partial differential equation. The equation, written in the form

Wave Equation

$$\boxed{\frac{\partial^2 y}{\partial t^2} = c^2\frac{\partial^2 y}{\partial x^2}, \quad c^2 = \frac{T}{m},} \tag{5}$$

is often called the one-dimensional **wave equation**, where the constant c^2 indicates that the coefficient is positive.

As yet we have not made use of the fact that the string is fixed at its endpoints. We may write these boundary conditions as

$$y(0,t) = y(L,t) = 0, \quad t \geq 0. \tag{6}$$

In addition, we have not taken into account the initial distortion of the string and the fact that it was at rest when released. These initial conditions can be written as

$$y(x,0) = f(x), \quad 0 \leq x \leq L, \tag{7}$$

$$\frac{\partial y}{\partial t}\bigg|_{t=0} = 0, \tag{8}$$

where $f(x)$ is a function depicting the original distortion (see Examples 1 and 2, following), and $(\partial y/\partial t)(x,t)$ is the velocity of the point x units away from the origin at time t.

The direct method we use to solve this problem is due to d'Alembert.[†] Since it is only rarely possible to apply this technique, we develop two more applicable methods in this chapter.

[†] The French mathematician Jean Le Rond d'Alembert (1717–1783) is known for his contributions in mechanics.

We begin by defining the variables
$$u = x + ct, \qquad v = x - ct, \tag{9}$$
and transforming the wave equation (5) into one involving variables u and v. By the chain rule, we have
$$\frac{\partial y}{\partial t} = \frac{\partial y}{\partial u}\frac{\partial u}{\partial t} + \frac{\partial y}{\partial v}\frac{\partial v}{\partial t} = c\left(\frac{\partial y}{\partial u} - \frac{\partial y}{\partial v}\right), \tag{10}$$
where we used (9) to obtain the partial derivatives $\partial u/\partial t = c = -\partial v/\partial t$. Taking the partial derivative with respect to t of (10), we obtain
$$\frac{\partial^2 y}{\partial t^2} = c\frac{\partial}{\partial t}\left(\frac{\partial y}{\partial u} - \frac{\partial y}{\partial v}\right) = c\left[\frac{\partial}{\partial u}\left(\frac{\partial y}{\partial u} - \frac{\partial y}{\partial v}\right)\frac{\partial u}{\partial t} + \frac{\partial}{\partial v}\left(\frac{\partial y}{\partial u} - \frac{\partial y}{\partial v}\right)\frac{\partial v}{\partial t}\right]$$
$$= c^2\left[\left(\frac{\partial^2 y}{\partial u^2} - \frac{\partial^2 y}{\partial u\,\partial v}\right) - \left(\frac{\partial^2 y}{\partial v\,\partial u} - \frac{\partial^2 y}{\partial v^2}\right)\right].$$

Since the mixed second partials are equal, we finally have
$$\frac{\partial^2 y}{\partial t^2} = c^2\left(\frac{\partial^2 y}{\partial u^2} - 2\frac{\partial^2 y}{\partial u\,\partial v} + \frac{\partial^2 y}{\partial v^2}\right).$$

Similarly, we find that
$$\frac{\partial^2 y}{\partial x^2} = \left(\frac{\partial^2 y}{\partial u^2} + 2\frac{\partial^2 y}{\partial u\,\partial v} + \frac{\partial^2 y}{\partial v^2}\right).$$

Substituting these results into (5) and canceling like terms yield the equation
$$\frac{\partial^2 y}{\partial u\,\partial v} = 0, \tag{11}$$
since $4c^2 \neq 0$. It is now easy to solve (11) by performing two successive integrations. Integrating with respect to u, we obtain
$$\frac{\partial y}{\partial v} = g'(v), \tag{12}$$
where g' is an unknown function in v. From (12) it is evident that g' is the partial derivative, with respect to v, of the solution y—which is the reason for using the prime in our notation. Integrating (12) with respect to v yields
$$y = g(v) + h(u), \tag{13}$$
where h is an unknown function in u. Substituting (9) into (13) yields **d'Alembert's solution,**
$$y(x, t) = g(x - ct) + h(x + ct), \tag{14}$$
of the wave equation (5).

Since h is arbitrary, if we ignore (7) we see that one possible solution of the wave equation is $g(x - ct)$. If we set $t = 0$, the initial distortion of the string is given by $y(x, 0) = g(x)$. At a fixed time later, its distortion is $y(x, t) = g(x - ct)$. Thus the shape of the string can be obtained from the graph of the function $g(x)$ by moving an interval of length L to the right with velocity c. For this reason, $g(x - ct)$ is called a **traveling wave.** Similarly, $h(x + ct)$ is a traveling wave that moves to the left with velocity c. Hence d'Alembert's solution is a superposition of two traveling waves, one moving to the right, the other to the left, with the same velocity c.

The functions g and h can be determined from the initial conditions. Setting $t = 0$, we find that

$$f(x) = y(x,0) = g(x) + h(x). \tag{15}$$

Noting that g and h are functions of one variable, we may use the chain rule to differentiate (14) with respect to t. Setting $t = 0$, we obtain

$$0 = \frac{\partial y}{\partial t}\bigg|_{t=0} = -cg'(x-ct) + ch'(x+ct)\bigg|_{t=0} = c[h'(x) - g'(x)].$$

From the last equation we know that $h' = g'$, so

$$h(x) = g(x) + k,$$

where k is a constant. Replacing this function for h in (15) yields $f(x) = 2g(x) + k$ or $g = (f-k)/2$. Hence $h = (f+k)/2$, so the solution of the problem finally becomes

$$y(x,t) = \tfrac{1}{2}[f(x-ct) + f(x+ct)]. \tag{16}$$

Note that since the boundary conditions (6) must be satisfied by (16), setting $x = 0$ in (16) yields

$$0 = \tfrac{1}{2}[f(-ct) + f(ct)],$$

or

$$f(-ct) = -f(ct).$$

Thus $f(x)$ is odd. This result may seem curious, since $f(x)$ has been defined only on the interval $0 \leq x \leq L$. However, it causes no difficulty, since we may extend f to the interval $-L \leq x \leq L$ by defining

$$f(-x) = -f(x) \quad \text{for} \quad -L \leq x \leq 0.$$

This extension is always possible as $f(0) = 0$. Similarly, setting $x = L$ and using the fact that f is odd, we find that

$$f(L + ct) = -f(L - ct)$$
$$= f(ct - L),$$

so that f has period $2L$. Again, it is clear that we can extend $f(x)$ to the entire real line by letting

$$f(x + 2L) = f(x),$$

for all x. This extension of $f(x)$ to the entire real line seems very reasonable if we return to the notion of traveling waves. It is clear, from equation (16), that we must immediately leave the interval $0 \leq x \leq L$ when $t > 0$. Now suppose we move $f(x)$ gradually to the right. Since $y(0, t) = y(L, t) = 0$ for all t, any positive contribution at 0 or L from the traveling wave $f(x - ct)$ must be offset by an equal negative contribution from $f(x + ct)$ [which is obtained by moving $f(x)$ the same distance to the left], and vice versa. In effect, this procedure amounts to a reflection of wave forms at the boundary.

By Theorem 10.4.2, the function $f(x)$ has a Fourier sine series—a fact we will refer to in the next section. (The continuity of f is guaranteed by the fact that the string is unbroken. We assume that f' is piecewise continuous, since that is what it is in most practical situations, such as in the example below.)

Example 1 Suppose that we pluck the string at its center (see Figure 10.11), a distance $y(L/2, 0) = y_0$ meters (y_0 is assumed to be small). We can assume that the distortion consists of two straight lines from $(0,0)$ to $(L/2, y_0)$ and from $(L/2, y_0)$ to $(L, 0)$. Hence

$$f(x) = \begin{cases} \dfrac{2y_0}{L} x, & 0 \le x \le \dfrac{L}{2}, \\ \dfrac{2y_0}{L}(L - x), & \dfrac{L}{2} \le x \le L, \end{cases}$$

Figure 10.11

which has the piecewise continuous derivative

$$f'(x) = \begin{cases} \dfrac{2y_0}{L}, & 0 \le x < \dfrac{L}{2}, \\ -\dfrac{2y_0}{L}, & \dfrac{L}{2} < x \le L. \end{cases}$$

If the string is 4 meters long, has a mass of 0.08 kilogram, and is subject to a constant tension of 2 newtons ($= \text{kg-m/s}^2$), we obtain the solution

$$y(x, t) = \tfrac{1}{2}[f(x - 10t) + f(x + 10t)],$$

since

$$c^2 = \frac{T}{m} = \frac{2}{0.08/4} = 100 \ (\text{meters/second})^2.$$

To discover the exact location of a point 1 meter from the end 0 after exactly 1 second, we set $x = t = 1$. Then

$$y(1, 1) = \tfrac{1}{2}[f(-9) + f(11)],$$

and since f has period $2L = 8$ and is odd, we get

$$y(1, 1) = \frac{1}{2}[f(-1) + f(3)] = \frac{1}{2}[f(3) - f(1)]$$

$$= \frac{1}{2}\left(\frac{2y_0}{L} - \frac{2y_0}{L}\right) = 0.$$

Example 2 Suppose that $f(x) = y_0 \sin(\pi x/L)$. By equation (16),

$$y(x, t) = \frac{y_0}{2}\left[\sin\frac{\pi}{L}(x - ct) + \sin\frac{\pi}{L}(x + ct)\right],$$

and by the addition formulas of trigonometry, this reduces to

$$y(x, t) = y_0 \sin\frac{\pi x}{L} \cos\frac{\pi c t}{L}.$$

Thus, if we fix our attention on the point x_0 units from the left, we see that it oscillates with a period $2L/c$ and an amplitude $y_0 \sin(\pi x_0/L)$.

PROBLEMS 10.6

In Problems 1–8 use d'Alembert's solution (16) to find the displacement $y(x, t)$ for each given function $f(x)$, point x, and time t. Assume that $c = 5$ and $L = 6$ meters.

1. $f(x) = 0.01 \sin(\pi x/L)$, $x = 2$, $t = 1$
2. $f(x) = 0.1 \sin(\pi x/L)$, $x = 3$, $t = 2$
3. $f(x) = 0.01 \sin(2\pi x/L)$, $x = 3$, $t = 1$
4. $f(x) = 0.1 \sin(2\pi x/L)$, $x = 2$, $t = 2$
5. $f(x) = \begin{cases} 0.1x, & 0 \le x \le 3, \\ 0.1(6-x), & 3 \le x \le 6 \end{cases}$ $x = 2$, $t = 2$
6. $f(x) = \begin{cases} 0.1x, & 0 \le x \le 2, \\ 0.2(3-x), & 2 \le x \le 4, \\ 0.1(x-6), & 4 \le x \le 6 \end{cases}$ $x = 3$, $t = 5$
7. $f(x) = \begin{cases} 0.1x, & 0 \le x \le 2, \\ 0.2, & 2 \le x \le 4, \\ 0.1(6-x), & 4 \le x \le 6 \end{cases}$ $x = 3$, $t = 5$
8. $f(x) = \begin{cases} 0.01x^2, & 0 \le x \le 3, \\ 0.01(6-x)^2, & 3 \le x \le 6 \end{cases}$ $x = 2$, $t = 2$

9. Assume, instead of the initial conditions (7) and (8), that a string is initially in its equilibrium position and is set vibrating by each of its points x being given an initial velocity $f'(x)$, where f is a differentiable function. Prove that the solution is given by

$$y(x, t) = \frac{1}{2c}[f(x + ct) - f(x - ct)].$$

Using the result of Problem 9 and setting $c = 5$ and $L = 6$ meters, do Problems 10–15 for the given functions $f'(x)$ and values x and t.

10. $f'(x) = 0.01 \sin(\pi x/L)$, $x = 2$ meters, $t = 3$ seconds
11. $f'(x) = 0.1 \sin(2\pi x/L)$, $x = 1$ meter, $t = 1$ second
12. $f'(x) = 3(6x - x^2)$, $x = 3$ meters, $t = 0.01$ second

13. $f'(x) = 0.1 \sin^2(\pi x/L)$, $x = 1$ meter, $t = 0.1$ second
14. $f'(x) = \begin{cases} 0.01x, & 0 \le x \le 3, \\ 0.01(6 - x), & 3 \le x \le 6 \end{cases}$ $x = 2$, $t = 1$
15. $f'(x) = \begin{cases} 0.01(3 - x), & 0 \le x \le 3, \\ 0.01(x - 3), & 3 \le x \le 6 \end{cases}$ $x = 2$, $t = 1$

16. Suppose that we subject a vibrating string to both an initial displacement $y(x, 0) = f(x)$, $0 \le x \le L$, and an initial velocity $y_t(x, 0) = g'(x)$, $0 \le x \le L$. Show that the solution to this initial-value problem is

$$y(x, t) = \frac{1}{2}[f(x - ct) + f(x + ct)]$$
$$+ \frac{1}{2c}[g(x + ct) - g(x - ct)].$$

17. Use the result of Problem 16 to solve the initial-value problem

$$y_{tt} = y_{xx}, \quad |x| < \infty, \quad t > 0$$

$$y(x, 0) = \sin\frac{\pi x}{2} = y_t(x, 0).$$

18. Suppose that we subject a vibrating string to a damping force that is proportional at each instance to the velocity at each point. Show that the resulting hyperbolic differential equation is of the form

$$\frac{\partial^2 y}{\partial t^2} - k\frac{\partial y}{\partial t} = c^2 \frac{\partial^2 y}{\partial x^2}.$$

*19. The distortionless solution of the transmission-line problem of a telegraph cable satisfies the partial differential equation

$$v_{xx} = \alpha v_{tt} + \beta v_t + \gamma v, \quad 0 < x < \infty, \quad t > 0,$$

where $\alpha = LC$, $\beta = RC + GL$, $\gamma = RG$, and v, L, R, C, and G are the voltage, inductance, resistance, capacitance, and conductance between the line and ground. What is the form of the solution?

10.7 Separation of Variables: The Wave Equation

One of the simplest techniques for solving ordinary differential equations consists of separating the variables. Although we now have *two* independent variables, we can nevertheless adapt the technique to all partial differential equations of the forms (10.5.10)–(10.5.13) whenever the function $F(u, v) \equiv 0$. The method is best explained by means of an example.

Suppose that we again consider the problem of a vibrating string given by the wave equation

$$\frac{\partial^2 y}{\partial t^2} = c^2 \frac{\partial^2 y}{\partial x^2}, \qquad c^2 = \frac{T}{m}, \tag{1}$$

with boundary conditions

$$y(0, t) = y(L, t) = 0, \qquad t \geq 0, \tag{2}$$

and initial conditions

$$y(x, 0) = f(x), \qquad \left.\frac{\partial y}{\partial t}\right|_{t=0} = 0, \qquad 0 \leq x \leq L. \tag{3}$$

We begin by seeking a solution of the form

$$y(x, t) = X(x)T(t), \tag{4}$$

where X is a function of x alone and T is a function involving only the independent variable t. Of course there is no guarantee that such a solution exists. In fact, as we will see later, separation is often not possible. Nevertheless, we can try it out and see if it works. Using primes to denote differentiation, we observe that

$$\frac{\partial^2 y}{\partial t^2} = \frac{\partial^2}{\partial t^2}[XT] = XT'' \qquad \text{and} \qquad \frac{\partial^2 y}{\partial x^2} = X''T,$$

so that equation (1) becomes

$$XT'' = c^2 X''T. \tag{5}$$

It is now possible to separate the variables and write equation (5) as

$$\frac{T''}{T} = c^2 \frac{X''}{X}. \tag{6}$$

Initially we had no assurance that it would be possible to separate the variables as we have done. Should such a separation prove impossible, this method would not apply and we would have to turn to other techniques.

Now suppose we fix t and allow x to vary. Since the left-hand side of equation (6) is constant, the right-hand side must be constant for all values of x. Consequently, the right-hand side is constant *regardless* of the value of t, and we have

$$\frac{T''}{T} = c^2 \frac{X''}{X} = k,$$

where k is constant. From this set of equalities we obtain two ordinary differential equations

$$T'' - kT = 0, \tag{7}$$

$$X'' - \frac{k}{c^2} X = 0. \tag{8}$$

Thus the problem of solving the partial differential equation (1) has been reduced to solving two ordinary differential equations, (7) and (8). The constant k is arbitrary, but its value must be the same for both equations. Using the characteristic equation (see Chapter 3) $\lambda^2 - k = 0$, we can easily obtain the general solution

$$T(t) = \begin{cases} a_1 e^{\sqrt{k}\,t} + a_2 e^{-\sqrt{k}\,t}, & \text{if } k > 0, \\ a_1 t + a_2, & \text{if } k = 0, \\ a_1 \cos\sqrt{-k}\,t + a_2 \sin\sqrt{-k}\,t, & \text{if } k < 0. \end{cases}$$

Similarly, we also have the general solution

$$X(x) = \begin{cases} b_1 e^{\sqrt{k}\,x/c} + b_2 e^{-\sqrt{k}\,x/c}, & \text{if } k > 0, \\ b_1 x + b_2, & \text{if } k = 0, \\ b_1 \cos\dfrac{\sqrt{-k}\,x}{c} + b_2 \sin\dfrac{\sqrt{-k}\,x}{c}, & \text{if } k < 0. \end{cases}$$

Since our solution must satisfy the boundary conditions (2), if a solution of the form (4) exists, we must have the equations

$$X(0)T(t) = X(L)T(t) = 0 \quad \text{for all } t.$$

If $T(t) \equiv 0$, then $y \equiv 0$ and the string is constantly at rest, which is impossible if $f(x) \not\equiv 0$. For $f \not\equiv 0$, T must be nonzero for at least one value of t, which implies that $X(0) = X(L) = 0$.

Setting $x = 0$ and $x = L$ in the general solution yields, for $k > 0$, the homogeneous system of equations

$$X(0) = b_1 + b_2 = 0,$$

$$X(L) = b_1 e^{\sqrt{k}\,L/c} + b_2 e^{-\sqrt{k}\,L/c} = 0.$$

This system has a nonzero solution if and only if the determinant of its coefficients is zero. But

$$\begin{vmatrix} 1 & 1 \\ e^{\sqrt{k}\,L/c} & e^{-\sqrt{k}\,L/c} \end{vmatrix} = e^{-\sqrt{k}\,L/c} - e^{\sqrt{k}\,L/c} = -2\sinh\frac{\sqrt{k}\,L}{c} \neq 0,$$

so that $X(x) \equiv 0$ for $k > 0$. Then $y \equiv 0$, which is impossible if $f(x) \not\equiv 0$. For $k = 0$, the condition $X(0) = 0$ implies that $b_2 = 0$, while $X(L) = 0$ requires that $b_1 = 0$, again leading to the zero solution. So we are left with the remaining possibility that k is negative, and we set $k = -r^2$. This yields the general solution

$$X(x) = b_1 \cos\frac{rx}{c} + b_2 \sin\frac{rx}{c}.$$

Letting $x = 0$ and $x = L$, we have

$$X(0) = b_1 = 0 \quad \text{and} \quad X(L) = b_2 \sin \frac{rL}{c} = 0.$$

To prevent our again having the zero solution, we must choose an r such that rL/c is a positive multiple of π:

$$\frac{rL}{c} = n\pi \quad \text{or} \quad r = \frac{n\pi c}{L}.$$

Thus we obtain an infinite set of solutions,

$$X_n(x) = b_2 \sin \frac{n\pi x}{L}, \tag{9}$$

each associated with the choice

$$k = -\frac{n^2 \pi^2 c^2}{L^2}. \tag{10}$$

Hence the corresponding solutions for T are

$$T_n(t) = a_1 \cos \frac{n\pi ct}{L} + a_2 \sin \frac{n\pi ct}{L}.$$

Finally, we obtain an infinite set of solutions,

$$y_n(x,t) = X_n(x)T_n(t) = \left(B_n \cos \frac{n\pi ct}{L} + B_n^* \sin \frac{n\pi ct}{L} \right) \sin \frac{n\pi x}{L}, \tag{11}$$

where B_n and B_n^* are constants, that satisfy the partial differential equation (1) and the boundary conditions (2). In the language of Section 10.1 we see that we have found the *eigenvalues* (10) and the corresponding *eigenfunctions* (11) of the boundary value problem (1), (2).

We now try to select a solution that satisfies the initial conditions (3). It is extremely likely that no one solution (11) will satisfy (3). However, any finite sum of solutions (11) is again a solution of the boundary value problem (by the *principle of superposition*), so we can try to find a finite linear combination of these solutions that satisfies the initial conditions (3). Generally, even this will fail, so as a last resort we try an infinite series of the solutions (11):

$$y(x,t) = \sum_{n=1}^{\infty} y_n(x,t) = \sum_{n=1}^{\infty} \left(B_n \cos \frac{n\pi ct}{L} + B_n^* \sin \frac{n\pi ct}{L} \right) \sin \frac{n\pi x}{L}. \tag{12}$$

There is no guarantee at this point that such a series converges, but since its form is very similar to that of the Fourier series (10.3.7), there is every reason to be optimistic. Furthermore, if (12) converges and the series obtained by the formal term-by-term partial differentiations $\partial/\partial t$, $\partial^2/\partial t^2$, $\partial/\partial x$, and $\partial^2/\partial x^2$ all converge uniformly, then (12) is again a solution of the boundary value problem (1), (2). If, in addition, we can also satisfy the initial conditions (3), then we have solved the vibrating string problem.

Setting $t = 0$ in equation (12) and using the first of equations (3), we have

$$y(x,0) = \sum_{n=1}^{\infty} B_n \sin \frac{n\pi x}{L} = f(x). \tag{13}$$

We again assume that $f(x)$ and $f'(x)$ are piecewise continuous (see Section

10.6). The infinite series (13) is a Fourier sine series, which requires that the function $f(x)$ be odd $[f(-x) = -f(x)]$, and periodic with period $2L$. As in Section 10.6, we can extend the function $f(x)$, defined on the interval $0 \leq x \leq L$, in such a way that it is odd and periodic with period $2L$. Since $f(0) = 0$, we let $f(-x) = -f(x)$ on $-L \leq x \leq 0$ and require that $f(x + 2L) = f(x)$ for all real numbers x. Then the coefficients B_n are given by the Euler formula

$$B_n = \frac{1}{L} \int_{-L}^{L} f(x) \sin \frac{n\pi x}{L} dx = \frac{2}{L} \int_0^L f(x) \sin \frac{n\pi x}{L} dx, \quad n = 1, 2, 3, \ldots. \quad (14)$$

The last equality holds because $f(x)$ and $\sin(n\pi x/L)$ are odd and their product is even [see equation (10.4.5), p. 433]. Taking the partial derivative of (12) with respect to t and using the second of the initial conditions (3), we have

$$\frac{\partial y}{\partial t} = \frac{\pi c}{L} \sum_{n=1}^{\infty} \left(nB_n^* \cos \frac{n\pi ct}{L} - nB_n \sin \frac{n\pi ct}{L} \right) \sin \frac{n\pi x}{L} \bigg|_{t=0} = 0,$$

or

$$\sum_{n=1}^{\infty} nB_n^* \sin \frac{n\pi x}{L} = 0.$$

Again we obtain an expression of the form (14) for B_n^*, but this time the function involved is zero. Hence all the coefficients B_n^* vanish. Thus solution (12) reduces to

$$y(x, t) = \sum_{n=1}^{\infty} B_n \cos \frac{n\pi ct}{L} \sin \frac{n\pi x}{L}, \quad (15)$$

where the coefficients B_n are given by (14). It is possible to write series (15) in closed form by using the trigonometric identity

$$2 \sin A \cos B = \sin(A + B) + \sin(A - B).$$

Then

$$y(x, t) = \frac{1}{2} \sum_{n=1}^{\infty} B_n \left[\sin \frac{n\pi}{L}(x + ct) + \sin \frac{n\pi}{L}(x - ct) \right]$$

$$= \frac{1}{2} \left[\sum_{n=1}^{\infty} B_n \sin \frac{n\pi}{L}(x + ct) + \sum_{n=1}^{\infty} B_n \sin \frac{n\pi}{L}(x - ct) \right],$$

and the series can be evaluated by substituting $x + ct$ and $x - ct$, respectively, for the variable x in (13). Hence

$$y(x, t) = \tfrac{1}{2}[f(x + ct) + f(x - ct)],$$

which is precisely the same result as we obtained directly in Section 10.6.

Example 1 Let a vibrating string have length π and let $c^2 = 1$. Suppose that the initial velocity is zero and the initial distortion $f(x) = x(\pi - x)$. Instead of using d'Alembert's solution, we here obtain an expansion of the form (15) by calculating the coefficients B_n.

From (14) we have

$$B_n = \frac{2}{\pi} \int_0^{\pi} x(\pi - x) \sin nx \, dx = \frac{2}{\pi} \int_0^{\pi} (\pi x - x^2) \sin nx \, dx.$$

We integrate by parts twice:

$$B_n = \frac{2}{\pi}\left[-(\pi x - x^2)\frac{\cos nx}{n}\bigg|_0^\pi + \frac{1}{n}\int_0^\pi (\pi - 2x)\cos nx\, dx\right]$$

$$= \frac{2}{n\pi}\left[(\pi - 2x)\frac{\sin nx}{n}\bigg|_0^\pi + \frac{2}{n}\int_0^\pi \sin nx\, dx\right]$$

$$= \frac{4}{n^2\pi}\left(-\frac{\cos nx}{n}\bigg|_0^\pi\right) = \frac{4}{n^3\pi}\left[1 - (-1)^n\right].$$

Hence the solution is

$$y(x,t) = \frac{4}{\pi}\sum_{n=1}^\infty \frac{[1-(-1)^n]}{n^3}\cos nt \sin nx$$

$$= \frac{8}{\pi}\left(\cos t \sin x + \frac{1}{3^3}\cos 3t \sin 3x + \frac{1}{5^3}\cos 5t \sin 5x + \cdots\right)$$

$$= \frac{8}{\pi}\sum_{n=0}^\infty \frac{\cos(2n+1)t \sin(2n+1)x}{(2n+1)^3}.$$

PROBLEMS 10.7

In Problems 1–6 use the method of separation of variables to find the displacement $y(x,t)$ in the wave equation for each initial condition $y(x,0) = f(x)$.

1. $f(x) = 0.1\sin(\pi x/L)$, $0 \leq x \leq L$
2. $f(x) = 0.01\sin(2\pi x/L)$, $0 \leq x \leq L$
3. $f(x) = 0.1x(L-x)$, $0 \leq x \leq L$
4. $f(x) = 0.1x^2(L-x)$, $0 \leq x \leq L$
5. $f(x) = \begin{cases} 0.1x, & 0 \leq x \leq 3 \\ 0.1(6-x), & 3 \leq x \leq 6 \end{cases}$
6. $f(x) = \begin{cases} 0.1x, & 0 \leq x \leq 2 \\ 0.2, & 2 \leq x \leq 4 \\ 0.1(6-x), & 4 \leq x \leq 6 \end{cases}$

In Problems 7–14 reduce the given equations to pairs of ordinary differential equations by the method of separation of variables.

7. $\dfrac{\partial^2 y}{\partial x^2} + \dfrac{\partial y}{\partial t} = 0$
8. $\dfrac{\partial^2 y}{\partial x^2} + x\dfrac{\partial y}{\partial t} = 0$
9. $\dfrac{\partial^2 y}{\partial x \partial t} + \dfrac{\partial y}{\partial t} = 0$
10. $z_{xx} = y^2 z_{yy}$
11. $z_{yy} + xyz_x = 0$
12. $z_{xx} - z_{yy} = z$
13. $u_{rr} + \dfrac{1}{r}u_r + \dfrac{1}{r^2}u_{\theta\theta} = 0$, where $u = u(r,\theta)$
14. $u_{rr} + \dfrac{1}{r}u_r + \dfrac{1}{r^2}u_{\theta\theta} + u_{zz} = 0$, where $u = u(r,\theta,z)$

For Problems 15–20 do the following:

a. Find every solution of the given equation by the technique of separation of variables.

b. Find the nontrivial solutions if we require the solutions to satisfy the boundary conditions

$$y(0,t) = y(\pi, t) = 0.$$

c. Using the nontrivial solutions in part (b), indicate whether it would be possible to find a solution that also satisfies the initial condition

$$y(x,0) = f(x), \quad 0 < x < \pi,$$

for $f(x) = x(\pi - x)$.

15. $\dfrac{\partial^2 y}{\partial x^2} + \dfrac{1}{x^2}\dfrac{\partial^2 y}{\partial t^2} = 0$

16. $\dfrac{\partial^2 y}{\partial x \partial t} = y$

17. $\dfrac{\partial^2 y}{\partial x^2} + \dfrac{\partial^2 y}{\partial t^2} = 0$

18. $\dfrac{\partial^2 y}{\partial x^2} - 2\dfrac{\partial y}{\partial x} = -\dfrac{\partial y}{\partial t}$

19. $\dfrac{\partial^2 y}{\partial t^2} = \dfrac{\partial^2 y}{\partial x^2} + y$

20. $\dfrac{\partial^2 y}{\partial t^2} = x^2 \dfrac{\partial^2 y}{\partial x^2}$

10.8 Heat Flow and the Heat Equation

Consider a cylindrical rod of length L and radius R (so that the cross-sectional area is $A = \pi R^2$) composed of a uniform heat-conducting material. We assume that heat can enter and leave the rod only through its ends; that is, the lateral surface of the rod is completely insulated. Let x measure the distance along the rod (see Figure 10.12) and let $T(x, t)$ denote the absolute temperature at time t at a point x units along the rod (we assume that the temperature is uniform across any cross-sectional area of the rod). Let ρ be the **density** of the rod (mass per unit volume), which is assumed to be constant. The **specific heat** c of the rod is defined as the amount of heat (joules, BTUs) that must be supplied to raise the temperature of one unit mass (kilograms, pounds) of the rod by one degree (Celsius, Fahrenheit).

Figure 10.12

Consider a section of the rod between x and $x + dx$. Since mass = volume × density, the mass between these two points is $\rho A \, dx$. In order to change the temperature of the rod between these two points from 0 to $T(x, t)$ degrees, we must supply $T(x, t) \cdot c\rho A \cdot dx$ units of heat. Thus, between any two points x_0 and x_1, the heat energy contained in the rod at time t is

$$Q(t) = \int_{x_0}^{x_1} T(x, t) c\rho A \, dx. \tag{1}$$

By the law of conservation of energy, if there are no heat sources within the rod, the heat energy in any part of the rod can increase or decrease only because of a lateral flow of heat through the boundaries of that part of the rod. This **heat flux**, written as $Q_F(x, t)$, is the quantity of heat energy per unit time passing through a unit area in the cross section x units from the left-hand end in a positive (rightward) direction. Thus, to find the rate of change of the heat energy between the x_0 and x_1 cross sections, we need only take the difference

$$\frac{dQ}{dt} = AQ_F(x_0, t) - AQ_F(x_1, t), \tag{2}$$

where the first term on the right-hand side is the heat energy flowing in and the second is that flowing out (see Figure 10.13).

It is an empirical law of physics that the heat flux at any point is proportional to the **temperature gradient** $\partial T(x, t)/\partial x$ at that point. The constant of proportionality is called the **thermal conductivity** of the rod and is denoted by κ. Since heat flows in the direction of decreasing temperature, we have

$$Q_F(x, t) = -\kappa \frac{\partial T}{\partial x}(x, t),$$

```
        Q_F(x_0,t) →   ← Q_F(x_1,t)
  0           x_0   x_1                L
```

Figure 10.13

where $\kappa > 0$. Since no heat sources are in this part of the rod, we have

$$AQ_F(x_0, t) - AQ_F(x_1, t) = A\left[\kappa\frac{\partial T}{\partial x}(x_1, t) - \kappa\frac{\partial T}{\partial x}(x_0, t)\right]$$

$$= \int_{x_0}^{x_1} \frac{\partial}{\partial x}\left[\kappa\frac{\partial T}{\partial x}(x, t)\right] A\, dx. \qquad (3)$$

Substituting the right-hand side of (3) for the right-hand side of (2) and differentiating (1), we have

$$\int_{x_0}^{x_1} \frac{\partial}{\partial x}\left(\kappa\frac{\partial T}{\partial x}\right) A\, dx = \frac{dQ}{dt}$$

$$= \int_{x_0}^{x_1} c\rho \frac{\partial T}{\partial t} A\, dx,$$

or

$$\int_{x_0}^{x_1}\left[c\rho\frac{\partial T}{\partial t} - \frac{\partial}{\partial x}\left(\kappa\frac{\partial T}{\partial x}\right)\right] A\, dx = 0. \qquad (4)$$

Equation (4) must hold in any interval $x_0 \leq x \leq x_1$, so if we assume that the functions involved are all continuous, the expression in brackets must vanish. To see this, note that if it were positive (negative) at some point, then by continuity it would be positive (negative) on some interval, in which case (4) would not hold. Thus, for all x in the interval $[0, L]$,

$$c\rho\frac{\partial T}{\partial t} = \frac{\partial}{\partial x}\left(\kappa\frac{\partial T}{\partial x}\right), \qquad (5)$$

and if we assume that κ is constant,

Heat Equation

$$\boxed{\frac{\partial T}{\partial t} = \delta\frac{\partial^2 T}{\partial x^2},} \qquad (6)$$

where $\delta = \kappa/c\rho$. Equation (6) is called the **heat (conduction) equation** and is clearly parabolic. The constant δ is positive and measures the **diffusivity** of the material of the rod. To completely specify our problem, we select the case in which the boundary conditions are

$$T(0, t) = T(L, t) = 0, \quad t \geq 0, \qquad (7)$$

and the initial condition is

$$T(x, 0) = f(x), \quad 0 \leq x \leq L, \qquad (8)$$

where f is a given function. In physical terms, we are keeping the ends of the rod at zero temperature and letting $f(x)$ denote the initial temperature at any point x of the rod. It is easy to modify the problem by selecting other, possibly time-dependent, boundary conditions (see Problems 7–9).

We begin to solve the heat equation by assuming the existence of a solution to the problem of the form

$$T(x,t) = X(x)\mathcal{T}(t).$$

Substituting this equation into (6) yields

$$X\mathcal{T}' = \delta X''\mathcal{T},$$

or

$$\frac{\mathcal{T}'}{\delta \mathcal{T}} = \frac{X''}{X}. \tag{9}$$

The function on the left-hand side depends only on t, so it is constant for fixed t and arbitrary x. On the other hand, the function on the right side of (9) depends only on x and is constant for fixed x and arbitrary t. The only way these situations can hold simultaneously is for each function to be constant, say, k. Then we obtain the pair of ordinary differential equations

$$\mathcal{T}' - k\delta\mathcal{T} = 0, \tag{10}$$

$$X'' - kX = 0. \tag{11}$$

The boundary conditions (7) may be written as

$$X(0)\mathcal{T}(t) = X(L)\mathcal{T}(t) = 0,$$

implying that $X(0) = X(L) = 0$ unless the rod has zero initial temperature at every point. If we ignore this uninteresting case, we note that the boundary value problem (11), $X(0) = X(L) = 0$, is almost identical to the boundary value problem for X in the case of the vibrating string [see equation (10.7.8)]. Paralleling the development of Section 10.7, we see that the only nonzero solutions for the boundary value problem (11) with $X(0) = X(L) = 0$ arise for negative values of k. Setting $k = -r^2$, we find that k is an eigenvalue of the problem only if r is a multiple of π/L, and the eigenfunctions corresponding to $-(n\pi/L)^2$ have the form

$$X_n(x) = A\sin\frac{n\pi x}{L}.$$

Setting $k = -(n\pi/L)^2$ in (10), we see that this first-order equation has the general solution

$$\mathcal{T}_n(t) = Be^{-(n\pi/L)^2\delta t}.$$

Hence we consider an infinite series of the form

$$T(x,t) = \sum_{n=1}^{\infty} X_n(x)\mathcal{T}_n(t) = \sum_{n=1}^{\infty} B_n \sin\frac{n\pi x}{L} e^{-(n\pi/L)^2\delta t}. \tag{12}$$

Setting $t = 0$ and making use of the initial condition (8), we have

$$T(x,0) = \sum_{n=1}^{\infty} B_n \sin\frac{n\pi x}{L} = f(x).$$

Thus, in order that separation of variables work, $f(x)$ must be representable as a Fourier sine series, implying that f must be extended as an odd, piecewise continuous function of period $2L$ with piecewise continuous derivatives. When we make this extension, the coefficients B_n are given by

$$B_n = \frac{1}{L}\int_{-L}^{L} f(x)\sin\frac{n\pi x}{L}\,dx = \frac{2}{L}\int_0^L f(x)\sin\frac{n\pi x}{L}\,dx, \qquad n = 1, 2, 3, \ldots. \tag{13}$$

The solution (12) is completely determined and the series must converge, since $T(x, 0)$ converges and the exponential factors are less than 1 for all $t > 0$. Observe that the exponential factors in (12) cause $T(x, t)$ to approach zero as t tends to infinity.

Example 1 Suppose that the rod has length π, $\delta = 1$, and initial temperature $f(x) = \sin 2x$. By the orthogonality of the set of functions $\{\sin nx\}$ (see Example 10.2.1), the coefficients B_n are all zero for $n \neq 2$. For $n = 2$, we have, from equation (13) and from Example 10.2.1,

$$B_2 = \frac{1}{\pi}\int_{-\pi}^{\pi} \sin^2 2x\,dx = 1.$$

Thus the solution of this problem is given by

$$T(x, t) = (\sin 2x)e^{-4t}.$$

Example 2 Let the rod have length 1, $\delta = 1$, and initial temperature $f(x) = x(1 - x^2)$. Observe that $f(x)$ is an odd function for all $|x| \leq 1$, since

$$f(-x) = -x\left[1 - (-x)^2\right] = -x(1 - x^2) = -f(x).$$

Extend f to the reals $\mathbb{R}$ periodically by letting $f(x + 2) = f(x)$. Then

$$B_n = 2\int_0^1 f(x)\sin n\pi x\,dx = 2\int_0^1 (x - x^3)\sin n\pi x\,dx.$$

We integrate by parts three times to obtain (the details should now be familiar)

$$B_n = \frac{12(-1)^{n+1}}{n^3\pi^3}.$$

Hence, by (12), the solution to this initial-value problem is

$$T(x, t) = \frac{12}{\pi^3}\sum_{n=1}^{\infty} \frac{(-1)^{n+1}\sin n\pi x\, e^{-n^2\pi^2 t}}{n^3}.$$

PROBLEMS 10.8

In Problems 1–6 find the temperature $T(x, t)$ in an insulated rod π units long whose ends are kept at 0°C, where the initial temperature is given by

1. $f(x) = x(\pi - x)$
2. $f(x) = x(\pi^2 - x^2)$
3. $f(x) = x^2(\pi - x)$
4. $f(x) = x\cos\dfrac{x}{2}$
5. $f(x) = \begin{cases} x, & 0 \leq x \leq \pi/2 \\ 0, & \pi/2 < x \leq \pi \end{cases}$
6. $f(x) = \begin{cases} 0, & 0 \leq x < \pi/3 \\ 1, & \pi/3 \leq x < 2\pi/3 \\ 0, & 2\pi/3 \leq x \leq \pi \end{cases}$

7. Suppose that the ends of the rod are kept at different constant temperatures:

$$T(0, t) = T_1, \quad T(L, t) = T_2.$$

Find the temperature at x as t tends to ∞; that is, find the **steady-state** temperature. [*Hint:* The steady-state temperature is *not* time dependent.]

8. Denote the steady-state temperature by $T_s(x)$ and define the **transient** temperature in the rod by
$$T_t(x,t) = T(x,t) - T_s(x).$$
Verify that $T_t(x,t)$ is given by equation (12). This shows that we need only superimpose the steady-state temperature on the solution (12) to obtain the solution for nonzero boundary conditions.

9. Suppose that the ends of the rod are kept at the temperatures
$$T(0,t) = T_1 e^{-c^2 \delta t}, \quad T(L,t) = T_2 e^{-c^2 \delta t}.$$
For what initial conditions $f(x)$ can a solution be found by the separation of variables? What happens if cL is a multiple of π?

*10. Consider the nonhomogeneous heat equation
$$\frac{\partial u}{\partial t} - k \frac{\partial^2 u}{\partial x^2} = A(x,t),$$
where $A(x,t)$ is given by
$$A(x,t) = \sum_{n=1}^{\infty} a_n(t) \sin \frac{n\pi x}{L}$$
for known functions $a_n(t)$. Suppose that initial and boundary conditions are given by
$$u(0,t) = u(L,t) = 0, \quad t \geq 0,$$
$$u(x,0) = f(x), \quad 0 \leq x \leq L,$$
with $f(x)$ known. Find a Fourier series for the solution $u(x,t)$. [*Hint*: Assume that B_n in (12) is a function of t.]

10.9 Two-Dimensional Heat Flow and Laplace's Equation

Consider a rectangular sheet of heat-conducting material of uniform thickness θ, density ρ, specific heat c, and thermal conductivity κ. We may suppose that the set of points (x, y) with $0 \leq x \leq L$ and $0 \leq y \leq M$ is a face of the sheet. We assume that the faces of the sheet are insulated and that heat can enter and leave only through the edges of the sheet.

Select any interior point (x, y) on the face and consider a rectangular region $ABCD$ whose corner coordinates are given in Figure 10.14. It is reasonable to assume the heat flow in each plane to be identical, so that the flow is two-dimensional. In the discussion below we assume that the distances Δx and Δy are very small.

The rate of change of heat energy in the sheet $ABCD$ at any time t is approximately (see Section 10.8).

$$c\rho\theta \Delta x \Delta y \frac{\partial T}{\partial t}, \tag{1}$$

where $T = T(x, y, t)$ is the temperature at any point (x, y) at time t. Since Δx and Δy are small, we assume that the heat flux along each edge is constant.

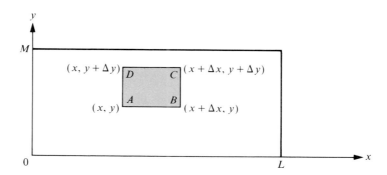

Figure 10.14

Then the heat energy passing through $ABCD$ in the vertical direction is

$$\theta \Delta x [Q_F(x, y, t) - Q_F(x, y + \Delta y, t)], \tag{2}$$

where Q_F is the heat flux. In the horizontal direction the heat energy is

$$\theta \Delta y [Q_F(x, y, t) - Q_F(x + \Delta x, y, t)]. \tag{3}$$

In Section 10.8 we mentioned that it has been shown empirically that the heat flux at any point is proportional to the temperature gradient at that point. Adding (2) and (3), we get another approximation of the rate of change of heat energy in the sheet $ABCD$:

$$\kappa \theta \Delta y \left[\frac{\partial T}{\partial x}(x + \Delta x, y, t) - \frac{\partial T}{\partial x}(x, y, t) \right]$$
$$+ \kappa \theta \Delta x \left[\frac{\partial T}{\partial y}(x, y + \Delta y, t) - \frac{\partial T}{\partial y}(x, y, t) \right]. \tag{4}$$

Equating (4) and (1) and dividing by $\Delta x \Delta y$ yield

$$c\rho\theta \frac{\partial T}{\partial t} = \kappa\theta \frac{\left[\frac{\partial T}{\partial x}(x + \Delta x, y, t) - \frac{\partial T}{\partial x}(x, y, t) \right]}{\Delta x}$$
$$+ \kappa\theta \frac{\left[\frac{\partial T}{\partial y}(x, y + \Delta y, t) - \frac{\partial T}{\partial y}(x, y, t) \right]}{\Delta y}.$$

Passing to the limit as Δx and Δy both approach zero, we obtain

$$c\rho\theta \frac{\partial T}{\partial t} = \kappa\theta \left[\frac{\partial^2 T}{\partial x^2} + \frac{\partial^2 T}{\partial y^2} \right],$$

which may be written in the form

$$\boxed{\frac{\partial^2 T}{\partial x^2} + \frac{\partial^2 T}{\partial y^2} = \frac{1}{\delta} \frac{\partial T}{\partial t},} \tag{5}$$

where $\delta = \kappa/c\rho$ is the **diffusivity**. Equation (5) is called the **two-dimensional heat equation**.

If the edges of the sheet are kept at constant temperatures, and an extremely long time is allowed to pass, the temperature at any given point stabilizes and remains almost constant for all larger values of t; that is, steady-state conditions are attained. The temperature T at a point on the sheet then depends only on its position and not on time. Thus $\partial T/\partial t$ is zero throughout the sheet and equation (5) becomes

Laplace's Equation

$$\boxed{\frac{\partial^2 T}{\partial x^2} + \frac{\partial^2 T}{\partial y^2} = 0.} \tag{6}$$

10.9/TWO-DIMENSIONAL HEAT FLOW AND LAPLACE'S EQUATION

This elliptic partial differential equation, commonly referred to as **Laplace's equation**, arises in many other problems of applied mathematics. We now give an example of the solution of Laplace's equation by the method of separation of variables.

Example 1   We begin by studying the steady-state heat flow problem in the rectangular sheet of size $L \times M$ in which Laplace's equation arose. Suppose that the upper horizontal edge is kept at 100°C, while the other three edges are kept at 0°C. We may write these boundary conditions for $T(x, y)$ as follows:

$$T(0, y) = T(L, y) = 0, \quad 0 < y < M, \tag{7}$$

$$T(x, 0) = 0, \quad T(x, M) = 100, \quad 0 < x < L. \tag{8}$$

Since we are seeking the steady-state solution, time is not a factor in this problem, and therefore no initial condition is needed.

Setting $T(x, y) = X(x)Y(y)$, we find that (6) becomes

$$X''Y + XY'' = 0,$$

which may be written in separated form as

$$\frac{X''}{X} = -\frac{Y''}{Y}. \tag{9}$$

The left-hand side of (9) is constant for fixed x and arbitrary y. Hence Y''/Y must be constant. Letting k be this constant, we obtain the pair of ordinary differential equations

$$\begin{aligned} X'' - kX &= 0, \\ Y'' + kY &= 0. \end{aligned} \tag{10}$$

From (7) we have

$$X(0)Y(y) = X(L)Y(y) = 0,$$

and $Y(y) \not\equiv 0$ [since otherwise $T(x, y) \equiv 0$, contradicting the second equation of (8)], so that $X(0) = X(L) = 0$. Thus again we have the situation encountered in Section 10.8, and the eigenvalues are all negative. Setting $k = -r^2$, we find that rL must be a multiple of π in order that we have a nonzero solution of the boundary value problem for X. Hence $r = n\pi/L$, and the functions

$$X_n(x) = A \sin \frac{n\pi x}{L}, \quad n = 1, 2, 3, \ldots, \tag{11}$$

are the eigenfunctions of the problem. Setting $k = -(n\pi/L)^2$ in the second equation of (10) yields the general solution

$$Y_n(y) = B_1 e^{n\pi y/L} + B_2 e^{-n\pi y/L}.$$

The first condition of (8) implies that $Y(0) = 0$, thus $B_1 = -B_2$, and we can rewrite Y_n as

$$Y_n(y) = B \sinh \frac{n\pi y}{L}. \tag{12}$$

As in Section 10.8, to enlarge the class of possible solutions, we consider an infinite sum of products of the terms (11) and (12):

$$T(x, y) = \sum_{n=1}^{\infty} c_n \sin \frac{n\pi x}{L} \sinh \frac{n\pi y}{L}. \tag{13}$$

Evaluating equation (13) at any point (x, M), we obtain, by (8),

$$100 = \sum_{n=1}^{\infty} \left(c_n \sinh \frac{n\pi M}{L} \right) \sin \frac{n\pi x}{L}, \qquad 0 < x < L,$$

which again is a Fourier sine series, requiring that we extend the boundary condition $T(x, M)$ to the interval $-L \leq x \leq L$ as an odd piecewise continuous periodic function of period $2L$ with a piecewise continuous derivative. Clearly the function

$$f(x) = \begin{cases} 100, & 0 < x < L, \\ -100, & -L < x < 0, \end{cases}$$

with $f(x + 2L) = f(x)$ satisfies these conditions. Hence $c_n \sinh(n\pi M/L)$ must equal the nth Fourier (sine) coefficient of $f(x)$:

$$c_n \sinh \frac{n\pi M}{L} = \frac{1}{L} \int_{-L}^{L} f(x) \sin \frac{n\pi x}{L} \, dx = \frac{200}{L} \int_{0}^{L} \sin \frac{n\pi x}{L} \, dx$$

$$= \frac{200}{n\pi} (1 - \cos n\pi) = \frac{200}{n\pi} \left[1 - (-1)^n \right].$$

Hence $c_{2k} = 0$, while

$$c_{2k+1} = \frac{400}{\pi(2k+1)\sinh\left[(2k+1)\pi M/L\right]},$$

and the solution is given by

$$T(x, y) = \frac{400}{\pi} \left[\frac{\sin(\pi x/L)\sinh(\pi y/L)}{\sinh(\pi M/L)} + \frac{\sin(3\pi x/L)\sinh(3\pi y/L)}{3\sinh(3\pi M/L)} + \cdots \right]$$

$$= \frac{400}{\pi} \sum_{n=1}^{\infty} \frac{\sin\left[(2n-1)\pi x/L\right] \sinh\left[(2n-1)\pi y/L\right]}{(2n-1)\sinh\left[(2n-1)\pi M/L\right]}.$$

We may use this formula to compute the steady-state temperature at any point in the rectangle. For example,

$$T\left(\frac{L}{2}, \frac{M}{2}\right) = \frac{400}{\pi} \sum_{n=1}^{\infty} \frac{\sin\left[(2n-1)\pi/2\right] \sinh\left[(2n-1)\pi M/2L\right]}{(2n-1)\sinh\left[(2n-1)\pi M/L\right]}.$$

Since

$$\frac{\sinh(A/2)}{\sinh A} = \frac{e^{A/2} - e^{-A/2}}{e^A - e^{-A}} = \frac{1}{e^{A/2} + e^{-A/2}} = \frac{1}{2\cosh(A/2)},$$

we obtain

$$T\left(\frac{L}{2}, \frac{M}{2}\right) = \frac{200}{\pi} \left[\frac{1}{\cosh(\pi M/2L)} - \frac{1}{3\cosh(3\pi M/2L)} + \frac{1}{5\cosh(5\pi M/2L)} - \cdots \right],$$

which, for given L and M, can be approximated to any desired degree of accuracy.

PROBLEMS 10.9

1. A square plate with sides of length L has both faces insulated. The upper horizontal edge is kept at $50°C$, while all the other edges are at $0°C$. Find the steady-state temperatures at the points $(L/2, L/2)$ and $(L/4, L/4)$.

2. If in Problem 1 the temperatures along the edge $y = L$ are given by

$$T(x, L) = x(L - x),$$

and all the other conditions remain the same, what is $T(x, y)$?

3. If in Problem 1 the temperatures along the upper horizontal edge are

$$T(x, L) = 100 \sin(\pi x/L),$$

with all the other conditions remaining unchanged, what is $T(x, y)$?

4. Suppose that the temperature along the edge $y = L$ in Problem 2 is changed to

$$T(x, L) = x(L^2 - x^2),$$

with all the other conditions remaining the same. Find the temperature at the points $(L/4, L/4)$ and $(L/4, 3L/4)$.

Assume in Problems 5–8 that all other previous conditions are unchanged.

5. Repeat Problem 4 with $T(x, L) = x^2(L - x)$.

6. Repeat Problem 2 with $T(x, L) = x \cos \dfrac{\pi x}{2L}$.

7. Repeat Problem 2 with $T(x, L) = 10x \sin \pi x/L$.

8. Repeat Problem 4 with $T(x, L) = 100 \sin^2(\pi x/L)$.

9. A rectangular plate of length $2L$ and height L has both faces insulated. The upper edge is kept at $50°C$ and the lower edge at $0°C$, while the vertical edges have temperature

$$T(-L, y) = T(L, y) = 50 \sin \dfrac{\pi y}{2L}, \quad 0 \le y \le L.$$

Find the temperature at any point in the plate.

10. Repeat Problem 9 with

$$T(-L, y) = T(L, y) = 50y/L, \quad 0 \le y \le L,$$

all other conditions remaining unchanged.

11. Repeat Problem 10 with the upper edge now changed to

$$T(x, L) = 50(x/L)^2,$$

all other conditions remaining unchanged.

*12. Extend the development in this section to a solid cube of side L.

   a. What is the three-dimensional heat equation for such a situation?

   b. What is the steady-state equation for part (a)?

   c. Assume that all faces, except $z = L$, of the cube are kept at $0°C$, while the $z = L$ face is kept at $100°C$. What is the steady-state solution in the form of a Fourier series?

10.10 Laplace's Equation in Polar and Spherical Coordinates

We now consider the steady-state heat flow problem in the disk D of radius 1 centered at the origin. It should be clear from the nature of the situation that the problem will be simplified if we can use polar coordinates (r, θ) instead of rectangular coordinates (x, y). Assume that the boundary of the disk is kept at the temperature

$$T(1, \theta) = f(\theta), \quad 0 \le \theta \le 2\pi, \tag{1}$$

where the function $f(\theta)$ is periodic with period 2π and has a continuous derivative (see Figure 10.15). Now we need only write Laplace's equation, $\partial^2 T/\partial x^2 + \partial^2 T/\partial y^2 = 0$, in polar coordinates. Recall the transformation formulas

$$r = \sqrt{x^2 + y^2}, \quad x = r \cos \theta,$$
$$\theta = \tan^{-1} \dfrac{y}{x}, \quad y = r \sin \theta.$$

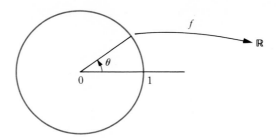

Figure 10.15

By the chain rule, we have

$$\frac{\partial T}{\partial x} = \frac{\partial T}{\partial r}\frac{\partial r}{\partial x} + \frac{\partial T}{\partial \theta}\frac{\partial \theta}{\partial x}$$

$$= \frac{\partial T}{\partial r}\cos\theta - \frac{\partial T}{\partial \theta}\frac{\sin\theta}{r},$$

since

$$\frac{\partial r}{\partial x} = \frac{\partial}{\partial x}\sqrt{x^2+y^2} = \frac{x}{\sqrt{x^2+y^2}} = \frac{x}{r} = \cos\theta,$$

and

$$\frac{\partial \theta}{\partial x} = \frac{-y/x^2}{1+(y/x)^2} = \frac{-y}{r^2} = \frac{-\sin\theta}{r}.$$

Then

$$\frac{\partial^2 T}{\partial x^2} = \frac{\partial}{\partial r}\left(\frac{\partial T}{\partial r}\frac{\partial r}{\partial x} + \frac{\partial T}{\partial \theta}\frac{\partial \theta}{\partial x}\right)\frac{\partial r}{\partial x} + \frac{\partial}{\partial \theta}\left(\frac{\partial T}{\partial r}\frac{\partial r}{\partial x} + \frac{\partial T}{\partial \theta}\frac{\partial \theta}{\partial x}\right)\frac{\partial \theta}{\partial x}$$

$$= \frac{\partial^2 T}{\partial r^2}\cos^2\theta - \frac{\partial^2 T}{\partial r\partial\theta}\frac{\sin\theta\cos\theta}{r} + \frac{\partial T}{\partial \theta}\frac{\sin\theta\cos\theta}{r^2}$$

$$- \frac{\partial^2 T}{\partial \theta\partial r}\frac{\sin\theta\cos\theta}{r} + \frac{\partial T}{\partial r}\frac{\sin^2\theta}{r} + \frac{\partial^2 T}{\partial \theta^2}\frac{\sin^2\theta}{r^2} + \frac{\partial T}{\partial \theta}\frac{\cos\theta\sin\theta}{r^2}.$$

Similarly,

$$\frac{\partial^2 T}{\partial y^2} = \frac{\partial}{\partial r}\left(\frac{\partial T}{\partial r}\sin\theta + \frac{\partial T}{\partial \theta}\frac{\cos\theta}{r}\right)\frac{\partial r}{\partial y} + \frac{\partial}{\partial \theta}\left(\frac{\partial T}{\partial r}\sin\theta + \frac{\partial T}{\partial \theta}\frac{\cos\theta}{r}\right)\frac{\partial \theta}{\partial y}$$

$$= \frac{\partial^2 T}{\partial r^2}\sin^2\theta + \frac{\partial^2 T}{\partial r\partial\theta}\frac{\sin\theta\cos\theta}{r} - \frac{\partial T}{\partial \theta}\frac{\cos\theta\sin\theta}{r^2}$$

$$+ \frac{\partial^2 T}{\partial \theta\partial r}\frac{\sin\theta\cos\theta}{r} + \frac{\partial T}{\partial r}\frac{\cos^2\theta}{r} + \frac{\partial^2 T}{\partial \theta^2}\frac{\cos^2\theta}{r^2} - \frac{\partial T}{\partial \theta}\frac{\sin\theta\cos\theta}{r^2}.$$

Laplace's Equation in Polar Coordinates

Adding the two equations, we obtain

$$\frac{\partial^2 T}{\partial x^2} + \frac{\partial^2 T}{\partial y^2} = \frac{\partial^2 T}{\partial r^2} + \frac{1}{r^2}\frac{\partial^2 T}{\partial \theta^2} + \frac{1}{r}\frac{\partial T}{\partial r} = 0, \qquad (2)$$

which is Laplace's equation in polar coordinates.

Now we let $T(r,\theta) = R(r)\Theta(\theta)$ and transform (2) into

$$R''\Theta + \frac{1}{r^2}R\Theta'' + \frac{1}{r}R'\Theta = 0,$$

or

$$\left(R'' + \frac{1}{r}R'\right)\Theta = -\frac{R}{r^2}\Theta''.$$

Separating the variables, we have

$$\frac{r^2 R'' + r R'}{R} = -\frac{\Theta''}{\Theta} = k,$$

since each side of the first equality is a function of only one variable. Hence we obtain two ordinary differential equations:

$$\Theta'' + k\Theta = 0, \qquad (3)$$

and

$$r^2 R'' + r R' - kR = 0. \qquad (4)$$

Since $f(\theta) = T(1,\theta) = R(1)\Theta(\theta)$, it is clear that $\Theta(\theta)$ is periodic with period 2π. We assume that f is not identically zero [see Problem 12 for the case where $f(\theta) \equiv 0$], so $R(1) \neq 0$. Observe that $\Theta(0) = \Theta(2\pi)$ and $\Theta'(0) = \Theta'(2\pi)$ so that the boundary value problem in Θ has periodic eigenfunctions of period 2π. Thus the eigenvalues must be $k = n^2$, with the corresponding eigenfunctions

$$\Theta_n(\theta) = A_1 \cos n\theta + A_2 \sin n\theta, \qquad n = 0, 1, 2, 3, \ldots .$$

Equation (4) is an Euler equation (see Section 3.7) and can be solved by letting $R = r^\lambda$. Letting $k = n^2$, we see that (4) becomes

$$r^\lambda [\lambda(\lambda - 1) + \lambda - n^2] = r^\lambda(\lambda^2 - n^2) = 0,$$

and the roots of the equation $\lambda^2 - n^2 = 0$ are $\lambda = \pm n$. Hence the general solution of (4) with $k = n^2$ is

$$R_n(r) = B_1 r^n + B_2 r^{-n}, \qquad n = 0, 1, 2, 3, \ldots .$$

Since we want R_n to exist for all values in the range $0 \leq r \leq 1$, B_2 must vanish and $R_n(r) = B r^n$, $n = 0, 1, 2, 3, \ldots$. Thus we seek a solution in the form of the infinite series

$$T(r,\theta) = \frac{A_0}{2} + \sum_{n=1}^{\infty} r^n (A_n \cos n\theta + B_n \sin n\theta). \qquad (5)$$

To evaluate the constants A_n, B_n, we note that

$$f(\theta) = T(1, \theta) = \frac{A_0}{2} + \sum_{n=1}^{\infty} A_n \cos n\theta + B_n \sin n\theta. \tag{6}$$

Hence (6) is the Fourier series of f, and the coefficients A_n and B_n are the Fourier coefficients of f given by the Euler formulas

$$A_n = \frac{1}{\pi} \int_{-\pi}^{\pi} f(\phi) \cos n\phi \, d\phi, \qquad B_n = \frac{1}{\pi} \int_{-\pi}^{\pi} f(\phi) \sin n\phi \, d\phi.$$

Since (6) converges and $r \leq 1$, series (5) also converges. For example, if $f(\theta) = \cos \theta$, all the coefficients except A_1 vanish, and $A_1 = 1$. Hence the temperature is given by

$$T(r, \theta) = r \cos \theta$$

at all points in the unit disk.

It is interesting to note what occurs if we replace the coefficients A_n and B_n in (5) by the Euler formulas and interchange the sums and integrals:

$$T(r, \theta) = \frac{1}{2\pi} \int_{-\pi}^{\pi} f(\phi) \, d\phi$$

$$+ \sum_{n=1}^{\infty} r^n \left[\frac{\cos n\theta}{\pi} \int_{-\pi}^{\pi} f(\phi) \cos n\phi \, d\phi + \frac{\sin n\theta}{\pi} \int_{-\pi}^{\pi} f(\phi) \sin n\phi \, d\phi \right]$$

$$= \frac{1}{\pi} \int_{-\pi}^{\pi} f(\phi) \left[\frac{1}{2} + \sum_{n=1}^{\infty} r^n (\cos n\theta \cos n\phi + \sin n\theta \sin n\phi) \right] d\phi.$$

Since

$$\cos n(\theta - \phi) = \cos n\theta \cos n\phi + \sin n\theta \sin n\phi,$$

we obtain

$$T(r, \theta) = \frac{1}{\pi} \int_{-\pi}^{\pi} f(\phi) \left[\frac{1}{2} + \sum_{n=1}^{\infty} r^n \cos n(\theta - \phi) \right] d\phi.$$

But

$$\cos n(\theta - \phi) = \frac{e^{in(\theta - \phi)} + e^{-in(\theta - \phi)}}{2},$$

so

$$\sum_{n=1}^{\infty} r^n \cos n(\theta - \phi) = \frac{1}{2} \sum_{n=1}^{\infty} r^n \left[e^{in(\theta - \phi)} + e^{-in(\theta - \phi)} \right]$$

$$= \frac{1}{2} \left\{ \sum_{n=1}^{\infty} [re^{i(\theta - \phi)}]^n + \sum_{n=1}^{\infty} [re^{-i(\theta - \phi)}]^n \right\}.$$

The complex number $e^{it} = \cos t + i \sin t$ can be represented as a vector of length $\sqrt{(\cos t)^2 + (\sin t)^2} = 1$ (see Appendix 5), so $re^{i(\theta - \phi)}$ and $re^{i(\phi - \theta)}$ both have length equal to $r(<1)$. Using the geometric series, we have

$$\sum_{n=1}^{\infty} r^n \cos n(\theta - \phi) = \frac{1}{2}\left[\frac{re^{i(\theta-\phi)}}{1-re^{i(\theta-\phi)}} + \frac{re^{-i(\theta-\phi)}}{1-re^{-i(\theta-\phi)}}\right]$$

$$= \frac{1}{2}\left\{\frac{r[e^{i(\theta-\phi)} + e^{-i(\theta-\phi)}] - 2r^2}{r^2 - r[e^{i(\theta-\phi)} + e^{-i(\theta-\phi)}] + 1}\right\}$$

$$= \frac{r\cos(\theta-\phi) - r^2}{r^2 - 2r\cos(\theta-\phi) + 1}.$$

Thus

$$\frac{1}{2} + \sum_{n=1}^{\infty} r^n \cos n(\theta-\phi) = \frac{1}{2} + \frac{r\cos(\theta-\phi) - r^2}{r^2 + 1 - 2r\cos(\theta-\phi)}$$

$$= \frac{1 - r^2}{2[r^2 + 1 - 2r\cos(\theta-\phi)]}$$

and

Poisson Integral Formula

$$\boxed{T(r,\theta) = \frac{1-r^2}{2\pi}\int_{-\pi}^{\pi} \frac{f(\phi)}{r^2 + 1 - 2r\cos(\theta-\phi)}\, d\phi.} \qquad (7)$$

Equation (7), the **Poisson integral formula**[†], is valid for all values $r < 1$ (see Problem 9). It indicates that the temperature at any interior point (r, θ) of the unit disk may be obtained by integrating the boundary temperatures according to formula (7). In particular, if $r = 0$, then the temperature at the center of the disk is

$$T(0,\theta) = \frac{1}{2\pi}\int_{-\pi}^{\pi} f(\phi)\, d\phi;$$

that is, the temperature at the center is the integral average of the boundary temperatures. This fact, often called the **mean value theorem**, holds for all functions that satisfy Laplace's equation on the unit disk $r \leq 1$.

Example 1 In this example we consider the three-dimensional Laplace equation

$$\boxed{\frac{\partial^2 T}{\partial x^2} + \frac{\partial^2 T}{\partial y^2} + \frac{\partial^2 T}{\partial z^2} = 0} \qquad (8)$$

associated with the steady-state heat flow problem in a solid.

Suppose that the solid is a rectangular cylinder of height π and radius 1 (Figure 10.16). Since such a solid is easily described in cylindrical coordinates,

---

[†] Named in honor of Siméon Denis Poisson (1781–1840), a french mathematician and physicist.

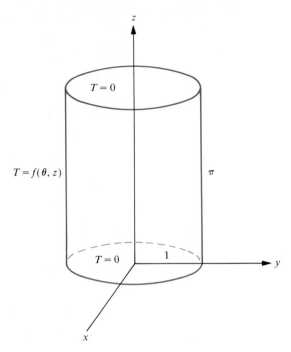

Figure 10.16

it is convenient to obtain Laplace's equation (8) in terms of the coordinates (r, θ, z), where

$$x = r\cos\theta, \qquad y = r\sin\theta, \qquad z = z.$$

Obviously we are simply writing the xy-coordinates in polar form. Thus the calculations we did earlier still apply, and Laplace's equation becomes [from equation (2)]

$$\boxed{\frac{\partial^2 T}{\partial r^2} + \frac{1}{r}\frac{\partial T}{\partial r} + \frac{1}{r^2}\frac{\partial^2 T}{\partial \theta^2} + \frac{\partial^2 T}{\partial z^2} = 0.} \tag{9}$$

We now apply the method of separation of variables to the function T and, accordingly, set $T(r, \theta, z) = R(r)\Theta(\theta)Z(z)$. Equation (9) becomes

$$R''\Theta Z + \frac{1}{r}R'\Theta Z + \frac{1}{r^2}R\Theta''Z + R\Theta Z'' = 0,$$

or

$$\left[(R'' + r^{-1}R')\Theta + r^{-2}R\Theta''\right]Z = -R\Theta Z''.$$

Separating out the z-variables, we have

$$\frac{R'' + r^{-1}R'}{R} + r^{-2}\frac{\Theta''}{\Theta} = -\frac{Z''}{Z} = k_1,$$

since the left-hand side of the equation is a function of r and θ, whereas the

right-hand side is a function of z alone. Thus
$$Z'' + k_1 Z = 0,$$
and
$$\frac{R'' + r^{-1}R'}{R} = k_1 - r^{-2}\frac{\Theta''}{\Theta}.$$
Multiplying both sides of the last equation by r^2, we have
$$\frac{r^2 R'' + rR'}{R} = k_1 r^2 - \frac{\Theta''}{\Theta},$$
or
$$\frac{r^2 R'' + rR'}{R} - k_1 r^2 = -\frac{\Theta''}{\Theta} = k_2,$$
since each side is a function of only one variable. Hence we also obtain the ordinary differential equations
$$\Theta'' + k_2 \Theta = 0$$
and
$$r^2 R'' + rR' - (k_2 + k_1 r^2) R = 0. \tag{10}$$

If we now assume the boundary conditions
$$T(r, \theta, 0) = T(r, \theta, \pi) = 0, \qquad T(1, \theta, z) = f(\theta, z),$$
where f is a continuously differentiable function in θ and z, we can translate the conditions into terms of R, Θ, and Z:
$$R(r)\Theta(\theta)Z(0) = R(r)\Theta(\theta)Z(\pi) = 0,$$
$$R(1)\Theta(\theta)Z(z) = f(\theta, z).$$

Since R and Θ are not constantly zero [assuming $f(\theta, z) \neq 0$], it follows that $Z(0) = Z(\pi) = 0$. Also, since f is continuous, it must be periodic with period 2π in θ (since f is the temperature on the surface of the cylinder). Thus $\Theta(\theta)$ is periodic with period 2π, so $\Theta(0) = \Theta(2\pi)$ and $\Theta'(0) = \Theta'(2\pi)$. Finally, we know that $R(1) \neq 0$ and that $R(0)$ is finite. As in (3), the eigenvalues of the boundary value problem
$$\Theta'' + k_2 \Theta = 0, \qquad \Theta(0) = \Theta(2\pi), \qquad \Theta'(0) = \Theta'(2\pi),$$
are $k_2 = m^2$, $m = 0, 1, 2, \ldots$, with the corresponding eigenfunctions
$$\Theta_m(\theta) = A_1 \cos m\theta + A_2 \sin m\theta.$$
Similarly, the boundary value problem
$$Z'' + k_1 Z = 0, \qquad Z(0) = Z(\pi) = 0,$$
has the eigenvalues $k_1 = n^2$, $n = 1, 2, 3, \ldots$, and the corresponding eigenfunctions
$$Z_n(z) = B \sin nz.$$
Setting $k_1 = n^2$ and $k_2 = m^2$ in (10), we obtain
$$r^2 R'' + rR' - (n^2 r^2 + m^2) R = 0, \tag{11}$$
which may be transformed into Bessel's equation by the substitution $\rho = inr$.

To verify this fact, we note that
$$\frac{dR}{d\rho} = \frac{dR}{dr}\frac{dr}{d\rho} = \frac{1}{in}R'(r),$$

and
$$\frac{d^2R}{d\rho^2} = \frac{d}{d\rho}\left(\frac{1}{in}R'\right) = \frac{1}{(in)^2}R''(r) = -\frac{1}{n^2}R''(r).$$

Substituting these values into equation (11), we obtain
$$-n^2 r^2 \frac{d^2R}{d\rho^2} + inr\frac{dR}{d\rho} - (n^2 r^2 + m^2)R = 0,$$

or
$$\rho^2 \frac{d^2R}{d\rho^2} + \rho\frac{dR}{d\rho} + (\rho^2 - m^2)R = 0. \tag{12}$$

By Theorem 5.4.2 (p. 190), equation (12) has the general solution
$$R(\rho) = C_1 J_m(\rho) + C_2 Y_m(\rho), \qquad \rho \neq 0.$$

And since $R(0)$ is finite, the constant $C_2 = 0$, since the Bessel function of the second kind, Y_m, involves a $\ln \rho$ term that approaches $-\infty$ as ρ approaches zero. We are, therefore, led to the double series
$$T(r, \theta, z) = \sum_{m=0}^{\infty} \sum_{n=1}^{\infty} (a_{mn} \cos m\theta + b_{mn} \sin m\theta) \sin nz\, J_m(inr).$$

The coefficients a_{mn} and b_{mn} can be determined by setting $r = 1$ and requiring that
$$f(\theta, z) = \sum_{m=0}^{\infty}\left[\left(\sum_{n=1}^{\infty} a_{mn} J_m(in) \sin nz\right)\cos m\theta \right.$$
$$\left. + \left(\sum_{n=1}^{\infty} b_{mn} J_m(in) \sin nz\right)\sin m\theta\right]. \tag{13}$$

If we define
$$a_m(z) = \sum_{n=1}^{\infty} a_{mn} J_m(in)\sin nz$$

and
$$b_m(z) = \sum_{n=1}^{\infty} b_{mn} J_m(in)\sin nz, \tag{14}$$

we see that (13) becomes
$$f(\theta, z) = \sum_{m=0}^{\infty} a_m(z)\cos m\theta + b_m(z)\sin m\theta,$$

which is a Fourier series in θ, where z is an arbitrary parameter. By the Euler formulas, it follows that
$$a_m(z) = \frac{1}{\pi}\int_{-\pi}^{\pi} f(\theta, z)\cos m\theta\, d\theta, \qquad b_m(z) = \frac{1}{\pi}\int_{-\pi}^{\pi} f(\theta, z)\sin m\theta\, d\theta.$$

If the functions $a_m(z)$ and $b_m(z)$ can be calculated, we can then use the Fourier sine series (14) in z to calculate the coefficients a_{mn} and b_{mn}:

$$a_{mn} = \frac{2}{\pi J_m(in)} \int_0^\pi a_m(z) \sin nz \, dz,$$

$$b_{mn} = \frac{2}{\pi J_m(in)} \int_0^\pi b_m(z) \sin nz \, dz.$$

Although the formal procedure delineated above is straightforward, the calculations involved are, to say the least, very tedious. In Problem 8 the reader is asked to develop a similar procedure that leads to the use of series of Legendre polynomials in place of Bessel functions.

PROBLEMS 10.10

1. Consider the steady-state heat flow problem on the unit disk D [see equation (2)] and suppose that the boundary condition (1) is given by $f(\theta) = |\theta - \pi|$. Find the Fourier series solution (5) of this problem.

2. Repeat Problem 1 with $f(\theta) = \cos 2\theta$.

3. Find the Poisson integral solution (7) of Problem 1. What is the temperature at the origin? What is the temperature at the point $(r, \theta) = (\frac{1}{2}, 0)$?

4. Find the Poisson integral solution for Problem 2. What is the temperature at the origin?

5. Consider the steady-state heat flow problem on the unit disk and suppose that $f(\theta) = \sin^2\theta$. Find the temperature at the point $r = \frac{1}{2}, \theta = 0°$ by
   a. the eigenfunction expansion method;
   b. the Poisson integral formula.

6. Repeat Problem 5 with $f(\theta) = \sin\theta - \cos\theta$.

7. Repeat Problem 5 with $f(\theta) = \theta(2\pi - \theta)$.

*8. a. Show that the three-dimensional Laplace equation (8) may be written in spherical coordinates r, θ, ϕ ($x = r\cos\theta\sin\phi$, $y = r\sin\theta\sin\phi$, $z = r\cos\phi$) in the form

$$\frac{\partial^2 T}{\partial r^2} + \frac{2}{r}\frac{\partial T}{\partial r} + \frac{1}{r^2}\frac{\partial^2 T}{\partial \phi^2} + \frac{\cot\phi}{r^2}\frac{\partial T}{\partial \phi}$$
$$+ \frac{1}{r^2 \sin^2\phi}\frac{\partial^2 T}{\partial \theta^2} = 0. \quad \text{(i)}$$

b. Separate the variables in equation (i) to obtain the ordinary differential equations

$$\Theta'' + k_1\Theta = 0,$$

$$R'' + \left(\frac{2}{r}\right)R' + \left(\frac{k_2}{r^2}\right)R = 0,$$

$$[(\sin\phi)\Phi']' - \left(\frac{k_2\sin\phi + k_1}{\sin\phi}\right)\Phi = 0.$$

c. Assume the boundary condition $T(1, \theta, \phi) = f(\theta, \phi)$ for a sphere of radius 1, where f is continuously differentiable in θ and ϕ and periodic with period 2π in θ. Find the eigenfunctions of the three boundary value problems thus defined. [*Hint*: Use Legendre polynomials.]

d. Obtain the eigenfunction expansion for $T(r, \theta, \phi)$.

*9. Justify the steps involved in obtaining the Poisson integral formula (7) from the solution (5) of the steady-state heat equation in the disk $r \leq 1$. [*Hint*: Show that series (5) converges uniformly in the disk $r \leq r_0 < 1$.]

*10. Generalize the problem in Figure 10.15 to obtain the steady-state temperature in a disk of radius R centered at the origin, and prove that the corresponding Poisson integral formula is

$$T(r, \theta) = \frac{R^2 - r^2}{2\pi} \int_{-\pi}^{\pi} \frac{f(\phi)\, d\phi}{R^2 + r^2 - 2rR\cos(\theta - \phi)},$$

$r < R$,

where the continuously differentiable function $f(\phi)$ describes the boundary temperature in the direction $0 \leq \phi < 2\pi$.

*11. Let D be any bounded polygon in the xy-plane, and let (x_0, y_0) be a point interior to D. Let u be a real-valued function defined on D that has continuous derivatives of at least order two.

a. Show that if u attains its maximal value M at (x_0, y_0), then

$$\left.\frac{\partial^2 u}{\partial x^2} + \frac{\partial^2 u}{\partial y^2}\right|_{(x_0, y_0)} \leq 0.$$

b. Show that if

$$\frac{\partial^2 u}{\partial x^2} + \frac{\partial^2 u}{\partial y^2} > 0$$

at all interior points of D, then u does not attain its maximal value at any interior point of D.

c. Suppose that

$$\frac{\partial^2 u}{\partial x^2} + \frac{\partial^2 u}{\partial y^2} \geq 0 \qquad \text{(ii)}$$

at all interior points of D. Show that the function

$$w(x, y) = u(x, y) + \epsilon\left[(x - x_0)^2 + (y - y_0)^2\right]$$

does not attain its maximal value in D at (x_0, y_0), for any value $\epsilon > 0$.

d. Conclude that any function satisfying equation (ii) at all interior points of D must attain its maximal value on the boundary of D. (This is called the **maximum principle**.)

*12. Use the maximum principle obtained in Problem 11 to show that any function u that satisfies Laplace's equation attains both its maximum and its minimum on the boundary of D. (Physically, this means that, in a steady-state heat flow problem, the maximum and minimum temperatures occur on the boundary of the region.)

Review Exercises for Chapter 10

In Exercises 1–4 show that the given set of functions is orthogonal on the given interval and determine the corresponding orthonormal set.

1. $\{\sin nx\}$, $n = 1, 2, 3, \ldots,$ $\quad 0 \leq x \leq \pi$
2. $\{\cos n\pi x\}$, $n = 0, 1, 2, \ldots,$ $\quad 0 \leq x \leq 2$
3. $\{\sin 2n\pi x, \cos 2n\pi x\}$, $n = 1, 2, \ldots,$ $\quad 0 \leq x \leq 1$
4. $\{\sin n\pi x\}$, $n = 1, 2, 3, \ldots,$ $\quad 0 \leq x \leq 1$

Expand each of the functions in Exercises 5–8 in a Fourier series. Examine each series at all the points of discontinuity.

5. $f(x) = x \sin x$, $\quad 0 < x < 2\pi$
6. $f(x) = \sqrt{1 - \cos x}$, $\quad |x| < \pi$
7. $f(x) = |\sin x|$, $\quad |x| < \pi$
8. $f(x) = \begin{cases} x, & 0 < x < \pi \\ 0, & \pi < x < 2\pi \end{cases}$

Find Fourier sine and cosine series for the functions in Exercises 9–12 by extending the given function on the interval $[0, a]$ to an odd or even function on $[-a, a]$.

9. $f(x) = x^2$, $\quad 0 < x < \pi$
10. $f(x) = e^x$, $\quad 0 < x < \pi$
11. $f(x) = x$, $\quad 0 < x < 1$
12. $f(x) = 2 - x$, $\quad 0 < x < 2$
13. Find a Fourier sine series for $f(x) = x^2 - 2$ in $0 < x < 2$ and use this series to obtain a series for π^3.

In Exercises 14–16 find the temperature in an insulated rod a units long whose ends are kept at 0°C, where the initial temperature is given by

14. $f(x) = x^2(a^2 - x^2)$

15. $f(x) = \begin{cases} x, & 0 \leq x \leq a/2 \\ a - x, & a/2 \leq x \leq a \end{cases}$

16. $f(x) = \begin{cases} x^2, & 0 \leq x \leq a/2 \\ 0, & a/2 < x \leq a \end{cases}$

Find solutions of the equations in Exercises 17–20 by separation of variables.

17. $\dfrac{\partial^2 y}{\partial x^2} = \dfrac{t^2}{x^2} \dfrac{\partial^2 y}{\partial t^2}$

18. $\dfrac{\partial^2 y}{\partial x^2} + \dfrac{\partial^2 y}{\partial x \partial t} = \dfrac{\partial y}{\partial t}$

19. $\dfrac{\partial^2 y}{\partial t^2} = \dfrac{\partial^2 y}{\partial x^2} - y$

20. $\dfrac{\partial^2 y}{\partial x^2} + \dfrac{\partial y}{\partial x} = \dfrac{\partial y}{\partial t}$

21. A tightly stretched string fixed at $x = 0$ and $x = a$ is initially displaced to the position $y(x, 0) = 2\sin^3(\pi x/a)$. At time $t = 0$, it is released from rest from this position. Find the displacement of any point on the string at any time t.

22. Repeat Exercise 21 if the initial displacement is $y(x, 0) = a^2 x - x^3$.

23. The vibrating string in Exercise 21 is subjected to a damping force that is proportional to the velocity at each point and instant in time.

    a. Find the differential equation that the damped string satisfies.

    b. Solve the equation by separation of variables.

    *c. If the initial displacement is $y(x, 0) = f(x)$, $0 \leq x \leq a$, express the solution as an infinite series.

24. A rectangular plate is bounded by the lines $x = 0$, $y = 0$, $x = 100$, $y = 50$. Its surfaces are insulated and the temperature along the upper edge is given by $T(x, 50) = x(100 - x)$, while the other edges are kept at 0°C. Find the steady-state temperature.

APPENDIX 1

Integral Tables†

STANDARD FORMS

1. $\int a\,dx = ax + C$

2. $\int af(x)\,dx = a\int f(x)\,dx + C$

3. $\int u\,dv = uv - \int v\,du$ (integration by parts)

4. $\int u^n\,du = \dfrac{u^{n+1}}{n+1} + C, \quad n \neq -1$

5. $\int \dfrac{du}{u} = \ln u \quad \text{if } u > 0 \quad \text{or} \quad \ln(-u) \quad \text{if } u < 0$
$= \ln|u| + C$

6. $\int e^u\,du = e^u + C$

7. $\int a^u\,du = \int e^{u \ln a}\,du$
$= \dfrac{e^{u \ln a}}{\ln a} = \dfrac{a^u}{\ln a} + C, \quad a > 0, \, a \neq 1$

8. $\int \sin u\,du = -\cos u + C$

9. $\int \cos u\,du = \sin u + C$

10. $\int \tan u\,du = \ln|\sec u| = -\ln|\cos u| + C$

11. $\int \cot u\,du = \ln|\sin u| + C$

12. $\int \sec u\,du = \ln|\sec u + \tan u|$
$= \ln\left|\tan\dfrac{u}{2} + \dfrac{\pi}{4}\right| + C$

13. $\int \csc u\,du = \ln|\csc u - \cot u| = \ln\left|\tan\dfrac{u}{2}\right| + C$

14. $\int \sec^2 u\,du = \tan u + C$

15. $\int \csc^2 u\,du = -\cot u + C$

16. $\int \sec u \tan u\,du = \sec u + C$

17. $\int \csc u \cot u\,du = -\csc u + C$

18. $\int \dfrac{du}{u^2 + a^2} = \dfrac{1}{a}\tan^{-1}\dfrac{u}{a} + C$

---

† All angles are measured in radians.

APPENDIX 1/INTEGRAL TABLES

19. $\displaystyle\int \frac{du}{u^2 - a^2} = \frac{1}{2a} \ln\left|\frac{u-a}{u+a}\right| + C$

$\displaystyle\qquad = -\frac{1}{a} \coth^{-1}\frac{u}{a} + C, \quad u^2 > a^2$

20. $\displaystyle\int \frac{du}{a^2 - u^2} = \frac{1}{2a} \ln\left|\frac{a+u}{a-u}\right| + C$

$\displaystyle\qquad = \frac{1}{a} \tanh^{-1}\frac{u}{a} + C, \quad u^2 < a^2$

21. $\displaystyle\int \frac{du}{\sqrt{a^2 - u^2}} = \sin^{-1}\frac{u}{|a|} + C$

22. $\displaystyle\int \frac{du}{\sqrt{u^2 + a^2}} = \ln\left(u + \sqrt{u^2 + a^2}\right) + C$

23. $\displaystyle\int \frac{du}{\sqrt{u^2 - a^2}} = \ln\left|u + \sqrt{u^2 - a^2}\right| + C$

24. $\displaystyle\int \frac{du}{u\sqrt{u^2 - a^2}} = \frac{1}{|a|} \sec^{-1}\left|\frac{u}{a}\right| + C$

25. $\displaystyle\int \frac{du}{u\sqrt{u^2 + a^2}} = -\frac{1}{a} \ln\left|\frac{a + \sqrt{u^2 + a^2}}{u}\right| + C$

26. $\displaystyle\int \frac{du}{u\sqrt{a^2 - u^2}} = -\frac{1}{a} \ln\left|\frac{a + \sqrt{a^2 - u^2}}{u}\right| + C$

INTEGRALS INVOLVING $au + b$

27. $\displaystyle\int \frac{du}{au+b} = \frac{1}{a}\ln|au+b| + C$

28. $\displaystyle\int \frac{u\,du}{au+b} = \frac{u}{a} - \frac{b}{a^2}\ln|au+b| + C$

29. $\displaystyle\int \frac{u^2\,du}{au+b} = \frac{(au+b)^2}{2a^3} - \frac{2b(au+b)}{a^3} + \frac{b^2}{a^3}\ln|au+b| + C$

30. $\displaystyle\int \frac{du}{u(au+b)} = \frac{1}{b}\ln\left|\frac{u}{au+b}\right| + C$

31. $\displaystyle\int \frac{du}{u^2(au+b)} = -\frac{1}{bu} + \frac{a}{b^2}\ln\left|\frac{au+b}{u}\right| + C$

32. $\displaystyle\int \frac{du}{(au+b)^2} = \frac{-1}{a(au+b)} + C$

33. $\displaystyle\int \frac{u\,du}{(au+b)^2} = \frac{b}{a^2(au+b)} + \frac{1}{a^2}\ln|au+b| + C$

34. $\displaystyle\int \frac{du}{u(au+b)^2} = \frac{1}{b(au+b)} + \frac{1}{b^2}\ln\left|\frac{u}{au+b}\right| + C$

35. $\displaystyle\int (au+b)^n\,du = \frac{(au+b)^{n+1}}{(n+1)a} + C, \quad n \neq -1$

36. $\displaystyle\int u(au+b)^n\,du = \frac{(au+b)^{n+2}}{(n+2)a^2} - \frac{b(au+b)^{n+1}}{(n+1)a^2} + C, \quad n \neq -1, -2$

37. $\displaystyle\int u^m(au+b)^n\,du = \begin{cases} \dfrac{u^{m+1}(au+b)^n}{m+n+1} + \dfrac{nb}{m+n+1}\displaystyle\int u^m(au+b)^{n-1}\,du \\[1em] \dfrac{u^m(au+b)^{n+1}}{(m+n+1)a} - \dfrac{mb}{(m+n+1)a}\displaystyle\int u^{m-1}(au+b)^n\,du \\[1em] \dfrac{-u^{m+1}(au+b)^{n+1}}{(n+1)b} + \dfrac{m+n+2}{(n+1)b}\displaystyle\int u^m(au+b)^{n+1}\,du \end{cases}$

INTEGRALS INVOLVING $\sqrt{au+b}$

38. $\displaystyle\int \frac{du}{\sqrt{au+b}} = \frac{2\sqrt{au+b}}{a} + C$

39. $\displaystyle\int \frac{u\,du}{\sqrt{au+b}} = \frac{2(au-2b)}{3a^2}\sqrt{au+b} + C$

40. $\displaystyle\int \frac{du}{u\sqrt{au+b}} = \begin{cases} \dfrac{1}{\sqrt{b}}\ln\left|\dfrac{\sqrt{au+b} - \sqrt{b}}{\sqrt{au+b} + \sqrt{b}}\right| + C, & b > 0 \\[1em] \dfrac{2}{\sqrt{-b}}\tan^{-1}\sqrt{\dfrac{au+b}{-b}} + C, & b < 0 \end{cases}$

41. $\int \sqrt{au+b}\, du = \dfrac{2\sqrt{(au+b)^3}}{3a} + C$

42. $\int u\sqrt{au+b}\, du = \dfrac{2(3au-2b)}{15a^2}\sqrt{(au+b)^3} + C$

43. $\int \dfrac{\sqrt{au+b}}{u}\, du = 2\sqrt{au+b} + b\int \dfrac{du}{u\sqrt{au+b}}$ (See 40.)

INTEGRALS INVOLVING $u^2 + a^2$

44. $\int \dfrac{du}{u^2+a^2} = \dfrac{1}{a}\tan^{-1}\dfrac{u}{a} + C$

45. $\int \dfrac{u\,du}{u^2+a^2} = \dfrac{1}{2}\ln(u^2+a^2) + C$

46. $\int \dfrac{u^2\,du}{u^2+a^2} = u - a\tan^{-1}\dfrac{u}{a} + C$

47. $\int \dfrac{du}{u(u^2+a^2)} = \dfrac{1}{2a^2}\ln\left(\dfrac{u^2}{u^2+a^2}\right) + C$

48. $\int \dfrac{du}{u^2(u^2+a^2)} = -\dfrac{1}{a^2 u} - \dfrac{1}{a^3}\tan^{-1}\dfrac{u}{a} + C$

49. $\int \dfrac{du}{(u^2+a^2)^n} = \dfrac{u}{2(n-1)a^2(u^2+a^2)^{n-1}} + \dfrac{2n-3}{(2n-2)a^2}\int \dfrac{du}{(u^2+a^2)^{n-1}}$

50. $\int \dfrac{u\,du}{(u^2+a^2)^n} = \dfrac{-1}{2(n-1)(u^2+a^2)^{n-1}} + C, \quad n \ne 1$

51. $\int \dfrac{du}{u(u^2+a^2)^n} = \dfrac{1}{2(n-1)a^2(u^2+a^2)^{n-1}} + \dfrac{1}{a^2}\int \dfrac{du}{u(u^2+a^2)^{n-1}}, \quad n \ne 1$

INTEGRALS INVOLVING $u^2 - a^2$, $u^2 > a^2$

52. $\int \dfrac{du}{u^2-a^2} = \dfrac{1}{2a}\ln\left|\dfrac{u-a}{u+a}\right| + C$

53. $\int \dfrac{u\,du}{u^2-a^2} = \dfrac{1}{2}\ln(u^2-a^2) + C$

54. $\int \dfrac{u^2\,du}{u^2-a^2} = u + \dfrac{a}{2}\ln\left|\dfrac{u-a}{u+a}\right| + C$

55. $\int \dfrac{du}{u(u^2-a^2)} = \dfrac{1}{2a^2}\ln\left|\dfrac{u^2-a^2}{u^2}\right| + C$

56. $\int \dfrac{du}{u^2(u^2-a^2)} = \dfrac{1}{a^2 u} + \dfrac{1}{2a^3}\ln\left|\dfrac{u-a}{u+a}\right| + C$

57. $\int \dfrac{du}{(u^2-a^2)^2} = \dfrac{-u}{2a^2(u^2-a^2)} - \dfrac{1}{4a^3}\ln\left|\dfrac{u-a}{u+a}\right| + C$

58. $\int \dfrac{du}{(u^2-a^2)^n} = \dfrac{-u}{2(n-1)a^2(u^2-a^2)^{n-1}} - \dfrac{2n-3}{(2n-2)a^2}\int \dfrac{du}{(u^2-a^2)^{n-1}}$

59. $\int \dfrac{u\,du}{(u^2-a^2)^n} = \dfrac{-1}{2(n-1)(u^2-a^2)^{n-1}} + C$

60. $\int \dfrac{du}{u(u^2-a^2)^n} = \dfrac{-1}{2(n-1)a^2(u^2-a^2)^{n-1}} - \dfrac{1}{a^2}\int \dfrac{du}{u(u^2-a^2)^{n-1}}$

INTEGRALS INVOLVING $a^2 - u^2$, $u^2 < a^2$

61. $\int \dfrac{du}{a^2-u^2} = \dfrac{1}{2a}\ln\left|\dfrac{a+u}{a-u}\right| + C = \dfrac{1}{a}\tanh^{-1}\dfrac{u}{a} + C$

62. $\int \dfrac{u\,du}{a^2-u^2} = -\dfrac{1}{2}\ln|a^2-u^2| + C$

63. $\int \dfrac{u^2\,du}{a^2-u^2} = -u + \dfrac{a}{2}\ln\left|\dfrac{a+u}{a-u}\right| + C$

64. $\int \dfrac{du}{u(a^2-u^2)} = \dfrac{1}{2a^2}\ln\left|\dfrac{u^2}{a^2-u^2}\right| + C$

65. $\int \dfrac{du}{(a^2-u^2)^2} = \dfrac{u}{2a^2(a^2-u^2)} + \dfrac{1}{4a^3}\ln\left|\dfrac{a+u}{a-u}\right| + C$

66. $\int \dfrac{u\,du}{(a^2-u^2)^2} = \dfrac{1}{2(a^2-u^2)} + C$

INTEGRALS INVOLVING $\sqrt{u^2+a^2}$

67. $\int \dfrac{du}{\sqrt{u^2+a^2}} = \ln(u+\sqrt{u^2+a^2}) + C = \sinh^{-1}\dfrac{u}{|a|} + C$

68. $\int \dfrac{u\,du}{\sqrt{u^2+a^2}} = \sqrt{u^2+a^2} + C$

69. $\int \dfrac{u^2\,du}{\sqrt{u^2+a^2}} = \dfrac{u\sqrt{u^2+a^2}}{2} - \dfrac{a^2}{2}\ln(u+\sqrt{u^2+a^2}) + C$

70. $\int \dfrac{du}{u\sqrt{u^2+a^2}} = -\dfrac{1}{a}\ln\left|\dfrac{a+\sqrt{u^2+a^2}}{u}\right| + C$

71. $\int \sqrt{u^2+a^2}\,du = \dfrac{u\sqrt{u^2+a^2}}{2} + \dfrac{a^2}{2}\ln(u+\sqrt{u^2+a^2}) + C$

72. $\int u\sqrt{u^2+a^2}\,du = \dfrac{(u^2+a^2)^{3/2}}{3} + C$

73. $\int u^2\sqrt{u^2+a^2}\,du = \dfrac{u(u^2+a^2)^{3/2}}{4} - \dfrac{a^2 u\sqrt{u^2+a^2}}{8} - \dfrac{a^4}{8}\ln(u+\sqrt{u^2+a^2}) + C$

74. $\int \dfrac{\sqrt{u^2+a^2}}{u}\,du = \sqrt{u^2+a^2} - a\ln\left|\dfrac{a+\sqrt{u^2+a^2}}{u}\right| + C$

75. $\int \dfrac{\sqrt{u^2+a^2}}{u^2}\,du = -\dfrac{\sqrt{u^2+a^2}}{u} + \ln(u+\sqrt{u^2+a^2}) + C$

INTEGRALS INVOLVING $\sqrt{u^2-a^2}$

76. $\int \dfrac{du}{\sqrt{u^2-a^2}} = \ln|u+\sqrt{u^2-a^2}| + C$

77. $\int \dfrac{u\,du}{\sqrt{u^2-a^2}} = \sqrt{u^2-a^2} + C$

78. $\int \dfrac{u^2\,du}{\sqrt{u^2-a^2}} = \dfrac{u\sqrt{u^2-a^2}}{2} + \dfrac{a^2}{2}\ln|u+\sqrt{u^2-a^2}| + C$

79. $\int \dfrac{du}{u\sqrt{u^2-a^2}} = \dfrac{1}{|a|}\sec^{-1}\left|\dfrac{u}{a}\right| + C$

80. $\int \sqrt{u^2-a^2}\,du = \dfrac{u\sqrt{u^2-a^2}}{2} - \dfrac{a^2}{2}\ln|u+\sqrt{u^2-a^2}| + C$

81. $\int u\sqrt{u^2-a^2}\,du = \dfrac{(u^2-a^2)^{3/2}}{3} + C$

82. $\int u^2\sqrt{u^2-a^2}\,du = \dfrac{u(u^2-a^2)^{3/2}}{4} + \dfrac{a^2 u\sqrt{u^2-a^2}}{8} - \dfrac{a^4}{8}\ln|u+\sqrt{u^2-a^2}| + C$

83. $\int \dfrac{\sqrt{u^2 - a^2}}{u}\, du = \sqrt{u^2 - a^2} - |a| \sec^{-1}\left|\dfrac{u}{a}\right| + C$

84. $\int \dfrac{\sqrt{u^2 - a^2}}{u^2}\, du = -\dfrac{\sqrt{u^2 - a^2}}{u} + \ln\left|u + \sqrt{u^2 - a^2}\right| + C$

85. $\int \dfrac{du}{(u^2 - a^2)^{3/2}} = -\dfrac{u}{a^2\sqrt{u^2 - a^2}} + C$

INTEGRALS INVOLVING $\sqrt{a^2 - u^2}$

86. $\int \dfrac{du}{\sqrt{a^2 - u^2}} = \sin^{-1}\dfrac{u}{|a|} + C$

87. $\int \dfrac{u\, du}{\sqrt{a^2 - u^2}} = -\sqrt{a^2 - u^2} + C$

88. $\int \dfrac{u^2\, du}{\sqrt{a^2 - u^2}} = -\dfrac{u\sqrt{a^2 - u^2}}{2} + \dfrac{a^2}{2}\sin^{-1}\dfrac{u}{|a|} + C$

89. $\int \dfrac{du}{u\sqrt{a^2 - u^2}} = -\dfrac{1}{a}\ln\left|\dfrac{a + \sqrt{a^2 - u^2}}{u}\right| + C$

90. $\int \dfrac{du}{u^2\sqrt{a^2 - u^2}} = -\dfrac{\sqrt{a^2 - u^2}}{a^2 u} + C$

91. $\int \sqrt{a^2 - u^2}\, du = \dfrac{u\sqrt{a^2 - u^2}}{2} + \dfrac{a^2}{2}\sin^{-1}\dfrac{u}{|a|} + C$

92. $\int u\sqrt{a^2 - u^2}\, du = -\dfrac{(a^2 - u^2)^{3/2}}{3} + C$

93. $\int u^2\sqrt{a^2 - u^2}\, du = -\dfrac{u(a^2 - u^2)^{3/2}}{4} + \dfrac{a^2 u\sqrt{a^2 - u^2}}{8} + \dfrac{a^4}{8}\sin^{-1}\dfrac{u}{|a|} + C$

94. $\int \dfrac{\sqrt{a^2 - u^2}}{u}\, du = \sqrt{a^2 - u^2} - a \ln\left|\dfrac{a + \sqrt{a^2 - u^2}}{u}\right| + C$

95. $\int \dfrac{\sqrt{a^2 - u^2}}{u^2}\, du = -\dfrac{\sqrt{a^2 - u^2}}{u} - \sin^{-1}\dfrac{u}{|a|} + C$

INTEGRALS INVOLVING THE TRIGONOMETRIC FUNCTIONS

96. $\int \sin au\, du = -\dfrac{\cos au}{a} + C$

97. $\int u \sin au\, du = \dfrac{\sin au}{a^2} - \dfrac{u \cos au}{a} + C$

98. $\int u^2 \sin au\, du = \dfrac{2u}{a^2}\sin au + \left(\dfrac{2}{a^3} - \dfrac{u^2}{a}\right)\cos au + C$

99. $\int \dfrac{du}{\sin au} = \dfrac{1}{a}\ln|\csc au - \cot au|$

$\qquad = \dfrac{1}{a}\ln\left|\tan\dfrac{au}{2}\right| + C$

100. $\int \sin^2 au\, du = \dfrac{u}{2} - \dfrac{\sin 2au}{4a} + C$

101. $\int u \sin^2 au\, du = \dfrac{u^2}{4} - \dfrac{u \sin 2au}{4a} - \dfrac{\cos 2au}{8a^2} + C$

102. $\int \dfrac{du}{\sin^2 au} = -\dfrac{1}{a}\cot au + C$

103. $\int \sin pu \sin qu\, du = \dfrac{\sin(p - q)u}{2(p - q)} - \dfrac{\sin(p + q)u}{2(p + q)} + C, \quad p \neq \pm q$

104. $\int \dfrac{du}{1 - \sin au} = \dfrac{1}{a}\tan\left(\dfrac{\pi}{4} + \dfrac{au}{2}\right) + C$

105. $\displaystyle\int \frac{u\,du}{1 - \sin au} = \frac{u}{a}\tan\left(\frac{\pi}{4} + \frac{au}{2}\right) + \frac{2}{a^2}\ln\left|\sin\left(\frac{\pi}{4} - \frac{au}{2}\right)\right| + C$

106. $\displaystyle\int \frac{du}{1 + \sin au} = -\frac{1}{a}\tan\left(\frac{\pi}{4} - \frac{au}{2}\right) + C$

107. $\displaystyle\int \frac{du}{p + q \sin au} = \begin{cases} \dfrac{2}{a\sqrt{p^2 - q^2}} \tan^{-1}\dfrac{p\tan\frac{1}{2}au + q}{\sqrt{p^2 - q^2}} + C, & |p| > |q| \\[2mm] \dfrac{1}{a\sqrt{q^2 - p^2}} \ln\left|\dfrac{p\tan\frac{1}{2}au + q - \sqrt{q^2 - p^2}}{p\tan\frac{1}{2}au + q + \sqrt{q^2 - p^2}}\right| + C, & |p| < |q| \end{cases}$

108. $\displaystyle\int u^m \sin au\,du = -\frac{u^m \cos au}{a} + \frac{mu^{m-1}\sin au}{a^2} - \frac{m(m-1)}{a^2}\int u^{m-2}\sin au\,du$

109. $\displaystyle\int \sin^n au\,du = -\frac{\sin^{n-1}au\cos au}{an} + \frac{n-1}{n}\int \sin^{n-2}au\,du$

110. $\displaystyle\int \frac{du}{\sin^n au} = \frac{-\cos au}{a(n-1)\sin^{n-1}au} + \frac{n-2}{n-1}\int \frac{du}{\sin^{n-2}au}, \quad n \ne 1$

111. $\displaystyle\int \cos au\,du = \frac{\sin au}{a} + C$

112. $\displaystyle\int u \cos au\,du = \frac{\cos au}{a^2} + \frac{u \sin au}{a} + C$

113. $\displaystyle\int u^2 \cos au\,du = \frac{2u}{a^2}\cos au + \left(\frac{u^2}{a} - \frac{2}{a^3}\right)\sin au + C$

114. $\displaystyle\int \frac{du}{\cos au} = \frac{1}{a}\ln|\sec au + \tan au| = \frac{1}{a}\ln\left|\tan\left(\frac{\pi}{4} + \frac{au}{2}\right)\right| + C$

115. $\displaystyle\int \cos^2 au\,du = \frac{u}{2} + \frac{\sin 2au}{4a} + C$

116. $\displaystyle\int u \cos^2 au\,du = \frac{u^2}{4} + \frac{u\sin 2au}{4a} + \frac{\cos 2au}{8a^2} + C$

117. $\displaystyle\int \frac{du}{\cos^2 au} = \frac{\tan au}{a} + C$

118. $\displaystyle\int \cos qu \cos pu\,du = \frac{\sin(q-p)u}{2(q-p)} + \frac{\sin(q+p)u}{2(q+p)} + C, \quad q \ne \pm p$

119. $\displaystyle\int \frac{du}{p + q\cos au} = \begin{cases} \dfrac{2}{a\sqrt{p^2 - q^2}}\tan^{-1}\left[\sqrt{(p-q)/(p+q)}\,\tan\tfrac{1}{2}au\right] + C, & |p| > |q| \\[2mm] \dfrac{1}{a\sqrt{q^2 - p^2}}\ln\left[\dfrac{\tan\frac{1}{2}au + \sqrt{(q+p)/(q-p)}}{\tan\frac{1}{2}au - \sqrt{(q+p)/(q-p)}}\right] + C, & |p| < |q| \end{cases}$

120. $\displaystyle\int u^m \cos au\,du = \frac{u^m \sin au}{a} + \frac{mu^{m-1}}{a^2}\cos au - \frac{m(m-1)}{a^2}\int u^{m-2}\cos au\,du$

121. $\displaystyle\int \cos^n au\,du = \frac{\sin au \cos^{n-1}au}{an} + \frac{n-1}{n}\int \cos^{n-2}au\,du$

122. $\displaystyle\int \frac{du}{\cos^n au} = \frac{\sin au}{a(n-1)\cos^{n-1}au} + \frac{n-2}{n-1}\int \frac{du}{\cos^{n-2}au}$

123. $\displaystyle\int \sin au \cos au\,du = \frac{\sin^2 au}{2a} + C$

124. $\displaystyle\int \sin pu \cos qu\,du = -\frac{\cos(p-q)u}{2(p-q)} - \frac{\cos(p+q)u}{2(p+q)} + C, \quad p \ne \pm q$

125. $\int \sin^n au \cos au \, du = \dfrac{\sin^{n+1} au}{(n+1)a} + C, \quad n \neq -1$

126. $\int \cos^n au \sin au \, du = -\dfrac{\cos^{n+1} au}{(n+1)a} + C, \quad n \neq -1$

127. $\int \sin^2 au \cos^2 au \, du = \dfrac{u}{8} - \dfrac{\sin 4au}{32a} + C$

128. $\int \dfrac{du}{\sin au \cos au} = \dfrac{1}{a} \ln|\tan au| + C$

129. $\int \dfrac{du}{\cos au(1 \pm \sin au)} = \mp \dfrac{1}{2a(1 \pm \sin au)} + \dfrac{1}{2a} \ln\left|\tan\left(\dfrac{au}{2} + \dfrac{\pi}{4}\right)\right| + C$

130. $\int \dfrac{du}{\sin au(1 \pm \cos au)} = \pm \dfrac{1}{2a(1 \pm \cos au)} + \dfrac{1}{2a} \ln\left|\tan\dfrac{au}{2}\right| + C$

131. $\int \dfrac{du}{\sin au \pm \cos au} = \dfrac{1}{a\sqrt{2}} \ln\left|\tan\left(\dfrac{au}{2} \pm \dfrac{\pi}{8}\right)\right| + C$

132. $\int \dfrac{\sin au \, du}{\sin au \pm \cos au} = \dfrac{u}{2} \mp \dfrac{1}{2a} \ln|\sin au \pm \cos au| + C$

133. $\int \dfrac{\cos au \, du}{\sin au \pm \cos au} = \pm \left[\dfrac{u}{2} \pm \dfrac{1}{2a} \ln|\sin au \pm \cos au|\right] + C$

134. $\int \dfrac{\sin au \, du}{p + q \cos au} = -\dfrac{1}{aq} \ln|p + q \cos au| + C$

135. $\int \dfrac{\cos au \, du}{p + q \sin au} = \dfrac{1}{aq} \ln|p + q \sin au| + C$

136. $\int \sin^m au \cos^n au \, du = \begin{cases} -\dfrac{\sin^{m-1} au \cos^{n+1} au}{a(m+n)} + \dfrac{m-1}{m+n} \int \sin^{m-2} au \cos^n au \, du, & m \neq -n \\ \dfrac{\sin^{m+1} au \cos^{n-1} au}{a(m+n)} + \dfrac{n-1}{m+n} \int \sin^m au \cos^{n-2} au \, du, & m \neq -n \end{cases}$

137. $\int \tan au \, du = -\dfrac{1}{a} \ln|\cos au| = \dfrac{1}{a} \ln|\sec au| + C$

138. $\int \tan^2 au \, du = \dfrac{\tan au}{a} - u + C$

139. $\int \tan^n au \sec^2 au \, du = \dfrac{\tan^{n+1} au}{(n+1)a} + C, \quad n \neq -1$

140. $\int \tan^n au \, du = \dfrac{\tan^{n-1} au}{(n-1)a} - \int \tan^{n-2} au \, du + C, \quad n \neq 1$

141. $\int \cot au \, du = \dfrac{1}{a} \ln|\sin au| + C$

142. $\int \cot^2 au \, du = -\dfrac{\cot au}{a} - u + C$

143. $\int \cot^n au \csc^2 au \, du = -\dfrac{\cot^{n+1} au}{(n+1)a} + C, \quad n \neq -1$

144. $\int \cot^n au \, du = -\dfrac{\cot^{n-1} au}{(n-1)a} - \int \cot^{n-2} au \, du, \quad n \neq 1$

145. $\int \sec au \, du = \dfrac{1}{a} \ln|\sec au + \tan au| = \dfrac{1}{a} \ln\left|\tan\left(\dfrac{au}{2} + \dfrac{\pi}{4}\right)\right| + C$

146. $\int \sec^2 au \, du = \dfrac{\tan au}{a} + C$

147. $\int \sec^3 au \, du = \dfrac{\sec au \tan au}{2a} + \dfrac{1}{2a} \ln|\sec au + \tan au| + C$

148. $\int \sec^n au \tan au \, du = \dfrac{\sec^n au}{na} + C$

149. $\int \sec^n au \, du = \dfrac{\sec^{n-2} au \tan au}{a(n-1)} + \dfrac{n-2}{n-1} \int \sec^{n-2} au \, du, \quad n \neq 1$

150. $\int \csc au\, du = \frac{1}{a}\ln|\csc au - \cot au| = \frac{1}{a}\ln\left|\tan\frac{au}{2}\right| + C$

151. $\int \csc^2 au\, du = -\frac{\cot au}{a} + C$

152. $\int \csc^n au \cot au\, du = -\frac{\csc^n au}{na} + C$

153. $\int \csc^n au\, du = -\frac{\csc^{n-2} au \cot au}{a(n-1)} + \frac{n-2}{n-1}\int \csc^{n-2} au\, du, \quad n \neq 1$

INTEGRALS INVOLVING INVERSE TRIGONOMETRIC FUNCTIONS

154. $\int \sin^{-1}\frac{u}{a}\, du = u\sin^{-1}\frac{u}{a} + \sqrt{a^2 - u^2} + C$

155. $\int u\sin^{-1}\frac{u}{a}\, du = \left(\frac{u^2}{2} - \frac{a^2}{4}\right)\sin^{-1}\frac{u}{a} + \frac{u\sqrt{a^2 - u^2}}{4} + C$

156. $\int \cos^{-1}\frac{u}{a}\, du = u\cos^{-1}\frac{u}{a} - \sqrt{a^2 - u^2} + C$

157. $\int u\cos^{-1}\frac{u}{a}\, du = \left(\frac{u^2}{2} - \frac{a^2}{4}\right)\cos^{-1}\frac{u}{a} - \frac{u\sqrt{a^2 - u^2}}{4} + C$

158. $\int \tan^{-1}\frac{u}{a}\, du = u\tan^{-1}\frac{u}{a} - \frac{a}{2}\ln(u^2 + a^2) + C$

159. $\int u\tan^{-1}\frac{u}{a}\, du = \frac{1}{2}(u^2 + a^2)\tan^{-1}\frac{u}{a} - \frac{au}{2} + C$

160. $\int u^m \sin^{-1}\frac{u}{a}\, du = \frac{u^{m+1}}{m+1}\sin^{-1}\frac{u}{a} - \frac{1}{m+1}\int \frac{u^{m+1}}{\sqrt{a^2 - u^2}}\, du$

161. $\int u^m \cos^{-1}\frac{u}{a}\, du = \frac{u^{m+1}}{m+1}\cos^{-1}\frac{u}{a} + \frac{1}{m+1}\int \frac{u^{m+1}}{\sqrt{a^2 - u^2}}\, du$

162. $\int u^m \tan^{-1}\frac{u}{a}\, du = \frac{u^{m+1}}{m+1}\tan^{-1}\frac{u}{a} - \frac{a}{m+1}\int \frac{u^{m+1}}{u^2 + a^2}\, du$

INTEGRALS INVOLVING e^{au}

163. $\int e^{au}\, du = \frac{e^{au}}{a} + C$

164. $\int ue^{au}\, du = \frac{e^{au}}{a}\left(u - \frac{1}{a}\right) + C$

165. $\int u^2 e^{au}\, du = \frac{e^{au}}{a}\left(u^2 - \frac{2u}{a} + \frac{2}{a^2}\right) + C$

166. $\int u^n e^{au}\, du = \frac{u^n e^{au}}{a} - \frac{n}{a}\int u^{n-1} e^{au}\, du$

$= \frac{e^{au}}{a}\left[u^n - \frac{nu^{n-1}}{a} + \frac{n(n-1)u^{n-2}}{a^2} - \cdots + \frac{(-1)^n n!}{a^n}\right]$, if n is a positive integer

167. $\int \frac{du}{p + qe^{au}} = \frac{u}{p} - \frac{1}{ap}\ln|p + qe^{au}| + C$

168. $\int e^{au}\sin bu\, du = \frac{e^{au}(a\sin bu - b\cos bu)}{a^2 + b^2} + C$

169. $\int e^{au}\cos bu\, du = \frac{e^{au}(a\cos bu + b\sin bu)}{a^2 + b^2} + C$

170. $\int ue^{au}\sin bu\, du = \dfrac{ue^{au}(a\sin bu - b\cos bu)}{a^2 + b^2} - \dfrac{e^{au}[(a^2 - b^2)\sin bu - 2ab\cos bu]}{(a^2 + b^2)^2} + C$

171. $\int ue^{au}\cos bu\, du = \dfrac{ue^{au}(a\cos bu + b\sin bu)}{a^2 + b^2} - \dfrac{e^{au}[(a^2 - b^2)\cos bu + 2ab\sin bu]}{(a^2 + b^2)^2} + C$

172. $\int e^{au}\sin^n bu\, du = \dfrac{e^{au}\sin^{n-1} bu}{a^2 + n^2 b^2}(a\sin bu - nb\cos bu) + \dfrac{n(n-1)b^2}{a^2 + n^2 b^2}\int e^{au}\sin^{n-2} bu\, du$

173. $\int e^{au}\cos^n bu\, du = \dfrac{e^{au}\cos^{n-1} bu}{a^2 + n^2 b^2}(a\cos bu + nb\sin bu) + \dfrac{n(n-1)b^2}{a^2 + n^2 b^2}\int e^{au}\cos^{n-2} bu\, du$

INTEGRALS INVOLVING ln u

174. $\int \ln u\, du = u\ln u - u + C$

175. $\int u\ln u\, du = \dfrac{u^2}{2}\left(\ln u - \dfrac{1}{2}\right) + C$

176. $\int u^m \ln u\, du = \dfrac{u^{m+1}}{m+1}\left(\ln u - \dfrac{1}{m+1}\right), \quad m \neq -1$

177. $\int \dfrac{\ln u}{u}\, du = \dfrac{1}{2}\ln^2 u + C$

178. $\int \dfrac{\ln^n u\, du}{u} = \dfrac{\ln^{n+1} u}{n+1} + C, \quad n \neq -1$

179. $\int \dfrac{du}{u\ln u} = \ln|\ln u| + C$

180. $\int \ln^n u\, du = u\ln^n u - n\int \ln^{n-1} u\, du + C, \quad n \neq -1$

181. $\int u^m \ln^n u\, du = \dfrac{u^{m+1}\ln^n u}{m+1} - \dfrac{n}{m+1}\int u^m \ln^{n-1} u\, du + C, \quad m, n \neq -1$

182. $\int \ln(u^2 + a^2)\, du = u\ln(u^2 + a^2) - 2u + 2a\tan^{-1}\dfrac{u}{a} + C$

183. $\int \ln|u^2 - a^2|\, du = u\ln|u^2 - a^2| - 2u + a\ln\left|\dfrac{u+a}{u-a}\right| + C$

INTEGRALS INVOLVING HYPERBOLIC FUNCTIONS

184. $\int \sinh au\, du = \dfrac{\cosh au}{a} + C$

185. $\int u\sinh au\, du = \dfrac{u\cosh au}{a} - \dfrac{\sinh au}{a^2} + C$

186. $\int \cosh au\, du = \dfrac{\sinh au}{a} + C$

187. $\int u\cosh au\, du = \dfrac{u\sinh au}{a} - \dfrac{\cosh au}{a^2} + C$

188. $\int \cosh^2 au\, du = \dfrac{u}{2} + \dfrac{\sinh au\cosh au}{2a} + C$

189. $\int \sinh^2 au\, du = \dfrac{\sinh au\cosh au}{2a} - \dfrac{u}{2} + C$

190. $\int \sinh^n au\, du = \dfrac{\sinh^{n-1} au\cosh au}{an} - \dfrac{n-1}{n}\int \sinh^{n-2} au\, du$

191. $\int \cosh^n au\, du = \dfrac{\cosh^{n-1} au\sinh au}{an} + \dfrac{n-1}{n}\int \cosh^{n-2} au\, du$

192. $\int \sinh au\cosh au\, du = \dfrac{\sinh^2 au}{2a} + C$

193. $\int \sinh pu\cosh qu\, du = \dfrac{\cosh(p+q)u}{2(p+q)} + \dfrac{\cosh(p-q)u}{2(p-q)} + C$

194. $\int \tanh au\,du = \dfrac{1}{a}\ln\cosh au + C$

195. $\int \tanh^2 au\,du = u - \dfrac{\tanh au}{a} + C$

196. $\int \tanh^n au\,du = \dfrac{-\tanh^{n-1} au}{a(n-1)} + \int \tanh^{n-2} au\,du$

197. $\int \coth au\,du = \dfrac{1}{a}\ln|\sinh au| + C$

198. $\int \coth^2 au\,du = u - \dfrac{\coth au}{a} + C$

199. $\int \operatorname{sech} au\,du = \dfrac{2}{a}\tan^{-1} e^{au} + C$

200. $\int \operatorname{sech}^2 au\,du = \dfrac{\tanh au}{a} + C$

201. $\int \operatorname{sech}^n au\,du = \dfrac{\operatorname{sech}^{n-2} au \tanh au}{a(n-1)} + \dfrac{n-2}{n-1}\int \operatorname{sech}^{n-2} au\,du$

202. $\int \operatorname{csch} au\,du = \dfrac{1}{a}\ln\left|\tanh\dfrac{au}{2}\right| + C$

203. $\int \operatorname{csch}^2 au\,du = -\dfrac{\coth au}{a} + C$

204. $\int \operatorname{sech} u \tanh u\,du = -\operatorname{sech} u + C$

205. $\int \operatorname{csch} u \coth u\,du = -\operatorname{csch} u + C$

SOME DEFINITE INTEGRALS

Unless otherwise stated, all letters stand for positive numbers.

206. $\displaystyle\int_0^\infty \dfrac{dx}{x^2 + a^2} = \dfrac{\pi}{2a}$

207. $\displaystyle\int_0^\infty \dfrac{x^{p-1}}{1+x}\,dx = \dfrac{\pi}{\sin p\pi}$

208. $\displaystyle\int_0^a \dfrac{dx}{\sqrt{a^2 - x^2}} = \dfrac{\pi}{2}$

209. $\displaystyle\int_0^a \sqrt{a^2 - x^2}\,dx = \dfrac{\pi a^2}{4}$

210. $\displaystyle\int_0^\pi \sin mx \sin nx\,dx = \begin{cases} 0, & \text{if } m, n \text{ integers and } m \neq n \\ \dfrac{\pi}{2}, & \text{if } m, n \text{ integers and } m = n \end{cases}$

211. $\displaystyle\int_0^\pi \cos mx \cos nx\,dx = \begin{cases} 0, & \text{if } m, n \text{ integers and } m \neq n \\ \dfrac{\pi}{2}, & \text{if } m, n \text{ integers and } m = n \end{cases}$

212. $\displaystyle\int_0^\pi \sin mx \cos nx\,dx = \begin{cases} 0, & \text{if } m, n \text{ integers and } m + n \text{ is even} \\ \dfrac{2m}{(m^2 - n^2)}, & \text{if } m, n \text{ integers and } m + n \text{ is odd} \end{cases}$

213. $\displaystyle\int_0^{\pi/2} \sin^2 x\,dx = \int_0^{\pi/2} \cos^2 x\,dx = \dfrac{\pi}{4}$

214. $\displaystyle\int_0^\infty e^{-ax}\cos bx\,dx = \dfrac{a}{a^2 + b^2}$

215. $\displaystyle\int_0^\infty e^{-ax}\sin bx\,dx = \dfrac{b}{a^2 + b^2}$

216. $\displaystyle\int_0^\infty e^{-a^2 x^2}\,dx = \dfrac{\sqrt{\pi}}{2a}$

217. $\displaystyle\int_0^{\pi/2} \sin^{2m} x\,dx = \int_0^{\pi/2} \cos^{2m} x\,dx = \dfrac{1\cdot 3\cdot 5\cdots(2m-1)}{2\cdot 4\cdot 6\cdots 2m}\cdot\dfrac{\pi}{2}, \quad m = 1, 2, 3, \ldots$

218. $\displaystyle\int_0^{\pi/2} \sin^{2m+1} x\,dx = \int_0^{\pi/2} \cos^{2m+1} x\,dx = \dfrac{2\cdot 4\cdot 6\cdots 2m}{1\cdot 3\cdot 5\cdots(2m+1)}, \quad m = 1, 2, 3, \ldots$

219. $\displaystyle\int_0^\infty \dfrac{e^{-x}}{\sqrt{x}}\,dx = \sqrt{\pi}$

220. $\displaystyle\int_0^1 x^m (\ln x)^n\,dx = \dfrac{(-1)^n n!}{(m+1)^{n+1}}$

APPENDIX 2

Table of Laplace Transforms[†]

$F(s) = \mathscr{L}\{f(t)\}$	$f(t)$
1. $\dfrac{1}{s^r}, \quad r > 0$	$\dfrac{t^{r-1}}{\Gamma(r)}, \quad \Gamma(n+1) = n!$
2. $\dfrac{1}{(s-a)^r}, \quad r > 0$	$\dfrac{t^{r-1}e^{at}}{\Gamma(r)}$
3. $\dfrac{1}{(s-a)(s-b)}, \quad a \neq b$	$\dfrac{e^{at} - e^{bt}}{a - b}$
4. $\dfrac{s}{(s-a)(s-b)}, \quad a \neq b$	$\dfrac{ae^{at} - be^{bt}}{a - b}$
5. $\dfrac{1}{s^2 + k^2}$	$\dfrac{1}{k} \sin kt$
6. $\dfrac{s}{s^2 + k^2}$	$\cos kt$
7. $\dfrac{1}{s^2 - k^2}$	$\dfrac{1}{k} \sinh kt$
8. $\dfrac{s}{s^2 - k^2}$	$\cosh kt$
9. $\dfrac{1}{(s-a)^2 + k^2}$	$\dfrac{1}{k} e^{at} \sin kt$

[†] For a more extensive list of Laplace transforms and their inverses, see A. Erdelyi et al., *Tables of Integral Transforms*, 2 vols. (New York: McGraw-Hill, 1954).

$F(s) = \mathscr{L}\{f(t)\}$	$f(t)$
10. $\dfrac{s-a}{(s-a)^2 + k^2}$	$e^{at}\cos kt$
11. $\dfrac{1}{(s-a)^2 - k^2}$	$\dfrac{1}{k}e^{at}\sinh kt$
12. $\dfrac{s-a}{(s-a)^2 - k^2}$	$e^{at}\cosh kt$
13. $\dfrac{1}{s(s^2 + k^2)}$	$\dfrac{1}{k^2}(1 - \cos kt)$
14. $\dfrac{1}{s^2(s^2 + k^2)}$	$\dfrac{1}{k^3}(kt - \sin kt)$
15. $\dfrac{1}{(s^2 + k^2)^2}$	$\dfrac{1}{2k^3}(\sin kt - kt\cos kt)$
16. $\dfrac{s}{(s^2 + k^2)^2}$	$\dfrac{t}{2k}\sin kt$
17. $\dfrac{s^2}{(s^2 + k^2)^2}$	$\dfrac{1}{2k}(\sin kt + kt\cos kt)$
18. $\dfrac{1}{(s^2 + a^2)(s^2 + b^2)}$, $a^2 \neq b^2$	$\dfrac{a\sin bt - b\sin at}{ab(a^2 - b^2)}$
19. $\dfrac{s}{(s^2 + a^2)(s^2 + b^2)}$, $a^2 \neq b^2$	$\dfrac{\cos bt - \cos at}{a^2 - b^2}$
20. $\dfrac{1}{s^4 - k^4}$	$\dfrac{1}{2k^3}(\sinh kt - \sin kt)$
21. $\dfrac{s}{s^4 - k^4}$	$\dfrac{1}{2k^2}(\cosh kt - \cos kt)$
22. $\dfrac{1}{s^4 + 4k^4}$	$\dfrac{1}{4k^3}(\sin kt\cosh kt - \cos kt\sinh kt)$
23. $\dfrac{s}{s^4 + 4k^4}$	$\dfrac{1}{2k^2}\sin kt\sinh kt$
24. $\sqrt{s-a} - \sqrt{s-b}$	$\dfrac{e^{bt} - e^{at}}{2\sqrt{\pi t^3}}$
25. $\dfrac{s}{(s-a)^{3/2}}$	$\dfrac{e^{at}}{\sqrt{\pi t}}(1 + 2at)$
26. $\dfrac{1}{\sqrt{s+a}\sqrt{s+b}}$	$e^{-(a+b)t/2} I_0\left(\dfrac{a-b}{2}t\right)$
27. $\dfrac{\left(\sqrt{s^2+k^2} - s\right)^r}{\sqrt{s^2+k^2}}$, $r > -1$	$k^r J_r(kt)$
28. $\dfrac{1}{(s^2+k^2)^r}$, $r > 0$	$\dfrac{\sqrt{\pi}}{\Gamma(r)}\left(\dfrac{t}{2k}\right)^{r-1/2} J_{r-1/2}(kt)$
29. $\left(\sqrt{s^2+k^2} - s\right)^r$, $r > 0$	$\dfrac{rk^r}{t}J_r(kt)$

$F(s) = \mathcal{L}\{f(t)\}$	$f(t)$
30. $\dfrac{(s - \sqrt{s^2 - k^2})^r}{\sqrt{s^2 - k^2}}, \quad r > -1$	$k^r I_r(kt)$
31. $\dfrac{1}{(s^2 - k^2)^r}, \quad r > 0$	$\dfrac{\sqrt{\pi}}{\Gamma(r)} \left(\dfrac{t}{2k}\right)^{r-1/2} I_{r-1/2}(kt)$
32. $\dfrac{e^{-k/s}}{s^r}, \quad r > 0$	$\left(\dfrac{t}{k}\right)^{(r-1)/2} J_{r-1}(2\sqrt{kt})$
33. $\dfrac{e^{-k/s}}{\sqrt{s}}$	$\dfrac{1}{\sqrt{\pi t}} \cos 2\sqrt{kt}$
34. $\dfrac{e^{k/s}}{s^r}, \quad r > 0$	$\left(\dfrac{t}{k}\right)^{(r-1)/2} I_{r-1}(2\sqrt{kt})$
35. $\dfrac{e^{k/s}}{\sqrt{s}}$	$\dfrac{1}{\sqrt{\pi t}} \cosh 2\sqrt{kt}$
36. $\dfrac{1}{s} \ln s$	$-\ln t - \gamma, \quad \gamma \approx 0.5772$
37. $\ln \dfrac{s-a}{s-b}$	$\dfrac{e^{bt} - e^{at}}{t}$
38. $\ln\left(1 + \dfrac{k^2}{s^2}\right)$	$\dfrac{2}{t}(1 - \cos kt)$
39. $\ln\left(1 - \dfrac{k^2}{s^2}\right)$	$\dfrac{2}{t}(1 - \cosh kt)$
40. $\arctan\left(\dfrac{k}{s}\right)$	$\dfrac{\sin kt}{t}$
41. $\dfrac{1}{s} \arctan\left(\dfrac{k}{s}\right)$	$\text{Si}(kt) = \displaystyle\int_0^{kt} \dfrac{\sin u}{u} \, du$
42. $e^{-r\sqrt{s}}, \quad r > 0$	$\dfrac{r}{2\sqrt{\pi t^3}} \exp\left(-\dfrac{r^2}{4t}\right)$
43. $\dfrac{e^{-r\sqrt{s}}}{s}, \quad r \geq 0$	$1 - \text{erf}\left(\dfrac{r}{2\sqrt{t}}\right) = 1 - \dfrac{2}{\sqrt{\pi}} \displaystyle\int_0^{r/2\sqrt{t}} e^{-u^2} \, du$
44. $e^{r^2 s^2}(1 - \text{erf}(rs)), \quad r > 0$	$\dfrac{1}{r\sqrt{\pi}} \exp\left(-\dfrac{t^2}{4r^2}\right)$
45. $\dfrac{1}{s} e^{r^2 s^2}(1 - \text{erf}(rs)), \quad r > 0$	$\text{erf}\left(\dfrac{t}{2r}\right)$
46. $\text{erf}\left(\dfrac{r}{\sqrt{s}}\right)$	$\dfrac{1}{\pi t} \sin(2k\sqrt{t})$
47. $e^{rs}(1 - \text{erf}\sqrt{rs}), \quad r > 0$	$\dfrac{\sqrt{r}}{\pi\sqrt{t}(t+r)}$
48. $\dfrac{1}{\sqrt{s}} e^{rs}(1 - \text{erf}\sqrt{rs}), \quad r > 0$	$\dfrac{1}{\sqrt{\pi(t+r)}}$

APPENDIX 3

The Existence and Uniqueness of Solutions

In this appendix we prove a number of results that guarantee the existence of unique solutions to first-order initial-value problems and first-order systems of differential equations. We begin with the first-order initial-value problem

$$x'(t) = f(t, x(t)), \qquad x(t_0) = x_0, \tag{1}$$

where t_0 and x_0 are real numbers. Equation (1) includes all the first-order equations we have discussed in this book. For example, for the linear nonhomogeneous equation $x' + a(t)x = b(t)$,

$$f(t, x) = -a(t)x + b(t).$$

We here show that if $f(t, x)$ and $(\partial f/\partial x)(t, x)$ are continuous in some region containing the point (t_0, x_0), then there is an interval (containing t_0) on which a unique solution of equation (1) exists.

Before continuing with our discussion, we advise the reader that in this appendix we use theoretical tools from calculus that have not been widely used earlier in the text. In particular, we need the following facts about continuity and convergence of functions. They are discussed in most intermediate and advanced calculus texts:[†]

a. Let $f(t, x)$ be a continuous function of the two variables t and x and let the closed, bounded region D be defined by

$$D = \{(t, x) : a \le t \le b, c \le x \le d\},$$

where a, b, c, and d are finite real numbers. Then $f(t, x)$ is bounded for (t, x) in D; that is, there is a number $M > 0$ such that $|f(t, x)| \le M$ for every pair (t, x) in D.

[†] See, for example, R. C. Buck, *Advanced Calculus*, 3d ed., (New York: McGraw-Hill, 1978).

b. Let $f(x)$ be continuous on the closed interval $a \le x \le b$ and differentiable on the open interval $a < x < b$. Then the **mean value theorem** of differential calculus states that there is a number ξ between a and b ($a < \xi < b$) such that
$$f(b) - f(a) = f'(\xi)(b-a).$$
This equation can be written as

$$\boxed{\dfrac{f(b) - f(a)}{b - a} = f'(\xi),}$$

which says, geometrically, that the slope of the tangent to the curve $y = f(x)$ at the point ξ between a and b is equal to the slope of the secant line passing through the points $(a, f(a))$ and $(b, f(b))$ (see Figure A3.1).

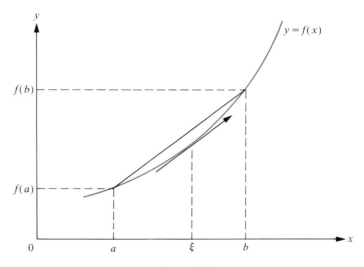

Figure A3.1

c. Let $\{x_n(t)\}$ be a sequence of functions. Then $x_n(t)$ is said to **converge uniformly** to a (limit) function $x(t)$ on the interval $a \le t \le b$ if for every real number $\epsilon > 0$, there exists an integer $N > 0$ such that whenever $n \ge N$, we have
$$|x_n(t) - x(t)| < \epsilon,$$
for every t, $a \le t \le b$.

d. If the functions $\{x_n(t)\}$ of statement (c) are continuous on the interval $a \le t \le b$, then the limit function $x(t)$ is also continuous there. This fact is often stated "the uniform limit of continuous functions is continuous."

e. Let $f(t, x)$ be a continuous function in the variable x and suppose that $x_n(t)$ converges to $x(t)$ uniformly as $n \to \infty$. Then
$$\lim_{n \to \infty} f(t, x_n(t)) = f(t, x(t)).$$

f. Let $f(t)$ be an integrable function on the interval $a \le t \le b$. Then
$$\left| \int_a^b f(t)\, dt \right| \le \int_a^b |f(t)|\, dt,$$

and if $|f(t)| \leq M$, then
$$\int_a^b |f(t)|\, dt \leq M \int_a^b dt = M(b-a).$$

g. Let $\{x_n(t)\}$ be a sequence of functions with $|x_n(t)| \leq M_n$ for $a \leq t \leq b$. Then, if $\sum_{n=0}^{\infty} |M_n| < \infty$ (that is, if $\sum_{n=0}^{\infty} M_n$ converges absolutely), then $\sum_{n=0}^{\infty} x_n(t)$ converges uniformly on the interval $a \leq t \leq b$ to a unique limit function $x(t)$. This is often called the **Weierstrass M-test** for uniform convergence.

h. Let $\{x_n(t)\}$ converge uniformly to $x(t)$ on the interval $a \leq t \leq b$ and let $f(t, x)$ be a continuous function of t and x in the region D defined in statement (a). Then

$$\boxed{\lim_{n \to \infty} \int_a^b f(s, x_n(s))\, ds = \int_a^b \lim_{n \to \infty} f(s, x_n(s))\, ds = \int_a^b f(s, x(s))\, ds.}$$

The first result we need was proved in Section 2.7 (see p. 74).

THEOREM 1

Let $f(t, x)$ be continuous for all values t and x. Then the initial-value problem (1) is equivalent to the integral equation

$$x(t) = x_0 + \int_{t_0}^t f(s, x(s))\, ds \qquad (2)$$

in the sense that $x(t)$ is a solution of equation (1) if and only if $x(t)$ is a solution of equation (2).

Let D denote the rectangular region in the tx-plane defined by

$$D: a \leq t \leq b, c \leq x \leq d, \qquad (3)$$

where $-\infty < a < b < +\infty$ and $-\infty < c < d < +\infty$ (see Figure A3.2). We say that the function $f(t, x)$ is **Lipschitz continuous** in x over D if there exists a constant k, $0 < k < \infty$, such that

$$\boxed{|f(t, x_1) - f(t, x_2)| \leq k|x_1 - x_2|,} \qquad (4)$$

whenever (t, x_1) and (t, x_2) belong to D. The constant k is called a **Lipschitz constant**. Clearly, according to equation (4), every Lipschitz continuous function is continuous in x for each fixed t. However, *not every continuous function is Lipschitz continuous.*

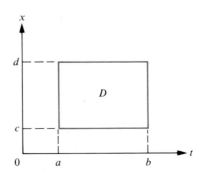

Figure A3.2

Example 1 Let $f(t, x) = \sqrt{x}$ on the set $0 \le t \le 1, 0 \le x \le 1$. Then $f(t, x)$ is certainly continuous on this region. But

$$|f(t, x) - f(t, 0)| = |\sqrt{x} - 0| = \frac{1}{\sqrt{x}}|x - 0|,$$

for all $0 < x < 1$, and $x^{-1/2}$ tends to infinity as x approaches zero. Thus no finite Lipschitz constant can be found to satisfy equation (4).

However, Lipschitz continuity is not a rare occurrence, as shown by the following theorem.

THEOREM 2

Let $f(t, x)$ and $(\partial f/\partial x)(t, x)$ be continuous on D. Then $f(t, x)$ is Lipschitz continuous in x over D.

Proof Let (t, x_1) and (t, x_2) be points in D. For fixed t, $(\partial f/\partial x)(t, x)$ is a function of x, and so we may apply the mean value theorem of differential calculus [statement (b)] to obtain

$$|f(t, x_1) - f(t, x_2)| = \left|\frac{\partial f}{\partial x}(t, \xi)\right||x_1 - x_2|,$$

where $x_1 < \xi < x_2$. But since $\partial f/\partial x$ is continuous in D, it is bounded there [according to statement (a)]. Hence there is a constant k, $0 < k < \infty$, such that

$$\left|\frac{\partial f}{\partial x}(t, x)\right| \le k,$$

for all (t, x) in D. ∎

Example 2 If $f(t, x) = tx^2$ on $0 \le t \le 1, 0 \le x \le 1$, then

$$\left|\frac{\partial f}{\partial x}\right| = |2tx| \le 2,$$

so that

$$|f(t, x_1) - f(t, x_2)| \le 2|x_1 - x_2|.$$

In Section 2.7 (p. 75) we defined the **Picard iterations**:

$$\begin{aligned} x_0(t) &= x_0, \\ x_1(t) &= x_0 + \int_{t_0}^{t} f(s, x_0(s))\, ds, \\ x_2(t) &= x_0 + \int_{t_0}^{t} f(s, x_1(s))\, ds, \\ &\vdots \\ x_n(t) &= x_0 + \int_{t_0}^{t} f(s, x_{n-1}(s))\, ds. \end{aligned} \tag{5}$$

We here show that under certain conditions the Picard iterations defined by (5) converge uniformly to a solution of equation (2). In Example 2.7.1 (p. 75) we showed that the Picard iterations converge to the unique solution of the initial-value problem

$$x'(t) = x(t), \qquad x(0) = 1. \tag{6}$$

We now state and prove the main result of this appendix.

APPENDIX 3/THE EXISTENCE AND UNIQUENESS OF SOLUTIONS

THEOREM 3   EXISTENCE THEOREM

Let $f(t, x)$ be Lipschitz continuous in x with the Lipschitz constant k on the region D of all points (t, x) satisfying the inequalities

$$|t - t_0| \leq a, \qquad |x - x_0| \leq b.$$

(See Figure A3.3.) Then there exists a number $\delta > 0$ with the property that the initial-value problem

$$x' = f(t, x), \qquad x(t_0) = x_0,$$

has a solution $x = x(t)$ on the interval $|t - t_0| \leq \delta$.

Proof The proof of this theorem is complicated and is done in several stages. However, the basic idea is simple: We need only justify that the Picard iterations converge uniformly and yield, in the limit, the solution of the integral equation (2).

Since f is continuous on D, it is bounded there [statement (a)] and we may begin by letting M be a finite upper bound for $|f(t, x)|$ on D. We then define

$$\delta = \min\{a, b/M\}. \tag{7}$$

1. We first show that the iterations $\{x_n(t)\}$ are continuous and satisfy the inequality

$$|x_n(t) - x_0| \leq b. \tag{8}$$

Inequality (8) is necessary in order that $f(t, x_n(t))$ be defined for $n = 0, 1, 2, \ldots$. To show the continuity of $x_n(t)$, we first note that $x_0(t) = x_0$ is continuous (a constant function is always continuous). Then

$$x_1(t) = x_0 + \int_{t_0}^{t} f(t, x_0(s))\, ds.$$

But $f(t, x_0)$ is continuous [since $f(t, x)$ is continuous in t and x], and the integral of a continuous function is continuous. Thus $x_1(t)$ is continuous. In a similar fashion, we can show that

$$x_2(t) = x_0 + \int_{t_0}^{t} f(t, x_1(s))\, ds$$

is continuous, and so on for $n = 3, 4, \ldots$.

Obviously inequality (8) holds when $n = 0$, because $x_0(t) = x_0$. For $n \neq 0$, we use definition (5) and equation (7) to obtain

$$|x_n(t) - x_0| = \left|\int_{t_0}^{t} f(s, x_{n-1}(s))\, ds\right| \leq \left|\int_{t_0}^{t} |f(s, x_{n-1}(s))|\, ds\right|$$

$$\leq M\left|\int_{t_0}^{t} ds\right| = M|t - t_0| \leq M\delta \leq b.$$

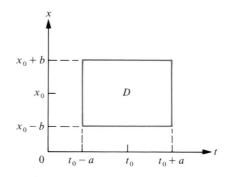

Figure A3.3

These inequalities follow from statement (f). Note that the last inequality helps explain the choice of δ in equation (7).

2. Next, we show by induction that

$$|x_n(t) - x_{n-1}(t)| \leq Mk^{n-1}\frac{|t-t_0|^n}{n!} \leq \frac{Mk^{n-1}\delta^n}{n!}. \tag{9}$$

If $n = 1$, we obtain

$$|x_1(t) - x_0(t)| \leq \left|\int_{t_0}^t f(s, x_0(s))\, ds\right| \leq M\left|\int_{t_0}^t ds\right|$$
$$= M|t - t_0| \leq M\delta.$$

Thus the result is true for $n = 1$.

We assume that the result is true for $n = m$ and prove that it holds for $n = m+1$; that is, we assume that

$$|x_m(t) - x_{m-1}(t)| \leq \frac{Mk^{m-1}|t-t_0|^m}{m!} \leq \frac{Mk^{m-1}\delta^m}{m!}.$$

Then, since $f(t, x)$ is Lipschitz continuous in x over D,

$$|x_{m+1}(t) - x_m(t)| = \left|\int_{t_0}^t f(s, x_m(s))\, ds - \int_{t_0}^t f(s, x_{m-1}(s))\, ds\right|$$
$$\leq \left|\int_{t_0}^t |f(s, x_m(s)) - f(s, x_{m-1}(s))|\, ds\right|$$
$$\leq k\left|\int_{t_0}^t |x_m(s) - x_{m-1}(s)|\, ds\right|$$
$$\leq \frac{Mk^m}{m!}\left|\int_{t_0}^t (s-t_0)^m\, ds\right|^\dagger = \frac{Mk^m|t-t_0|^{m+1}}{(m+1)!} \leq \frac{Mk^m\delta^{m+1}}{(m+1)!},$$

which is what we wanted to show.

3. We now show that $x_n(t)$ converges uniformly to a limit function $x(t)$ on the interval $|t - t_0| \leq \delta$. By statement (d), this shows that $x(t)$ is continuous.

We first note that

$$x_n(t) - x_0(t) = x_n(t) - x_{n-1}(t) + x_{n-1}(t) - x_{n-2}(t) + \cdots + x_1(t) - x_0(t)$$
$$= \sum_{k=0}^n [x_m(t) - x_{m-1}(t)]. \tag{10}$$

But by inequality (9),

$$|x_m(t) - x_{m-1}(t)| \leq \frac{Mk^{m-1}\delta^m}{m!} = \frac{M}{k}\frac{k^m\delta^m}{m!},$$

so

$$\sum_{m=1}^\infty |x_m(t) - x_{m-1}(t)| \leq \frac{M}{k}\sum_{m=1}^\infty \frac{(k\delta)^m}{m!} = \frac{M}{k}(e^{k\delta} - 1),$$

since

$$e^{k\delta} = \sum_{m=0}^\infty \frac{(k\delta)^m}{m!} = 1 + \sum_{m=1}^\infty \frac{(k\delta)^m}{m!}.$$

---

† This inequality follows from the induction assumption that inequality (9) holds for $n = m$.

By the Weierstrass M-test [statement (g)], we conclude that the series

$$\sum_{m=1}^{\infty} [x_m(t) - x_{m-1}(t)]$$

converges absolutely and uniformly on $|t - t_0| \leq \delta$ to a unique limit function $y(t)$. But

$$y(t) = \lim_{n \to \infty} \sum_{m=1}^{n} [x_m(t) - x_{m-1}(t)]$$

$$= \lim_{n \to \infty} [x_n(t) - x_0(t)] = \lim_{n \to \infty} x_n(t) - x_0(t),$$

or

$$\lim_{n \to \infty} x_n(t) = y(t) + x_0(t).$$

We denote the right-hand side of this equation by $x(t)$. Thus the limit of the Picard iterations $x_n(t)$ exists and the convergence $x_n(t) \to x(t)$ is uniform for all t in the interval $|t - t_0| \leq \delta$.

4. It remains to be shown that $x(t)$ is a solution to equation (2) for $|t - t_0| < \delta$. Since $f(t, x)$ is a continuous function of x and $x_n(t) \to x(t)$ as $n \to \infty$, we have, by statement (e),

$$\lim_{n \to \infty} f(t, x_n(t)) = f(t, x(t)).$$

Hence, by equation (5),

$$x(t) = \lim_{n \to \infty} x_{n+1}(t) = x_0 + \lim_{n \to \infty} \int_{t_0}^{t} f(s, x_n(s)) \, ds$$

$$= x_0 + \int_{t_0}^{t} \lim_{n \to \infty} f(s, x_n(s)) \, ds = x_0 + \int_{t_0}^{t} f(s, x(s)) \, ds.$$

The step in which we interchange the limit and integral is justified by statement (h). Thus $x(t)$ solves equation (2), and therefore it solves the initial-value problem (1). ∎

It turns out that the solution obtained in Theorem 3 is unique. Before proving this, however, we derive a simple version of a very useful result known as **Gronwall's inequality**.

THEOREM 4 GRONWALL'S INEQUALITY

Let $x(t)$ be a continuous nonnegative function and suppose that

$$x(t) \leq A + B \left| \int_{t_0}^{t} x(s) \, ds \right|, \tag{11}$$

where A and B are positive constants, for all values of t such that $|t - t_0| \leq \delta$. Then

$$\boxed{x(t) \leq A e^{B|t - t_0|}} \tag{12}$$

for all t in the interval $|t - t_0| \leq \delta$.

Proof We prove this result for $t_0 \leq t \leq t_0 + \delta$. The proof for $t_0 - \delta \leq t \leq t_0$ is similar (see Problem 14). Since $x(t) \geq 0$ and $t > t_0$,

$$\left| \int_{t_0}^{t} x(s) \, ds \right| = \int_{t_0}^{t} x(s) \, ds$$

and we define
$$y(t) = B\int_{t_0}^{t} x(s)\,ds.$$

Then
$$y'(t) = Bx(t) \leq B\left[A + B\int_{t_0}^{t} x(s)\,ds\right] = AB + By,$$

or
$$y'(t) - By(t) \leq AB. \tag{13}$$

We note that
$$\frac{d}{dt}\left[y(t)e^{-B(t-t_0)}\right] = e^{-B(t-t_0)}[y'(t) - By(t)].$$

Therefore, multiplying both sides of equation (13) by the integrating factor $e^{-B(t-t_0)}$ (which is greater than zero), we have
$$\frac{d}{dt}\left[y(t)e^{-B(t-t_0)}\right] \leq ABe^{-B(t-t_0)}.$$

An integration of both sides of the inequality from t_0 to t yields
$$y(s)e^{-B(s-t_0)}\Big|_{t_0}^{t} \leq AB\int_{t_0}^{t} e^{-B(s-t_0)}\,ds = -Ae^{-B(s-t_0)}\Big|_{t_0}^{t}.$$

But $y(t_0) = 0$, so
$$y(t)e^{-B(t-t_0)} \leq A(1 - e^{-B(t-t_0)}),$$

from which, after multiplying both sides by $e^{B(t-t_0)}$, we obtain
$$y(t) \leq A[e^{B(t-t_0)} - 1].$$

Then by equation (11),
$$x(t) \leq A + y(t) \leq Ae^{B(t-t_0)}. \quad\blacksquare$$

THEOREM 5   UNIQUENESS THEOREM

Let the conditions of Theorem 3 (existence theorem) hold. Then $x(t) = \lim_{n\to\infty} x_n(t)$ is the only continuous solution of the initial-value problem (1) in $|t - t_0| \leq \delta$.

Proof   Let $x(t)$ and $y(t)$ be two continuous solutions of equation (2) in the interval $|t - t_0| \leq \delta$ and suppose that $(t, y(t))$ belongs to the region D for all t in that interval.[†] Define $v(t) = |x(t) - y(t)|$. Then $v(t) \geq 0$ and $v(t)$ is continuous. Since $f(t,x)$ is Lipschitz continuous in x over D,

$$v(t) = \left\|\left[x_0 + \int_{t_0}^{t} f(s, x(s))\,ds\right] - \left[x_0 + \int_{t_0}^{t} f(s, y(s))\,ds\right]\right\|$$
$$\leq k\left|\int_{t_0}^{t} |x(s) - y(s)|\,ds\right| = k\left|\int_{t_0}^{t} v(s)\,ds\right|$$
$$\leq \epsilon + k\left|\int_{t_0}^{t} v(s)\,ds\right|$$

[†] Note that without this assumption, the function $f(t, y(t))$ may not even be defined at points where $(t, y(t))$ is not in D.

for every $\epsilon > 0$. By Gronwall's inequality, we have

$$v(t) \leq \epsilon e^{k|t-t_0|}.$$

But $\epsilon > 0$ can be chosen arbitrarily close to zero, so that $v(t) \leq 0$. Since $v(t) \geq 0$, it follows that $v(t) \equiv 0$, implying that $x(t)$ and $y(t)$ are identical. Hence the limit of the Picard iterations is the only continuous solution. ∎

THEOREM 6 LOCAL EXISTENCE-UNIQUENESS THEOREM

Let $f(t, x)$ and $(\partial f/\partial x)$ be continuous on D. Then there exists a constant $\delta > 0$ such that the Picard iterations $\{x_n(t)\}$ converge to a unique continuous solution of the initial-value problem (1) on $|t - t_0| \leq \delta$.

Proof This theorem follows directly from Theorem 2 and the existence and uniqueness theorems. ∎

We note that Theorems 3, 5, and 6 are *local* results. By this we mean that unique solutions are guaranteed to exist only "near" the initial point (t_0, x_0).

Example 3 Let

$$x'(t) = x^2(t), \qquad x(1) = 2.$$

Without solving this equation, we can show that there is a unique solution in some interval $|t - t_0| = |t - 1| \leq \delta$. Let $a = b = 1$. Then $|f(t, x)| = x^2 \leq 9$ ($= M$) for all $|x - x_0| \leq 1$, $x_0 = x(1) = 2$. Therefore, $\delta = \min\{a, b/M\} = \frac{1}{9}$, and Theorem 6 guarantees the existence of a unique solution on the interval $|t - 1| \leq \frac{1}{9}$. The solution of this initial-value problem is easily found by a separation of variables to be $x(t) = 2/(3 - 2t)$. This solution exists as long as $t \neq \frac{3}{2}$. Starting at $t_0 = 1$, we see that the maximum interval of existences is $|t - t_0| < \frac{1}{2}$. Hence the value $\delta = \frac{1}{9}$ is not the best possible.

Example 4 Consider the initial-value problem

$$x' = \sqrt{x}, \qquad x(0) = 0.$$

As we saw in Example 1, $f(t, x) = \sqrt{x}$ does *not* satisfy a Lipschitz condition in any region containing the point $(0, 0)$. By a separation of variables, it is easy to calculate the solution

$$x(t) = \left(\frac{t}{2}\right)^2.$$

However $y(t) = 0$ is also a solution. Hence, without a Lipschitz condition the solution to an initial-value problem (if one exists) may fail to be unique.

The last two examples illustrate the local nature of our existence-uniqueness theorem. We now show that it is possible to extend our local results to systems and derive *global* existence-uniqueness results for certain linear differential equations; that is, we show that a unique solution exists for every real number t.

Consider the initial-value system

$$\begin{aligned}
x_1' &= f_1(t, x_1, x_2, \ldots, x_n), & x_1(t_0) &= x_{10}, \\
x_2' &= f_2(t, x_1, x_2, \ldots, x_n), & x_2(t_0) &= x_{20}, \\
&\vdots & &\vdots \\
x_n' &= f_n(t, x_1, x_2, \ldots, x_n), & x_n(t_0) &= x_{n0}.
\end{aligned} \qquad (14)$$

As in Section 7.6, we write the vectors

$$\mathbf{x} = \begin{pmatrix} x_1 \\ x_2 \\ \vdots \\ x_n \end{pmatrix}, \quad \mathbf{f}(t, \mathbf{x}) = \begin{pmatrix} f_1(t, x_1, \ldots, x_n) \\ f_2(t, x_1, \ldots, x_n) \\ \vdots \\ f_n(t, x_1, \ldots, x_n) \end{pmatrix}, \quad \mathbf{x}_0 = \begin{pmatrix} x_{10} \\ x_{20} \\ \vdots \\ x_{n0} \end{pmatrix}.$$

Then we may write (14) in the compact form

$$\mathbf{x}' = \mathbf{f}(t, \mathbf{x}), \quad \mathbf{x}(t_0) = \mathbf{x}_0. \tag{15}$$

In order to generalize the results of Theorems 3 and 5, it is necessary to generalize the notion of the absolute value of a number to the absolute value or *norm* of a vector $\mathbf{x}$ or matrix A. If $\mathbf{x} = (x_1, x_2)$ is a two-vector, then the distance from (x_1, x_2) to the origin is given by the Pythagorean theorem (see Figure A3.4):

$$\|\mathbf{x}\| = \sqrt{(x_1^2 + x_2^2)}. \tag{16}$$

Therefore, it is natural to define the length or **norm** of an n-vector $\mathbf{x} = (x_1, x_2, \ldots, x_n)$ by

$$\|\mathbf{x}\| = \sqrt{(x_1^2 + x_2^2 + \cdots + x_n^2)}. \tag{17}$$

If A is an $n \times n$ matrix, there are several ways to define its norm. For our purposes, the simplest choice for the norm of A is

$$\|A\| = \sum_{i=1}^{n} \sum_{j=1}^{n} |a_{ij}|; \tag{18}$$

that is, the norm of A is the sum of the absolute values of the components of A.

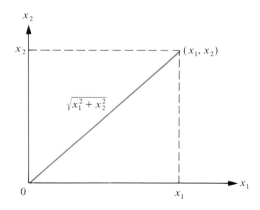

Figure A3.4

Using equations (17) and (18), we may prove that

a. $\|\mathbf{x}\| \geq 0$, where the equality holds only if $\mathbf{x}$ is the zero vector; (19)

b. $\|\alpha \mathbf{x}\| = |\alpha| \|\mathbf{x}\|$ when α is a scalar; (20)

c. $\|\mathbf{x} + \mathbf{y}\| \leq \|\mathbf{x}\| + \|\mathbf{y}\|$. (21)

Moreover, if A is an $n \times n$ matrix, it is not difficult to show (see Problems 17–20) that

$$\|A\mathbf{x}\| \leq \|A\| \|\mathbf{x}\|. \tag{22}$$

Now using this notation and a norm $\|\cdot\|$ for every absolute value $|\cdot|$ that appeared in the proofs of Theorems 3 and 5, we obtain the existence and uniqueness of a local vector solution $\mathbf{x}(t)$ of the initial-value problem (15). The verification of these details is left as an exercise. In fact, all the steps of the proof of Theorem 3 are as before with $\|\cdot\|$ replacing $|\cdot|$. The notions of continuity and uniform convergence apply to the component functions of $\mathbf{x}$ and $\mathbf{f}$. We have the following result:

THEOREM 7 EXISTENCE-UNIQUENESS THEOREM FOR SYSTEMS

Let D denote the region [in $(n+1)$-dimensional space, one dimension for t and n dimensions for the vector $\mathbf{x}$]

$$|t - t_0| \leq a, \qquad \|\mathbf{x} - \mathbf{x}_0\| \leq b, \tag{23}$$

and suppose that $\mathbf{f}(t, \mathbf{x})$ satisfies the Lipschitz condition

$$\|\mathbf{f}(t, \mathbf{x}_1) - \mathbf{f}(t, \mathbf{x}_2)\| \leq k \|\mathbf{x}_1 - \mathbf{x}_2\|, \tag{24}$$

whenever the pairs $(t, \mathbf{x}_1)$ and $(t, \mathbf{x}_2)$ belong to D, where k is a positive constant. Then there is a constant $\delta > 0$ such that there exists a unique continuous vector solution $\mathbf{x}(t)$ of system (15) in the interval $|t - t_0| \leq \delta$.

Before leaving this theorem, we should briefly discuss condition (24). It is easy to see that (24) is implied by the inequalities

$$|f_i(t, x_{11}, \ldots, x_{1n}) - f_i(t, x_{21}, \ldots, x_{2n})| \leq k_i \sum_{j=1}^{n} |x_{1j} - x_{2j}|, \tag{25}$$

for $i = 1, \ldots, n$ with $k = n\sqrt{\sum_{i=1}^{n} k_i^2}$. This fact follows from the double inequality

$$\frac{1}{n} \sum_{j=1}^{n} |x_j| \leq \|\mathbf{x}\| \leq \sum_{j=1}^{n} |x_j|, \tag{26}$$

which is an immediate consequence of the definition of $\|\mathbf{x}\|$ (see Problem 21).

Another condition that implies inequality (24) is

$$|f_i(t, x_{11}, \ldots, x_{1n}) - f_i(t, x_{21}, \ldots, x_{2n})| \leq k_2 \max_j |x_{1j} - x_{2j}|, \tag{27}$$

for $i = 1, 2, \ldots, n$. The two inequalities (25) and (27) are useful, since it is often very difficult to verify inequality (24) directly.

Finally, if the partial derivatives $\partial f_i / \partial x_j$, $i, j = 1, 2, \ldots, n$, are continuous in D, then they are bounded on D and conditions (25) and (27) both follow from the mean value theorem of differential calculus.

GLOBAL EXISTENCE AND UNIQUENESS FOR A LINEAR SYSTEM

Consider the system

$$\mathbf{x}' = A(t)\mathbf{x} + \mathbf{f}(t), \qquad \mathbf{x}(t_0) = \mathbf{x}_0, \tag{28}$$

where t_0 is a point in the interval $\alpha \leq t \leq \beta$ and $A(t)$ is an $n \times n$ matrix.

THEOREM 8 GLOBAL EXISTENCE-UNIQUENESS THEOREM

Let $A(t)$ and $\mathbf{f}(t)$ be a continuous matrix and vector function, respectively, on the interval $\alpha \le t \le \beta$. Then there exists a unique vector function $\mathbf{x}(t)$ that is a solution of equation (28) on the entire interval $\alpha \le t \le \beta$.

Remark We proved this theorem for the scalar case ($n=1$) in Section 2.7 (see p. 76).

Proof We define the Picard iterations

$$\mathbf{x}_0(t) = \mathbf{x}_0,$$
$$\mathbf{x}_1(t) = \mathbf{x}_0 + \int_{t_0}^{t} [A(s)\mathbf{x}_0(s) + \mathbf{f}(s)]\, ds,$$
$$\vdots \qquad \vdots \qquad \vdots \qquad \qquad (29)$$
$$\mathbf{x}_{n+1}(t) = \mathbf{x}_0 + \int_{t_0}^{t} [A(s)\mathbf{x}_n(s) + \mathbf{f}(s)]\, ds,$$
$$\vdots \qquad \vdots \qquad \vdots$$

Clearly these iterations are continuous on $\alpha \le t \le \beta$ for $n=1, 2, \ldots$. Since $A(t)$ is a continuous matrix function, we have

$$\sup_{\alpha \le t \le \beta} \|A(t)\| = \sup_{\alpha \le t \le \beta} \sum_{i=1}^{n} \sum_{j=1}^{n} |a_{ij}(t)| = k < \infty.$$

Let
$$M = \sup_{\alpha \le t \le \beta} \|A(t)\mathbf{x}_0 + \mathbf{f}(t)\|,$$

which is finite [according to statement (a)] since $A(t)$ and $\mathbf{f}(t)$ are continuous. Observe that we are unable to define the value δ of equation (7), since the domain of the vector function $\mathbf{x}(t)$ is *unrestricted*. Thus a slight modification of the proof of Theorem 3 is necessary:

a. Note that for any vectors $\mathbf{x}_1$ and $\mathbf{x}_2$ and for $\alpha \le t \le \beta$, we have

$$\|[A(t)\mathbf{x}_1 + \mathbf{f}(t)] - [A(t)\mathbf{x}_2 + \mathbf{f}(t)]\| \le \|A(t)\| \|\mathbf{x}_1 - \mathbf{x}_2\| \le k\|\mathbf{x}_1 - \mathbf{x}_2\|.$$

Thus the vector function $A(t)\mathbf{x} + \mathbf{f}$ satisfies a Lipschitz condition for any $\mathbf{x}$ and $\alpha \le t \le \beta$.

b. By induction, we now show that the Picard iterations (29) satisfy the inequality

$$\|\mathbf{x}_n(t) - \mathbf{x}_{n-1}(t)\| \le Mk^{n-1} \frac{|t - t_0|^n}{n!} \le Mk^{n-1} \frac{(\beta - \alpha)^n}{n!}. \qquad (30)$$

We prove the first inequality. The second follows easily. First, if $n=1$, then

$$\|\mathbf{x}_1(t) - \mathbf{x}_0(t)\| = \left\| \int_{t_0}^{t} [A(s)\mathbf{x}_0 + \mathbf{f}(s)]\, ds \right\|$$
$$\le \left| \int_{t_0}^{t} \|A(s)\mathbf{x}_0 + \mathbf{f}(s)\|\, ds \right|$$
$$\le M \left| \int_{t_0}^{t} ds \right| = M|t - t_0| \le M(\beta - \alpha).^{\dagger}$$

---

† The first of these inequalities is called the **Cauchy-Schwarz inequality**. Its proof may be found in any advanced calculus text. See, for example, R. Buck, *Advanced Calculus*, 3d ed., (New York: McGraw-Hill, 1978).

Assume that the result holds for $n = m$. Then by part (a),

$$\|\mathbf{x}_{m+1}(t) - \mathbf{x}_m(t)\| \le \left| \int_{t_0}^{t} \|[A(s)\mathbf{x}_m(s) + \mathbf{f}(s)] - [A(s)\mathbf{x}_{m-1}(s) + \mathbf{f}(s)]\| \, ds \right|$$

$$\le k\left| \int_{t_0}^{t} \|\mathbf{x}_m(s) - \mathbf{x}_{m-1}(s)\| \, ds \right| \le k\left| \int_{t_0}^{t} Mk^{m-1} \frac{|s - t_0|^m}{m!} \, ds \right|$$

$$= Mk^m \frac{|t - t_0|^{m+1}}{(m+1)!} \le Mk^m \frac{(\beta - \alpha)^{m+1}}{(m+1)!},$$

which is the desired result.

The major difference between this proof and the proof of Theorem 3 is that, in the linear case, the bound M can be defined independently of x, whereas in Theorem 3 we must restrict x as well as t. This difference enables us to prove the global result that solutions exist and are unique on the entire interval $\alpha \le t \le \beta$.

With estimate (30) [which is identical to estimate (9)], the proof proceeds exactly as those for Theorems 3 and 5. ∎

COROLLARY

Let $A(t)$ and $\mathbf{f}(t)$ be continuous on the interval $-\infty < t < \infty$. Then there exists a unique continuous solution $\mathbf{x}(t)$ of (28) defined for *all* t on $-\infty < t < \infty$.

Proof Suppose $|t_0| \le n$. We define $\mathbf{x}_n(t)$ to be the unique solution of (28) on the interval $|t| \le n$ which is guaranteed by Theorem 8. We observe that $\mathbf{x}_n(t)$ coincides with $\mathbf{x}_{n+k}(t)$ on the interval $|t| \le n$ for $k = 1, 2, 3, \ldots$, since both are solutions of (28) and Theorem 8 requires any solution on $|t| \le n$ to be unique. Thus the vector function

$$\mathbf{x}(t) = \lim_{n \to \infty} \mathbf{x}_n(t)$$

is defined for all real values t, since $|t| \le n$ for sufficiently large n. Clearly $\mathbf{x}(t)$ is a solution of (28) on $-\infty < t < \infty$, and such a solution must be unique since it is uniquely defined on each finite interval containing t_0. ∎

PROBLEMS APPENDIX 3

For each initial-value problem of Problems 1–10 determine whether a unique solution can be guaranteed. If so, let $a = b = 1$ if possible, and find the number δ as given by equation (7). When possible, solve the equation and find a better value for δ, as in Example 3.

1. $x' = x^3$, $x(2) = 5$
2. $x' = x^3$, $x(1) = 2$
3. $x' = \dfrac{x}{t - x}$, $x(0) = 1$
4. $x' = x^{1/3}$, $x(1) = 0$
5. $x' = \sin x$, $x(1) = \pi/2$
6. $x' = \sqrt{x(x-1)}$, $x(1) = 2$
7. $x' = \ln|\sin x|$, $x(\pi/2) = 1$
8. $x' = \sqrt{x(x-1)}$, $x(2) = 3$
9. $x' = |x|$, $x(0) = 1$
10. $x' = tx$, $x(5) = 10$
11. Compute a Lipschitz constant for each of the following functions on the indicated region D:
    a. $f(t, x) = te^{-2x/t}$, $t > 0$, $x > 0$
    b. $f(t, x) = \sin tx$, $|t| \le 1$, $|x| \le 2$
    c. $f(t, x) = e^{-t^2} x^2 \sin\dfrac{1}{x}$, all t, $-1 \le x < 1$, $x \ne 0$

    [Part (c) shows that a Lipschitz constant may exist even when $\partial f/\partial x$ is not bounded in D.]
    d. $f(t, x) = (t^2 x^3)^{3/2}$, $|t| \le 2$, $|x| \le 3$

12. Consider the initial-value problem
$$x' = x^2, \quad x(1) = 3.$$
Show that the Picard iterations converge to the unique solution of this problem.

13. Construct the sequence $\{x_n(t)\}$ of Picard iterations for the initial-value problem
$$x' = -x, \quad x(0) = 3,$$
and show that it converges to the unique solution $x(t) = 3e^{-t}$.

14. Prove Gronwall's inequality (Theorem 4) for $t_0 - \delta \le t \le t_0$. [*Hint*: For $t < t_0$, $y(t) \le 0$.]

15. Let $v(t)$ be a positive function that satisfies the inequality
$$v(t) \le A + \int_{t_0}^{t} r(s) v(s) \, ds,$$
where $A \ge 0$, $r(t)$ is a continuous, positive function, and $t \ge t_0$. Prove that
$$v(t) \le A \exp\left[\int_{t_0}^{t} r(s) \, ds\right],$$
for all $t \ge t_0$. What kind of result holds for $t < t_0$? This is a general form of Gronwall's inequality. [*Hint*: Define $y(t) \equiv \int_{t_0}^{t} r(s)v(s)\, ds$ and show that $y'(t) = r(t)v(t) \le r(t)[A + y(t)]$. Finish the proof by following the steps of the proof of Theorem 4, using the integrating factor $\exp[-\int_{t_0}^{t} r(s)\, ds]$.]

16. Consider the initial-value problem
$$x'' = f(t, x), \quad x(0) = x_0, \quad x'(0) = x_1, \quad (i)$$
where f is defined on the rectangle D: $|t| \le a$, $|x - x_0| \le b$. Prove under appropriate hypotheses that if a solution exists, then it must be unique. [*Hint*: Let $x(t)$ and $y(t)$ be continuous solutions of (i). Verify by differentiation that
$$x(t) = x_0 + x_1 t - \int_0^t (t-s) f(s, x(s))\, ds,$$
$$y(t) = x_0 + x_1 t - \int_0^t (t-s) f(s, y(s))\, ds,$$
in some interval $|t| \le \delta$, $\delta > 0$. Then subtract these two expressions, use an appropriate Lipschitz condition, and apply the Gronwall inequality of Problem 15.]

17. Prove inequality (19).
18. Prove equation (20).
*19. Prove inequality (21). [*Hint*: (a) First show that the **inner product**
$$\mathbf{x} \cdot \mathbf{y} = x_1 y_1 + x_2 y_2 + \cdots + x_n y_n$$
satisfies the condition $|\mathbf{x} \cdot \mathbf{y}| \le \|\mathbf{x}\| \|\mathbf{y}\|$, where $\mathbf{x} = (x_1, x_2, \ldots, x_n)$ and $\mathbf{y} = (y_1, y_2, \ldots, y_n)$. (b) Show that $\mathbf{x} \cdot \mathbf{x} = \|\mathbf{x}\|^2$. (c) Apply (b) to $\|\mathbf{x} + \mathbf{y}\|^2$ and use the inequality in (a) to verify (21).]

*20. Prove inequality (22).
21. Prove inequality (26).
22. Using inequality (26), show that inequalities (24), (25), and (27) are all equivalent.
23. Show that, if the partial derivatives $\partial f_i / \partial x_j$ are continuous on the region D defined by (23), then there exists a constant $k_i > 0$ such that inequality (24) holds for $i = 1, \ldots, n$.
*24. Prove Theorem 7 by carrying out the following steps:

a. Show that system (15) is equivalent to the integral equation
$$\mathbf{x}(t) = \mathbf{x}_0 + \int_{t_0}^{t} \mathbf{f}(s, \mathbf{x}(s))\, ds. \quad \text{(ii)}$$

b. Define the vector-valued Picard iterates $\{\mathbf{x}_n\}$ as in equations (5).

c. Prove that
$$M = \sup_{(t, \mathbf{x}) \in D} \|\mathbf{f}(t, \mathbf{x})\|$$
is finite.

d. Prove that if $\delta = \min\{a, b/M\}$, then the Picard iterates defined in (b) satisfy the condition $\|\mathbf{x}_n(t) - \mathbf{x}_0\| \le b$, for $|t - t_0| \le a$.

e. Show by induction that
$$\|\mathbf{x}_n(t) - \mathbf{x}_{n-1}(t)\| \le Mk^{n-1} \frac{|t - t_0|^n}{n!}$$
$$\le Mk^{n-1} \frac{\delta^n}{n!}.$$

f. Prove that the series
$$\sum_{m=1}^{\infty} \|\mathbf{x}_m(t) - \mathbf{x}_{m-1}(t)\|$$
converges uniformly for $|t - t_0| < \delta$ and conclude that $\mathbf{x}_n(t)$ converges uniformly to a vector function $\mathbf{x}(t)$ as n tends to ∞.

g. Show that the $\mathbf{x}(t)$ defined in part (f) is a continuous solution to (ii) on the interval $|t - t_0| \le \delta$.

h. Use Gronwall's inequality to show that this solution is unique.

25. Consider the system
$$x_1' = t x_1 x_2 + t^2, \quad x_1(0) = 1,$$
$$x_2' = x_1^2 + x_2^2 + t, \quad x_2(0) = 2.$$

a. Write the system in form (15).

b. Find a Lipschitz constant for $\mathbf{f}$ in the region
$$D: |t| \le 1, \quad (x_1 - 1)^2 + (x_2 - 2)^2 \le 1.$$

c. Find $\delta = \min\{a, b/M\}$ as in Problem 24.

26. Follow the steps of Problem 25 for the system
$$x_1' = x_2^2 + 1, \quad x_1(1) = 0,$$
$$x_2' = x_1^2 + t, \quad x_2(1) = 1,$$
in the region D: $|t - 1| \leq 3$, $x_1^2 + (x_2 - 1)^2 \leq 4$.

27. Follow the steps of Problem 25 for the system
$$x_1' = x_1^2 x_2,$$
$$x_2' = x_3 + t, \quad \mathbf{x}(0) = \mathbf{c},$$
$$x_3' = x_3^2,$$
in the region D: $|t| \leq a$, $\|\mathbf{x} - \mathbf{c}\| \leq b$.

*28. Under the assumption of Theorem 7, consider the systems
$$\mathbf{x}' = \mathbf{f}(t, \mathbf{x}), \quad \mathbf{x}(t_0) = \mathbf{x}_{10},$$
$$\mathbf{x}' = \mathbf{f}(t, \mathbf{x}), \quad \mathbf{x}(t_0) = \mathbf{x}_{20}.$$
Let $\mathbf{x}_1$ and $\mathbf{x}_2$ be solutions of these two systems, respectively. Prove that
$$\|\mathbf{x}_1(t) - \mathbf{x}_2(t)\| \leq \|\mathbf{x}_{10} - \mathbf{x}_{20}\| e^{k\delta}.$$
This shows that the solutions of Theorem 7 vary continuously with the initial vector $\mathbf{x}(t_0)$.

*29. Consider the system
$$\mathbf{x}' = \mathbf{f}(t, \mathbf{x}). \qquad \text{(iii)}$$
Let D denote the region $a \leq t \leq b$, $\mathbf{x} \in R^n$, and assume that
$$\|\mathbf{f}(t, \mathbf{x}_1) - \mathbf{f}(t, \mathbf{x}_2)\| \leq k\|\mathbf{x}_1 - \mathbf{x}_2\|,$$
for any pair of points $(t, \mathbf{x}_1)$ and $(t, \mathbf{x}_2)$ in D. Prove that for any pair $(t_0, \mathbf{x}_0)$ in D there exists a unique solution $\mathbf{x}(t)$ of equation (iii) defined on the entire interval $a \leq t \leq b$, which satisfies $\mathbf{x}(t_0) = \mathbf{x}_0$. [*Hint*: Proceed as in the proof of Theorem 8. This is a more general global existence-uniqueness result.]

APPENDIX 4

Determinants

In several parts of this book we make use of determinants. In this appendix we show how determinants arise and discuss their uses.

We begin by considering the system of two linear equations in two unknowns:

$$a_{11}x_1 + a_{12}x_2 = b_1,$$
$$a_{21}x_1 + a_{22}x_2 = b_2. \quad (1)$$

For simplicity we assume that the constants a_{11}, a_{12}, a_{21}, and a_{22} are all nonzero (otherwise, the system can be solved directly). To solve system (1), we reduce it to one equation in one unknown. To accomplish this, we multiply the first equation by a_{22} and the second by a_{12} to obtain

$$a_{11}a_{22}x_1 + a_{22}a_{12}x_2 = a_{22}b_1,$$
$$a_{12}a_{21}x_1 + a_{22}a_{12}x_2 = a_{12}b_2. \quad (2)$$

Subtracting the second equation from the first, we have

$$(a_{11}a_{22} - a_{12}a_{21})x_1 = a_{22}b_1 - a_{12}b_2. \quad (3)$$

Now we define the quantity

$$\boxed{D = a_{11}a_{22} - a_{12}a_{21}.} \quad (4)$$

If $D \neq 0$, then (3) yields

$$x_1 = \frac{a_{22}b_1 - a_{12}b_2}{D}, \quad (5)$$

and x_2 may be obtained by substituting this value of x_1 into either of the equations of (1). *Thus, if $D \neq 0$, system (1) has a unique solution.*

On the other hand, suppose that $D = 0$. Then $a_{11}a_{22} = a_{12}a_{21}$, and equation (3) leads to the equation

$$0 = a_{22}b_1 - a_{12}b_2.$$

Either this equation is true or it is false. If it is false, that is, if $a_{22}b_1 - a_{12}b_2 \neq 0$, then system (1) has *no* solution. If the equation is true, that is, $a_{22}b_1 - a_{12}b_2 = 0$, then the second equation of (2) is a multiple of the first and (1) consists essentially of only one equation. In this case we may choose x_1 arbitrarily and calculate the corresponding value of x_2, which means that there are an *infinite* number of solutions. In sum, we have shown that *if $D = 0$, then system (1) has either no solution or an infinite number of solutions.*

These facts are easily visualized geometrically by noting that (1) consists of the equations of two straight lines. A solution of the system is a point of intersection of the two lines. If the slopes are different, then $D \neq 0$ and the two lines intersect at a single point, which is the unique solution. It is easy to show (see Problem 21) that $D = 0$ if and only if the slopes of the two lines are the same. If $D = 0$, either we have two parallel lines and no solution, since the lines never intersect, or both equations yield the same line and every point on this line is a solution. These results are illustrated in Figure A4.1.

(a) Unique solution

(b) No solution

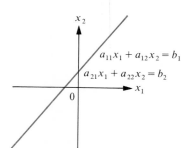

(c) Infinitely many solutions

Figure A4.1

Example 1 Consider the following systems of equations:

(i) $2x_1 + 3x_2 = 12$ (ii) $x_1 + 3x_2 = 3$ (iii) $x_1 + 3x_2 = 3$
$x_1 + x_2 = 5$ $3x_1 + 9x_2 = 8$ $3x_1 + 9x_2 = 9$

In system (i), $D = 2 \cdot 1 - 3 \cdot 1 = -1 \neq 0$, so there is a unique solution, which is easily found to be $x_1 = 3$, $x_2 = 2$. In system (ii), $D = 1 \cdot 9 - 3 \cdot 3 = 0$. Multiplying the first equation by 3 and then subtracting this from the second equation, we obtain the equation $0 = -1$, which is impossible. Thus there is no solution. In (iii), $D = 1 \cdot 9 - 3 \cdot 3 = 0$. But now the second equation is simply three times the first equation. If x_2 is arbitrary, then $x_1 = 3 - 3x_2$, and there are an infinite number of solutions.

Returning again to system (1), we define the **determinant of the system** as

$$D = a_{11}a_{22} - a_{12}a_{21}. \tag{6}$$

For convenience of notation, we denote the determinant by writing the coefficients of the system in a square array:

$$D = \begin{vmatrix} a_{11} & a_{12} \\ a_{21} & a_{22} \end{vmatrix} = a_{11}a_{22} - a_{12}a_{21}. \tag{7}$$

Therefore, a 2×2 determinant is the product of the two components in the upper-left-to-lower-right diagonal minus the product of the other two components.

We have proved the following theorem:

THEOREM 1

For the 2×2 system (1), there is a unique solution if and only if the determinant D is not equal to zero. If $D = 0$, then there is either no solution or an infinite number of solutions.

Let us now consider the general system of n equations in n unknowns,

$$\begin{aligned} a_{11}x_1 + a_{12}x_2 + \cdots + a_{1n}x_n &= b_1, \\ a_{21}x_1 + a_{22}x_2 + \cdots + a_{2n}x_n &= b_2, \\ \vdots \qquad \vdots \qquad \qquad \vdots \quad &\vdots \\ a_{n1}x_1 + a_{n2}x_2 + \cdots + a_{nn}x_n &= b_n, \end{aligned} \tag{8}$$

and define the determinant of such a system in order to obtain a theorem like the one above for $n \times n$ systems. We begin by defining the determinant of a 3×3 system:

$$D = \begin{vmatrix} a_{11} & a_{12} & a_{13} \\ a_{21} & a_{22} & a_{23} \\ a_{31} & a_{32} & a_{33} \end{vmatrix} = a_{11}\begin{vmatrix} a_{22} & a_{23} \\ a_{32} & a_{33} \end{vmatrix} - a_{12}\begin{vmatrix} a_{21} & a_{23} \\ a_{31} & a_{33} \end{vmatrix} + a_{13}\begin{vmatrix} a_{21} & a_{22} \\ a_{31} & a_{32} \end{vmatrix}. \tag{9}$$

We see that to calculate a 3×3 determinant, it is necessary to calculate three 2×2 determinants.

Example 2

$$\begin{vmatrix} 3 & 5 & 2 \\ 4 & 2 & 3 \\ -1 & 2 & 4 \end{vmatrix} = 3 \begin{vmatrix} 2 & 3 \\ 2 & 4 \end{vmatrix} - 5 \begin{vmatrix} 4 & 3 \\ -1 & 4 \end{vmatrix} + 2 \begin{vmatrix} 4 & 2 \\ -1 & 2 \end{vmatrix}$$

$$= 3 \cdot 2 - 5 \cdot 19 + 2 \cdot 10 = -69.$$

The general definition of the determinant of the $n \times n$ system of equations (8) is simply an extension of this procedure:

$$D = \begin{vmatrix} a_{11} & a_{12} & \cdots & a_{1n} \\ a_{21} & a_{22} & \cdots & a_{2n} \\ \vdots & \vdots & & \vdots \\ a_{n1} & a_{n2} & \cdots & a_{nn} \end{vmatrix} = a_{11}A_{11} - a_{12}A_{12} + \cdots + (-1)^{n+1} a_{1n}A_{1n}, \quad (10)$$

where A_{1j} is the $(n-1) \times (n-1)$ determinant obtained by crossing out the first row and jth column of the original $n \times n$ determinant. Thus an $n \times n$ determinant can be obtained by calculating n $(n-1) \times (n-1)$ determinants. Note that in definition (10) the signs alternate. The signs of the n^2 $(n-1) \times (n-1)$ determinants can easily be illustrated by the following schematic diagram:

$$\begin{vmatrix} + & - & + & - & + & - & \cdots \\ - & + & - & + & - & + & \cdots \\ + & - & + & - & + & - & \cdots \\ - & + & - & + & - & + & \cdots \\ + & - & + & - & + & - & \cdots \\ \vdots & \vdots & \vdots & \vdots & \vdots & \vdots & \ddots \end{vmatrix}$$

Example 3

$$\begin{vmatrix} 1 & 3 & 5 & 2 \\ 0 & -1 & 3 & 4 \\ 2 & 1 & 9 & 6 \\ 3 & 2 & 4 & 8 \end{vmatrix} = 1 \begin{vmatrix} -1 & 3 & 4 \\ 1 & 9 & 6 \\ 2 & 4 & 8 \end{vmatrix} - 3 \begin{vmatrix} 0 & 3 & 4 \\ 2 & 9 & 6 \\ 3 & 4 & 8 \end{vmatrix} + 5 \begin{vmatrix} 0 & -1 & 4 \\ 2 & 1 & 6 \\ 3 & 2 & 8 \end{vmatrix} - 2 \begin{vmatrix} 0 & -1 & 3 \\ 2 & 1 & 9 \\ 3 & 2 & 4 \end{vmatrix}$$

$$= 1(-92) - 3(-70) + 5(2) - 2(-16) = 160.$$

(The values in parentheses are obtained by calculating the four 3×3 determinants.)

The reason for considering determinants of systems of n equations in n unknowns is that Theorem 1 also holds for these systems (this fact is not proven here).[†]

THEOREM 2

For system (8) there is a unique solution if and only if the determinant D, defined by (10), is not zero. If $D = 0$, then there is either no solution or an infinite number of solutions.

It is clear that calculating determinants by formula (10) can be extremely tedious, especially if $n \geq 5$. For that reason techniques are available for greatly simplifying these calculations. Some of these techniques are described in the theorems below. The proofs of these theorems can be found in the text cited in the footnote.

We begin with the result that states that the determinant can be obtained by expanding in any row.

[†] For a proof, see S. I. Grossman, *Elementary Linear Algebra*, 3d ed., (Belmont, Calif.: Wadsworth, 1987), Chapter 2.

THEOREM 3

For any i, $i = 1, 2, \ldots, n$,

$$D = \begin{vmatrix} a_{11} & a_{12} & \cdots & a_{1n} \\ a_{21} & a_{22} & \cdots & a_{2n} \\ \vdots & \vdots & & \vdots \\ a_{n1} & a_{n2} & \cdots & a_{nn} \end{vmatrix}$$

$$= (-1)^{i+1} a_{i1} A_{i1} + (-1)^{i+2} a_{i2} A_{i2} + \cdots + (-1)^{i+n} a_{in} A_{in},$$

where A_{ij} is the $(n-1) \times (n-1)$ determinant obtained by crossing off the ith row and jth column of D. Notice that the signs in the expansion of a determinant alternate.

Example 4 Calculate

$$\begin{vmatrix} 3 & 5 & 2 \\ 4 & 2 & 3 \\ -1 & 2 & 4 \end{vmatrix}$$

by expanding in the second row (see Example 2).

Solution

$$\begin{vmatrix} 3 & 5 & 2 \\ 4 & 2 & 3 \\ -1 & 2 & 4 \end{vmatrix} = (-1)^{2+1}(4) \begin{vmatrix} 5 & 2 \\ 2 & 4 \end{vmatrix} + (-1)^{2+2}(2) \begin{vmatrix} 3 & 2 \\ -1 & 4 \end{vmatrix} + (-1)^{2+3}(3) \begin{vmatrix} 3 & 5 \\ -1 & 2 \end{vmatrix}$$

$$= -4(16) + 2(14) - 3(11) = -69.$$

We remark that we can also get the same result by expanding in the third row of D.

THEOREM 4

For any j, $j = 1, 2, \ldots, n$,

$$D = \begin{vmatrix} a_{11} & a_{12} & \cdots & a_{1n} \\ a_{21} & a_{22} & \cdots & a_{2n} \\ \vdots & \vdots & & \vdots \\ a_{n1} & a_{n2} & \cdots & a_{nn} \end{vmatrix}$$

$$= (-1)^{1+j} a_{1j} A_{1j} + (-1)^{2+j} a_{2j} A_{2j} + \cdots + (-1)^{n+j} a_{nj} A_{nj},$$

where A_{ij} is as defined in Theorem 3.

Example 5 Calculate

$$D = \begin{vmatrix} 3 & 5 & 2 \\ 4 & 2 & 3 \\ -1 & 2 & 4 \end{vmatrix}$$

by expanding in the third column.

Solution

$$\begin{vmatrix} 3 & 5 & 2 \\ 4 & 2 & 3 \\ -1 & 2 & 4 \end{vmatrix} = (-1)^{1+3}(2)\begin{vmatrix} 4 & 2 \\ -1 & 2 \end{vmatrix} + (-1)^{2+3}(3)\begin{vmatrix} 3 & 5 \\ -1 & 2 \end{vmatrix} + (-1)^{3+3}(4)\begin{vmatrix} 3 & 5 \\ 4 & 2 \end{vmatrix}$$

$$= 2(10) - 3(11) + 4(-14) = -69.$$

THEOREM 5

Let

$$D = \begin{vmatrix} a_{11} & a_{12} & \cdots & a_{1n} \\ a_{21} & a_{22} & \cdots & a_{2n} \\ \vdots & \vdots & & \vdots \\ a_{n1} & a_{n2} & \cdots & a_{nn} \end{vmatrix}.$$

> i. If any row or column of D is a zero vector, then $D = 0$.
>
> ii. If any row (column) is a multiple of any other row (column), then $D = 0$.
>
> iii. Interchanging any two rows (columns) of D has the effect of multiplying D by -1.
>
> iv. Multiplying a row (column) of D by a constant α has the effect of multiplying D by α.
>
> v. If any row (column) of D is multiplied by a constant and added to a different row (column) of D, then D is unchanged.

Example 6 Calculate

$$D = \begin{vmatrix} 2 & 1 & 4 & 3 \\ 3 & 1 & -2 & -1 \\ 14 & -2 & 0 & 6 \\ 6 & 2 & -4 & -2 \end{vmatrix}.$$

Solution $D = 0$ according to (ii) since the fourth row is twice the second row. This can easily be verified.

Example 7 Calculate

$$D = \begin{vmatrix} 1 & 14 & 3 \\ 2 & 28 & -2 \\ 0 & -42 & 1 \end{vmatrix}.$$

Solution According to (iv), we may divide the second column by 14, which has the effect of dividing D by 14. Then

$$\frac{D}{14} = \begin{vmatrix} 1 & 1 & 3 \\ 2 & 2 & -2 \\ 0 & -3 & 1 \end{vmatrix},$$

or

$$D = 14\begin{vmatrix} 1 & 1 & 3 \\ 2 & 2 & -2 \\ 0 & -3 & 1 \end{vmatrix} = 14(-24) = -336.$$

Example 8 Calculate

$$D = \begin{vmatrix} 0 & -42 & 1 \\ 2 & 28 & -2 \\ 1 & 14 & 3 \end{vmatrix}.$$

Solution D is obtained from the determinant of Example 7 by interchanging the first and third rows. Hence, according to (iii), $D = -(-336) = 336$.

The results in Theorem 5 can be used to simplify the calculation of determinants.

Example 9 Calculate

$$D = \begin{vmatrix} 1 & 3 & 5 & 2 \\ 0 & -1 & 3 & 4 \\ 2 & 1 & 9 & 6 \\ 3 & 2 & 4 & 8 \end{vmatrix}.$$

Solution This determinant was calculated in Example 3. The idea is to use Theorem 5 to make the evaluation of the determinant almost trivial. We begin by multiplying the first row by -2 and adding it to the third row. By (v), this manipulation leaves the determinant unchanged:

$$D = \begin{vmatrix} 1 & 3 & 5 & 2 \\ 0 & -1 & 3 & 4 \\ 2+(-2)1 & 1+(-2)3 & 9+(-2)5 & 6+(-2)2 \\ 3 & 2 & 4 & 8 \end{vmatrix}$$

$$= \begin{vmatrix} 1 & 3 & 5 & 2 \\ 0 & -1 & 3 & 4 \\ 0 & -5 & -1 & 2 \\ 3 & 2 & 4 & 8 \end{vmatrix}.$$

Now we multiply the first row by -3 and add it to the fourth row:

$$D = \begin{vmatrix} 1 & 3 & 5 & 2 \\ 0 & -1 & 3 & 4 \\ 0 & -5 & -1 & 2 \\ 0 & -7 & -11 & 2 \end{vmatrix}.$$

We now expand D by its first column:

$$D = 1\begin{vmatrix} -1 & 3 & 4 \\ -5 & -1 & 2 \\ -7 & -11 & 2 \end{vmatrix} - 0\begin{vmatrix} 3 & 5 & 2 \\ -5 & -1 & 2 \\ -7 & -11 & 2 \end{vmatrix} + 0\begin{vmatrix} 3 & 5 & 2 \\ -1 & 3 & 4 \\ -7 & -11 & 2 \end{vmatrix} + 0\begin{vmatrix} 3 & 5 & 2 \\ -1 & 3 & 4 \\ -5 & -1 & 2 \end{vmatrix}$$

$$= \begin{vmatrix} -1 & 3 & 4 \\ -5 & -1 & 2 \\ -7 & -11 & 2 \end{vmatrix},$$

which is a 3×3 determinant. We can calculate it by expansion or we can reduce further. By Theorem 5, parts (iv) and (v),

$$\begin{vmatrix} -1 & 3 & 4 \\ -5 & -1 & 2 \\ -7 & -11 & 2 \end{vmatrix} = -\begin{vmatrix} 1 & -3 & -4 \\ -5 & -1 & 2 \\ -7 & -11 & 2 \end{vmatrix} = -\begin{vmatrix} 1 & -3 & -4 \\ 0 & -16 & -18 \\ 0 & -32 & -26 \end{vmatrix} = -\begin{vmatrix} -16 & -18 \\ -32 & -26 \end{vmatrix}$$

$$= -[(-16)(-26) - (-18)(-32)] = 160.$$

In the second step we multiplied the first row by 5 and added it to the second and multiplied the first row by 7 and added it to the third. Note how we were able to reduce the problem to the calculation of a single 2×2 determinant.

There is one further useful result about determinants.

THEOREM 6

Let

$$D = \begin{vmatrix} a_{11} & a_{12} & \cdots & a_{1n} \\ a_{21} & a_{22} & \cdots & a_{2n} \\ \vdots & \vdots & & \vdots \\ a_{i1}+b_{i1} & a_{i2}+b_{i2} & \cdots & a_{in}+b_{in} \\ \vdots & \vdots & & \vdots \\ a_{n1} & a_{n2} & \cdots & a_{nn} \end{vmatrix}.$$

Then

$$D = \begin{vmatrix} a_{11} & a_{12} & \cdots & a_{1n} \\ a_{21} & a_{22} & \cdots & a_{2n} \\ \vdots & \vdots & & \vdots \\ a_{i1} & a_{i2} & \cdots & a_{in} \\ \vdots & \vdots & & \vdots \\ a_{n1} & a_{n2} & \cdots & a_{nn} \end{vmatrix} + \begin{vmatrix} a_{11} & a_{12} & \cdots & a_{1n} \\ a_{21} & a_{22} & \cdots & a_{2n} \\ \vdots & \vdots & & \vdots \\ b_{i1} & b_{i2} & \cdots & b_{in} \\ \vdots & \vdots & & \vdots \\ a_{n1} & a_{n2} & \cdots & a_{nn} \end{vmatrix}. \tag{11}$$

Example 10 To illustrate Theorem 6, we note that

$$\begin{vmatrix} 2 & 1 & 4 \\ 3+5 & 2-3 & 1+2 \\ 0 & -4 & 2 \end{vmatrix} = \begin{vmatrix} 2 & 1 & 4 \\ 3 & 2 & 1 \\ 0 & -4 & 2 \end{vmatrix} + \begin{vmatrix} 2 & 1 & 4 \\ 5 & -3 & 2 \\ 0 & -4 & 2 \end{vmatrix}$$
$$= -38 - 86 = -124.$$

We conclude this appendix by showing how determinants can be used to obtain the unique solution (if one exists) of system (8) of n equations in n unknowns. We define the determinants

$$D_1 = \begin{vmatrix} b_1 & a_{12} & \cdots & a_{1n} \\ b_2 & a_{22} & \cdots & a_{2n} \\ \vdots & \vdots & & \vdots \\ b_n & a_{n2} & \cdots & a_{nn} \end{vmatrix},$$

$$D_2 = \begin{vmatrix} a_{11} & b_1 & a_{13} & \cdots & a_{1n} \\ a_{21} & b_2 & a_{23} & \cdots & a_{2n} \\ \vdots & \vdots & \vdots & & \vdots \\ a_{n1} & b_n & a_{n3} & \cdots & a_{nn} \end{vmatrix}, \ldots,$$

$$D_k = \begin{vmatrix} a_{11} & a_{12} & \cdots & a_{1,k-1} & b_1 & a_{1,k+1} & \cdots & a_{1n} \\ a_{21} & a_{22} & \cdots & a_{2,k-1} & b_2 & a_{2,k+1} & \cdots & a_{2n} \\ \vdots & \vdots & & \vdots & \vdots & \vdots & & \vdots \\ a_{n1} & a_{n2} & \cdots & a_{n,k-1} & b_n & a_{n,k+1} & \cdots & a_{nn} \end{vmatrix}, \ldots, \tag{12}$$

$$D_n = \begin{vmatrix} a_{11} & a_{12} & \cdots & a_{1,n-1} & b_1 \\ a_{21} & a_{22} & \cdots & a_{2,n-1} & b_2 \\ \vdots & \vdots & & \vdots & \vdots \\ a_{n1} & a_{n2} & \cdots & a_{n,n-1} & b_n \end{vmatrix},$$

obtained by replacing the kth column of D by the column

$$\begin{pmatrix} b_1 \\ b_2 \\ \vdots \\ b_n \end{pmatrix}.$$

Then we have the following theorem, known as **Cramer's rule**.

THEOREM 7   CRAMER'S RULE

Let D and D_k, $k = 1, 2, \ldots, n$, be given as in (10) and (12). If $D \neq 0$, then the unique solution to system (8) is given by the values

$$\boxed{x_1 = \frac{D_1}{D}, \quad x_2 = \frac{D_2}{D}, \ldots, x_n = \frac{D_n}{D}.} \tag{13}$$

Example 11 Consider the system

$$2x_1 + 4x_2 - x_3 = -5,$$
$$-4x_1 + 3x_2 + 5x_3 = 14,$$
$$6x_1 - 3x_2 - 2x_3 = 5.$$

We have

$$D = \begin{vmatrix} 2 & 4 & -1 \\ -4 & 3 & 5 \\ 6 & -3 & -2 \end{vmatrix} = 112, \quad D_1 = \begin{vmatrix} -5 & 4 & -1 \\ 14 & 3 & 5 \\ 5 & -3 & -2 \end{vmatrix} = 224,$$

$$D_2 = \begin{vmatrix} 2 & -5 & -1 \\ -4 & 14 & 5 \\ 6 & 5 & -2 \end{vmatrix} = -112, \quad D_3 = \begin{vmatrix} 2 & 4 & -5 \\ -4 & 3 & 14 \\ 6 & -3 & 5 \end{vmatrix} = 560.$$

Therefore

$$x_1 = \frac{D_1}{D} = 2, \quad x_2 = \frac{D_2}{D} = -1, \quad x_3 = \frac{D_3}{D} = 5.$$

Note As a general rule, avoid Cramer's rule if $n > 3$. There is too much work involved.

PROBLEMS APPENDIX 4

For each of the 2×2 systems in Problems 1–8 calculate the determinant D. If $D \neq 0$, find the unique solution. If $D = 0$, determine whether there is no solution or an infinite number of solutions.

1. $2x_1 + 4x_2 = 6$
   $x_1 + x_2 = 3$
2. $2x_1 + 4x_2 = 6$
   $x_1 + 2x_2 = 5$
3. $2x_1 + 4x_2 = 6$
   $x_1 + 2x_2 = 3$
4. $6x_1 - 3x_2 = 3$
   $-2x_1 + x_2 = -1$
5. $6x_1 - 3x_2 = 3$
   $-2x_1 + x_2 = 1$
6. $6x_1 - 3x_2 = 3$
   $-2x_1 + 2x_2 = -1$
7. $2x_1 + 5x_2 = 0$
   $3x_1 - 7x_2 = 0$
8. $2x_1 - 3x_2 = 0$
   $-4x_1 + 6x_2 = 0$

In Problems 9–20 calculate the determinant.

9. $\begin{vmatrix} 1 & 2 & 3 \\ 6 & -1 & 4 \\ 2 & 0 & 6 \end{vmatrix}$
10. $\begin{vmatrix} 4 & -1 & 0 \\ 2 & 1 & 7 \\ -2 & 3 & 4 \end{vmatrix}$
11. $\begin{vmatrix} 7 & 2 & 3 \\ 0 & 4 & 1 \\ 0 & 0 & 5 \end{vmatrix}$
12. $\begin{vmatrix} 1 & -1 & 4 \\ 3 & -2 & 1 \\ 5 & 1 & 7 \end{vmatrix}$

13. $\begin{vmatrix} 4 & 2 & 7 \\ 1 & 5 & 3 \\ -1 & 1 & 4 \end{vmatrix}$ 14. $\begin{vmatrix} -1 & 0 & 4 \\ 7 & 3 & 2 \\ 4 & 1 & 5 \end{vmatrix}$

15. $\begin{vmatrix} 1 & 4 & 7 & 2 \\ 0 & 5 & 8 & 1 \\ 0 & 0 & -3 & 4 \\ 0 & 0 & 0 & 8 \end{vmatrix}$

16. $\begin{vmatrix} a_1 & a_2 & a_3 & a_4 \\ 0 & b_1 & b_2 & b_3 \\ 0 & 0 & c_1 & c_2 \\ 0 & 0 & 0 & d_1 \end{vmatrix}$

17. $\begin{vmatrix} 2 & 1 & 3 & 4 \\ 3 & -2 & 5 & 1 \\ 4 & 0 & 4 & 5 \\ 2 & 1 & 7 & -4 \end{vmatrix}$

18. $\begin{vmatrix} 1 & 3 & -1 & 7 \\ -2 & 5 & 2 & 8 \\ -3 & 7 & 3 & 3 \\ 5 & 0 & -5 & 11 \end{vmatrix}$

19. $\begin{vmatrix} 2 & 3 & 1 & 4 \\ 2 & 2 & 4 & 6 \\ 3 & -1 & -2 & 4 \\ 4 & 2 & -3 & -5 \end{vmatrix}$

20. $\begin{vmatrix} 1 & 0 & 2 & 3 & 1 \\ 0 & 4 & -1 & -2 & 3 \\ 2 & 1 & 0 & -1 & 1 \\ -3 & 2 & 2 & 0 & 5 \\ 0 & 3 & 6 & 1 & -3 \end{vmatrix}$

21. Show that two lines in system (1) have the same slope if and only if the determinant of the system is zero.

In Problems 22–28 solve the system by using Cramer's rule.

22. $3x_1 - x_2 = 13$
    $-4x_1 + 6x_2 = -8$

23. $2x_1 + 6x_2 + 3x_3 = 9$
    $-3x_1 - 17x_2 - x_3 = 4$
    $4x_1 + 3x_2 + x_3 = -7$

24. $2x_1 + x_3 = 0$
    $3x_1 - 2x_2 + 2x_3 = -4$
    $4x_1 - 5x_2 = 3$

25. $x_1 - 8x_2 - x_3 = -1$
    $-x_1 + 4x_2 + x_3 = 3$
    $3x_1 - 2x_2 + 6x_3 = 5$

26. $x_1 + 2x_2 - x_3 - 4x_4 = 1$
    $-x_1 + 2x_3 + 6x_4 = 5$
    $ -4x_2 - 2x_3 - 8x_4 = -8$
    $3x_1 - 2x_2 + 5x_4 = 3$

27. $2x_1 + 5x_2 - 3x_3 + 2x_4 = 3$
    $-x_1 - 3x_2 + 2x_3 - x_4 = -1$
    $-3x_1 + 4x_2 + 8x_3 - 2x_4 = 4$
    $6x_1 - x_2 - 6x_3 + 4x_4 = 2$

28. $x_1 - 2x_2 + x_3 = 0$
    $2x_1 - x_2 - 3x_3 = 0$
    $5x_1 + 7x_2 - 8x_3 = 0$

APPENDIX 5

Complex Numbers

In algebra we encounter the problem of finding the roots of the polynomial

$$\lambda^2 + a\lambda + b = 0. \tag{1}$$

To find the roots, we use the quadratic formula to obtain

$$\lambda = \frac{-a \pm \sqrt{a^2 - 4b}}{2}. \tag{2}$$

If $a^2 - 4b > 0$, there are two real roots. If $a^2 - 4b = 0$, we obtain the single root (of multiplicity 2) $\lambda = -a/2$. To deal with the case $a^2 - 4b < 0$, we introduce the **imaginary number**†

$$i = \sqrt{-1}. \tag{3}$$

---

† You should not be troubled by the term "imaginary." It is just a name. The British mathematician Alfred North Whitehead, in the chapter on imaginary numbers in his *Introduction to Mathematics*, wrote:

> At this point it may be useful to observe that a certain type of intellect is always worrying itself and others by discussion as to the applicability of technical terms. Are the incommensurable numbers properly called numbers? Are the positive and negative numbers really numbers? Are the imaginary numbers imaginary, and are they numbers?—are types of such futile questions. Now, it cannot be too clearly understood that, in science, technical terms are names arbitrarily assigned, like Christian names to children. There can be no question of the names being right or wrong. They may be judicious or injudicious; for they can sometimes be so arranged as to be easy to remember, or so as to suggest relevant and important ideas. But the essential principle involved was quite clearly enunciated in Wonderland to Alice by Humpty Dumpty, when he told her, apropos of his use of words, 'I pay them extra and make them mean what I like'. So we will not bother as to whether imaginary numbers are imaginary, or as to whether they are numbers, but will take the phrase as the arbitrary name of a certain mathematical idea, which we will now endeavour to make plain.

Then for $a^2 - 4b < 0$,

$$\sqrt{a^2 - 4b} = \sqrt{(4b - a^2)(-1)} = \sqrt{4b - a^2}\sqrt{-1} = \sqrt{4b - a^2}\, i,$$

and the two roots of (1) are given by

$$\lambda_1 = -\frac{a}{2} + \frac{\sqrt{4b - a^2}}{2}i \quad \text{and} \quad \lambda_2 = -\frac{a}{2} - \frac{\sqrt{4b - a^2}}{2}i.$$

Example 1 Find the roots of the quadratic equation $\lambda^2 + 2\lambda + 5 = 0$.

Solution We have $a = 2$, $b = 5$, and $a^2 - 4b = -16$. Thus $\sqrt{a^2 - 4b} = \sqrt{-16} = \sqrt{16}\sqrt{-1} = 4i$, and the roots are

$$\lambda_1 = \frac{-2 + 4i}{2} = -1 + 2i \quad \text{and} \quad \lambda_2 = -1 - 2i.$$

DEFINITION 1 COMPLEX NUMBER

A **complex number** is a number of the form

$$z = \alpha + i\beta, \tag{4}$$

where α and β are real numbers. α is called the **real part** of z and is denoted Re z. β is called the **imaginary part** of z and is denoted Im z. Representation (4) is sometimes called the **Cartesian form** of the complex number z.

Remark If $\beta = 0$ in equation (4), then $z = \alpha$ is a real number. In this context we can regard the set of real numbers as a subset of the set of complex numbers.

Example 2 In Example 1, Re $\lambda_1 = -1$ and Im $\lambda_1 = 2$.

We can add and multiply complex numbers by using the standard rules of algebra.

Example 3 Let $z = 2 + 3i$ and $w = 5 - 4i$. Calculate (a) $z + w$, (b) $3w - 5z$, and (c) zw.

Solution

(a) $z + w = (2 + 3i) + (5 - 4i) = (2 + 5) + (3 - 4)i = 7 - i$.

(b) $3w = 3(5 - 4i) = 15 - 12i$, $5z = 10 + 15i$, and $3w - 5z = (15 - 12i) - (10 + 15i) = (15 - 10) + i(-12 - 15) = 5 - 27i$.

(c) $zw = (2 + 3i)(5 - 4i) = (2)(5) + 2(-4i) + (3i)(5) + (3i)(-4i) = 10 - 8i + 15i - 12i^2 = 10 + 7i + 12 = 22 + 7i$. Here we use the fact that $i^2 = -1$.

We can plot a complex number z in the xy-plane by plotting Re z along the x-axis and Im z along the y-axis. Thus each complex number can be thought of as a point in the xy-plane. With this representation the xy-plane is called the **complex plane**. Some representative points are plotted in Figure A5.1.

If $z = \alpha + i\beta$, then we define the **conjugate** of z, denoted $\bar{z}$, by

$$\bar{z} = \alpha - i\beta. \tag{5}$$

Figure A5.2 depicts a representative value of z and $\bar{z}$.

Figure A5.1

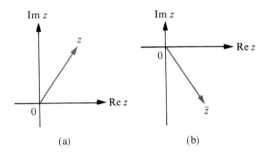

Figure A5.2

Example 4 Compute the conjugate of (a) $1 + i$, (b) $3 - 4i$, (c) $-7 + 5i$, and (d) -3.

Solution (a) $\overline{1 + i} = 1 - i$. (b) $\overline{3 - 4i} = 3 + 4i$. (c) $\overline{-7 + 5i} = -7 - 5i$. (d) $\overline{-3} = -3$.

It is not difficult to show (see Problem 35) that

$$\bar{z} = z, \quad \text{if and only if } z \text{ is real.} \tag{6}$$

If $z = \beta i$ with β real, then z is said to be **pure imaginary**. We can then show (see Problem 36) that

$$\bar{z} = -z, \quad \text{if and only if } z \text{ is pure imaginary.} \tag{7}$$

Let $p_n(x) = a_0 + a_1 x + a_2 x^2 + \cdots + a_n x^n$ be a polynomial with real coefficients. Then it can be shown (see Problem 41) that the complex roots of the equation $p_n(x) = 0$ occur in complex conjugate pairs; that is, if z is a root of $p_n(x) = 0$, then so is $\bar{z}$. We saw this fact illustrated in Example 1 in the case in which $n = 2$.

For $z = \alpha + i\beta$ we define the **magnitude** of z, denoted $|z|$, by

$$|z| = \sqrt{\alpha^2 + \beta^2}, \tag{8}$$

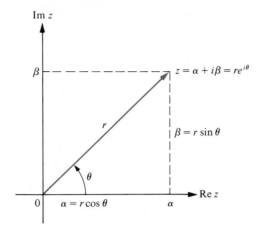

Figure A5.3

and we define the **argument** of z, denoted arg z, as the angle θ between the line $0z$ and the positive x-axis. From Figure A5.3 we see that $r = |z|$ is the distance from z to the origin, and, if $-\pi/2 < \theta < \pi/2$,

$$\theta = \arg z = \tan^{-1}\frac{\beta}{\alpha}. \tag{9}$$

By convention, we always choose values of arg z that lie in the interval

$$-\pi < \theta \leq \pi. \tag{10}$$

From Figure A5.4 we see that

$$|\bar{z}| = |z| \tag{11}$$

and

$$\arg \bar{z} = -\arg z. \tag{12}$$

We can use $|z|$ and arg z to describe what is often a more convenient way to represent complex numbers. From Figure A5.3 it is evident that, if $z = \alpha + i\beta$, $r = |z|$,

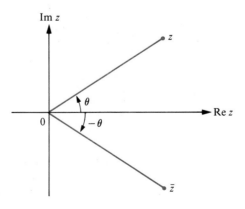

Figure A5.4

and $\theta = \arg z$, then

$$\alpha = r\cos\theta \quad \text{and} \quad \beta = r\sin\theta. \tag{13}$$

We see at the end of this appendix that

$$e^{i\theta} = \cos\theta + i\sin\theta. \tag{14}$$

Since $\cos(-\theta) = \cos\theta$ and $\sin(-\theta) = -\sin\theta$, we also have

$$e^{-i\theta} = \cos(-\theta) + i\sin(-\theta) = \cos\theta - i\sin\theta. \tag{14'}$$

Formula (14) is called the **Euler formula**. Using the Euler formula and equation (13), we have

$$z = \alpha + i\beta = r\cos\theta + ir\sin\theta = r(\cos\theta + i\sin\theta),$$

or

$$z = re^{i\theta}. \tag{15}$$

Representation (15) is called the **polar form** of the complex number z.

Example 5 Determine the polar forms of the following complex numbers: (a) 1, (b) -1, (c) i, (d) $1 + i$, (e) $-1 - \sqrt{3}\,i$, and (f) $-2 + 7i$.

Solution The six points are plotted in Figure A5.5.

(a) From Figure A5.5(a) it is clear that $\arg 1 = 0$. Since $\operatorname{Re} 1 = 1$, we see that, in polar form,

$$1 = 1e^{i0} = 1e^0 = e^0.$$

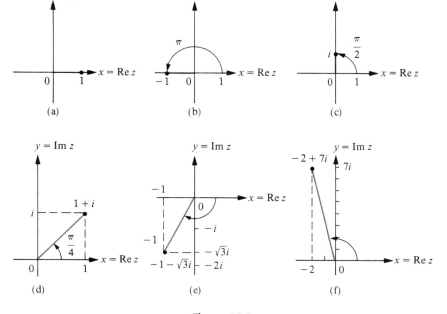

Figure A5.5

(b) Since $\arg(-1) = \pi$ [Figure A5.5(b)] and $|-1| = 1$, we have
$$-1 = 1e^{\pi i} = e^{i\pi}.$$

(c) From Figure A5.5(c) we see that $\arg i = \pi/2$. Since $|i| = \sqrt{0^2 + 1^2} = 1$, it follows that
$$i = e^{i\pi/2}.$$

(d) $\arg(1 + i) = \tan^{-1}(1/1) = \pi/4$, and $|1 + i| = \sqrt{1^2 + 1^2} = \sqrt{2}$, so
$$1 + i = \sqrt{2}\, e^{i\pi/4}.$$

(e) Here $\tan^{-1}(\beta/\alpha) = \tan^{-1}\sqrt{3} = \pi/3$. However, $\arg z$ is in the third quadrant, so $\arg z = (\pi/3) - \pi = -2\pi/3$. Also, $|-1 - \sqrt{3}\, i| = \sqrt{1^2 + (\sqrt{3})^2} = \sqrt{1 + 3} = 2$, so
$$-1 - \sqrt{3}\, i = 2e^{-2\pi i/3}.$$

(f) To compute this complex number, we need a calculator. A calculator indicates that
$$\tan^{-1}\left(-\tfrac{7}{2}\right) = \tan^{-1}(-3.5) \approx -1.2925.$$

But $\tan^{-1} x$ is defined as a number in the interval $(-\pi/2, \pi/2)$. Since from Figure A5.5(f) θ is in the second quadrant, we see that $\arg z = \tan^{-1}(-3.5) + \pi \approx 1.8491$. Next, we see that
$$|-2 + 7i| = \sqrt{(-2)^2 + 7^2} = \sqrt{53}.$$

Hence
$$-2 + 7i \approx \sqrt{53}\, e^{1.8491 i}.$$

Example 6 Convert the following complex numbers from polar to Cartesian form: (a) $2e^{i\pi/3}$, (b) $4e^{3\pi i/2}$.

Solution

(a) $e^{i\pi/3} = \cos(\pi/3) + i \sin(\pi/3) = 1/2 + (\sqrt{3}/2)i$. Thus $2e^{i\pi/3} = 1 + \sqrt{3}\, i$.

(b) $e^{3\pi i/2} = \cos(3\pi/2) + i \sin(3\pi/2) = 0 + i(-1) = -i$. Thus $4e^{3\pi i/2} = -4i$.

If $\theta = \arg z$, then by equation (12), $\arg \bar{z} = -\theta$. Thus, since $|\bar{z}| = |z|$, we have the following:

$$\text{If } z = re^{i\theta}, \quad \text{then } \bar{z} = re^{-i\theta}. \tag{16}$$

Suppose we write a complex number in its polar form $z = re^{i\theta}$. Then
$$z^n = (re^{i\theta})^n = r^n(e^{i\theta})^n = r^n e^{in\theta} = r^n(\cos n\theta + i \sin n\theta). \tag{17}$$

Formula (17) is useful for a variety of computations. In particular, when $r = |z| = 1$, we obtain the **De Moivre formula**:[†]

$$(\cos \theta + i \sin \theta)^n = \cos n\theta + i \sin n\theta. \tag{18}$$

[†] Abraham De Moivre (1667–1754) was a French mathematician well known for his work in probability theory, infinite series, and trigonometry. He was so highly regarded that Newton often told those who came to him with questions on mathematics, "Go to M. De Moivre; he knows these things better than I do."

Example 7 Compute $(1+i)^5$.

Solution In Example 5(d) we showed that $1 + i = \sqrt{2}\, e^{\pi i/4}$. Then

$$(1+i)^5 = (\sqrt{2}\,e^{\pi i/4})^5 = (\sqrt{2})^5 e^{5\pi i/4} = 4\sqrt{2}\left(\cos\frac{5\pi}{4} + i\sin\frac{5\pi}{4}\right)$$

$$= 4\sqrt{2}\left(-\frac{1}{\sqrt{2}} - \frac{1}{\sqrt{2}}i\right) = -4 - 4i.$$

This can be checked by direct calculation. If the direct calculation seems no more difficult, then try to compute $(1+i)^{20}$ directly. Proceeding as above, we obtain

$$(1+i)^{20} = (\sqrt{2})^{20} e^{20\pi i/4} = 2^{10}(\cos 5\pi + i\sin 5\pi)$$
$$= 2^{10}(-1+0) = -1024.$$

PROOF OF EULER'S FORMULA

We now show that

$$e^{i\theta} = \cos\theta + i + \sin\theta \tag{19}$$

by using power series. We have

$$e^x = 1 + x + \frac{x^2}{2!} + \frac{x^3}{3!} + \cdots,^\dagger \tag{20}$$

$$\sin x = x - \frac{x^3}{3!} + \frac{x^5}{5!} - \cdots, \tag{21}$$

$$\cos x = 1 - \frac{x^2}{2!} + \frac{x^4}{4!} - \cdots. \tag{22}$$

Then

$$e^{i\theta} = 1 + (i\theta) + \frac{(i\theta)^2}{2!} + \frac{(i\theta)^3}{3!} + \frac{(i\theta)^4}{4!} + \frac{(i\theta)^5}{5!} + \cdots. \tag{23}$$

Now $i^2 = -1$, $i^3 = -i$, $i^4 = 1$, $i^5 = i$, and so on. Thus (23) can be written

$$e^{i\theta} = 1 + i\theta - \frac{\theta^2}{2!} - \frac{i\theta^3}{3!} + \frac{\theta^4}{4!} + \frac{i\theta^5}{5!} - \cdots$$

$$= \left(1 - \frac{\theta^2}{2!} + \frac{\theta^4}{4!} - \cdots\right) + i\left(\theta - \frac{\theta^3}{3!} + \frac{\theta^5}{5!} - \cdots\right)$$

$$= \cos\theta + i\sin\theta.$$

This completes the proof. ∎

---

†Although we do not prove it here, these series expansions are also valid when x is a complex number.

PROBLEMS APPENDIX 5

In Problems 1–5 perform the indicated operation.
1. $(2-3i)+(7-4i)$
2. $3(4+i)-5(-3+6i)$
3. $(1+i)(1-i)$
4. $(2-3i)(4+7i)$
5. $(-3+2i)(7+3i)$

In Problems 6–15 convert the complex number to its polar form.
6. $5i$
7. $5+5i$
8. $-2-2i$
9. $3-3i$
10. $2+2\sqrt{3}\,i$
11. $3\sqrt{3}+3i$
12. $1-\sqrt{3}\,i$
13. $4\sqrt{3}-4i$
14. $-6\sqrt{3}-6i$
15. $-1-\sqrt{3}\,i$

In Problems 16–25 convert from polar to Cartesian form.
16. $e^{3\pi i}$
17. $2e^{-7\pi i}$
18. $\frac{1}{2}e^{3\pi i/4}$
19. $\frac{1}{2}e^{-3\pi i/4}$
20. $6e^{\pi i/6}$
21. $4e^{5\pi i/6}$
22. $4e^{-5\pi i/6}$
23. $3e^{-2\pi i/3}$
24. $\sqrt{3}\,e^{23\pi i/4}$
25. e^{i}

In Problems 26–34 compute the conjugate of the given number.
26. $3-4i$
27. $4+6i$
28. $-3+8i$
29. $-7i$
30. 16
31. $2e^{\pi i/7}$
32. $4e^{3\pi i/5}$
33. $3e^{-4\pi i/11}$
34. $e^{0.012i}$

35. Show that $z=\alpha+i\beta$ is real if and only if $z=\bar{z}$. [*Hint*: If $z=\bar{z}$, show that $\beta=0$.]

36. Show that $z=\alpha+i\beta$ is pure imaginary if and only if $z=-\bar{z}$. [*Hint*: If $z=-\bar{z}$, show that $\alpha=0$.]

37. For any complex number z, show that $z\bar{z}=|z|^2$.

38. Show that the circle of radius 1 centered at the origin (the **unit circle**) is the set of points in the complex plane that satisfy $|z|=1$.

39. For any complex number z_0 and real number a, describe $\{z:|z-z_0|=a\}$.

40. Describe $\{z:|z-z_0|\le a\}$, where z_0 and a are as in Problem 39.

*41. Let $p(\lambda)=\lambda^n+a_{n-1}\lambda^{n-1}+a_{n-2}\lambda^{n-2}+\cdots+a_1\lambda+a_0$ with $a_0,a_1,\ldots,a_{n-1}$ real numbers. Show that if $p(z)=0$, then $p(\bar{z})=0$; that is, *that the roots of polynomials with real coefficients occur in complex conjugate pairs.*

42. Derive expressions for $\cos 4\theta$ and $\sin 4\theta$ by comparing the De Moivre formula and the expansion of $(\cos\theta+i\sin\theta)^4$.

*43. Prove the De Moivre formula by mathematical induction. [*Hint*: Recall the trigonometric identities $\cos(x+y)=\cos x\cos y-\sin x\sin y$ and $\sin(x+y)=\sin x\cos y+\cos x\sin y$.]

Answers to Odd-Numbered Problems

CHAPTER 1

Problems 1.1, page 9

1. 62,500 after 20 days; 156,250 after 30 days
3. $250{,}000(1.06)^{10} \approx 447{,}712$ in 1980; $250{,}000(1.06)^{30} \approx 1{,}435{,}873$ in 2000
5. **a.** $150(0.16)^2 \approx 3.84°C$ **b.** $-\ln 15/\ln 0.16 \approx 1.48$ minutes
7. $5580 \ln 0.7/\ln 0.5 \approx 2871$ years
9. **a.** $25(0.6)^{2.4} \approx 7.34$ kg **b.** $10 \ln 0.02/\ln 0.6 \approx 76.6$ hours
11. $\ln 0.5/(-\alpha) \approx 4.62 \times 10^6$ years
13. $\beta = (1/1500) \ln 845.6/1013.25 \approx -1.205812 \times 10^{-4}$ **a.** 625.526 mbar **b.** 303.416 mbar **c.** 429.033 mbar **d.** 348.632 mbar **e.** $-(1/\beta)\ln 1013.25 \approx 57.4$ km
15. **a.** $\approx 10{,}000$ **b.** 7 **c.** $214.20 \approx 8.9$ days
17. 55.1 minutes 19. 38.7 minutes

Problems 1.2, page 18

1. first 3. third 5. second
7. third 9. initial-value
11. initial-value 13. boundary value
19. $y_1 = e^{2x} \cos x$, $y_2 = e^{2x} \sin x$
25. $\phi''(-1) = -3$, $\phi'''(-1) = 2$

27.

29.

31.

33.

35. a.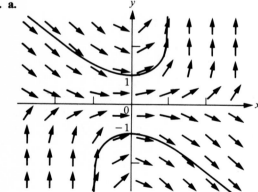

  b. It is the constant solution $y = 0$.

  c. Their graphs are symmetric with respect to the origin.

37. All points except those on the line $x + y = 0$

39. All points in the horizontal strip $-1 < y < 1$

41. Valid for all $x < 1$

Problems 1.3, page 25

1. 2.98   **3.** 0.71   **5.** 8.31   **7.** 0.34   **9.** 156.45

11. $y_1 = 1.02$, $y_2 = 1.04$, $y_3 = 1.07$, $y_4 = 1.09$, $y_5 = 1.12$

13. $y_1 = 0.40$, $y_2 = 0.68$, $y_3 = 0.79$, $y_4 = 0.95$, $y_5 = 1.20$

15. $y_1 = 1.00$, $y_2 = 1.56$, $y_3 = 2.06$, $y_4 = 1.28$, $y_5 = 0.67$, $y_6 = 1.05$, $y_7 = 0.28$, $y_8 = 1.07$

17. $y_1 = 1.10$, $y_2 = 1.21$, $y_3 = 1.33$, $y_4 = 1.46$, $y_5 = 1.60$, $y_6 = 1.75$, $y_7 = 1.91$, $y_8 = 2.10$, $y_9 = 2.29$, $y_{10} = 2.50$

19. $y_1 = 1.55$, $y_2 = 1.19$, $y_3 = 0.91$, $y_4 = 0.69$, $y_5 = 0.52$

Review Exercises for Chapter 1, page 26

1. $y = \frac{3}{2}x^2 + C$ **3.** $x = -3e^{t/2}$

5. $P(5) = 10{,}000(1.15)^5 \approx 20{,}114$; $P(10) = 10{,}000(1.15)^{10} \approx 40.456$

7. a. $T(20) = 23 + 102\left(\frac{57}{102}\right)^2 \approx 54.9°C$

  b. $t = \dfrac{10 \ln(2/102)}{\ln(57/102)} \approx 67.6$ minutes

9. $h = \ln 0.5 / \ln 0.8 \approx 3.11$ weeks

11.

13. $y(3) \approx 6.076$ **15.** $y(3) \approx 5.445$

17. $y(3) \approx 0.625$

19. a. $y(1) \approx 25.32$ **c.** $(1-x)^{-1/2}$ is not defined at $x = 1$

CHAPTER 2

Problems 2.1, page 39

1. $y = \sqrt{e^x + c}$ **3.** $y = \frac{1}{2}\ln(x^2 + 1) + c$

5. $z = \tan\left(\dfrac{r^3}{3} + c\right)$ **7.** $P = ce^{\sin Q - \cos Q}$

9. $s = e^{(t^3/3) - 2t}$ **11.** $y = \dfrac{2401}{(1+x)^3}$

13. $y = c \cos x - 3$

15. If $|x| < 1$ and $|y| < 1$, $y = \sin(c - \sin^{-1} x)$
    If $|x| > 1$ and $|y| > 1$,
    $y = \pm \cosh(c - \cosh^{-1} x)$
17. $y = \dfrac{3x}{4x-3}$ 19. $x = \ln\left(1 - \dfrac{1-e}{e^t}\right)$
21. $y = \pm\sqrt{2e^x + c}$ 23. $y = \pm\sqrt{x^2 - 2x + c}$
25. At $t = 9$, predicted mass is 447.2; error is 1.4%.
    At $t = 18$, predicted mass is 660.4; error is 0.2%.
27. $x = x_0 + v_0(\cos\theta)t$, $y = y_0 + v_0(\sin\theta)t - \tfrac{1}{2}gt^2$
29. $t = 3\ln 10/\ln 2 \approx 9.97$ hr
31. $19{,}825{,}000\left(\dfrac{23{,}210}{19{,}825}\right)^{32/5} \approx 54{,}371{,}290$
33. $715{,}718{,}491$
35. $t = \dfrac{\pi d}{r}\left[\dfrac{\sqrt{i}}{2(k\pi)^{3/2}} \ln\left|\dfrac{\sqrt{i}+\sqrt{k\pi}\,u}{\sqrt{i}-\sqrt{k\pi}\,u}\right| - \dfrac{u}{k\pi}\right]$
    where $u = \left(\dfrac{3rv}{\pi d}\right)^{1/3}$; to prevent overflow,
    $i = k\pi u^2$
37. Revolve $y = kx^4$ about the y-axis to get the cistern's shape.
39. 12:19:31.7 PM.

Problems 2.2, page 45

1. $x^2 y + y = c$ 3. $x^4 y^3 + \ln\left|\dfrac{x}{y}\right| = e^3 - 1$
5. $x^2 - 2y\sin x = \dfrac{\pi^2}{4} - 2$
7. $e^{xy} + 4xy^3 - y^2 + 3 = 0$
9. $x^2 y + xe^y = c$ 11. $\tan^{-1}\dfrac{y}{x} + \ln\left|\dfrac{x}{y}\right| = c$
13. $2x + \ln(x^2 + y^2) = c$ and $x^2 + y^2 = 0$
15. $x^2 \ln|x| - y = cx^2$ and $x = 0$
17. $y^2(x^2 + y^2 + 2) = c$ 19. $x^2 y = c$

Problems 2.3, page 51

1. $x = ce^{3t}$ 3. $x = 2e^t - 1$
5. $x = y^y(1 + ce^{-y})$ 7. $y = x^4 + 3x^3$
9. $x = ce^{2t} + \dfrac{t^3 e^{2t}}{3}$ 11. $s = \left(c + \dfrac{u^2}{2}\right)e^{-u} + 1$
13. $x = e^{-y}(c + y)$ 15. $x = \dfrac{1}{k} + ce^{-kt}$
17. $y = B + ce^{-kt}$ 19. $P = \dfrac{\beta}{\delta + (\beta/P_0 - \delta)e^{-\beta t}}$
21. a. $\dfrac{dN}{dt} = rN\left(1 - \dfrac{N}{K}\right) - h$
       $= -\dfrac{r}{K}\left(N^2 - KN + \dfrac{hK}{r}\right)$

b. $N = \tfrac{1}{2}K + \tfrac{1}{2}D \tanh\left[\tfrac{1}{2}D\dfrac{r}{K}t + c\right]$ where
   $D = \sqrt{K^2 - \dfrac{4Kh}{r}}$ and
   $c = \tanh^{-1}[2N_0 - K/D]$
23. Graph $y = rN(1 - N/K)$ and observe maximum
    $rK/4$ occurs at $N = K/2$
25. The object satisfies $m\dot{v} = -mg + kv^2$. Thus
    $\lim_{t\to\infty} v = -\sqrt{mg/k}$
27. $y = \pm\sqrt{\dfrac{x}{6 + ce^{-x}}}$
29. $y = \dfrac{1}{x\sqrt{2-x^2}}$, $0 < x < \sqrt{2}$
31. $y = \dfrac{1}{x^3(\tfrac{1}{2} - \ln x)}$, $0 < x < \sqrt{e}$
33. $y = \left[\int g(x) e^{\int f(x)\,dx}dx + k\right]e^{-\int f(x)\,dx}$
35. b. $y = -ce^{-3x} + ke^{-2x}$
37. $y = (cx + k)e^{-ax}$
39. $y = \begin{cases} e^{-x}, & \text{if } 0 \le x \le 1 \\ e^{-1}, & \text{if } x > 1 \end{cases}$
41. See the reference.

Problems 2.4, page 59

1. $y = \tfrac{1}{4}x(c + \operatorname{sgn} x \ln|x|)^2$; $y \equiv 0$ is also a solution
3. $y = -x\ln(1 - \ln x)$ 5. $y = x/(1 - \ln x)$
7. $y = \tfrac{1}{2}\left(x + \sqrt{x^2 + 8/x}\right)$ 9. $y = (c^2 x^2 - 1)/2c$
11. $y = \tan(2x + c) - \tfrac{1}{2}x$
13. $3x + c = x + y + \ln(x + y - 1)$
15. $y = \tan^{-1}(x + c) - x$ 17. $y = \pm 1/x\sqrt{c - 2x}$
19. $y = -\ln|1 + e^{-1} - x|/x$ 21. $y = e^{ce^x} - x$
25. $y = 2x \pm \sqrt{5(x + \tfrac{2}{5})^2 + c}$
27. $(x + 1)e^{(1-t)/(x+1)} = c$
29. $y = 5 - x \pm \sqrt{4(x-2)^2 + c}$
31. $xe^{tx} = ct$, by setting $v = tx$
33. General solution: $y = cx + c^3$; singular solution:
    $y = \pm\tfrac{2}{3}x\sqrt{-\tfrac{1}{3}x}$
35. General solution: $y = cx + \tfrac{1}{4}c^4$; singular solution: $y = -\tfrac{3}{4}x^{4/3}$

Problems 2.5, page 64

1. $I = \tfrac{6}{5}(1 - e^{-10t})$ 3. $I = 2(1 - e^{-25t})$
5. $I = \tfrac{1}{20}(e^t - e^{-t})$
7. $Q = \tfrac{1}{680}[3\sin 60t + 5\cos 60t - 5e^{-100t}]$

9. $Q = \frac{1}{100}(1 + 99e^{-100t})$

11. $1000\frac{dQ}{dt} + 10^6 Q = 0$; $Q(0) = 10$, $Q = 10e^{-1000t}$

13. $Q(60) = \frac{1}{2000}[1 - e^{-10^5(600 + 3600)}]$

15. $I(t) = \frac{E_0 C}{1 + (\omega RC)^2}$
$\times \left[RC\omega^2 \cos \omega t - \omega \sin \omega t + \frac{1}{RC} e^{-t/RC} \right]$;

$Q(t) = \frac{E_0 C}{1 + (\omega RC)^2}$
$\times [(\cos \omega t - e^{-t/RC}) + \omega RC \sin \omega t]$

17. $I_{\text{transient}}(0) = \frac{E_0}{R[1 + (\omega RC)^2]} \approx \frac{E_0}{R}$ for very small R

19. $I(t)$
$= \begin{cases} \frac{3}{50}(1 - e^{-50t}), & 0 \le t \le 10 \\ \left[\frac{e^{500}}{(70)^2} - \frac{3}{50}\right]e^{-50t} \\ \quad + \frac{7}{100} - \frac{e^{10-t}}{98}, & t \ge 10 \end{cases}$

Problems 2.6, page 72

1. 2000-year-old wood has $(\frac{1}{2})^{20/57} \approx 0.784$ the C^{14}-concentration of freshly cut wood.

3. $I(t) = \frac{1 + N}{1 + Ne^{-k(1+N)t}}$

5. $t = 25 \ln 2 \approx 17.3$ min   7. $47\frac{43}{91}$ g/liter

9. $p(t) = 400 + \frac{2}{\pi} - \frac{1}{3}t + \frac{2}{\pi} \sin \frac{\pi}{12}(t - 6)$

11. $x(t) = \frac{A}{k} + \frac{B}{k^2 + \omega^2}(k \sin \omega t - \omega \cos \omega t)$
$+ \left(x_0 + \frac{B\omega}{\omega^2 + k^2} - \frac{A}{k} \right)e^{-kt}$

13. distance is $\sqrt{x^2 + (vt - y)^2} = x\sqrt{1 + (y')^2}$
$= \frac{x}{2}\left[\left(\frac{x}{a}\right)^{v/w} + \left(\frac{a}{x}\right)^{v/w} \right]$

15. At the point $(0, -b/2)$

17. $r = 2e^{(\theta + \pi/2)/\sqrt{8}}$ 19. $r = ae^{-\sqrt{3}\theta}/\sqrt{3}$

21. The eagle's speed is $\frac{3 + \sqrt{73}}{8}$ times the pigeon's. The pigeon flies $\frac{200}{3}$ ft, the hawk flies $\frac{400}{3}$ ft, and the eagle flies $\frac{25}{3}(3 + \sqrt{73})$ ft

Problems 2.7, page 77

1. $y = e^{-x}$ 3. $y = e^x - 1 - x$

Problems 2.8, page 80

1. $y_n = y_0 + 2\left(1 - \frac{1}{2^n}\right)$ 3. $y_n = (-1)^n y_0$

5. $y_n = n + 1$ 7. $y_n = \frac{5^{n+1} - 1}{2}$

9. $y_n = e^{-n^2 + n}\left(y_0 + \frac{e^n - 1}{e - 1}\right)$

11. $P_n = \frac{2}{3} + \frac{1}{3}(-\frac{1}{2})^n$

13. $P_n = P_0/n!$, $1 = P_0\left(1 + \frac{1}{1!} + \frac{1}{2!} + \cdots\right) = P_0 e$

15. $x_k = \frac{n - (k - 1)}{k} \cdot \frac{n - (k - 2)}{k}$
$\cdots \cdot \frac{n - 1}{2} \cdot x_1$

Review Exercises for Chapter 2, page 81

1. $y = \frac{1}{1 - \ln|x|}$ 3. $y = \sin(\ln|x| + c)$

5. $y = \frac{2}{c - x^2}$ 7. $y = \sqrt{2} \sin x$

9. $y = cx + \frac{1}{c}$ and $y = 2\sqrt{x}$

11. $y = e^{(x^2 - 1)/2}$ 13. $y = ce^{-\cos x} - 1$

15. $y = \frac{x + c}{\sqrt{1 + x^2}}$ 17. $y = x^2(1 + \ln|x|)$

19. $y = \begin{cases} e^{x^2/2}\left(c - \int_x^0 e^{-t^2/2}\, dt\right), & x \le 0 \\ ce^{x^2/2}, & x > 0 \end{cases}$

21. $y = \frac{1}{2}(\sqrt{3x^2 + 4} - x)$ 23. $y = \frac{(x + c)^2 - 4}{4x}$

25. $y = \frac{1}{1 - x + ce^{-x}}$ 27. $xy + c = e^y \tan x$

29. $\frac{y}{x^3} + \frac{3}{x} + \frac{x}{y^2} = c$

31. $x = \frac{1 - e^{-bt}}{1001}$ where $b = 1.001 \times 10^{-6}$;

$x(36{,}000) \approx 3.536 \times 10^{-5}$; $x(360{,}000) \approx 3.023 \times 10^{-4}$; $x(31{,}536{,}000) \approx 9.99 \times 10^{-4}$

33. Approximate by assuming that pollutants are dumped continuously at the rate $\frac{1}{24}$ m³/sec; $x \approx (1 - e^{-bt})/24{,}024$; maximum concentration is $1/24{,}024$

35. 1.18 hr   37. $P + \exp \frac{1}{b}[a - e^{-b(t+c)}]$;
$\lim_{t \to \infty} P(t) = e^{a/b}$

39. 3.125 ft.   41. $6 - \cosh(\sqrt{g/6}\, t)$ as long as this is nonnegative

45. $y = e^x - 1$ 47. $y = x$

CHAPTER 3

Problems 3.1, page 96

1. Linear, homogeneous, variable coefficients
3. Nonlinear
5. Linear, nonhomogeneous, constant coefficients
7. Linear, nonhomogeneous, variable coefficients
9. Linear, homogeneous, constant coefficients
11. 3   13. 3   15. $y = c_1 + c_2 x$
17. $y = c_1 + c_2 x^2$
19. At that point $y_1(x_1) = y_1'(x_1) = 0$. But $y \equiv 0$ also satisfies that initial condition.
21. b. No; $a(x) = -3/x$ is not continuous on $|x| < 1$
23. $y_3 = 2y_1 - \tfrac{1}{2}y_2$ 27. $W = -2/x$

Problems 3.2, page 99

1. $y_2(x) = xe^x$ 3. $y_2(x) = e^{-2x}$
5. $y_2(x) = e^{-3x}$ 7. $y_2(x) = e^{-x/2} \sin \tfrac{3\sqrt{3}}{2} x$
9. $y_2(x) = e^{10x}$ 11. $y_2(x) = x^{-2}$
13. $y_2(x) = x \cos x$ 15. $y_2(x) = (7x + 1)e^{-4x}$
17. $y_2(x) = e^x \int x^n e^{-x}\, dx = -(x^n + nx^{n-1} + n(n-1)x^{n-2} + \cdots + n!)$
19. $y_2(x) = \cos x / \sqrt{x}$

Problems 3.3, page 102

1. $y = c_1 e^{2x} + c_2 e^{-2x}$ 3. $y = c_1 e^x + c_2 e^{2x}$
5. $x = (1 + \tfrac{9}{2}t)e^{-5t/2}$ 7. $x = -\tfrac{1}{5}(e^{3t} + 4e^{-2t})$
9. $y = c_1 + c_2 e^{5x}$ 11. $y = (c_1 + c_2 x)e^{-\pi x}$
13. $z = c_1 e^{-5x} + c_2 e^{3x}$ 15. $y = (1 + 2x)e^{4x}$
17. $y = c_1 e^{\sqrt{2}x} + c_2 e^{-\sqrt{2}x}$ 19. $y = e^{\sqrt{5}x} + 2e^{-\sqrt{5}x}$
23. $y = 1 - \dfrac{4}{e^{2x} + 2}$ 25. $y = 1 + \dfrac{1}{c + x}$
27. $y = 2 - \dfrac{3c}{e^{3x} + c}$

Problems 3.4, page 105

1. $y = e^{-x}(c_1 \cos x + c_2 \sin x)$
3. $x = e^{-t/2}\left(c_1 \cos \dfrac{3\sqrt{3}}{2} t + c_2 \sin \dfrac{3\sqrt{3}}{2} t\right)$
5. $x = -\tfrac{3}{2}\cos 2\theta + \sin 2\theta$ 7. $y = \sin \dfrac{x}{2} + 2\cos \dfrac{x}{2}$
9. $y = e^{-x}(c_1 \cos 2x + c_2 \sin 2x)$
11. $y = e^{-x}(\sin x - \cos x)$
13. $y = c_1 \sin x$ is zero at $x = \pi$.

Problems 3.5, page 111

1. $y = c_1 \sin 2x + c_2 \cos 2x + \sin x$

3. $y = c_1 e^x + c_2 e^{2x} + 3e^{3x}$
5. $y = e^x(c_1 + c_2 x - 2x^2)$
7. $y = 3e^{5x} - 10e^{2x} + 10x + 7$
9. $y = c_1 + c_2 e^{-x} + \dfrac{x^4}{4} - \dfrac{4}{3}x^3 + 4x^2 - 8x$
11. $y = e^{2x}(c_1 \cos x + c_2 \sin x) + xe^{2x} \sin x$
13. $y = e^{-3x}(c_1 + c_2 x + 5x^2)$
15. $y = c_1 e^{3x} + c_2 e^{-x} + \tfrac{20}{27} - \tfrac{7}{9}x + \tfrac{1}{3}x^2 - \tfrac{1}{4}e^x$
17. $y = (c_1 + c_2 x)e^{-2x} + (\tfrac{1}{9}x - \tfrac{2}{27})e^x + \tfrac{3}{25} \sin x - \tfrac{4}{25} \cos x$
21. $y = \left(-\tfrac{1}{6}x^3 - \tfrac{1}{4}x^2 + \tfrac{1}{4}x + c_1\right)\cos x + \left(\tfrac{1}{4}x^2 + \tfrac{1}{4}x + c_2\right)\sin x$
23. $y = \left(\tfrac{1}{12}x^4 + c_1 x + c_2\right)e^x$
25. a. $y = \begin{cases} a \sin x + b \cos x + x, & 0 \le x \le 1 \\ a \sin x + b \cos x + 1, & x \ge 1 \end{cases}$
    b. $y = \begin{cases} x, & 0 \le x \le 1 \\ 1, & x \ge 1 \end{cases}$

Problems 3.6, page 117

1. $y = c_1 e^x + c_2 e^{2x} + 5$
3. $y = c_1 \cos 2x + c_2 \sin 2x + \sin x$
5. $y = c_1 e^x + c_2 e^{2x} + 3e^{3x}$
7. $y = (c_1 + c_2 x - 2x^2)e^x$
9. $y = c_1 e^{2x} + c_2 e^{5x} + 10x + 7$
11. $y = c_1 + c_2 e^x - \ln|\cos x|$
13. $y = c_1 \cos 2x + c_2 \sin 2x + \tfrac{1}{2}x \sin 2x + \tfrac{1}{4}\cos 2x \ln|\cos 2x|$
15. $y = e^x[c_1 + c_2 x - \ln|1 - x|]$
17. $y = c_1 e^x + c_2 e^{-x} + e^x \ln|x|$
19. $y = e^{2x}[c_1 + c_2 x + (x + 1)\ln|x + 1|]$
21. $y_p = \tfrac{1}{2}\left[-1 - x \ln x + \left(x - \dfrac{1}{x}\right)\ln(1 + x)\right]$

Problems 3.7, page 122

1. $y = c_1 x + c_2 x^{-1}$
3. $y = x[c_1 \cos(\ln x) + c_2 \sin(\ln x)]$
5. $y = x^{3/2} - x^{1/2}$ 7. $y = c_1 x + c_2 x^3$
9. $y = x^{-2}[c_1 \cos(\ln x) + c_2 \sin(\ln x)]$
11. $y = c_1 x^3 + c_2 x^{-4}$
13. $y = c_1 x^3 + c_2 x^{-5} - \dfrac{1}{16x}$
15. $y = x^3(c_1 + c_2 \ln x + \tfrac{1}{2} \ln^2 x)$
17. $y = c_1 \cos(\ln x) + c_2 \sin(\ln x) + 10$
19. $y = c_1 x^{-5} + c_2 x^{-1} + x/12$
21. $y = c_1 x^2 + c_2 x^{-1} + \tfrac{1}{4} - \tfrac{1}{2}\ln x$
23. $y = c_1 + c_2 x^{-3}$

Problems 3.8, page 126

1. $y = (c_1 + c_2 x)\cos x + (c_3 + c_4 x)\sin x$

3. $y = (1+x)e^x$ 5. $y = 1 + e^{3x} + e^{-3x}$
7. $y = c_1 + c_2 x + c_3 x^2 + c_4 x^3$
9. $y = c_1 e^x + c_2 e^{-x} + c_3 e^{2x} + c_4 e^{-2x}$
11. $y = 1 - x + e^{2x} - e^{-2x}$
13. $y = c_1 e^x + c_2 \cos x + c_3 \sin x$
15. $y = c_1 e^{3x} + e^{-3x/2}\left(c_2 \cos\frac{3\sqrt{3}}{2}x + c_3 \sin\frac{3\sqrt{3}}{2}x\right)$
17. If $c_1 y_1 + c_2 y_2 + c_3 y_3 \equiv 0$, then we also must have
$$c_1 y_1' + c_2 y_2' + c_3 y_3' \equiv 0,$$
$$c_1 y_1'' + c_2 y_2'' + c_3 y_3'' \equiv 0.$$
In order for all three of these identities to hold at x_0, it must be that $c_1 = c_2 = c_3 = 0$.
19. a. $y_1(v')'' + (3y_1' + ay_1)(v')'$
$+ (3y_1'' + 2ay_1' + by_1)v' = 0$
c. $v' = \frac{W(y_1, y_2)}{y_1^2} \int \frac{y_1 e^{-\int a(x)dx}}{W^2(y_1, y_2)} dx$
25. $y = c_1 e^x + c_2 x e^x + c_3 e^{-x} + \frac{1}{4} x^2 e^x$
27. $y = c_1 e^{2x} + c_2 e^{-3x} + c_3 e^{4x} + \frac{1}{24}x + \frac{41}{288}$
29. b. $y_p = \frac{1}{10}e^{2x}\int e^{-3x} \tan x\, dx + \frac{1}{10}e^{-x}[(3\cos x - \sin x)\ln|\sec x + \tan x| + 1]$
31. $y = c_1 x^{-1} + (c_2 + c_3 \ln|x|)x$
33. $y = c_1 x^{-1} + c_2 \cos(\ln|x|) + c_3 \sin(\ln|x|)$

Problems 3.9, page 135
1. $y_n = c_1 4^n + c_2 (-1)^n$
3. $y_n = 3(-\frac{1}{3})^n - 2(-\frac{1}{2})^n$
5. $y_n = (c_1 + c_2 n)(-1)^n$
7. $y_n = 2^{n/2}\left(\cos\frac{n\pi}{4} + \sin\frac{n\pi}{4}\right)$
9. $y_n = \frac{2^n}{\sqrt{3}}\sin\frac{n\pi}{3}$
11. $P_n = \sqrt{5}\left[\left(\frac{1}{2} + \frac{1}{2\sqrt{5}}\right)^n - \left(\frac{1}{2} - \frac{1}{2\sqrt{5}}\right)^n\right]$;
$P_{10} \approx 0.088$
13. If there are at least 125,000 pairs in any one year, there is no danger of extinction
15. a. $4; $18
b. Even if B has unlimited resources, the probability of B bankrupting A is only $8/27$

Review Exercises for Chapter 3, page 135
1. $y_2 = \cos 2x$ 3. $y_2 = x^{-2}$
5. $y_2 = -2 + x \ln\left|\frac{1+x}{1-x}\right|$ 7. $y = 13e^{4x} - 10e^{5x}$
9. $y = \frac{2}{\sqrt{7}}e^{(3/2)x}\sin\frac{\sqrt{7}}{2}x$
11. $y = (c_1 + c_2 x)e^{-x/2}$

13. $y = e^x(c_1 \cos\sqrt{6}\,x + c_2 \sin\sqrt{6}\,x)$
15. $y = c_1 e^x + c_2 e^{2x} + c_3 e^{3x}$
17. $y = -3e^x + x^2 + 4x + 5$
19. $y = e^x(c_1 + c_2 x + x^{-1})$
21. $y = (c_1 + c_2 \ln x)x^{-2}$
23. $y = (c_1 + c_2 x)e^{-2x} + c_3 e^{3x}$

CHAPTER 4

Problems 4.1, page 141
1. $x = \cos 10t$ 3. $x = 3\cos t + 4\sin t$
5. $x = \frac{3}{5}\sin 5t$
7. $v = -0.2\sqrt{g}$ m/sec
9. $\sqrt{\frac{8\pi}{g}}$ seconds; $x(t) = \left(1 - \frac{2}{\pi}\right)\cos\left(\sqrt{\frac{\pi g}{2}}\,t\right)$ m
11. $\theta = \frac{1}{2}\cos\sqrt{2g}\,t$, $f = \frac{\sqrt{2g}}{2\pi} \approx 1.28$ Hz
13. $f = \frac{\sqrt{96g}}{2\pi} \approx 8.85$ Hz
15. 844 sec = 14.1 min
17. $x'' + \frac{T_0 g}{L\omega} x = 0$, $2\pi\sqrt{\frac{L\omega}{T_0 g}}$

Problems 4.2, page 146
1. $x(t) = (1 + 10t)e^{-10t}$
3. $x(t) = e^{-\sqrt{5}\,t/2}[3\cosh\frac{1}{2}t + (8 + 3\sqrt{5})\sinh\frac{1}{2}t]$
$= \left(\frac{3\sqrt{5} + 11}{2}\right)e^{[(1-\sqrt{5})t/2]} - \left(\frac{3\sqrt{5} + 5}{2}\right)$
$\times e^{[(-1-\sqrt{5})t/2]}$
5. $x(t) = e^{-4t}\sin 3t$
7. $k = \frac{25}{4}\left(1 + \frac{\pi^2}{10000}\right)$
In Problems 9–13, $p = \sqrt{g/96} \approx 0.579$
9. $x(t) = e^{-3pt}\left(\frac{1}{6}\cos\sqrt{91}\,pt - \frac{1-\frac{1}{2}p}{\sqrt{91}\,p}\sin\sqrt{91}\,pt\right)$
11. $x(t) = e^{-10pt}[\frac{1}{6} - (1 - \frac{5}{3}p)t]$
13. $x(t) = e^{-3pt/\sqrt{2}}$
$\times\left(-\frac{24}{25}\cos\frac{1}{2}\sqrt{182}\,pt - \frac{72}{25\sqrt{91}}\sin\frac{1}{2}\sqrt{182}\,pt\right)$
15. No

Problems 4.3, page 151
1. $x(t) = \frac{2001}{2000}(1 + 10t)e^{-10t} - \frac{1}{2000}\cos 10t$

3. $x(t) = \frac{1}{50}\left\{e^{-\sqrt{5}t/2}\left[(150+\sqrt{5})\cosh\frac{1}{2}t\right.\right.$
$\left.\left.+(405+150\sqrt{5})\sinh\frac{1}{2}t\right] - \sqrt{5}\cos t\right\}$
$= \frac{1}{100}\left[(555+151\sqrt{5})e^{(1-\sqrt{5})t/2}\right.$
$\left. - (255+149\sqrt{5})e^{(-1-\sqrt{5})t/2} - 2\sqrt{5}\cos t\right]$

5. $x(t) = \frac{1}{104}\left[e^{-4t}(3\cos 3t + 106\sin 3t)\right.$
$\left. - 3\cos 3t + 2\sin 3t\right]$

7. $x(t) = -0.1419e^{-2t} + 0.0029e^{-19.62t}$
$- 0.1390\cos 2t + 0.1133\sin 2t$

9. Use $c^2 = \dfrac{144(240)}{g(4\pi^2 + 0.36)}$ and solve the system

$\left(\dfrac{240}{100}g - 144\right)A + \dfrac{12}{100}cgB = \dfrac{g}{5}$

$-\dfrac{12}{100}cgA + \left(\dfrac{240}{100}g - 144\right)B = 0.$

11. b. Let $A = \dfrac{F_0}{k - m\omega^2}$. With $x(0) = x'(0) = 0$,

$x = -\dfrac{A\omega}{\omega_0}\sin\omega_0 t + A\sin\omega t.$

If $\omega \approx \omega_0$, $x \approx A(\sin\omega t - \sin\omega_0 t)$
$= 2A\sin\frac{1}{2}(\omega - \omega_0)t\cos\frac{1}{2}(\omega + \omega_0)t,$
the product of a slowly varying component and a rapidly varying one.

Problems 4.4, page 155

1. $I(t) = 6(e^{-5t} - e^{-20t})$; $Q(t) = \frac{1}{10}(9 - 12e^{-5t} + 3e^{-20t})$.

3. $I_{\text{steady state}} = \cos t + 2\sin t$

5. $I_{\text{steady state}} = \frac{1}{13}(70\sin 10t - 90\cos 10t)$

7. $I_{\text{transient}} = e^{-t}(-\cos t - 3\sin t)$

9. $I_{\text{transient}} = \frac{40}{3}e^{-5t} - \frac{250}{39}e^{-2t}$

11. $I_{\text{transient}} = \dfrac{e^{-t}}{481}\left(-245\cos 7t - \frac{255}{7}\sin 7t\right)$

13. $I_{\text{transient}} = \dfrac{e^{-600t}}{2320}[-153\sin 800t - 96\cos 800t];$

$I_{\text{steady state}} = \dfrac{1}{580}[24\cos 600t + 27\sin 600t]$

15. $Q = \begin{cases} c_1\cos\dfrac{t}{\sqrt{LC}} + c_2\sin\dfrac{t}{\sqrt{LC}} \\ \quad + \left(\dfrac{CE_0}{1 - CL\omega^2}\right)\sin\omega t, \quad \omega \neq \dfrac{1}{\sqrt{LC}}, \\ \left(c_1 - \dfrac{E_0 t}{2\omega L}\right)\cos\omega t \\ \quad + c_2\sin\omega t, \quad \omega = \dfrac{1}{\sqrt{LC}} \end{cases}$

17. b. $\omega = \dfrac{\sqrt{(4L/C) - R^2}}{2L}$

19. b. No ω produces resonance.

Review Exercises for Chapter 4, page 155

1. a. $x = \cos\sqrt{50}\,t$
   b. $x = e^{-5t}(\cos 5t + \sin 5t)$
   c. $x = \frac{1}{1000}[1000\cos\sqrt{50}\,t + \sqrt{2}\sin\sqrt{50}\,t - \sin 10t]$
   d. $x = \frac{1}{2500}\left[e^{-5t}(2501\cos 5t + 2502\sin 5t)\right.$
   $\left. - \cos 10t - \frac{1}{2}\sin 10t\right]$

3. a. $x = 3\cos\dfrac{2\sqrt{10}}{5}t + \sqrt{10}\sin\dfrac{2\sqrt{10}}{5}t$

   b. $x = e^{-t/\sqrt{5}}\left(3\cos\sqrt{\dfrac{7}{5}}t + \dfrac{3 + 4\sqrt{5}}{\sqrt{7}}\sin\sqrt{\dfrac{7}{5}}t\right)$

   c. $x = 3\cos\dfrac{2\sqrt{10}}{5}t + \dfrac{59\sqrt{10}}{60}\sin\dfrac{2\sqrt{10}}{5}t + \dfrac{1}{15}\sin t$

   d. $x = \dfrac{1}{145}\left[e^{-t/\sqrt{5}}\left((435 + 2\sqrt{5})\cos\sqrt{\dfrac{7}{5}}t\right.\right.$
   $\left.\left. + \left(579\sqrt{\dfrac{5}{7}} + \dfrac{435}{\sqrt{7}}\right)\sin\sqrt{\dfrac{7}{5}}t\right)\right.$
   $\left. - 2\sqrt{5}\cos t + 3\sin t\right]$

5. a. $x = \frac{6}{5}\sin\frac{5}{2}t$

   b. $x = \frac{2}{7}\sqrt{21}\,e^{-t}\sin\dfrac{\sqrt{21}}{2}t$

   c. $x = \frac{72}{55}\sin\frac{5}{2}t - \frac{1}{11}\sin 3t$

   d. $x = \dfrac{1}{697}\left[e^{-t}\left(24\cos\dfrac{\sqrt{21}}{2}t + \dfrac{4296}{\sqrt{21}}\sin\dfrac{\sqrt{21}}{2}t\right)\right.$
   $\left. - 24\cos 3t - 11\sin 3t\right]$

7. $Q_{ss} = \frac{1}{10}$,

$Q_{\text{tran}} = -\dfrac{1}{10}e^{-25t/2}\left[\cos\dfrac{25\sqrt{31}}{2}t + \dfrac{1}{\sqrt{31}}\sin\dfrac{25\sqrt{31}}{2}t\right]$

9. $I_{ss} = \frac{50}{14841}[120\cos 5t + 21\sin 5t]$

11. $I_{ss} = 4\cos 5t + 2\sin 5t$

13. $x = c_1\cos\sqrt{2g}\,t + c_2\sin\sqrt{2g}\,t$
$+ \dfrac{g}{10(g - 32)}\cos 8t$

15. $x = e^{-gt/2000}(c_1 \cos rt + c_2 \sin rt)$
$- \dfrac{13g^2}{8000D} \cos \dfrac{65}{8} t$
$+ \left(\dfrac{2g^2 - 4225g/64}{5D} \right) \sin \dfrac{65t}{8}$
where $r^2 = 2g - g^2/4 \times 10^6$ and
$D = (2g - 4225/64)^2 + (65g/8000)^2$.

CHAPTER 5

Problems 5.1, page 165

1. $e^x = e\left[1 + (x-1) + \dfrac{(x-1)^2}{2!} + \dfrac{(x-1)^3}{3!} + \cdots \right]$

3. $\cos x = \dfrac{\sqrt{2}}{2}\left[1 + \left(x - \dfrac{\pi}{4}\right) - \dfrac{\left(x - \dfrac{\pi}{4}\right)^2}{2!} - \dfrac{\left(x - \dfrac{\pi}{4}\right)^3}{3!} + \cdots \right]$

5. $e^{bx} = e^{-b}\left[1 + b(x+1) + \dfrac{b^2}{2!}(x+1)^2 + \dfrac{b^3}{3!}(x+1)^3 + \cdots \right]$

7. $x^2 e^{-x^2} = \sum\limits_{n=0}^{\infty} \dfrac{(-1)^n x^{2n+2}}{n!}$

9. $(x-1)\ln x = (x-1)^2 - \tfrac{1}{2}(x-1)^3 + \tfrac{1}{3}(x-1)^4 - \cdots$

19. $y = e^x + x + 1$
21. $y = c_0 \cos x + (c_1 - 1)\sin x + x$
23. $y = c_0(1 + x \tan^{-1} x) + c_1 x$ 25. $y = xe^x$
27. $y = 1 - 2x^2$ 29. $y = x + \sin x$
31. $y = \dfrac{1}{1-x}$ 33. $y = e^{x^2} - \dfrac{x}{4}$
35. $y = c_0\left[1 + \sum\limits_{n=1}^{\infty} \dfrac{1 \cdot 4 \cdots (3n-2)}{(3n)!} x^{3n} \right]$
$+ c_1\left[x + \sum\limits_{n=1}^{\infty} \dfrac{2 \cdot 5 \cdots (3n-1)}{(3n+1)!} x^{3n+1} \right]$

37. $y = c_1 x +$
$c_0\left[1 - x^2 - \sum\limits_{n=2}^{\infty} \dfrac{1 \cdot 3 \cdot 5 \cdots (2n-3)}{(2n)!}(2x^2)^n \right]$

39. $y = c_0\left[1 + 2(x-1)^2 + \tfrac{1}{2} \sum\limits_{n=3}^{\infty} (n+1)(1-x)^n \right]$
$+ c_1\left[(x-1) - \tfrac{1}{2}(x-1)^2 - \tfrac{1}{4}\sum\limits_{n=3}^{\infty} (n+1)(1-x)^n \right]$

41. $y = c_0(1 - 3x^2) + c_1\left[x - \tfrac{2}{3}x^3 \right.$
$\left. -2 \sum\limits_{n=2}^{\infty} \dfrac{2^n[3 \cdot 12 \cdot 25 \cdots (n+1)(2n-3)]}{(2n+1)!} x^{2n+1} \right]$

Problems 5.2, page 172

1. $y_1 = x$;
$y_2 = 1 - \dfrac{x^2}{2!} - \dfrac{x^4}{4!} - \dfrac{3x^6}{6!} - \dfrac{3 \cdot 5}{8!}x^8 - \cdots$

3. $y_1 = 1 + \dfrac{x^3}{2} + \dfrac{1}{2!(5 \cdot 2)}x^6 + \dfrac{1}{3!(8 \cdot 5 \cdot 2)}x^9 + \cdots$
$= 1 + \sum\limits_{n=1}^{\infty} \dfrac{x^{3n}}{n![2 \cdot 5 \cdots (3n-1)]}$
$y_2 = x + \dfrac{1}{4}x^4 + \dfrac{1}{2!(7 \cdot 4)}x^7$
$+ \dfrac{1}{3!(10 \cdot 7 \cdot 4)}x^{10} + \cdots$
$= \sum\limits_{n=0}^{\infty} \dfrac{x^{3n+1}}{n![1 \cdot 4 \cdots (3n+1)]}$

5. $y_1 = (x-2)e^x$; $y_2 = (x-2)e^{x-2}\int \dfrac{e^{(x-2)^2/2}}{(x-2)^2}\, dx$

7. $y_1 = e^{-x^3/3}$; $y_2 = e^{-x^3/3}\int e^{x^3/3}\, dx$

9. $y_1 = x$, $y_2 = 1 + x \tan^{-1} x$

11. $y_1 = 1 - x$, $y_2 = \dfrac{1}{1-x}$

13. $y_1 = x + \tfrac{8}{3}x^3$,
$y_2 = 1 + 9x^2 + \tfrac{15}{2}x^4 - \tfrac{7}{2}x^6 + \tfrac{27}{8}x^8 - \cdots$

15. $y = c_1 x + c_0\left(1 - \dfrac{1}{2}x^2 - \dfrac{1}{4!}x^4 - \dfrac{3}{6!}x^6 - \cdots \right)$
$+ \sin x = c_1 x - c_0 x \int \dfrac{e^{x^2/2}}{x^2}\, dx + \sin x$

17. $y = 1 + \dfrac{x^4}{4!} + \dfrac{1 \cdot 5 x^8}{8!} + \dfrac{1 \cdot 5 \cdot 9}{12!}x^{12} + \cdots$

19. $y = e^{x^2/2}$

21. $y = 1 + x + \dfrac{x^2}{2!} + \dfrac{2}{3!}x^3 + \dfrac{4}{4!}x^4 + \cdots$

23. $y = x - \dfrac{x^3}{3!} + \dfrac{4x^5}{5!} - \dfrac{19x^7}{7!} + \cdots$

25. $y = e^x(c_1 + c_2 \ln|x|)$

27. $y = c_1 x + c_2 x \int \dfrac{e^{-x}}{x} dx$

$= c_1 x + c_2 \left(x \ln|x| + \sum\limits_{n=1}^{\infty} \dfrac{(-1)^n x^{n+1}}{(n!)n} \right)$

29. $y = c_1 x^2 + c_2 x^2 \int \dfrac{e^{-x}}{x^3} dx = c_1 x^2$

$+ c_2 \left(-\tfrac{1}{2} + x + \tfrac{1}{2} x^2 \ln|x| + \sum\limits_{n=3}^{\infty} \dfrac{(-1)^n x^n}{n!(n-2)} \right)$

31. $y = c_1 (x^2 + 2x + 3) + \dfrac{c_2 x^4}{(1-x)^2}$

33. No; the solutions $y = ce^{-1/x}$ and $y = ce^{-1/2x^2}$ do not have power series about $x = 0$

Problems 5.3, page 184

1. $y_0 = c_0 \cos\sqrt{x} + c_1 \sin\sqrt{x}$ if $x > 0$; $c_0 \cosh\sqrt{-x} + c_1 \sinh\sqrt{-x}$ if $x < 0$

3. $y = c_0 \dfrac{\sinh x}{x^3} + c_1 \dfrac{\cosh x}{x^3}$

5. $y = c_0 \dfrac{\sin x^2}{x^2} + c_1 \dfrac{\cos x^2}{x^2}$ 7. $y = c_0 x + \dfrac{c_1}{x^2}$

9. $y = c_0 \dfrac{\sin x}{x} + c_1 \dfrac{\cos x}{x}$

11. $y = c_0 e^x + c_1 e^x \ln|x|$

13. $y = \sqrt{|x|}\, e^x (c_0 + c_1 \ln|x|)$

15. $y = \dfrac{c_0}{1-x} + c_1 \dfrac{\ln|x|}{1-x}$

17. $y = (c_0 + c_1 \ln|x|)\left(1 + \dfrac{x^2}{2^2} + \dfrac{x^4}{(2\cdot 4)^2} + \cdots\right)$

$- c_1 \left(\dfrac{x^2}{4} + \dfrac{3x^4}{8\cdot 16} + \cdots \right)$

19. $y = c_0 \sqrt{x}\left(1 - \tfrac{7}{6}x + \tfrac{21}{40}x^2 + \cdots\right)$

$+ c_1 (1 - 3x + 2x^2 + \cdots)$

21. $y = c_0 x + c_1 \left(x \ln|x| + \sum\limits_{n=1}^{\infty} \dfrac{(-1)^n}{n!n} x^{n+1} \right)$

23. $y = \sqrt{|x|}\, (c_0 + c_1 \ln|x|)$
25. b. $y_2 = xe^{-1/x}$

Problems 5.4, page 194

1. $\left(\dfrac{384}{x^4} - \dfrac{72}{x^2} + 1 \right) J_1(x) - \left(\dfrac{192}{x^3} - \dfrac{12}{x} \right) J_0(x)$

13. Use equation (17)
15. Set $z = \sqrt{x}$, then $y = AJ_p(\sqrt{x}) + BY_p(\sqrt{x})$
17. Set $y = xu$, then $y = x[AJ_1(x) + BY_1(x)]$
19. Set $y = ux^{-k}$, then $y = x^{-k}(AJ_k(x) + BY_k(x))$
21. Set $y = \sqrt{x}\, u$, $z = kx^3/3$,
then $y = \sqrt{x}\, (AJ_{1/6}(kx^3/3) + BY_{1/6}(kx^3/3))$

33. b. $u^2 \dfrac{d^2 x}{du^2} + \left(u^2 + \dfrac{4k_1}{ma^2} \right) x = 0$

c. Substitute $x = \sqrt{u}\, z$ getting $u^2 z'' + uz'$

$+ \left(u^2 - \left(\dfrac{1}{4} - \dfrac{4k_1}{ma^2} \right) \right) z = 0.$

Problems 5.5, page 200

1. $P_5 = \dfrac{63x^5 - 70x^3 + 15x}{8}$,

$P_6 = \dfrac{231x^6 - 315x^4 + 105x^2 - 5}{16}$,

$P_7 = \dfrac{429x^7 - 693x^5 + 315x^3 - 35x}{16}$,

$P_8 = \dfrac{429(15x^8 - 28x^6) + 630(11x^4 - 2x^2) + 35}{128}$

9. $H_0(x) = 1$, $H_1(x) = 2x$, $H_2(x) = 4x^2 - 2$,
$H_3 = 8x^3 - 12x$, $H_4(x) = 16x^4 - 48x^2 + 12$

Review Exercises for Chapter 5, page 207

7. $y = x^2 + e^x$ 9. $y = x(\sin x + 1)$
11. $y = e^{-x^2}$ 13. $y = c_0 e^x + c_1 \sqrt{x}$
15. $y = \dfrac{c_0}{x} + \dfrac{c_1}{1-x}$

17. $y = c_1 \dfrac{e^{x/2}}{\sqrt{x}}$

$+ c_2 x \left(1 + \sum\limits_{n=1}^{\infty} \dfrac{x^n}{5\cdot 7 \cdots (2n+3)} \right)$

23. $y = c_0 x J_0(x) + c_1 x Y_0(x)$ 25. $y = P_n(\cos x)$

CHAPTER 6

Problems 6.1, page 219

1. $\dfrac{5}{s^2} + \dfrac{2}{s}$, $s > 0$ 3. $\dfrac{18}{s^3} - \dfrac{7}{s}$, $s > 0$

5. $\dfrac{2}{s^3} + \dfrac{8}{s^2} - \dfrac{16}{s}$, $s > 0$

7. $\dfrac{3}{4s^4} + \dfrac{1}{2s^3} + \dfrac{1}{2s^2} + \dfrac{1}{s}$, $s > 0$

9. $\dfrac{a}{s^2} + \dfrac{b}{s}$, $s > 0$ 11. $\dfrac{e^2}{s-5}$, $s > 5$

13. $\dfrac{1}{s - 1/2}$, $s > 1/2$ 15. $\dfrac{e^{-1/2}}{s+1}$, $s > -1$

17. $\dfrac{3}{s^2 + 9}$, $s > 0$ 19. $\dfrac{s}{s^2 + 49}$, $s > 0$

21. $\dfrac{5\cos 2}{s^2 + 25} + \dfrac{s \sin 2}{s^2 + 25}$, $s > 0$

23. $\dfrac{s\cos b}{s^2+a^2} - \dfrac{a\sin b}{s^2+a^2},\ s>0$

25. $\dfrac{s}{s^2-1/4},\ s>1/2$

27. $\dfrac{s\cosh 2}{s^2-25} - \dfrac{5\sinh 2}{s^2-25},\ s>5$

29. $\dfrac{a\cosh b}{s^2-a^2} + \dfrac{s\sinh b}{s^2-a^2},\ s>|a|$

31. $\dfrac{1}{(s-1)^2},\ s>1$

33. $\dfrac{6}{(s+1)^4} - \dfrac{1}{s+1},\ s>-1$

35. $\dfrac{1}{(s-1)^2+1},\ s>1$

37. $\dfrac{s-4}{(s-4)^2+4},\ s>4$ 39. $\dfrac{s+2}{(s+1)^2+1},\ s>-1$

43. $9t^2+7$ 45. $\cos t + \sin t$

47. $\cosh\sqrt{2}\,t - \sqrt{2}\,\sinh\sqrt{2}\,t$ 49. te^t

51. $\dfrac{3}{\sqrt{5}}e^{-2t}\sin\sqrt{5}\,t$

53. $2e^{-t}\cos\sqrt{7}\,t - \dfrac{3}{\sqrt{7}}e^{-t}\sin\sqrt{7}\,t$

55. $ce^{-at}\cos\sqrt{b-a^2}\,t + \dfrac{d-ac}{\sqrt{b-a^2}}e^{-at}\sin\sqrt{b-a^2}\,t$

63. $\dfrac{2abs}{(s^2-a^2-b^2)^2 - 4a^2b^2}$

65. $\dfrac{s(s^2-a^2+b^2)}{(s^2+a^2+b^2)^2 - 4a^2s^2}$

67. $\dfrac{a(s^2-a^2-b^2)}{(s^2+a^2+b^2)^2 - 4a^2s^2}$ 71. $\dfrac{6}{(s+1)^4}$

73. $\dfrac{2s^3 - 54s}{(s^2+9)^3}$ 75. $\dfrac{(s-a)^2 - b^2}{\left[(s-a)^2+b^2\right]^2}$

77. $\dfrac{3(s+1)^2 + 3}{\left[(s+1)^2-1\right]^2}$

81. a. $\sqrt{\pi/s}$ b. $\dfrac{\sqrt{\pi}}{2s^{3/2}}$ c. $\dfrac{15\sqrt{\pi}}{8s^{7/2}}$

Problems 6.2, page 233

1. $\cos t$ 3. $A\cosh at + \dfrac{B}{a}\sinh at$

5. $e^{-t}(\cos 2t + \sin 2t)$ 7. $\tfrac{1}{3} - e^t + \tfrac{5}{3}e^{3t}$

9. $-\dfrac{t}{9} + \dfrac{23}{27}e^{3t} + \dfrac{4}{27}e^{-3t}$ 11. e^{-t}

13. $\sinh t$ 15. $\left(1+\dfrac{t}{2k}\right)\sin kt$

17. $\left(a+\dfrac{1}{2a^2}\right)\sin at + \left(a-\dfrac{t}{2a}\right)\cos at$

19. $\tfrac{5}{4}\cosh t - \tfrac{1}{4}(\cos t + t\sin t)$

21. $\dfrac{s^2+2a^2}{s(s^2+4a^2)}$ 23. $\dfrac{2a(3s^2-a^2)}{(s^2+a^2)^3}$

25. $\dfrac{8+6s^2}{s^2(s^2+4)^2}$ 27. $\dfrac{2a^2(4a^2+3s^2)}{s^2(s^2+4a^2)^2}$

29. $\dfrac{16}{s^2+16}\left[\dfrac{1}{s^3} + \dfrac{1}{s(s^2+16)} + \dfrac{4s}{(s^2+16)^2}\right]$

33. $\dfrac{\pi}{2} - \tan^{-1}s = \tan^{-1}(1/s)$

35. $\dfrac{\pi}{2} - \tan^{-1}(s/3) = \tan^{-1}(3/s)$ 37. $\tan^{-1}(k/s)$

39. $\dfrac{1}{2}\ln\dfrac{s^2-a^2}{s^2}$ 41. $\dfrac{1}{2s}\ln\dfrac{s^2-a^2}{s^2}$

43. $\dfrac{e^{s^2/4}}{s}\left(1-\mathrm{erf}\left(\dfrac{s}{2}\right)\right)$ 45. $\dfrac{2}{t}(1-\cos at)$

47. $\dfrac{\sin t}{t}$

49. a. $Y' - s^2 Y = -1$

b. $Y(s) = \left[Y(0) - \displaystyle\int_0^s e^{-u^3/3}\,du\right]e^{s^3/3}$

51. $I(t) = \dfrac{1}{676}[12\cos t + 5\sin t - (12+65t)e^{-5t}]$

53. $Q(t) = \dfrac{1}{125}[-4\cos 10t - 3\sin 10t + (4+50t)e^{-5t}]$

Problems 6.3, page 245

9. $\dfrac{b}{s(e^{as}+1)}$ 11. $\dfrac{b(1-e^{-as})}{s(1+e^{-2as})}$

13. $\dfrac{1}{s^2(1+e^{-as})}$ 15. $\dfrac{1}{(s^2+1)(1-e^{-\pi s})}$

17. $\dfrac{1+se^{-4\pi s}}{s^2+1}$ 19. $\sin(t-\pi)H(t-\pi)$

21. $\cos t(1+H(t-\pi))$

23. $\dfrac{1}{4!}[t^4 + (t-4)^4 H(t-4)]$

25. $f(t) = 1 + 2H(t-1) + 2H(t-7)$;
$\mathscr{L}\{f\} = \dfrac{1}{s}[1 + 2e^{-s} + 2e^{-7s}]$

27. $y = t - \sin t - H(t-\pi)(t + \sin t + \pi\cos t)$

29. $y = \tfrac{1}{10}[1 - e^{-t}(\cos 3t + \tfrac{1}{3}\sin 3t)]$
$\quad - \tfrac{1}{10}[1 - e^{-(t-1)}(\cos 3(t-1)$
$\quad + \tfrac{1}{3}\sin 3(t-1))]H(t-1)$

31. $y = \tfrac{1}{10} + \tfrac{9}{10}e^{-t}\cos 3t + \tfrac{3}{10}e^{-t}\sin 3t$
$\quad - \tfrac{1}{10}H(t-1)[1 - e^{-(t-1)}\cos 3(t-1)$
$\quad\quad - \tfrac{1}{3}e^{-(t-1)}\sin 3(t-1)]$

33. $y = e^{2t}\sin 3t + e^{2(t-1)}\sin 3(t-1)H(t-1)$

Problems 6.4, page 251

1. $I(t) = \dfrac{10}{\sqrt{6}} e^{-20(t-30)} \sin\sqrt{600}\,(t-30)\,H(t-30)$
3. $I(t) = 1 - \cos 4t + H(t-5)[\cos 4(t-5) - 1]$
5. $I(t) = 1 - \cos\sqrt{10}\,t - H(t-2)$
   $\times [1 - \cos\sqrt{10}\,(t-2)] + 2H(t-4)$
   $\times [1 - \cos\sqrt{10}\,(t-4)]$
   $+ \dfrac{60}{\sqrt{10}} H(t-4) \sin\sqrt{10}\,(t-4)$
7. a. $x(t) = \dfrac{t}{2} - \dfrac{\sin 2t}{4} - \dfrac{1}{4} H\!\left(t - \dfrac{\pi}{2}\right)$
   $\times [2t - \pi\cos(2t-\pi) - \sin(2t-\pi)]$
   b. $x(t) = \dfrac{t}{2} - \dfrac{\sin 2t}{4} + \cos 2t$
   $- \dfrac{1}{4} H\!\left(t - \dfrac{\pi}{2}\right)$
   $\times [2t - \pi\cos(2t-\pi) - \sin(2t-\pi)]$

Problems 6.5, page 257

1. $\dfrac{3!}{s^4(s^2+1)}$ 3. $\dfrac{3!5!}{s^{10}}$
5. $\dfrac{19!}{s^{20}(s-17)}$
7. $\dfrac{1}{2}\int_0^t (t-u)^3 \sin u\,du = \dfrac{t^3}{2} - 3t + 3\sin t$
9. $\dfrac{1}{a^2}(1 - \cos at)$ 11. $\dfrac{1}{2}(t-3)^2 H(t-3)$
13. $e^{-t}(1-t)^2$
15. $\tfrac{1}{13}[1 - e^{-2t}(\cos 3t + \tfrac{2}{3}\sin 3t)]$
    $-\tfrac{1}{13}[1 - e^{-2(t-\pi)}(\cos 3(t-\pi)$
    $+ \tfrac{2}{3}\sin 3(t-\pi)] H(t-\pi)$

Review Exercises for Chapter 6, page 258

1. $\dfrac{3}{s^2} - \dfrac{2}{s}$ 3. $\dfrac{1}{e(s-2)}$
5. $\dfrac{1}{(s+1)^2}$ 7. $\dfrac{3\cosh 4 - s(\sinh 4)}{s^2 - 9}$
9. $\dfrac{(s-1)^2 - 1}{[(s-1)^2 + 1]^2}$ 11. $\dfrac{s^2 + 2}{s(s^2+4)}$
13. $\dfrac{2s(s^2-12)}{(s^2+4)^3}$ 15. e^{-3s}
17. $\dfrac{1 + se^{-2\pi s}}{s^2 + 1}$ 19. -14
21. te^{2t} 23. $e^{-2t}\sin t$
25. $2e^{2t} - e^t$ 27. $-2e^{-t} - 5te^{-t} + 2\cos t + 4\sin t$
29. $\dfrac{2}{t}(1 - \cos 2t)$
31. $\cos t - \cos\!\left(t - \dfrac{\pi}{2}\right) H\!\left(t - \dfrac{\pi}{2}\right)$
    $= \begin{cases} \cos t, & t < \dfrac{\pi}{2} \\ \cos t - \sin t, & t > \dfrac{\pi}{2} \end{cases}$
33. $y = \cos t + 3\sin t$ 35. $y = 5e^{2t} - 3e^{3t}$
37. $y = \dfrac{e^{2t}}{9}(3t - 4) + e^t - \dfrac{5}{9}e^{-t}$
39. $y = \tfrac{1}{2}[1 - e^{-(t-3)}(\cos(t-3) - \sin(t-3))]$
    $\times H(t-3)$
41. $y = te^{2t} - (t-1)e^{2(t-1)} H(t-1)$ 43. $\dfrac{4!7!}{s^{13}}$
45. a. $\displaystyle\int_0^t \sin(t-u)u^2\,du = t^2 + 2\cos t - 2$;
    b. $\displaystyle\int_0^t \tfrac{1}{2}\sin 2u\,du = \tfrac{1}{4}(1 - \cos 2t)$

CHAPTER 7

Problems 7.1, page 266

1. $x_1' = x_2$
   $x_2' = -3x_1 - 2x_2$
3. $x_1' = x_2$
   $x_2' = x_3$
   $x_3' = x_1^3 - x_2^2 + x_3 + t$
5. $x_1' = x_2$
   $x_2' = x_3$
   $x_3' = -x_1 x_3 + x_1^4 x_2 + \sin t$
7. $x_1' = x_2$
   $x_2' = x_3$
   $x_3' = x_1 - 4x_2 + 3x_3$
9. a. $x_1 = c_1 e^t,\ x_2 = c_2 e^t$

Problems 7.2, page 270

1. b. $W = e^{-6t}$
   c. $([c_1 + c_2(1-t)]e^{-3t},\ (-c_1 + c_2 t)e^{-3t})$
3. d. System is not defined at $t = 0$.

Problems 7.3, page 279

1. $x = c_1 e^{-t} + c_2 e^{4t},\ y = -c_1 e^{-t} + \tfrac{3}{2} c_2 e^{4t}$
3. $x = (c_1 + c_2 t) e^{-3t},\ y = -(c_1 + c_2 + c_2 t) e^{-3t}$
5. $x = (c_1 + c_2 t) e^{10t},\ y = -(2c_1 + c_2 + 2c_2 t) e^{10t}$
7. $x = c_1 + c_2 e^{2t} - \tfrac{1}{4}(t^2 + 9t)$,
   $y = -c_1 - \tfrac{1}{3}c_2 e^{2t} + \tfrac{1}{4}(t^2 + 7t - 3)$
9. $x = c_1 e^{(1+\sqrt{3})t} + c_2 e^{(1-\sqrt{3})t}$
   $y = \dfrac{\sqrt{3}}{2}\left[c_1 e^{(1+\sqrt{3})t} - c_2 e^{(1-\sqrt{3})t} \right]$
   $- \dfrac{1}{4}\sin 2\!\left(c_1 e^{(1+\sqrt{3})t} + c_2 e^{(1-\sqrt{3})t} \right)$

11. $x_1 = 3c_1 e^{-t} + 4c_2 e^{-2t}$,
    $x_2 = -4c_1 e^{-t} - 5c_2 e^{-2t} - c_3 e^{2t}$,
    $x_3 = -2c_1 e^{-t} - 7c_2 e^{-2t} + c_3 e^{2t}$
13. $(3e^{3t}, -2e^{3t}), (2e^{10t}, e^{10t})$
15. $(e^{2t}, e^{2t}), (te^{2t}, (1+t)e^{2t})$
17. $(e^{3t} \cos 3t, \tfrac{1}{2}e^{3t}(3 \sin 3t + \cos 3t));$
    $(e^{3t} \sin 3t, \tfrac{1}{2}e^{3t}(\sin 3t - 3 \cos 3t))$
19. $(\tfrac{1}{4}te^{2t}, -\tfrac{11}{4}e^{2t})$
21. $(-\sin t, -\cos t - 2 \sin t)$
23. $(-t + 3e^t, \tfrac{1}{2} - t + 3e^t)$
25. $\dfrac{dx}{dt} = -\dfrac{3x}{100+t} + \dfrac{2y}{100-t}, \quad x(0) = 100$
    $\dfrac{dy}{dt} = \dfrac{3x}{100+t} - \dfrac{4y}{100-t}, \quad y(0) = 0$
27. $y_{\max} = \dfrac{500}{\sqrt{3}}\left[(2+\sqrt{3})^{(1-\sqrt{3})/2} - (2+\sqrt{3})^{(-1-\sqrt{3})/2}\right] = 500\sqrt{\tfrac{2}{3}}(2+\sqrt{3})^{-\sqrt{3}/2}$
    $t_{\max} = \dfrac{25}{\sqrt{3}} \ln(2+\sqrt{3})$ min.
29. $x_1 = c_1 \cos \sqrt{10}\,t + c_2 \sin \sqrt{10}\,t + c_3 \cos t + c_4 \sin t$
    $x_2 = -\tfrac{1}{4}c_1 \cos \sqrt{10}\,t - \tfrac{1}{4}c_2 \sin \sqrt{10}\,t + 2c_3 \cos t + 2c_4 \sin t$
31. The concentration of H_2S approaches $\dfrac{\gamma}{\alpha}$;
    the concentration of SO_2 approaches $\dfrac{\gamma+\delta}{\beta}$

Problems 7.4, page 285

1. $x = \cosh t, \; y = \sinh t$
3. $x = e^{3t}(2 \cos 3t + 2 \sin 3t)$,
   $y = e^{3t}(-2 \cos 3t + 4 \sin 3t)$
5. $x = \tfrac{3}{2} \cos t + \tfrac{1}{2} \sin t + \tfrac{7}{2}e^t - 5e^{-t} - 4t$,
   $y = \cos t + \tfrac{7}{2}e^t - \tfrac{5}{2}e^{-t} - 1 - 3t$
7. $x = \dfrac{e^{3t}}{4(13)^3}(6887 \cos 2t + 2637 \sin 2t) + \dfrac{e^t}{4}$
   $- \dfrac{5t^2}{13} - \dfrac{34t}{(13)^2} - \dfrac{74}{(13)^3}$,
   $y = \dfrac{e^{3t}}{4(13)^3}(-9524 \cos 2t + 4250 \sin 2t) + \dfrac{4}{13}t^2$
   $+ \dfrac{48t}{(13)^2} + \dfrac{184}{(13)^3}$
9. $x = \cos t, \; y = -\cos t - \sin t$
11. $x = \cos t - \sin t, \; y \equiv 1, \; z = \sin t + \cos t$

Problems 7.5, page 297

1. $\begin{pmatrix} 3 & 9 \\ 6 & 15 \\ -3 & 6 \end{pmatrix}$
3. $\begin{pmatrix} 2 & 2 \\ -2 & -1 \\ 6 & -1 \end{pmatrix}$
5. $\begin{pmatrix} 0 & 0 \\ 0 & 0 \\ 0 & 0 \end{pmatrix}$
7. $\begin{pmatrix} -2 & 4 \\ 7 & 15 \\ -15 & 10 \end{pmatrix}$
9. $\begin{pmatrix} 4 & 10 \\ 17 & 22 \\ -9 & 1 \end{pmatrix}$
11. $\begin{pmatrix} 0 & 6 \\ 5 & 14 \\ -9 & 9 \end{pmatrix}$
13. $\begin{pmatrix} 1 & -5 & 0 \\ -3 & 4 & -5 \\ -14 & 13 & -1 \end{pmatrix}$
15. $\begin{pmatrix} 1 & 1 & 5 \\ 9 & 5 & 10 \\ 7 & -7 & 3 \end{pmatrix}$
17. $\begin{pmatrix} -1 & -1 & -1 \\ -3 & -3 & -10 \\ -7 & 3 & 5 \end{pmatrix}$
19. $\begin{pmatrix} -1 & -1 & -5 \\ -9 & -5 & -10 \\ -7 & 7 & -3 \end{pmatrix}$
21. $\begin{pmatrix} 8 & 20 \\ -4 & 11 \end{pmatrix}$
23. $\begin{pmatrix} -3 & -3 \\ 1 & 3 \end{pmatrix}$
25. $\begin{pmatrix} 13 & 35 & 18 \\ 20 & 26 & 20 \end{pmatrix}$
27. $\begin{pmatrix} 19 & -17 & 34 \\ 8 & -12 & 20 \\ -8 & -11 & 7 \end{pmatrix}$
29. $\begin{pmatrix} 18 & 15 & 35 \\ 9 & 21 & 13 \\ 10 & 9 & 9 \end{pmatrix}$
31. $(7, 16)$
33. $\begin{pmatrix} 3 & -2 & 1 \\ 4 & 0 & 6 \\ 5 & 1 & 9 \end{pmatrix}$
35. $\begin{pmatrix} 0 & -8 \\ 32 & 32 \end{pmatrix}$
37. $\begin{pmatrix} 11 & 38 \\ 57 & 106 \end{pmatrix}$
39. $\begin{pmatrix} \tfrac{1}{2} & -\tfrac{1}{2} \\ -\tfrac{1}{4} & \tfrac{3}{4} \end{pmatrix}$
41. $\begin{pmatrix} 0 & 1 \\ 1 & 0 \end{pmatrix}$
43. $\begin{pmatrix} \tfrac{1}{3} & -\tfrac{1}{4} & -\tfrac{1}{6} \\ 0 & \tfrac{1}{4} & \tfrac{1}{2} \\ 0 & \tfrac{1}{4} & -\tfrac{1}{2} \end{pmatrix}$
45. $\begin{pmatrix} 0 & 1 & -1 \\ 2 & -2 & -1 \\ -1 & 1 & 1 \end{pmatrix}$
47. not invertible
49. a. $\begin{pmatrix} i & 2 \\ 1 & -i \end{pmatrix}$ b. $\begin{pmatrix} \tfrac{1}{2}+\tfrac{1}{2}i & 0 \\ 0 & \tfrac{1}{2}-\tfrac{1}{2}i \end{pmatrix}$
    c. $\dfrac{1}{2}\begin{pmatrix} 1+i & 1+i & -i \\ -1-i & -1+i & 1 \\ -1+i & 1+i & 1 \end{pmatrix}$
51. $\begin{pmatrix} \tfrac{1}{2} & 0 & 0 \\ 0 & \tfrac{1}{3} & 0 \\ 0 & 0 & \tfrac{1}{4} \end{pmatrix}$
53. $\begin{pmatrix} 1/a_{11} & 0 & \cdots & 0 \\ 0 & 1/a_{22} & \cdots & 0 \\ & & \vdots & \\ 0 & 0 & \cdots & 1/a_{nn} \end{pmatrix}$
55. $A_{1,2}(2)$ yields a row of zeros

57. $\mathbf{x}'(t) = (1, \cos t)$; $\int \mathbf{x}(t)\,dt = (\tfrac{1}{2}t^2, -\cos t) + \mathbf{c}$

59. $\mathbf{y}'(t) = \begin{pmatrix} e^t \\ -\sin t \\ \sec^2 t \end{pmatrix}$; $\int \mathbf{y}(t)\,dt = \begin{pmatrix} e^t \\ \sin t \\ -\ln|\cos t| \end{pmatrix} + \mathbf{c}$

Problems 7.6, page 303

1. $\mathbf{x}' = \begin{pmatrix} 2 & 3 \\ 4 & -6 \end{pmatrix} \mathbf{x}$

3. $\mathbf{x}' = \begin{pmatrix} 0 & 1 & 0 \\ 0 & 0 & 1 \\ 1 & -4t & 2 \end{pmatrix} \mathbf{x} + \begin{pmatrix} 0 \\ 0 \\ \sin t \end{pmatrix}$

5. $\mathbf{x}' = \begin{pmatrix} 2t & -3t^2 & \sin t \\ 2 & 0 & -4 \\ 0 & 17 & 4t \end{pmatrix} \mathbf{x} + \begin{pmatrix} 0 \\ -\sin t \\ e^t \end{pmatrix}$

Problems 7.7, page 313

1. fundamental   3. fundamental
5. not fundamental

7. $C = \begin{pmatrix} 1 & -2 \\ -1 & 1 \end{pmatrix}$ 9. $C = \begin{pmatrix} 1 & -1 & 1 \\ 0 & 1 & 1 \\ 1 & 0 & 0 \end{pmatrix}$

13. $\Psi(t) = e^{6t} \begin{pmatrix} \cos 2t - \sin 2t & \tfrac{1}{2}\sin 2t \\ -4\sin 2t & \cos 2t + \sin 2t \end{pmatrix}$

15. $\Psi(t) =$

$\dfrac{1}{6} \begin{pmatrix} -e^{-t} + 9e^t - 2e^{2t} & -2e^{-t} + 2e^{2t} & 7e^{-t} - 9e^t + 2e^{2t} \\ 6e^t - 6e^{2t} & 6e^{2t} & -6e^t + 6e^{2t} \\ -e^{-t} + 3e^t - 2e^{2t} & -2e^{-t} + 2e^{2t} & 7e^{-t} - 3e^t + 2e^{2t} \end{pmatrix}$

17. a. $e^t \begin{pmatrix} 1 - 2t \\ 2 - 2t \end{pmatrix}$ b. $e^t \begin{pmatrix} -10t - 2 \\ -10t + 3 \end{pmatrix}$

c. $e^{t-1} \begin{pmatrix} 2 - 2t \\ 3 - 2t \end{pmatrix}$ d. $e^{t+1} \begin{pmatrix} 4 + 2t \\ 3 + 2t \end{pmatrix}$

e. $e^{t-3} \begin{pmatrix} 3 \\ 3 \end{pmatrix}$ f. $e^{t-a} \begin{pmatrix} 2(c-b)(a-t) + b \\ 2(c-b)(a-t) + c \end{pmatrix}$

19. a. $\mathbf{x}' = A(t)\mathbf{x}$,

where $A(t) = \begin{pmatrix} 0 & 1 \\ -b(t) & -a(t) \end{pmatrix}$

21. $\varphi_2(t) = e^{t^2} \int e^{-t^2}\,dt$

Problems 7.8, page 326

In the following answers, each eigenvalue λ is followed by linearly independent eigenvectors corresponding to λ.

1. $\lambda_1 = -4, \begin{pmatrix} 1 \\ 1 \end{pmatrix}$; $\lambda_2 = 3, \begin{pmatrix} 2 \\ -5 \end{pmatrix}$

3. $\lambda_1 = i, \begin{pmatrix} 2+i \\ 5 \end{pmatrix}$; $\lambda_2 = -i, \begin{pmatrix} 2-i \\ 5 \end{pmatrix}$

5. $\lambda_1 = -3$ (alg. mult. 2), $\begin{pmatrix} 1 \\ 0 \end{pmatrix}$

7. $\lambda_1 = 0, \begin{pmatrix} 1 \\ 1 \\ 1 \end{pmatrix}$; $\lambda_2 = 1, \begin{pmatrix} -1 \\ 0 \\ 1 \end{pmatrix}$; $\lambda_3 = 3, \begin{pmatrix} 1 \\ -2 \\ 1 \end{pmatrix}$

9. $\lambda_1 = 1$ (alg. mult. 2), $\begin{pmatrix} 1 \\ 0 \\ -2 \end{pmatrix} \begin{pmatrix} 0 \\ 1 \\ -2 \end{pmatrix}$;

$\lambda_2 = 10, \begin{pmatrix} 2 \\ 2 \\ 1 \end{pmatrix}$

11. $\lambda_1 = 1$ (alg. mult. 3), $\begin{pmatrix} 1 \\ 1 \\ 1 \end{pmatrix}$

13. $\lambda_1 = -1, \begin{pmatrix} 0 \\ -1 \\ 1 \end{pmatrix}$; $\lambda_2 = i, \begin{pmatrix} 1+i \\ 1 \\ 1 \end{pmatrix}$;

$\lambda_3 = -i, \begin{pmatrix} 1-i \\ 1 \\ 1 \end{pmatrix}$

15. $\lambda_1 = 1, \begin{pmatrix} 4 \\ 1 \\ -3 \end{pmatrix}$, $\lambda_2 = 2$ (alg. mult. 2), $\begin{pmatrix} 3 \\ 1 \\ -2 \end{pmatrix}$

17. $\lambda_1 = a$ (alg. mult. 4), $\begin{pmatrix} 1 \\ 0 \\ 0 \\ 0 \end{pmatrix}, \begin{pmatrix} 0 \\ 1 \\ 0 \\ 0 \end{pmatrix}, \begin{pmatrix} 0 \\ 0 \\ 1 \\ 0 \end{pmatrix}, \begin{pmatrix} 0 \\ 0 \\ 0 \\ 1 \end{pmatrix}$

19. $\lambda_1 = a$ (alg. mult. 4), $\begin{pmatrix} 1 \\ 0 \\ 0 \\ 0 \end{pmatrix}, \begin{pmatrix} 0 \\ 0 \\ 0 \\ 1 \end{pmatrix}$

Problems 7.9, page 337

1. $\dfrac{1}{7} \begin{pmatrix} 5e^{-4t} + 2e^{3t} & 2e^{-4t} - 2e^{3t} \\ 5e^{-4t} - 5e^{3t} & 2e^{-4t} + 5e^{3t} \end{pmatrix}$

3. $\begin{pmatrix} 2\sin t + \cos t & -\sin t \\ 5\sin t & -2\sin t + \cos t \end{pmatrix}$

5. $e^{-t} \begin{pmatrix} 1 + 4t & -2t \\ 8t & -4t + 1 \end{pmatrix}$

7. $e^{-t} \begin{pmatrix} \tfrac{3}{4} + t \\ 1 + 2t \end{pmatrix}$ (see Problem 5)

9. $e^{-t} \begin{pmatrix} 2 - 2t \\ 5 - 4t \end{pmatrix}$ (see Problem 5)

11. $\dfrac{1}{6} \begin{pmatrix} 2 + 3e^t + e^{3t} & 2 - 2e^{3t} & 2 - 3e^t + e^{3t} \\ 2 - 2e^{3t} & 2 + 4e^{3t} & 2 - 2e^{3t} \\ 2 - 3e^t + e^{3t} & 2 - 2e^{3t} & 2 + 3e^t + e^{3t} \end{pmatrix}$;

$\dfrac{1}{6} \begin{pmatrix} 4 - 6e^t + 8e^{3t} \\ 4 - 16e^{3t} \\ 4 + 6e^t + 8e^{3t} \end{pmatrix}$

13.

$\begin{pmatrix} 4e^t - 3e^{2t} + 6te^{2t} & -12e^t + 12e^{2t} - 6te^{2t} & 6te^{2t} \\ e^t - e^{2t} + 2te^{2t} & -3e^t + 4e^{2t} - 2te^{2t} & 2te^{2t} \\ -3e^t + 3e^{2t} - 4te^{2t} & 9e^t - 9e^{2t} + 4te^{2t} & -4te^{2t} + e^{2t} \end{pmatrix}$

15.
$$\frac{1}{9}\begin{pmatrix} 5e^t + 4e^{10t} & -4e^t + 4e^{10t} & -2e^t + 2e^{10t} \\ -4e^t + 4e^{10t} & 5e^t + 4e^{10t} & -2e^t + 2e^{10t} \\ -2e^t + 2e^{10t} & -2e^t + 2e^{10t} & 8e^t + e^{10t} \end{pmatrix}$$

17.
$$\frac{1}{2}\begin{pmatrix} e^{2t} + e^{6t} & -e^{2t} + e^{4t} & 0 & -e^{4t} + e^{6t} \\ -e^{2t} + e^{6t} & e^{2t} + e^{4t} & 0 & -e^{4t} + e^{6t} \\ e^{2t} - e^{6t} & -e^{2t} + e^{4t} & 2e^{2t} & 2e^{2t} - e^{4t} - e^{6t} \\ -e^{2t} + e^{6t} & e^{2t} - e^{4t} & 0 & e^{4t} + e^{6t} \end{pmatrix}$$

19. $\frac{1}{6}\begin{pmatrix} 4 + 2e^{3t} \\ 4 - 4e^{3t} \\ 4 + 2e^{3t} \end{pmatrix}$ (see Problem 11)

21. $\begin{pmatrix} -4e^t + 3e^{2t} + 6te^{2t} \\ -e^t + e^{2t} + 2te^{2t} \\ 3e^t - e^{2t} - 4te^{2t} \end{pmatrix}$ (see Problem 13)

Problems 7.10, page 343

1. $\phi_p(t) = \begin{pmatrix} \frac{1}{3}e^{2t} \\ \frac{1}{2}e^t - \frac{2}{3}e^{2t} \end{pmatrix}$

3. $\phi(t) = \begin{pmatrix} -t\cos t \\ -t\sin t - 2t\cos t + \cos t + \sin t \end{pmatrix}$

5. $\phi(t)$
$$= \begin{pmatrix} \frac{3}{2}te^t + \frac{1}{3}te^{2t} + \frac{5}{2}e^t - \frac{115}{72}e^{-t} - \frac{7}{9}e^{2t} - \frac{1}{8}e^{3t} \\ te^t + te^{2t} + \frac{5}{2}e^t - 2e^{2t} + \frac{1}{2}e^{3t} \\ \frac{1}{2}te^t + \frac{1}{3}te^{2t} + e^t - \frac{115}{72}e^{-t} - \frac{7}{9}e^{2t} + \frac{3}{8}e^{3t} \end{pmatrix}$$

7. $\phi_p(t) = \begin{pmatrix} t \\ t^2 \\ t^4 \end{pmatrix}$

9. $\phi_p(t) = \begin{pmatrix} 5\sin t \ln|\csc t + \cot t| \\ 1 + (2\sin t - \cos t)\ln|\csc t + \cot t| \end{pmatrix}$

11. $\Phi(t) = \begin{pmatrix} 1/t & \ln t/t \\ -1/t^2 & (1 - \ln t)/t^2 \end{pmatrix}$

$\Phi^{-1}(t) = \begin{pmatrix} t(1 - \ln t) & -t^2 \ln t \\ t & t^2 \end{pmatrix}$

Problems 7.11, page 351

1. $\begin{pmatrix} e^t & 0 \\ 0 & e^{3t} \end{pmatrix}$ **3.** See answer to Problem 7.9.1.

5. See answer to Problem 7.9.3.

7. See answer to Problem 7.9.11.

9. See answer to Problem 7.9.15.   **11.** $\begin{pmatrix} e^{4t} & te^{4t} \\ 0 & e^{4t} \end{pmatrix}$

13. $\begin{pmatrix} e^t & 2te^t \\ 0 & e^t \end{pmatrix}$ **15.** See answer to Problem 7.9.5.

17. See answer to Problem 7.9.11.

19. See answer to Problem 7.9.21.

Review Exercises for Chapter 7, page 352

1. $x_1' = x_2,\ x_2' = x_3,\ x_3' = 6x_3 - 2x_2 + 5x_1$

3. $x_1' = x_2,\ x_2' = x_3,\ x_3' = (\ln t - x_1 x_3)/x_2$

5. $x = c_1 e^{3t} + c_2(1 + t)e^{3t},\ y = (c_1 + c_2 t)e^{3t}$

7. $x = c_1 e^{5t} + c_2 e^{-t},\ y = 2c_1 e^{5t} - c_2 e^{-t}$

9. $x = c_1 e^{-t} + 3e^{-2t},\ y = (c_2 - 2c_1 t)e^{-t} + 12e^{-2t}$

11. $\begin{pmatrix} x_1 \\ x_2 \end{pmatrix}' = \begin{pmatrix} 3 & -4 \\ -2 & 7 \end{pmatrix}\begin{pmatrix} x_1 \\ x_2 \end{pmatrix}$

13. $\begin{pmatrix} x_1 \\ x_2 \end{pmatrix}' = \begin{pmatrix} 1 & 1 \\ -3 & 2 \end{pmatrix}\begin{pmatrix} x_1 \\ x_2 \end{pmatrix} + \begin{pmatrix} e^t \\ e^{2t} \end{pmatrix}$

15. a. $\frac{1}{5}\begin{pmatrix} -2e^t + 12e^{6t} \\ 3e^t + 12e^{6t} \end{pmatrix}$ **b.** $\frac{1}{5}\begin{pmatrix} -2e^t - 3e^{6t} \\ 3e^t - 3e^{6t} \end{pmatrix}$

c. $\begin{pmatrix} 0 \\ 0 \end{pmatrix}$ **d.** $\frac{1}{5}\begin{pmatrix} 18e^t + 17e^{6t} \\ -27e^t + 17e^{6t} \end{pmatrix}$

e. $\frac{1}{5}\begin{pmatrix} 2(a-b)e^t + (3a+2b)e^{6t} \\ -3(a-b)e^t + (3a+2b)e^{6t} \end{pmatrix}$

17. $\lambda_1 = 2$ (alg. mult. 2), $\begin{pmatrix} 1 \\ 0 \end{pmatrix}$

19. $\lambda_1 = 1,\ \begin{pmatrix} 1 \\ 0 \\ -1 \end{pmatrix},\ \lambda_2 = 2,\ \begin{pmatrix} -1 \\ 1 \\ 1 \end{pmatrix},$

$\lambda_3 = -1,\ \begin{pmatrix} -1 \\ -2 \\ 7 \end{pmatrix}$

21. $\lambda_1 = -2$ (alg. mult. 3), $\begin{pmatrix} 1 \\ 0 \\ 0 \end{pmatrix}$

23. $\frac{1}{3}\begin{pmatrix} 2e^{2t} + e^{5t} & e^{2t} - e^{5t} \\ 2e^{2t} - 2e^{5t} & e^{2t} + 2e^{5t} \end{pmatrix}$

25.
$$e^{-2t}\begin{pmatrix} -5t^2 + t + 1 & \frac{5}{2}t^2 - 18t & -\frac{5}{2}t^2 - 7t \\ -3t^2 + t & \frac{3}{2}t^2 - 11t + 1 & -\frac{3}{2}t^2 - 4t \\ 7t^2 - t & -\frac{7}{2}t^2 + 25t & \frac{7}{2}t^2 + 10t + 1 \end{pmatrix}$$

27. $\phi(t) = e^{2t}\begin{pmatrix} 3\cos 2t + \frac{19}{8}\sin 2t + \frac{1}{4}t \\ -6\sin 2t + \frac{19}{4}\cos 2t - \frac{11}{4} \end{pmatrix}$

CHAPTER 8

Problems 8.1, page 359

1. $(I_L, I_R) = (1 - 1.025e^{-0.05}, 1 - 1.05e^{-0.05})$

3. $(100(k_1 e^{\lambda_1 t} + k_2 e^{\lambda_2 t}) + 1$,
   $-8(k_1 \lambda_1 e^{\lambda_1 t} + k_2 \lambda_2 e^{\lambda_2 t}) + 1)$
   where $\begin{Bmatrix} k_1 \\ k_2 \end{Bmatrix} = \dfrac{\mp 3\sqrt{2} - 4}{800}$, $\begin{Bmatrix} \lambda_1 \\ \lambda_2 \end{Bmatrix} = -50 \pm 25\sqrt{2}$

5. The general solution $(I_L, I_R) = (I_L, I_R)_h + (I_L, I_R)_p$ is given by
   $(I_L, I_R)_p = (A \sin \omega t + B \cos \omega t,$
   $C \sin \omega t + D \cos \omega t)$,
   where $\omega = 60\pi$, $\Delta = \omega^2 + 2500$, $A = (2500/\Delta)^2$,
   $B = -25\omega(\omega^2 + 7500)/\Delta^2$, $C = -2500(\omega^2 - 2500)/\Delta^2$, $D = -250{,}000\, \omega/\Delta^2$, and

   a. $(I_L, I_R)_h = \left(\dfrac{e^{-50t}}{2} \left[k_1 + k_2 \left(t + \dfrac{1}{25} \right) \right], \right.$
   $\left. e^{-50t} \left[k_1 + k_2 \left(t + \dfrac{1}{25} \right) \right] \right)$

   b. $(I_L, I_R)_h$
   $= (e^{-50t}[k_1 \cos 50\sqrt{3}\, t + k_2 \sin 50\sqrt{3}\, t],$
   $e^{-50t}[k_3 \cos 50\sqrt{3}\, t + k_4 \sin 50\sqrt{3}\, t])$

   c. $(I_L, I_R)_h = (e^{-50t}(k_1 e^{25\sqrt{2}\, t} + k_2 e^{-25\sqrt{2}\, t}),$
   $e^{-50t}[(2 - \sqrt{2})k_1 e^{25\sqrt{2}\, t} + (2 + \sqrt{2})k_2 e^{-25\sqrt{2}\, t}])$

7. $I_1 = \frac{1}{20}(8 \sin t - 6 \cos t + 5e^{-t} + e^{-3t})$;
   $I_2 = \frac{1}{20}(2 \sin t - 4 \cos t + 5e^{-t} - e^{-3t})$

9. $I = 50t$

11. $E_k = A \dfrac{\sinh a(n-k)}{\sinh an} \cos \omega t$

13. Natural frequencies are
    $\omega_N = \left[2\sqrt{CL} \sin \dfrac{N\pi}{2(n+1)} \right]^{-1}$.
    Cut-off frequency is $\omega_n \approx \dfrac{1}{2\sqrt{CL}}$.

15. See the reference.

Problems 8.2, page 365

1. When $t = \sqrt{5} \ln\left(\dfrac{3 + \sqrt{5}}{3 - \sqrt{5}} \right)$,
   $y_{\max} = \dfrac{40}{3 - \sqrt{5}} \left(\dfrac{3 + \sqrt{5}}{3 - \sqrt{5}} \right)^{(-5 - 3\sqrt{5})/10}$

3. When $t = \dfrac{150}{\sqrt{3}} \ln \dfrac{7 + \sqrt{13}}{7 - \sqrt{13}}$,
   $y_{\max} = \dfrac{3000}{7 - \sqrt{13}} \left(\dfrac{7 - \sqrt{13}}{7 + \sqrt{13}} \right)^{(7 + \sqrt{13})/(2\sqrt{13})}$

5. Let $k = AB - (a+b)^2/4$. If $k > 0$
   $x = e^{(a-b)t/2}[x_0 \cos \sqrt{k}\, t + (\tfrac{1}{2}x_0(a+b) - Ay_0) \sin \sqrt{k}\, t / \sqrt{k}\,]$
   $y = e^{(a-b)t/2}[y_0 \cos \sqrt{k}\, t - (\tfrac{1}{2}y_0(a+b) - Bx_0) \sin \sqrt{k}\, t / \sqrt{k}\,]$.
   If $k < 0$, replace cos, sin, $\sqrt{k}$ with cosh, sinh, $\sqrt{-k}$.

7. Let $D = a_{11}a_{22} - a_{12}a_{21} \neq 0$, λ_1 and λ_2 be the roots of $\lambda^2 - (a_{11} + a_{22})\lambda + D = 0$. Then
   $x = c_1 e^{\lambda_1 t} + c_2 e^{\lambda_2 t} + (b_2 a_{12} - b_1 a_{22})/D$
   $y = c_1(\lambda_1 - a_{11})e^{\lambda_1 t}/a_{12} + c_2(\lambda_2 - a_{11})e^{\lambda_2 t}/a_{12}$
   $+ (b_1 a_{21} - b_2 a_{11})/D$.

9. Let $D = a_1 b_2 - a_2 b_1 \neq 0$, $\omega = \pi/4$, λ_1 and λ_2 be the roots of $\lambda^2 + (a_1 + b_2)\lambda + D = 0$, and A_1, A_2, A_3, A_4 be the solution of the system
   $a_1 A_1 + \omega A_2 - a_2 A_3 \qquad = -a$
   $-\omega A_1 + a_1 A_2 \qquad - a_2 A_4 = 0$
   $-b_1 A_1 \qquad + b_2 A_3 + \omega A_4 = 0$
   $\qquad b_1 A_2 + \omega A_3 - b_2 A_4 = 0$.
   $x = c_1 e^{\lambda_1 t} + c_2 e^{\lambda_2 t} + ab_2/D$
   $+ A_1 \cos \omega t + A_2 \sin \omega t$
   $y = (c_1 \lambda_1 + a_1) e^{\lambda_1 t}/a_2 + (c_2 \lambda_2 + a_2) e^{\lambda_2 t}/a_2 +$
   $ab_1/D + A_3 \cos \omega t + A_4 \sin \omega t$.

Problems 8.3, page 368

1. If $\tan \theta = B/A$, then $A \cos \omega t + B \sin \omega t$
   $= \sqrt{A^2 + B^2} \cos(\omega t - \theta)$.

3. $x_1 = (\tfrac{1}{2}x_{10} + \tfrac{1}{4}x_{20}) \cos t + (\tfrac{1}{2}x_{30} + \tfrac{1}{4}x_{40}) \sin t$
   $+ (\tfrac{1}{2}x_{10} - \tfrac{1}{4}x_{20}) \cos \sqrt{3}\, t$
   $+ (\tfrac{1}{2}x_{30} - \tfrac{1}{4}x_{40}) \sin \sqrt{3}\, t / \sqrt{3}$
   $x_2 = (x_{10} + \tfrac{1}{2}x_{20}) \cos t + (x_{30} + \tfrac{1}{2}x_{40}) \sin t$
   $- (x_{10} - \tfrac{1}{2}x_{20}) \cos \sqrt{3}\, t$
   $- (x_{30} - \tfrac{1}{2}x_{40}) \sin \sqrt{3}\, t / \sqrt{3}$

5. $\omega_n = \dfrac{n^2 \pi^2}{64} \sqrt{\dfrac{4.32 \times 10^9}{24(8)^4/\pi g}} \approx 325 n^2$ rev/sec. The lowest frequency is 325 rev/sec.

7. $l_1(m_1 + m_2)\theta'' + m_2 l_2 \phi'' \cos(\theta - \phi)$
   $+ m_2 l_2 \phi'^2 \sin(\theta - \phi) + g(m_1 + m_2) \sin \theta = 0$
   $l_1 \theta'' \cos(\theta - \phi) \quad + \quad l_2 \phi'' - l_1 \theta'^2 \sin(\theta - \phi)$
   $+ g \sin \phi = 0$

9. The characteristic equation is $m_1 l_1 l_2 \lambda^4 + (m_1 + m_2)(l_1 + l_2)g\lambda^2 + (m_1 + m_2)g^2 = 0$. The discriminant as a quadratic in λ^2 is positive, so the roots are pure imaginary, and the solution is a superposition of two simple harmonics.

Problems 8.4, page 374

1. yes; $W = 907$ 3. no epidemic
5. yes; $W = 22{,}994$
7. From (11) $x(t) = \beta/\alpha$ at $y_{\max}$. Substitute this into (13). Then use $N \geq x(0)$ and a power series for the logarithm.

Review Exercises for Chapter 8, page 374

1. a. $\dfrac{dE}{dx} = -IR$, $\dfrac{dI}{dx} = -EG$

   b. $E = E_0 e^{-\sqrt{RG}\,x}$, $I = \sqrt{\dfrac{G}{R}}\, E_0 e^{-\sqrt{RG}\,x}$

3. a. $x = 30(1 - e^{-9t/75})$, $y = 30(1 + e^{-9t/75})$

   b. y decreases steadily to 30 lb, x increases steadily to 30 lb.

   c. There is no maximum.

5. $y = \dfrac{bd}{(a+c)(a+d)}[a + ce^{-(a+c)t}]$

CHAPTER 9

Problems 9.1, page 378

1. 0.33333333×10^0 3. -0.35×10^{-4}
5. 0.77777777×10^0
7. 0.77272727×10^1 9. -0.18833333×10^2
11. 0.23705963×10^9 13. 0.83742×10^{-20}
15. $\epsilon_a = 0.1$; $\epsilon_r = 0.0002$
17. $\epsilon_a = 0.005$; $\epsilon_r = 0.04$
19. $\epsilon_a = 0.333\cdots \times 10^{-2}$; $\epsilon_r = 0.571428 \times 10^{-3}$
21. $\epsilon_a = 1$, $\epsilon_r = 0.14191 \times 10^{-4}$

Problems 9.2, page 385

1. $y = 2e^x - x - 1$, $y(1) = 2(e-1) \approx 3.4366$;
   a. $y_E = 2.98$; b. $y_{IE} = 3.405$

3. $y = \sqrt{2x^2 + 1} - x$; $y(1) = \sqrt{3} - 1 \approx 0.73205$
   a. $y_E = 0.71$; b. $y_{IE} = 0.73207$

5. $y = \sinh(\tfrac{1}{2}x^2 - \tfrac{1}{2})$; $y(3) \approx 27.2899$
   a. $y_E = 8.31$; b. $y_{IE} = 21.671$

7. $y = (x^3 + x^{-2})^{-1/2}$, $y(2) = 2/\sqrt{33} \approx 0.3481553$;
   a. $y_E = 0.343$ b. $y_{IE} = 0.34939$

9. $y = 2e^{e^x - 1}$, $y(2) = 2e^{e^2 - 1} \approx 1190.59$;
   a. $y_E = 156$; b. $y_{IE} = 781.56$

In 11–19 answers are given for the improved Euler method. Part (a) is given in the answers for Section 1.3.

11. $y_1 = 1.02$, $y_2 = 1.04$, $y_3 = 1.07$, $y_4 = 1.09$, $y_5 = 1.12$

13. $y_1 = 0.34$, $y_2 = 0.56$, $y_3 = 0.76$, $y_4 = 0.98$, $y_5 = 1.34$

15. $y_1 = 1.28$, $y_2 = 1.65$, $y_3 = 1.46$, $y_4 = 1.09$, $y_5 = 0.87$, $y_6 = 0.79$, $y_7 = 0.93$, $y_8 = 0.95$

17. $y_1 = 1.10$, $y_2 = 1.22$, $y_3 = 1.35$, $y_4 = 1.48$, $y_5 = 1.63$, $y_6 = 1.79$, $y_7 = 1.97$, $y_8 = 2.16$, $y_9 = 2.37$, $y_{10} = 2.60$

19. $y_1 = 1.60$, $y_2 = 1.27$, $y_3 = 1.00$, $y_4 = 0.80$, $y_5 = 0.64$

21. No method will provide a correct answer since the solution to the differential equation is the hyperbola $x^2 - 2xy - y^2 = 4$ or $y = \sqrt{2x^2 - 4} - x$, which is not defined if $x = 1$.

23. The solution to the differential equation is $y = [x^2(2 - x^2)]^{-1/2}$, which is not defined at $x = 3$.

25. Using $h = 0.2$, $y(1) = 3.1700$; using $h = 0.1$, $y(1) = 3.1586$; exact value is 3.15484548.

Problems 9.3, page 392

1. $y = 2e^x - x - 1$; $y(1) = 2(e - 1) \approx 3.4365637$;
   $y_{RK} = 3.436502$

3. $y = \sqrt{2x^2 + 1} - x$; $y(1) = \sqrt{3} - 1 \approx 0.732050807$; $y_{RK} = 0.73205044$

5. $y = \sinh(\tfrac{1}{2}x^2 - \tfrac{1}{2})$; $y(3) \approx 27.2899$;
   $y_{RK} = 27.0275$

7. $y = (x^3 + x^{-2})^{-1/2}$; $y(2) = 2/\sqrt{33} \approx 0.3481553$;
   $y_{RK} = 0.348161$

9. $y = 2e^{e^x - 1}$, $y(2) = 2e^{e^2 - 1} \approx 1190.59$;
   $y_{RK} = 1164.76$

11. $y_1 = 1.02$, $y_2 = 1.04$, $y_3 = 1.07$, $y_4 = 1.10$, $y_5 = 1.12$.

13. $y_1 = 0.38$, $y_2 = 0.61$, $y_3 = 0.78$, $y_4 = 0.98$, $y_5 = 1.35$.

15. $y_1 = 1.33$, $y_2 = 1.91$, $y_3 = 1.61$, $y_4 = 1.16$, $y_5 = 0.92$, $y_6 = 0.79$, $y_7 = 0.54$, $y_8 = 0.77$.

17. $y_1 = 1.11$, $y_2 = 1.22$, $y_3 = 1.35$, $y_4 = 1.48$, $y_5 = 1.63$, $y_6 = 1.79$, $y_7 = 1.97$, $y_8 = 2.16$, $y_9 = 2.37$, $y_{10} = 2.60$.

19. $y_1 = 1.59$, $y_2 = 1.26$, $y_3 = 1.00$, $y_4 = 0.79$, $y_5 = 0.64$.

21. See answer to Problem 9.2.21.
23. See answer to Problem 9.2.23.
25. $a = \tfrac{1}{4}$, $b = 0$, $c = \tfrac{3}{4}$, $d = 0$ 27. $n = p = \tfrac{2}{3}$
29. Because the determinant of the resulting 4×4 system is zero. There are an infinite number of solutions; no.

Problems 9.4, page 397

1. $y = 2e^x - x - 1$, $y(1) = 2(e - 1) \approx 3.43656$;
   a. $y = 3.43761$; b. $y = 3.43022$.

3. $y = \sqrt{2x^2 + 1} - x$, $y(1) = \sqrt{3} - 1 \approx 0.73205$;
   a. $y = 0.74146$; b. $y = 0.74566$.

5. $y = \sinh(\tfrac{1}{2}x^2 - \tfrac{1}{2})$,
   $y(3) = (e^4 - e^{-4})/2 \approx 27.28992$;
   a. $y = 26.19221$; b. $y = 26.55498$.

7. $y = (x^3 + x^{-2})^{-1/2}$, $y(2) = 2/\sqrt{33} \approx 0.34816$;
   a. $y = 0.34840$; b. $y = 0.34465$.

9. $y = 2e^{e^x - 1}$, $y(2) = 2e^{e^2 - 1} \approx 1190.58883$;
   a. $y = 1082.39224$; b. $y = 1086.87818$.

19. $y = 3.41076$

Problems 9.5, page 401

1. a. $|e_n| \leq 0.859h$; **b.** $h = 0.1$, $|e_n| \leq 0.086$; $h = 0.2$, $|e_n| \leq 0.172$.

3. $|e_n| \leq 6.06h$; **a.** 1,212,000; **b.** 12,120,000.

Problems 9.7, page 409

1. Exact solution: $x = e^{3t}(\cos 3t - \sin 3t)$, $y = e^{3t}(2\cos 3t + \sin 3t)$, $x(1) = -22.719$, $y(1) = -36.935$
  a. $x = -17.43$, **b.** $x = -30.09$,
  $y = 7.76$ $y = -31.31$
  c. $x = -22.907$,
  $y = -37.412$

3. Exact solution: $x = (2/5)(e^{4t} - e^{-t})$, $y = (1/5)(3e^{4t} + 2e^{-t})$, $x(1) = 21.692$, $y(1) = 32.906$
  a. $x = 7.43$, **b.** $x = 16.98$, **c.** $x = 21.54$,
  $y = 11.47$ $y = 25.84$ $y = 32.68$

5. Exact solution:
$x = (1/27)(16e^{3t} + 9t^2 + 6t + 11)$,
$y = (1/27)(16e^{3t} - 18t^2 - 21t - 16)$,
$x(1) = 12.866$, $y(1) = 9.866$
  a. $x = 7.13$, **b.** $x = 11.58$, **c.** $x = 12.84$,
  $y = 4.43$ $y = 8.61$ $y = 9.84$

7. Exact solution:
$x = (33/8)(e^{2t} - 1) - (1/4)(t^2 + 9t)$,
$y = (1/8)(27 - 11e^{2t}) + (1/4)(t^2 + 7t)$,
$x(1) = 23.855$, $y(1) = -4.785$
  a. $x = 15.61$, **b.** $x = 22.67$, **c.** $x = 23.85$,
  $y = -2.07$ $y = -4.39$ $y = -4.78$

9. $y(1) = 1.09$

11. Maximum satisfies $y(x) = \sqrt{2(y'(0) - x)}$. Exact solution can be obtained by transforming the Riccati equation $y' + y^2/2 = y'(0) - x$ and using Example 5.4.4. Numerically with $h = 0.1$ use $y'(0) = 0.504$ and get $y_{\max} \approx 0.125$.

Review Exercises for Chapter 9, page 410

1. $y^2 = 2(e^x + 1)$, $y(3) = \sqrt{2(e^3 + 1)} \approx 6.4939$;
  a. $y_{IE} = 6.56$; **b.** $y_{RK} = 6.4942$.

3. $y = x + \sqrt{1 + x^2}$, $y(3) = 3 + \sqrt{10} \approx 6.1623$;
  a. $y_{IE} = 5.96$; **b.** $y_{RK} = 6.1605$.

5. $y^{-2} = (2x - 1 + 3e^{-2x})/2$, $y(3) = 0.6320$;
  a. $y_{IE} = 0.6356$; **b.** $y_{RK} = 0.6329$.

7. The exact solution is $y^2(y + 1) = x^2$. If we graph this curve and begin moving to the right from the initial point, we cannot proceed *on that branch* beyond $x = 2/3\sqrt{3}$.

9. Solution $y = \tan x$ becomes infinite at $x = \pi/2$, so no numerical method will yield $y(2)$.

11. $y = \cos(\ln|x|) + \sin(\ln|x|)$, $y(2) = \cos(\ln 2) + \sin(\ln 2) \approx 1.4082$;
  a. $y_E = 1.42163$; **b.** $y_{RK} = 1.40820$;
  c. $y_{PC} = 1.39531$.

13. $x = \left(\dfrac{2 - \sqrt{2}}{4}\right)e^{(3+\sqrt{8})t} + \left(\dfrac{2 + \sqrt{2}}{4}\right)e^{(3-\sqrt{8})t}$,

$y = \dfrac{\sqrt{2}}{4}e^{(3+\sqrt{8})t} - \dfrac{\sqrt{2}}{4}e^{(3-\sqrt{8})t}$,

$x(2) \approx 16912.88$, $y(2) \approx 40827.91$.
  a. $x_E = 333.57$, **b.** $x_{RK} = 15783.44$,
  $y_E = 801.94$; $y_{RK} = 38101.20$;
  c. $x_{PC} = 5090.404$,
  $y_{PC} = 12285.920$.

CHAPTER 10

Problems 10.1, page 418

1. eigenvalues: $\left(\dfrac{n\pi}{T}\right)^2$;

eigenfunctions: $A \sin \dfrac{n\pi t}{T}$, $n = 1, 2, 3, \ldots$

3. eigenvalues: $-\left(\dfrac{n\pi}{T}\right)^2$;

eigenfunctions: $A \sin \dfrac{n\pi t}{T}$, $n = 1, 2, 3, \ldots$

5. eigenvalues: $\left(\dfrac{n\pi}{2T}\right)^2$;

eigenfunctions; $A \sin \dfrac{n\pi(t + T)}{2T}$, $n = 1, 2, 3, \ldots$

7. eigenvalues: $\left[\left(n + \dfrac{1}{2}\right)\dfrac{\pi}{T}\right]^2$; eigenfunctions:

$A \sin\left(n + \dfrac{1}{2}\right)\dfrac{\pi t}{T}$, $n = 0, 1, 2, 3, \ldots$

9. eigenvalues: $(n\pi)^2$; eigenfunctions: $A \cos n\pi t$, $n = 0, 1, 2, 3, \ldots$

11. eigenvalues: all values of λ such that if $\lambda > \dfrac{1}{4}$

and $\beta = \dfrac{\sqrt{4\lambda - 1}}{2}$, then β satisfies $2\beta + \tan \beta = 0$; eigenfunctions: $Ax^{-1/2}[2\beta \cos(\beta \ln|x|) + \sin(\beta \ln|x|)]$.

Problems 10.2, page 422

1. $\dfrac{1}{\sqrt{2\pi}}, \dfrac{\cos x}{\sqrt{\pi}}, \dfrac{\cos 2x}{\sqrt{\pi}}, \ldots$

3. $\dfrac{1}{\sqrt{T}}, \sqrt{\dfrac{2}{T}} \cos \dfrac{2\pi x}{T}, \sqrt{\dfrac{2}{T}} \cos \dfrac{4\pi x}{T}, \ldots$

5. $\sqrt{\dfrac{1}{\pi}}, \sqrt{\dfrac{2}{\pi}} \cos 2x, \sqrt{\dfrac{2}{\pi}} \cos 4x, \ldots$

7. $\dfrac{1}{\sqrt{2\pi}}, \dfrac{1}{\sqrt{\pi}}\cos 3x, \dfrac{1}{\sqrt{\pi}}\cos 6x, \dfrac{1}{\sqrt{\pi}}\cos 9x, \ldots$

Problems 10.3, page 431

1. $\pi, 2, T/n, 1/k$ 3. $2bT/a$

5. $2(\sin x - \tfrac{1}{2}\sin 2x + \tfrac{1}{3}\sin 3x - \tfrac{1}{4}\sin 4x + \cdots)$

7. $\dfrac{\pi^2}{3} - 4(\cos x - \tfrac{1}{4}\cos 2x + \tfrac{1}{9}\cos 3x - \tfrac{1}{16}\cos 4x + \cdots)$

9. $\dfrac{\pi}{2} - \dfrac{4}{\pi}\left[\cos x + \dfrac{1}{3^2}\cos 3x + \dfrac{1}{5^2}\cos 5x + \cdots\right]$

11. $\dfrac{2}{\pi}[(1 + \sin(\pi - 1))\sin x$
$\quad - \tfrac{1}{2}(1 + \tfrac{1}{2}\sin 2(\pi - 1))\sin 2x$
$\quad + \tfrac{1}{3}(1 + \tfrac{1}{3}\sin 3(\pi - 1))\sin 3x - \cdots]$

13. $\dfrac{\sinh \pi}{\pi}\left[1 + 2\sum_{n=1}^{\infty}\dfrac{(-1)^n}{1+n^2}\cos nx \right.$
$\left.\quad - 2\sum_{n=1}^{\infty}\dfrac{n(-1)^n}{1+n^2}\sin nx\right]$

15. $1 - \dfrac{2}{\pi}\sum_{n=1}^{\infty}\dfrac{\sin n\pi x}{n}$

17. $\dfrac{1}{3} + \dfrac{4}{\pi^2}\sum_{n=1}^{\infty}\dfrac{(-1)^n}{n^2}\cos n\pi x$

19. $\dfrac{1}{2} - \dfrac{2}{\pi}\left(\sin \pi x + \dfrac{1}{3}\sin 3\pi x + \dfrac{1}{5}\sin 5\pi x + \cdots\right)$

21. $\dfrac{3}{4} - \dfrac{1}{\pi}\sum_{n=1}^{\infty}\dfrac{1}{n}\sin n\pi x$
$\quad - \dfrac{2}{\pi^2}\left(\cos \pi x + \dfrac{1}{3^2}\cos 3\pi x + \dfrac{1}{5^2}\cos 5\pi x + \cdots\right)$

Problems 10.4, page 437

1. even   3. even   5. odd   7. neither   9. neither

11. $\dfrac{\pi^2}{3} - 4(\cos x - \tfrac{1}{4}\cos 2x + \tfrac{1}{9}\cos 3x - \tfrac{1}{16}\cos 4x + \cdots)$

13. $\dfrac{\pi^3}{4} - 6\pi\left(\cos x - \dfrac{\cos 2x}{4} + \dfrac{\cos 3x}{9} - \dfrac{\cos 4x}{16} + \cdots\right)$
$\quad + \dfrac{24}{\pi}\left(\cos x + \dfrac{\cos 3x}{3^4} + \dfrac{\cos 5x}{5^4} + \cdots\right)$

15. $\dfrac{T}{2}$
$\quad + \dfrac{4T}{\pi^2}\left(\cos \dfrac{\pi x}{T} + \dfrac{1}{9}\cos \dfrac{3\pi x}{T} + \dfrac{1}{25}\cos \dfrac{5\pi x}{T} + \cdots\right)$

17. $\dfrac{2T}{\pi}\left(\sin \dfrac{\pi x}{T} + \dfrac{1}{2}\sin \dfrac{2\pi x}{T}\right.$
$\left.\quad + \dfrac{1}{3}\sin \dfrac{3\pi x}{T} + \dfrac{1}{4}\sin \dfrac{4\pi x}{T} + \cdots\right)$

19. $\dfrac{1}{3} + \dfrac{4}{\pi^2}(\cos \pi x + \tfrac{1}{4}\cos 2\pi x$
$\quad + \tfrac{1}{9}\cos 3\pi x + \tfrac{1}{16}\cos 4\pi x + \cdots)$

21. even: k
$\quad$ odd: $\dfrac{4k}{\pi}(\sin \pi x + \tfrac{1}{3}\sin 3\pi x + \tfrac{1}{5}\sin 5\pi x + \cdots)$

23. even: $\dfrac{1}{4} - \dfrac{6}{\pi^2}(\cos \pi x - \tfrac{1}{4}\cos 2\pi x$
$\quad + \tfrac{1}{9}\cos 3\pi x - \tfrac{1}{16}\cos 4\pi x + \cdots)$
$\quad + \dfrac{24}{\pi^4}\left(\cos \pi x + \dfrac{\cos 3\pi x}{3^4} + \dfrac{\cos 5\pi x}{5^4} + \cdots\right)$

$\quad$ odd: $\dfrac{2}{\pi}[\sin \pi x - \tfrac{1}{2}\sin 2\pi x$
$\quad + \tfrac{1}{3}\sin 3\pi x - \tfrac{1}{4}\sin 4\pi x + \cdots]$
$\quad - \dfrac{12}{\pi^3}\left[\sin \pi x - \dfrac{\sin 2\pi x}{2^3}\right.$
$\left.\quad + \dfrac{\sin 3\pi x}{3^3} - \dfrac{\sin 4\pi x}{4^3} + \cdots\right]$

25. even: $\dfrac{1}{2} + \dfrac{2}{\pi}\left(\cos \dfrac{\pi x}{2} - \dfrac{1}{3}\cos \dfrac{3\pi x}{2}\right.$
$\left.\quad + \dfrac{1}{5}\cos \dfrac{5\pi x}{2} - \dfrac{1}{7}\cos \dfrac{7\pi x}{2} + \cdots\right)$

$\quad$ odd: $\dfrac{2}{\pi}\left(\sin \dfrac{\pi x}{2} + \sin \pi x + \dfrac{1}{3}\sin \dfrac{3\pi x}{2}\right.$
$\quad + \dfrac{1}{5}\sin \dfrac{5\pi x}{2} + \dfrac{1}{3}\sin 3\pi x + \dfrac{1}{7}\sin \dfrac{7\pi x}{2}$
$\left.\quad + \dfrac{1}{9}\sin \dfrac{9\pi x}{2} + \dfrac{1}{5}\sin 5\pi x + \cdots\right)$

27. even: $\dfrac{3}{4} + \dfrac{4}{\pi^2}\left(\cos \dfrac{\pi x}{2} - \dfrac{2}{2^2}\cos \pi x\right.$
$\quad + \dfrac{1}{3^2}\cos \dfrac{3\pi x}{2} + \dfrac{1}{5^2}\cos \dfrac{5\pi x}{2} - \dfrac{2}{6^2}\cos 3\pi x$
$\left.\quad + \dfrac{1}{7^2}\cos \dfrac{7\pi x}{2} + \dfrac{1}{9^2}\cos \dfrac{9\pi x}{2} - \dfrac{2}{10^2}\cos 5\pi x + \cdots\right)$

$\quad$ odd: $\dfrac{2}{\pi}\left(\sin \dfrac{\pi x}{2} + \dfrac{1}{2}\sin \pi x + \dfrac{1}{3}\sin \dfrac{3\pi x}{2}\right.$
$\quad + \dfrac{1}{4}\sin 2\pi x + \dfrac{1}{5}\sin \dfrac{5\pi x}{2} + \cdots\Big)$
$\quad + \dfrac{4}{\pi^2}\left(\sin \dfrac{\pi x}{2} - \dfrac{1}{3^2}\sin \dfrac{3\pi x}{2} + \dfrac{1}{5^2}\sin \dfrac{5\pi x}{2} - \cdots\right)$

29. even:

$$\frac{2}{3} - \frac{16}{\pi^3}\left(\cos\frac{\pi x}{2} - \frac{1}{3^3}\cos\frac{3\pi x}{2} + \frac{1}{5^3}\cos\frac{5\pi x}{2} - \cdots\right)$$
$$-\frac{8}{\pi^2}\left(\frac{1}{2^2}\cos\pi x - \frac{1}{4^2}\cos 2\pi x + \frac{1}{6^2}\cos 3\pi x - \cdots\right)$$

odd: $\frac{2}{\pi}\left(\sin\frac{\pi x}{2} - \frac{1}{2}\sin\pi x + \frac{1}{3}\sin\frac{3\pi x}{2}\right.$
$-\frac{1}{4}\sin 2\pi x + \frac{1}{5}\sin\frac{5\pi x}{2} - \cdots\Big)$
$+\frac{8}{\pi^2}\left(\sin\frac{\pi x}{2} - \frac{1}{3^2}\sin\frac{3\pi x}{2} + \frac{1}{5^2}\sin\frac{5\pi x}{2} - \cdots\right)$
$-\frac{16}{\pi^3}\left(\sin\frac{\pi x}{2} + \frac{2}{2^3}\sin\pi x + \frac{1}{3^3}\sin\frac{3\pi x}{2} + \frac{1}{5^3}\sin\frac{5\pi x}{2}\right.$
$+\frac{2}{6^3}\sin 3\pi x + \frac{1}{7^3}\sin\frac{7\pi x}{2} + \frac{1}{9^3}\sin\frac{9\pi x}{2} + \cdots\Big)$

Problems 10.5, page 443
1. hyperbolic 3. parabolic 5. elliptic
7. elliptic 9. elliptic
11. $\dfrac{\partial^2 z}{\partial x^2} + \dfrac{\partial^2 z}{\partial y^2} - z = 0$
13. $\dfrac{\partial^2 z}{\partial x^2} - \dfrac{\partial^2 z}{\partial y^2} + \dfrac{57}{4}z = 0$
15. $\dfrac{\partial^2 z}{\partial x^2} + \dfrac{z}{2} = 0$
17. $\dfrac{\partial^2 z}{\partial x^2} + \dfrac{\partial^2 z}{\partial y^2} + 5z = 0$
19. $\dfrac{\partial^2 z}{\partial x^2} - \dfrac{\partial^2 z}{\partial y^2} - \dfrac{1}{12}z = 0$
21. Multiply $-AC$ to get $\left[b^2 - \dfrac{(a-c)^2}{4}\right]\sin^2 2\alpha$
$+ b(a-c)\sin 2\alpha \cos 2\alpha - ac$. Then substitute $a - c = 2b/\tan 2\alpha$ to obtain $b^2 - ac$.
23. Elliptic in second and fourth quadrants, hyperbolic in first and third quadrants, and parabolic on axes.
25. Elliptic for $|y| < |x|$, hyperbolic for $|y| > |x|$, and parabolic for $|y| = |x|$.
27. Elliptic for $y > e^{-x}$, hyperbolic for $y < e^{-x}$, and parabolic for $y = e^{-x}$.

Problems 10.6, page 449
1. -0.0075 m 3. 0.0 m
5. 0.1 m 7. 0.2 m.
9. By (14) $y(x, t) = h(x + ct) + g(x - ct)$. Since $y(x, 0) = 0$ we get $h(x) = -g(x)$. Thus $f'(x) =$
$y_t(x, 0) = c[h'(x + ct) + h'(x - ct)]|_{t=0} = 2ch'(x)$.
Thus $h(x) = \dfrac{1}{2c}[f(x) + k]$ from which the result follows. Furthermore, since $y(0, t) = 0$, it follows that f is even with period $2L$.

11. $-\dfrac{0.045}{\pi}$ m   13. $\dfrac{0.03}{\pi}\left[\dfrac{\pi}{6} - \dfrac{1}{4}\right]$ m

15. 0.002 m

17. $y(x, t) = \sin\dfrac{\pi x}{2}\left[\cos\dfrac{\pi t}{2} + \dfrac{2}{\pi}\sin\dfrac{\pi t}{2}\right]$

19. If $RC = LG$, let $v(x, t) = u(x, t)e^{-Rt/L}$. Then u satisfies the wave equation $u_{tt} = (1/LC)u_{xx}$, so that v has the form

$$v(x, t) = e^{-Rt/L}\left[g\left(x - \dfrac{1}{\sqrt{LC}}t\right) + h\left(x + \dfrac{1}{\sqrt{LC}}t\right)\right].$$

A much more complicated solution involving integral equations exists when $RC \ne LG$ (see Zachmanoglou and Thoe, *Introduction to Partial Differential Equations*, Williams & Wilkins, Baltimore, MD (1976), p. 375).

Problems 10.7, page 454

1. $y(x, t) = 0.1\cos\dfrac{\pi ct}{L}\sin\dfrac{\pi x}{L}$

3. $y(x, t) = \dfrac{0.8L^2}{\pi^3}\sum_{n=0}^{\infty}\dfrac{1}{(2n+1)^3}$
$\times \cos\dfrac{(2n+1)\pi ct}{L}\sin\dfrac{(2n+1)\pi x}{L}$

5. $y(x, t) = \dfrac{2.4}{\pi^2}\sum_{n=1}^{\infty}\dfrac{1}{n^2}$
$\times \sin\left(\dfrac{n\pi}{2}\right)\cos\left(\dfrac{n\pi ct}{6}\right)\sin\left(\dfrac{n\pi x}{6}\right)$

7. $T' = kT$ and $X'' + kX = 0$
9. $X' + X = 0$ or $T' = 0$; hence $y(x, t) = c_1 te^{-x} + c_2$
11. $Y'' - kyY = 0$ and $xX' + kX = 0$
13. $\Theta'' + k\Theta = 0$ and $r^2R'' + rR' - kR = 0$
15. **a.** $y(x, t) = \sqrt{x}\left(b_1 e^{\sqrt{k}\,t} + b_2 e^{-\sqrt{k}\,t}\right)$
$\times \left(c_1 x^{\sqrt{1-4k}/2} + c_2 x^{-\sqrt{1-4k}/2}\right)$, all k
b. $y(x, t) = c_1\sqrt{x}\left(x^{\sqrt{1-4k}/2} - \dfrac{\pi^{\sqrt{1-4k}}}{x^{\sqrt{1-4k}/2}}\right)$
$\times \left(b_1 e^{\sqrt{k}\,t} + b_2 e^{-\sqrt{k}\,t}\right)$, $k > 0$
17. **a.** $y(x, t) = \left(b_1 e^{\sqrt{k}\,t} + b_2 e^{-\sqrt{k}\,t}\right)$
$\times \left(c_1 \cos\sqrt{k}\,x + c_2 \sin\sqrt{k}\,x\right)$, all k
b. $c_1 = 0$ and $k = 1^2, 2^2, 3^2, \ldots$, so $y(x, t) = \sin nx(b_1 e^{nt} + b_2 e^{-nt})$
c. Use a Fourier sine series.

19. a. $y(x,t) = \left(b_1 e^{\sqrt{k}\,x} + b_2 e^{-\sqrt{k}\,x}\right)$
$\times \left(c_1 e^{\sqrt{k-1}\,t} + c_2 e^{-\sqrt{k-1}\,t}\right)$
b. Let $k < 0$, then $b_1 = 0$ and $k = -1^2, -2^2, \ldots$
so that $y(x,t) = \sin nx (c_1 \cos\sqrt{n^2+1}\,t + c_2 \sin\sqrt{n^2+1}\,t)$.
c. Use a Fourier sine series.

Problems 10.8, page 458

1. $T(x,t) = \dfrac{4}{\pi} \sum\limits_{n=1}^{\infty} \dfrac{[1-(-1)^n]}{n^3}(\sin nx)e^{-n^2\delta t}$

3. $T(x,t) = 4 \sum\limits_{n=1}^{\infty} \dfrac{[2(-1)^{n+1}-1]}{n^3}(\sin nx)e^{-n^2\delta t}$

5. $T(x,t) = \dfrac{2}{\pi} \sum\limits_{n=1}^{\infty} \dfrac{1}{n^2}$
$\times \left[\sin\dfrac{n\pi}{2} - \dfrac{n\pi}{2}\cos\dfrac{n\pi}{2}\right](\sin nx)\,e^{-n^2\delta t}$

7. $[(L-x)T_1 + xT_2]/L$

9. $f(x) = \dfrac{c_0^2}{T_1}\cos cx + c_0 c_2 \sin cx$, where
$c_2 = \dfrac{c_0(1-\cos cL)}{T\sin cL}$. If cL is a multiple of π, then c_2 is arbitrary.

Problems 10.9, page 463

1. $T(x,y) = \dfrac{100}{\pi} \sum\limits_{n=1}^{\infty} \dfrac{[1-(-1)^n]}{n\sinh n\pi}$
$\times \sin\dfrac{n\pi x}{L} \sinh\dfrac{n\pi y}{L}$
Evaluate at $\left(\dfrac{L}{2}, \dfrac{L}{2}\right)$ and $\left(\dfrac{L}{4}, \dfrac{L}{4}\right)$.

3. $T(x,y) = \dfrac{100}{\sinh \pi} \sin\dfrac{\pi x}{L} \sinh\dfrac{\pi y}{L}$

5. $T(x,y) =$
$\dfrac{-4L^3}{\pi^3} \sum\limits_{n=1}^{\infty} \dfrac{[1+2(-1)^n]}{n^3 \sinh n\pi}$
$\times \sin\dfrac{n\pi x}{L} \sinh\dfrac{n\pi y}{L}$.
Evaluate at $\left(\dfrac{L}{4}, \dfrac{L}{4}\right)$ and $\left(\dfrac{L}{4}, \dfrac{3L}{4}\right)$.

7. $T(x,y) = \dfrac{-160L}{\pi^2} \sum\limits_{n=1}^{\infty} \dfrac{n \sin\dfrac{2n\pi x}{L} \sinh\dfrac{2n\pi y}{L}}{(4n^2-1)^2 \sinh(2n\pi)}$

9. The solution is the superposition of the solutions T_1 and T_2 of two problems: first where $T_1 = 50°$ on the top edge and $0°$ on the other edges; second where $T_2 = 50\sin \pi y/2L°$ on the side edges and $0°$ on the horizontal edges. Thus
$T(x,y) = T_1(x,y) + T_2(x,y)$
$= \dfrac{200}{\pi} \sum\limits_{n=0}^{\infty} \dfrac{(-1)^n}{(2n+1)\sinh[(2n+1)\pi/2]}$
$\times \cos\left(\dfrac{2n+1}{2}\right)\dfrac{\pi x}{L} \sinh\left(\dfrac{2n+1}{2}\right)\dfrac{\pi y}{L}$
$- \dfrac{400}{\pi} \sum\limits_{n=1}^{\infty} \dfrac{(-1)^n n}{(4n^2-1)\cosh n\pi}$
$\times \cosh\dfrac{n\pi x}{L} \sin\dfrac{n\pi y}{L}$.

11. $T(x,y) = 200 \sum\limits_{n=0}^{\infty} \dfrac{(-1)^n}{\sinh[(2n+1)\pi/2]}$
$\times \left[\dfrac{1}{(2n+1)\pi} - \left(\dfrac{2}{(2n+1)\pi}\right)^3\right]$
$\times \cos\left(\dfrac{2n+1}{2}\right)\dfrac{\pi x}{L} \sinh\left(\dfrac{2n+1}{2}\right)\dfrac{\pi y}{L}$
$- \dfrac{100}{\pi} \sum\limits_{n=1}^{\infty} \dfrac{(-1)^n}{n\cosh n\pi} \cosh\dfrac{n\pi x}{L} \sin\dfrac{n\pi y}{L}$

Problems 10.10, page 471

1. $T(r,\theta) = 2 \sum\limits_{n=1}^{\infty} \dfrac{r^n(-1)^n}{n} \sin n\theta$

3. $T(r,\theta) = \dfrac{1-r^2}{2\pi}$
$\times \displaystyle\int_{-\pi}^{\pi} \dfrac{(\pi-\phi)\,d\phi}{r^2+1-2r\cos(\theta-\phi)} \bigg|_{(\frac{1}{2},0)}$
$= \dfrac{3}{8\pi} \displaystyle\int_{-\pi}^{\pi} \dfrac{(\pi-\phi)\,d\phi}{\frac{5}{4}-\cos\phi}$

5. a. $A_0 = 1$, $A_2 = -\dfrac{1}{2}$; all other coefficients are zero, so that $T(r,\theta) = \dfrac{1}{2}(1-r^2\cos 2\theta)$. Hence $T(\dfrac{1}{2},0) = \dfrac{3}{8}°$.

b. $T(\dfrac{1}{2},0) = \dfrac{3}{8\pi} \displaystyle\int_{-\pi}^{\pi} \dfrac{\sin^2\phi}{\frac{5}{4}-\cos\phi}\,d\phi = \dfrac{3}{8}°$. (The last equality can be proved using the calculus of residues.)

7. a. $T(r,\theta) = -\dfrac{\pi^2}{3}$
$- 4\sum\limits_{n=1}^{\infty}(-r)^n\left(\dfrac{\cos n\theta}{n^2} + \dfrac{\pi \sin n\theta}{n}\right)$

b. $T(r, \theta) = \dfrac{1-r^2}{2\pi} \displaystyle\int_{-\pi}^{\pi} \dfrac{\phi(2\pi - \phi)\, d\phi}{r^2 + 1 - 2r\cos(\theta - \phi)}$

9. Integrating by parts, we get
$$|a_n| \le \dfrac{T}{n^2 \pi^2} \left| \int_{-T}^{T} F''(x) \cos \dfrac{n\pi x}{T}\, dx \right| < \dfrac{2MT^2}{n^2 \pi^2}$$
and compare with $\sum 1/n^2$.

11. a. If u attains a maximal value at (x_0, y_0), the cross sections of the surface with the planes $x = x_0$ and $y = y_0$ have slope zero at (x_0, y_0) and are concave down. Thus $u_{xx} \le 0$ and $u_{yy} \le 0$ at (x_0, y_0).

c. $w_{xx} + w_{yy} = u_{xx} + u_{yy} + 4\varepsilon > 0$

Review Exercises for Chapter 10, page 472

1. $\left\{\sqrt{\dfrac{2}{\pi}} \sin nx\right\}$ **3.** $\{\sqrt{2} \sin 2n\pi x, \sqrt{2} \cos 2n\pi x\}$

5. $-1 + \pi \sin x - \dfrac{1}{2} \cos x + 2 \displaystyle\sum_{n=2}^{\infty} \dfrac{\cos nx}{n^2 - 1}$

7. $\dfrac{2}{\pi}\left[1 - 2\displaystyle\sum_{n=1}^{\infty} \dfrac{\cos 2nx}{4n^2 - 1}\right]$

9. $2\pi \displaystyle\sum_{n=1}^{\infty} \left\{\left(\dfrac{1}{2n-1} - \dfrac{4}{(2n-1)^{2n-1}\pi^2}\right) \right.$
$\left. \times \sin(2n-1)x - \dfrac{1}{2^n} \sin 2nx\right\};$

$\dfrac{\pi^2}{3} + 4 \displaystyle\sum_{n=1}^{\infty} \dfrac{(-1)^n}{n^2} \cos nx$

11. $\dfrac{2}{\pi} \displaystyle\sum_{n=1}^{\infty} \dfrac{(-1)^{n-1}}{n} \sin n\pi x$;

$\dfrac{-4}{\pi^2} \displaystyle\sum_{n=1}^{\infty} \dfrac{1}{(2n-1)^2} \cos(2n-1)\pi x$

13. $\dfrac{-32}{\pi^3} \left(\sin \dfrac{\pi x}{2} + \dfrac{\pi^2}{8} \sin \dfrac{2\pi x}{2} \right.$
$+ \dfrac{1}{3^3} \sin \dfrac{3\pi x}{2} + \dfrac{\pi^2}{16} \sin \dfrac{4\pi x}{2} + \cdots\bigg);$

$\pi^3 = 32\left(1 - \dfrac{1}{3^3} + \dfrac{1}{5^3} - \dfrac{1}{7^3} + \cdots\right)$

15. $T(x, t) = \dfrac{4a}{\pi^2} \displaystyle\sum_{n=0}^{\infty} \dfrac{(-1)^n}{(2n+1)^2}$

$\times \sin \dfrac{(2n+1)\pi x}{a} e^{-[(2n+1)\pi/a]^2 \delta t}$

17. $y = X(x)T(t)$ where the general solutions to X and T are

$$X(x) = \begin{cases} c_1 x^{\lambda_1} + c_2 x^{\lambda_2}, & k > -\tfrac{1}{4} \\ c_1 \sqrt{x} + c_2 \sqrt{x} \ln x, & k = -\tfrac{1}{4} \\ \sqrt{x}\,[c_1 \cos(\ln|x|^\beta) \\ \quad + c_2 \sin(\ln|x|^\beta)], & k < -\tfrac{1}{4}, \end{cases}$$

$$T(t) = \begin{cases} b_1 t^{\lambda_1} + b_2 t^{\lambda_2}, & k > -\tfrac{1}{4} \\ b_1 \sqrt{t} + b_2 \sqrt{t} \ln t, & k = -\tfrac{1}{4} \\ \sqrt{t}\,[b_1 \cos(\ln|t|^\beta) \\ \quad + b_2 \sin(\ln|t|^\beta)], & k < -\tfrac{1}{4}, \end{cases}$$

where $\left.\begin{array}{l}\lambda_1 \\ \lambda_2\end{array}\right\} = \dfrac{1 \pm \sqrt{1+4k}}{2}$, respectively, and

$\beta = \dfrac{\sqrt{-(1+4k)}}{2}$.

19. $y = X(x)T(t)$ where the general solutions to X and T are

$$X(x) = \begin{cases} c_1 e^{\sqrt{k+1}\,x} + c_2 e^{-\sqrt{k+1}\,x}, & k > -1 \\ c_1 + c_2 x, & k = -1 \\ c_1 \cos\sqrt{-(k+1)}\,x \\ \quad + c_2 \sin\sqrt{-(k+1)}\,x, & k < -1, \end{cases}$$

$$T(t) = \begin{cases} b_1 e^{\sqrt{k}\,t} + b_2 e^{-\sqrt{k}\,t}, & k > 0 \\ b_1 + b_2 t, & k = 0 \\ b_1 \cos\sqrt{-k}\,t + b_2 \sin\sqrt{-k}\,t, & k < 0. \end{cases}$$

21. $y(x, t) = \dfrac{1}{2}\left(3 \sin \dfrac{\pi x}{a} \cos \dfrac{c\pi t}{a} - \sin \dfrac{3\pi x}{a} \cos \dfrac{3c\pi t}{a}\right)$

23. a. $\dfrac{\partial^2 y}{\partial t^2} = c^2 \dfrac{\partial^2 y}{\partial x^2} - 2b \dfrac{\partial y}{\partial t}$, where b is a constant of proportionality;

b. $y(x, t) = \dfrac{e^{-bt}}{4}\left[3\left(\cos k_1 t + \dfrac{b}{k_1} \sin k_1 t\right) \sin \dfrac{\pi x}{a}\right.$
$\left. - \left(\cos k_3 t + \dfrac{b}{k_3} \sin k_3 t\right) \sin \dfrac{3\pi x}{a}\right];$

c. $y(x, t)$
$= e^{-bt} \displaystyle\sum_{n=1}^{\infty} A_n \left(\cos k_n t + \dfrac{b}{k_n} \sin k_n t\right) \sin \dfrac{n\pi x}{a},$

where $k_n = \sqrt{n^2 \pi^2 c^2 - b^2 a^2}/a$

and $A_n = \dfrac{2}{a} \displaystyle\int_0^a f(x) \sin \dfrac{n\pi x}{a}\, dx$.

APPENDIXES

Problems Appendix 3, page A26

1. yes; $\delta = \tfrac{1}{216}$ (better: $\delta = \tfrac{1}{50}$)

3. Yes, but $a = b = 1$ is not possible. You need $a + b < 1$. If $a = b = \frac{1}{3}$, then $\delta = \frac{1}{4}$.
5. yes; $\delta = 1$
7. Yes; but you need $b < 1$. If $b = \frac{1}{2}$, then $\delta = 1$.
9. yes; $\delta = \frac{1}{2}$ (better: $\delta = +\infty$; there is a unique solution defined for $-\infty < t < \infty$)
11. a. $k = 1$;   b. $k = 1$;

   c. $k = 3$

   $\left(\text{a smaller constant is } \sup_{|x| \leq 1} \left| 2x \sin\frac{1}{x} - \cos\frac{1}{x} \right| \right);$

   d. $972\sqrt{3} \approx 1683$.

13. $x_n(t) = 3 \sum_{k=0}^{n} \frac{(t-3)^k}{k!}$

25. a. $\begin{pmatrix} x_1 \\ x_2 \end{pmatrix}' = \begin{pmatrix} tx_1 x_2 + t^2 \\ x_1^2 + x_2^2 + t \end{pmatrix}; \quad x(0) = \begin{pmatrix} 1 \\ 2 \end{pmatrix}$

   b. $k = 10$;   c. $\delta = \frac{1}{7\sqrt{5}} \approx 0.064$.

27. a. $\begin{pmatrix} x_1 \\ x_2 \\ x_3 \end{pmatrix}' = \begin{pmatrix} x_1^2 x_2 \\ x_3 + t \\ x_3^2 \end{pmatrix}; \quad x(0) = c;$

   b. $k \leq \max_{\substack{1 \leq i \leq 3 \\ (t,x) \in D}} \left| \frac{\partial f}{\partial x_i} \right| \leq \max\{1 + a, 2(b + |c|)^2\};$

   c. $\delta = \min\left\{ a, \frac{b}{M} \right\}$ where $M \leq ((b + |c|)^6 + (b + |c| + a)^2 + (b + |c|)^4)$

Problems Appendix 4, page A37

1. $D = -2;\ x_1 = 3,\ x_2 = 0$
3. $D = 0$; infinite number—the entire line $x_1 + 2x_2 = 3$
5. $D = 0$; no solution   7. $D = -29;\ x_1 = x_2 = 0$
9. -56 11. 140 13. 96
15. -120 17. -132 19. -398
23. $x_1 = -3,\ x_2 = 0,\ x_3 = 5$
25. $x_1 = -\frac{26}{9},\ x_2 = -\frac{1}{2},\ x_3 = \frac{19}{9}$
27. $x_1 = -2,\ x_2 = 0,\ x_3 = 1,\ x_4 = 5$

Problems Appendix 5, page A46

1. $9 - 7i$ 3. 2 5. $-27 + 5i$
7. $5\sqrt{2}\, e^{\pi i/4}$ 9. $3\sqrt{2}\, e^{-\pi i/4}$ 11. $6e^{\pi i/6}$
13. $8e^{-\pi i/6}$ 15. $2e^{-2\pi i/3}$ 17. -2
19. $-(\sqrt{2}/4) - (\sqrt{2}/4)i$
21. $-2\sqrt{3} + 2i$ 23. $-\frac{3}{2} - (3\sqrt{3}/2)i$
25. $\cos 1 + i \sin 1 \approx 0.54 + 0.84i$
27. $4 - 6i$ 29. $7i$ 31. $2e^{-\pi i/7}$ 33. $3e^{4\pi i/11}$
39. circle of radius a centered at z_0 if $a > 0$, single point z_0 if $a = 0$, empty set if $a < 0$

Index

Abel's formula, 91
Absolute error, 377
Acceleration, 2
Adams-Bashforth formula, 394
Airy, G. B., 166
Airy equation, 166, 173
Algebraic multiplicities, 317
Amplitude, 140
Analytic function, 159
Argument, A42
Associative law, 289
Augmented matrix, 292

Beats, 152
Bernoulli, J., 50
Bernoulli's equation, 49–51
Bessel, F. W., 186
Bessel equation, 99, 180, 182, 186
Bessel function, 186–195
    first kind, 188
    generating function, 194
    integrals of, 192
    modified, 194–195
    order 0, 99
    order $\frac{1}{2}$, 180
    order 1, 182
    oscillation, 205
    second kind, 189
    zeros, 205
Binomial coefficient, 81

Binomial series, 166
Binomial theorem, 166
Birth rate, 34
Black box, 231
Block diagram, 65
Boundary conditions, 85
Boundary value problems, 12, 85, 411
    numerical solution of, 407

Capacitance, 61
Carbon dating, 7
Carrying capacity, 35
Cayley-Hamilton theorem, 347
Characteristic equation, 100, 118, 124, 275, 317, 402
    of a difference equation, 128
    polynomial, 317
    of a system, 275, 317
Chebyshëv, P. L., 208
Chebyshëv's equation, 208
Clairaut's equation, 60
Closed system, 65
Coefficient, Fourier, 423
Commutative, 289
Comparison theorem, Sturm, 204
Compartmental analysis, 65
Competition model, 362–363
Complex number, A39–A46
Conjugate, A40
Constant coefficients, 84, 124, 261
Continuous dependence on initial conditions, A28

INDEX

Convergence
  absolute, 158
  of an improper integral, 209
  of a numerical method, 378
  radius of, 158
  of series, 158
  uniform, A15
Convolution, 252
Cramer's rule, A37
Critically damped, 144
Cross-derivative test, 41
Curves of pursuit, 69–74

d'Alembert, J., 445
  solution of wave equation, 446–447
Damped vibration, 143–146
Damping factor, 144
Damping force, 143
Damping ratio, 148
Dead space, 77
Death rate, 34
De Moivre, A., A44
De Moivre formula, A44
Density, 455
Determinant, A29–A37
Difference equation, 22, 77–81, 128–135
  homogeneous, 78, 402
  linear, 78–81, 128–135
  nonhomogeneous, 78, 402
  order of, 78
  solution of, 78
Differential equation, 1, 11
  constant coefficients, 84
  direction field of, 15, 19
  exact, 41
  homogeneous, 46, 83, 100–105
  linear, 46–53, 83–127
  nonhomogeneous, 46, 83, 105–116
  order of, 11, 83
  ordinary, 11
  ordinary point of, 169
  partial, 11
  singular point of, 169
    irregular, 174
    regular, 173
  singular solution of, 44, 60
  solution of, 7, 11
  system of, 260
  vector, 300
  vector solution of, 301
Diffusivity, 456, 460
Dirac delta function, 238
Direction field, 15, 19
Discontinuity, jump, 212
Discrete harmonic motion, 130
Discretization error, 25
Divergence, 158
Dot product, 288
Drag force, 32

Ecosystem, 362
Eigenfunction, 413, 452
Eigenvalue, 315, 413, 452
Eigenvector, 315
  generalized, 334
Electric circuit, 61–65, 152–155, 230, 247–251, 354
Electromotive force, 61
Elliptic equation, 439, 440, 442
Epidemic, 369–374
Error
  absolute, 377
  analysis, 398–401
  discretization, 25
  relative, 377
  round-off, 25, 402
Escape velocity, 31–32
Euler, L., 21
Euler constant, 189
Euler equation, 118–122
Euler formula, 103, A43, A45
  for Fourier coefficients, 423, 425
Euler's method, 20, 379
  improved, 381–385
  for systems, 405
Even extension, 436
Even functions, 429, 432
Exact differential, 41
Exact equation, 40–44
Existence-uniqueness theorems, 13, 84, 85, 263, 266, 301, 307, A17, A21, A22, A24, A25
Exponential, matrix, 334
Exponential decay, 4
Exponential growth, 4
Exponents of a differential equation, 175

Feedback, 271, 361
Fibonacci numbers, 135
Filter
  band-pass, 360
  high-pass, 360
  low-pass, 360
Floating-point
  arithmetic, 376
  number, 376
Force
  damping, 143
  electromotive, 61
  gravitational, 2
Forced vibrations, 146–152
Fourier, J., 423
Fourier convergence theorem, 425
Fourier series, 416, 423–431
  coefficients, 423
  complex, 438
  cosine, 434
  generalized, 431
  sine, 434
Fractional transfer coefficient, 66
Free fall, 1
Free vibration, 139–142

Frequency, natural, 140, 155
Frequency ratio, 148
Frobenius, G. F., 174
Frobenius, method of, 173–184
Fundamental matrix solution, 307
Fundamental set of solutions, 304

Games and quality control, 131–135
Gamma function, 187, 220
Gauss, C. F., 207
Gauss-Jordan elimination, 293
Gauss's hypergeometric equation, 207
General solution, 6, 7, 44, 89, 93–94, 124, 312
Generalized eigenvector, 334
Golden ratio, 135
Gompertz equation, 82
Gravitational acceleration, 2
Green's function, 116
Gronwall's inequality, A20
Growth rate, 34

Half-life, 7
Half-range expansion, 436–437
Half-wave rectifier, 246
Harmonic motion, 104, 130, 137–142
Heat conductivity, 455
Heat equation, 439, 456, 460
Heat flux, 455
Heaviside, O., 234
Heaviside function, 234
Hermite, Charles, 166, 422
Hermite equation, 166, 200, 422
Hermite function, 166, 201, 422
  orthogonality, 422
Hermite polynomials, 166, 201, 422
Heun's formula, 392
Homogeneous, 54, 90
Homogeneous equation, 46, 54–55
Homogeneous systems, 261
Hooke's law, 138
Hyperbolic equation, 439, 440, 442
Hypergeometric series, 207

Imaginary number, A39
Imaginary part, A40
Improper integral, 209
  convergence of, 209
  divergence of, 209
Improved Euler's method, 381, 405
Indicial equation, 175
Inductance, 61
Infection rate, 370
Infectives, 369
Initial condition, 4, 12
Initial point, 12
Initial value problem, 4, 12, 262
Inner product, 418, A27
Integral equation, Volterra, 256
Integrating factor, 42, 47
Interval of convergence, 158

Intravenous feeding, 363
Inverse Laplace transform, 211
Inverse square law, 138
Invertible matrix, 290
Irregular singular point, 174

Jump discontinuity, 212

Kirchhoff's current law, 354
Kirchhoff's voltage law, 61, 354
Kutta-Simpson $\frac{3}{8}$ rule, 391

Lagrange, J. L., 112
Lagrange's method, 53, 112
Laguerre, E., 201
Laguerre equation, 201, 421
Laguerre polynomials, 201, 421
Laplace, P. S., 210
Laplace equation, 439, 460, 463–471
Laplace transforms, 209–258, A11–A13
  of convolution integrals, 252
  differentiation property, 221, 222
  existence theorem, 213
  first shifting property, 218
  inverse, 211
  linearity property, 214, 215
  periodicity property, 241
  second shifting property, 235
  for systems, 280–285
  uniqueness, 214
Law of mass action, 370
Law of reflection, 56
Legendre, A. M., 196
Legendre equation, 98, 195–200
Legendre function, 195
Legendre polynomials, 197–198
  generating function, 201
  Rodrigues's formula, 198
Legendre series, 471
Linear combination, 87, 123, 267
Linear differential equation, 46
Linear system, 266–270
Linearly dependent, 87, 123
Linearly independent, 87, 123, 267, 304
Lipschitz constant, A16
Lipschitz continuous function, A16
Logistic equation, 34
Logistic growth, 34

Maclaurin series, 158
Magnitude, A41
Main diagonal, 289
Malthus, T. R., 5
Malthusian law, 6
Matrix, 285
  addition, 287
  augmented, 292
  equality, 286
  exponential, 334
  function, 296
  identity, 289

inverse, 290
multiplication, 288
scalar multiplication, 287
solution, 306
square, 286
zero, 286
Maximum principle, 472
Mean value theorem, 467, A15
Method of determinants, 275
Method of elimination, 273
Method of Frobenius, 173–184
Method of Newton, 373
Method of separation of variables, 28–40
Milne's predictor-corrector method, 397
Motion, Newton's second law of, 2
Multiplicity, 124
Multi-step, 402
Mutualism, 365

Natural frequency, 140
Natural vibration, 146
Neumann, C. G., 189
Neumann function, 189
Newton's law of cooling, 6
Newton's method, 373
Newton's second law of motion, 2
Nonhomogeneous boundary value problem, 415
Nonhomogeneous equation, 46
Nonhomogeneous system, 261
Nonlinear equation, 46
Nonoscillatory, 203
Norm, 418–419
matrix, A23
weighted, 421
Normal form, 136
Numerical instability, 402

Obsidian dating, 39
Odd function, 428, 432
Odd extension, 436
Ohm's law, 61
Open system, 65
Order, 11
Ordinary differential equation. *See* Differential equation.
Ordinary point, 166–172
Orthogonal function, 418–421
weighted, 421
Orthogonality, 416
Hermite polynomials, 422
Laguerre polynomials, 421
Orthonormal functions, 419
weighted, 421
Oscillation, 202, 203
Oscillation transformer, 375
Overdamped, 144

Parabolic equation, 439, 440, 442
Pareto, V., 40
Pareto's law, 40

Partial differential equation, 11, 411, 438–472
degenerate, 439, 442
elliptic, 439, 440, 442
hyperbolic, 439, 440, 442
linear, 438
parabolic, 439, 440, 442
second-order, 438–443
Partial fractions, 225–227
Partial sum, 165
Particular solution, 7, 44, 94
Pendulum, 142
Periodic function, 423
Phase angle, 148
Picard, E., 75
Picard iterations, 75, A17
Piecewise continuous, 213
Planck, B. M., 282
Point
ordinary, 169
singular, 169
Poisson, S. D., 467
Poisson integral formula, 467
Polar form, A43
Polynomials
Hermite, 166, 201, 422
Laguerre, 201, 421
Legendre, 197–198
Power series, 157
center of, 158
interval of convergence, 158
method, 157–163
radius of convergence, 158
Predator-prey equations, 365
Predictor-corrector formulas, 393–397
for systems, 406
Principal matrix solution, 307, 327
Principle of superposition, 105, 266, 452
Propagation of error, 402
Pure imaginary number, A41

Quadrature formulas, 393–397

Radioactive decay, 7
Radius of convergence, 158
Ratio test, 158
Real part, A40
Recursion formula, 162
Reduction of order, 97–99
Reentry to atmosphere, 32
Regular singular point, 173
Remainder term, 159
Removal rate, 370
Resistance, 61
Resonance, 149, 155
Ricatti, J. F., 102
Riccati equation, 102
Rodrigues, O., 198
Rodrigues formula, 198
Root test, 158
Rounding, 377

Round-off error, 25
Row operations, 292
Row reduction, 292
Runge-Kutta method, 386–392
   for systems, 406–407

Schaefer model, 52
Schrödinger, E., 166
Separation of variables, method of, 4, 28–40, 450–454
Series
   binomial, 166
   Fourier cosine, 434
   Fourier sine, 434
   Frobenius, 173–184
   Legendre, 471
   Maclaurin, 158
   power, 157
   Taylor, 158
Shifting theorem
   first, 218
   second, 235
Shooting method, 407–408
Significant digits, 376
Simple harmonic motion, 137–141
Simpson's rule, 386, 393
Singular point, 166–172
   irregular, 174
   regular, 173
Singular solution, 44, 60
SIR model, 370
Snowplow problem, 40
Solution, matrix, 306
Solution of a differential equation, 7, 11, 93
Solution of a system of differential equations, 268, 269
Solution of an initial-value problem, 12
Specific heat, 455
Specific infection rate, 73
Spring constant, 138
Stability of a numerical method, 378
Steady-state current, 63, 154, 458
Stefan's law, 82
Step function, 234
Stiffness, 408
String insulator, 360
Sturm, J. C. F., 202
   comparison theorem, 204
Substitution technique, 55–59
Successive approximations, 74–77
Superposition, principle, 105, 266, 452
Susceptibles, 369
System, 259–375
   characteristic equation of, 275, 317
   closed, 65
   first order, 261
   linear homogeneous, 261, 299
   linear nonhomogeneous, 261, 299, 337
   numerical solution of, 404–407
   open, 65
   solution of, 260, 301, 312

Taylor series, 158, 164–165
Taylor's formula, 159
Temperature gradient, 455
Thermal conductivity, 455
Torricelli, E., 40
Torricelli's law, 40
Total differential, 40
Transfer function, 231
Transform
   inverse Laplace, 211
   Laplace, 211
Transient current, 63, 154, 459
Trapezoidal rule, 381, 393
Triangular matrix, 299
   upper, 299
   lower, 299
Trigonometric series, 423
Truncation, 377

Uncoupled, 279
Underdamped, 144
Undetermined coefficients, 105–111
Uniform convergence, A15
Uniqueness theorem, 76, A21
Unit circle, A46
Unit impulse function, 238

Variable coefficients, 84
Variation of parameters, 53, 112–118
   for linear systems, 279, 338–339
Vector, 286
   column, 286
   function, 296
   multiplication, 288
   product, 288
   row, 286
   solution, 301
Velocity, 2
Verhulst, P. F., 34
Vibrating string, 444–454
Vibrational motion, 137–152
   damped, 143–146
   forced, 146–152
   free, 139–142
   natural, 146
Volterra, V., 255
Volterra integral equation, 256

Wave equation, 439, 445
   traveling, 446
Weierstrass M-test, A16
Weight function, 421, 431
Whirl, 368
Wronsky, J. M., 90
Wronskian, 90, 123
   of a system, 267, 306